Advances in Big Data Analytics

Yong Shi

Advances in Big Data Analytics

Theory, Algorithms and Practices

Yong Shi
University of Chinese Academy of Sciences
Beijing, China

University of Nebraska at Omaha
Omaha, Nebraska, USA

ISBN 978-981-16-3609-7 ISBN 978-981-16-3607-3 (eBook)
https://doi.org/10.1007/978-981-16-3607-3

This Springer imprint is published by the registered company Springer Nature Singapore Pte Ltd.
The registered company address is: 152 Beach Road, #21-01/04 Gateway East, Singapore 189721, Singapore

Preface

Today, we are in the big data era. Big data has become a reality that no one can ignore. Big data is our environment whenever we need to make decision. Big data is a buzz word that makes everyone understand how important it is. Big data shows a big opportunity for academia, industry, and government. Big data then is a big challenge for all parties. The meaning of big data contains both data science and applications, where big data analysis, across data science and applications, is also a subset of big data.

This book is called *Advances in Big Data Analytics: Theory, Algorithms, and Practices*. Based on more than 80 published papers and reports, it provides the reader an up-to-date research progress and application findings of my students and colleagues and that of mine in big data analytics and related areas in the last decade (2010–2020). Since the contents of big data vary from the application domains, the book summarizes the algorithms, procedures, analyses, and empirical studies as a general picture of big data analytics development.

This book is organized into three parts and each part contains several related chapters. Part I addresses the basic concepts of big data and theoretical foundations. It contains Chaps. 1–3. Chapter 1, based on three published papers and a big data world report, first discusses the big data evolution and challenges and then presents the concepts of big data, big data analytics, data science, and application as well as some open research problems. Chapter 2 outlines the recent advances of multiple criteria linear programming classification in ten related published papers. It has three sections. They are multi-criteria linear programming for supervised learning, multi-criteria linear programming with expert, and rule-based knowledge and multi-criteria decision-making-based data analytics. Chapter 3 presents research findings in support vector machine classification based on 15 published papers. There are five sections in this chapter. They are support vector machine in data science, twin support vector machine in classification, nonparallel support vector machine classifiers, Laplacian support vector machine classifiers, and loss functions of support vector machine classification.

Part II mainly presents different functional research in big data analytics. It consists of Chaps. 4–9. Among them, Chap. 4 is about feature selection with three sections. They are feature selection in classification, feature selection with regularization programming, and feature selection with knowledge functions. Chapter 5 shows data stream analysis. Its first section is application-driven classification of data streams while the second section is robust ensemble learning for mining noisy data streams. Chapter 6 discusses learning analysis with three sections. They are concept learning, label proportion-based learning, and enlarged learning models. Chapter 7 presents sentiment analysis in three sections. They are development of sentiment analysis, word embedding analysis, and domain-based sentiment analysis. Chapter 8 is about link analysis with two sections. They are market-oriented link analysis and variation of link analyses. Chapter 9 is called evaluation analysis. It has three sections. They are evaluation for methodologies, evaluation for software, and evaluation for sociology and economics.

Part III presents different applications and future analysis. It contains Chaps. 10–12. Chapter 10 is about business and engineering applications with three sections. They are banking and financial market analysis, agriculture classification, and engineering problems. Chapter 11 deals with healthcare applications with two sections. They are the underlying transmission patterns of COVID-19 outbreak—an age-specific social contact characterization and evaluating doctor performance. Finally, Chap. 12 is the ongoing research for Artificial Intelligence test problems in three sections. They are basic concepts of Artificial Intelligence (AI) Intelligence-Quotient (IQ) tests, laws of intelligence based on AI-IQ research, and a fuzzy cognitive map approach to characteristics AI-IQ test.

I would like to express my sincere thanks to my colleagues and graduate students who have been co-authors of the papers and reports that formed the basis of this book. They are Prof. Gang Kou and Prof. Yi Peng, Prof. Daji Ergu, Prof. Wikil Kwak, Prof. Zhengxin Chen, Prof. Yingjie Tian, Prof. Lingling Zhang, Prof. Xiaohui Liu, Prof. Xianhua Wei, Prof. Xiaofei Zhou, Prof. Jiming Liu, Dr. Zhiquan Qi, Dr. Peng Zhang, Dr. Bo Wang, Dr. Huimin Tang, Dr. Jianyu Miao, Dr. Yunlong Mi, Mr. Wei Li and Miss Luyao Zhu, Dr. Xi Zhao, Dr. Zhuofan Yang, Dr. Guangli Nie and Dr. Yibing Chen, Dr. Limeng Cui, Dr. Fan Meng, Dr. Zhensong Chen, Dr. Peijia Li, Dr. Fangyao Liu, and Dr. Feng Liu.

I am also indebted to my graduate students who helped me organize this book with different chapters for the consistency of formation, notation, figures, and mathematical symbols. They are Dr. Bo Wang for Chap. 2, Miss Jiayu Xue, Mr. Yi Qu, Miss Mengyu Shang, and Miss Linzi Zhang for Chap. 3, Dr. Jianyu Miao for Chaps. 4 and 5, Dr. Yunlong Mi for Chap. 6, Mr. Wei Li and Miss Luyao Zhu for Chap. 7, Dr. Xi Zhao and Dr. Wei Dai for Chap. 8, Dr. Yang Xiao and Dr. Huimin Tang for Chap. 9, Dr. Fangyao Liu and Dr. Bo Li for Chaps. 10 and 11, and Dr. Wei Dai for Chap. 12.

In addition, I would like to thank several individuals who have encouraged me to prepare and publish this book. They are Prof. Daniel Berg (University of Miami, USA), Prof. James M. Tien (University of Miami, USA), Prof. Zhengxin Chen (University of Nebraska at Omaha, USA), Prof. Florin G. Filip (Romania Academy of Sciences, Romania), Prof. Svetozar D. Margenov (Bulgarian Academy of Sciences, Bulgaria), Prof. Enrique Herrera-Viedma (University of Granada,

Spain), and Prof. Fuad Aleskerov (National Research University Higher School of Economics, Russia).

Finally, we would like to acknowledge the funding agencies who provided their generous support to our research activities on this book. They are Nebraska Furniture Market—a unit of Berkshire Hathaway Investment Co., Omaha, USA for credit scoring and recommendation system fund (2008–2009, 2016–2017); Nebraska EPScOR, the National Science Foundation of USA for industrial partnership fund (2009–2010); BHP Billiton Co., Australia for the research fund (2005–2010); the CAS/SAFEA International Partnership Program for Creative Research Teams (2010–2012); Sojern, Inc., Omaha, USA for Business Intelligence (2012–2013); National Science Foundation of China (Key Project: #71932008, 2020–2024, Key Project: #91546201, 2016–2020, Key Project: #71331005, 2014–2018, International Collaboration Project: #71110107026, 2012–2016).

The first draft of this book was compiled during the COVID-19 pandemic from February to June 2020 in Omaha, Nebraska when I was in the self-quarantine. The book was finally completed in December 2020, Qingshuiwan, Hannan, China.

Beijing, China
February to December 2020

Yong Shi

Contents

Part I
Concept and Theoretical Foundation

Chapter 1
Big Data and Big Data Analytics

Big data now is a common term. However, the evolution of big data comes from twofold. The creation of the computer in the 1940s gradually provides tools for human beings to collect massive data, while the term "big data" becomes a popular slogan to represent the collection, processing, and analysis of various data [1]. The data has been exponentially growing for the last 70 decades. EMC2 [2] estimated that the world generated 1.8 zettabytes of data (1.8 multiple 21 zeros) by 2011. In fact, this figure has grown to 44 zettabytes, about 24 times in 2020. Big Data Analytics has arisen as the technical means dealing with both theory and application of big data. This chapter elaborates on the understanding of big data and its analytics. Section 1.1 briefly describes big data evolution and challenges. Section 1.2 is about big data's current status, including its development in the world as well as in China. Section 1.3 explores big data analysis and data science problems.

1.1 Big Data Evolution and Challenges

Nowadays, in human society, big data is the environment that we cannot ignore in our daily activities. Big data occurs as a phenomenon. Whenever we make a decision, the impact of big data must be considered. Big data is a "buzz" word that is a better capture about the name of data collection and analysis which went through the stages of database management in 1960s, data warehouse in 1970s, knowledge discovery in databases (KDD) in 1980s, enterprise resource planning (ERP) and data mining in 1990s, customer relationship management (CRM) and business analytics (BA) in 2000s. Big data, as a good term, unifies all of the above concepts so that the majority of people know what it means. Big data is also a big challenge for those people who analyze the data due to its complex structure and lack of available technology. Furthermore, big data provides a big opportunity to the business world for increasing productivity [3].

Y. Shi, *Advances in Big Data Analytics*,
https://doi.org/10.1007/978-981-16-3607-3_1

The common concept of big data contains the applications, engineering, and scientific issues of big data. The definition of big data varies from academic and business communities and there is no unified definition about big data. In some professional communities, the terms of business intelligence and business analytics are also used to represent big data analytics [4].

In 2012, the National Science Foundation of USA [5] provided the following definition:

Definition 1.1 Big data is “large, diverse, complex, longitudinal, and/or distributed datasets generated from instruments, sensors, internet transactions, email, video, click streams, and/or all other digital sources available today and in the future”.

In May 2013, a group of international scholars has brain-stormed two versions of big data definitions at the 462nd Session: Data Science and Big Data in Xiangshan Science Conferences [6] at Beijing, China, where the author served as one of co-chairs. The first version of big data was given for academic and business communities as:

Definition 1.2 Big data is a collection of data with complexity, diversity, heterogeneity and high potential value, which are difficult to process and analyze in reasonable time.

The second version is for organizations and governmental policy making as:

Definition 1.3 Big data is a new type of strategic resource in digital era and the key factor to drive innovation, which is changing the way of human being’s current production and living.

In addition, “4V’s” have been commonly used to capture the main characteristics of big data: Volume, Velocity, Variety and Veracity [7, 8].

The history of data analytics can be traced back to more than 200 years ago when people used statistics to solve real-life problems. In the area of statistics, Bayes’ Theorem has been playing a key role in developing probability theory and statistical applications. However, it was Richard Price (1723–1791), a famous statistician, edited Bayes’ Theorem after Thomas Bayes’ death [9]. Richard Price was also one of the scientists who initiated the use of statistics in analyzing social and economic datasets. In 1783, Price published “Northampton table”, which collected observations for calculating of the probability of the duration of human life in England. In this work, Price showed the observations via tables with rows for records and columns for attributes as the basis of statistical analysis. Such tables now are commonly used in data mining as multi-dimensional tables. Therefore, from the historical point of view, the multi-dimensional table should be called as “Richard Price Table” and Price should be honored as the father of data analytics, later called data mining. Since the 1950s, as computing technology has been gradually used in commercial applications, many corporations have developed databases to store and analyze collected datasets. Mathematical tools employed to handle datasets revolute from statistics to methods of artificial intelligence, including neural networks and decision trees. In the 1990s, the database community started using the term "data mining", which is interchangeable with the term "knowledge discovery in databases (KDD)" [10]. Now data mining becomes the common technology of data analytics over the intersection of human intervention, machine learning, mathematical modeling, and databases.

In recent years, many authors published their opinions about how big data is deeply impacting the evolution of science and engineering as well as the development of society. One of the popular books, written by Mayer-Schönberger and Cukier [11] showed three advantages of big data: (1) access to all data undermines the sampling; (2) rough measurements for big data replace the requirement of high-quality data preparation; and (3) decision making is based on correlations of big data, instead of the reasoning. Although such advantages represent the importance of big data analytics, they cannot change the fundamentals of data analytics, which are still sampling, accuracy and reasoning (see Sect. 1.3). Big data does not mean the entire data. It is impossible to collect the entire data, which is a relative concept. However, big data, comparing with small data, can provide a very large sample. The larger the sample is, the more robust the results are. Big data may lead a better learning result, but the sampling process is needed to test and predict. With a rough data preparation, big data may produce a quick response or rough knowledge for people to make decision. However, such a decision could be good for a short run, not for a long run. The long run decision requires the solid and high-quality data preparation. In addition, for decision makers, seeking the reasoning of big data is more important than finding the correlations. Using of big data in engineering practice or business actions is not only for what we can do, but more on what we should do for the future. Therefore, big data needs data mining techniques to discover knowledge and predict the future. Based on known data mining methods, big data analytics should consider the large sample from all available data (structure or non-structured data); look for the precise solution based on the rough solutions; and identify the reasoning from the correlations. Big data analysis does not remove the fundamentals of data analysis or data mining. Instead, it improves the analytic methodologies since all data are supposedly are available.

Among many challenges of the big data problems, it is believed that the following three problems are urgent to solve in order to gain benefits from big data in science, engineering and business applications:

Challenge 1.1 Transforming Semi-structured and Non-structured Data into "Structured Format"

In the academic field of big data, it is not clear about the principle, basic rules and properties of data, especially semi-structured and non-structured data due to complexity of such data. The data complexity reflects not only the variety of the objects that data represents, but also a partial image that each dataset can present for a given object. The relationship of data representation and a real object just likes that of "the blind men and an elephant" [12]. Even though each data set truly represents an angle of the object, it cannot be its whole picture. The investigation of theoretical components of big data, which can be viewed as "data science" deserves the interdisciplinary efforts from mathematics, sociology, economics, computational science and management science. However, the term of data science is still under discussion among the research communities. Thanks to the advancement of information technology in recent years, the techniques, such Hadoop and MapReduce allow us to collect a large amount of semi-structured and non-structured data in a reasonable amount of time. Now, the key engineering challenge is how to effectively analyze these data and discover knowledge from them in an expected time. The answer could be that first transform the given semi-structured and/or non-structured data into a structured data-like format (or pseudo multi-dimensional table), and then conduct a data mining process by taking advantage of the existing data mining algorithms that are mainly developed for the structured data. Note that, the transformation from semi-structured and non-structured data into structured format should be subject-oriented. Once the structured data-like format is built up, the "first-order mining" by using data mining tools can result "rough knowledge" (called hidden patterns in data mining). To upgrade such knowledge into the "intelligent knowledge" that can be used for decision support, the analysts should combine some sort of human knowledge, such as experience, common sense, domain preference with the rough knowledge. This is viewed as the "second-order mining" [13].

Since most big data is based on semi-structured and/or non-structured representations, the "structured rough knowledge" from big data may reflect new properties, which can be captured by decision makers when it is upgraded as intelligent knowledge. The key value of big data analytics or data mining is to obtain intelligent knowledge.

Challenge 1.2 Exploring the Complexity, Uncertainty and Systematic Modeling of Big Data

As mentioned as in the above, any data representation of a given object is its partial picture of the facts. The complexity of big data comes from the coherent of data representation while the uncertainty of big data causes from the changes of the objects in the nature as well as the variety of data representations duo to measurements. Although a certain data analytic method is applied on big data, the knowledge discovered from the analysis is just knowledge from that particular angle of the real object. Once the angle is changed by the way of collecting or viewing the data from the object, the knowledge is no longer to be useful. For example, in petroleum exploration engineering, which can be viewed as a big data problem, the data mining has been done on spatial database generated from seismic tests and well log data collection. The underground geological structure itself is complicated. The non-linear patterns of data are changeable via different dimension and angles. Any results of data mining or data analysis could be knowledge that is only true for the given surface. If the surface is changed, the result is changes as well. Therefore, how can one derive the intelligent knowledge from knowledge from a surface and knowledge from a surface turning around 90° is challengeable [14]. The breakthrough to a systematic modeling on complexity and uncertainty of big data analysis and mining is needed for gaining knowledge from big data. Form a long-run point of view, it could be not easy for us to establish a comprehensive mathematic system design about big data as a whole. However, through the understanding of particular complexity or uncertainty in given subjects or domain of fields, it is possible to build a domain-based systematic modeling for the specific big data. As long as a series of such modeling structures are founded, the collection of them can be viewed as a systematic modeling of the big data. From a short-run point of view, if the engineers can find out some general approaches to deal with complexity and uncertainty of big data in a certain field, say in financial market (with data stream and media news) or internet retails (images and media evaluations), it will bring added value to social and economic development. In addition, the formats of complexity and uncertainty of big data result in the measurement and evaluation on the rough knowledge from big data mining. Many known techniques in engineering, such as optimization, utility theorem, expectation analysis, can be used to measure how the rough knowledge gaining from big data should be better combined with human judgment into the "second-order mining" process to effectively elicit the intelligent knowledge for decision support. Note that since the knowledge changes with individual and situation, the machine-man (big data mining vs. human knowledge) is still playing a key role in big data modeling.

Challenge 1.3 Exploring the Relationship of Data Heterogeneity, Knowledge Heterogeneity and Decision Heterogeneity
At the big data environment, decision makers face three heterogeneous problems: data heterogeneity, knowledge heterogeneity and decision heterogeneity. Traditionally speaking, decision making depends on the learning of knowledge from others and accumulation of experience. Learning of knowledge now is more based on the data analysis and data mining. In a theory of management information system, decision making can be classified as three levels: structured decision, semi-structured decision and non-structured decision depending on the responsibilities of individuals in an organization [15]. The operational staffs handling routine works relate to structured decision. The managers' decision is based on subordinates' reports (almost of them are structured) and their own judgments and refers as semi-structured. The top-managers or chief executive officer (CEO) make a final decision is non-structured, which is most likely text or voice. The demand of decision makers for data or information (quantitative forms) and knowledge (qualitative forms) are different according to different levels of the responsibilities. However, big data is disruptively changing the decision-making process. Based on big data analysis or mining, the functions of business operation (structured decision), manager (semi-structured decision), and CEO (non-structured decision) can be combined as a whole picture for decision making. For instance, a marketing person may use a real-time credit card approval system based on big data mining technology to quickly approve a credit limit to a customer without reporting to a supervisor. Such a decision has almost zero risk. He or she is a final decision maker, representing both manager and CEO.

In a data mining process using structured data, the rough knowledge normally is structured knowledge due to its numerical formats. In big data mining, although rough knowledge in the "first-order mining" is derived from heterogeneous data, it can be still reviewed as structure knowledge since the data mining is carried out on structured data-like format or pseudo multi-dimensional table. When the "second-order mining" is used, the structured knowledge is combined with domain knowledge of managers or CEO that are semi-structured or non-structured and gradually upgraded into intelligent knowledge [16]. Therefore, intelligent knowledge may be the representation of non-structured knowledge. Note that if the business operations only involve with semi-structured data and/or non-structured data, either it results in non-structured knowledge without data analysis (mining) or structured knowledge which is from data mining. Such structured or non-structured knowledge can impact semi-structured decision or non-structured decision depending on the levels of management involvements. Big data, nevertheless, creates a challenge to traditional decision-making process. Research on how the impact of big data on decision making is complicate and perhaps philosophy oriented. An observation is that no matter which kind of data heterogeneity is presented by big data, rough knowledge is in the domain of "first-order mining" and searching intelligent knowledge by the "second-order mining" is a key to study the relationship between data heterogeneity, knowledge heterogeneity and decision heterogeneity. Exploring how decision making can be changed in big data environment is equivalent to investigating the relationships of processing heterogeneous data, big data mining, domain knowledge of decision makers and involvement in decision making.

It can be predicted that any of theoretical contribution and engineering technologic breakthrough on the above three challenges can enhance the applications of big data in our society. It will start from the field of information technology, and then widely spreads to multi-media, finance, insurance, education, etc. for the formulation of new business models, boosting investment, driving the consumption, improving production, increasing productivity. In a word, it generates the big data revolution.

1.2 Big Data Development

It is not easy to describe how big data deeply and quickly influences the world. However, four big data events in academic community should be first mentioned. They are the big data associations, big data conferences, big data journals and big data sources opened by governments.

1.2.1 Big Data in Academic Community

Recent years, both academic and professional communities built various big data related non-profit organizations to exchange and disseminate theoretical findings, practical experience and case studies about big data as well as data science. Some of them are the National Consortium for Data Science (NCDS), the Big Data Institute

(BDI), Data Science Association (DSA), Institute for Big Data Analytics (IBDA), Institute for Data Science, Institute for Data Sciences & Engineering (IDSE), Data & Society Research Institute (DSRI), the Data Warehousing Institute (TDWI), Global Association for Research Methods and Data Science, ACM Special Interest Group on Knowledge Discovery and Data Mining (SIGKDD), SNIA—Analytics and Big Data Committee (ABDC), Association of BIG DATA professionals (aBIGDATAp), The Big Data Alliance (BDA), Digital Analytics Association (DAA), and Data Science Consortium. It can be observed that, in some academic communities, the term "big data" means the information technology business applications of dealing with massive data problems while the scientific components or research aspects of big data is called data science. This is why data science somehow is interchangeable with big data. In some professional communities, terms "business intelligent and business analytics" are used to describe big data analysis or big data mining [4].

In 2013–2020, for example, numerous big data conferences have been held around the world. Some of them are International Conference on Big Data and Cloud Computing, IEEE International Conference on Big Data, ISC Big Data Conference, Big Data Technology Conference, CCF Big Data, IEEE International Congress on Big Data, the series of Big Data Conferences (Stanford, Beijing and Cambridge), International Conference on Algorithms for Big Data, International Conference on Data Science (ICDS, which was founded by the author and his institutions), the Conference on Nonparametric Statistics for Big Data and INFORMS Conference on the Business of Big Data. Most of the above conferences have been held annually since 2014. And these conferences have attracted thousands of scholars, engineers and practitioners for their common interests in big data problems.

There two categories of big data related academic journals. One is under the name of big data and another is under names of data science. The big data journals are Journal of Big Data, Big Data Research, International Journal of Big Data Intelligence, International Journal of Big Data, Big Data Journal and Big Data & Society Journal. The data science journals are Annals of Data Science, The Data Science Journal, Journal of Data Science, EPJ Data Science, Data Science Journal, and International Journal of Data Science. Most of these journals are newly established in recent years and need to demonstrate the reputations by publish cutting-edge research findings and technological advances in big data related areas.

1.2.2 Big Data in the World

It has been recognized that most data of "big data" come from three sources: The large amount of data generated from social and economic activities are controlled by governments. The enterprises, especially the well-known big data companies such as Google in U.S., Baidu in China and Yandex in Russia, own their business data as the important assets. The rest of open data accesses from online are free for anyone to download or use. Therefore, governments play a key role in making policies and promoting big data applications. Governmental actions on big data can be categorized as two stages.

1.2.2.1 Stage 1 (2009–2012): Start-Up

In 2009 the U.S. government launched its Data.gov website to offer governmental datasets to the public, and later in 2011 the Open Government Partnership (OGP) initiated by a United Nation General Assembly meeting. The United States and the United Kingdom, both are founders of OGP, delivered their national action plans for the first time, in order to open the government data as their main priority and promise to accomplish this goal. In 2011 the McKinsey Global Institute (MGI) published a special report called "Big Data: The next frontier for innovation, competition, and productivity", which was for the first-time a thorough introduction and prospection of big data released from a distinguished institute. Around that time, big data was widely discussed among both academic circles and economic circles. The Obama Administration launched the "Big Data Research and Development Initiative" on the White House official website in 2012, demonstrating that big data technologies evolved from early commercial operations to national scientific and technological strategies. Meanwhile, as the ongoing development of Internet and mobile communication technologies are increasing exponentially, (including intelligent terminal devices, semi structures and unstructured data), mathematical tools used to process datasets were shifted from statistics to artificial intelligence. Hence, the Age of Big data began.

1.2.2.2 Stage 2 (2013–Today): High-Speed Developing

Along with matured big data fundamental technologies and techniques, academic and business domains steered to applications research accordingly.

Big data technologies started to infiltrate into all society sectors, such as government administration, finance, science and technology, health care, education, transportation, industry. These sectors formed a complete big data industrial chain and developed amounts of applications in diverse fields: smart government, smart city, intelligent manufacturing, new retailing, etc. And this is when Hereupon big data entered its high-speed development stage. In 2013 at the G8 Summit, eight G8 members signed an Open Data Charter, which established the basic principles and standards for members to improve transparency of government information [17]. It encourages the governments open their data to public on five principles: Open Data by Default, Quality and Quantity, Usable by All, Releasing Data for Improved Governance, and Releasing Data for Innovation. So far, a number of countries have set up their data.gov style websites. The released big data covers broad categories, including Agriculture, Climate & Weather, Infrastructure, Energy, Finance & Economy, Environment, Health, Crime & Justice, Government &

Policy, Law, Job & Employment, Public Safety & Security, Science & Technology, Education, Society & Culture, Tourism and Transportations [18].

In terms of the subject of "Open Data", Europe is at the forefront. In 2014, the European Commission adopted the "Towards a Thriving Data-driven Economy" strategy and advocated European countries to seize the opportunities in data economy development. In March 2018, news of Facebook data leakage scandal started spreading and it is now still heating up. In the Age of Big Data (data sharing and data safety), individual privacy balancing and protection became a worldwide problem. In 2016, General Data Protection Regulation (GDPR) was approved by the European Parliament and came into effect on May 25 2018. On April 25 the same year, the European Commission released the policy document "Towards a Common European Data Space", addressing principles on how public sectors open datasets, retain and collect research data, and how private companies are processing and opening data. On October 4 2018, the European Parliament voted through the Regulation on "The Free Flow of Non-personal Data". Henceforth, the European Union has built a systematical legal system for individuals' privacy protection, as well as data opening and sharing.

In 2019, the Tianfu Institute of International Big Data Strategy and Technology (TIBD, which was founded by the author and local government and institutions), Chengdu Government Service Management & Network Administration Office, the Research Center on Fictitious Economy and Data Science Chinese Academy of Science, and the Key Laboratory of Big Data Mining and Knowledge Management Chinese Academy of Sciences jointly released the first "Annual Big Data World Report", called "Global Big Data Development Analysis Report 2018" [19]. This report was based on the source data of 79 OGP-membership countries as well as that of China. It produced many interesting findings regarding how big data developed in the world. For example, the proportion of OGP membership countries by continents making commitments to open government data in November 2018 is distributed as the following: European countries account for 36.5%, African 19.2%, North American 13.5%, Asian 13.5%, Latin American 13.5%, and Oceania 3.8% (see Fig. 1.1).

Another example is in terms of correlation between per capita GDP growth and government's efforts in opening-up data in major countries, the former is proportional to the latter with an exception of India (see Fig. 1.2). Though the per capita GDP growth rate in India is comparatively low, India is also a pioneering country in government open data efforts.

In 2021, the second "Annual Big Data World Report" named "Global Big Data Development Analysis Report 2020" was released by the TIBD and these agencies mentioned above. This report introduces the COVID-19 pandemic and its acceleration of big data development. It also incorporates the current status of data opening efforts around the world with its promotional effect on digitalization and "High-quality Development".

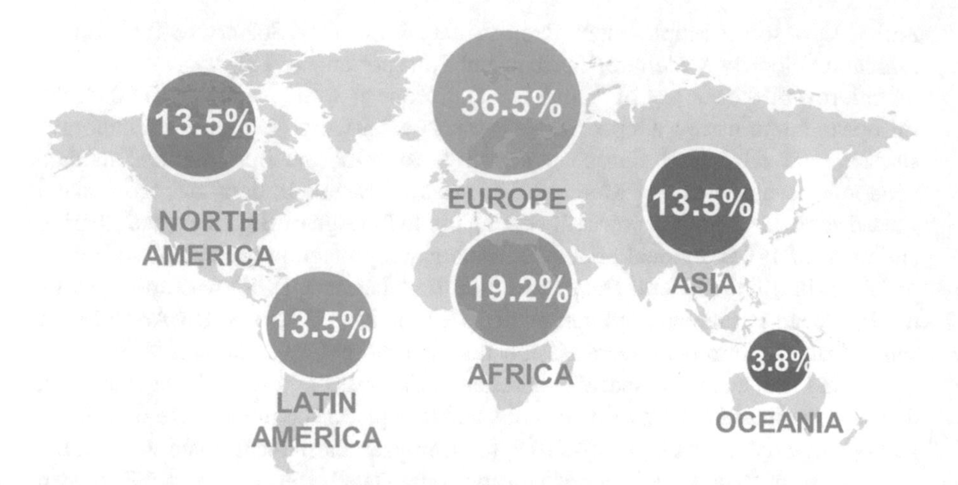

Fig. 1.1 OGP participating countries making commitments of open data (2018)

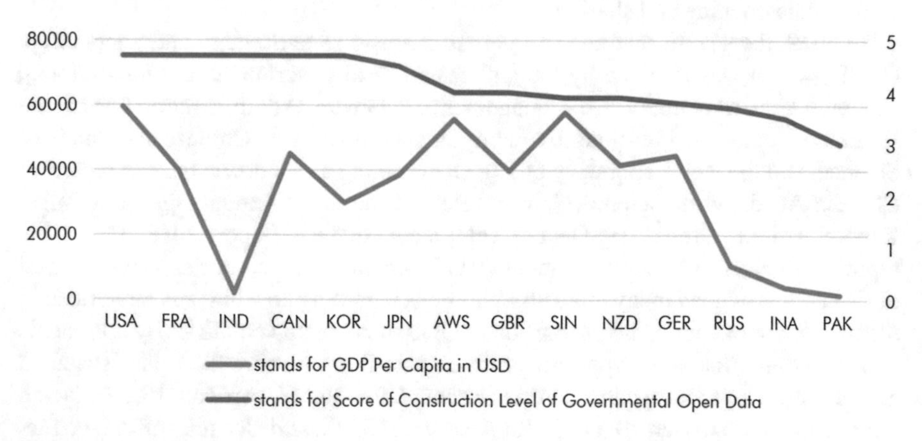

Fig. 1.2 The Relationship between Per Capita GDP and the score of construction level of Governmental Big Data Open in world's main countries (2018)

1.2.3 Big Data in China

In 2015, the State Council of the People's Republic of China issued the "Action Plan for Promoting the Development of Big Data" with the purpose of comprehensively promoting the development and application of big data in China and accelerating the construction of a powerful data nation [20]. Big data industry in China comes into flourish in all fields. Later, Big data becomes one of China's national strategies of economic development in it's the 13th Five-year plan (2016–2020).

The action framework can be interpreted as a Top-Level Design, which consists of three national platforms: National Data Opening Platform, Trans-Departmental Data Sharing Platform and Internet based National Data Service Platform (see Fig.

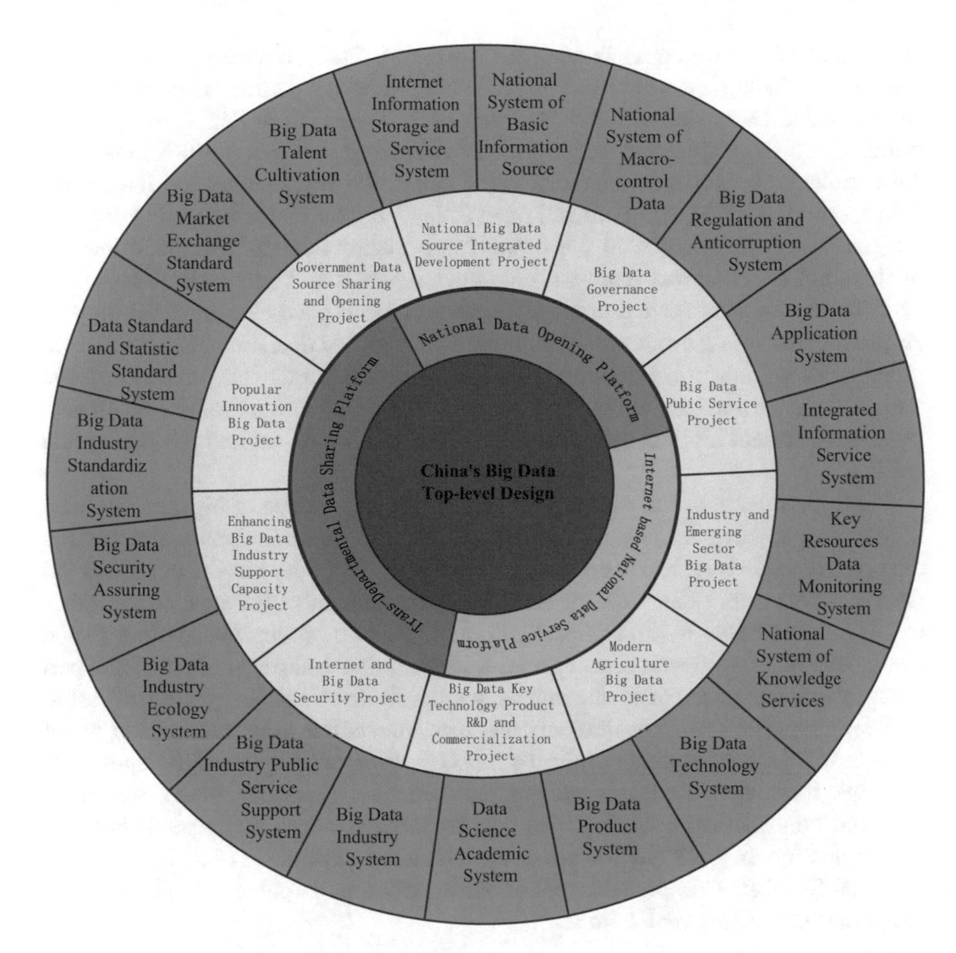

Fig. 1.3 China's big data top-down design

1.3). By 2020, the Chinese government commits to complete ten key big data projects for three platforms so as to provide big data applications in a number of public areas, including credit, transportation, healthcare, employment, social security, geography, culture, education, science and technology, agriculture, environment, safety and security, product quality, statistics, meteorology and ocean service. These projects will finally assemble a numbers of big data systems across a wide range of governmental departments, industries, academic and education institutes. After the announcement of the Action Framework, the Chinese government has published a planning program to further nail down key tasks for related departments to carry out its respective potions of Fig. 1.3 in terms of the responsibilities, road maps and target dates into 2020. This program offers a strong support for the completion of the Action Framework [21].

Importantly, the Chinese government tries to use the implementation of the framework to influence the Chinese culture and social value towards data and builds social data awareness. In May 2013, a group of the Chinese and international scholars, including the author of this book, brain-stormed the meanings of big data and provided Definition 1.2 and Definition 1.3. These definitions had a positive influence on the Chinese leadership in the following events to build China's big data strategy. A Chinese version of big data can be regarded as the large-scale data set produced and being utilized from China's modern informatization process, the totality of data source in the current information society and the whole set of data, not only internet data, but also governmental and commercial data [22]. The framework calls for the entire Chinese society to have big data thinking and emphases social data awareness. The traditional Chinese decision making relies on qualitative thinking, not quantitative thinking. Such a cultural behavior has burdened nation's science and technology development. Thus, the Action Framework aims to change the current Chinese culture by enhancing data awareness and promoting data spirit.

Since the release of the Action Framework, the importance of big data has been highly recognized by leaders. An integrated national big data center was proposed as the country seeks to enhance governance capability. Not only has the national level paid attention to big data development, local governments also attach importance to big data increasingly. Up to June 2017, more than 40 provinces and cities issued nearly 100 big data development policies and big data industrial plans. An expert committee is composed of academicians, scholars from scientific research institutes, and representatives of industrial circles. The innovation alliance is made up of more than 70 related entities of the 14 Big Data National Engineering Laboratory. Through these two mechanisms, big data scientists and companies would be gathered contributing to policy making, technology consulting and technology transformation. In addition, a series of open data competitions will be hold in eight National Big Data Comprehensive Test Areas, to promote pubic data opening and encourage innovative applications.

1.3 Big Data Analytics

This subsection presents some fundamental scientific problems in big data analytics. Section 1.3.1 describes an overview of big data analytics based on multiple domains. They include the influences of Management Science on data acquisition and data management, Information Science on data access and processing, Mathematics and Statistics on data understanding, and Engineering on data applications. Section 1.3.2 outlines six open research problems in big data analytics.

1.3.1 Overview of Big Data Analytics

Although there are many different interpretations of big data, big data analytics and data science, one can view that big data is consist of both data science and applications, where big data analytics is an intersection of both. To distinguish these three concepts, the following definition of data science is used in the book:

Definition 1.4 Data Science is mathematical means and algorithm to extract knowledge from big data.

The definition above is very rough, not precise at this point due to the complexity of big data. The boundary of data science for a given filed can be change because the nature of big data in the field differs from others.

With Definitions 1.1–1.4, it can be viewed that if the common-known big data is represented as a set, then data science and application are two subsets while big data analytics is also a subset of big data, across both data science and application since big data analytics is used data science to deal with some specific application problems. A relationship of big data, big data analysis, data science and application can be shown as Fig. 1.4.

In general, the process of big data analytics, as a subset of big data can be shown in Fig. 1.5. It is consisting of several steps, including data acquisition and management, data access and processing, data mining and interpretation, and data applications [23]. However, due to the "4Vs" characteristics of big data, the activities of each step in the process also face Challenges 1.1–1.3. The techniques of multidisciplinary fields need to apply in addressing such challenges.

For Challenge 1.1, majority of big data are represented as semi-structured and non-structured formats. Even though the technologies of MapReduce (Hadoop) can be used to acquire big data, the traditional data acquisition and management of Computer Science should be reinforced by the knowledge of Management Science. For example, the organizational strategy of using big data must be considered before performing the big data acquisition. The basic design of big data base and management should be built up in terms of data capabilities, value, ethic, ownership, policy, quality assurance etc. [15] With help of Management Science, big data can play as an important role for us to make effective decision.

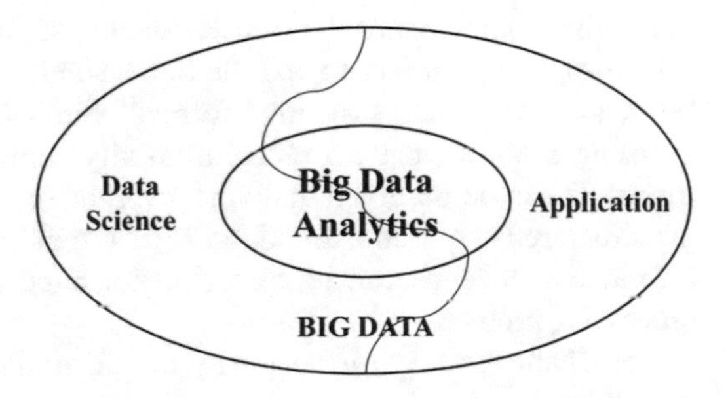

Fig. 1.4 Relationship of big data, big data analytics, data science and application

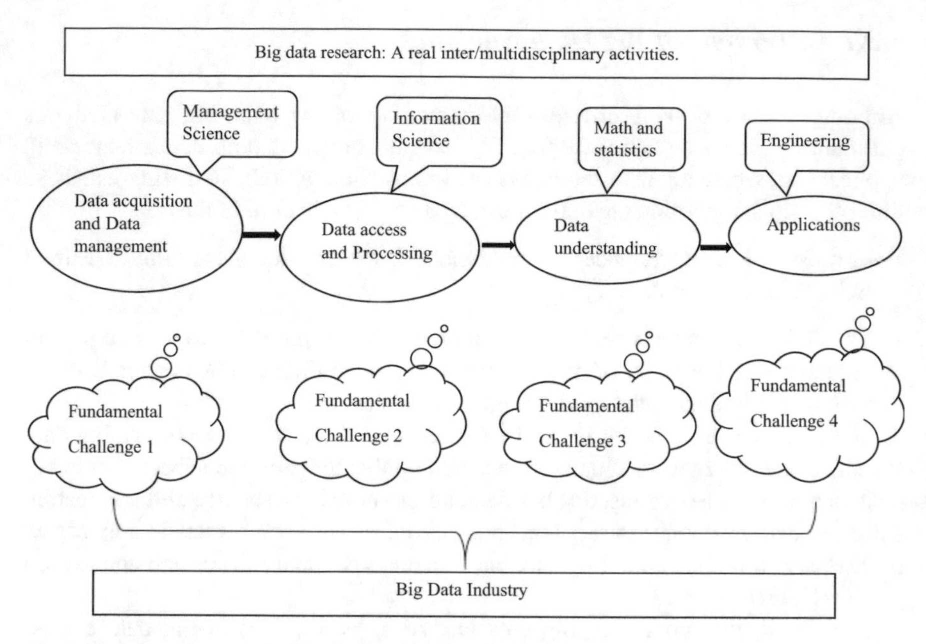

Fig. 1.5 An overview of big data analytics

For Challenge 1.2, the complex formats and features of big data lead the difficulty of assessing, especially processing the data for data mining and interpretation. Many existing techniques of Information Science are ready to respond this challenge. Since most data mining or machine leaning algorithms are constructed to handle structured data, they cannot be used directly analyze a large-scale of semi-structured and non-structured data. Note that the current information technology still lacks the ability of computing big volume of semi-structured and non-structured data, such as clustering millions of text files, images or both in a reasonable time. To do this, we must find a way to transform the semi-structured and non-structured data to structured data or pseudo structured formats which can be analyzed by many known data mining or machine leaning algorithms [18]. This transformation process can be done by using the existing information retrieval algorithms in documents such as for information within documents and metadata about documents and web page. For a given objective of the transformation, some information retrieval algorithm can be applied to turn each text file into a single record with many attributes into a "structured or pseudo structured format". Similarly, an image can be transformed by using a known pattern recognition algorithm as a record of the transformed format. It can be observed that whenever the transformation objective is changed, the structured or pseudo structured format will vary. Therefore, the knowledge of Information Science can be effectively applied to treat the big data access and processing problem.

For Challenge 1.3, it is necessary to utilize rules and principles of Mathematics and Statistics in big data analytics and its interpretation. With the analyzable big data formats, all possible methods in Mathematics and Statistics may be used to conduct Big data analysis. For instance, the modeling methods can include parent space identification and sampling; clustering, classification, regression, prediction and variable selection in data mining methods; relevance analysis, latent variable analytics and statistical inference in analytical methods; and sub sampling, complexity and distributed computation in computation methods. The challenge reflects when and which method is appropriate to be used in a particular case of big data analysis. Because the transformation of Big data is subject to the pre-determined objective, it can be useful to choose a method for data mining or knowledge discovery. Like traditional data mining procedure, experimental design for method choice should be conducted in such Big data mining for most of cases. However, the results of big data mining should be interacted with the user's judgment for the reason which knowledge changes with the individual and situation [16]. In order to let the user has a better understanding of knowledge from big data analysis, different representation or visualization methods, like uniform scheme can be employed to show the simple versions of big data complexity.

In addition, how to use knowledge from big data analysis in the real-life applications is not easy. This perhaps turns to an Engineering problem. Engineering is generally defined as "the application of scientific, economic, social, and practical knowledge in order to invent, design, build, maintain, research, and improve structures, machines, devices, systems, materials and processes" [24]. Use of big data knowledge in most of situations has to do with enhancing the current stages of either scientific, economic, or social conditions. Nowadays every corner and event of our human society depends on big data. Data-driven decision eventually becomes the most reliable approach to any problem. A good engineering design for Big data application will naturally yield the better way to achieve scientific, social and/or economic benefits.

Variety of big data applications can form a new industry, which can be called big data Industry. In such an industry, big data is the input, through big data analytic process mentioned in the above, the output will be data generated knowledge that can be easily turned into products, such as value chain management, business pattern, etc., to create a remarkable productivity.

1.3.2 Some Open Big Data Research Problems

In the notion of the theoretical and technical components of big data, big data analytics has the following open problems.

Problem 1 High Dimensionality
Given a database, when the number of features (p) is far larger than the sample size (n), and n varies with p ($n = n(p)$), the situation is called high dimensionality (HD) problem. When the problem occurs at Big data, $p \gg n(p)$. HD frequently appears in

medical science, such as DNA scanning. In the linear case a basic solution can be shown as:

Consider a linear model as $y = \beta_1 x_1 + \beta_2 x_2 + , \cdots, \beta_p x_p$ for dataset, $D = \{(x_1, y_1), (x_2, y_2), \cdots, (x_n, y_n)\}$. Then, the matrix format can be represented as $Y = X_{n \times p} \beta_{p \times 1}$ and the solution is $\hat{\beta} = (X'X)^{-1} X' Y$.

An asymptotical normality of this is:

$$\sqrt{n}\left(\hat{\beta} - \beta\right) \sim N\left(0, \frac{1}{n}(X'X)^{-1}\sigma^2\right) \xrightarrow{d} N\left(0, \sigma^2 I_{p \times p}\right)$$

There are a number of recent approaches that may categorized as sparse modeling, including compressed sensing, low rank decomposition of matrix and sparse learning to deal with HD problems (for instance, see [25–28]). Some of these developed algorithms are available to be used to handle HD problems in big data. The open research questions for HD problems are how to add priors so that a HD problem can be well defined; and how to find effective sparse modeling, etc. Eventually, systematically solving HD problems need a build-up on the theory and methodology of either HD statistics or HD data mining.

Problem 2 Sub-sampling

The current technologies, like Hadoop system of processing Big data are some types of "divide-and-conquer" schemes, where sub-sampling techniques have been employed. For example, in MapReduce, Map is designed as random sub-sampling of sub datasets with intermediate solutions from given large database, where Reduce is an aggregation process of intermediate solutions for the final estimation of a given database [29]. Although sub-sample is one of the key concepts in big data processing, there are many open questions that need to address so that the more advanced technologies on big data can be developed. For example, how to sub-sampling/aggregate so that the final estimation model of the given original database is properly representing the database? Is the distributed processing feasible? How about traditional sub-sampling/re-sampling technologies working? Are there sub-sampling axioms, such as similarity and transitivity?

Problem 3 Computational Complexity

Traditionally, computational complexity concerns with how difficult a problem can be solved, or how much computation cost must be paid if an algorithm is used to solve a problem.

As an illustration, if a traditional setting can be $R = A(P) := A(D)$, where D is database, A is computation and R is the complexity. Then a Big data setting should be $R_t = A_t(D_t)$, where all D, A and R are changed with time associated with the cost. In this case, the core open questions are how to properly define complexity in big data setting? Is the complexity easy or difficult to measure for a given big data problem? How to establish complexity theory for some specific types of big data problems?

Problem 4 Real and Distributed Computation

Parallel and distributed processing become necessary, perhaps is a unique way of processing for Big data [30]. The main challenges of such a real and distributed

(R/D) computation in handling Big data come from the relationship of three components of Hadoop system: the Hadoop Distributed File System (HDFS) which is a distributed file system designed to run on commodity hardware, HBase which is an open source, non-relational, distributed database, and MapReduce. The quality measurements of Hadoop system for a real and distributed computation include real time, feasibility, efficiency, scalability, etc. It should be noted that some of the measures are conflicted each other and a compromised standard among them is a way to look for a good computational result. There are some open questions in this area. For example, does the R/D computation support fast storage/reading/ranking? For problem of decomposability, can a data modeling problem be decomposed into a series of sub-data set dependent problems? For solution assemblies, how can the solution of a problem be assembled with its sub-solution (component solutions)? When the distributed process is conducted, can the forward and incremental steps be performed by on-line computation?

Problem 5 Unstructured Processing
It has been commonly recognized that structured data are those that can be represented with finite number of rules and can be processed within acceptable time. Otherwise, the data are unstructured (some of them are also called semi-structured), which are difficulty to process (for example, thousands of images or text files). The main challenge of processing unstructured data is that they are multi-sourced and heterogeneous. In most of cases, the understanding of the data is cognition dependent. In this area, the core open questions are how to build a uniform platform on which different types of unstructured data (e.g., mixture of images, text, video and audio) can be processed simultaneously? How to develop the cognition consistent approaches for unstructured data modeling?

Problem 6 Visualization
Using visual-consistent figures or graphics to exhibit the intrinsic structure and patterns in HD big data is challenge visualization analysis. This requires building a basic tool for human-machine interface and expanding applications. For example, by using feature extraction, a HD data space can be transformed into feature space with low dimension (LD), and then by using to visualization techniques, the latter can be turned into visualized space with 2-dimension or 3-dimension. The key concept of judging a good visualization tool is that the end user can easily understand the meaning of big data results without knowing any technical analysis behind. Some current visualization techniques used in showcases, such as The Second Life (http://secondlife.com/) and video games, can be effectively applied to Big data visualization. The core open questions are: is there essential feature extraction of HD data (say, dimension-reduction)? What is structured representation of imaginable thinking? How to construct appropriate visualized space? How to map a problem in feature space (or data space) to a representation problem in visualized space? [31].

Big data analytics is still a very pre-mature field at this point. Fundamentally speaking, in order to conduct an applicable big data analysis, one should think about how to design big data analytic algorithm structure. Here are some ideas open to be discussed. First, a big data analytic algorithm should be an algorithm that can process and analyze big data under available computational resource and complete in a reasonable time. The big data can be handled by it has at least one of following characteristics: large-size, heterogeneous, distributed, multi-sources, data steam, high-dimension, and high-uncertainty. The algorithm can be performed at appropriate degree of time, storage and communication complexity. It also has some unique properties, such as highly fault toleration, solution integration and assembled capability. Second, the key ideas of designing a big data analytic algorithm could include maintaining the proper ratio of data sample and population; simple modeling and simple procedure; inferior preciseness, complex inherence and theory based. Finally, in addition to well-known statistics or data mining methods, other computational methods, such as set-based processing, stochastic computing, online computing, distributed/parallel computing, cloud computing may be employed to construct a high-efficient big data analytic algorithm. These concepts and discussions about big data and big data analytics have been implemented in the following chapters of this book.

Looking around the world, big data development is just at the beginning. Big data is treasure created by the people and should be used to benefit the people. Even the precise meaning of big data analytics is not clear yet, the data scientists and engineers should figure out the fundamental issues of big data which may lead a context of data science. The advancement of data science will provide more theoretical findings and creative or innovative techniques to support the big data development into the future.

References

1. Tuitt, D.: A history of big data. HCL Technologies Blogs. http://www.hcltech.com/blogs/transformation-through-technology/history-big-data (2012)
2. EMC2: Digital universe study: extracting value from chaos. http://www.emc.com/leadership/programs/digital-universe.htm (2011)
3. Shaw, J.: Why "big data" is a big deal. Harvard Business Review, March–April. http://harvardmagazine.com/2014/03/why-big-data-is-a-big-deal (2014)
4. Chen, H., Chiang, R.H.L., Storey, V.: Business intelligence and analytics: from big data to big import. MIS Q. **36**(4), 1165–1188 (2012)
5. NSF: Core Techniques and Technologies for Advancing Big Data Science & Engineering (BIGDATA). National Science Foundation. http://www.nsf.gov/pubs/2012/nsf12499/nsf12499.htm (2012)
6. XSSC The 462nd Session: Data Science and Big Data of Xiangshan Science Conferences. http://www.xssc.ac.cn/xs/showconf.asp?tid=4&pid=342 (2013)
7. Laney, D.: The Importance of "Big Data": A Definition. A Gartner Co. Report (2012)
8. Villanova University: What is big data? http://www.villanovau.com/university-online-programs/what-is-big-data/ (2014)
9. Bayes, T., Price, R.: An essay towards solving a problem in the doctrine of chances. Philos. Trans. R. Soc. Lond. **53**, 370–418 (1763)
10. Fayyad, U.M., Piatetsky, S.G., Smyth, P.: From data mining to knowledge discovery: an overview. In: Fayyad, U.M., Piatetsky, S.G., Smyth, P., Uthurusamy, R. (eds.) Advances in

Knowledge Discovery and Data Mining, pp. 1–34. AAAI Press/The MIT Press, Menlo Park (1996)
11. Mayer-Schönberger, V., Cukier, K.: Big Data: A Revolution That Will Transform How We Live, Work, and Think. Houghton Mifflin Harcourt, New York, NY (2013)
12. Blind men and an elephant. http://en.wikipedia.org/wiki/Blind_men_and_an_elephant (2014)
13. Zhang, L., Li, J., Shi, Y., Liu, X.: Foundations of intelligent knowledge management. J. Human Syst. Manag. **28**(4), 145–161 (2009)
14. Ouyang, Z.B., Shi, Y.: A fuzzy clustering algorithm for petroleum data. In: WI-IAT '11 Proceedings of the 2011 IEEE/WIC/ACM International Conferences on Web Intelligence and Intelligent Agent Technology, vol. 03, pp. 233–236 (2011)
15. Laudon, K.C., Laudon, J.P.: Management Information Systems. Pearson, Upper Saddle River, NJ (2012)
16. Shi, Y., Zhang, L.L., Tian, Y.J., Li, X.S.: Intelligent Knowledge: A Study beyond Data Mining. Springer, New York (2015)
17. G8 report on open data charter and technical annex. https://www.gov.uk/government/publications/open-data-charter/g8-open-data-charter-and-technical-annex (2014)
18. Shi, Y.: Big data: history, current status, and challenges going forward. The Bridge. **44**(4), 6–11 (2014)
19. Zhong, Y., Shi, Y., Jing, X.: Global big data development analysis report 2018. A report by Tianfu Institute of International Strategy and Technology, Chengdu, China (2019)
20. Action Framework for Promoting the Development of Big Data (AFPDBD). www.gov.cn (in Chinese) (September 2015)
21. Shi, Y., Shan, Z., Li, J., Fang, Y.: How China deals with big data. Ann. Data Sci. **4**, 433–440 (2017)
22. Shan, Z.: Interpretation on action framework for promoting big data. http://news.xinhuanet.com/info/2015-09/17/c_134632375.htm (2015) (in Chinese)
23. Xu, Z., Shi, Y.: Exploring big data analysis: fundamental scientific problems. Ann. Data Sci. **2**, 363–372 (2015)
24. Wikipedia: Engineering. http://en.wikipedia.org/wiki/Engineering (2020)
25. Chang, X., Wang, Y., Li, R., Xu, Z.: Sparse K-means with L∞/Lo penalty for high-dimensional data clustering. arxiv.org/pdf/1403.7890 (2014)
26. Donoho, D.L.: For most large underdetermined systems of linear equations the minimal L1-norm solution is also the sparsest solution. Commun Pure Appl Math. **56**(6), 797–829 (2006)
27. Kriegel, H.P., Kröger, P., Renz, M., Wurst, S.: A generic framework for efficient subspace clustering of high-dimensional data. In: IEEE International Conference on Data Mining (ICDM), Houston, Texas, USA, pp. 205–257 (2005)
28. Wang, Y., Chang, X., Li, R., Xu, Z.: Sparse K-means with Lq (0<=q<1) constrain for high-dimensional data clustering. In: IEEE International Conference on Data Mining, Dallas, USA (2013)
29. Kleiner, A., Talwalkar, A., Sarkar, P.: The big data bootstrap. In: The 29th International Conference on Machine Learning, Edinburgh, Scotland, UK (2012)
30. Xu, Z., Leung, K.S., Liang, Y., Leung, Y.: Efficiency speed-up strategies for evolutionary computation: fundamentals and fast-Gas. Appl. Math. Comput. **142**(2, 3), 341–388 (2003)
31. Zheng, Y., Zhang, C., Xie, W., Xie, X., Sun, G., Huang, Y.: T-Drive: driving directions based on taxi trajectories. In: ACM SIGSPATIAL GIS (2010)

Chapter 2
Multiple Criteria Optimization Classification

As the increasingly strong computational power of computers fills the shortage of human brain at calculating, data mining, a major component of data science, has emerged as the times require due to its merit of being capable of extracting novel and useful knowledge which has potential value from large scale of complex data. However, from the mathematical perspective, some data mining methods, such as decision tree, genetic algorithm, and association rules could be considered as heuristic algorithms: which means to select a "better solution" from several alternative solutions as the criterion of classification. These methods lack of exploring how to locate the "best solution" systematically.

Based on [1] and [2], this chapter describes the advanced techniques of applying multi-criteria decision making methods and multi-criteria mathematical programming to conducting data mining process for selecting the "best solution" from multiple alternative solutions, instead of using heuristic algorithms. Section 2.1 is Multi-Criteria Linear Programming (MCLP) for supervised learning, which includes error correction method in classification by using Multiple-Criteria and Multiple-Constraint Levels Linear Programming (MC2LP) [3], Multi-Instance classification based on regularized Multiple Criteria Linear Programming (RMCLP) [4], supportive instances for RMCLP classification [5], and kernel based simple RMCLP for binary classification and regression [6]. Then, Sect. 2.2 describes a group of knowledge-incorporated MCLP classifier [7] and decision rule extraction for RMCLP model [1]. Finally, Sect. 2.3 summarizes three methods of MCDM based data analytics. They are a MCDM approach for estimating the number of clusters [8], parallel RMCLP classification algorithm [9], and an effective intrusion detection framework based on MCLP and support vector machine.

Y. Shi, *Advances in Big Data Analytics*,
https://doi.org/10.1007/978-981-16-3607-3_2

2.1 Multi-criteria Linear Programming for Supervised Learning

2.1.1 Error Correction Method in Classification by Using Multiple-Criteria and Multiple-Constraint Levels Linear Programming

First, the MCLP model for classification is outlined as below [10, 11]:

Given a set of n variables about the records $X^T = (x_1, x_2, \ldots, x_l)$, and then let $x_i = (x_{i1}, x_{i2}, \ldots, x_{in})^T$ be one sample of data, where $i = 1, 2, \ldots, l$ and l is the sample size. In linear discriminant analysis, data separation can be achieved by two opposite objectives, that is, minimizing the sum of the deviations (MSD) and maximizing the minimum distances (MMD) of observations from the critical value. That is to say, in order to solve classification problem, we need to minimize the overlapping of data, i.e. α, at the same time, to maximize the distances from the well classified points to the hyperplane, i.e. β.

However, it is difficult for traditional linear programming to optimize MMD and MSD simultaneously. According to the concept of Pareto optimality, we can check all the possible trade-offs between the objective functions by using multiple-criteria linear programming algorithm. The MCLP model can be described by Fig. 2.1.

Moreover, the first Multiple Criteria Linear Programming (MCLP) model can be described as follows:

$$
\begin{aligned}
&\min \sum_i \alpha_i \\
&\min \sum_i \beta_i \\
s.t.\ &A_i X = b + \alpha_i - \beta_i,\ A_i \in Bad, \\
&A_i X = b + \alpha_i - \beta_i,\ A_i \in Good, \\
&\alpha_i, \beta_i \geq 0, i = 1, 2, \ldots, l
\end{aligned}
$$

Fig. 2.1 MCLP model

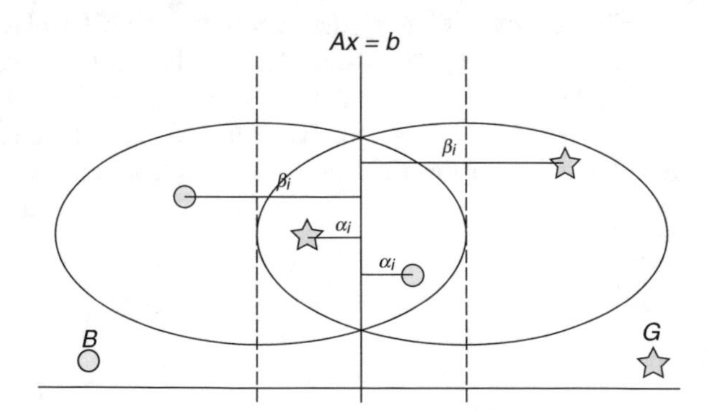

Here, α_i is the overlapping and β_i is the distance from the training sample xi to the discriminator (w $\cdot$ x_i) $= b$ (classification separating hyperplane).

Then, the MC2LP model for classification is introduced in [10].

According to the discussion above, a non-fixed b is very important to our problem. At the same time, for the simplicity and existence of the solution, b should be fixed in some interval.

As a result, for different data, we fix b in different pairs of intervals $[b_l, b_u]$, where b_l and b_u are two fixed numbers. Now our problem is to search the best cutoff between b_l and b_u at every level of their tradeoffs, that is to say, to test every point in the interval $[b_l, b_u]$. We keep the multiple-criteria the same as MCLP, which is, MMD and MSD. And then, the following model is posed:

$$\begin{aligned} &\min \sum_i \alpha_i \\ &\min \sum_i \beta_i \\ s.t.\, &A_i X = [b_l, b_u] + \alpha_i - \beta_i,\ A_i \in Bad, \\ &A_i X = [b_l, b_u] + \alpha_i - \beta_i,\ A_i \in Good, \\ &\alpha_i, \beta_i \geq 0, i = 1, 2, \ldots, l \end{aligned}$$

where A_i, b_l and b_u are given, and X is unrestricted.

In the model, $[b_l, b_u]$ represents a certain tradeoff in the interval. By virtue of the technical of Multiple-criteria and multiple-constraint levels linear programming (MC2LP), we can test each tradeoff between the multiple-criteria and multiple-constraint levels as follows:

$$\begin{aligned} &\min \lambda_1 \sum_i \alpha_i - \lambda_2 \sum_i \beta_i \\ s.t.\, &A_i X = \gamma_1 b_l + \gamma_2 b_u + \alpha_i - \beta_i,\ A_i \in Bad, \\ &A_i X = \gamma_1 b_l + \gamma_2 b_u + \alpha_i - \beta_i,\ A_i \in Good, \\ &\alpha_i, \beta_i \geq 0, i = 1, 2, \ldots, l \end{aligned}$$

Here, the parameters of $\lambda \times \gamma$ are fixed for each programming problem. Moreover, the advantage of MC2LP is that it can find the potential solutions for all possible trade-offs in the parameter space systematically [12, 13] where the parameter space is

$$\{(\lambda, \gamma) \,|\lambda_1 + \lambda_2 = 1, \gamma_1 + \gamma_2 = 1\}.$$

Of course, in this model, choosing a suitable pair for the goal problem is a key issue and needs domain knowledge. Consequently, a non-parameter choosing MC2LP method should be posed.

For the original MCLP model, one cutoff b is used to predict a new sample's class, that is to say, there is only one hyperplane. The former MC2LP model points out that we can define two cutoffs b_l and b_u instead of the original single cutoff. And then a systematical method can be used to solve this problem. Consequently, all potential solutions at each constrain level tradeoff can be acquired. However, one problem is how to find the cutoffs b_l and b_u.

On one hand, we utilize two cutoffs to discover the solution of higher accuracy; on the other hand, we hope the cutoffs can be obtained from the system directly. Inspired by the idea above, we address our first MC2LP model, which solves the classification problem twice.

For the first step, MCLP model is used to find the vector of external deviations α. It is a function of λ. For simplicity, we set $b = 1$. And then, we fix the parameter of λ to get one potential solution. Now a non-parameter vector of external deviations α is acquired. The component $(\alpha_i > 0)$ means the corresponding sample in the training set is misclassified. In other words, Type I and Type II errors occur. According to the idea of MC2LP, we can detect the result of every single MCLP by fixing the parameter of γ at each level in the interval $[b_l, b_u]$. Now, we find the maximal component of α:

$$\alpha_{\max} = \max\{\alpha_i, 1 \le i \le l\}. \tag{2.1}$$

Indeed, the smaller the weight of external deviations is, the bigger α_{max} is.

The misclassified samples are all projected into the interval $[1 - \alpha_{max}, 1 + \alpha_{max}]$ according to the weight vector X obtained from the MCLP model. In this way, we define b_l and b_u as $1 - \alpha_{max}$ and $1 + \alpha_{max}$, respectively. It is easy to see, if we want to lessen the number of two types of error, in effect, we just need to inspect the cutoffs by altering the cutoff in the interval

$$[1 - \alpha_{max}, 1 + \alpha_{max}].$$

Moreover, for the second step, a new MC2LP classification model can be stated as follows:

$$\begin{aligned}
&\min \lambda_1 \sum_i \alpha_i - \lambda_2 \sum_i \beta_i \\
&s.t. A_i X = [1 - \alpha_{\max}, 1 + \alpha_{\max}] + \alpha_i - \beta_i,\ A_i \in Bad, \\
&\quad A_i X = [1 - \alpha_{\max}, 1 + \alpha_{\max}] + \alpha_i - \beta_i,\ A_i \in Good, \\
&\quad \alpha_i, \beta_i \ge 0, i = 1, 2, \ldots, l
\end{aligned}$$

where A_i, α_{max} are given, and X is unrestricted, $[1 - \alpha_{max}, 1 + \alpha_{max}]$ means a certain tradeoff in the interval. At the same time, $\lambda = (\lambda_1, \lambda_2)$ is the parameter chosen in the first step.

The most direct modification of the new MC2LP model is to transfer the single objective function to be a multiple-criteria one. Because the vector of external deviations is a function of λ, it is easy to observe that if the weight between external deviations and internal deviations changes, α changes. Consequently, α_{max} alters. And the ideal α is the one that makes α_{max} not too huge. In other words, we do not hope to check the weight that satisfies λ_1 not too small. Actually, some papers have proved that only if $\lambda_1 > \lambda_2$, then $\alpha \cdot \beta = 0$, which makes the model meaningful [14]. As a result, we only need to check the parameters of objective functions that make α_{max} not too big, in short, not too far away from the original one.

On the other hand, we expect α_{max} not too small. That is to say, we hope the model has some generalization. Hence, two small positive numbers ϵ_1 and ϵ_2 are chosen manually. And then, the interval is built as $[(1 - \alpha_{max} - \epsilon_1, 1 - \alpha_{max} + \epsilon_1),$ $(1 + \alpha_{max} - \epsilon_2, 1 + \alpha_{max} + \epsilon_2)]$. This means that the lower and the upper bound of the interval should be trade-off of some intervals, i.e. the multiple-constrained levels are actually multiple-constrained intervals. Indeed, checking every tradeoff of the intervals is the same as checking every tradeoff of $1 - \alpha_{max} - \epsilon_1$ and $1 + \alpha_{max} + \epsilon_2$. In this case, we can consider the objective function as a multiple-criteria one. It can be stated as follows:

$$
\begin{aligned}
&\min \sum_i \alpha_i \\
&\min \sum_i \beta_i \\
s.t.\, &A_i X = [1 - \alpha_{\max} - \varepsilon_1, 1 + \alpha_{\max} + \varepsilon_2] + \alpha_i - \beta_i,\ A_i \in Bad, \\
&A_i X = [1 - \alpha_{\max} - \varepsilon_1, 1 + \alpha_{\max} + \varepsilon_2] - \alpha_i + \beta_i,\ A_i \in Bad, \\
&\alpha_i, \beta_i \geq 0, i = 1, 2, \ldots, l
\end{aligned}
\tag{2.2}
$$

where A_i, α_{max}, ϵ_1 and ϵ_2 are given, and X is unrestricted. Here, ϵ_1 and ϵ_2 are two nonnegative numbers.

Lemma 2.1 *For certain trade-off between the objective functions, if b is maintained to be the same sign, then hyperplanes, which are obtained in the MCLP model, keep the same. Furthermore, different signs result in different hyperplanes.*

Proof Assume that the tradeoff between the objective functions is $\lambda = (\lambda_1, \lambda_2)$ and X_1 is the solution obtained by fixing b to be 1. Then, set b_1 as an arbitrary positive number. The MCLP model can be transformed as follows:

$$
\begin{aligned}
&\min \lambda_1 \sum_i \alpha_i - \lambda_2 \sum_i \beta_i \\
s.t.\, &A_i X = b_1 + \alpha_i - \beta_i,\ A_i \in Bad, \\
&A_i X = b_1 - \alpha_i + \beta_i,\ A_i \in Good, \\
&\alpha_i, \beta_i \geq 0, i = 1, 2, \ldots, l
\end{aligned}
$$

The problem above is the same as:

$$\min \lambda_1 \frac{\sum_i \alpha_i}{b_1} - \lambda_2 \frac{\sum_i \beta_i}{b_1}$$
$$s.t. A_i \frac{X}{b_1} = 1 + \frac{\alpha_i}{b_1} - \frac{\beta_i}{b_1}, A_i \in Bad,$$
$$A_i \frac{X}{b_1} = 1 - \frac{\alpha_i}{b_1} + \frac{\beta_i}{b_1}, A_i \in Good,$$
$$\alpha_i, \beta_i \geq 0, i = 1, 2, \ldots, l$$

And then, we let $\alpha i' = \frac{\alpha_i}{b_1}$, $\beta i' = \frac{\beta_i}{b_1}$, $X' = \frac{X}{b_1}$. It is obvious that the solution is $X' = \frac{X}{b_1}$ and the hyperplane $AX' = b_1$ is the same as $AX_1 = 1$.

Similarly, we can prove that when b is a negative number, the solution is the same as the one that is obtained from $b = -1$.

As a result, we just need to compare the solutions (hyperplanes) resulted from $b = 1$ and $b = -1$. For this case, it is easy to see that the signs before α_i and β_i swap when we transform $b = 1$ into $b = 1$. If this happens, then the objective function changes into $-\lambda_1 \sum_i \alpha_i + \lambda_2 \sum_i \beta_i$. This means that the solutions will be different.

According to the lemma, we have the theorem below:

Theorem 2.1 *For our MC2LP model (2.2) above, according to the solutions (hyperplanes), space γ is divided into two non-intersect parts.*

Remark 2.1 When $[1 - \alpha_{max}, 1 + \alpha_{max}]$ is achieved, ϵ_1 and ϵ_2 are chosen to satisfy that 0 is contained by the interval $[1 - \alpha_{max} - \epsilon_1, 1 + \alpha_{max} + \epsilon_2]$. In this case, for any λ, the solutions belong to the trade-offs with same sign will result in the same hyperplane. In other words, there are only two different hyperplanes corresponding to model (2.2). In short, the flexibility of model (2.2) is limited.

In many classification models, including original MCLP model, two types of error is a big issue. In credit card account classification, to correct two types of error can not only improve the accuracy of classification but also help to find some important accounts.

Accordingly, many researchers have focused on this topic. Based on this consideration, more attention should be paid to the samples that locate between two hyperplanes acquired by the original MCLP model, that is, the points in the grey zone [15]. Consequently, we define the external deviations and internal deviations related to two different hyperplanes, the left one and the right one, that is, α^l, α^r, β^l and β^r.

Definition 2.1 The conditions the deviations should satisfy are stated as follows:

$$\alpha_i^l = \begin{cases} 0, & A_i X < 1 - \alpha_{max} and A_i \in Bad; \\ A_i X - (1 - \alpha_{max}), & A_i X \geq 1 - \alpha_{max} and A_i \in Bad; \\ 0, & A_i X \geq 1 - \alpha_{max} and A_i \in Good; \\ (1 - \alpha_{max}) - A_i X, & A_i X < 1 - \alpha_{max} and A_i \in Good. \end{cases}$$

$$\alpha_i^r = \begin{cases} 0, & A_i X < 1 + \alpha_{max} and A_i \in Bad; \\ A_i X - (1 + \alpha_{max}), & A_i X \geq 1 + \alpha_{max} and A_i \in Bad; \\ 0, & A_i X \geq 1 + \alpha_{max} and A_i \in Good; \\ (1 + \alpha_{max}) - A_i X, & A_i X < 1 + \alpha_{max} and A_i \in Good. \end{cases}$$

$$\beta_i^l = \begin{cases} (1 - \alpha_{max}) - A_i X, & A_i X < 1 - \alpha_{max} and A_i \in Bad; \\ 0, & A_i X \geq 1 - \alpha_{max} and A_i \in Bad; \\ A_i X - (1 - \alpha_{max}), & A_i X \geq 1 - \alpha_{max} and A_i \in Good; \\ 0, & A_i X < 1 - \alpha_{max} and A_i \in Good. \end{cases}$$

$$\beta_i^r = \begin{cases} (1 + \alpha_{max}) - A_i X, & A_i X < 1 + \alpha_{max} and A_i \in Bad; \\ 0, & A_i X \geq 1 + \alpha_{max} and A_i \in Bad; \\ A_i X - (1 + \alpha_{max}), & A_i X \geq 1 + \alpha_{max} and A_i \in Good; \\ 0, & A_i X < 1 + \alpha_{max} and A_i \in Good. \end{cases}$$

Figure 2.2 is a sketch for the model. In the graph, the green and the red lines are the left and right hyperplane, b^l and b^r respectively, which are some trade-offs in two intervals, i.e. $[1 - \alpha_{max} - \epsilon_2, 1]$ and $[1, 1 + \alpha_{max} + \epsilon_1]$. And all

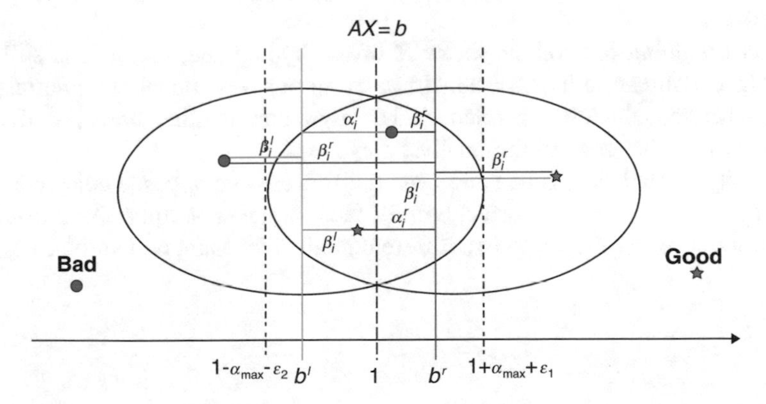

Fig. 2.2 MC2LP model

the deviations are measured according to them in different colors. For instance, if a sample in "Good" class is misclassified as "Bad" class, it means ${\alpha_i}^r > {\beta_i}^l \geq 0$ and ${\alpha_i}^l = {\beta_i}^r = 0$. And then, if a sample in "Bad" class is misclassified as "Good" class, it means ${\alpha_i}^l > {\beta_i}^r \geq 0$ and ${\alpha_i}^r = {\beta_i}^l = 0$. Thus, for the misclassified ones, ${\alpha_i}^r + {\alpha_i}^l - {\beta_i}^r - {\beta_i}^l$ should be minimized.

As a result, a more meticulous model could be stated as follows:

$$
\begin{aligned}
&\min \sum_i (\alpha_i^r + \alpha_i^l)\\
&\min \sum_i (\alpha_i^l - \beta_i^r)\\
&\min \sum_i (\alpha_i^r - \beta_i^l)\\
&\max \sum_i (\beta_i^r + \beta_i^l)\\
s.t.\, &A_i X = 1 + [0, \alpha_{\max} + \varepsilon_1] + \alpha_i^r - \beta_i^r,\ A_i \in Bad,\\
&A_i X = 1 - [0, \alpha_{\max} + \varepsilon_2] + \alpha_i^l - \beta_i^l,\ A_i \in Bad,\\
&A_i X = 1 + [0, \alpha_{\max} + \varepsilon_1] - \alpha_i^r + \beta_i^r,\ A_i \in Good,\\
&A_i X = 1 - [0, \alpha_{\max} + \varepsilon_2] - \alpha_i^l + \beta_i^l,\ A_i \in Good,\\
&\alpha_i^r, \alpha_i^l \beta_i^r, \beta_i^l \geq 0,\ i = 1, 2, \ldots, l.
\end{aligned}
$$

where A_i, α_{max}, $\epsilon_1 > 0$, $\epsilon_2 > 0$ are given, and X is unrestricted.

In Fig. 2.2, for each point, there are at most two kinds of deviations nonzero. The objective functions appear to deal with the deviations according to the position shown in Fig. 2.2, respectively, whereas they have their own special meaning. That is to say, it measures two types of error in some degree by means of the second and third objective functions. As a result, in this new version of MC2LP, we not only consider the deviations respectively, but also take the relationship of the deviations based on two types of error into account in the objective functions. By virtue of MC2LP method, each tradeoff between $1 - \alpha_{max} - \epsilon_2$ and 1 for the left hyperplane as well as each tradeoff between 1 and $1 + \alpha_{max} + \epsilon_1$ for the right hyperplane can be checked.

After obtaining the weight vector X of the hyperplane, $AX = 1$ is still used to be the classification hyperplane. However, in our new model, we minimize the distance between the left hyperplane and the right one. In other words, we discover the hyperplane that genders the smallest grey area.

Actually, in statistics, Type I and Type II errors are two opposite objectives. That is to say, it is very hard to correct both of them at the same time. As a result, we modify the former model into two different models focusing on two types of error respectively as follows:

$$\begin{aligned}
&\min \sum_i (\alpha_i^r + \alpha_i^l)\\
&\min \sum_i (\alpha_i^l - \beta_i^r)\\
&\max \sum_i (\beta_i^r + \beta_i^l)\\
s.t.A_iX &= 1 + [0, \alpha_{\max} + \varepsilon] + \alpha_i^r - \beta_i^r, A_i \in Bad,\\
A_iX &= 1 + \alpha_i^l - \beta_i^l, A_i \in Bad,\\
A_iX &= 1 + [0, \alpha_{\max} + \varepsilon] - \alpha_i^r + \beta_i^r, A_i \in Good,\\
A_iX &= 1 - \alpha_i^l + \beta_i^l, A_i \in Good,\\
&\alpha_i^r, \alpha_i^l \beta_i^r, \beta_i^l \geq 0, i = 1, 2, \ldots, l.
\end{aligned} \tag{2.3}$$

where A_i, α_{max} and $\epsilon > 0$ are given, and X is unrestricted. In this model, $\sum_i \alpha_i^r - \beta_i^l$ is not contained in the objective functions. This model can deal with Type II error, that is, classifying a "Good" point to be a "Bad" one. Now we provide an example to illustrate the effect of model (2.2).

As the result shown above, model (2.3) can correct Type II error in some degree. We conclude this in the proposition below.

Proposition 2.1 *Model (2.3) can correct Type II error by moving the right hyperplane to the right based on the concept of multiple-constraint levels.*

Note that the second objective function in model (2.3) is nonzero for the samples in class "Bad" and getting negative when the right hyperplane moving to the right. That is to say, we tolerate some Type I errors. At the same time, the first objective function in model (2.3) renders Type II errors an increasing punishment with moving the right hyperplane to the right. As a result, it can correct Type II error in some degree.

Similar to model (2.3), (2.4) is posed to deal with Type I error as follows:

$$\begin{aligned}
&\min \sum_i (\alpha_i^r + \alpha_i^l)\\
&\min \sum_i (\alpha_i^l - \beta_i^r)\\
&\min \sum_i (\beta_i^r + \beta_i^l)\\
s.t. \quad A_iX &= 1 + \alpha_i^r - \beta_i^r, A_i \in Bad,\\
A_iX &= 1 - [0, \alpha_{\max} + \varepsilon_2] + \alpha_i^l - \beta_i^l, A_i \in Bad,\\
A_iX &= 1 - \alpha_i^r + \beta_i^r, A_i \in Good,\\
A_iX &= 1 - [0, \alpha_{\max} + \varepsilon_2] - \alpha_i^l + \beta_i^l, A_i \in Good,\\
&\alpha_i^r, \alpha_i^l \beta_i^r, \beta_i^l \geq 0, i = 1, 2, \ldots, l.
\end{aligned} \tag{2.4}$$

where A_i, α_{max} and $\epsilon > 0$ are given, and X is unrestricted. In this model, $\sum_i \alpha_i^l - \beta_i^r$ is not contained in the objective functions. This model focuses on Type I error, that is, classifying a "Bad" point to be a "Good" one.

The numerical examples to illustrate the theoretical results of this section can be found in [3].

2.1.2 Multi-instance Classification Based on Regularized Multiple Criteria Linear Programming

Multi-instance learning (MIL) has received intense interest recently in the field of machine learning. This idea was originally proposed for handwritten digit recognition by [16]. The term multi-instance learning was first introduced by [17] when they were investigating the problem of binding ability of a drug activity prediction. In MIL framework, the training set consists of positive and negative bags of points in the n-dimensional real-space R^n, and each bag contains a number of points (instances). A positive training bag contains at least one positive instance, whereas a negative bag contains only negative instances. The aim of MIL is to construct a learned classifier from the training set for correctly labeling unseen bags. Multi-instance learning has been found useful in diverse domains such as object detection, text categorization, image categorization, image retrieval, web mining, computer-aided medical diagnosis, etc. [12–14, 18].

In this subsection, we propose a novel Multi-instance Learning method based on Regularized Multiple Criteria Linear Programming (called MI-RMCLP), which includes two algorithms for linear and nonlinear cases separately. To our knowledge, MI-RMCLP is the first RMCLP implementation based on MIL, which is a useful extension of RMCLP. The original MI-RMCLP model proposed itself is a nonconvex optimization problem. By an appropriate modification, we will the model to derive two quadratic programming subproblems, which can arrive at the optimal value by an iterative strategy solving these sequential subproblems. All preliminary numerical experiments show that our approach is competitive with other multiple learning formulations.

We first give a brief introduction of RMCLP in the following. For classification about the training data:

$$T = \{(x_1, y_1), \cdots, (x_l, y_l)\} \in \left(R^n \times y\right)^l,$$

where $x_i \in R^n$, $y_i \in \ = \{1, -1\}$, $i = 1, \cdots, l$, data separation can be achieved by two opposite objectives. The first objective separates the observations by minimizing the sum of the deviations (MSD) among the observations. The second maximizes the minimum distances (MMD) of observations from the critical value [19]. The overlapping of data u should be minimized, while the distance v has to be maximized. However, it is difficult for traditional linear programming to optimize MMD and MSD simultaneously. According to the concept of Pareto optimality, we can seek the best trade-off of the two measurements [2, 20]. So MCLP model can be described as follows:

$$\min_{u} e^T u \& \max_{v} e^T v, \tag{2.5}$$

$$s.t. (w \cdot x_i) + (u_i - v_i) = b, \text{ for } \{i \,|y_i = 1\}, \tag{2.6}$$

$$(w \cdot x_i) - (u_i - v_i) = b, \text{ for } \{i \,|y_i = -1\}, \tag{2.7}$$

$$u, v \geq 0, \tag{2.8}$$

where $e \in R^l$ be vector whose all elements are 1, w and b are unrestricted, u_i is the overlapping, and v_i the distance from the training sample x_i to the discriminator $(w \cdot x_i) = b$ (classification separating hyperplane). By introducing penalty parameter $c, d > 0$, MCLP has the following version

$$\min_{u,v} c e^T u - d e^T v, \tag{2.9}$$

$$s.t. (w \cdot x_i) + (u_i - v_i) = b, \text{ for } \{i \,|y_i = 1\}, \tag{2.10}$$

$$(w \cdot x_i) - (u_i - v_i) = b, \text{ for } \{i \,|y_i = -1\}, \tag{2.11}$$

$$u, v \geq 0, \tag{2.12}$$

The geometric meaning of the model is shown in Fig. 2.3.

A lot of empirical studies have shown that MCLP is a powerful tool for classification. However, we cannot ensure this model always has a solution under different kinds of training samples. To ensure the existence of solution, recently, Shi et al. proposed a RMCLP model by adding two regularized items $\frac{1}{2} w^T H w$ and

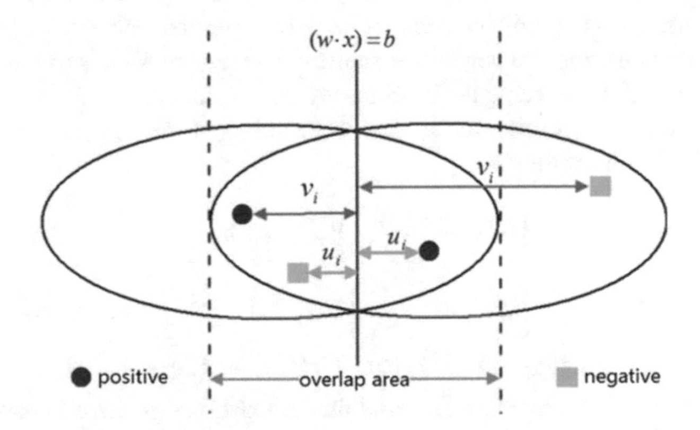

Fig. 2.3 Geometric meaning of MCLP

$\frac{1}{2}u^T Qu$ on MCLP as follows (more theoretical explanation of this model can be found in [2]):

$$\min_z \frac{1}{2} w^T Hw + \frac{1}{2} u^T Qu + de^T u - ce^T v, \tag{2.13}$$

$$s.t. (w \cdot x_i) + (u_i - v_i) = b, \text{for } \{i \,|y_i = 1\}, \tag{2.14}$$

$$(w \cdot x_i) - (u_i - v_i) = b, \text{for } \{i \,|y_i = -1\}, \tag{2.15}$$

$$u, v \geq 0, \tag{2.16}$$

where $z = (w^T, u^T, v^T, b)^T \in R^{n+l+l+1}$, $H \in R^n \times n$, $Q \in R^{l \times l}$ are symmetric positive definite matrices. Obviously, the regularized MCLP is a convex quadratic programming.

Compared with traditional SVM, we can find that the RMCLP model is similar to the Support Vector Machine model in terms of the formation by considering minimization of overlapping of the data. However, RMCLP tries to measure all possible distances v from the training samples x_i to separating hyperplane, while SVM fixes the distance as 1 (through bounding planes $(w \cdot x) = b \pm 1$) from the support vectors. Although the interpretation can vary, RMCLP addresses more control parameters than the SVM, which may provide more flexibility for better separation of data under the framework of the mathematical programming. In addition, different with SVM, RMCLP considers all the samples to solve classification problem. These make RMCLP have stronger insensitivity to outliers.

One of the drawbacks of applying the supervised learning model is that it is not always possible for a teacher to provide labeled examples for training. Multiple instance learning (MIL) provides a new way of modeling the teachers' weakness. MIL considers a particular form of weak supervision in which training class labels are associated with sets of patterns, or bags, instead of individual patterns. A negative bag only consists of negative instances, whereas a positive bag comprises both positive and negative instances. The goal of MIL is to find a separate hyperplane, which can decide the label of any new instance.

In the following, we give the formal description of multiple instance learning problem. Given a training set

$$\{\mathbf{B}_1^+, \cdots, \mathbf{B}_{m^+}^+, \mathbf{B}_1^-, \cdots, \mathbf{B}_{m^-}^-\} \tag{2.17}$$

where a bag $\mathbf{B}_i^+ = \left\{x_{i1}, \cdots, x_{im_i^+}\right\}, x_{ij} \in R^n, j = 1, \cdots, m_i^+, i = 1, \cdots, m^+; \mathbf{B}_i^- = \left\{x_{i1}, \cdots, x_{im_i^-}\right\}, x_{ij} \in R^n, j = 1, \cdots, m_i^-, i = 1, \cdots, m^-$; $\mathbf{B}^+$ means that the positive bag $\mathbf{B}^+$ contains at least one positive instance x_{ij}; $\mathbf{B}^-$ means that all instance x_{ij} of the negative bag $\mathbf{B}^-$ are negative. The goal is to induce a real-valued function

$$y = \text{sgn}\left(\mathbf{g}(x)\right) \tag{2.18}$$

such that the label of any instance x in R^n space can be predicted.

Now we rewrite the training set (2.17) as

$$\text{Train} = \left\{\mathbf{B}_1^+, \cdots, \mathbf{B}_{m^+}^+, \mathbf{B}_{m^++1}^-, \cdots, \mathbf{B}_{m^++m^-}^-, \right\} = \left\{\mathbf{B}_1^+, \cdots, \mathbf{B}_{m^+}^+, x_{z+1}, \cdots, x_{z+f}\right\} \tag{2.19}$$

where z is the number of the instances in all positive bags and f the number of the instances in negative bags.

The set consisting of subscripts of B_i is expressed as:

$$\Im(i) = \{i \,|x_i \in \mathbf{B}_i\,\} \tag{2.20}$$

For a separable multi-instance classification problem, if a positive bag can be correctly classified, it should satisfy the following constraint:

$$\max_{j\in\Im(i)} \left(w \cdot x_j\right) - b > 0. \tag{2.21}$$

In RMCLP, v_i means the distance from the training sample x_i to the separating hyperplane and be a nonnegative number. Thus, we can always find an appropriate v_i such that

$$\max_{j\in\Im(i)} \left(w \cdot x_j\right) - b = v_i. \tag{2.22}$$

For nonseparable multi-instance classification, we need to add corresponding slack variable $u_i \geq 0$. Finally, the (2.22) is expressed by

$$\max_{j\in\Im(i)} \left(w \cdot x_j\right) - b = v_i - u_i. \tag{2.23}$$

Similar to [21], it is equivalent to the fact that there exist convex combination coefficients set $\left\{\lambda_j^i \,|j \in \Im(i), i = 1, \cdots, m^+\right\}$, such that

$$\left(w \cdot \sum_{j\in\Im(i)} \lambda_j^i x_j\right) + u_i - v_i = b, \tag{2.24}$$

$$\lambda_j^i \geq 0, \sum_{j\in\Im(i)} \lambda_j^i = 1. \tag{2.25}$$

For solving multi-instance classification, so (2.6–2.9) can be converted as:

$$\min_z \frac{1}{2}\|w\|^2 + \frac{1}{2}\|u\|^2 + d\sum_{i=1}^{m^+} u_i + d\sum_{i=z+1}^{z+f} u_i - c\sum_{i=1}^{m^+} v_i - c\sum_{i=z+1}^{z+f} v_i, \tag{2.26}$$

$$s.t. \left(w \cdot \sum_{j\in\Im(i)} \lambda_j^i x_j \right) + (u_i - v_i) = b, i = 1, \cdots, m^+, \tag{2.27}$$

$$(w \cdot x_i) - (u_i - v_i) = b, i = z+1, \cdots, z+f, \tag{2.28}$$

$$\lambda_j^i \geq 0, j \in \Im(i), i = 1, \cdots, m^+, \tag{2.29}$$

$$\sum_{j\in\Im(i)} \lambda_j^i = 1, i = 1, \cdots, m^+, \tag{2.30}$$

$$u, v \geq 0, \tag{2.31}$$

where $z = \left(w^T, u^T, v^T, b, \lambda^T\right)^T, \lambda = \left\{\lambda_j^i \,|j \in \Im(i)\,, i = 1, \cdots, m^+\right\}, \Im(i) = \left\{i \,|x_i \in \mathbf{B}_i^+\right\}$.

As both λ_j^i and w are variables, the constraint (2.27) is no longer a linear constraint and (2.26–2.31) becomes a nonlinear optimization problem.

In the following, we give an approximate iterative solution via solving successive quadratic programming problem. Firstly, we fix λ, and solve a quadratic programming with respect to w, u, v, b; then fix w, solve a quadratic programming with respect to u, v, b, λ.

1. For fixed λ_j^i, $i = 1, \cdots, m^+$, $j \in \Im(i)$, we can obtain

$$\hat{x}_i = \sum_{j\in\Im(i)} \lambda_j^i x_j, i = 1, \cdots, m^+, \tag{2.32}$$

So the problem (2.26–2.31) can be written as

$$\min_z \frac{1}{2} w^T H w + \frac{1}{2} u^T Q u + d e^T u - c e^T v, \tag{2.33}$$

$$s.t. \left(w \cdot \hat{x}_i\right) + (u_i - v_i) = b, i = 1, \cdots, m^+, \tag{2.34}$$

$$\left(w \cdot \hat{x}_i\right) - (u_i - v_i) = b, i = z+1, \cdots, z+f, \tag{2.35}$$

$$u, v \geq 0, \tag{2.36}$$

The problem (2.33–2.36) is a standard quadratic programming problem and as same as RMCLP. We choose H and Q to be identity matrix. Its dual problem can be formulated as

$$\max_{\alpha,u} -\frac{1}{2}\sum_{i=1}^{m^+}\sum_{j=1}^{m^+}\left((\hat{x}_i\cdot\hat{x}_j)+1\right)\alpha_i\alpha_j$$
$$-\frac{1}{2}\sum_{i=1}^{m^+}\sum_{j=z+1}^{z+f}\left((\hat{x}_i\cdot\hat{x}_j)+1\right)\alpha_i\alpha_j$$
$$-\frac{1}{2}\sum_{i=2+1}^{z+f}\sum_{j=1}^{m^+}\left((\hat{x}_i\cdot\hat{x}_j)+1\right)\alpha_i\alpha_j \tag{2.37}$$
$$-\frac{1}{2}\sum_{i=2+1}^{z+f}\sum_{j=z+1}^{z+f}\left((\hat{x}_i\cdot\hat{x}_j)+1\right)\alpha_i\alpha_j$$
$$-\frac{1}{2}\sum_{i=1}^{m^+}\sum_{j=1}^{m^+}u_iu_j-\frac{1}{2}\sum_{i=1}^{m^+}\sum_{j=z+1}^{z+f}u_iu_j$$
$$-\frac{1}{2}\sum_{i=z+1}^{z+f}\sum_{j=1}^{m^+}u_iu_j\frac{1}{2}\sum_{i=z+1}^{z+f}\sum_{j=z+1}^{z+f}u_iu_j$$

$$s.t. -u_i-d\leq\alpha_i\leq -c, i=1,\cdots,m^+, \tag{2.38}$$

$$-u_i-d\leq -\alpha_i\leq -c, i=z+1,\cdots,z+f, \tag{2.39}$$

where $c, d > 0$. We can compute: $\hat{\alpha}=\left(\hat{\alpha}_1,\cdots,\hat{\alpha}_{m^+},\hat{\alpha}_{z+1},\cdots,\hat{\alpha}_{z+f}\right)^T$ by solving the problem of (2.37–2.39), and (w, b) can be expressed as

$$\hat{w}=-\sum_{i=1}^{m^+}\hat{\alpha}_i\hat{x}_i-\sum_{i=z+1}^{z+f}\hat{\alpha}_i\hat{x}_i, \tag{2.40}$$
$$\hat{b}=\sum_{i=1}^{m^+}\hat{\alpha}_i+\sum_{i=z+1}^{z+f}\hat{\alpha}_i, \tag{2.41}$$

$\hat{w}, \hat{b}$ is the updating value of (w, b).

2. For fixed w, the formula (2.26–2.31) can be substituted as:

$$\min_{\lambda,u,v,b}\frac{1}{2}\sum_{i=1}^{m^+}\sum_{j=1}^{m^+}u_iu_j+\frac{1}{2}\sum_{i=1}^{m^+}\sum_{j=z+1}^{z+f}u_iu_j+\frac{1}{2}\sum_{i=z+1}^{z+f}\sum_{j=1}^{m^+}u_iu_j+\frac{1}{2}\sum_{i=z+1}^{z+f}\sum_{j=z+1}^{z+f}u_iu_j$$
$$+d\sum_{i=1}^{m^+}u_i+d\sum_{i=z+1}^{z+f}u_i-c\sum_{i=1}^{m^+}v_i-c\sum_{i=z+1}^{z+f}v_i \tag{2.42}$$

$$s.t. \left(w \cdot \sum_{j \in \Im(i)} \lambda_j^i x_j \right) + (u_i - v_i) = b, i = 1, \cdots, m^+, \tag{2.43}$$

$$(w \cdot x_i) - (u_i - v_i) = b, i = z + 1, \cdots, z + f, \tag{2.44}$$

$$\lambda_j^i \geq 0, j \in \Im(i), i = 1, \cdots, m^+, \tag{2.45}$$

$$\sum_{j \in \Im(i)} \lambda_j^i = 1, i = 1, \cdots, m^+, \tag{2.46}$$

$$u, v \geq 0, \tag{2.47}$$

thus we are able to establish the following Algorithm 2.1 based on the formulas above.

Algorithm 2.1 Linear MI-RMCLP

Initialize: Given a training set (see (2.19));
Choose appropriate penalty parameters $c, d > 0$;
Choose Q and H to be identity matrixes;
Setting initial values for λ ($k = 1$), where $\left\{\lambda_j^i(1) \mid j \in \Im(i), i = 1, \cdots, m^+\right\}$;
Process: **1**. For fixed $\lambda(k) = \left\{\lambda_j^i(k)\right\}$, the goal is to compute $w(k)$:
1.1. Compute $\left\{\hat{x}_1, \cdots, \hat{x}_{m^+}, \hat{x}_{r_1}, \cdots, \hat{x}_{z+f}\right\}$ by (2.32);
1.2. Solve quadratic programming (2.38) ~ (2.39),
obtaining the solution $\hat{\alpha} = \left(\hat{\alpha}_1, \cdots, \hat{\alpha}_p, \hat{\alpha}_{z+1}, \cdots, \hat{\alpha}_{z+f}\right)^T$;
1.2. Compute $\hat{w}$ from (2.40);
1.4. Set $w(k) = \hat{w}$.
2. For fixed $w(k)$, the goal is to compute $\lambda(k + 1)$:
2.1. Solve quadratic programming (2.42) ~ (2.47) with the
variables λ, u, v, b, obtaining the solution $\hat{\lambda}$, $\hat{b}$.
2.2. Set λ $(k + 1) = \hat{\lambda}$, b $(k + 1) = \hat{b}$;
2. If $|\lambda(k + 1) - \lambda(k)| < \varepsilon$, goto **Output**:; otherwise,
goto the step 1, setting $k = k + 1$.
Output: Obtain the decision function $f(x) = \text{sgn}\left((w^* \cdot x) + b^*\right)$,
where $w^* = w(k)$, $b^* = b(k)$.

For nonlinear MI-RMCLP, we firstly introduce the kernel function $K(x, x') = (\Phi(x) \cdot \Phi(x'))$ to replace (x, x'), where $\Phi(x)$ is a mapping from the input space R^n to some Hilbert space $\mathbb{H}$:

$$\Phi : R^n \rightarrow \mathbb{H}$$

$$x \rightarrow \mathrm{x} = \Phi(x) \tag{2.48}$$

Therefore, the problem (2.26–2.31) can be expressed as

$$\min_{z}\frac{1}{2}\|w\|^2+\frac{1}{2}\|u\|^2+d\sum_{i=1}^{m^+}u_i+d\sum_{i=z+1}^{z+f}u_i-c\sum_{i=1}^{m^+}v_i-c\sum_{i=z+1}^{z+f}v_i \tag{2.49}$$

$$s.t.\left(\mathrm{w}\cdot\sum_{j\in\Im(i)}\lambda_j^i\Phi\left(x_j\right)\right)+(u_i-v_i)=b, i=1,\cdots,m^+, \tag{2.50}$$

$$(\mathrm{w}\cdot\Phi\left(x_i\right))-(u_i-v_i)=b, i=z+1,\cdots,z+f, \tag{2.51}$$

$$\lambda_j^i\geq 0, j\in\Im(i), i=1,\cdots,m^+, \tag{2.52}$$

$$\sum_{j\in\Im(i)}\lambda_j^i=1, i=1,\cdots,m^+, \tag{2.53}$$

$$u,v\geq 0, \tag{2.54}$$

Similar to Algorithm 2.1, as a given λ, the current problem can be solved by the following quadratic programming problem:

$$\begin{aligned}\max_{\alpha,u}&-\frac{1}{2}\sum_{i=1}^{m^+}\sum_{j=1}^{m^+}\left(\sum_{k\in\Im(i)}\lambda_k^i\sum_{l\in I(j)}\lambda_l^jK\left(x_k\cdot x_l\right)+1\right)\alpha_i\alpha_j\\&-\frac{1}{2}\sum_{i=1}^{m^+}\sum_{j=z+1}^{z+f}\left(\sum_{k\in\Im(i)}\lambda_k^iK\left(x_k\cdot x_j\right)+1\right)\alpha_i\alpha_j\\&-\frac{1}{2}\sum_{i=z+1}^{z+f}\sum_{j=1}^{m^+}\left(\sum_{l\in I(j)}\lambda_l^jK\left(x_i\cdot x_l\right)+1\right)\alpha_i\alpha_j\\&-\frac{1}{2}\sum_{i=z+1}^{z+f}\sum_{j=z+1}^{z+f}\left(K\left(x_i\cdot x_j\right)+1\right)\alpha_i\alpha_j\end{aligned} \tag{2.55}$$

$$\begin{aligned}&-\frac{1}{2}\sum_{i=1}^{m^+}\sum_{j=1}^{m^+}u_iu_j-\frac{1}{2}\sum_{i=1}^{m^+}\sum_{j=z+1}^{z+f}u_iu_j\\&-\frac{1}{2}\sum_{i=z+1}^{z+f}\sum_{j=1}^{m^+}u_iu_j-\frac{1}{2}\sum_{i=z+1}^{z+f}\sum_{j=z+1}^{z+f}u_iu_j\end{aligned} \tag{2.56}$$

$$s.t.-u_i-d\leq\alpha_i\leq -c, i=1,\cdots,m^+, \tag{2.57}$$

$$-u_i-d\leq -\alpha_i\leq -c, i=z+1,\cdots,z+f, \tag{2.58}$$

We can obtain a solution of $\left(\hat{w}, \hat{b}\right)$ by computing

$$\hat{\mathrm{w}} = -\sum_{i=1}^{m^+} \hat{\alpha}_i \sum_{j \in \Im(i)} \lambda_j^i \Phi(x_i) - \sum_{i=z+1}^{z+f} \hat{\alpha}_i \Phi(x_i), \tag{2.59}$$

$$\hat{b} = \sum_{i=1}^{m^+} \hat{\alpha}_i + \sum_{i=z+1}^{z+f} \hat{\alpha}_i, \tag{2.60}$$

where $\hat{\alpha} = \left(\hat{\alpha}_1, \cdots, \hat{\alpha}_p, \hat{\alpha}_{z+1}, \cdots, \hat{\alpha}_{z+f}\right)^T$ is a solution of the problem (2.56)–(2.58).

For fixed w, the problem (2.49–2.54) can be written as

$$\min_{\lambda,u,v,b} \frac{1}{2} \sum_{i=1}^{m^+} \sum_{j=1}^{m^+} u_i u_j + \frac{1}{2} \sum_{i=1}^{m^+} \sum_{j=z+1}^{z+f} u_i u_j + \frac{1}{2} \sum_{i=z+1}^{z+f} \sum_{j=1}^{m^+} u_i u_j + \frac{1}{2} \sum_{i=z+1}^{z+f} \sum_{j=z+1}^{z+f} u_i u_j + d \sum_{i=1}^{m^+} u_i + d \sum_{i=z+1}^{z+f} u_i - c \sum_{i=1}^{m^+} v_i - c \sum_{i=z+1}^{z+f} v_i \tag{2.61}$$

$$s.t. - \sum_{j=1}^{m^+} \hat{\alpha}_j \sum_{k \in I(j)} \tilde{\lambda}_k^j \sum_{l \in \Im(i)} \lambda_l^i K(x_k, x_l) - \sum_{j=z+1}^{z+f} \hat{\alpha}_j \sum_{l \in \Im(i)} \lambda_l^i K\left(x_j, x_l\right) - (u_i - v_i) = b, \quad i = 1, \cdots, m^+, \tag{2.62}$$

$$-\sum_{j=1}^{m^+} \hat{\alpha}_j \sum_{k \in I(j)} \tilde{\lambda}_k^j K(x_k, x_i) - \sum_{j=z+1}^{z+f} \hat{\alpha}_j K\left(x_j, x_i\right) + (u_i - v_i) = b, i = z+1, \cdots, z+f \tag{2.63}$$

$$\lambda_j^i \geq 0,\ j \in \Im(i), i = 1, \cdots, m^+, \tag{2.64}$$

$$\sum_{j \in \Im(i)} \lambda_j^i = 1, i = 1, \cdots, m^+, \tag{2.65}$$

$$u, v \geq 0, \tag{2.66}$$

where $\tilde{\lambda} = \left(\tilde{\lambda}_i^i \,|\, j \in \Im(i),\ i = 1, \ldots, m^+\right)$ and $\hat{\alpha} = \left(\hat{\alpha}_1, \cdots, \hat{\alpha}_{z+1}, \cdots, \hat{\alpha}_{z+f}\right)^T$ are known.

The ultimate separating hypersurface can be expressed as

$$g(x) = -\sum_{j=1}^{m^+} \hat{\alpha}_j \sum_{k \in I(j)} \tilde{\lambda}_k^j K(x_k, x) - \sum_{j=z+1}^{z+f} \hat{\alpha}_j K\left(x_j, x\right) + \hat{b}, \tag{2.67}$$

In the following, we give out Algorithm 2.2 for nonlinear MI-RMCLP.

Algorithm 2.2 Nonlinear MI-RMCLP

Initialize: Given a training set (see (2.19));
Choose appropriate penalty parameters c, $d > 0$;
Choose Q and H to be identity matrixes;
Choose appropriate
Setting initial values for λ $(k = 1)$, where $\left\{\lambda_j^i(1) \,|\, j \in \Im(i), i = 1, \cdots, m^+\right\}$;
Process: **1**. For fixed $\lambda(k) = \left\{\lambda_j^i(k)\right\}$, the goal is to compute $w(k)$:
1.1. Solve quadratic programming (2.56) ~ (2.58), obtaining the solution.
$\hat{\alpha} = (\hat{\alpha}_1, \cdots, \hat{\alpha}_p, \hat{\alpha}_{z+1}, \cdots, \hat{\alpha}_{z+f})^T$;
1.2. Set $\tilde{\lambda} = \lambda(k)$;
2. For fixed $\hat{\alpha}$, $\tilde{\lambda}$, the goal is to compute $\hat{\lambda} = \left\{\lambda_j^i\right\}$:
2.1. Solve quadratic programming (2.61) ~ (2.66) with the variables (λ, u, v, b), obtaining the solution $\hat{\lambda} = \left\{\lambda_j^i\right\}$.
2.2. Set λ $(k + 1) = \hat{\lambda}$, b $(k + 1) = \hat{b}$;
2. If $|\lambda(k + 1) - \lambda(k)| < \varepsilon$, goto **Output**:; otherwise, goto the step 1, setting $k = k + 1$.
Output: Obtain the decision function $f(x) =$ sgn $(g(x))$, where $g(x)$ by (2.18).

To demonstrate the capabilities of our algorithm, we report results on 12 data sets, 2 from the UCI machine learning repository [22], and 10 from [23]. "Elephant," "Fox" and "Tiger" data sets are from an image annotation task in which the goal is to determine whether or not a given animal is present in an image. The other seven data sets are from the OHSUMED data, and the task is to learn binary concepts associated with the Medical Subject Headings of MEDLINE documents. The "Musk1" and "Musk2" data sets from the UCI machine learning repository are used to test our nonlinear multi-instance RMCLP, which involves bags of molecules and their activity levels and is commonly used in multi-instance classification. Detailed information about these data sets can be found in [21].

Our algorithm code was written in MATLAB 2010. The experiment environment is Intel Core i5 CPU, 2 GB memory. The "quadprog" function with MATLAB is employed to solve quadratic programming problem related to this section. The testing accuracies for our method are computed using standard tenfold cross-validation [24]. The RBF kernel parameter σ is selected from the set $\{2^i | i = -7, \cdots, 7\}$ by tenfold cross-validation on the tuning set comprising of random 10% of the training data. Once the parameters are selected, the tuning set was returned to the training set to learn the final decision function. The (c, d) are set 1. If the difference between 2 is less than 10^{-4} or the iterations $K > 100$, our algorithms will be stopped.

We compare our results with MICA [21], mi-SVM [23], MI-SVM [25], EM-DD [25] and SVM-CC [26]. MI-RMCLP is our method in Table 2.1 and Fig. 2.4. The results of tenfold cross-validation accuracy are listed in Table 2.1 and Fig. 2.4. The results for mi-SVM, MI-SVM and EM-DD are taken from [21].

Table 2.1 Results of all methods in the case of rbf kernel

Data Sets	MICA (%)	mi-SVM (%)	MI-SVM (%)	EM-DD (%)	SVM-CC (%)	MI-RMCLP (%)
Elephant	80.5	**82.2**	81.4	78.3	81.5	79.3
Fox	**58.7**	58.2	57.8	56.1	57.3	57.6
TST1	94.5	92.6	92.9	85.8	**95.0**	91.2
TST2	85.0	78.2	84.5	84.0	82.7	**86.0**
TST3	86.0	**87.0**	82.2	69.0	86.4	85.1
TST4	**87.7**	82.8	82.4	80.5	82.1	81.4
TST7	78.9	81.3	78.0	75.4	77.4	**82.7**
TST9	61.4	**67.5**	60.2	65.5	62.0	62.9
TST10	**82.3**	79.6	79.5	78.5	81.5	77.6
Musk-1	84.4	87.4	77.9	84.8	**88.9**	85.8
Musk-2	90.5	82.6	84.3	84.9	89.6	**91.7**

Note: Best accuracy is in bold

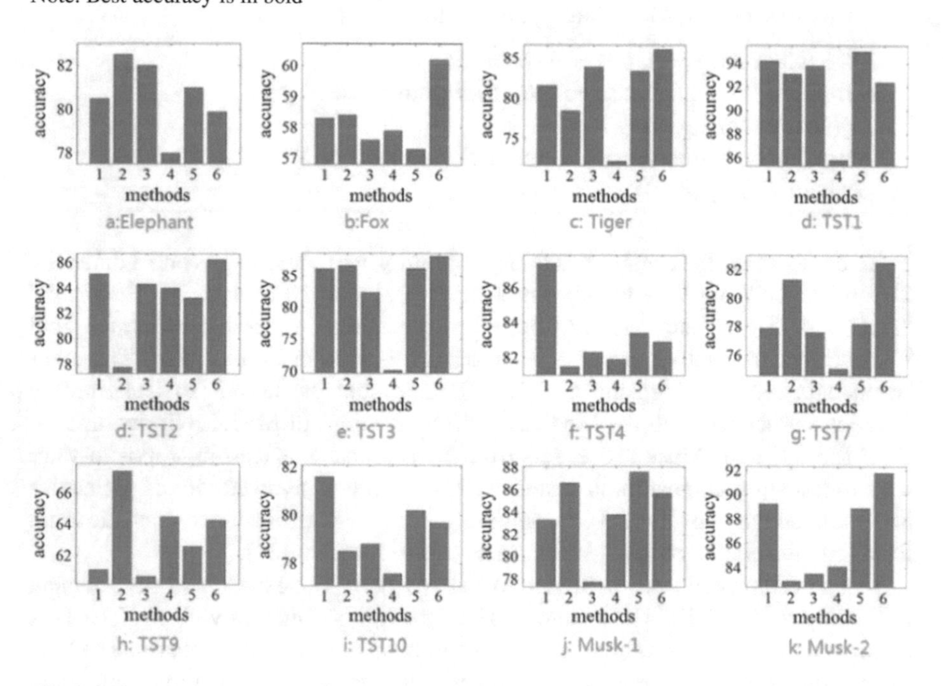

Fig. 2.4 Results of all methods in the case of linear kernel. X-axis represents different methods: 1: MICA; 2: mi-SVM; 3: MI-SVM; 4: EM-DD; 5: SVM-CC; 6: Mi-RMCLP. Y-axis represents the accuracy

2.1.3 Supportive Instances for Regularized Multiple Criteria Linear Programming Classification

Although RMCLP performs excellently in classifying lots of benchmark datasets, its shortage is also obvious. By taking account of every training instances into consideration, RMCLP is sensitive to noisy and imbalanced training samples. In other words, the classification boundary may shift significantly even if there is merely a slight change of training samples. This difficulty can be described in Fig. 2.5, assume there is a two groups classification problem, the first group is denoted by "." and the second group is denoted by "☆" . We can observe that it is a linear-separable dataset and the classification boundary is denoted by a line "/". Figure 2.5a shows that on an ideal training sample, RMCLP successfully classify all the instances. In Fig. 2.5b, when we add some noisy instances into the first group, the classification boundary shifts towards the first group, making more instances in the first group misclassified. In Fig. 2.5c, we can observe that when we add instances into the second group to make the number of instances in two groups imbalanced, the classification boundary also changes significantly, causing a great number of misclassifications. In Fig. 2.5d, we can see that if we choose some representative instances (also called supportive instances) for RMCLP, which locate inside the blue circle, then although more noisy and imbalanced instances are added into the training sample, the classification boundary always keeps unchanged and will have a good ability to do prediction. That is to say, building RMCLP model only on supportive instances can improve its accuracy and stability.

According to the above observation, in this subsection, we propose a clustering-based sample selection method, which chooses the instances in the clustering center as the supportive samples (just as SVM [27] chooses the support vectors to draw a classification boundary). Experimental results on synthetic and real-life datasets show that our new method not only can significantly improve the prediction accuracy, but also can dramatically reduce the number of training instances.

Lots of empirical studies have shown that MCLP is a powerful tool for classification. However, there is no theoretical work on whether MCLP always can

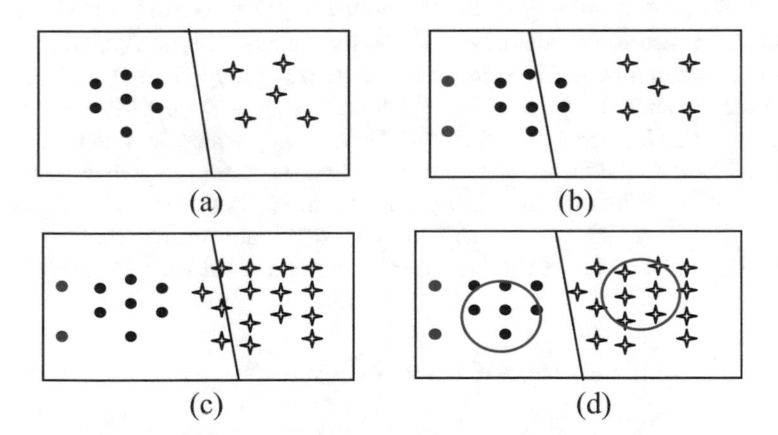

Fig. 2.5 (**a**) The original RMCLP model built on an ideal training sample; (**b**) when adding two noisy instances in the left side, the classification boundary shifts towards the left side; (**c**) when the training sample is imbalanced, the boundary also shifts significantly; (**d**) if we select representative training instances which locate around the distribution centers (inside the circle), the classification boundary becomes satisfactory

find an optimal solution under different kinds of training samples. To go over this difficulty, recently, [2] proposed a RMCLP model by adding two regularized items $\frac{1}{2}x^T Hx$ and $\frac{1}{2}\alpha^T Q\alpha$ on MCLP as follows:

$$\textbf{Minimize}\frac{1}{2}x^T Hx + \frac{1}{2}\alpha^T Q\alpha + d^T\alpha - c^T\beta \tag{2.68}$$

$$\textbf{\textit{Subject to}}: \begin{aligned} &A_i x - \alpha_i + \beta_i = b, \forall A_i \in G_1; \\ &A_i x + \alpha_i - \beta_i = b, \forall A_i \in G_2; \\ &\alpha_i, \beta_i \geq 0. \end{aligned}$$

where $H \in R^{r*r}$, $Q \in R^{n*n}$ are symmetric positive definite matrices. $d^T, c^T \in R^n$. The RMCLP model is a convex quadratic program. Theoretically studies [2] have shown that RMCLP can always find a global optimal solution.

Besides two groups classification problem, a recent work [28] also introduced a multiple groups RMCLP model. As far as three groups classification problem be considered, we first find a projection direction x and a group of hyper planes (b_1, b_2), to an arbitrary training instance A_i, if $A_i x < b_1$, then $A_i \in G_1$; if $b_1 \leq A_i x < b_2$ then $A_i \in G_2$; and if $A_i x \geq b_2$, then $A_i \in G_3$. Extending this method to n group classification, we can also find a direction x and $n - 1$ dimension vector $b = [b_1, b_2, \ldots, b_{n-1}] \in R^{n-1}$, to make sure that to any training instance A_i:

$$\begin{aligned} &A_i x < b_1, \forall A_i \in G_1; \\ &b_{j-1} \leq A_i x < b_j, \forall A_i \in G_i, 1 < i < n; \\ &A_i x \geq b_{n-1}, \forall A_i \in G_n; \end{aligned} \tag{2.69}$$

We first define $c_i = \frac{b_{i-1}+b_i}{2}$ as the midline in group $i(1 < i < n)$. Then, to the misclassified records, we define α_i^+ as the distance from c_i to $A_i x$, which equals $(c_i - A_i x)$, when misclassify a group i's record into group j $(j < i)$, and we define α_i^- as the distance from $A_i x$ to c_i, which equals $(c_i - A_i x)$, when misclassify a group i's record into group j $(j > i)$. Similarly, to the correct classified records, we define β_i^- when A_i is in the left side of c_i, and we define β_i^+ when A_i is in the right side of c_i. When we have a n groups training sample with size m, we have $\alpha = \{\alpha_i^+, \alpha_i^-\} \in R^{m*2}$, $\beta = \{\beta_i^+, \beta_i^-\} \in R^{m*2}$, and we can build a multiple groups Regularized Multi-Criteria Linear Programming (SRMCLP) as follows:

$$\textbf{Minimize } \frac{1}{2}x^T Hx + \frac{1}{2}\alpha^T Q\alpha + d^T\alpha + c^T\beta$$

$$\textbf{\textit{Subject to}}: \begin{aligned} &A_i x - \alpha_i^- - \beta_i^- + \beta_i^+ = \tfrac{1}{2}b_1, \forall A_i \in G_1; \\ &A_i x - \alpha_i^- + \alpha_i^+ - \beta_i^- + \beta_i^+ = \tfrac{1}{2}(b_{i-1} + b_i), \forall A_i \in G_i, 1 < i < n; \\ &A_i x + \alpha_i^+ - \beta_i^- + \beta_i^+ = 2b_{n-1}, \forall A_i \in G_n; \\ &\alpha_i^-, \alpha_i^+, \beta_i^-, \beta_i^+ \geq 0. \end{aligned} \tag{2.70}$$

Input: training sample *Tr*, testing sample *Ts*, parameter ε, exclusion percentage *s*
Output: selected sample *Tr'*
Begin
1.Set *Tr'=Tr*
2. While (|PrevClusteringCenter-CurrClusteringCenter| < ε)
{
2.1. Calculate current clustering center; $cent = \frac{1}{|T_{r'}|}\sum_i x_i$

2.2. For each instances $i \in T'$ do
{
2.2.1 Calculate the Euclidean distance of the clustering center,

$$dis_i = \sqrt{\sum_{r \in r} \left| cent_r - Tr_r^i \right|^2}$$

2.2.2 get *s%* of the instances which are farthest from the clustering center, denoted as the subset *{P}*
2.2.3 exclude *{P}* from the training, *Tr'=Tr\{P}*.
}
}
3. Return the selected sample $T_{r'}$
End

Fig. 2.6 Clustering method to get the supportive method

Since this multiple groups RMCLP model is mainly designed to solve the ordinal separable dataset, we also call it Ordinal RMCLP model [28].

Figure 2.6 gives the whole procedure of the sample selection algorithm. The main idea of our algorithm is that it iteratively discards training instances in each group which are far away from the clustering center until the clustering center for each group is stable (with the given threshold ε), then the remained instances will be taken as the supportive instances (just as the support vectors to SVM) and used to build a classifier. From Fig. 2.6, We can observe that this algorithm is similar to the well-known k-means algorithm. However, the main difference between them is that, our algorithm is based on supervised learning framework, while k-means is an unsupervised learning algorithm. In our algorithm, although the clustering centers shift in each iteration, each instance keeps a constant class label. But in k-means, the class label of each instance may change frequently. An important issue of k-means clustering is how to choose the initial points, if we choose a good initial point, we can get a global optimal solution; otherwise, we may only get a local optimal solution. On the contrast, our sample selection method can avoid this problem. It always leads to a global minimal solution.

There are some important parameters in our algorithm. The first important parameter is ε, which determinates when the algorithm stops. The second parameter is the exclusion percentage s, which indicates how many instances that are far away from the clustering center should be discarded in each iteration. This parameter, in fact, determines the convergence speed. The larger value of s, the faster algorithm converges. To analyze the computation complexity of our new algorithm, we take an extremely bad situation into consideration. Assume there are n instances in the training sample, we assign the values $s = 1$ and $\varepsilon = 0$. Then, the algorithm will discard only one instances in each iteration. To the worst case, after n times iterations, the algorithm converges to the clustering center. In the ith iteration, it needs to calculate the $(n - i)$ instances to get the clustering center, so we can roughly infer that the computation complexity is about $O(n^2)$.

To investigate whether our new algorithm works, we use two synthetic datasets and a well-known US bank's real-life credit card dataset for testing. In our experiments, the RMCLP is implemented by Visual Fortran 6.5.

The 6000 credit card records are randomly selected from 25,000 real-life credit card records of a major US bank. Each record has 113 variables, with 38 original variables and 65 derived variables. The 38 original variables are balance, purchase, payment, cash advance, and related variables, with the former 5 items each have six variables that represent raw data of six consecutive months and the last item includes interest charges, data of last payment, times of cash advance, account open data and so on. The 65 derived variables (CHAR01–CHAR65) are derived from original 38 variables using simple arithmetic methods to reinforce the comprehension of cardholders' behaviors. In this section, we use the derived 65 variables. We then define five classes for this dataset using a label variable: The Number of Over-limits. The five classes are defined as Bankrupt charge-off accounts (THE NUMBER OF OVER-LIMITS$\geq$13), Non-bankrupt charge-off accounts (7$\leq$THE NUMBER OF OVER-LIMITS$\leq$12), Delinquent accounts (3$\leq$THE NUMBER OF OVER-LIMITS$\leq$6), Current accounts (1$\leq$THE NUMBER OF OVER-LIMITS$\leq$2), and Outstanding accounts (no over limit). Bankrupt charge-off accounts are accounts that have been written off by credit card issuers due to reasons other than bankrupt claims. The charge-off policy may vary among authorized institutions. Delinquent accounts are accounts that haven't paid the minimum balances for more than 90 days. Current accounts are accounts that have paid the minimum balances. The outstanding accounts are accounts that have not balances. In our randomly selected 6000 records, there are 72 Bankrupt charge-off accounts, 205 Non-bankrupt charge-off accounts, 454 Delinquent accounts, 575 Current accounts and 4694 outstanding accounts.

Two groups credit card dataset To acquire a two groups training sample, we combine the Bankrupt charge-off accounts, Non-bankrupt charge-off accounts and Delinquent accounts together to form a "bad" group. And then we combine the current accounts and the outstanding accounts into a "good" group. According to the previous research work on this dataset, we first randomly select a benchmark training size of 700 bad records and 700 good records, and the remained 4600

Table 2.2 Comparison of different percentage of training instances

Percent. of training	Training sample		Testing sample (4600 instances)	
	Right instances	Accuracy (%)	Right instances	Accuracy (%)
100 (1400)	1096	78.29	3394	72.78
90 (1260)	998	79.20	3295	71.63
80 (1120)	912	81.43	3292	71.57
70 (980)	789	80.51	3571	77.63
60 (840)	667	79.40	3761	81.76
50 (700)	559	79.86	3881	84.37
40 (560)	449	80.18	3964	86.17
30 (420)	331	78.81	4050	88.04
20 (280)	232	82.86	4073	88.54
10 (140)	116	82.86	1971	42.85

records are combined to test the performance. Now what we need to do is to examine three assumptions: first, is the randomly selected 1400 points are suitable to build model? second, are there any noisy instances in this randomly selected dataset? third, can we reduce the 1400 points in a much smaller size and improve the accuracy synchronously? Experimental results in Table 2.2 tell us the answers. The first column of Table 2.2 is the current training sample's size, from the 1400 instances to 140 instances, the second and the third columns list the performance on different training samples and the fourth and the fifth columns exhibit the performance on the same 4600 testing instances. The experiment is conducted as follows: firstly, we build a RMCLP model on all the 1400 training instances, and we get a benchmark accuracy as 72.78%. Then we call our sample selection algorithm with parameter $s = 1$ and $\varepsilon = 0.1$. We do experiments on night special datasets, 10%, 20%, ..., 90% of the original 1400 training sample. We finally list the performance of RMCLP in Table 2.2. Intuitionally, we though the larger the training sample, the more information we could get, and thus the model would be more accurate when do prediction. However, Table 2.2, we can see that the 1400 randomly selected instances is not the best training set for RMCLP model, there exist noisy and useless instances which deteriorate its performance. Our new sample selection method reduces the training samples continuously. When get 20% of the original training sample (that is 280 instances), we can build a RMCLP with the highest accuracy of 88.54% on the testing set.

Multiple Groups credit card dataset Besides two groups RMCLP model, in this part, we also study the performance of our new algorithm on multiple groups RMCLP model. For three groups classification, we choose the Bankrupt charge-off accounts as the first group, the Non-bankrupt charge-off as the second group and the Delinquent as the third group. Based upon the three groups dataset, we construct the four groups dataset by adding the Current account as the fourth group. At last, we construct a five groups dataset by adding the Outstanding accounts as the fifth group.

Table 2.3 Comparison on three groups credit card dataset

3 Groups (22 + 155 + 404)	Original RMCLP		RMCLP After sample selection	
	Corrected Rec.	Accuracy (%)	Corrected Rec.	Accuracy (%)
Group1	12	54.5	19	86.36
Group2	12	7.7	89	57.42
Group3	402	99.5	390	96.53
Average	426	72.32	481	85.71

Table 2.4 Comparison on four groups credit card dataset

4 Groups (22 + 155 + 404 + 525)	Original RMCLP		RMCLP After sample selection	
	Corrected Rec.	Accuracy (%)	Corrected Rec.	Accuracy (%)
Group1	16	72.7	19	86.36
Group2	52	32.5	122	78.71
Group3	38	9.4	267	66.09
Group4	525	100.0	510	97.14
Average	631	57.05	918	82.00

Table 2.5 Comparison on five groups credit card dataset

5 Groups (22 + 155 + 404 + 525 + 4644)	Original RMCLP		RMCLP After sample selection	
	Corrected Rec.	Accuracy (%)	Corrected Rec.	Accuracy (%)
Group1	13	59.1	16	72.73
Group2	130	82.9	130	82.87
Group3	273	67.6	365	90.35
Group4	161	30.7	438	82.43
Group5	4644	100.0	4520	97.33
Average	5221	90.80	5469	95.11

Tables 2.3, 2.4 and 2.5 list the results of comparisons. The second and the third columns list the results of the original RMCLP method, the fourth and the fifth columns list the results of RMCLP after selecting the supportive instances. We can observe that in three groups classification, the original RMCLP's average accuracy is 72.32%, while that of the supportive instances is 85.71%. The improvement of accuracy is as large as 12.39%. In four groups classification, the average accuracy of the original RMCLP is 57.05%, on the contrast, after selecting the supportive instances, the accuracy improves to 82.00%, as high as 25.95% improvement. To the five groups classification, the improvement after selecting supportive instances is 4.31%. From these compressive results, we can validate our former conclusion that selecting supportive instances for RMCLP can significantly improve its accuracy.

2.1.4 Kernel Based Simple Regularized Multiple Criteria Linear Programming for Binary Classification and Regression

In this section, a novel kernel based regularized multiple criteria linear program are proposed for both classification and regression scenarios.

Given an observed dataset $T = \{(x_1, y_1), (x_2, y_2), \ldots, (x_l, y_l)\}$ with l instances. Each instance x_i belongs to the category y_i. $x_i \in \chi \subseteq R^n$ and $y_i \in y$ are the n attributes values and corresponding label for the instance i. The goal of classification problem is to predict the corresponding label $y_i \in y$ when new instance $x_j \in \chi$ arrives. When $Card(y) = 2$, the issue is binary classification problem. In order to facilitate description, here we let $y = \{-1, 1\}$ for following introduction. Under this binary classification problem, supposed we have positive instances number is l1, negative instances number is l_2, where $l_1 + l_2 = l$. $\xi_A = 0$, $\xi_B = 0$ which are not marked in the picture.

In contrast to points A and B, points C and D are improperly predicted. Hence their distance could be constructed as $\beta_C = 0$, $\beta_D = 0$ and $\xi_C > 0$, $\xi_D > 0$. In summary, following the idea described above the basic MCLP model [29] for classification could be written as this:

$$\begin{aligned}
&\min_{w,b,\xi,\beta} \sum_{i=1}^{l} \xi_i \\
&\max_{w,b,\xi,\beta} \sum_{i=1}^{l} \beta_i \\
&s.t. y_i \left(x_i^T w + b\right) = \beta_i - \xi_i, \\
&\xi_i \geq 0, \beta_i \geq 0, i = 1, \cdots, l;
\end{aligned} \tag{2.71}$$

Here w and b could be seem as the slope and intercept of the discriminant hyperplane. One of the objectives $\sum \xi_i$ could be considered as the measure of misclassification, thus we minimized it to avoid the inappropriate model construction.

And the other goal $\sum \beta_i$ is to maximize the generalization capability of the chosen classification function. As we introduced before, there exist no single solution that could make the both these two goals in conflict optimal at the same time. In [30, 31], compromise solution is introduced and analyzed for this multiple objective model Eq. (2.71). However, the algorithm that obtained compromise solution were usually time consuming and not suitable for real world application.

As a result, many methods convert model Eq. (2.71) into single objective linear program:

$$
\begin{aligned}
&\min_{w,b,\xi,\beta} \sum_{i=1}^{l} \xi_i - \gamma \sum_{i=1}^{l} \beta_i \\
&s.t. y_i \left(x_i^T w + b \right) = \beta_i - \xi_i, \\
&\xi_i \geq 0, \beta_i \geq 0, i = 1, \cdots, l;
\end{aligned}
\tag{2.72}
$$

Unfortunately, naive model Eq. (2.72) confronts the unsolvable defect because of the nature of linear programming. More sophisticated approaches need to be investigated. Therefore, an improved model would be illustrated in next section.

Although model Eq. (2.72) avoided the computational cost of multiple objectives, it had a fatal solvability problem. Therefore, we added new quadratic term to the objective function and proposed a new simple regularized MCLP model showed as below:

$$
\begin{aligned}
&\min_{w,b,\xi,\beta} \sum_{i=1}^{l} \xi_i - \gamma \sum_{i=1}^{l} \beta_i + \frac{1}{2} \tau \beta^T H \beta \\
&s.t. y_i \left(x_i^T w + b \right) = \beta_i - \xi_i, \\
&\xi_i \geq 0, \beta_i \geq 0, i = 1, \cdots, l; \\
&b \in \{-1, 1\}.
\end{aligned}
\tag{2.73}
$$

Furthermore, users want to guarantee the slope of the hyperplane not too large. Then, we made the regularization term $w^T K w$ as a part of the goal and obtained the following model:

$$
\begin{aligned}
&\min_{w,b,\xi,\beta} \sum_{i=1}^{l} \xi_i - \gamma \sum_{i=1}^{l} \beta_i + \frac{1}{2} \tau \beta^T H \beta + \frac{1}{2} \kappa w^T K w \\
&s.t. y_i \left(x_i^T w + b \right) = \beta_i - \xi_i, \\
&\xi_i \geq 0, \beta_i \geq 0, i = 1, \cdots, l; \\
&b \in \{-1, 1\};
\end{aligned}
\tag{2.74}
$$

In order to write the formulas in matrix form, we let

$$A = \begin{bmatrix} x_1^T \\ x_2^T \\ \vdots \\ x_l^T \end{bmatrix}_{l*n}, Y = \begin{bmatrix} y_1 & 0 & \cdots & 0 \\ 0 & y_2 & \cdots & 0 \\ \cdots & \cdots & \cdots & \cdots \\ 0 & \cdots & 0 & y_l \end{bmatrix}_{l*l} \tag{2.75}$$

So model Eq. (2.74) could be rewritten as this:

$$\begin{aligned} &\min_{w,\beta,\xi} \frac{1}{2} w^T H w + \frac{1}{2}\lambda_1 \beta^T K \beta - \lambda_2 e^T \beta + \lambda_3 e^T \xi \\ &s.t. Y\left(Aw + be\right) - \beta + \xi = 0, \\ &b \in \{-1, 1\}, \beta \geq 0, \xi \geq 0 \end{aligned} \tag{2.76}$$

Where $w \in R^n$, $\beta \in R^l$, $\xi \in R^l$, $e = [1, \cdots, 1]_l^T$ is the vector of all ones. K and H are $n \times n$ and $l \times l$ positive matrix, respectively. We simply set positive matrix H, K in model Eq. (2.76) as identity matrix. And to solve the problem with inequality type constraints, we have to find the saddle point of the Lagrangian function for model Eq. (2.76)

$$\begin{aligned} L\left(w, \beta, \xi, \alpha_{equ}, \alpha_\beta, \alpha_\xi\right) = &\left(\frac{1}{2} w^T w + \frac{1}{2}\lambda_1 \beta^T \beta - \lambda_2 e^T \beta + \lambda_3 e^T \xi\right) \\ &+ \alpha_{equ}^T \left(Y\left(Aw + be\right) - \beta + \xi\right) - \alpha_\beta^T \beta - \alpha_\xi^T \xi \end{aligned} \tag{2.77}$$

where α_{equ} is free, $\alpha_\beta \geq 0$, $\alpha_\xi \geq 0$ are Lagrangian multipliers. Minimization with respect to w, β, ξ implies the following

$$\nabla_w L\left(w, \beta, \xi, \alpha_{equ}, \alpha_\beta, \alpha_\xi\right) = w + A^T Y \alpha_{equ} = 0 \tag{2.78}$$

$$\nabla_\beta L\left(w, \beta, \xi, \alpha_{equ}, \alpha_\beta, \alpha_\xi\right) = \lambda_1 \beta - \lambda_2 e - \alpha_{equ} - \alpha_\beta = 0 \tag{2.79}$$

$$\nabla_\xi L\left(w, \beta, \xi, \alpha_{equ}, \alpha_\beta, \alpha_\xi\right) = \lambda_3 e + \alpha_{equ} - \alpha_\xi = 0 \tag{2.80}$$

Sustaining Eq. (2.78) into function Eq. (2.77), we get

$$L\left(w, \beta, \xi, \alpha_{equ}, \alpha_\beta, \alpha_\xi\right) = -\frac{1}{2}\alpha_{equ}^T Y A A^T Y \alpha_{equ} - \frac{1}{2}\lambda_1 \beta^T \beta + b e^T Y \alpha_{equ}$$

Therefore, the dual problem for model Eq. (2.76) is obtained as

$$\begin{aligned} \max -\frac{1}{2}\alpha_{equ}^{T} Y A A^{T} Y \alpha_{equ} - \frac{1}{2}\lambda_1 \beta^T \beta + b e^T Y \alpha_{equ} \\ s.t. \lambda_1 \beta - \lambda_2 e - \alpha_{equ} \geq 0, \\ \lambda_3 e + \alpha_{equ} \geq 0, \\ \beta \geq 0, \\ b \in \{-1, 1\} \end{aligned} \tag{2.81}$$

According to the Eq. (2.78), the decision function is

$$f(x) = \text{sign}\,(w \cdot x + b) = \text{sign}\left(-Y A^T \alpha_{equ} x + b\right).$$

When introduce kernel functions

$$\begin{aligned} R^n \to H \\ x \to \Phi(x) \end{aligned} \tag{2.82}$$

We have $K(x_i, x_j) = \Phi(x_i) \cdot \Phi(x_j)$. Therefore, the dual problem Eq. (2.81) could be rewritten as

$$\begin{aligned} \min \frac{1}{2}\alpha_{equ}^{T} Y K\,(A, A)\, Y \alpha_{equ} + \frac{1}{2}\lambda_1 \beta^T \beta - b e^T Y \alpha_{equ} \\ s.t. \lambda_1 \beta - \lambda_2 e - \alpha_{equ} \geq 0, \\ \lambda_3 e + \alpha_{equ} \geq 0, \\ \beta \geq 0, \\ b \in \{-1, 1\} \end{aligned} \tag{2.83}$$

Furthermore, the decision boundary turns into

$$f(x) = \text{sign}\,(w \cdot \Phi(x) + b) = \text{sign}\left(-Y K\,(A, x)\, \alpha_{equ} + b\right).$$

Theorem 2.2 *Given the solution of the dual problem Eq. (2.83) as* $\left(\alpha_{equ}^*, \beta^*\right)$, *the solution of its corresponding primal problem w.r.t. H space can be obtained as* *below:*

$$w^* = -Y\Phi(A)^T\alpha_{equ}^* \tag{2.84}$$

Proof From dual problem Eq. (2.83), we can get its Lagrangian function as:

$$L\left(\alpha_{equ}, \beta, \alpha_1, \alpha_2\right) = \frac{1}{2}\alpha_{equ}^T YK\left(A, A\right)Y\alpha_{equ} + \frac{1}{2}\lambda_1\beta^T\beta - be^TY\alpha_{equ}$$
$$-\alpha_1^T\left(\lambda_1\beta - \lambda_2 e - \alpha_{equ}\right) - \alpha_2^T\left(\lambda_3 e + \alpha_{equ}\right) - \alpha_3^T\beta \tag{2.85}$$

Where $\alpha_1 \geq 0, \alpha_2 \geq 0, \alpha_3 \geq 0$. From the KTT condition, we have the equations below:

$$\lambda_1\beta - \lambda_2 e - \alpha_{equ} \geq 0 \tag{2.86}$$

$$\lambda_3 e + \alpha_{equ} \geq 0 \tag{2.87}$$

$$\beta \geq 0 \tag{2.88}$$

$$\left(\lambda_1\beta - \lambda_2 e - \alpha_{equ}\right)^T\alpha_1 = 0 \tag{2.89}$$

$$\left(\lambda_3 e + \alpha_{equ}\right)^T\alpha_2 = 0 \tag{2.90}$$

$$\beta^T\alpha_3 = 0 \tag{2.91}$$

$$\nabla_{\alpha_{equ}}L\left(\alpha_{equ}, \beta, \alpha_1, \alpha_2\right) = YK\left(A, A\right)Y\alpha_{equ} - bYe + \alpha_1 - \alpha_2 = 0 \tag{2.92}$$

$$\nabla_\beta L\left(\alpha_{equ}, \beta, \alpha_1, \alpha_2\right) = \lambda_1\beta - \lambda_1\alpha_1 - \alpha_3 = 0 \tag{2.93}$$

Sustaining Eq. (2.84) into Eq. (2.92), so

$$\nabla_{\alpha_{equ}}L\left(\alpha_{equ}, \beta, \alpha_1, \alpha_2\right) = YK\left(A, A\right)Y\alpha_{equ} - bYe + \alpha_1 - \alpha_2$$
$$= -Y\left(w^* \cdot \Phi(A) + be\right) + \alpha_1^* - \alpha_2^* = 0 \tag{2.94}$$

This satisfies the constraint of problem Eq. (2.76) when $\beta = \alpha_1^*$, $\xi = \alpha_2^*$. Therefore, $\left(w^*, \alpha_1^*, \alpha_2^*\right)$ is the feasible solution of primal problem Eq. (2.76) w.r.t. H space. Furthermore, introducing Eqs. (2.89), (2.90) and (2.92), the objective function of primal problem Eq. (2.76) turns into:

$$\frac{1}{2}w^{*T}w^* + \frac{1}{2}\lambda_1\beta^{*T}\beta^* - \lambda_2 e^T\beta^* + \lambda_3 e^T\xi^*$$
$$= -\frac{1}{2}\alpha_{equ}^{*T}YK\left(A, A\right)Y\alpha_{equ}^* - \frac{1}{2}\lambda_1\beta^{*T}\beta^* + be^TY\alpha_{equ}^* \tag{2.95}$$

As a result, the object value of the primal problem at points (w^*, β^*, ξ^*) is the optimal value of its dual problem at points (α_{equ}, β^*) w.r.t. H space.

Base on the Theorem 2.2, we introduced Algorithm 2.3 using kernel based simple regular multiple constraint linear program (KSRMCLP) for binary classification problem.

Given a training set $\{(x_1, y_1), \cdots, (x_l, y_l)\}$, being different from classification problem, regression is not to give a new arrival instance x_i a category label but a real number value, $y_i \in R$. That is mean the possible set of y_i has been changed from finite labels set y to infinite R. Following the idea of $\epsilon - tube$, a model for regression problem could be constructed from a binary classification model [32]. Given a real number ϵ, two different category points could be generated when we add and minus ϵ on the regression output y_i. When we have l instances $\{(x_1, y_1), \cdots, (x_l, y_l)\}$ for regression, $2 \times l$ instances $\{(x_1, y_1 + \epsilon)_{pos}, \cdots, (x_l, y_l + \epsilon)_{pos}, (x_1, y_1 - \epsilon)_{neg}, \cdots, (x_\ell, y_\ell - \epsilon)_{neg}\}$ could be constructed. According to the binary classification model we propose in the last section, a model for regression problem could be given as:

$\min \frac{1}{2}w^T Hw + \frac{1}{2}\lambda_1\beta^T K\beta - \lambda_2 e^T\beta + \lambda_3 e^T\xi$

$$s.t. Y\left(A_{reg}w + be\right) = \beta - \xi, \tag{2.96}$$

$$\beta \geq 0, \xi \geq 0$$

Algorithm 2.3 KSRMCLP Algorithm for Binary Classification

Input:

Training dataset $S = \{(x_1, y_1), (x_2, y_2), \cdots, (x_l, y_l)\}$ with l instances, $x_i \in R^n$ and$y_i \in \{-1, 1\}$, kernel function $K_\theta(x_i, x_j)$ and its parameters θ, model parameters $\lambda_1 \geq 0$, $\lambda_2 \geq 0$, $\lambda_3 \geq 0$.

Output:

Binary classification discriminate function $f(x)$.

1: **Begin**

2: Construct data matrix A, label matrix Y according to Eq. (2.75).

$$A = \begin{bmatrix} x_1^T \\ x_2^T \\ \vdots \\ x_l^T \end{bmatrix}_{l*n}, Y = \begin{bmatrix} y_1 & 0 & \cdots & 0 \\ 0 & y_2 & \cdots & 0 \\ \cdots & \cdots & \cdots & \cdots \\ 0 & \cdots & 0 & y_l \end{bmatrix}_{l*l}$$

3: Construct and solve the optimization problem according to model Eq. (2.83).

$\min \frac{1}{2}\alpha_{equ}^T Y K_\theta(A, A) Y\alpha_{equ} + \frac{1}{2}\lambda_1\beta^T\beta - be^T Y\alpha_{equ},$

$s.t.\ \lambda_1\beta - \lambda_2 e - \alpha_{equ} \geq 0,$

$\lambda_3 e + \alpha_{equ} \geq 0,$

$\beta \geq 0,$

$$b \in \{-1, 1\}$$

4: Obtain the decision function $f(x) = sign\,(-YK_\theta(A,x)\alpha_{equ} + b)$.
5: **End**

where $w \in R^{n+1}$, $\beta, \xi \in R^{2l}$, and

$$A_{reg} = \begin{bmatrix} x_1^T, y_1 + \epsilon \\ \vdots \\ x_l^T, y_l + \epsilon \\ x_1^T, y_1 - \epsilon \\ \vdots \\ x_l^T, y_l - \epsilon \end{bmatrix}_{2l\times(n+1)}, Y = \begin{bmatrix} I_{l\times l} & O \\ O & -I_{l\times l} \end{bmatrix}_{2l\times 2l} \tag{2.97}$$

The constraint of Eq. (2.96) could be divided into two parts, the positive and the negative. For positive points, the corresponding target value is $y_i + \epsilon$, for negative points is $y_i - \epsilon$. Thus matrix Y is useless and variables β, ξ, also change into Boos, $\beta_{pos}, \beta_{neg}, \xi_{pos}, \xi_{neg}$. Then, model Eq. (2.96) could be written as,
$\min \frac{1}{2}w^T Hw + \frac{1}{2}\lambda_1\beta_{pos}^T K\beta_{pos} + \frac{1}{2}\lambda_1\beta_{neg}^T K\beta_{neg} - \lambda_2 e^T\left(\beta_{pos} + \beta_{neg}\right) + \lambda_3 e^T\left(\xi_{pos} + \xi_{neg}\right)$

$$s.t. Aw + be + \eta\,(y + \epsilon e) = \beta_{pos} - \xi_{pos} \tag{2.98}$$

$$Aw + be + \eta\,(y - \epsilon e) = -\left(\beta_{neg} - \xi_{neg}\right)$$

$$\beta_{pos} \geq 0, \beta_{neg} \geq 0, \xi_{pos} \geq 0, \xi_{neg} \geq 0$$

where $w \in R^n$, $\beta_{pos}, \beta_{neg}, \xi_{pos}, \xi_{neg} \in R^l$, $b \in R$ are variables.

$$A = \begin{bmatrix} x_1^T \\ x_2^T \\ \vdots \\ x_l^T \end{bmatrix}_{l\times n}, y = \begin{bmatrix} y_1 \\ y_2 \\ \vdots \\ y_l \end{bmatrix}_{l\times 1} \tag{2.99}$$

We know $\eta \neq 0$, w, b, β_{pos}, β_{neg}, ξ_{pos}, ξ_{neg} are all variables, so η could be removed from the expression. Model Eq. (2.98) turns into:
$\min \frac{1}{2}w^T Hw + \frac{1}{2}\lambda_1\beta_{pos}^T K\beta_{pos} + \frac{1}{2}\lambda_1\beta_{neg}^T K\beta_{neg} - \lambda_2 e^T\left(\beta_{pos} + \beta_{neg}\right) + \lambda_3 e^T\left(\xi_{pos} + \xi_{neg}\right)$

$$s.t. Aw + be + (y + \epsilon e) = \beta_{pos} - \xi_{pos} \tag{2.100}$$

$$Aw + be + (y - \epsilon e) = -\left(\beta_{neg} - \xi_{neg}\right),$$

$$\beta_{pos} \geq 0, \beta_{neg} \geq 0, \xi_{pos} \geq 0, \xi_{neg} \geq 0$$

where $w \in R^n$, $\beta_{pos}, \beta_{neg}, \xi_{pos}, \xi_{neg} \in R^l$, $b \in R$ are variables. And $\epsilon, \lambda_1, \lambda_2, \lambda_3 \in R$, positive matrices H, K are given in advance. Similar to the procedure last part, we set K, H as identity matrix, the Lagrangian function of model Eq. (2.100) is derived as

$$L\left(w, \beta_{pos}, \beta_{neg}, \xi_{pos}, \xi_{neg}\right) = -\tfrac{1}{2}\left(\alpha_{pos} + \alpha_{neg}\right)^T AA^T\left(\alpha_{pos} + \alpha_{neg}\right) - \tfrac{1}{2}\lambda_1 \beta_{pos}^T \beta_{pos}$$
$$-\tfrac{1}{2}\lambda_1 \beta_{neg}^T \beta_{neg} + (be + y)^T\left(\alpha_{pos} + \alpha_{neg}\right) + \epsilon e^T\left(\alpha_{pos} - \alpha_{neg}\right) \tag{2.101}$$

where α_{pos}, α_{neg} are free variables, $\alpha_{\beta_{pos}} \geq 0, \alpha_{\beta_{neg}} \geq 0, \alpha_{\xi_{pos}} \geq 0, \alpha_{\xi_{neg}} \geq 0$ are corresponding Lagrangian multipliers. Also, from KKT condition, we have

$$\nabla_w L\left(w, \beta_{pos}, \beta_{neg}, \xi_{pos}, \xi_{neg}\right) = w + A^T\left(\alpha_{pos} + \alpha_{neg}\right) = 0 \tag{2.102}$$

$$\nabla_{\beta_{pos}} L\left(w, \beta_{pos}, \beta_{neg}, \xi_{pos}, \xi_{neg}\right) = \lambda_1 \beta_{pos} - \lambda_2 e - \alpha_{pos} - \alpha_{\beta_{pos}} = 0 \tag{2.103}$$

$$\nabla_{\beta_{neg}} L\left(w, \beta_{pos}, \beta_{neg}, \xi_{pos}, \xi_{neg}\right) = \lambda_1 \beta_{neg} - \lambda_2 e + \alpha_{neg} - \alpha_{\beta_{neg}} = 0 \tag{2.104}$$

$$\nabla_{\xi_{pos}} L\left(w, \beta_{pos}, \beta_{neg}, \xi_{pos}, \xi_{neg}\right) = \lambda_3 e + \alpha_{pos} - \alpha_{\xi_{pos}} = 0 \tag{2.105}$$

$$\nabla_{\xi_{neg}} L\left(w, \beta_{pos}, \beta_{neg}, \xi_{pos}, \xi_{neg}\right) = \lambda_3 e - \alpha_{neg} - \alpha_{\xi_{neg}} = 0 \tag{2.106}$$

Therefore, the dual problem for model Eq. (2.100) is obtained:

$$\begin{aligned}
\max -\frac{1}{2}\left(\alpha_{pos} + \alpha_{neg}\right)^T AA^T\left(\alpha_{pos} + \alpha_{neg}\right) - \frac{1}{2}\lambda_1\left(\beta_{pos}^T \beta_{pos} + \beta_{neg}^T \beta_{neg}\right) + \\
(be + y)^T\left(\alpha_{pos} + \alpha_{neg}\right) + \epsilon e^T\left(\alpha_{pos} - \alpha_{neg}\right) \\
s.t. \lambda_1 \beta_{pos} - \lambda_2 e - \alpha_{pos} \geq 0, \\
\lambda_1 \beta_{neg} - \lambda_2 e + \alpha_{neg} \geq 0, \\
\lambda_3 e + \alpha_{pos} \geq 0, \\
\lambda_3 e - \alpha_{neg} \geq 0, \\
\beta_{pos} \geq 0, \\
\beta_{neg} \geq 0, \\
b \in \{-1, 1\}
\end{aligned} \tag{2.107}$$

where $\alpha_{pos}, \alpha_{neg}, \beta_{pos}, \beta_{neg} \in R^l$, $b \in R$ are variables. And $\epsilon \geq 0$, $\lambda_1 \geq 0$, $\lambda_2 \geq 0$, $\lambda_3 \geq 0$ are given in advance.

When introducing kernel function Eq. (2.82), model Eq. (2.107) turns into

$$\begin{aligned}
\min \tfrac{1}{2}\left(\alpha_{pos} + \alpha_{neg}\right)^T K(A, A)\left(\alpha_{pos} + \alpha_{neg}\right) + \tfrac{1}{2}\lambda_1\left(\beta_{pos}^T\beta_{pos} + \beta_{neg}^T\beta_{neg}\right) \\
-(be + y)^T\left(\alpha_{pos} + \alpha_{neg}\right) - \epsilon e^T\left(\alpha_{pos} - \alpha_{neg}\right) \\
s.t. \lambda_1\beta_{pos} - \lambda_2 e - \alpha_{pos} \geq 0, \\
\lambda_3 e + \alpha_{pos} \geq 0, \\
\lambda_3 e - \alpha_{neg} \geq 0, \\
\beta_{pos} \geq 0, \\
\beta_{neg} \geq 0, \\
b \in \{-1, 1\}
\end{aligned} \tag{2.108}$$

From the decision hyperplane $w \cdot x + b + y = 0$, the regression function could be obtained as

$$f(x) = -(w \cdot x + b) = A^T\left(\alpha_{pos} + \alpha_{neg}\right) \cdot x - b$$

With kernel function, regression function could be derived from

$$f(x) = \Phi(A)^T\left(\alpha_{pos} + \alpha_{neg}\right)\Phi(x) - b = K(A, x)\left(\alpha_{pos} + \alpha_{neg}\right) - b$$

Theorem 2.3 *Given the solution of Dual Problem Eq. (2.108) $\left(\alpha_{pos}^*, \alpha_{neg}^*, \beta_{pos}^*, \beta_{neg}^*\right)$, the solution of its corresponding primal problem w.r.t. H space can be obtained as below:*

$$w^* = -\Phi\left(A^T\right)\left(\alpha_{pos}^* + \alpha_{neg}^*\right) \tag{2.109}$$

Proof From dual problem Eq. (2.108), we can get its Lagrangian function as:

$$\begin{aligned}
L\left(\alpha_{pos}, \alpha_{neg}, \beta_{pos}, \beta_{neg}\right) = \frac{1}{2}\left(\alpha_{pos} + \alpha_{neg}\right)^T K(A, A)\left(\alpha_{pos} + \alpha_{neg}\right) \\
+\frac{1}{2}\lambda_1\left(\beta_{pos}^T\beta_{pos} + \beta_{neg}^T\beta_{neg}\right) \\
-(be + y)^T\left(\alpha_{pos} + \alpha_{neg}\right) - \epsilon e^T\left(\alpha_{pos} - \alpha_{neg}\right) \\
-\alpha_1^T\left(\lambda_1\beta_{pos} - \lambda_2 e - \alpha_{pos}\right)
\end{aligned} \tag{2.110}$$

$$-\alpha_2^T\left(\lambda_1\beta_{neg}-\lambda_2 e+\alpha_{neg}\right)$$

$$-\alpha_3^T\left(\lambda_3 e+\alpha_{pos}\right)$$

$$-\alpha_4^T\left(\lambda_3 e-\alpha_{neg}\right)$$

$$-\alpha_5^T\beta_{pos}$$

$$-\alpha_6^T\beta_{neg}$$

where $\alpha_1 \geq 0$, $\alpha_2 \geq 0$, $\alpha_3 \geq 0$, $\alpha_4 \geq 0$, $\alpha_5 \geq 0$, $\alpha_6 \geq 0$. from the KTT condition, we have the equation below:

$$\lambda_1\beta_{pos}-\lambda_2 e-\alpha_{pos}\geq 0 \tag{2.111}$$

$$\lambda_1\beta_{neg}-\lambda_2 e+\alpha_{neg}\geq 0 \tag{2.112}$$

$$\lambda_3 e+\alpha_{pos}\geq 0 \tag{2.113}$$

$$\lambda_3 e-\alpha_{neg}\geq 0 \tag{2.114}$$

$$\beta_{pos}\geq 0 \tag{2.115}$$

$$\beta_{neg}\geq 0 \tag{2.116}$$

$$\alpha_1^T\left(\lambda_1\beta_{pos}-\lambda_2 e-\alpha_{pos}\right)=0 \tag{2.117}$$

$$\alpha_2^T\left(\lambda_1\beta_{neg}-\lambda_2 e+\alpha_{neg}\right)=0 \tag{2.118}$$

$$\alpha_3^T\left(\lambda_3 e+\alpha_{pos}\right)=0 \tag{2.119}$$

$$\alpha_4^T\left(\lambda_3 e-\alpha_{neg}\right)=0 \tag{2.120}$$

$$\alpha_5^T\beta_{pos}=0 \tag{2.121}$$

$$\alpha_6^T\beta_{neg}=0 \tag{2.122}$$

$$\nabla_{\alpha_{pos}}L=K\left(A,A\right)\left(\alpha_{pos}+\alpha_{neg}\right)-\left(be+y\right)-\epsilon e+\alpha_1-\alpha_3=0 \tag{2.123}$$

$$\nabla_{\alpha_{neg}}L=K\left(A,A\right)\left(\alpha_{pos}+\alpha_{neg}\right)-\left(be+y\right)+\epsilon e-\alpha_2+\alpha_4=0 \tag{2.124}$$

$$\nabla_{\beta_{pos}}L=\lambda_1\beta_{pos}-\lambda_1\alpha_1-\alpha_5=0 \tag{2.125}$$

$$\nabla_{\beta_{neg}}L=\lambda_1\beta_{neg}-\lambda_1\alpha_2-\alpha_6=0 \tag{2.126}$$

Sustaining Eq. (2.109) into Eqs. (2.123) and (2.124), we have

$$\begin{aligned} &\nabla_{\alpha_{pos}} L\left(\alpha_{pos}, \alpha_{neg}, \beta_{pos}, \beta_{neg}\right) \\ &= K(A, A)\left(\alpha_{pos}+\alpha_{neg}\right)-(be+y)-\epsilon e+\alpha_1-\alpha_3 \\ &= -\left(w^* \cdot \Phi(A)+be+(y+\epsilon e)-\alpha_1+\alpha_3\right)=0 \end{aligned} \tag{2.127}$$

$$\begin{aligned} &\nabla_{\alpha_{pos}} L\left(\alpha_{pos}, \alpha_{neg}, \beta_{pos}, \beta_{neg}\right) \\ &= K(A, A)\left(\alpha_{pos}+\alpha_{neg}\right)-(be+y)+\epsilon e-\alpha_1+\alpha_3 \\ &= -\left(w^* \cdot \Phi(A)+be+(y-\epsilon e)+\alpha_1-\alpha_3\right)=0 \end{aligned} \tag{2.128}$$

This satisfies the constraint of primal problem Eq. (2.100), so $\left(w^*, \alpha_1^*, \alpha_2^*, \alpha_3^*, \alpha_4^*\right)$ is the feasible solution of primal problem Eq. (2.100) w.r.t. H space. Furthermore, introducing Eqs. (2.117)–(2.122), the objective function of primal problem Eq. (2.100) turns into:

$$\begin{aligned} &\frac{1}{2}w^T w+\frac{1}{2}\lambda_1\left(\beta_{pos}^T\beta_{pos}+\beta_{neg}^T\beta_{neg}\right)-\lambda_2 e^T\left(\beta_{pos}+\beta_{neg}\right)+\lambda_3 e^T\left(\xi_{pos}+\xi_{neg}\right) \\ &= -\frac{1}{2}\left(\alpha_{pos}+\alpha_{neg}\right)^T K(A, A)\left(\alpha_{pos}+\alpha_{neg}\right)-\frac{1}{2}\lambda_1\left(\beta_{pos}^T\beta_{pos}+\beta_{neg}^T\beta_{neg}\right) \\ &\quad +(be+y)^T\left(\alpha_{pos}+\alpha_{neg}\right)+\epsilon e^T\left(\alpha_{pos}-\alpha_{neg}\right) \end{aligned}$$

As a result, the object value of the primal problem at points $\left(w^*, \beta_{pos}^*, \beta_{neg}^*, \xi_{pos}^*, \xi_{neg}^*\right)$ is the optimal value of its dual problem at points $\left(\alpha_{pos}^*, \alpha_{neg}^*, \beta_{pos}^*, \beta_{neq}^*\right)$.

Base on the Theorem 2.3, we introduced Algorithm 2.4 from kernel based simple regular multiple constraint linear programming (KSRMCLP) for regression problem.

Algorithm 2.4 KSRMCLP Algorithm for Regression
Input:
Training dataset $S=\{(x_1, y_1), (x_2, y_2), \cdots, (x_l, y_l)\}$, $x_i \in R^n$ and $y_i \in R$. Kernel function $K_\theta(x_i, x_j)$ and its parameters θ, model parameters $\epsilon \geq 0$, $\lambda_1 \geq 0$, $\lambda_2 \geq 0$, $\lambda_3 \geq 0$.
Output:
Regression estimated function $f(x)$.
1: **Begin**
2: Construct data matrix A, target value vector y according to equation formula

below:

$$A = \begin{bmatrix} x_1^T \\ x_2^T \\ \vdots \\ x_l^T \end{bmatrix}_{l*n}, y = \begin{bmatrix} y_1 \\ y_2 \\ \vdots \\ y_l \end{bmatrix}_{l\times 1}$$

3: Construct and solve the optimization problem according to Eq. (2.108).

$$\min \tfrac{1}{2}(\alpha_{pos} + \alpha_{neg})^T K_\theta (A, A) (\alpha_{pos} + \alpha_{neg}) + \tfrac{1}{2}\lambda_1 \left(\beta_{pos}^T \beta_{pos} + \beta_{neg}^T \beta_{neg}\right)$$
$$-(be + y)^T (\alpha_{pos} + \alpha_{neg}) - \epsilon e^T (\alpha_{pos} - \alpha_{neg}),$$
$$s.t.\ \lambda_1 \beta_{pos} - \lambda_2 e - \alpha_{pos} \geq 0,$$
$$\lambda_1 \beta_{neg} - \lambda_2 e + \alpha_{neg} \geq 0,$$
$$\lambda_3 e + \alpha_{pos} \geq 0,$$
$$\lambda_3 e - \alpha_{neg} \geq 0,$$
$$\beta_{pos} \geq 0,$$
$$\beta_{neg} \geq 0,$$
$$b \in \{-1, 1\}$$

4: Obtain the decision function $f(x) = K_\theta(A, x)(\alpha_{pos} + \alpha_{neg}) - b$.

5: **End**

2.2 Multiple Criteria Linear Programming with Expert and Rule Based Knowledge

2.2.1 A Group of Knowledge-Incorporated Multiple Criteria Linear Programming Classifier

Prior knowledge in some classifiers usually consists of a set of rules, such as, if A then $x \in G$ (or $x \in B$), where condition A is relevant to the attributes of the input data. One example of such form of knowledge can be seen in the breast cancer recurrence or nonrecurrence prediction. Usually, doctors can judge if the cancer recur or not in terms of some measured attributes of the patients. The prior knowledge used by doctors in the breast cancer dataset includes two rules which depend on two features of the total 32 attributes: tumor size (T) and lymph node status (L). The rules are [33]:

If $L \geq 5$ *and* $T \geq 4$ *Then RECUR and If* $L = 0$ *and* $T \leq 1.9$ *Then NONRECUR*

The conditions $L \geq 5$ *and* $T \geq 4$ ($L = 0$ *and* $T \leq 1.9$) in the above rules can be written into such inequality as $Cx \leq c$, where C is a matrix driven from the condition, x represents each individual sample, c is a vector. For example, if each sample x is expressed by a vector $[x_1, \ldots, x_L, \ldots, x_T, \ldots, x_r]^T$, for the rule: *if* $L \geq 5$ *and* $T \geq 4$ *then RECUR*, it also means: *if* $x_L \geq 5$ *and* $x_T \geq 4$, *then* $x \in$ *RECUR*, where x_L and x_T are the corresponding values of attributes L and T of a certain sample data, r is the number of attributes. Then its corresponding inequality $Cx \leq c$ can be written as:

$$\begin{bmatrix} 0 \ldots -1 \ldots \;0\; \ldots 0 \\ 0 \ldots \;0\; \ldots -1 \ldots 0 \end{bmatrix} x \le \begin{bmatrix} -5 \\ -4 \end{bmatrix}.$$

where x is the vector with r attributes include two features relevant to prior knowledge.

Similarly, the condition $L = 0$ *and* $T \le 1.9$ can also be reformulated to be inequalities. With regard to the condition $L = 0$, in order to express it into the formulation of $Cx \le c$, we must replace it with the condition $L \ge 0$ *and* $L \le 0$. Then the condition $L = 0$ *and* $T \le 1.9$ can be represented by two inequalities: $C^1 x \le c^1$ and $C^2 x \le c^2$, as follows:

$$\begin{bmatrix} 0 \ldots -1 \ldots 0 \ldots 0 \\ 0 \ldots \;0\; \ldots 1 \ldots 0 \end{bmatrix} x \le \begin{bmatrix} 0 \\ 1.9 \end{bmatrix} \text{and} \begin{bmatrix} 0 \ldots 1 \ldots 0 \ldots 0 \\ 0 \ldots 0 \ldots 1 \ldots 0 \end{bmatrix} x \le \begin{bmatrix} 0 \\ 1.9 \end{bmatrix}$$

We notice the fact that the set $\{x| \; Cx \le c\}$ can be regarded as polyhedral convex set. In Fig. 2.7, the triangle and rectangle are such sets.

In two-class classification problem, the result RECUR or NONRECUR is equal to the expression $x \in B$ or $x \in G$. So according to the above rules, we have:

$$Cx \le c \Rightarrow x \in G \quad (or \quad x \in B) \tag{2.129}$$

In MCLP classifier, if the classes are linearly separable, then $x \in G$ is equal to $x^T w \ge b$, similarly, $x \in B$ is equal to $x^T w \le b$. That is, the following implication must hold:

$$Cx \le c \Rightarrow x^T w \ge b \quad \left(or \quad x^T w \le b\right) \tag{2.130}$$

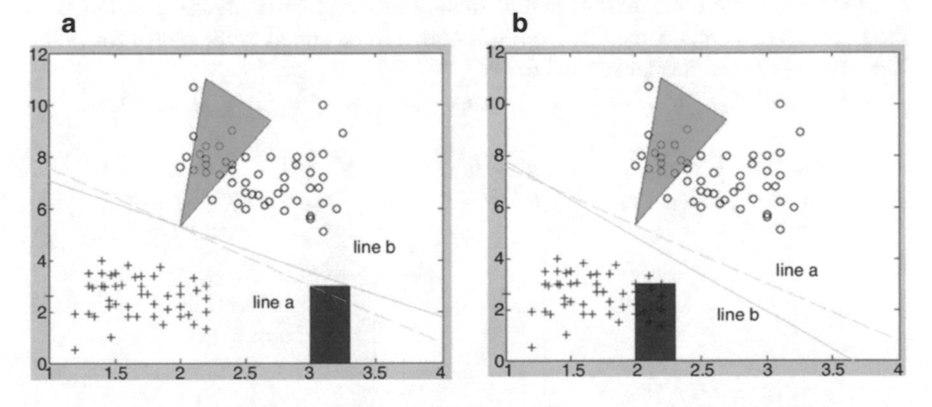

Fig. 2.7 The classification result by MCLP (line a) and knowledge-incorporated MCLP (line b)

For a given (w, b), the implication $Cx \leq c \Rightarrow x^T w \geq b$ holds, this also means that $Cx \leq c, x^T w < b$ has no solution x. According to nonhomogeneous Farkas theorem, we can conclude that $C^T u + w = 0$, $c^T u + b \leq 0$, $u \geq 0$, has a solution (u, w) [33].

The above statement is able to be added to constraints of an optimization problem. In this way, the prior knowledge in the form of some equalities and inequalities in constraints is embedded to the original multiple linear programming (MCLP) model. The knowledge-incorporated MCLP model is described in the following.

Knowledge-incorporated MCLP model Now, we are to explain the knowledge-incorporated MCLP model. This model is to deal with linear knowledge and linear separable data. The combination of the two kinds of input can help to improve the performances of both methods.

Suppose there are a series of knowledge sets as follows:

$$\begin{aligned} &\text{If } C^i x \leq c^i, i = 1, \ldots, k \quad \text{Then } x \in G \\ &\text{If } D^j x \leq d^j, j = 1, \ldots, l \quad \text{Then } x \in B \end{aligned}$$

This knowledge also means the convex sets $\{x| C^i x \leq c^i\}$, $i = 1, \ldots, k$ lie on the G side of the bounding plane, the convex sets $\{x| D^j x \leq d^j\}$, $j = 1, \ldots, l$ on the B side.

Based on the above theory in the last section, we converted the knowledge to the following constraints:

There exist u^i, $i = 1, \ldots, k$, $v^j, j = 1, \ldots, l$, such that:

$$\begin{array}{llll} C^{iT} u^i + w = 0, & c^{iT} u^i + b \leq 0, & u^i \geq 0, & i = 1, \ldots, k \\ D^{jT} v^j - w = 0, & d^{jT} v^j - b \leq 0, & v^j \geq 0, & j = 1, \ldots, l \end{array} \tag{2.131}$$

However, there is no guarantee that such bounding planes precisely separate all the points. Therefore, some error variables need to be added to the above formulas. The constraints are further revised to be:

There exist u^i, r^i, ρ^i, $i = 1, \ldots, k$ *and* $v^j, s^j, \sigma^j, j = 1, \ldots, l$, such that:

$$\begin{array}{llll} -r^i \leq C^{iT} u^i + w \leq r^i, & c^{iT} u^i + b \leq \rho^i, & u^i \geq 0, & i = 1, \ldots, k \\ -s^j \leq D^{jT} v^j - w \leq s^j, & d^{jT} v^j - b \leq \sigma^j, & v^j \geq 0, & j = 1, \ldots, l \end{array} \tag{2.132}$$

After that, we embed the above constraints to the MCLP classifier, and obtained the knowledge-incorporated MCLP classifier:

$$
\begin{aligned}
&\text{Minimize} \quad d_\alpha^+ + d_\alpha^- + d_\beta^+ + d_\beta^- + C\left(\sum\left(r_{\mathrm{i}} + \rho^{\mathrm{i}}\right) + \sum\left(s^{\mathrm{j}} + \sigma^{\mathrm{j}}\right)\right)\\
&\text{Subject to :}\\
&\quad \alpha^* + \sum_{i=1}^{n} \alpha_i = d_\alpha^- - d_\alpha^+\\
&\quad \beta^* - \sum_{i=1}^{n} \beta_i = d_\beta^- - d_\beta^+\\
&\quad x_{11}w_1 + \cdots + x_{1r}w_r = \mathrm{b} + \alpha_1 - \beta_1, \qquad \text{for } \mathrm{A}_1 \in \mathrm{B},\\
&\quad \vdots\\
&\quad x_{n1}w_1 + \cdots + x_{nr}w_r = \mathrm{b} - \alpha_n + \beta_n, \qquad \text{for } \mathrm{A}_n \in \mathrm{G},\\
&\quad -r^{\mathrm{i}} \le \mathrm{C}^{\mathrm{i}'} u^{\mathrm{i}} + w \le r^{\mathrm{i}}, \qquad \mathrm{i} = 1, \ldots, \mathrm{k}\\
&\quad \mathrm{c}^{\mathrm{i}'} u^{\mathrm{i}} + b \le \rho^{\mathrm{i}}\\
&\quad -s^{\mathrm{j}} \le \mathrm{D}^{\mathrm{j}'} v^{\mathrm{j}} - w \le s^{\mathrm{j}}, \qquad \mathrm{j} = 1, \ldots, l\\
&\quad \mathrm{d}^{\mathrm{j}'} v^{\mathrm{j}} - b \le \sigma^{\mathrm{j}}\\
&\quad \alpha_1, \ldots, \alpha_n \ge 0, \quad \beta_1, \ldots, \beta_n \ge 0, \quad \left(u^{\mathrm{i}}, v^{\mathrm{j}}, r^{\mathrm{i}}, \rho^{\mathrm{i}}, s^{\mathrm{j}}, \sigma^{\mathrm{j}}\right) \ge 0
\end{aligned}
\tag{2.133}
$$

In this model, all the inequality constraints are derived from the prior knowledge. The last objective $\mathrm{C}(\sum(r_{\mathrm{i}} + \rho^{\mathrm{i}}) + \sum(s^{\mathrm{j}} + \sigma^{\mathrm{j}}))$ is about the slack error variables added to the original knowledge equality constraints. The last objective attempts to drive the error variables to zero. We want to get the best bounding plane (w, b) by solving this model to separate the two classes.

We notice the fact that if we set the value of parameter C to be zero, this means to take no account of knowledge. Then this model will be equal to the original MCLP model. Theoretically, the larger the value of C, the greater impact on the classification result of the knowledge sets.

Knowledge-incorporated KMCLP Model If the data set is nonlinear separable, the above model will be inapplicable. We need to figure out how to embed prior knowledge into the KMCLP model, which can solve nonlinear separable problem.

As is shown in the above part, in generating KMCLP model, we suppose:

$$
w = \sum_{i=1}^{n} \lambda_i y_i X_i \tag{2.134}
$$

If expressed by matrix, the above formulation will be:

$$w = X^T Y\lambda \tag{2.135}$$

where Y is $n*n$ diagonal matrix, the value of each diagonal element depends on the class label of the corresponding sample data, which can be $+1$ or -1. X is the $n*r$ input matrix with n samples, r attributes. λ is a n-dimensional vector $\lambda = (\lambda_1, \lambda_2, \ldots, \lambda_n)^T$.

$$Y = \begin{bmatrix} y_1 & 0 & \ldots & 0 \\ 0 & y_2 & \ldots & 0 \\ \vdots & \vdots & \ddots & \vdots \\ 0 & 0 & \ldots & y_n \end{bmatrix}, \qquad X = \begin{bmatrix} x_{11} & x_{12} & \ldots & x_{1r} \\ x_{21} & x_{22} & \ldots & x_{2r} \\ \vdots & \vdots & \ddots & \vdots \\ x_{n1} & x_{n2} & \ldots & x_{nr} \end{bmatrix}$$

Therefore, w in the original MCLP model is replaced by $X^T Y\lambda$, thus forming the KMCLP model. And in this new model, the value of each λ_i is to be worked out by the optimization model.

In order to incorporate prior knowledge into KMCLP model, the inequalities about the knowledge must be transformed to be the form with λ_i instead of w. Enlightened by the KMCLP model, we also introduce kernel to the expressions of knowledge. Firstly, the equalities in (2.131) are multiplied by input matrix X [34]. Then replacing w with $X^T Y\lambda$, (2.131) will be:

$$\begin{array}{llll} XC^{iT}u^i + XX^T Y\lambda = 0, & c^{iT}u^i + b \le 0, & u^i \ge 0, & i = 1, \ldots, k \\ XD^{jT}v^j - XX^T Y\lambda = 0, & d^{jT}v^j - b \le 0, & v^j \ge 0, & j = 1, \ldots, l \end{array} \tag{2.136}$$

Kernel function is introduced here to replace XC^{iT} and XX^T. Also slack errors are added to the expressions, then such kind of constraints are formulated:

$$\begin{array}{ll} -r^i \le K\left(X, C^{iT}\right)u^i + K\left(X, X^T\right)Y\lambda \le r^i, & i = 1, \ldots, k \\ c^{iT}u^i + b \le \rho^i & \\ -s^j \le K\left(X, D^{jT}\right)v^j - K\left(X, X^T\right)Y\lambda \le s^j, & j = 1, \ldots, l \\ d^{jT}v^j - b \le \sigma^j & \end{array} \tag{2.137}$$

These constraints can be easily embedded to KMCLP model as the constraints acquired from prior knowledge.

Knowledge-incorporated KMCLP classifier:

$$
\begin{aligned}
&\text{Min}\left(d_\alpha^+ + d_\alpha^- + d_\beta^+ + d_\beta^-\right) + C\left(\sum_{i=1}^{k}\left(r_i + \rho^i\right) + \sum_{j=1}^{l}\left(s^j + \sigma^j\right)\right)\\
\text{s.t.}\quad &\lambda_1 y_1 K(X_1, X_1) + \cdots + \lambda_n y_n K(X_n, X_1) = b + \alpha_1 - \beta_1,, \quad \text{for } X_1 \in \text{B},\\
&\vdots\\
&\lambda_1 y_1 K(X_1, X_n) + \cdots + \lambda_n y_n K(X_n, X_n) = b - \alpha_n + \beta_n, \quad \text{for } X_n \in \text{G},\\
&\alpha^* + \sum_{i=1}^{n}\alpha_i = d_\alpha^- - d_\alpha^+,\\
&\beta^* - \sum_{i=1}^{n}\beta_i = d_\beta^- - d_\beta^+,\\
&-r^i \le K\left(X, C^{iT}\right)u^i + K\left(X, X^T\right)Y\lambda \le r^i, \quad i = 1, \ldots, k\\
&c^{iT}u^i + b \le \rho^i\\
&-s^j \le K\left(X, D^{jT}\right)v^j - K\left(X, X^T\right)Y\lambda \le s^j, \quad j = 1, \ldots, l\\
&d^{jT}v^j - b \le \sigma^j\\
&\alpha_1, \ldots, \alpha_n \ge 0, \quad \beta_1, \ldots, \beta_n \ge 0, \quad \lambda_1, \ldots, \lambda_n \ge 0,\\
&\left(u^i, v^j, r^i, \rho^i, s^j, \sigma^j\right) \ge 0\\
&d_\alpha^-, d_\alpha^+, d_\beta^-, d_\beta^+ \ge 0
\end{aligned}
\tag{2.138}
$$

In this model, all the inequality constraints are derived from prior knowledge. u^i, $v^i \in R^p$, where p is the number of conditions in one knowledge. For example, in the knowledge *if* $x_L \ge 5$ *and* $x_T \ge 4$*, then* $x \in RECUR$, the value of p is 2. r^i, ρ^i, s^j and σ^j are all real numbers. And the last objective Min $\sum\left(r_i + \rho^i\right) + \sum\left(s^j + \sigma^j\right)$ is about the slack error variables added to the original knowledge equality constraints. As we talked in last section, the larger the value of C, the greater impact on the classification result of the knowledge sets.

In this model, several parameters need to be set before optimization process. Apart from C we talked about above, the others are parameter of kernel function q (if we choose RBF kernel) and the ideal compromise solution α^* and β^*. We want to get the best bounding plane (λ, b) by solving this model to separate the two classes. And the discrimination function of the two classes is:

$$
\begin{aligned}
&\lambda_1 y_1 K(X_1, z) + \cdots + \lambda_n y_n K(X_n, z) \le b, \quad \textit{then} \quad z \in B\\
&\lambda_1 y_1 K(X_1, z) + \cdots + \lambda_n y_n K(X_n, z) \ge b, \quad \textit{then} \quad z \in G
\end{aligned}
\tag{2.139}
$$

where z is the input data which is the evaluated target with r attributes. X_i represents each training sample. y_i is the class label of *ith* sample.

In the above models, the prior knowledge we deal with is linear. That means the conditions in the above rules can be written into such inequality as $Cx \le c$, where C is a matrix driven from the condition, x represents each individual sample, c is a vector. The set $\{x|\ Cx \le c\}$ can be viewed as polyhedral convex set, which is a linear geometry in input space. But, if the shape of the region which consists of knowledge is nonlinear, for example, $\{x|\ ||x||^2 \leq c\}$, how to deal with such kind of knowledge?

Suppose the region is nonlinear convex set, we describe the region by $g(x) \leq 0$. If the data is in this region, it must belong to class B. Then, such kind of nonlinear knowledge may take the form of:

$$\begin{array}{lll} g(x) \leq 0 & \Rightarrow & x \in B \\ h(x) \leq 0 & \Rightarrow & x \in G \end{array} \tag{2.140}$$

Here $g(x)$: $R^r \rightarrow R^p$ $(x \in \Gamma)$ and $h(x)$: $R^r \rightarrow R^q$ $(x \in \Delta)$ are functions defined on a subset Γ and Δ of R^r which determine the regions in the input space. All the data satisfied $g(x) \leq 0$ must belong to the class B and $h(x) \leq 0$ to the class G.

With KMCLP classifier, this knowledge equals to:

$$\begin{array}{lll} g(x) \leq 0 & \Rightarrow & \lambda_1 y_1 K(X_1, x) + \cdots + \lambda_n y_n K(X_n, x) \leq b, \quad (x \in \Gamma) \\ h(x) \leq 0 & \Rightarrow & \lambda_1 y_1 K(X_1, x) + \cdots + \lambda_n y_n K(X_n, x) \geq b, \quad (x \in \Delta) \end{array} \tag{2.141}$$

This implication can be written in the following equivalent logical form [35]:

$$\begin{array}{ll} g(x) \leq 0 & , \ \lambda_1 y_1 K(X_1, x) + \cdots + \lambda_n y_n K(X_n, x) - b > 0, \text{ has no solution } x \in \Gamma. \\ h(x) \leq 0 & , \ \lambda_1 y_1 K(X_1, x) + \cdots + \lambda_n y_n K(X_n, x) - b < 0, \text{ has no solution } x \in \Delta. \end{array} \tag{2.142}$$

The above expressions hold, then there exist $v \in R^p$, $r \in R^q$, $v,r \geq 0$ such that:

$$\begin{array}{l} -\lambda_1 y_1 K(X_1, x) - \cdots - \lambda_n y_n K(X_n, x) + b + v^T g(x) \geq 0, \quad (x \in \Gamma) \\ \lambda_1 y_1 K(X_1, x) + \cdots + \lambda_n y_n K(X_n, x) - b + r^T h(x) \geq 0, \quad (x \in \Delta) \end{array} \tag{2.143}$$

Add some slack variables on the above two inequalities, then they are converted to:

$$\begin{array}{l} -\lambda_1 y_1 K(X_1, x) - \cdots - \lambda_n y_n K(X_n, x) + b + v^T g(x) + s \geq 0, \quad (x \in \Gamma) \\ \lambda_1 y_1 K(X_1, x) + \cdots + \lambda_n y_n K(X_n, x) - b + r^T h(x) + t \geq 0, \quad (x \in \Delta) \end{array} \tag{2.144}$$

The above statement is able to be added to constraints of an optimization problem.

Suppose there are a series of knowledge sets as follows:

$$\text{If } g_i(x) \le 0, \text{ Then } x \in B \quad \left(g_i(x) : R^r \to R^{p}{}_i \ (x \in \Gamma_i)\,, i = 1, \ldots, k\right)$$

$$\text{If } h_j(x) \le 0, \text{ Then } x \in G \quad \left(h_j(x) : R^r \to R^{q}{}_j \left(x \in \Delta_j\right), j = 1, \ldots, l\right)$$

Based on the above theory in last section, we converted the knowledge to the following constraints:

There exist $v_i \in R^{p}{}_i$, $i = 1, \ldots, k$, $r_j \in R^{q}{}_j$, $j = 1, \ldots, l$, $v_i, r_j \ge 0$ such that:

$$\begin{aligned} &-\lambda_1 y_1 K(X_1, x) - \cdots - \lambda_n y_n K(X_n, x) + b + {v_i}^T g_i(x) + s_i \ge 0, \quad (x \in \Gamma) \\ &\lambda_1 y_1 K(X_1, x) + \cdots + \lambda_n y_n K(X_n, x) - b + {r_j}^T h_j(x) + t_j \ge 0, \quad (x \in \Delta) \end{aligned} \tag{2.145}$$

These constraints can be easily imposed to KMCLP model as the constraints acquired from prior knowledge.

Nonlinear knowledge in KMCLP classifier [36]:

$$\begin{aligned} &\text{Min} \left(d_\alpha^+ + d_\alpha^- + d_\beta^+ + d_\beta^-\right) + C\left(\sum_{i=1}^{k} s_i + \sum_{j=1}^{l} t^j\right) \\ \text{s.t.} \quad &\lambda_1 y_1 K(X_1, X_1) + \cdots + \lambda_n y_n K(X_n, X_1) = b + \alpha_1 - \beta_1,, \quad \text{for } X_1 \in \text{B}, \\ &\quad \cdot \\ &\quad \cdot \\ &\quad \cdot \\ &\lambda_1 y_1 K(X_1, X_n) + \cdots + \lambda_n y_n K(X_n, X_n) = b - \alpha_n + \beta_n, \quad \text{for } X_\text{n} \in \text{G}, \\ &\alpha^* + \sum_{i=1}^{n} \alpha_i = d_\alpha^- - d_\alpha^+, \\ &\beta^* - \sum_{i=1}^{n} \beta_i = d_\beta^- - d_\beta^+, \\ &-\lambda_1 y_1 K(X_1, x) - \cdots - \lambda_n y_n K(X_n, x) + b + {v_i}^T g_i(x) + s_i \ge 0, \qquad \text{i} = 1, \ldots, \text{k} \\ &s_i \ge 0, \qquad \text{i} = 1, \ldots, \text{k} \\ &\lambda_1 y_1 K(X_1, x) + \cdots + \lambda_n y_n K(X_n, x) - b + {r_j}^T h_j(x) + t_j \ge 0, \qquad \text{j} = 1, \ldots, l \\ &t_j \ge 0, \qquad \text{j} = 1, \ldots, l \\ &\alpha_1, \ldots, \alpha_\text{n} \ge 0, \quad \beta_1, \ldots, \beta_\text{n} \ge 0, \quad \lambda_1, \ldots, \lambda_\text{n} \ge 0, \\ &(v_i, r_j) \ge 0 \\ &d_\alpha^-, d_\alpha^+, d_\beta^-, d_\beta^+ \ge 0 \end{aligned} \tag{2.146}$$

In this model, all the inequality constraints are derived from the prior knowledge. The last objective $C\left(\sum_{i=1}^{k} s_i + \sum_{j=1}^{l} t^j\right)$ is about the slack error. Theoretically, the larger the value of C, the greater impact on the classification result of the knowledge sets.

The parameters need to be set before optimization process are C, q (if we choose RBF kernel), α^* and β^*. The best bounding plane of this model decided by (λ, b) of the two classes is the same with formula (2.139).

2.2.2 Decision Rule Extraction for Regularized Multiple Criteria Linear Programming Model

In this section, we present a clustering-based rule extraction method to generate decision rules from the black box RCMLP model. Our method can improve the interpretability of the RMCLP model by using explicit and explainable decision rules. To achieve this goal, a clustering algorithm will first be used to generate *prototypes* (which are the clustering centers) for each group of examples identified by the RMCLP model. Then, hyper cubes (whose edges are parallel to the axes) will be extracted around each prototype. This procedure will be repeated until all the training examples are covered by a hyper cube. Finally, the hyper cubes will be translated to a set of *if-then* decision rules. Experiments on both synthetic and real-world data sets have demonstrate the effectiveness of our rule extraction method.

For ease of description, we introduce some notations first. Assume a r-dimensional space, the coordinate of the clustering center p is $p = (p_1, \ldots, p_r)$, and the classification hyper plane is $\sum_{i=1}^{r} a_i x_i = b$ (where x_i is the direction of the hyper plane). For each class, we prefer hyper cubes which cover as many examples as possible. Intuitively, if we pick a point u on the classification boundary and then draw cubes based on both clustering center p and u, then the generated hyper cube will cover the largest area with respect to the current prototype p. The distance from p to the hyper plane can be calculated by Eq. (2.147) as follows:

$$d = Distance\,(f, p_i) = \frac{\sum_{i=1}^{r} p_i x_i - b}{\sqrt{x_i^2}} \tag{2.147}$$

After computing d, Step 2.3 draws hyper cubes $H = DrawHC(d, P_i)$ by using the prototype point P_i as the central point, and each edge has a length of $\sqrt{2}d$ meanwhile parallel with the axis. By so doing, we can get *if-then* rules which are easily understood. For example, for a specific example $a_1 \in G_1$, a decision rule can be described in the following form:

$$\begin{aligned} &if \;\; (l_1 \le a_{11} \le u_1) \;\; and \;\; (l_2 \le a_{12} \le u_2) \ldots\ldots and \;\; (l_r \le a_{1r} \le u_r) \\ &then \; a_1 \; belongs \; to \; class \; 1 \end{aligned} \tag{2.148}$$

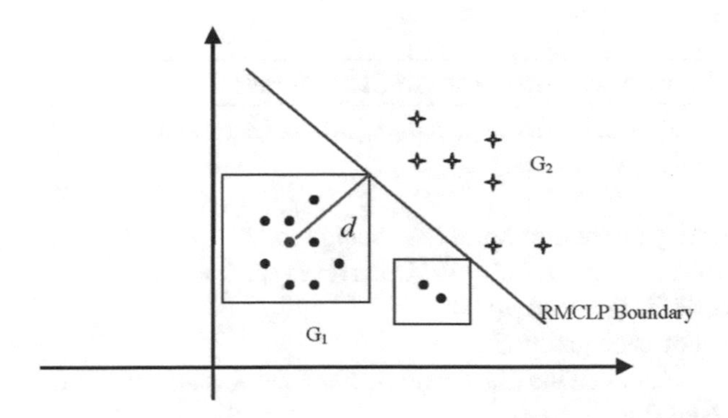

Fig. 2.8 An illustration of Algorithm 2.5 which generates hyper cubes from RMCLP models. Based on the RMCLP model's decision boundary (the red line), Algorithm 2.5 first calculates several clustering centers for each class (e.g., the red circle in Group 1), then it calculates the distance d from the classification boundary to the clustering center (the blue line). After that, it generates a series of hyper cubes. Each hyper cube's edge is parallel to the axes and the length is $\sqrt{2}d$. Finally, the hyper cubes can be easily translated into rules that are explainable and understandable

Figure 2.8 illustrates an example with two dimensions. Examples in G_1 ($a_i \in G_1$) are covered by hyper cubes with a central point as its clustering center and a vertex on the hyper plane $\sum_{i=1}^{r} a_i x_i = b$.

The main computational cost of Algorithm 2.5 is from *Steps 2.1~2.3*, where a K-Means clustering model and two distance functions are calculated. Assume there are l iterations of K-Means. In each iteration, there are k clusters. Therefore, the total time complexity of K-Means will be O($lknr$), where n is the number of training examples, r is the number of dimensions.

On the other hand, calculating distance d for each clustering center by (2.147) will take a linear time complexity, so the computational cost of *Step 2.2* will be $O(k)$ for k clustering centers. Finally, the time cost of extracting hyper cubes in *Step 2.3* will be $O(kr)$ for k clustering centers in r dimensional space. To sum up, the total computational complexity of Algorithm 2.5 can be denoted by (2.149),

$$O(lknr) + O(k) + O(kr) = O(lknr) \tag{2.149}$$

The above analysis indicates that the hyper cube extracting method in *Steps 2.2* and *2.3* is dominated by the K-Means clustering model in *Step 2.1*. It is in linear time complexity with respect to training example size.

Algorithm 2.5 Extract Rules from MCLP Models

Input: The data set $A = \{a_1, a_2, \ldots, a_n\}$, RMCLP model f
Output: Rule Set $\{w\}$
Begin
Step 1. Classify all the examples in A using model f;
Step 2. Define Covered set $C = \Phi$, Uncovered set $U = A$;
Step 3. **While** (*U is not empty*) do
Step 3.1 **For** each group G_i,
Calculate the clustering center $P_i =$ *K-means*$(G_i \cap U)$;
End for
Step 3.2 Calculate distances between each P_i and boundary $d =$ *Distance*(f, P_i);
Step 3.3 Draw a new hypercube $H =$ *DrawHC*(d, P_i);
Step 3.4 **For** all the examples$a_i \in U$,
If a_i is covered by H
$U = U \backslash a_i$, $C = C \cup a_i$;
End If
End For
End While
Step 4 Translate each hypercube H into rule;
Step 5 Return the rule set $\{w\}$
End

To demonstrate the effectiveness of the proposed rules extraction method, we will test our method on both synthetic and real-world data sets. The whole testing system is implemented in a Java environment by integrating WEKA data mining tools [37]. The clustering method used in our experiments is the *simple K-Means* package in WEKA.

As shown in Fig. 2.9a, we generate a 2-dimensional 2-class data set containing 60 examples, with 30 examples for each class. In each class, we use 50% of the examples to train a RMCLP model. That is, 30 training examples in total are used to train the RMCLP model. All examples comply with Gaussian distribution $x \sim N(\mu, \Sigma)$, where μ is mean vector and Σ is covariate matrix. The first group is generated by a mean vector $\mu_1 = [1,1]$ with a covariance matrix $\Sigma_1 = \begin{bmatrix} 0.1 & 0 \\ 0 & 0.1 \end{bmatrix}$. The second group is generated by a mean vector $\mu_2 = [2,2]$ with a covariance matrix $\Sigma_2 = \Sigma_1$.

Here we only discuss the two-group classification problem. It is not difficult to extend to multiple-group classification applications. It is expected to extract knowledge from the RMCLP model in the form of:

$$\textbf{\textit{if}}\ (a \le x1 \le b, c \le x2 \le d)\ \textbf{\textit{then}}\ Definition\ 1 \tag{2.150}$$

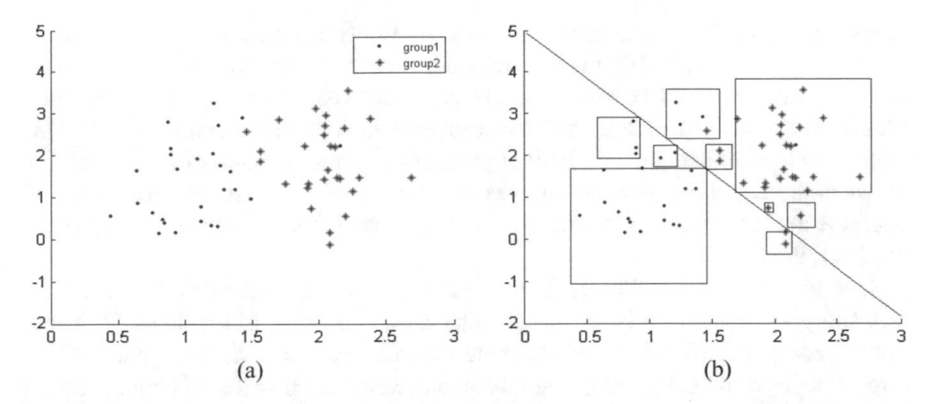

Fig. 2.9 (**a**) The synthetic dataset; (**b**) Experimental results. The straight line is the RMCLP model's classification boundary, and the squares are hyper cubes generated by using Algorithm 2.5. All the examples are covered by the squares whose edges are parallel to the axes

The result is shown in Fig. 2.9b; we can observe that for the total of 60 examples, three examples in group 1, and one example in group 2 are misclassified by the RMCLP model. That is to say, the accuracy of RMCLP on this synthetic dataset is $56/60 = 92.3\%$. By using our rule extraction algorithm, we can generate nine squares, four squares for group 1, and five squares for group 2. All the squares can be translated to explainable rules in the form of (6) as follows:

K_1: if $0.6 \le x_1 \le 0.8$ and $2 \le x_2 \le 2.8$, then $x \in G_1$;
K_2: if $1.1 \le x_1 \le 1.3$ and $1.8 \le x_2 \le 2.1$, then $x \in G_1$;
K_3: if $0.4 \le x_1 \le 1.5$ and $-1 \le x_2 \le 1.6$, then $x \in G_1$;
K_4: if $0.9 \le x_1 \le 2.2$ and $-0.8 \le x_2 \le 0$, then $x \in G_1$;
K_5: if $1.2 \le x_1 \le 1.6$ and $2.2 \le x_2 \le 3.2$, then $x \in G_2$;
K_6: if $1.4 \le x_1 \le 1.6$ and $1.8 \le x_2 \le 2.0$, then $x \in G_2$;
K_7: if $1.7 \le x_1 \le 2.8$ and $1.0 \le x_2 \le 4.0$, then $x \in G_2$;
K_8: if $1.9 \le x_1 \le 2.0$ and $0.7 \le x_2 \le 0.8$, then $x \in G_2$;
K_9: if $2.1 \le x_1 \le 2.4$ and $0.1 \le x_2 \le 0.5$, then $x \in G_2$;

where k_i $(i = 1, \ldots, 9)$ denotes the ith rule. From the results on this synthetic data set, we can observe that by using the proposed rule extraction method, we can not only obtain prediction results from RMCLP, but also comprehensible rule.

As one of the basic services offered by the Internet, E-Mail usage is becoming increasingly widely adopted. Along with constant global network expansion and network technology improvement, people's expectations of an E-Mail service are increasingly demanding. E-Mail is no longer merely a communication tool for people to share their ideas and information; its wide acceptance and technological advancement has given it the characteristics of a business service [38], and it is being commercialized as a technological product.

At the same time, many business and specialized personal users of E-Mail want an E-Mail account that is safe, reliable, and equipped with a first-class customer support service. Therefore, many websites have developed their own user-pays E-mail service to satisfy this market demand. According to statistics, the Chinese network has advanced so much in the past few years that, by 2005, the total market size of Chinese VIP E-mail services reached 6.4 hundred million RMB. This enormous market demand and market prospect also means increasing competition between the suppliers. How to analyze the pattern of lost customer accounts and decrease the customer loss rate have become a focal point of competition in today's market [39, 40].

Our partner company's VIP E-Mail data are mainly stored in two kinds of repository systems; one is customer databases, the other is log files. They are mainly composed of automated machine recorded customer activity journals and large amount of manually recorded tables; these data are distributed among servers located in different departments of our partnering companies, coving more than 30 kinds of transaction data charts and journal documents, with over 600 attributes.

If we were to directly analysis these data, it would lead to a "course of dimensionality", that is to say, a drastic rise in computational complexity and classification error with data of large dimensions. Hence, the dimensionality of the feature space must be reduced before classification is undertaken. According to the accumulated experience functions, we eventually selected 230 attributes from the original 600 attributes.

Figure 2.10 displays the procedure of feature selection of the VIP E-Mail dataset. We selected a part of the data charts and journal documents from the VIP E-Mail System. The left upper part of Fig. 2.10 displays the three logging journal documents and two email transaction journal documents; when the user logs into the pop3 server, the machine will record the user's login into the log file *pop3login*; similarly when the user logs into the smtp server, the machine will record this into the log file *smtplogin*; when the user logs into the E-Mail system through http protocol, the machine will record it into the log file *weblogin*; when the user successfully sends an E-Mail by smtp protocol, the system will record it into the log file *smtprcptlog*; when receiving a letter, it will be recorded into the log file *mx_rcptlog*.

We extracted 37 attributes from these five log files, that is, 184 attributes in total, to describe user logins and transactions. From the databases, shown in the left lower section of Fig. 2.8, we extracted six features about "customer complaint about the VIP E-Mail Service", 24 features about "customer payment" and 16 features about "customer's personal information" (for example, age, gender, occupation, income etc.) to form the operational table. Thus, 185 features from log files and 65 features from databases eventually formed the Large Table, and the 230 attributes depicted the features of the customers. The accumulated experience functions used in the feature selection are confidential, and further discussion of them exceeds the range of this section.

Considering the integrality of the customer records, we eventually extracted two groups from a huge number of data: the current and the lost. Ten thousand nine hundred and ninety-six customers, 5498 for each class, were chosen from the dataset. Combining the 10,996 SSN with the 230 features, we eventually acquired the Large Table with 5498 current records and 5498 lost records, which became the dataset for data mining.

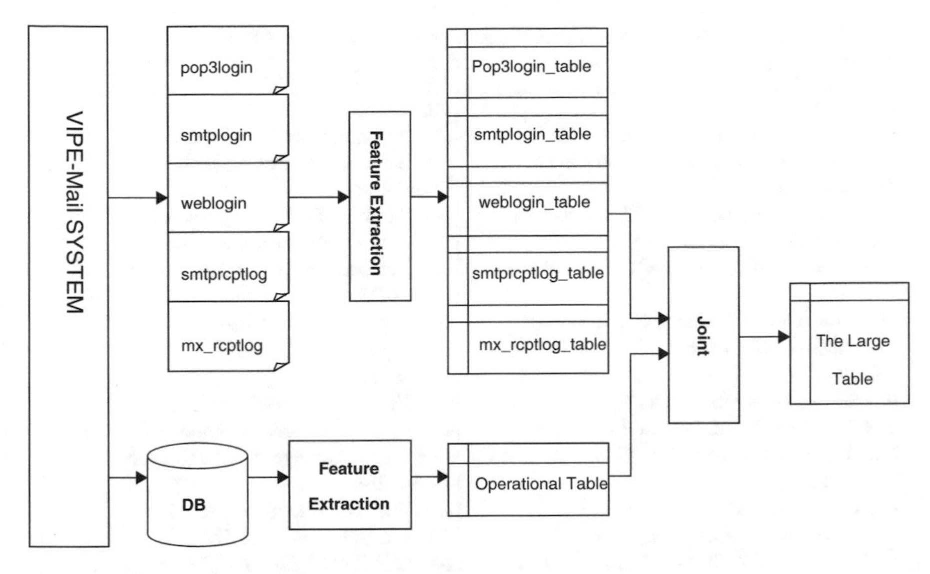

Fig. 2.10 The roadmap of the VIP Email Dataset

Table 2.6 Ten Folder Cross Validation on VIP Email Dataset

Cross validation	Training set (500 Bad data + 500 Good data)				Testing set (4998 Bad data + 4998 Good data)			
	LOST	Accuracy (%)	CURRENT	Accuracy (%)	LOST	Accuracy (%)	CURRENT	Accuracy (%)
DataSet 1	444	88.80	455	91.00	4048	80.99	4311	86.25
DataSet 2	447	89.40	459	91.80	4081	81.65	4355	87.13
DataSet 3	449	89.80	465	92.00	4079	81.61	4362	87.27
DataSet 4	440	88.00	467	92.40	4006	80.15	4286	85.75
DataSet 5	435	87.00	460	92.00	4010	80.23	4420	88.44
DataSet 6	436	87.20	460	92.00	3995	79.93	4340	86.83
DataSet 7	445	89.00	464	92.80	4008	80.19	4403	88.10
DataSet 8	443	88.60	455	91.00	4052	81.07	4292	85.87
DataSet 9	429	85.80	457	91.40	3955	79.13	4436	88.76
DataSet10	440	88.00	456	91.20	4087	81.77	4355	87.13

Table 2.6 lists the ten-folder cross validation results of the RMCLP model's performance on the VIP Email Dataset. The columns "LOST" and "CURRENT" refer to the number of records that were correctly classified as "lost" and "current" respectively. The column "Accuracy" was calculated using correctly classified records divided by the total records in that class. From Table 2.6, we can observe that the average prediction accuracy of the RMLCP on this data set is 80.67% on the first class and 87.15% on the second class. That is, on the whole 10,996 test examples, the average accuracy of RMCLP is 82.91%.

Table 2.7 Comparisons between RMCLP's Rule and Decision Tree's Rule

RMCLP's Rule	Decision Tree's Rule
RULE 1: if 0 <= The number of emails <= 3 *and* 0 <= the number of POP3 login on Tuesday <= 6 *and* 0 <= the number of HTTP login <= 1 *and* 0 <= Free Email Service <= 1 *and* 0 <= The percentage of Charge Type 7 <= 0.3 *and* 0 <= The total Charge Fee <= 45 ... then *class LOST* [0.816]	**RULE** $\mathbf{1'}$: if The number of emails <= 1 *and* the number of POP3 login on Tuesday <= 3 *and* number of HTTP login <= 1 *and* Free Email Service = 1 *and* The percentage of Charge Type 7 <= 0.25 *and* The total Charge Fee <= 50 ... then *class LOST* [0.746]
RULE 6: if 0 <= The number of HTTP Login <= 5 *and* 0 <= Free Email Service Status <= 1 *and* 0.2 <= The percentage of Charge Type 11 <= 0.5 *and* 0 <= The total Charge Fee <= 4 *and* 0 <= The number of emails <= 3 *and* 0 <= CONTACT_NUMBER <= 1 *and* 0 <= IDNUM <= 1 ... then *class CURRENT* [0.802]	**RULE** $\mathbf{6'}$: if The number of HTTP Login <= 3 *and* Free Email Service Status = 0 *and* The percentage of Charge Type 11 > 0.294 *and* The total Charge Fee <= 5 *and* The number of Emails <= 1 *and* CONTACT_NUMBER = 1 *and* IDNUM = 0 ... then *class CURRENT* [0.739]
Average Accuracy: 80.90%	Average Accuracy: 74.25%

As discussed above, a decision tree is widely used to extract rules from training examples. In the following experiments, we will compare our method with a decision tree (which is implemented by the WEKA *J48* package).

Table 2.7 shows the comparison results between our method and the decision tree. By using our rule extraction method, we obtain more than 20 hyper cubes. Due to space limitation, we only list the two most representative rules (i.e., Rule 1 for class "LOST" and Rule 6 for class "CURRENT") in the left side of Table 2.7. Then we find the corresponding rules from the decision tree (i.e., Rule $1'$ for class "LOST" and Rule $6'$ for class "CURRENT"), and list them in the right side of Table 2.7.

From these results, we can observe that our rule extraction method acquires much more accurate rules than the decision tree method. For example, when comparing Rule 1 with Rule $1'$, we can safely say that Rule 1 is supported by 81.6% examples in the "LOST" class; by contrast, rules from decision tree only get 74.6% supportive examples. Similarly, when comparing Rule 6 with Rule $6'$, our method also achieves better support than the decision tree.

At the bottom of Table 2.7, we list the average accuracy of the two methods. It is obvious that the average accuracy of rules extracted from RMCLP is 80.90%. This is better than the decision tree's accuracy of 74.25%. Moreover, compared to the RMCLP's performance in Table 2.6 (which equals 82.91%), we can say that the average accuracy of the extracted rules (i.e., 80.90%) suffers only a little loss in performance. Therefore, our rule extraction method from the RMCLP model can effectively extract comprehensible rules from the RMCLP model.

2.3 Multiple-Criteria Decision Making Based Data Analysis

2.3.1 A Multicriteria Decision Making Approach for Estimating the Number of Clusters

Estimating the number of clusters for a given data set is closely related to the validity measures and the data set structures. Many validity measures have been proposed and can be classified into three categories: external, internal, and relative [41]. External measures use predefined class labels to examine the clustering results. Because external validation uses the true class labels in the comparison, it is an objective indicator of the true error rate of a clustering algorithm. Internal measures evaluate clustering algorithms by measuring intra- and inter-cluster similarity. An algorithm is regarded as good if the resulting clusters have high intra-class similarities and low inter-class similarities. Relative measures try to find the best clustering structure generated by a clustering algorithm using different parameter values. Extensive reviews of cluster validation techniques can be found in [41] and [42, 43].

Although external measures perform well in predicting the clustering error in previous studies, they require a priori structure of a data set and can only be applied to data sets with class labels. Since this study concentrates on data sets without class labels, it utilizes relative validity measures. The proposed approach can be applied to a wide variety of clustering algorithms. For simplicity, this study chooses the well-known k-means clustering algorithm. Figure 2.11 describes the MCDM-based approach for determining the number of clusters in a data set. For a given data set, different numbers of clusters are considered as alternatives and the performances of k-means clustering algorithm on the relative measures with different numbers of clusters represent criteria by MCDM methods. The output is a ranking of numbers of clusters, which evaluates the appropriateness of different numbers of clusters for a given data set based on their overall performances for multiple criteria (i.e., selected relative measures).

2.3.1.1 MCDM Methods

This study chooses three MCDM methods for estimating the number of clusters for a data set. This section introduces the selected MCDM methods (i.e., WSM, PROMETHEE, and TOPSIS) and explains how they are used to estimate the optimal number of clusters for a given data set.

MCDM Method 1: Weighted Sum Method (WSM)

The weighted sum method (WSM) was introduced by Zadeh [44]. It is the most straightforward and widely-used MCDM method for evaluating alternatives. When

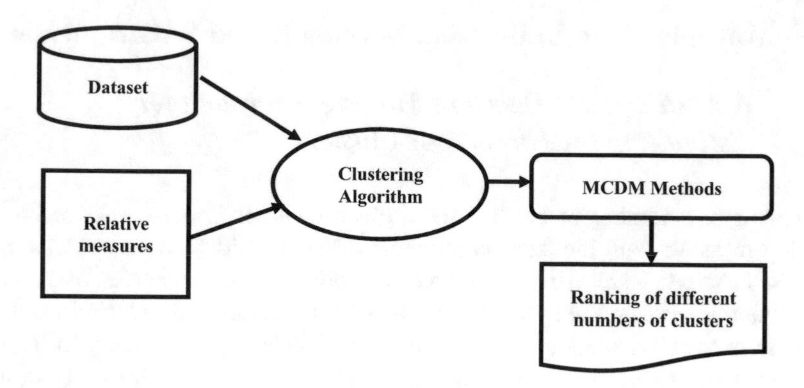

Fig. 2.11 A MCDM-based approach for determining the number of clusters in a dataset

an MCDM problem involves both benefit and cost criteria, two approaches can be used to deal with conflicting criteria. One is the benefit to cost ratio and the other is the benefit minus cost [45]. For the estimation of optimal number of clusters for a data set, the relative indices Dunn, silhouette, and PBM are benefit criteria and have to be maximized, while Hubert, normalized Hubert, Davies-Bouldin index, SD, S_Dbw, CS, and C-index are cost criteria and have to be minimized. This study chooses the benefit minus cost approach and applies the following formulations to rank different numbers of clusters.

Suppose there are m alternatives, k benefit criteria, and n cost criteria. The total benefit of alternative $A_i^{benefit}$ is defined as follows:

$$A_i^{benefit} = \sum_{j=1}^{k} w_j a_{ij},\ for\ \ i = 1, 2, 3, \ldots, m$$

where a_{ij} represents the performance measure of the jth criterion for alternative Ai. Similarly, the total cost of alternative $A_i^{\cos t}$ is defined as follows:

$$A_i^{\cos t} = \sum_{j=1}^{n} w_j a_{ij},\ for\ \ i = 1, 2, 3, \ldots, m$$

$where \sum_{j=1}^{k} w_j + \sum_{j=1}^{n} w_j = 1; 0 < w_j \leq 1$. Then the importance of alternative $A_i^{WSM-score}$ is defined as follows:

$$A_i^{WSM-score} = A_i^{benefit} - A_i^{\cos t},\ for\ \ i = 1, 2, 3, \ldots, m$$

The best alternative is the one has the largest WSM score [45].

MCDM Method 2: Preference Ranking Organization Method for Enrichment of Evaluations (PROMETHEE)

Brans proposed the PROMETHEE I and PROMETHEE II, which use pairwise comparisons and outranking relationships to choose the best alternative [46]. The final selection is based on the positive and negative preference flows of each alternative. The positive preference flow indicates how an alternative is outranking all the other alternatives and the negative preference flow indicates how an alternative is outranked by all the other alternatives [47]. While PROMETHEE I obtains partial ranking because it does not compare conflicting actions [48], PROMETHEE II ranks alternatives according to the net flow which equals to the balance of the positive and the negative preference flows. An alternative with a higher net flow is better [47]. Since the goal of this study is to provide a complete ranking of different numbers of clusters, PROMETHEE II is utilized. The following procedure presented by Brans and Mareschal [47] is used in the experimental study:

Step 1. Define aggregated preference indices.

Let a,b ∈A, and let

$$\begin{cases} \pi(a,b) = \sum\limits_{j=1}^{k} p_j(a,b)\, w_j, \\ \pi(a,b) = \sum\limits_{j=1}^{k} p_j(b,a)\, w_j. \end{cases}$$

where A is a finite set of possible alternatives {a_1, $a_2, \ldots,$ a_n}, k represents the number of evaluation criteria, and w_j is the weight of each criterion. For estimating the number of clusters for a given data set, the alternatives are different numbers of clusters and the criteria are relative indices. Arbitrary numbers for the weights can be assigned by decision-makers. The weights are then normalized to ensure that $\sum_{j=1}^{k} w_j = 1$. $\pi(a,b)$ indicates how a is preferred to b over all the criteria and $\pi(b,a)$ indicates how b is preferred to a over all the criteria. $P_j(a,b)$ and $P_j(b,a)$ are the preference functions for alternatives a and b. The relative indices Dunn, silhouette, and PBM have to be maximized, and Hubert, normalized Hubert, DB, SD, S_Dbw, CS, and C-index have to be minimized.

Step 2. Calculate $\pi(a,b)$ and $\pi(b,a)$ for each pair of alternatives of A. There are six types of preference functions and the decision-maker needs to choose one type of the preference functions for each criterion and the values of the corresponding parameters [49]. The usual preference function, which requires no input parameter, is used for all criteria in the experiment.

Step 3. Define the positive and the negative outranking flow as follows:
The positive outranking flow:

$$\phi^{+}(a) = \frac{1}{n-1}\sum_{x \in A} \pi\,(a, x)\,,$$

The negative outranking flow:

$$\phi^{-}(a) = \frac{1}{n-1}\sum_{x \in A} \pi\,(x, a)\,,$$

Step 4. Compute the net outranking flow for each alternative as follows:

$$\phi(a) = \phi^{+}(a) - \phi^{-}(a).$$

When $\phi(a) > 0$, a is more outranking all the alternatives on all the evaluation criteria. When $\phi(a)<0$, a is more outranked.

MCDM Method 3: Technique for Order Preference by Similarity to Ideal Solution (TOPSIS)

The Technique for order preference by similarity to ideal solution (TOPSIS) method was proposed by Hwang and Yoon [50] to rank alternatives over multiple criteria. It finds the best alternatives by minimizing the distance to the ideal solution and maximizing the distance to the nadir or negative-ideal solution [37]. This section uses the following TOPSIS procedure, which was adopted from [51] and [37], in the empirical study:

Step 1. Calculate the normalized decision matrix. The normalized value r_{ij} is calculated as

$$r_{ij} = x_{ij} / \sqrt{\sum_{i=1}^{J} x_{ij}^{2}},\ j = 1, \ldots, J; i = 1, .., n.$$

Step 2. Develop a set of weights wi for each criterion and calculate the weighted normalized decision matrix. The weighted normalized value vij is calculated as:

$$v_{ij} = w_i r_{ij},\ j = 1, .., J; i = 1, .., n.$$

Weight of the ith criterion, and $\sum_{i=1}^{n} w_i = 1$.

Step 3. Find the ideal alternative solution S+, which is calculated as:

$$S^{+} = \{v_1^{+}, \ldots, v_n^{+}\} = \left\{ \left(\max_j v_{ij} | i \in I' \right), \left(\min_j v_{ij} | i \in I'' \right) \right\}$$

where I′ is associated with benefit criteria and I″ is associated with cost criteria. In this study, benefit and cost criteria of TOPSIS are defined the same as the benefit and cost criteria in WSM.

Step 4. Find the negative-ideal alternative solution S2, which is calculated as:

$$S^{-} = \{v_1^{-}, \ldots, v_n^{-}\} = \left\{ \left(\min_j v_{ij} | i \in I' \right), \left(\max_j v_{ij} | i \in I'' \right) \right\}$$

Step 5. Calculate the separation measures, using the n-dimensional Euclidean distance. The separation of each alternative from the ideal solution is calculated as:

$$D_j^{+} = \sqrt{\sum_{i=1}^{n} \left(v_{ij} - v_i^{+}\right)^2}, \; j = 1, \ldots, J.$$

The separation of each alternative from the negative-ideal solution is calculated as:

$$D_j^{-} = \sqrt{\sum_{i=1}^{n} \left(v_{ij} - v_i^{-}\right)^2}, \; j = 1, \ldots, J.$$

Step 6. Calculate a ratio R_j^{+} that measures the relative closeness to the ideal solution and is calculated as:

$$R_j^{+} = D_j^{-} / \left(D_j^{+} + D_j^{-}\right), \; j = 1, \ldots, J.$$

Step 7. Rank alternatives by maximizing the ratio R_j^{+}.

2.3.1.2 Clustering Algorithm

The k-means algorithm, the most well-known partitioning method, is an iterative distance-based technique [32]. The input parameter k predefines the number of clusters. First, k objects are randomly chosen to be the centers of these clusters. All objects are then partitioned into k clusters based on the minimum squared-error criterion, which measures the distance between an object and the cluster center. The new mean of each cluster is calculated and the whole process iterates until the cluster centers remain the same [11, 52]. Let X = $\{x_i\}$ be the n objects to be clustered, C = $\{C_1, C_2, \ldots, C_k\}$ is the set of clusters. Let mi be the mean of cluster C_i. The squared-error between μ_i and the objects in cluster C_i is defined as.

$$WCSS\,(C_i) = \sum_{x_j \in C_i} \left\| x_j - \mu_i \right\|^2$$

Then the aim of k-means algorithm is to minimize the sum of the squared error over all *k* clusters, that is

$$\min\left(WCSS(C) = \arg\min_{C} \sum_{x_j \in C_i} \left\| x_j - \mu_i \right\|^2\right.$$

where WCSS denotes the sum of the squared error in the inner-cluster.

Two critical steps of *k*-means algorithm have impact on the sum of squared error. First, generate a new partition by assigning each observed point to its closest cluster center, the formula is as follows:

$$C_i{}^{(t)} = \left\{x_j : \left\| x_j - m_i{}^{(t)} \right\| \le \left\| x_j - m_{i*}{}^{(t)} \right\| foralli* = 1, .., k\right\}$$

where $m_i{}^{(t)}$ denotes the mean of the ith cluster in t^{th} times clustering, while $C_i{}^{(t)}$ represents all sets contained in the ith cluster in t^{th} times clustering. Second, compute new cluster mean centers using the following formula.

$$m_i{}^{(t+1)} = \frac{1}{\mid C_i{}^{(t+1)} \mid} \sum_{x_j \in C_i^{(t)}} x_j$$

where $m_i{}^{(t+1)}$ denotes the mean of the ith cluster in $(t + 1)^{th}$ times clustering while $C_i{}^{(t+1)}$ represents all sets contained in the ith cluster in $(t + 1)^{th}$ times clustering. The algorithm is implemented using WEKA (Waikato Environment for Knowledge Analysis), a free machine learning software [53].

2.3.1.3 Clustering Validity Measures

Ten relative measures are selected for the experiment, namely, the Hubert Γ statistic, the normalized Hubert Γ, the Dunn's index, the Davies-Bouldin index, the CS measure, the SD index, the S_Dbw index, the silhouette index, PBM, and the C-index. Relative measures can also be used to identify the optimal number of clusters in a data set and some of them, such as the C-index and silhouette, have exhibited good performance in previous studies. The following paragraphs define these relative measures.

- **Hubert Γ statistic** [54]:

$$\Gamma = (1/M) \sum_{i=1}^{n-1} \sum_{j=i+1}^{n} P(i, j) \cdot Q(i, j)$$

where n is the number of objects in a data set, M = n(n − 1)/2, P is the proximity matrix of the data set, and Q is an n*n matrix whose (i, j) element is equal to the distance between the representative points (v_{ci},v_{cj}) of the clusters where the objects xi and x_j belong [42]. C indicates the agreement between P and Q.

- **Normalized Hubert Γ**:

$$\hat{\Gamma} = \frac{\left[(1/M) \sum_{i=1}^{n-1} \sum_{j=i+1}^{n} (P(i,j) - \mu_P) \cdot (Q(i,j) - \mu_Q) \right]}{\sigma_P \sigma_Q}$$

where μ_P, μ_Q, σ_p, and σ_Q represent the respective means and variances of P and Q matrices [43].

Dunn's index [55] evaluates the quality of clusters by measuring inter cluster distance and intra cluster diameter.

$$D = \min_{i=1,\ldots,K} \left\{ \min_{j=i+1,\ldots,K} \left[\frac{d(C_i, C_j)}{\max_{l=1,\ldots,K} diam(C_l)} \right] \right\}$$

where K is the number of clusters, C_i is the ith cluster, d(C_i,C_j) is the distance between cluster C_i and C_j, and $diam(C_l)$ is the diameter of the lth cluster. Larger values of D suggest good clusters, and a D larger than 1 indicates compact separated clusters.

- **Davies-Bouldin index is defined as** [56]:

$$DB_K = \frac{1}{K} \sum_{i=1}^{K} R_i, R_i \max_{i=1,\ldots,K, i \neq j} R_{ij}, R_{ij} = \frac{s_i + s_j}{d_{ij}}, i = 1, \ldots, K$$

where K is the number of clusters, s_i and sj represent the respective dispersion of clusters i and j, dij measures the dissimilarity between two clusters, and R_{ij} measures the similarity between two clusters [42, 43]. It is the average similarity between each cluster and its most similar one.

- **The CS measure is proposed to evaluate clusters with different densities and/or sizes** [57]. It is computed as:

$$CS = \frac{\sum_{i=1}^{K} \left\{ \frac{1}{N_i} \sum_{x_j \in C_i} \max_{x_k \in C_i} \left\{ d\left(x_j, x_k\right) \right\} \right\}}{\sum_{i=1}^{K} \left\{ \min_{j \in \{1,2,\ldots,K\}, j \neq i} \left\{ d\left(v_i, v_j\right) \right\} \right\}}, v_i = \frac{1}{N_i} \sum_{x_j \in C_i} x_j$$

where N_i is the number of objects in cluster i and d is a distance function. The smallest CS measure indicates a valid optimal clustering.

- **SD index combines the measurements of average scattering for clusters and total separation between clusters** [42]:

$$SD(K) = Dis\left(c_{\max}\right) \times Scat(K) + Dis(K)$$

where c_{max} is the maximum number of input clusters,

$$Scat(K) = \frac{1}{K} \sum_{i=1}^{K} \|\sigma\left(v_i\right)\| / \|\sigma(X)\| \quad \text{and}$$

$$Dis(K) = \frac{D_{\max}}{D_{\min}} \sum_{k=1}^{K} \left(\sum_{z=1}^{K} \|v_k - v_z\| \right)^{-1}$$

D_{max} is the maximum distance between cluster centers and the D_{min} is the minimum distance between cluster centers.

S_Dbw index is similar to SD index and is defined as [42]:

$$SDbw(K) = Scat(K) + Densbw(K),$$

$$Densbw(K) = \frac{1}{K \cdot (K-1)} \sum_{i=1}^{K} \left(\sum_{\substack{j = 1 \\ j \neq i}}^{K} \frac{density(u_{ij})}{\max\{density(v_i), density(v_j)\}} \right),$$

$$density(u) = \sum_{l=1}^{N_{ij}} f\left(x_l, u\right)$$

where N_{ij} is the number of objects that belong to the cluster C_i and C_j, and function f(x,u) is defined as:

$$f(x,u) = \begin{cases} 0, if \ \ d(x,u) > stdev \\ 1, otherwise \end{cases}, stdev = \frac{1}{K}\sqrt{\sum_{i=1}^{K} \|\sigma(v_i)\|}$$

Silhouette is an internal graphic display for clustering methods evaluation. It represents each cluster by a silhouette, which shows how well objects lie within their clusters. It is defined as [58]:

$$s(i) = \frac{b(i) - a(i)}{\max\{a(i), b(i)\}}$$

where i represents any object in the data set, a(i) is the average dissimilarity of i to all other objects in the same cluster A, and b(i) is the average dissimilarity of i to all objects in the neighboring cluster B, which is defined as the cluster that has the smallest average dissimilarity of i to all objects in it. Note that A ≠B and the dissimilarity is computed using distance measures. Since a(i) measures how dissimilar i is to its own cluster and b(i) measures how dissimilar i is to its neighboring cluster, an s(i) close to one indicates a good clustering method. The average s(i) of the whole data set measures the quality of clusters.

- **PBM is developed by [40] and it is based on the intra-cluster and inter-cluster distances**:

$$PBM = \left(\frac{1}{K}\frac{E_1}{E_K}D_K\right)^2$$
$$\text{where } E_1 = \sum_{i=1}^{N} \|x_i - \overline{x}\|, E_K = \sum_{l=1}^{N}\sum_{x_i \in C_l} \|x_i - \overline{x}_l\|,$$
$$D_K = \max_{l,m=1,\ldots,K} \|\overline{x}_l - \overline{x}_m\|$$

The C-index [59] is based on intra-cluster distances and their maximum and minimum possible values [60]:

$$CI = \frac{\theta - \min\theta}{\max\theta - \min\theta}, \theta = \sum_{i=1}^{n-1}\sum_{j=i+1}^{n} q_{i,j} \|x_i - x_j\|$$

2.3.2 *Parallel Regularized Multiple Criteria Linear Programming Classification Algorithm*

In this section, the focus is on the RMCLP, and the designed and proposed Parallel version of RMCLP algorithm (PRMCLP). In order to overcome the compute and storage requirements that increase rapidly with the number of training sample, the second strategy is adopted, inspire by some findings in [61].

Let us give a brief introduction of MCLP as follows. For classification of the training data:

$$T = \{(x_1, y_1), \ldots, (x_l, y_l)\} \in \left(\Re^n \times y\right)^l \tag{2.151}$$

where $x_i \in \Re^n$, $y_i \in y = \{1, \text{-}1\}$, $i = 1, \ldots, l$, data separation can be achieved by two opposite objectives. The first objective separates the observations by minimizing the sum of the deviations (MSD) among the observations. The second maximizes the minimum distances (MMD) of observations from the critical value [62]. The overlapping of data $\xi^{(1)}$ should be minimized while the distance $\xi^{(2)}$ has to be maximized. However, it is difficult for traditional linear programming to optimize MMD and MSD simultaneously. According to the concept of Pareto optimality, we can seek the best trade-off between the two measurements [10, 63]. So MCLP model can be described as follows:

$$\min e^{\mathrm{T}}\xi^{(1)} \& \max e^{\mathrm{T}}\xi^{(2)} \tag{2.152}$$

$$s.t. (w \cdot x_i) + \left(\xi_i{}^{(1)} - \xi_i{}^{(2)}\right) = b,\ for\ \{i | y_i = 1\}, \tag{2.153}$$

$$(w \cdot x_i) - \left(\xi_i{}^{(1)} - \xi_i{}^{(2)}\right) = b,\ for\ \{i | y_i = -1\}, \tag{2.154}$$

$$\xi^{(1)}, \xi^{(2)} \geq 0 \tag{2.155}$$

where $e \in \mathrm{R}^l$ be vector whose all elements are 1, w and b are unrestricted, $\xi_i^{(1)}$ is the overlapping and $\xi_i^{(2)}$ the distance from the training sample x_i to the discriminator $(w \cdot x_i) = b$ (classification separating hyperplane). By introducing penalty parameter C,D > 0, MCLP has the following version

$$\min_{\xi_i{}^{(1)}, \xi_i{}^{(2)}} Ce^{\mathrm{T}}\xi^{(1)} - De^{\mathrm{T}}\xi^{(2)}, <?pag? > \tag{2.156}$$

$$s.t. (w \cdot x_i) + \left(\xi_i{}^{(1)} - \xi_i{}^{(2)}\right) = b,\ for\ \{i | y_i = 1\}, \tag{2.157}$$

$$(w \cdot x_i) - \left(\xi_i{}^{(1)} - \xi_i{}^{(2)}\right) = b,\ for\ \{i | y_i = -1\}, \tag{2.158}$$

$$\xi^{(1)}, \xi^{(2)} \geq 0 \tag{2.159}$$

A lot of empirical studies have shown that MCLP is a powerful tool for classification. However, we cannot ensure that this model always has a solution under different kinds of training samples. To ensure the existence of solution, recently, Shi et al. proposed a RMCLP model by adding two regularized items $\frac{1}{2}\omega^{\mathrm{T}}H\omega$ and $\frac{1}{2}\xi^{(1)\mathrm{T}}Q\xi^{(1)}$ in MCLP as follows (more theoretical explanation of this model can be found in [63]):

$$\min_z \frac{1}{2} w^{\mathrm{T}} H w + \frac{1}{2} \xi^{(1)\mathrm{T}} Q \xi^{(1)} + \frac{1}{2} b^2 + C e^{\mathrm{T}} \xi^{(1)} - D e^{\mathrm{T}} \xi^{(2)}, \tag{2.160}$$

$$s.t. \, (w \cdot x_i) + \left({\xi_i}^{(1)} - {\xi_i}^{(2)} \right) = b, \, for \, \{i | y_i = 1\}, \tag{2.161}$$

$$(w \cdot x_i) - \left({\xi_i}^{(1)} - {\xi_i}^{(2)} \right) = b, \, for \, \{i | y_i = -1\}, \tag{2.162}$$

$$\xi^{(1)}, \xi^{(2)} \geq 0 \tag{2.163}$$

where $z = (w^{\mathrm{T}}, \xi^{(1)T}, \xi^{(2)T}, b)^{\mathrm{T}} \in R^{n+l+l+1}$, $H \in R^{n \times n}$ is symmetric positive definite matrices. Obviously, the regularized MCLP is a convex quadratic programming. According to the dual theorem, (2.160)–(2.163) can be formulated as:

$$\min_{\alpha, \xi^{(1)}} \frac{1}{2} \alpha^{\mathrm{T}} \left(K \left(A, A^{\mathrm{T}} \right) + e e^{\mathrm{T}} \right) \alpha + \frac{1}{2} \xi^{(1)\mathrm{T}} Q \xi^{(1)}, \tag{2.164}$$

$$s.t. - Q \xi^{(1)} - C e \leq E \alpha \leq -D e, \tag{2.165}$$

where $A = \left[x_1^{\mathrm{T}}, \ldots, x_l^{\mathrm{T}} \right]^{\mathrm{T}} \in R^{l \times n}$, $E = \operatorname{diag} \{y_1, \ldots, y_l\}$
and
$K \left(A, A^{\mathrm{T}} \right) = \Phi(A) \Phi(A)^{\mathrm{T}} = \left(\Phi(A) \cdot \Phi(A)^{\mathrm{T}} \right)_{l \times l}$

and Φ is a mapping from the input space Rn to some Hilbert space H [64].

In order to realize the parallelization of RMCLP, we firstly translate RMCLP into a unconstrained optimization problem. To simplify, (2.164) can be rewritten as

$$\begin{aligned} & \min_{\pi} \tfrac{1}{2} \pi^{\mathrm{T}} \Lambda \pi, \\ & s.t. G\pi - Ce \leq 0, \\ & H\pi + De \leq 0, \end{aligned} \tag{2.166}$$

where $\pi = [\alpha^{\mathrm{T}}, \xi^{(1)\mathrm{T}}]^{\mathrm{T}}$, and $G = [-Q, -E]$, $H = [E, O]$, $O \in R^{l \times l}$ is a null matrix, Λ is written as

$$\begin{pmatrix} K \left(A, A^{\mathrm{T}} \right) + e e^{\mathrm{T}} & 0 \\ 0 & Q \end{pmatrix} \tag{2.167}$$

Next, we represent the objective (2.164) as the following unconstrained optimization problem

$$\min_{\pi} f(\pi) = \frac{1}{2} \pi^{\mathrm{T}} \Lambda \pi + \lambda^{\mathrm{T}} \max \{G\pi - Ce, 0\}^2 + \mu \max \{H\pi + De, 0\}^2 \tag{2.168}$$

where C,D $\in$R are the artificial parameters, and $\lambda = \{\lambda_1, \ldots, \lambda_l\}$, $\mu = \{\mu_1, \ldots, \mu_l\}$.

Define d is the search direction of the optimization problem (2.168), here, we choose the negative gradient direction as the feasible direction:

$$d = -\nabla f(\pi) / \|\nabla f(\pi)\| \tag{2.169}$$

where

$$\nabla f(\pi) = \Lambda\pi + 2\lambda^{\mathrm{T}}\mathrm{diag}\left(G^{\mathrm{T}} \max\{G\pi - Ce, 0\}\right) + 2\mu^{\mathrm{T}}\mathrm{diag}\left(H^{\mathrm{T}} \max\{H\pi + De, 0\}\right) \tag{2.170}$$

Now, we use PVD idea to split our model [61]. Suppose we can use p processors, the variable of the unconstrained optimization problem (2.168) can be divided into p chunks: {1, ..., p}, where the dimension of the ith chunk is mi

$$\pi = \{\pi_1, \ldots, \pi_m\}, \pi_i \in R^{m_i}, i = 1, \ldots, p, \sum_{i=1}^{p} m_i = 2l \tag{2.171}$$

In the next step, we allocate the p-th variable to p-th processor, and decompose the problem (2.168) into the subproblem with mi dimensions. Each processor solves one corresponding subproblem, which update other variables on the basis of some rules except for computing the mi variables itself. After each processor finishes updating, a quick synchronous step is performed: searching the results obtained by each processor and computing the current solution. Repeating then this, our algorithm can be described as

Theorem 2.4 *The sequence generated by* $\{\pi^k\}$ *of Algorithm 2.4 either terminates at a stationary point* $\{\pi^k\}$*, or is an infinite sequence, whose accumulation point is stationary and* $\lim_{k\to\infty} \nabla f(\pi^k) = 0$.

Proof

$$\begin{gathered}
For\ \forall\pi, \pi' \in R^{2l}, \text{we have} \\
\nabla f(\pi) = \Lambda\pi + 2\lambda^{\mathrm{T}}\mathrm{diag}\left(G^{\mathrm{T}} \max\{G\pi - Ce, 0\}\right) + 2\mu^{\mathrm{T}}\mathrm{diag}\left(H^{\mathrm{T}} \max\{H\pi + De, 0\}\right) \\
So
\end{gathered} \tag{2.172}$$

$$\begin{aligned}
&\left\|\nabla f(\pi) - \nabla f\left(\pi'\right)\right\| = \left\|\Lambda\left(\pi - \pi'\right) + 2\lambda^{\mathrm{T}}\mathrm{diag}\left(G^{\mathrm{T}}(\max\{G\pi - Ce, 0\} - \max\{G\pi - Ce, 0\})\right)\right. \\
&\left. + 2\mu^{\mathrm{T}}\mathrm{diag}\left(H^{\mathrm{T}} \max\{H\pi + De, 0\} - \max\{H\pi + De, 0\}\right)\right)\| \\
&\le \|\Lambda\| \left\|\pi - \pi'\right\| + 2\left\|\lambda^{\mathrm{T}}\right\| \left\|\mathrm{diag}\left(G^{\mathrm{T}}(\max\{G\pi - Ce, 0\} - \max\{G\pi - Ce, 0\})\right)\right\| \\
&+ 2\left\|\mu^{\mathrm{T}}\right\| \left\|\mathrm{diag}\left(H^{\mathrm{T}} \max\{H\pi + De, 0\} - \max\{H\pi + De, 0\}\right)\right)\|
\end{aligned} \tag{2.173}$$

i) For any $G\pi_i, G\pi'_i \leq Ce, where i = 1, \ldots, m$, we have

$$\left\|\mathrm{diag}\left(G^{\mathrm{T}}\left(\max\left\{G\pi - Ce, 0\right\} - \max\left\{G\pi - Ce, 0\right\}\right)\right)\right\| = 0 \leq \left\|G^{\mathrm{T}}G\left(\pi_i - \pi_i'\right)\right\| \tag{2.174}$$

ii) For any $G\pi_i, G\pi'_i > Ce, where i = 1, \ldots, m$, we have

$$\left\|\mathrm{diag}\left(G^{\mathrm{T}}\left(\max\left\{G\pi - Ce, 0\right\} - \max\left\{G\pi - Ce, 0\right\}\right)\right)\right\| = \left\|G^{\mathrm{T}}G\left(\pi_i - \pi_i'\right)\right\| \tag{2.175}$$

Taken together, we can obtain

$$\begin{aligned}&\left\|\mathrm{diag}\left(G^{\mathrm{T}}\left(\max\left\{G\pi - Ce, 0\right\} - \max\left\{G\pi' - Ce, 0\right\}\right)\right)\right\| \\ &\leq \left\|G^{\mathrm{T}}G\left(\pi - \pi'\right)\right\| \leq \left\|G^{\mathrm{T}}\right\| \, \|G\| \left(\pi - \pi'\right) \|\end{aligned} \tag{2.176}$$

Similarly, we have

$$\begin{aligned}&\left\|\mathrm{diag}\left(H^{\mathrm{T}}\left(\max\left\{H\pi - De, 0\right\} - \max\left\{H\pi' - De, 0\right\}\right)\right)\right\| \\ &\leq \left\|H^{\mathrm{T}}\left(\pi - \pi'\right)\right\| \leq \left\|H^{\mathrm{T}}\right\| \, \|H\| \left(\pi - \pi'\right) \|\end{aligned} \tag{2.177}$$

As the result, let $\|\Lambda\| + 2\,\|\Lambda\|\,\left\|G^{\mathrm{T}}\right\|\,\|G\| + 2\,\|\mu\|\,\left\|H^{\mathrm{T}}\right\|\,\|\mathrm{H}\| = K$,
we can obtain

$$\left\|\nabla f(\pi) - \nabla f\left(\pi'\right)\right\| \leq \left\|\pi - \pi'\right\| \tag{2.178}$$

According to the Theorem 2.2 in [19], $\{\pi^k\}$ either terminates at a stationary point $\left\{\pi^{\bar{k}}\right\}$, or is an infinite sequence, whose accumulation point is stationary and $\lim_{k\to\infty} \nabla f\left(\pi^k\right) = 0$.

Theorem 2.5 *If A of Algorithm 2.4 is positive definite, then the sequence of iterates* $\{\pi^k\}$ *generated by the subproblem of* (2.168) *converges linearly to the unique solution* $\overline{\pi}$, *and the rate of convergence is*

$$\left\|\pi^k - \overline{\pi}\right\| \leq \left(\frac{2}{\gamma}\left(f\left(\pi^k\right) - f\left(\overline{\pi}\right)\right)\right)^{\frac{1}{2}}\left(1 - \frac{1}{p}\left(\frac{\gamma}{K}\right)^2\right)^{\frac{1}{2}}, \tag{2.179}$$

where $\gamma, K > 0$ *are constants.*

Proof For

$$\begin{aligned}&\forall\pi,\pi^{'}\in R^{2l}\\&\left(\nabla f(\pi)-\nabla f\left(\pi^{'}\right)\right)\left(\pi-\pi^{'}\right)=(\pi-\pi^{'})^{\mathrm{T}}\Lambda\left(\pi-\pi^{'}\right)+(2\lambda^{\mathrm{T}}\mathrm{diag}(G^{\mathrm{T}}\Big(\max\{G\pi-Ce,0\}\\&-\max\{G\pi-Ce,0\}))+2\mu^{\mathrm{T}}\mathrm{diag}\Big(H^{\mathrm{T}}\max\{H\pi+De,0\}\\&-\max\{H\pi+De,0\}))))\left(\pi-\pi^{'}\right)\end{aligned} \tag{2.180}$$

It is known that

$$\begin{aligned}&\mathrm{diag}\left(G^{\mathrm{T}}(\max\{G\pi-Ce,0\}-\max\{G\pi-Ce,0\})\right)\left(\pi-\pi^{'}\right)\geq 0,\\&\mathrm{diag}\left(G^{\mathrm{T}}(\max\{G\pi-Ce,0\}-\max\{G\pi-Ce,0\})\right)\left(\pi-\pi^{'}\right)\geq 0\end{aligned} \tag{2.181}$$

Since Λ is a positive definite matrix, we have

$$\begin{aligned}&\left(\nabla f(\pi)-\nabla f\left(\pi^{'}\right)\right)\left(\pi-\pi^{'}\right)\geq(\pi-\pi^{'})^{\mathrm{T}}\Lambda\left(\pi-\pi^{'}\right)\geq\tfrac{\gamma}{2}\|\pi-\pi^{'}\|^2,\\&\forall\pi\in R^{2l}\end{aligned} \tag{2.182}$$

where γ is a constant. As a result, subproblem of (2.168) converges linearly to the unique solution $\overline{\pi}$, and the rate of convergence is

$$\left\|\pi^k-\overline{\pi}\right\|\leq\left(\frac{2}{\gamma}\left(f\left(\pi^k\right)-f\left(\overline{\pi}\right)\right)\right)^{\frac{1}{2}}\left(1-\frac{1}{p}\left(\frac{\gamma}{K}\right)^2\right)^{\frac{1}{2}} \tag{2.183}$$

2.3.3 An Effective Intrusion Detection Framework Based on Multiple Criteria Linear Programming and Support Vector Machine

The main contributions of this section include the following:

(a) Modifications to the chaos particle swarm optimization have been proposed by adopting the time-varying inertia weight factor (TVIW) and time-varying acceleration coefficients (TVAC), namely TVCPSO, to make it faster in searching for the optimum and avoid the search being trapped into local optimum.
(b) A weighted objective function that simultaneously takes into account trade-off between the maximizing the detection rate and minimizing the false alarm rate, along with considering the number of features is proposed to eliminate the redundant and irrelevant features, as long as increase the attacks' detection rate.

(c) An extended version of multiple criteria linear programming, namely PMCLP, has been adopted to increase the performance of this classifier in dealing with the unbalance intrusion detection dataset.
(d) The proposed TVCPSO has been adopted to provide an effective IDS framework by determining parameters and selecting a subset of features for multiple criteria linear programming and support vector machines.

In the recent years, biology inspired approaches has been used to solve complex problems in a variety of domains such as computer science, medicine, finance and engineering [65]. Swarm intelligence considered as an artificial intelligence techniques which inspired from a flock of birds, a school of fish swims or a colony of ants and their unique capability to solve complex problems [65]. Briefly, swarm intelligence (SI) considered as some methodologies, techniques and algorithms inspired by study of collective behaviors in decentralized systems [66]. Particle swarm optimization is one of these techniques, which introduced by Eberhart and Kennedy in 1995 [67]. Particle swarm optimization is a population based metaheuristic optimization technique that simulates the social behavior of individuals, namely, particles. This technique, compare with the other algorithms in this group has several advantages such as simple to implement, scalability, robustness, quick in finding approximately optimal solutions and flexibility [39].

In particle swarm optimization, each individual of a population that considered as a representative of the potential solution move through an n-dimensional search space. After the initialization of the population, at each iteration particle seeks the optimal solution by changing its direction which consists of its velocity and position according to two factors, its own best previous experience (*pbest*) and the best experience of all particles (*gbest*). Equations (2.184) and (2.185), respectively represents updating the velocity and position of each percale at iteration $[t + 1]$. At the end of each iteration the performance of all particles will be evaluated by predefined fitness functions.

$$\begin{aligned} v^{id}[t+1] = w.v^{id}[t] + c_1 r_1\left(p^{id,best}[t] - x^{id}[t]\right) \\ + c_2 r_2\left(p^{gd,best}[t] - x^{id}[t]\right) \quad \mathrm{d} = 1, 2, \ldots, \mathrm{D} \end{aligned} \tag{2.184}$$

$$x^{id}[t+1] = p^{id}[t] + v^{id}[t+1] \quad \mathrm{d} = 1, 2, \ldots, \mathrm{D} \tag{2.185}$$

Where, $i = 1, 2, \ldots, N$, N is the number of swarm population. In D-dimensional search space, $x^i[t] = \{x^{i1}[t], x^{i2}[t], \ldots, x^{iD}[t]\}$ represent the current position of the ith particle at iteration [t]. Likewise, the velocity vector of each particle at iteration [t] represented by $v^i[t] = \{v^{i1}[t], v^{i2}[t], \ldots, v^{iD}[t]\}$. $p^{i,\,best}[t] = \{p^{i1}[t], p^{i2}[t], \ldots, p^{iD}[t]\}$ represent the best position that particle i has obtained until iteration t, and $p^{g,\,best}[t] = \{p^{g1}[t], p^{g2}[t], \ldots, p^{gD}[t]\}$ represent the previous best position of whole particle until iteration t.

To control the pressure of local and global search, the concept of an inertia weight w was introduced in the PSO algorithm by [68]. r_1 *and* r_2 are two D-dimensional vectors with random number between 0 and 1. c_1 *and* c_2 are positive acceleration coefficients which respectively called cognitive parameter and social parameter. In fact, these two parameters control the importance of particles' self-learning versus learning from all the swarm's population.

In this research, in order to balance the global exploration and local exploitation, time-varying acceleration coefficients (TVAC) [68, 69] and time-varying inertia weight (TVIW) [69, 70] is adopted to justify the acceleration coefficients and inertia weight, respectively. Both of these concepts help PSO algorithm to have better performance to find the region of global optimum and do not trap in local minima [68, 69, 71].

In TVAC, the acceleration coefficients adjusted by decreasing the value of c_1 from initial value of c_{1i} to c_{1f}, while the value of c_2 is increasing from its initial value of c_{2i} to c_{2f} as shown in Eqs. (2.186) and (2.187). Moreover, in TVIW, the inertia weight w is updated according to the Eq. (2.188), which means a large inertia weight makes PSO has more global search ability at the beginning of the run and by a linearly decreasing the inertia weight makes PSO has better local search.

$$c_1 = c_{1i} + \frac{t}{t_{max}}\left(c_{1f} - c_{1i}\right) \tag{2.186}$$

$$c_2 = c_{2i} + \frac{t}{t_{max}}\left(c_{2f} - c_{2i}\right) \tag{2.187}$$

$$w = w_{max} - \frac{t}{t_{max}}\left(w_{max} - w_{min}\right) \tag{2.188}$$

Here, t represents the current iteration and t_{max} means the maximum number of iterations, c_{1i}, c_{1f},c_{2i}, c_{2f} are the constant values and w_{max}, w_{min} are the predefined maximum and minimum inertia weight.

2.3.3.1 Discrete Binary PSO

Although the original PSO was proposed to act in continuing space, Kennedy and Eberhart [67] proposed the discrete binary version of PSO. In this model particle moves in a state space restricted to zero and one on each dimension, in terms of the changes in probabilities that a bit will be in one state or the other. The formula proposed in Eq. (2.8) remains unchanged except that $x^{id}[t]$, $p^{gd,\,best}[t]$ *and* $p^{id,\,best}[t] \in \{0, 1\}$ and v^{id} restricted to the [0.0, 1.0] [15, 65]. By introducing the sigmoid function, the velocity mapped from a continuous space to probability space as following:

$$sig\left(v^{id}\right) = \frac{1}{1 + e^{\left(-v^{id}\right)}} \quad \mathrm{d} = 1, 2, \ldots, \mathrm{D} \tag{2.189}$$

The new particle position calculated by using the following rule:

$$x^{id}\left[t+1\right] = \begin{cases} 1, & if\ rnd\left(\right) < sig\left(v^{id}\right) \\ 0 & if\ rnd\left(\right) \geq sig\left(v^{id}\right) \end{cases}, \mathrm{d} = 1, 2, \ldots, \mathrm{D} \tag{2.190}$$

Where, $sig(v^{id})$ is a sigmoid function and *rnd( )* is a random number in range [0.0, 1.0].

Although traditional PSO gains considerable results in different fields, however, the performance of the PSO depends on the preset parameters and it often suffers the problem of being trapped in local optima. In order to further enhance the search ability of swarm in PSO and avoids the search being trapped in local optimum, chaotic concept has been introduced by [68, 69, 71]. Here, chaos is characterized as ergodicity, randomicity and regularity.

In this section, Logistic equation which is a typical chaotic system adopted to make the chaotic local search as represented in the following:

$$z_{j+1} = \mu z_j \left(1 - z_j\right) \quad j = 1, 2, \ldots m \tag{2.191}$$

Here, by considering n-dimensional vector $z_j = (z_{j1}, z_{j2}, \ldots, z_{jn})$, each component of this system is a random value in the range [0, 1], μ is the control parameter and the system of Eq. (2.15) has been proved to be completely chaotic when $0 \leq z_0 \leq 1$ and $\mu = 4$. Chaos queues $z_1, z_2, z_3, \ldots, z_m$ are generated by iteration of Logistic equation.

In fact, the basic ideas of chaotic are adopted in this section are described as follows:

Chaos initialization: In spite of standard PSO, which particle's position in the search space initialized randomly, here chaos initialization is adopted to better initialize the position of each particle and to increase the diversity of the population.

Chaotic local search (CLS): By using the chaos queues, it helps PSO to does not trapped in a local optimum besides it can cause to search the optimum quickly. It will happen by generating the chaos queues based on the optimal position ($p^{g,\,best}$), and then replace the position of one particle of the population with the best position of the chaos queues.

Although different performance metrics has been proposed to evaluate the effectiveness of IDSs, the most two popular of these metrics are detection rate (DR) and false alarm rate (FAR). By comparing the actual nature of a given record which here "Positive" means an "attack classes" and "Negative" means a "normal record" to the prediction ones, it's possible to consider four outcomes for this situation as shown in Table 2.8, which known as the confusion matrix.

Table 2.8 Confusion matrix

	Test Result Positive (Predicted as an attack)	Test Result Negative (Predicted as a normal record)
Actual Positive Class (Attack record)	True positive (TP)	False negative (FN)
Actual Negative Class (Normal record)	False positive (FP)	True negative (TN)

Here, true positive and true negative means correctly labeled the records as an attack and normal, respectively, that is, IDSs predict the labels perfectly. False positive (FP), refer to normal record is considered as an attack and False negative (FN) means those attack records falsely considered as a normal one.

A well performed IDS should has a high detection rate (DR) as well as low false positive rate. In intrusion detection domain false positive rate typically named false alarm rate (FAR). Thus, the particles with higher detection rate, lower false positive rate and the small number of selected features can produce a high objective function value. Hence, in this research a weighted objective function that simultaneously takes into account trade-off between the maximizing the detection rate and minimizing the false alarm rate, along with considering the number of features is proposed according to the following equation:

$$\text{Objective function } \left(\mathrm{F}_{fit}\right) = w_{DR}\cdot\left[\frac{TP}{(\mathrm{TP+FN})}\right] + w_{FAR}\cdot\left[1-\frac{FP}{(\mathrm{FP+TN})}\right] + w_{F}\cdot\left[1-\frac{\sum_{i=1}^{nF}\mathrm{f}_i}{n_F}\right] \tag{2.192}$$

Since any of these three elements of objective function have different effect on the performance of IDS, we convert this multiple criteria problem to a single weighted fitness function that combines the three goals linearly into one. Where w_{DR}, w_{FAR} and w_F represents the importance of detection rate, false alarm rate and number of selected features in the objective function. Detection rate or sensitivity in biomedical informatics terms, known as a true positive rate (TPR), which means the ratio of true positive recognition to the total actual positive class; $\frac{TP}{(\mathrm{TP+FN})}$. False alarm rate (FAR) or false positive rate (FPR) defined as: $\frac{FP}{(\mathrm{FP+TP})}$. f_i represents the value of feature mask ("1" represents that feature i is selected and "0" represents that feature i is not selected), and n_F indicates the number of features.

The specific steps of TVCPSO–MCLP and TVCPSO–SVM are described as follows:

Step 1: Chaotic initialization for n + 2 particle, for the MCLP algorithm, the first two parameters are α^* and β^* and for SVM algorithm the first two parameters are c and γ. The rest of n particle is binary features mask of feature sets which here is 41 features of NSL-KDD cup 99 datasets. Here in binary features mask, 1 and 0 adopted to present as selected features and discarded features, respectively.

(a) Initialize a vector $z_0 = (z_{01}, z_{02}, \ldots, z_{0n})$, each component of it is set as a random value in the range [0, 1], and by iteration of Logistic equation a chaos queue $z_1, z_2, \ldots, z_n$ is obtained.

(b) In order to transfer the chaos queue z_j into the parameter's range the following equation is used:

$$\hat{Z}_{jk} = a_k + (b_k - a_k)\,.z_{jk} \qquad (k = 1, 2, \ldots, n) \tag{2.193}$$

Where the value range of each particle defined by $[a_k, b_k]$.

Step 2: Compute the fitness value of the initial vector $\hat{Z}_j$ $(j = 1, 2, \ldots, m)$ and then choose the best M solutions as the initial positions of M particles.

Step 3: Randomly initialize the velocity of M particles, here, $v_j = (v_{j1}, v_{j2}, \ldots, v_{jn})$ $j = (1, 2, \ldots, M.)$

Step 4: Update the velocity and position of each classifier's parameters (α*, β* in MCLP and c, γ in SVM) according to Eqs. (2.184) and (2.185), and in order to update the velocity and position of the features in each particle Eqs. (2.184) and (2.190) have been used, respectively.

Step 5: Evaluate the fitness of each particle according to Eq. (2.192) and then compare the evaluated fitness value of each particle (personal optimal fitness (pfit)) to its personal best position ($p^{i,\,best}$):

(a) If the pfit is better than $p^{i,\,best}$ then update the $p^{i,\,best}$ as the current position, otherwise keep the previous ones in memory.
(b) If the pfit is better than $p^{g,\,best}$ then update the $p^{g,\,best}$ as the current position, otherwise keep the previous $p^{g,\,best}$.

Step 6: Optimize $p^{g,\,best}$ by chaos local search according to the following steps:

(a) Consider $T = 0$, scale the $p^{gk,\,best}$ into the range of [0,1] by $z_k^T = \frac{p^{gk,best} - a_k}{b_k - a_k}$ $(k = 1, 2, \ldots, n)$.
(b) Generate the chaos queues Z_j^T $(T = 1, 2, \ldots, m)$ by iteration of Logistic equation.
(c) Obtain the solution set $p = (p^1, p^2, \ldots, p^m)$ by scale the chaotic variables Z_j^T into the decision variable according to the $p_k^T = a_k + (b_k - a_k).z_k^T$.
(d) Evaluate the fitness value of each feasible solution $p = (p^1, p^2, \ldots, p^m)$, and get the best solution $\hat{p}^{g,best}$.

Step 7: If the stopping criteria are satisfied, then stop the algorithms and get the global optimum that are the optimal value of (α*, β* in MCLP and c, γ in SVM) and the most appropriate subset of features. Otherwise, go to step 5.

References

1. Sun, D., Liu, L., Zhang, P., Zhu, X., Shi, Y.: Decision rule extraction for regularized multiple criteria linear programming model. Int. J. Data Warehousing Mining. **7**(3), 88–101 (2011)
2. Shi, Y., Tian, Y., Chen, X., Zhang, P.: Regularized multiple criteria linear programs for classification. Sci. China Ser. F Inf. Sci. **52**(10), 1812–1820 (2009)
3. Wang, B., Shi, Y.: Error correction method in classification by using multiple-criteria and multiple-constraint levels linear programming. Int. J. Comput. Commun. Contr. **7**(5), 976–989 (2014)
4. Qi, Z., Tian, Y., Shi, Y.: Multi-instance classification based on regularized multiple criteria linear programming. Neural Comput. Applic. **23**(3), 857–863 (2013)
5. Zhang, P., Tian, Y., Zhang, Z., Shi, Y., Li, X.: Supportive instances for regularized multiple criteria linear programming classification. Int. J. Oper. Quant. Manag. **14**(4), 249–263 (2008)

6. Zhao, X., Shi, Y., Niu, L.: Kernel based simple regularized multiple criteria linear program for binary classification and regression. Intellig. Data Anal. **19**(3), 505–527 (2015)
7. Zhang, D., Tian, Y., Shi, Y.: A group of knowledge-incorporated multiple criteria linear programming classifiers. J. Comput. Appl. Math. **235**(13), 3705–3717 (2011)
8. Peng, Y., Zhang, Y., Kou, G., Shi, Y.: A multicriteria decision making approach for estimating the number of clusters in a data set. PLoS One. **7**(7), e41713 (2012)
9. Qi, Z., Tian, Y., Shi, Y., Alexandrov, V.: Parallel rmclp classification algorithm and its application on the medical data. IEEE Trans. Cloud Comput. (2015). https://doi.org/10.1109/TCC.2015.2481381
10. Shi, Y., Wise, W., Lou, M., et al.: Multiple criteria decision making in credit card portfolio management. In: Multiple Criteria Decision Making in New Millennium, pp. 427–436 (2001)
11. Witten, I.H., Frank, E., Hall, M.A., Pal, C.J.: Practical Machine Learning Tools and Techniques, p. 578. Morgan Kaufmann, Burlington, MA (2005)
12. Qi, Z., Xu, Y., Wang, L., Song, Y.: Online multiple instance boosting for object detection. Neurocomputing. **74**(10), 1769–1775 (2011)
13. Shao, Y., Yang, Z., Wang, X., Deng, N.: Multiple instance twin support vector machines. Lect. Note Oper. Res. **12**, 433–442 (2010)
14. Zhou, Z.: Multi-instance learning: a survey. Department of Computer Science & Technology, Nanjing University, Tech. Rep 2 (2004)
15. Chen, Y., Zhang, L., Shi, Y.: Post mining of multiple criteria linear programming classification model for actionable knowledge in credit card churning management. In: 2011 IEEE 11th International Conference on Data Mining Workshops, pp. 204–211. IEEE, New York (2011)
16. Keeler, J.D., Rumelhart, D.E., Leow, W.K.: Integrated segmentation and recognition of hand-printed numerals. In: Proceedings of the NIPS, pp. 557–563 (1990)
17. Dietterich, T.G., Lathrop, R.H., Lozano-Pérez, T.: Solving the multiple instance problem with axis-parallel rectangles. Artif. Intell. **89**(1–2), 31–71 (1997)
18. Viola, P., Platt, J., Zhang, C.: Multiple instances boosting for object detection. In: Proceedings of the NIPS, pp. 1417–1424 (2006)
19. Ferris, M.C., Mangasarian, O.L.: Parallel variable distribution. SIAM J. Optim. **4**(4), 815–832 (1994)
20. Shi, Y., Liu, R., Yan, N., Chen, Z.: Multiple criteria mathematical programming and data mining. In: International Conference on Computational Science, pp. 7–17. Springer, New York (2008)
21. Mangasarian, O.L., Wild, E.W.: Multiple instance classification via successive linear programming. J. Optim. Theory Appl. **137**(3), 555–568 (2008)
22. Murphy, P.M., Aha, D.W.: UCI machine learning repository (1992)
23. Andrews, S., Tsochantaridis, I., Hofmann, T.: Support vector machines for multiple instance learning. In: NIPS, vol. 2, pp. 561–568 (2002)
24. Deng, N., Tian, Y.: Support vector machines: theory, algorithms and extensions. Science Press, Beijing (2009)
25. Zhang, Q., Goldman, S.A.: Em-dd: an improved multiple-instance learning technique. In: Advances in Neural Information Processing Systems, pp. 1073–1080 (2001)
26. Yang, Z.X., Deng, N.: Multi-instance support vector machine based on convex combination. In: The Eighth International Symposium on Operations Research and Its Applications, vol. 481, p. 487. Citeseer (2009)
27. Vapnik, V.N.: The Nature of Statistical Learning Theory, 2nd edn. Springer, New York (2000)
28. Zhang, J., Shi, Y., Zhang, P.: Several multi-criteria programming methods for classification. Comput. Oper. Res. **36**(3), 823–836 (2009)
29. He, J., Shi, Y., Xu, W.: Classifications of credit cardholder behavior by using multiple criteria non-linear programming. In: CASDMKM, pp. 154–163 (2004)
30. Kou, G., Liu, X., Peng, Y., Shi, Y., Wise, M., Xu, W.: Multiple criteria linear programming approach to data mining: models, algorithm designs and software development. Optim. Methods Softw. **18**(4), 453–473 (2003)
31. Yu, P.L.: A class of solutions for group decision problems. Manag. Sci. **19**(8), 936–946 (1973)

32. MacQueen, J., et al.: Some methods for classification and analysis of multivariate observations. In: Proceedings of the Fifth Berkeley Symposium on Mathematical Statistics and Probability, Oakland, CA, USA, vol. 1, pp. 281–297 (1967)
33. Fung, G., Mangasarian, O.L., Shavlik, J.W.: Knowledge-based support vector machine classifiers. In: NIPS, pp. 521–528. Citeseer (2002)
34. Fung, G., Mangasarian, O.L., Shavlik, J.W.: Knowledge-based nonlinear kernel classifiers. In: Learning Theory and Kernel Machines, pp. 102–113. Springer, New York (2003)
35. Mangasarian, O.L., Wild, E.W.: Nonlinear knowledge in kernel machines. In: Data Mining and Mathematical Programming. Centre de Recherches Mathématiques Montréal Proceedings and & Lecture Notes, pp. 181–198 (2008)
36. Zhang, D., Tian, Y., Shi, Y.: Nonlinear knowledge in kernel-based multiple criteria linear programming classifier. In: Proceedings of the MCDM, pp. 622–629 (2009)
37. Olson, D.L.: Comparison of weights in topsis models. Math. Comput. Model. **40**(7–8), 721–727 (2004)
38. Thomsen, C., Pedersen, T.B.: A survey of open source tools for business intelligence. Int. J. Data Warehousing Mining. **5**(3), 56–75 (2009)
39. Olariu, S., Zomaya, A.Y.: Handbook of Bioinspired Algorithms and Applications. CRC Press, Boca Raton, FL (2005)
40. Pakhira, M.K., Bandyopadhyay, S., Maulik, U.: Validity index for crisp and fuzzy clusters. Pattern Recogn. **37**(3), 487–501 (2004)
41. Jain, A.K., Murty, M.N., Flynn, P.J.: Data clustering: a review. ACM Comput. Surv. (CSUR). **31**(3), 264–323 (1999)
42. Halkidi, M., Batistakis, Y., Vazirgiannis, M.: Cluster validity methods: Part I. ACM SIGMOD Rec. **31**(2), 40–45 (2002)
43. Halkidi, M., Batistakis, Y., Vazirgiannis, M.: Clustering validity checking methods: Part II. ACM SIGMOD Rec. **31**(3), 19–27 (2002)
44. Zadeh, L.: Optimality and non-scalar-valued performance criteria. IEEE Trans. Autom. Control. **8**(1), 59–60 (1963)
45. Triantaphyllou, E.: Multi-criteria Decision Making: A Comparative Study. Kluwer Academic Publishers, Dordrecht, The Netherlands (2000)
46. Brans, J.P.: L'ingénierie de la décision: l'élaboration d'instruments d'aide a la décision. Université Laval, Faculté des sciences de l'administration (1982)
47. Brans, J.P., Mareschal, B.: Promethee methods. In: Multiple Criteria Decision Analysis: State of the Art Surveys, pp. 163–186. Springer, New York (2005)
48. Brans, J.: How to decide with promethee. http://www.visualdecision.com/Pdf/How%20to%20use%20PROMETHEE.pdf (1994)
49. Brans, J.P., Vincke, P.: Note—a preference ranking organisation method: (the promethee method for multiple criteria decision-making). Manag. Sci. **31**(6), 647–656 (1985)
50. Hwang, C.L., Yoon, K.: Multiple attribute decision making methods and applications. Springer, Berlin (1981)
51. Opricovic, S., Tzeng, G.H.: Compromise solution by mcdm methods: a comparative analysis of vikor and topsis. Eur. J. Oper. Res. **156**(2), 445–455 (2004)
52. Han, J., Kamber, M.: Data Mining: Concepts and Techniques, 2nd edn. Morgan Kaufmann, Burlington, MA (2006)
53. Hall, M., Frank, E., Holmes, G., Pfahringer, B., Reutemann, P., Witten, I.H.: The weka data mining software: an update. ACM SIGKDD Explor. Newsl. **11**(1), 10–18 (2009)
54. Theodoridis, S., Koutroumbas, K.: Pattern Recognition, 4th edn. Academic Press, Cambridge (2008)
55. Dunn, J.C.: A fuzzy relative of the isodata process and its use in detecting compact well-separated clusters. J. Cybernetics. **3**, 32–57 (1973)
56. Davies, D.L., Bouldin, D.W.: A cluster separation measure. IEEE Trans. Pattern Anal. Mach. Intell. **1**(2), 224–227 (1979)
57. Chou, C., Su, M., Lai, E.: A new cluster validity measure and its application to image compression. Pattern. Anal. Applic. **7**(2), 205–220 (2004)

58. Rousseeuw, P.J.: Silhouettes: a graphical aid to the interpretation and validation of cluster analysis. J. Comput. Appl. Math. **20**, 53–65 (1987)
59. Hubert, L.J., Levin, J.R.: A general statistical framework for assessing categorical clustering in free recall. Psychol. Bull. **83**(6), 1072–1080 (1976)
60. Vendramin, L., Campello, R.J., Hruschka, E.R.: Relative clustering validity criteria: a comparative overview. Statist. Anal. Data Mining. **3**(4), 209–235 (2010)
61. Mangasarian, L.: Parallel gradient distribution in unconstrained optimization. SIAM J. Control. Optim. **33**(6), 1916–1925 (1995)
62. Freed, N., Glover, F.: Evaluating alternative linear programming models to solve the two-group discriminant problem. Decis. Sci. **17**(2), 151–162 (1986)
63. Shi, Y., Peng, Y., Xu, W., Tang, X.: Data mining via multiple criteria linear programming: applications in credit card portfolio management. Int. J. Inf. Technol. Decis. Making. **1**(01), 131–151 (2002)
64. Chen, W., Tian, Y.: Kernel regularized multiple criteria linear programming. In: 3rd International Symposium on Optimization and Systems Biology, pp. 345–352. Citeseer (2009)
65. Kolias, C., Kambourakis, G., Maragoudakis, M.: Swarm intelligence in intrusion detection: a survey. Comput. Secur. **30**(8), 625–642 (2011)
66. Wu, S.X., Banzhaf, W.: The use of computational intelligence in intrusion detection systems: a review. Appl. Soft Comput. **10**(1), 1–35 (2010)
67. Kennedy, J., Eberhart, R.: Particle swarm optimization. In: Proceedings of ICNN'95-International Conference on Neural Networks, vol. 4, pp. 1942–1948. IEEE, New York (1995)
68. Shi, Y., Eberhart, R.: A modified particle swarm optimizer. In: 1998 IEEE International Conference on Evolutionary Computation Proceedings. IEEE World Congress on Computational Intelligence (Cat. No. 98TH8360), pp. 69–73. IEEE, New York (1998)
69. Chen, H., Yang, B., Wang, S., Wang, G., Liu, D., Li, H., Liu, W.: Towards an optimal support vector machine classifier using a parallel particle swarm optimization strategy. Appl. Math. Comput. **239**, 180–197 (2014)
70. Huang, C.L., Dun, J.F.: A distributed pso–svm hybrid system with feature selection and parameter optimization. Appl. Soft Comput. **8**(4), 1381–1391 (2008)
71. Ratnaweera, A., Halgamuge, S.K., Watson, H.C.: Self-organizing hierarchical particle swarm optimizer with time-varying acceleration coefficients. IEEE Trans. Evol. Comput. **8**(3), 240–255 (2004)

Chapter 3
Support Vector Machine Classification

Support vector machine (SVM) has been a popular technique in data analytics. Shi et al. [1] has reported some SVM algorithms. They vary from leave-one-out (LOO) bounds approaches, multi-class, unsupervised, semi-supervised and robust SVMs. Following the direction of the research afterwards, this Chapter provides five sections about advances of SVM in big data analytics. Section 3.1 has two subsections. The first one outlines the recent findings of the author's research team on SVM [2] while the second one is about two new decomposition algorithms for training bound-constrained SVM [3]. Section 3.2 describes different twin SVM in classification with four subsections. The first one explores the improved twin SVM [4]. The second one is extending twin SVM for multi-category classification problems [5]. The third one provides robust twin SVM for pattern classification [6]. The fourth one elaborates structural twin SVM for classification [7]. Section 3.3 shows nonparallel SVM with four subsections. The first one is about a nonparallel SVM for a classification problem with universum learning [8]. The second one is about a divide-and-combine method for large scale nonparallel SVM [9]. The third one explores nonparallel SVM for pattern classification [4]. The fourth one is a multi-instance learning algorithm based on nonparallel classifier [10]. Section 3.4 shows Laplacian SVM classifiers with two subsections. One is about successive overrelaxation for Laplacian SVM [11] while another one is about Laplacian twin SVM for semi-supervised classification [12]. Finally, Sect. 3.5 discusses loss functions of SVM classification with three subsections. The first one is about the ramp loss least squares SVM [13]. The second is about the ramp loss nonparallel SVM for pattern classification [14]. The third one is about a classification model using privileged information and its application [10].

Y. Shi, *Advances in Big Data Analytics*,
https://doi.org/10.1007/978-981-16-3607-3_3

3.1 Support Vector Machine in Data Analytics

3.1.1 Recent Advances on Support Vector Machines Research

3.1.1.1 The Nature of C-Support Vector Machines

In this section, standard C-SVM [15–18] for binary classification is briefly summarized and understood from several points of view.

Definition 3.1 (Binary classification). For the given training set

$$T = \{(x_1, y_1), \ldots, (x_l, y_l)\} \in \left(R^n \times y\right)^l \tag{3.1}$$

where, the goal is to find a real function in and derive the value of y for any x by the decision function:

$$f(x) = sgn\,(g(x)) \tag{3.2}$$

C-SVM formulates the problem as a convex quadratic programming

$$min\ \frac{1}{2}\|w\|^2 + C\sum\nolimits_{i=1}^{l} \xi_j \tag{3.3}$$

$$s.t. y_j\left(\left(w \cdot x_j\right) + b\right) \geq 1 - \xi_j, i = 1, \cdots, l, \tag{3.4}$$

$$\xi_j \geq 0, i = 1, \cdots, l, \tag{3.5}$$

where and C is a penalty parameter. For this primal problem, C-SVM solves its Lagrangian dual problem

$$min\ \frac{1}{2}\sum\nolimits_{i=1}^{l}\sum\nolimits_{j=1}^{l} \alpha_i \alpha_j y_j y_j K\left(x_j, x_j\right) - \sum\nolimits_{j=1}^{l} \alpha_j \tag{3.6}$$

$$s.t.\sum\nolimits_{i=1}^{l} y_i \alpha_i = 0, \tag{3.7}$$

$$0 \leq \alpha_j \leq C, i = 1, \cdots, l, \tag{3.8}$$

where $K(x, x^{'})$ is the kernel function, which is also a convex quadratic problem and then construct the decision function.

As we all know, the principal of Structural Risk Minimization (SRM) is embodied in SVM, the confidential interval and the empirical risk should be considered at the same time. The two terms in the objective function (3.3) indicate that we not only minimize $\|w\|^2$ (maximize the margin), but also minimize $\sum_{i=1}^{l} \xi_j$, which is a measurement of violation of the constraints $y_j((w \cdot x_i) + b) \geq 1, i = 1, \cdots, l$.

Here the parameter C determines the weighting between the two terms, the larger the value of C, the larger the punishment on empirical risk.

In fact, the parameter C has another meaningful interpretation [16, 17]. Consider the binary classification problem, select a decision function candidate set F(t) depending on a real parameter t:

$$F(t) = \{f(x) = sgn((w \cdot x) + b) \,|\, ||w|| \leq t, t \in [0, \infty)\}, \tag{3.9}$$

and suppose that the loss function to be the soft margin loss function defined by

$$c(x, y, f(x)) = max\ \{0, 1 - yg(x)\}, \text{ where } g(x) = (w \cdot x) + b. \tag{3.10}$$

Thus, structural risk minimization is implemented by solving the following convex programming for an appropriate parameter t:

$$min \sum_{i=1}^{l} \xi_i \tag{3.11}$$

$$s.t.\ \ y_i\left(\left(w \cdot x_j\right) + b\right) \geq 1 - \xi_i,\ \ i = 1, \cdots, l, \tag{3.12}$$

$$\xi_i \geq 0, i = 1, \cdots, l, \tag{3.13}$$

$$\|w\| \leq t. \tag{3.14}$$

An interesting point is proved that when the parameters C and t are chosen satisfying t = ψ(C), where ψ is nondecreasing in the interval, problem (3.3)–(3.5) and problem (3.11)–(3.14) will get the same decision function [19]. Hence the very interesting and important meaning of the parameter C is proposed: C corresponds to the size of the decision function candidate set in the principle of SRM: the larger the value of C, the larger the decision function candidate set.

Now we can summarize and understand C-SVM from following points of view: (1) Construct a decision function by selecting a proper size of the decision function candidate set via adjusting the parameter C; (2) Construct a decision function by selecting the weighting between the margin of the decision function and the deviation of the decision function measured by the soft-margin loss function via adjusting the parameter C; (3) Another understanding about C-SVM can also be seen in the literatures [17]: Construct a decision function by selecting the weighting between flatness of the decision function and the deviation of the decision function measured by the soft-margin loss function via adjusting the parameter C.

3.1.1.2 Optimization Models of Support Vector Machines

In this section, several representative and important SVM optimization models with different variations are described and analyzed. These models can be divided into three categories: models for standard problems, models for nonstandard learning problems, and models combining SVMs with other issues in machine learning.

Models for Standard Problems

For the standard classification or regression problems, a lot of methods are developed based on standard SVM models to be the powerful new algorithms. Here we briefly introduce several basic and efficient models, lots of developments of these models are omitted here.

Least Squares Support Vector Machine

Just like the standard C-SVM the starting point of least squares SVM (LSSVM) [20] is also to find a separating hyperplane, but with different primal problem. In fact, introducing the transformation x = Φ(x) and the corresponding kernel K, the primal problem becomes the convex quadratic programming

$$min\ \frac{1}{2}\|w\|^2 + \frac{C}{2}\sum\nolimits_{i=1}^{l}\eta_i^2 \tag{3.15}$$

$$s.t.\ \ y_i\left(\left(w \cdot \Phi\left(x_j\right)\right) + b\right) = 1 - \eta_i, i = 1, \ldots, l. \tag{3.16}$$

The geometric interpretation of the above problem with x is shown in Fig. 3.1, where minimizing w realizes the maximal margin between the straight lines

$$(w \cdot x) + b = 1\ and\ (w \cdot x) + b = -1, \tag{3.17}$$

while minimizing implies making the straight lines (3.17) be proximal to all inputs of positive points and negative points respectively.

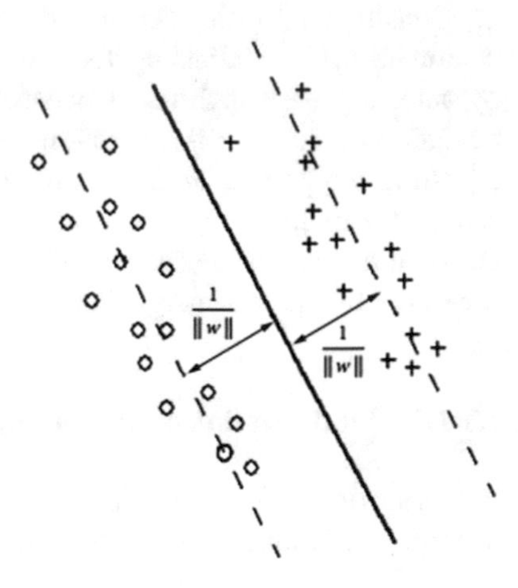

Fig. 3.1 Geometric interpretation of LSSVM

Its dual problem to be solved in LSSVM is also a convex quadratic programming

$$max -\frac{1}{2}\sum_{i=1}^{l}\sum_{j=1}^{l}\alpha_j\alpha_j y_j y_j \left(K\left(x_j, x_j\right)+\frac{6_{ij}}{C}\right)+\sum_{i=1}^{l}\alpha_i \tag{3.18}$$

$$s.t. \sum_{i=1}^{l}\alpha_i y_i = 0, \tag{3.19}$$

where

$$\delta_{ij} = \begin{cases} 1, i = j \\ 0, i \neq j \end{cases} \tag{3.20}$$

In C-SVM, the error is measured by the soft margin loss function, this leads to the fact that the decision function is decided only by the support vectors. While in LSSVM, almost all training points contribute to the decision function, which makes it lose the sparseness. However, LSSVM needs to solve a quadratic programming with only equality constraints, or equivalently a linear system of equations. Therefore, it is simpler and faster than C-SVM.

Twin Support Vector Machine

Twin support vector machine (TWSVM) is a binary classifier that perform classification using two nonparallel hyperplanes instead of a single hyperplane as in the case of conventional SVMs [21]. Suppose the two non-parallel hyperplanes are the positive hyperplane

$$(w_+ \cdot x) + b_+ = 0 \tag{3.21}$$

and the negative hyperplane

$$(w_- \cdot x) + b_- = 0 \tag{3.22}$$

The primal problems for finding these two hyperplanes are two convex quadratic programming problems [21]

$$min \frac{1}{2}c_1\left(\|w_+\|^2 + b_+^2\right) + \frac{1}{2}\sum_{i=1}^{p}\left(\left(w_+ \cdot x_j\right) + b_+\right)^2 + c_2\sum_{j=p+1}^{p+q}\xi_j \tag{3.23}$$

$$s.t. \ \left(w_+ \cdot x_j\right) + b_+ \leq -1 + \xi_j, \ \ j = p+1, \ldots, p+q, \tag{3.24}$$

$$\xi_j \geq 0, j = p+1, \cdots, p+q \tag{3.25}$$

and

$$min\ \frac{1}{2}c_3\left(\|w_-\|^2+b_-^2\right)+\frac{1}{2}\sum\nolimits_{i=p+1}^{p+q}((w_-\cdot x_i)+b_-)^2+c_4\sum\nolimits_{j=1}^{p}\xi_j \tag{3.26}$$

$$s.t.\ (w_-\cdot x_j)+b_-\geq 1-\xi_j,\ j=1,\ldots,p \tag{3.27}$$

$$\xi_j\geq 0,\ j=1,\cdots,p, \tag{3.28}$$

where x_i, $i=1,\ \ldots,p$ are positive inputs, and x_j, $i=p+1,\ \ldots,p+q$ are negative inputs, $c_1>0$, $c_2>0$, $c_3>0$, $c_4>0$ are parameters, $\xi_-=(\xi_{p+1},\ \ldots,\xi_{p+q})^{\mathrm{T}}$, $\xi_+=(\xi_1,\ \ldots,\xi_p)^{\mathrm{T}}$.

For both of the above primal problems an interpretation can be offered in the same way. The geometric interpretation of the problem (3.23)–(3.25) with shown in Fig. 3.2, where minimizing the second term $\sum_{i=1}^{p}\left((w_+\cdot x_j)+b_+\right)^2$ makes the positive hyperplane (blue solid line in Fig. 3.2) to be proximal to all positive inputs, minimizing the third term with the constraints (3.24) and (3.25) requires the positive hyperplane to be at a distance from the negative inputs by pushing the negative inputs to the other side of the bounding hyperplane (blue dotted line in Fig. 3.2), where a set ξ of variables is used to measure the error whenever the positive hyperplane is close to the negative inputs. Minimizing the first term $\frac{1}{2}\left(\|w_+\|^2+b_+^2\right)$ realizes the maximal margin between the positive hyperplane $(w_+\cdot x)+b_+=0$ and the bounding hyperplane $(w_+\cdot x)+b_+=-1$ in R^{n+1}space.

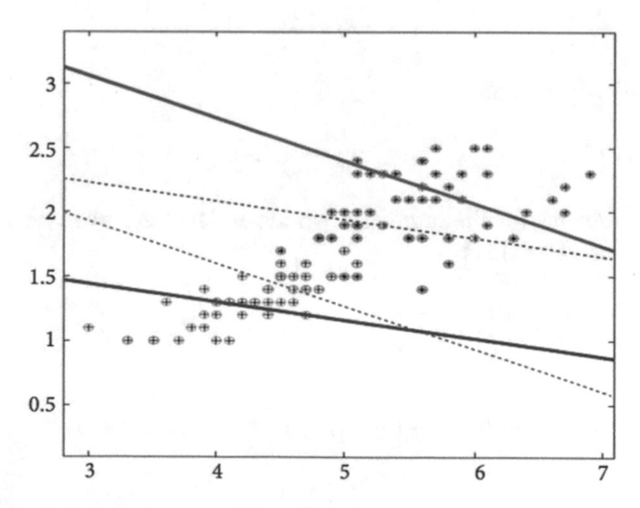

Fig. 3.2 Geometric interpretation of TWSVM

TWSVM is established based on solving two dual problems of the above primal problems separately. The generalization of TWSVM has been shown to be significantly better than standard SVM for both linear and nonlinear kernels. It has become one of the popular methods in machine learning because of its low computational complexity, since it solves above two smaller sized convex quadratic programming problems. On average, it is about four times faster than the standard SVMs.

AUC Maximizing Support Vector Machine

Nowadays the area under the receiver operating characteristics (ROC) curve, which corresponds to the Wilcoxon-Mann-Whitney test statistic, is increasingly used as a performance measure for classification systems, especially when one often has to deal with imbalanced class priors or misclassification costs. The area of that curve is the probability that a randomly drawn positive example has a higher decision function value than a random negative example; it is called the AUC (area under ROC curve). When the goal of a learning problem is to find a decision function with high AUC value, then it is natural to use a learning algorithm that directly maximizes this criterion. Over the last years, AUC maximizing SVMs (AUCSVM) have been developed [22, 23], in which one kind of primary problem to be solved is a convex problem.

$$min\ \frac{1}{2}\|w\|^2 + C\sum_{i=1}^{l^+}\sum_{i=1}^{l^-}\xi_{ij}, \tag{3.29}$$

$$s.t. \left(w \cdot \left(x_i^+ - x_j^-\right)\right) \geq 1 - \xi_{ij}, i = 1, \cdots, l^+, j = 1, \cdots, l^-, \tag{3.30}$$

$$\zeta_{ij} \geq 0, i = 1, \cdots, l^+, j = 1, \cdots, l^- \tag{3.31}$$

where x_i^+, $i = 1, \ldots, l^+$ and x_j^-, $j = 1$ are positive and negative inputs separately. Its dual problem is also a convex quadratic programming problem.

However, the existing algorithms all have the serious drawback that the number of constraints is quadratic in the number of training points, so they become very large even for small training set. To cope with this, different strategies can be constructed, in one of which a Fast and Exact k-Means (FEKM) [24] algorithm is applied to approximate the problem by representing the l^+l^- many pairs $\left(x_i^+ - x_{\overline{j}}\right)$ by only $l^+ - l^-$ cluster centers and thereby reduce the number of constraints and parameters. The approximate k-Means AUCSVM is more effective at maximizing the AUC than the SVM for linear kernels. Its execution time is quadratic in the sample size.

Fuzzy Support Vector Machine

In standard SVMs, each sample is treated equally; i.e., each input point is fully assigned to one of the two classes. However, in many applications, some input points, such as the outliers, may not be exactly assigned to one of these two classes, and each point does not have the same meaning to the decision surface. To solve this problem, each data point in the training dataset is assigned with a membership, if one data point is detected as an outlier, it is assigned with a low membership, so its contribution to total error term decreases. Unlike the equal treatment in standard SVMs, this kind of SVM fuzzifiers the penalty term in order to reduce the sensitivity of less important data points. Fuzzy SVM (FSVM) construct its primal problem as [25]

$$min \frac{1}{2}|w|^2 + C\sum_{i=1}^{l} s_j \xi_i \tag{3.32}$$

$$s.t. y_j \left((w \cdot x_j) + b \right) \geq 1 - \xi_j, i = 1, \ldots, l, \tag{3.33}$$

$$\xi_i \geq 0, i = 1, \cdots, l, \tag{3.34}$$

where s is the membership generalized by some outlier-detecting methods. Its dual problem is similarly deduced as C-SVM to be a convex quadratic programming

$$min \frac{1}{2}\sum_{i=1}^{l}\sum_{j=1}^{l} \alpha_i \alpha_j y_j y_j K\left(x_j, x_j\right) - \sum_{j=1}^{l} \alpha_j \tag{3.35}$$

$$s.t. \sum_{i=1}^{1} y_i \, \alpha_j = 0 \tag{3.36}$$

$$0 \leq \alpha_j \leq C s_j, i = 1, \cdots, l \tag{3.37}$$

Model (3.32)–(3.34) is also the general formulation of the cost sensitive SVM [26] solving the imbalanced problem, in which different error costs are used for the positive (C_+) and negative (C_-) classes

$$min \frac{1}{2}\|w\|^2 + C_+ \sum_{y_i=1} \xi_i + C_- \sum_{y_i=-1} x_i \tag{3.38}$$

$$s.t. y_i \left((w \cdot x_i) + b \right) \geq 1 - \xi_j, i = 1, \ldots, l, \tag{3.39}$$

$$\xi_j \geq 0, i = 1, \cdots, l. \tag{3.40}$$

Models for Nonstandard Problems

For the nonstandard problems appeared in different practical applications, a wide range of programming methods are used to build novel optimization models. Here we present several important and interesting models to show the interplay of SVMs and optimization.

Support Vector Ordinal Regression

Support vector ordinal regression (SVOR) [27] is a method to solve a specialization of the multi-class classification problem: ordinal regression problem. The problem of ordinal regression arises in many fields, e.g., information retrieval, econometric models, and classical statistics. It is complementary to the classification problem and metric regression problem due to its discrete and ordered outcome space.

Definition 3.2 (Ordinal regression problem). Given a training set

$$T = \left\{x_i^j\right\}_{i=1,,lj}^{j=1,\cdot\cdot\cdot,M} \tag{3.41}$$

where x_i^j is an input of a training point, the subscript $j = 1, \ldots,$ M denotes the corresponding class number, is the index within each class, and is the number of the training points in class. Find M − 1 parallel hyperplanes in R^n

$$(w \cdot x) - b_r = 0, r = 1, M - 1 \tag{3.42}$$

where $w \in R^n$, $b_1 \leq b_2 \leq \cdots \leq b_{M-1}$, $b_0 = -\infty$, $b_M = +\infty$, such that the class number for any x can be predicted by

$$f(x) = arg \min_{r \in \{1,\ldots,M\}} \{r : (w \cdot x) - b_r < 0\} \tag{3.43}$$

SVOR constructs the primal problem as

$$min \frac{1}{2}\|w\|^2 + C\sum\nolimits_{j=1}^{M}\sum\nolimits_{i=1}^{l^j} \left(\xi_i^j + \xi_i^{\star j}\right) \tag{3.44}$$

$$s.t. \left(w \cdot x_i^j\right) - b_j \leq -1 + \zeta_i^j, j = 1, \cdots, M, i = 1, \cdots, l^i \tag{3.45}$$

$$\left(w \cdot x_i^j\right) - b_{j-1} \geq 1 - \xi_i^j\star, j = 1, \cdots, M, i = 1, \cdots, l^i \tag{3.46}$$

$$\xi_i^j \geq 0, \xi_i^j \geq 0*, j = 1, \cdots, M, i = 1, \cdots, l^j \tag{3.47}$$

Where $b = (b_1, b_{M-1})^{\mathrm{T}}, b_0 = -\infty, b_M = +\infty$. Its dual problem is the following convex quadratic programming

$$\min_{\alpha^{(*)}} \frac{1}{2}\sum_{j,i}\sum_{j',i'}\left(\alpha_i^{*j} - \alpha_i^j\right)\left(\alpha_{i'}^{*j'} - \alpha_{i'}^{j'}\right)\left(x_i^j \cdot x_{i'}^{j'}\right) - \sum_{j,i}\left(\alpha_i^j + \alpha_i^{*j}\right) \tag{3.48}$$

$$s.t. \sum_{i=1}^{lj} \alpha_i^j = \sum_{i=1}^{l^{j+1}} \alpha_i^{*j+1}, j = 1, \cdots, M-1 \tag{3.49}$$

$$0 \le \alpha_i^j, \alpha_i^j \star \le C, j = 1, \cdots, M, i = 1, \ldots, l^j \tag{3.50}$$

$$\alpha_i 1 = 0, i = 1, \cdots, l^1 \tag{3.51}$$

$$\alpha_i^M = 0, i = 1, \cdots, l^M \tag{3.52}$$

Though SVOR is a method to solve a specialization of the multi-class classification problem and has many applications itself [27], it is also used in the context of solving general multi-class classification problem [16, 17, 28, 29] in which the SVOR is used as a basic classifier and used several times instead of only once, just as the binary classifiers for multi-class classification. There are many choices since any p-class SVOR with different order can be candidate, where. When p = 2, this approach reduces to the approach based on binary classifiers.

Semi-supervised Support Vector Machine

In practice, labeled instances are often difficult, expensive, or time consuming to obtain, meanwhile unlabeled instance may be relatively easy to collect. Different with standard SVMs using only labeled training points, lots of semi-supervised SVMs (S^3VM) use large amount of unlabeled data, together with the labeled data, to build better classifiers. Transductive support vector machine (TSVM) [30] is such an efficient method finding a labeling of the unlabeled data, so that a linear boundary has the maximum margin on both the original labeled data and the (now labeled) unlabeled data. The decision function has the smallest generalization error bound on unlabeled data.

For a training set given by

$$T = \{(x_1, y_1), \cdots, (x_l y_l)\} \cup \left\{x_{l+1}, \cdots, x_{l+q}\right\}, \tag{3.53}$$

where $x_j \in R^n$, $y_j \in \{-1, 1\}$, $i = 1, \cdots, l$, $x_i \in R^n$, $i = l + 1, \ldots, l + q$, and the set $\{x_{l+1}, \cdots, x_{l+q}\}$ is a collection of unlabeled inputs. The primal problem in TSVM is constructed as the following (partly) combinational optimization problem

$$min\ \frac{1}{2}\|w\|^2 + C\sum\nolimits_{i=1}^{l}\xi_j + C^*\sum\nolimits_{i=1}^{l}\xi_i^\star \tag{3.54}$$

$$s.t.\ \ y_i\left((w\cdot x_j)+b\right) \geq 1-\xi_j, i=1,\cdots,l, \tag{3.55}$$

$$y_i^*\left(w\cdot x_i^*\right)+b\Big) \geq 1-\xi_i^*,\ \ i=l+1,\cdots,l+q, \tag{3.56}$$

$$\xi_j \geq 0, i=1,\cdots,l \tag{3.57}$$

$$\xi_i^* \geq 0, i=l+1,\cdots,l+q, \tag{3.58}$$

where $y^*=\left(y_{l+1}*, y_{l+q}^\star\right)$, $C>0$, $C^*>0$ are parameters. However, finding the exact solution to this problem is NP-hard. Major effort has focused on efficient approximation algorithms. The SVM-light is the first widely used software [30]. In the approximation algorithms, several relax the above TSVM training problem to semi-definite programming (SDP) [31–33]. The basic idea is to work with the binary label matrix of rank 1, and relax it by a positive semi-definite matrix without the rank constraint. However, the computational cost of SDP is still expensive for large scale problems.

3.1.1.3 Universum Support Vector Machine

Different with semi-supervised SVM leveraging unlabeled data from the same distribution, Universum support vector machine (USVM) uses the additional data not belonging to either class of interest. Universum contains data belonging to the same domain as the problems of interest and is expected to represent meaningful information related to the pattern recognition task at hand. Universum classification problem can be formulated as follows:

Definition 3.3 (Universum classification problem). Given a training set

$$T=\{(x_1,y_1),\cdots,(x_l y_l)\}\cup\left\{x_1^\star,\cdots,x_u^\star\right\} \tag{3.59}$$

where $x_j\in R^n$, $y_j\in\{-1,1\}$, $i=1,\cdots,l$, $x_j^\star\in R^n$, $j=1,\ \ldots,u$, and the set

$$U=\left\{x_1^\star,\cdots,x_u^\star\right\} \tag{3.60}$$

is a collection of unlabeled inputs known not to belong to either class, find a real function g(x) in R such that the value of y for any x can be predicted by the decision function

$$f(x) = sgn\,(g(x)) \tag{3.61}$$

Universum SVM constructs the following primal problem

$$min\ \frac{1}{2}\|w\|_2^2 + C_t\sum\nolimits_{i=1}^{l}\xi_j + C_u\sum\nolimits_{s=1}^{u}\left(\psi_s + \psi_s^*\right) \tag{3.62}$$

$$s.t. y_j\left(\left(w\cdot x_j\right)+b\right) \geq 1-\xi_i, \xi_i \geq 0, i = 1, \cdots, l, \tag{3.63}$$

$$-\varepsilon - \psi_s^\star \leq \left(w\cdot x_s^\star\right)+b \leq \varepsilon + \psi_s, s = 1, \cdots, u, \tag{3.64}$$

$$\psi_s, \psi_s^\star \geq 0, s = 1, \cdots, u, \tag{3.65}$$

where $\psi^{(*)} = \left(\psi_1, \psi_1^*, \cdots, \psi_u, \psi_u^\star\right)^T$ and $C_t > 0$, $C_u > 0$, $\varepsilon > 0$ are parameters. Its goal is to find a separating hyperplane $(w \cdot x) + b = 0$ such that, on the one hand, it separates the inputs $\{x_1, \ldots, x_l\}$ with maximal margin, and on the other hand, it approximates to the inputs $\left\{x_1^\star, \ldots, x_u^\star\right\}$. We can also get its dual problem and introduce kernel function for dealing with nonlinear classification.

It is natural to consider the relationship between USVM and some 3-class classification. In fact, it can be shown that, under some assumptions, USVM is equivalent to K-SVCR [34] and is also equivalent to the SVOR with M = 3 with slight modification [35]. USVM's performance depends on the quality of the Universum, methodology of choosing the appropriate Universum is the subject of future research.

3.1.1.4 Robust Support Vector Machine

In standard SVMs, the parameters in the optimization problems are implicitly assumed to be known exactly. However, in practice, some uncertainty is often resent in many real-world problems, these parameters have perturbations since they are estimated from the training data which are usually corrupted by measurement noise. The solutions to the optimization problems are sensitive to parameter perturbations. So, it is useful to explore formulations that can yield discriminants robust to such measurement errors. For example, when the inputs are subjected to measurement errors, it would be better to describe the inputs by uncertainty sets $X_i \in R^n$,

$i = 1, \ldots, l$, since all we know is that the input belongs to the set X_i. Therefore, the standard problem turns to be the following robust classification problem.

Definition 3.4 (Robust classification problem). Given a training set

$$T = \{(X_1, Y_1), \cdots, (X_l, Y_l)\} \quad (3.66)$$

where Xi is a set in. Find a real function g(x) in R, such that the value of y for any x can be predicted by the decision function

$$f(x) = sgn(g(x)) \quad (3.67)$$

The geometric interpretation of the robust problem with circle perturbations is shown in Fig. 3.3, where the circles with "+" and "o" are positive and negative input sets respectively, the optimal separating hyperplane by the principle of maximal margin is constructed by robust SVM (RSVM). Now, the primal problem of RSVM for such case is a semi-infinite programming problem

$$min \frac{1}{2}\|w\|^2 + C\sum_{i=1}^{l} \xi_j \quad (3.68)$$

$$s.t. y_i\left(\left(w \cdot \left(x_i \cdot r_i u_j\right)\right) + b\right) \geq 1 - \xi_j, \forall \left\|u_j\right\| \leq 1, i = 1, \ldots, l \quad (3.69)$$

$$\xi_i \geq 0, i = 1, \ldots, l, \quad (3.70)$$

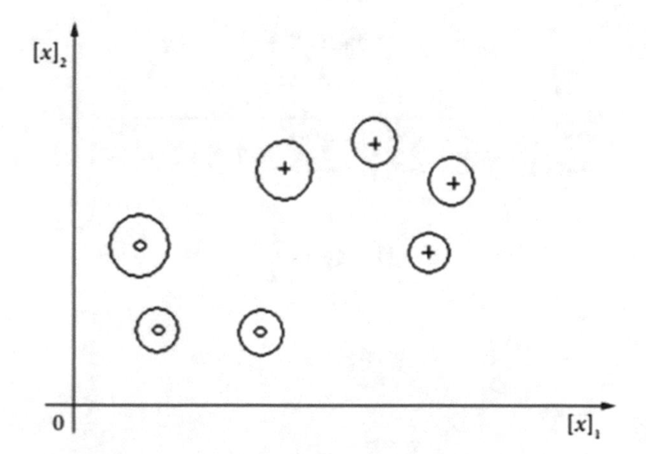

Fig. 3.3 Geometric interpretation of robust classification problem

where the set X_i is a super sphere obtained from perturbation of a point i

$$X_i = \{x \mid \| x - x_i \| \leq r_i\} \tag{3.71}$$

This semi-infinite programming problem can be proved to be equivalent to the following second order cone programming [28, 36]

$$min \ \frac{1}{2}(u - \nu) + C\sum\nolimits_{i=1}^{l}\xi_j \tag{3.72}$$

$$s.t. y_i \left(\left(w \cdot x_j \right) + b \right) - r_i t \geq 1 - \xi_i, i = 1, \cdots, l, \tag{3.73}$$

$$\xi_i \geq 0, i = 1, \cdots, l, \tag{3.74}$$

$$u + v = 1 \tag{3.75}$$

$$\begin{pmatrix} u \\ t \\ \nu \end{pmatrix} \in L^3 \tag{3.76}$$

$$\begin{pmatrix} t \\ w \end{pmatrix} \in L^{n+1}, \tag{3.77}$$

its dual problem is also a second order cone programming

$$\alpha, \beta, \gamma, z_u z_v max, \beta + \sum\nolimits_{i=1}^{l}\alpha_i, \tag{3.78}$$

$$s.t. Y \leq \sum\nolimits_{i=1}^{l} r_i \alpha_j - \sqrt{\sum\nolimits_{i--1}^{l}\sum\nolimits_{j--1}^{l}\alpha_i\alpha_j y_i y_j K\left(x_i, x_j\right)} \tag{3.79}$$

$$\beta + z_u = \frac{1}{2}, \tag{3.80}$$

$$\beta + z_v = -\frac{1}{2}, \tag{3.81}$$

$$\sum\nolimits_{i=1}^{l} y_j \alpha_j = 0, \tag{3.82}$$

$$0 \le \alpha_i \le C, i = 1, \cdots, l, \tag{3.83}$$

$$\sqrt{\gamma^2 + z_v^2} \le z_u, \tag{3.84}$$

which can be efficiently solved by Self-Dual-Minimization (SeDuMi). SeDuMi is a tool for solving optimization problems. It can be used to solve linear programming, second-order cone programming and semi-definite programming, and is available at the web site http://sedumi.mcmaster.ca.

3.1.1.5 Knowledge Based Support Vector Machine

In many real-world problems, we are given not only the traditional training set, but also prior knowledge such as some advised classification rules. If appropriately used, prior knowledge can significantly improve the predictive accuracy of learning algorithms or reduce the amount of training data needed. Now the problem can be extended in the following way: the single input points in the training points are extended to input sets, called knowledge sets. If we consider the input sets restricted as polyhedrons, the problem is formulated mathematically as follows:

Definition 3.5 (Knowledge-based classification problem). Given a training set

$$T = \{(X_1, y_1), \cdots, (X_p, y_p), (X_{p+1}, y_{p+1}), \cdots, (X_{p+q}, y_{p+q})\} \tag{3.85}$$

where X_i is a polyhedron in R_n defined by $X_i = \{x| Q_j x \le d_i\}$, and $Q_i \in R^{l_i \times n}$, $d_i \in R^{l_i}$, $y_1 = \cdots = y_p = 1$, $y_{p+1} = \cdots = y_{p+q} = -1$. Find a real valued function $g(x)$ in R_n such that the value of y for any x can be predicted by the decision function

$$f(x) = sgn\,(g(x)) \tag{3.86}$$

Of course, we can construct the primal problem to be the following semi-infinite programming problem

$$min\ \frac{1}{2}\|w\|^2 + C\sum_{i=1}^{p+q} \xi_i, \tag{3.87}$$

$$s.t.\ \ (w \cdot x) + b \ge 1, for\ x \in X_i.i = 1, \cdots, p, \tag{3.88}$$

$$(w \cdot x) + b \le -1, for \in X_i.i = p+1, \cdots, p+q, \tag{3.89}$$

$$\xi_i \ge 0, i = 1, \cdots, p+q \tag{3.90}$$

However, it was shown that the constraints (3.88)–(3.90) can be converted into a set of limited constraints and then the problem becomes a quadratic programming [37]

$$min\ \frac{1}{2}\|w\|^2 + C\sum_{i=1}^{p+q}\left(\left(\sum_{j=1}^{n}\xi_{i_j}\right) + \eta_i\right) \tag{3.91}$$

$$s.t. -\xi_j \leq Q_i^T u_i + w \leq \xi_j, i = 1, \cdots, p, \tag{3.92}$$

$$d_i^T u_i - b + 1 \leq \eta_j, i = 1, \cdots, p, \tag{3.93}$$

$$-\xi_j \leq Q_i^T u_i - w \leq \xi_j, i = p+1, \cdots, p+q, \tag{3.94}$$

$$d_i^T u_j + b + 1 \leq \eta_i, i = p+1, \cdots, p+q, \tag{3.95}$$

$$\xi, \eta, u \geq 0. \tag{3.96}$$

This model considered the linear knowledge incorporated to linear SVM, while linear knowledge based nonlinear SVM and nonlinear knowledge based SVM were also proposed by Mangasarian and his co-workers [37, 38]. Handling prior knowledge is worthy of further study, especially when the training data may not be easily available whereas expert knowledge may be readily available in the form of knowledge sets. Another prior information such as some additional descriptions of the training points was also considered and a method called privileged SVM was proposed [39], which allows one to introduce human elements of teaching: teacher's remarks, explanations, analogy, and so on in the machine learning process.

3.1.1.6 Multi-instance Support Vector Machine

Multi-instance problem was proposed in the application domain of drug activity prediction, and similar to both the robust and knowledge-based classification problems, it can be formulated as follows.

Definition 3.6 (Multi-instance classification problem). Suppose that there is a training set

$$T = \{(x_1, u), \cdots, (x_l, u)\} \tag{3.97}$$

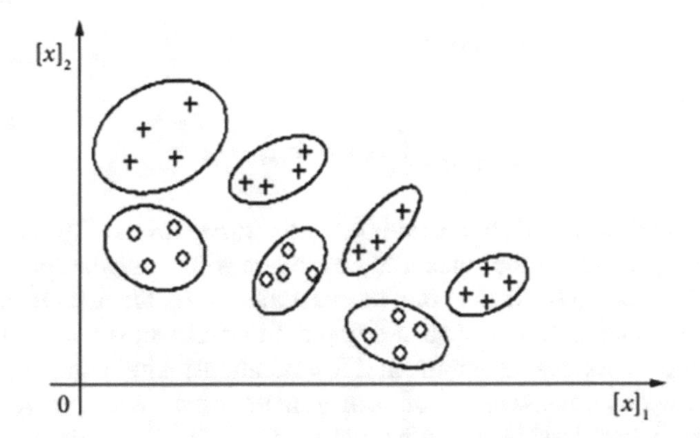

Fig. 3.4 Geometric interpretation of multi-instance classification problem

Where. Find a real function g(x) in R, such that the label y for any instance x can be predicted by the decision function

$$f(x) = sgn\,(g(x)) \tag{3.98}$$

The set X_i is called a bag containing a number of instances. Note that the interesting point of this problem is that: the label of a bag is related with the labels of the instances in the bag and decided by the following way: a bag is positive if and only if there is at least one instance in the bag is positive; a bag is negative if and only if all instances in the bag are negative. A geometric interpretation of multi-instance classification problem is shown in Fig. 3.4, where every enclosure stands for a bag; a bag with "+" is positive and a bag with "o" is negative, and both "+" and "o" stand for instances.

For a linear classifier, a positive bag is classified correctly if and only if some convex combination of points in the bag lies on the positive side of a separating plane. Thus, the primal problem in the multi-instance SVM (MISVM) is constructed as the following nonlinear programming problem [40]

$$min\ \frac{1}{2}\|w\|^2 + C_1\sum\nolimits_{i=1}^{p}\xi_i + C_2\sum\nolimits_{i=r+1}^{r+s}\xi_j, \tag{3.99}$$

$$s.t. \left(w \cdot \sum\nolimits_{j\in I(i)} v_j^i x_j\right) + b \geq 1 - \xi_j, i = 1, \cdots, p, \tag{3.100}$$

$$(w \cdot x_i) + b \leq -1 + \xi_i,\ \ j = r+1, r+s, \tag{3.101}$$

$$\xi_j \geq 0, i = 1, \cdots, p, r+1, \cdots, r+s, \tag{3.102}$$

$$v_j^i \geq 0, j \in I(i), i = 1, \ldots, p, \tag{3.103}$$

$$\sum\nolimits_{j \in I(i)} v_j^i = 1, i = 1, \cdots, p. \tag{3.104}$$

where r and s are respectively the number of the instances in all positive bags and all negative bags, and p is the number of positive bags. Though the above problem is nonlinear, it is easy to see that among its constraints, only the first one is nonlinear, and in fact is bilinear. Then a local solution to this problem is obtained by solving a succession of fast linear programs in a few iterations: Alternatively, hold one set of variables which constitute the bilinear terms constant while varying the other set. For a nonlinear classifier, a similar statement applies to the higher dimensional space induced by the kernel.

3.1.2 Two New Decomposition Algorithms for Training Bound-Constrained Support Vector Machines

In this section, we consider a simple modification model of the standard SVM as follow

$$\min \ \frac{1}{2}\left(\|\mathrm{w}\|^2 + b^2\right) + C\sum\nolimits_{i=1}^{l} \xi_i \tag{3.105}$$

$$s.t. \ \ y_i\left(\mathrm{w}^T \varphi\left(\mathrm{x}_i\right) + b\right) \geq 1 - \xi_i, \ \ \forall i = 1, \ldots, l, \tag{3.106}$$

$$\xi_i \geq 0, \quad \forall i = 1, \cdots, l, \tag{3.107}$$

The dual form of the above problem is the following QP, which contains bound constraints only:

$$\min \ f(\alpha) = \frac{1}{2}\alpha^T Q \alpha^T - 1^T \alpha \tag{3.108}$$

$$s.t. 0 \leq \alpha \leq \mathrm{C} \tag{3.109}$$

Where $Q_{i,j} \equiv y_i y_j (K(\mathrm{x}_i, \mathrm{x}_j) + 1)$. Suppose the optimal solution of problem above is. Then the classifier can be written as

$$H(\mathrm{x}) = \mathrm{sign}\left\{\sum\nolimits_{i=1}^{l} \alpha_i^* y_i \left(K\left(\mathrm{x}, \mathrm{x}_i\right) + 1\right)\right\} \tag{3.110}$$

Similar to the standard SVM, if $\alpha_i^* \neq 0$, the corresponding sample i is called a Support Vector (SV). Furthermore, if $\alpha_i^* = C$, it is called a Bounded Support Vector (BSV). To show the difference with the standard SVM, (3.105)–(3.107) and (3.108)–(3.109) are usually called the bound-constrained SVMs.

Bound-constrained SVMs are once proposed independently by Fireß, Cristianini and Campbell in [41]. Mangasarian and Musicant [42] proves that for the linear kernel case, there exist some kind of equivalence between bound-constrained SVMs and the standard SVMs. The result can be concluded in the following proposition.

Proposition 3.1 [42] *For the bound-constrained SVMs with linear kernel function* $k(\mathrm{x}, \mathrm{y})$, *Suppose* (w, b, ξ) *is an arbitrary solution of problem* (3.105)–(3.107). *If any solution* $\hat{\mathrm{v}}$ *of the following system:*

$$A^T v = 0, 1^T v = b, \mathrm{v} \geq 0 \tag{3.111}$$

satisfies

$$1^T v \left(1^T \xi - 1\right) \leq b^2 \tag{3.112}$$

There must exists a sufficient large C, such that (w, b, ξ) *is also a solution of the standard primal SVM model* (1), *where* $A^T = (x_1, \cdots, x_l) \in \Re^{n \times l}$, *i.e., each row of A is the feature vector of a sample.*

Because of the simple formulation and good classification performance of bound-constrained SVM, it draws a lot of attention in the recent years. For the linear kernel case, Mangasarian and Musicant proposed to solve the model (3.108)–(3.109) by the over-relaxation method in [42]. They proved the global convergence and the linear convergent rate of the algorithm at the same time. Hsieh et al. gave a novel dual coordinate descent method and proved the algorithm reaches an ε-accurate solution in $O(\log(1/\varepsilon))$ iterations in [43]. Several researchers also explore how to train the primal form of (3.108)–(3.109) and the extended models fast. The existing algorithms can be broadly categorized into two categories: the cutting-plane methods [44–46], and subgradient methods [47]. For example, in [48], Shalev-Shwartz et al. described and analyzed a simple and effective stochastic sub-gradient descent algorithm and prove that the number of iterations required to obtain a solution of accuracy ε is $(1/\varepsilon)$. Generally speaking, without counting the loading time, these recent advances on linear classification have shown that training one million instances takes only a few seconds [49].

For the general nonlinear kernel case, the scale of the problems grows quadratically with the number of samples l due to the appearance of kernel matrix. Therefore, for the middle and large-scale problem, decomposition techniques are needed to handle the problems arising from the limitation of the memory. Hsu and Lin analyzed in thorough how to design a decomposition algorithm for problem (4) in [50], which mainly discussed how to design an effective working set selection rule based on the existing successful working set selection rule for standard SVMs.

They started from applying the working set selection strategy of SVM^{light}(Joachims) to model (4) directly. Step by step, a series of heuristic techniques were added to improve the performance of the selection rule. Different form the designation idea in [50], we will discuss how to derive effective working set selection rule from the optimization algorithm design point of view. Although the derivation is much more concise than [50], the numerical experiments in subsection show that the computation efficiency of our new methods is at least as good as Hsu and Lin's method.

3.1.2.1 The Decomposition Algorithm Framework

Because Q_{ij} is in general not zero, Q becomes a fully dense matrix. Due to the density of Q, a prohibitive amount of memory is required to store the matrix. Thus, traditional optimization algorithms, which needs the whole Hessian matrix of objective function, cannot be directly used. To conquer this difficulty, several researchers [51–54] have proposed decomposition methods. The key idea of decomposition is to update a small part of variables at each iteration, and to solve a sequence of constant-size problems. Then, the solution of a large-scale dense problem can be found by solving a number of small problems.

Now we denote the gradient of objective function (3.105) as:

$$F(\alpha) = \left(\frac{\partial f(\alpha)}{\partial \alpha_i} \right) = Q\alpha - 1, \tag{3.113}$$

for any $i \in \{1, \cdots, l\}$, the i-th element of $F(\alpha)$ is:

$$F_i(\alpha) = \frac{\partial f(\alpha)}{\partial \alpha_i} = \sum\nolimits_{j=1}^{l} Q_{ij}\alpha_j - 1. \tag{3.114}$$

The index set of variables optimized at a current iteration is called the working set and denoted as $\mathcal{B}$. Let $\mathcal{N} \equiv \{1, \cdots l\}, /\mathcal{B}$, and superscript record the iteration number. Using the notation above, the general decomposition method for problem (3.108)–(3.109) is described in Algorithm 3.1. Notice that Algorithm 3.1 is just a framework, to make it be implemented in practice, the specific working set selection rule and QP subproblem solver should be given. In the rest part of this section, we will discuss how to select the working set effectively from the optimization design point of view.

Denote $Qsub\left(\alpha^{(k)}, \mathcal{B}\right)$ as the optimal objective function value of QP

$$\min \ f\left(\alpha^{(k)} + d\right) - f^{(k)} \tag{3.115}$$

$$s.t. \ -\alpha^{(k)} \le d_i \le C - \alpha_i^{(k)} \forall i \in \mathcal{B}, \tag{3.116}$$

$$d_i = 0, \forall i \notin \mathcal{B}. \tag{3.117}$$

If only one iteration is considered, the best working set is $\text{argmin}_{\mathcal{B}:|\mathcal{B}|=n_B}\ Qsub\left(\alpha^{(k)}, \mathcal{B}\right)$. However, this choice requires solving $\begin{pmatrix} l \\ n_B \end{pmatrix}$ QPs (7), which is too expensive to manipulate in practice. To decrease the computational cost, we will present two new working set selection methods in the next two subsections.

Algorithm 3.1 Decomposition Algorithm Framework for Bound-Constrained SVMs

Step 0. Initialization. Given the upper bound for the size of the workings set $n_B \geq 1$, the initial working set $\mathcal{B}^{(0)}$ and the initial point $\alpha^{(0)} = 0 \in \Re^l$. Set $k = 0$.
Step 1. Subproblem Solution Construct and solve the quadratic subproblem

$$\min \frac{1}{2}\sum\nolimits_{i,j\in\mathcal{B}^{(k)}} Q_{ij}\alpha_i\alpha_j + \sum\nolimits_{i\in\mathcal{B}^{(k)}}\left(F_i^{(k)} - \sum\nolimits_{j\in\mathcal{B}^{(k)}} Q_{ij}\alpha_j^{(k)}\right)\alpha_i \tag{3.118}$$

$$s.t.\, 0 \leq \alpha_i \leq C, \forall i \in \mathcal{B}^{(k)}. \tag{3.119}$$

Denote the optimal solution as $\{\overline{\alpha}_i\}_{i\in\mathcal{B}(k)}$. Update the iteration point as

$$\alpha_i^{(k+1)} = \begin{cases} \overline{\alpha}_i & \text{if } i \in \mathcal{B}^{(k)}, \\ \alpha_i^{(k)} & \text{if } i \notin \mathcal{B}^{(k)}. \end{cases} \tag{3.120}$$

Step 2. Gradient Update. Update gradient based on formula (5). Set $k := k + 1$.
Step 3. Working Set Selection. Test whether the iterates should be terminated according to some stopping criteria. If not stop, select at most n_B indices to form a new working set $\mathcal{B}^{(k)}$, and go back to Step 1. Otherwise, output $\alpha^{(k)}$ as the optimal solution and stop.

3.1.2.2 Using First Order Information for Working Set Selection

A straightforward way of simplifying problem (3.115)–(3.117) is dropping the second order information in the objective function, and confine the incremental variables between 1 and -1, which results in the following Linear Programming (LP):

$$\min \sum\nolimits_{i\in\mathcal{B}} F_i^{(k)} d_i \tag{3.121}$$

$$-\alpha_i^{(k)} \leq d_i \leq C - \alpha_i^{(k)}, \forall i \in \mathcal{B}, \tag{3.122}$$

$$-1 \leq d_i \leq 1, \forall i \in \mathcal{B}. \tag{3.123}$$

Denote the optimal objective function value of problem (3.133)–(3.135) as *Lsub* $(\alpha^{(k)}, \mathrm{B})$. Then, argminB: $|\mathrm{B}| = n_{\mathrm{B}}$ *Lsub* $(\alpha^{(k)}, \mathrm{B})$ should be a good choice as the working set $\mathrm{B}^{(k)}$. Dividing the constraints (3.134) and (3.135) by samples, problem (3.133)–(3.135) can be solved as $|\mathrm{B}|$ independent LPs. In details, for any $i \in \mathrm{B}$,

$$\min \ d_i F_i^{(k)} \tag{3.124}$$

$$s.t. \max\left\{-1, -\alpha^{(k)}\right\} \leq d_i \leq \min\left\{1, C - \alpha_i^{(k)}\right\}. \tag{3.125}$$

which solution can be represented as

$$\hat{d}_i = \begin{cases} \max\left\{-\alpha_i^{(k)}, -1\right\}, \text{if } F_i^{(k)} \geq 0, \\ \min\left\{C - \alpha_i^{(k)}, 1\right\}, \text{if } F_i^{(k)} < 0, \end{cases} \tag{3.126}$$

The above formula can be rewritten as a more compact form:

$$\hat{d}_i = -\operatorname{sign}\left(F_i^{(k)}\right) \min\left\{\frac{1}{2}C\left(1 - \operatorname{sign}\left(F_i^{(k)}\right)\right) - \alpha_i^{(k)}, 1\right\}. \tag{3.127}$$

Furthermore, the corresponding optimal objective function value of LP (10) can be represented as

$$- \mid F_i^{(k)} \mid \min\left\{\frac{1}{2}C\left(1 - \operatorname{sign}\left(F_i^{(k)}\right)\right) - \alpha_i^{(k)}, 1\right\}. \tag{3.128}$$

Based on these discussions, for $\forall i = 1, \cdots, l$, define

$$\ell_i(\alpha) = -\left|F^{(k)}\right| \min\left\{\frac{1}{2}C\left(1 - \operatorname{sign}\left(F_i^{(k)}\right)\right) - \alpha_i^{(k)}, 1\right\}, \tag{3.129}$$

namely,

$$\ell_i(\alpha) = \begin{cases} \max\left\{-\alpha_i^{(k)}, -1\right\} F_i(\alpha), \text{if } F_i^{(k)} \geq 0, \\ \min\left\{C - \alpha_i^{(k)}, 1\right\} F_i(\alpha), \text{if } F_i^{(k)} < 0, \end{cases} \tag{3.130}$$

Then,

$$Lsub\left(\alpha^{(k)}, \mathrm{B}\right) = \sum\nolimits_{i \in \mathrm{B}} \ell_i^{(k)}. \tag{3.131}$$

Therefore, instead of solving $\binom{l}{n_{\mathrm{B}}}$ LPs (3.124)–(3.125), we can compute $\ell^{(k)}$, and select the indices corresponding to the n_B smallest elements of $\ell_i^{(k)}$ directly to get the value of

$$\min \ Lsub\left(\alpha^{(k)}, \mathrm{B}\right) \tag{3.132}$$

Especially, when $\ell^{(k)} = 0$, no new working set can be selected, and the iterates can be terminated naturally. To summarize, we describe this simple working set selection rule in

---

Algorithm 3.2 The First-Order Working Set Selection Rule

Compute $\ell^{(k)}$ by formula (3.129).
 if $l^{(k)} \neq 0$ then
 Sort the indices set $\{1, \ldots, l\}$ by the values of $\ell^{(k)}$ in increasing order. Choose the first n_B indices to form the working set $\mathcal{B}^{(k)}$
 else
 Output $\boldsymbol{\alpha}^{(k)}$ as the optimal solution and terminate the iterates.
 end if

In practice, we find that some indices enter and leave the working set lots of times when applying working set selection rule Algorithm 3.2 directly to Algorithm 3.1. This causes the decomposition algorithm converging very slow. In order to avoid the zigzagging phenomenon, we keep part of indices from previous working sets at each iteration. To be more precise, at most n_N new indices are allowed to enter into the working set, where $1 \leq n_N \leq n_B$. Other indices are taken from the working set of last iteration. One thing we want to stress is that different techniques of inheriting indices has been used by several state-of-the-art solvers [55]. In this work, we use the inheriting strategy used in [55] to revise our working set selection rules we proposed before. Details are given in

---

Algorithm 3.3 The Practical First-Order Working Set Selection Rule

---

Compute $\ell^{(k)}$ by (11)
 if $\ell^{(k)} \neq 0$ then
 i) Set $\hat{\mathcal{B}} = B^{(k)}$ and $\mathcal{B}^{(k+1)} = \varnothing$.
 ii) Sort the indices set $\{1_{\backslash}. \backslash' l\}b_3$' the values of $P^{(k)}$ in increasing order; Add the first n_N indices to form the working set $\mathcal{B}^{(k+1)}$ Set $\hat{\mathcal{B}} := \hat{\mathcal{B}} \backslash \mathcal{B}^{(k+1)}$
 iii) Fill $j \in \left\{i | i \in \hat{\mathcal{B}}, 0 < \alpha_i^{(k)} < C\right\}$, which has the lowest number of consecutive iterations in working set, to $\mathrm{B}^{(k+1)}$ and remove j from $\hat{\mathcal{B}}$ until $| \mathcal{B}^{(k+1)} | = n_B$ or there is no such element.

iv) Fill $j \in \left\{i | i \in \hat{\mathcal{B}}, \alpha_i^{(k)} = 0\right\}$, which has the lowest number of consecutive iterations in working set, to $\mathcal{B}^{(k+1)}$ and remove it from $\hat{\mathcal{B}}$ until $| \mathcal{B}^{(k+1)} |= n_B$ or there is no such element.

v) Fill $j \in \left\{i | i \in \hat{\mathcal{B}}, \alpha_i^{(k)} = C\right\}$, which has the lowest number of consecutive iterations in working set to $\mathcal{B}^{(k+1)}$ and remove it from $\hat{\mathcal{B}}$ until $| \mathcal{B}^{(k+1)} |= n_B$ or there is no such element.

else

Output $\alpha^{(k)}$ as the optimal solution and terminate the iterations.

end if

3.1.2.3 Using Second Order Information for Working Set Selection

In the algorithm described in the last subsection, only first-order information of the objective function is used for simplicity. However, this does not mean that there is no simple and feasible method which uses second order information for choosing working set. In this subsection, we will consider how to include second order information into the process of working set selection.

Besides keeping both the first order information, we incorporate the diagonal second order information, and get the following revised QP:

$$\min \ \sum\nolimits_{i\in\mathcal{B}} \left(F_i^{(k)} d_i + \frac{1}{2} Q_{ii} d_i^2\right) \tag{3.133}$$

$$-\alpha_i^{(k)} \le d_i \le C - \alpha_i^{(k)}, \forall i \in \mathcal{B}, \tag{3.134}$$

$$-1 \le d_i \le 1, \forall i \in \mathcal{B}. \tag{3.135}$$

Denote the optimal objective function value of problem (3.133)–(3.135) as DQsub $\left(\alpha^{(k)}, \mathcal{B}\right)$. Since more information of the objective model is used, arg $\min_{\mathcal{B}:|\mathcal{B}|=n_B}$ DQsub $\left(\alpha^{(k)}, \mathcal{B}\right)$ should be a better choice of $\mathcal{B}^{(k)}$ than the set selection rule given in the last subsection.

Similar to the discussion of Sect. 2.2, dividing the constraints (3.134) and (3.135) by samples, problem (3.133)–(3.135) can be solved as $| \ \mathcal{B} \ |$ independent QPs. In details, for any $i \in \mathcal{B}$,

$$\min \ d_i F_i^{(k)} + \frac{1}{2} Q_{ii} d_i^2 \tag{3.136}$$

$$s.t. \ \max \left\{-1, -\alpha^{(k)}\right\} \le d_i \le \min \left\{1, C - \alpha_i^{(k)}\right\}. \tag{3.137}$$

which solution can be represented as

$$\hat{d}_i = \begin{cases} \max\left\{-\frac{F_i^{(k)}}{Q_{ii}}, \max\left\{-\alpha_i^{(k)}, -1\right\}\right\}, & \text{if } F_i^{(k)} \geq 0 \\ \min\left\{-\frac{F_i^{(k)}}{Q_{ii}}, \min\left\{C - \alpha_i^{(k)}, 1\right\}\right\}, & \text{if } F_i^{(k)} < 0 \end{cases} \tag{3.138}$$

The above formula can be rewritten as a more compact form:

$$\hat{d}_i = mid\left\{-\frac{F_i^{(k)}}{Q_{\text{ii}}}, \ \max\left\{-\alpha_i^{(k)}, -1\right\}, \ \min\left\{C - \alpha_i^{(k)}, 1\right\}\right\}. \tag{3.139}$$

Furthermore, the corresponding optimal objective function value of problem (3.136)–(3.137) can be represented as $F_i^{(k)}\hat{d}_i + \frac{1}{2}Q_{ii}\hat{d}_i^2$. Based on these discussions, for $\forall i = 1, \ldots, l$, define

$$\dot{d}_i\ (\alpha) = mid\left\{-\frac{F_i\ (\alpha)}{Q_{i_i}}, \ \max\ \{-\alpha_i, -1\}, \ \min\ \{C - \alpha_i, 1\}\right\}, \tag{3.140}$$

and

$$q_i(\alpha) = F(\alpha)_i\breve{d}_i(\alpha) + \tfrac{1}{2}Q_{ii}\breve{d}_i(\alpha)^2. \tag{3.141}$$

Then, $DQsub\ \left(\alpha^{(k)}, \mathcal{B}\right) = \sum_{i\in\mathcal{B}} q_i^{(k)}$. Therefore, instead of solving $\binom{l}{n_{\text{B}}}$ QPs (3.141), we can compute $q^{(k)}$, and select the indices corresponding to the n_B smallest elements of $q_i^{(k)}$ to form arg $min_{B:|B|=n_{\text{B}}}\ DQsub\ \left(\alpha^{(k)}, \mathcal{B}\right)$ directly. Especially, when $q^{(k)} = 0$, no new working set can be selected, and the iterates terminate naturally. Similar practical working set selection strategy can also be used here to avoid the zigzagging phenomenon. To summarize, we describe the new working set selection rule in Algorithm 3.4.

3.1.2.4 Global Convergence Analysis

In this section, we will prove that the decomposition algorithms based on our new working set selection rules (Algorithm 3.3 or Algorithm 3.4) are globally convergent.

Algorithm 3.4 The Practical Second-Order Working Set Selection Rule

Compute $q^{(k)}$ by (14).

if $q^{(k)} \neq 0$ then

i) Set $\mathcal{B} = \mathcal{B}^{(k)}$ and $\mathcal{B}^{(k+1)} = \varnothing$.

ii) Sort the indices set $\{1, \cdots, l\}$ by the values of $q^{(k)}$ in increasing order; Add the first n_N indices to form the working set $\mathcal{B}^{(k+1)}$; Set $\hat{\mathcal{B}} := \hat{\mathcal{B}} \backslash \mathcal{B}^{(k+1)}$.

iii) Fill $j \in \left\{i | i \in \hat{\mathcal{B}}, 0 < \alpha_i^{(k)} < C\right\}$, which has the lowest number of consecutive iterations in working set, to $\mathcal{B}^{(k+1)}$ and remove j from $\hat{\mathcal{B}}$ until $| \mathcal{B}^{(k+1)} | = n_B$ or there is no such element.

iv) Fill $j \in \left\{i | i \in \hat{\mathcal{B}}, \alpha_i^{(k)} = 0\right\}$, which has the lowest number of consecutive iterations in working set, to $\mathcal{B}^{(k+1)}$ and remove it from $\hat{\mathcal{B}}$ until $| \mathcal{B}^{(k+1)} | = n_B$ or there is no such element.

v) Fill $j \in \left\{i | i \in \hat{\mathcal{B}}, \alpha_i^{(k)} = C\right\}$, which has the lowest number of consecutive iterations in working set to $\mathcal{B}^{(k+1)}$ and remove it from $\hat{\mathcal{B}}$ until $| \mathcal{B}^{(k+1)} | = n_B$ or there is no such element.

else

Output $\alpha^{(k)}$ as the optimal solution and terminate the iterations.

end if

Lemma 3.1 *Suppose α is a feasible point of problem* (3.108)–(3.109). *$\boldsymbol{\alpha}$ is the KKT point of problem* (4) *if and only if $\ell(\boldsymbol{\alpha}) = \mathbf{0}$.*

Proof Because α is the KKT point of problem (3.108)–(3.109), from Kuhn-Tucker theorem [56], we know that there exists Lagrange multipliers u and v which satisfies KKT condition:

$$F_i(\alpha) - v_i + u_i = 0, \forall i = 1, \ldots, l, \tag{3.142}$$

$$v_i \alpha_i = 0, \forall i = 1, \cdots, l, \tag{3.143}$$

$$u_i (C - \alpha_i) = 0, \forall i = 1, \ldots, l, \tag{3.144}$$

$$v_i, u_i \geq 0, \forall i = 1, \cdots, l. \tag{3.145}$$

Because α is a feasible point of problem (3.108)–(3.109), there are only the following three kinds of possible value for α_i:

1. If $\alpha_i \in (0, C)$, according to the complementary condition (3.157) and (3.158), we have $v_i = u_i = 0$. Furthermore, from (3.156), we know $F_i(\alpha) = 0$. Therefore, $\ell_i(\alpha) = d_i 0 = 0$.
2. If $\alpha_i = 0$, based on the complementary condition (3.158), we have $u_i = 0$. Furthermore, according to (3.156) and (3.159), we have $F_i(\alpha) = v_i \geq 0$. Since the feasible set of problem (3.124)–(3.125) becomes $\{d_i | 0 \leq d_i \leq \min\{C, 1\}\}$

when $\alpha_i = 0$, the corresponding optimal solution is $d_i = 0$. Therefore, $\ell_i(\alpha) = 0F_i(\alpha) = 0$.

3. If $\alpha_i = C$, based on the complementary condition (3.158), we have $v_i = 0$. Furthermore, according to (3.156)–(3.159), we have $F_i(\alpha) = -u_i \leq 0$. Since the feasible set of problem (3.124)–(3.125) becomes $\{d_i| - \min\{C, 1\} \leq d_i \leq 0\}$ when $\alpha_i = 0$, the corresponding optimal solution is $d_i = 0$. Therefore, $\ell_i(\alpha) = 0F_i(\alpha) = 0$.

Now we prove that if the feasible point of problem (3.108)–(3.109) α satisfies $\ell(\alpha) = 0$, α must be the KKT point.

For any $i \in \{1, \cdots, l\}$, let d_i^* denote the optimal solution of problem (3.124)–(3.125). Since $\ell_i(\alpha) = F_i(\alpha)\, d_i^* = 0$, we know $F_i(\alpha) = 0$ or $d_i^* = 0$.

1. If $F_i(\alpha) = 0$, we can choose $u_i = v_i = 0$;
2. If $F_i(\alpha) > 0$, we have $d_i^* = -\alpha_i = 0$, namely, $\alpha_i = 0$. Choose $v_i = F_i(\alpha)$ and $u_i = 0$;
3. If $F_i(\alpha) < 0$, we have $d_i^* = C - \alpha_i = 0$, namely, $\alpha_i = C$. Choose $u_i = -F_i(\alpha)$ and $v_i = 0$.

It is not difficult to verify that the above chosen value of u, v and α satisfy the KKT condition (3.156)–(3.159).

Lemma 3.2 *Suppose the kernel unction $K(\cdot, \cdot)$ satisfies Mercer condition* [57], *α is a feasible point of problem* (4), *then for any $p \in \{1, \cdots, l\}$, we have*

$$Qsub\,(\alpha, p) \leq \frac{\ell_p\,(\alpha)}{2} \min\left\{1, -\frac{\ell_p\,(\alpha)}{2MC^2}\right\}, \tag{3.146}$$

where $M = max_{i=1}^{l}\left\{K_{i,i} + 1\right\}$.

Proof For any index $p \in \{1, \cdots, l\}$, let $\overline{d}_p$ denote the solution of problem (3.124)–(3.125) when $i = p$. For any $t \in [0, 1]$, $t\overline{d}_p$ stay in the feasible region and $|\ \overline{d}_p\ | \leq C$. Furthermore, since kernel function $K(\cdot, \cdot)$ satisfies Mercer condition, we have $K_{p,p} \geq 0$, and $M = max_{i=1}^{l}\left\{K_{i,i} + 1\right\} \geq 1$. In all,

$$\begin{aligned} Qsub\left(\boldsymbol{\alpha}^{(k)}, p\right) &\leq \min_{t\in[0,1]} \left\{\overline{d}_p F_p\,(\boldsymbol{\alpha})\, t + \tfrac{1}{2}K_{p,p}\overline{d}_p^2 t^2\right\} \\ &\leq \min_{t\in[0,1]} \left\{\ell_p\,(\boldsymbol{\alpha})\, t + \tfrac{1}{2}K_{p,p}\overline{d}_p^2 t^2\right\} \\ &\leq \min_{t\in[0,1]} \left\{\ell_p\,(\boldsymbol{\alpha})\, t + \tfrac{1}{2}\left(K_{p,p} + 1\right)\overline{d}_p^2 t^2\right\} \\ &\leq \min_{t\in[0,1]} \left\{\ell_p\,(\boldsymbol{\alpha})\, t + \tfrac{1}{2}MC^2 t^2\right\} \\ &\leq \tfrac{\ell_p(\boldsymbol{\alpha})}{2} t^* \end{aligned} \tag{3.147}$$

where $t^* = \min\left\{1, -\dfrac{\ell_p\,(\alpha)}{2MC^2}\right\}$.

Theorem 3.1 *Suppose the kernel function $K(\cdot, \cdot)$ satisfies Mercer condition* [57], $n_B \geq n_N \geq 1$. *Let* $\{\alpha^{(k)}\}$ *denote the iterates generated by Algorithm* 3.1 *with the working selection rule in Algorithm* 3.3. *If* $\{\alpha^{(k)}\}$ *contains only finite elements, the last iteration point must be the global optima of problem* (3.108)–(3.109). *If* $\{\alpha^{(k)}\}$ *contains infinite elements, any accumulation point is a global optima of problem* (3.108)–(3.109).

Proof If $\{\alpha^{(k)}\}$ contains finite elements, based on the working selection rule in Algorithm 3.2, we know the last iteration point must satisfy $\ell(\alpha^{(k)}) = 0$. From Lemma 3.1, it must be a KKT point. Hence, we know that α is a KKT point. Hence, we only discuss the situation of infinite iteration points.

Let $\overline{\alpha}$ be any accumulation of the sequence $\{\alpha^{(k)}\}$. Without loss of generalization, we can assume $\{\alpha^{(k)}\}$ converge to $\overline{\alpha}$ (this requirement always can be obtained by the proper relabeled of the order of iteration points). Because the feasible region of problem (4) is a bounded closed set in $\mathfrak{R}^l$ and the iterates generated by the decomposition algorithm are always feasible, we know that $\overline{\alpha}$ is also a feasible point. Furthermore, the value of $f(\overline{\alpha})$ is a finite number.

Let the index p satisfy

$$\ell_p^{(k)} = \min \ \ell_i^{(k)} = -\left\| \ell^{(k)} \right\|_\infty. \tag{3.148}$$

Since $n_N \geq 1$, index p must be contained in the selected working set $\mathrm{B}^{(k)}$. Hence,

$$f^{(k+1)} - f^{(k)} \leq Qsub\left(\alpha^{(k)}, p\right). \tag{3.149}$$

From Lemma 3.2, we have

$$\begin{aligned} f^{(k+1)} - f^{(k)} &\leq \frac{1}{2} \ell_p^{(k)} \min \left\{ 1, -\frac{\ell_p^{(k)}}{2MC^2} \right\} \\ &= -\frac{1}{2} \left\| \ell^{(k)} \right\|_\infty \ \min \left\{ 1, \frac{\left\| \ell^{(k)} \right\|_\infty}{2MC^2} \right\}. \end{aligned} \tag{3.150}$$

Sum the above formulae from 0 to s, we get

$$\sum\nolimits_{k=0}^{s} \left\| \ell^{(k)} \right\|_\infty \ \min \left\{ 1, \frac{\left\| \ell^{(k)} \right\|_\infty}{2MC^2} \right\} \leq f^{(0)} - f^{(s+1)}. \tag{3.151}$$

Let $s \to \infty$, we have

$$\sum\nolimits_{k=0}^{\infty} \left\| \ell^{(k)} \right\|_\infty \ \min \left\{ 1, \frac{\left\| \ell^{(k)} \right\|_\infty}{2MC^2} \right\} \leq f^{(0)} - f(\overline{\alpha}) < +\infty. \tag{3.152}$$

Therefore,

$$\|\ell(\overline{\alpha})\|_{\infty} = \lim_{k\to+\infty} \left\|\ell\left(\alpha^{(k)}\right)\right\|_{\infty} = 0. \tag{3.153}$$

From Lemma 3.1, we know $\overline{\alpha}$ is the KKT point of problem (3.108)–(3.109).

Since the kernel function $K(\cdot,\cdot)$ satisfies Mercer condition, problem (3.108)–(3.109) is a convex problem. Therefore, $\overline{\alpha}$ is a global optima of problem (3.108)–(3.109).

Lemma 3.3 *Suppose kernel function $K(\cdot,\cdot)$ satisfies Mercer condition* [57], *a feasible point of problem* (3.108)–(3.109) *α is a KKT point if and only if $q(\alpha) = 0$.*

Proof Because kernel function $K(\cdot,\cdot)$ satisfies Mercer condition, for all $i \in \{1,\cdots,l\}$, $Q_{ii} > 0$. From the definition of (α), we know that $\forall i \in \{1,\cdots,l\}$, $q_i(\alpha)$ is the optimal objective function value of QP

$$\min \ F_i(\alpha)\, d_i + \frac{1}{2} Q_{i,i} d_i^2 \tag{3.154}$$

$$s.t. \max\{-1, -\alpha_i\} \leq d_i \leq \min\{1, C - \alpha_i\}. \tag{3.155}$$

Let d_i^* denote the optimal solution of (3.154)–(3.155), we have $q_i(\alpha) = d_i^* F_i(\alpha) + \frac{1}{2} Q_{ii}\left(d_i^*\right)^2$.

Firstly, let us prove that any KKT point α of problem (4) satisfies $q(\alpha) = 0$. Because α is the KKT point of problem (3.108)–(3.109), from Kuhn-Tucker theorem [56], we know that there exists Lagrange multipliers u and v which satisfies KKT condition:

$$F_i(\alpha) - v_i + u_i = 0, \forall i = 1,\cdots,l, \tag{3.156}$$

$$v_i \alpha_i = 0, \forall i = 1,\cdots,l, \tag{3.157}$$

$$u_i(C - \alpha_i) = 0, \forall i = 1,\cdots,l, \tag{3.158}$$

$$v_i, u_i \geq 0, \forall i = 1,\cdots,l. \tag{3.159}$$

Because α is a feasible point of problem (3.108)–(3.109), there are only the following three kinds of possible value for α_i:

1. If $\alpha_i \in (0, C)$, according to the complementary condition (3.157) and (3.158), we have $v_i = u_i = 0$. Furthermore, from (3.156), we know $F_i(\alpha) = 0$. Through simple computation, we get that the optimal objective function value of problem (3.154)–(3.155) is 0. Therefore, $q_i(\alpha) = 0$.
2. If $\alpha_i = 0$, based on the complementary condition (3.158), we have $u_i = 0$. Furthermore, according to (3.156) and (3.159), we have $F_i(\alpha) = v_i \geq 0$. Then,

the feasible set of problem (3.154)–(3.155) becomes $\{d_i | 0 \le d_i \le \min\{C, 1\}\}$. Since the symmetric axis of quadratic function (3.154) is $-\frac{F_i(\alpha)}{2Q_{ii}} \le 0$, we know the optimal solution is obtained at zero, namely $q_i(\alpha) = 0$.
3. Similar to the discussion in (2), we can prove that $d_i^* = 0$ and $q_i(\alpha) = 0$.

Now we prove that if a feasible point of problem (3.108)–(3.109) α satisfies $q(\alpha) = 0$, α is the KKT point of problem (3.108)–(3.109).

For any $i \in \{1, \cdots, l\}$, if $q_i(\alpha) = 0$, we know $d_i^* = 0$ or $d_i^* = -\frac{2F_i(\alpha)}{Q_{ii}}$. In fact, if $d_i^* = -\frac{2F_i(\alpha)}{Q_{\dot{x}i}} \ne 0$, because of the convexity of the feasible region, it is easy to prove that $q_i(\alpha) < 0$, which is contradict to the known fact of $q_i(\alpha) = 0$. Therefore, $d_i^* = 0$.

1. If the symmetric axis of (3.154) is $-\frac{F_i(\alpha)}{2Q_{ii}} < 0$, because $d_i^* = 0$, there is $\alpha_i = 0$. Set $v_i = F_i(\alpha) > 0$ and $u_i = 0$.
2. If the symmetric axis of (3.154) is $-\frac{F_\lambda(\alpha)}{2Q_{ii}} > 0$, because $d_i^* = 0$, there is $\alpha_i = C$. Set $u_i = -F_i(\alpha) > 0$ and $v_i = 0$.
3. If the symmetric axis of (3.154) is $-\frac{F_i(\alpha)}{2Q_{ii}} = 0$. We have $F_i(\alpha) = 0$. Set $u_i = v_i = 0$. It can be checked easily that the above chosen u and v, together with α satisfy the KKT condition (3.156)–(3.159).

Theorem 3.2 *Suppose the kernel function $K(\cdot,\cdot)$ satisfies Mercer condition, $n_B \ge n_N \ge 1$. Let $\{\alpha^{(k)}\}$ denote the iterates generated by Algorithm* 3.1 *with working selection rule in Algorithm* 3.4. *If $\{\alpha^{(k)}\}$ contains only finite elements, the last iteration point must be the global optima of problem* (4). *If $\{\alpha^{(k)}\}$ contains infinite elements, any accumulation point is a global optima of problem* (3.108)–(3.109).

Proof If $\{\alpha^{(k)}\}$ contains finite elements, the last iteration point must satisfy $q(\alpha^{(k)}) = 0$. From Lemma 3.3, it knows that α is a KKT point. Hence, we only discuss the situation of infinite iteration points.

Let $\overline{\alpha}$ be any accumulation point of $\{\alpha^{(k)}\}$. Without loss of generalization, we can assume $\{\alpha^{(k)}\}$ converge to $\overline{\alpha}$ (this requirement always can be obtained by the proper relabeled of the order of iteration points). Because the feasible region of problem (3.108)–(3.109) is a bounded closed set in $\Re^l$ and the iterates generated by the decomposition algorithm are always feasible, we know that $\overline{\alpha}$ is also a feasible point. Furthermore, the value of $f(\overline{\alpha})$ is a finite number.

Let the index p satisfy $q_p^{(k)} = \min_{1 \le i \le l} q_i^{(k)} = -\|q^{(k)}\|_\infty$, where the last equality is based on the definition of ℓ_∞ and the fact of vector $q \le 0$. Since $n_N \ge 1$, index p must be contained in the selected working set $\mathrm{B}^{(k)}$. On the other hand, according to the definition of (α), we know $Qsub(\alpha, p) = q_p(\alpha)$. Hence,

$$f^{(k+1)} - f^{(k)} \le Qsub\left(\alpha^{(k)}, p\right) = -\left\|q^{(k)}\right\|_\infty. \tag{3.160}$$

Sum the above formulae from 0 to s,

$$\sum\nolimits_{k=0}^{s} \left\|q^{(k)}\right\|_\infty \le f^{(0)} - f^{(s+1)}. \tag{3.161}$$

Let $s \to \infty$, we have

$$\sum_{k=0}^{\infty} \left\| q^{(k)} \right\|_{\infty} \leq f^{(0)} - f(\overline{\alpha}) < +\infty. \tag{3.162}$$

Therefore,

$$\| q(\overline{\alpha}) \|_{\infty} = \lim_{k \to +\infty.} q(\alpha)^k = 0 \tag{3.163}$$

From Lemma 3.3, we know $\overline{\alpha}$ is the KKT point of problem (3.108)–(3.109).

Since the kernel function K(,.,) satisfies Mercer condition, problem (3.108)–(3.109) is a convex problem. Therefore, $\overline{\alpha}$ is a global optimum.

3.2 Twin Support Vector Machine in Classification

3.2.1 Improved Twin Support Vector Machine

3.2.1.1 TBSVM (Twin Bounded Support Vector Machine)

In this section, we introduce the so-called improved TWSVM, TBSVM [21], and also point out its drawbacks.

Linear TBSVM

For the linear case, two primal problems solved in TBSVM are

$$\min \frac{1}{2} c_3 \left(\left\| w_+^2 \right\| + b_+^2 \right) + \frac{1}{2} (A w_+ + e_+ b_+)^{\mathrm{T}} (A w_+ + e_+ b_+) + c_1 e_-^{\mathrm{T}} \xi_- \tag{3.164}$$

$$s.t. \ -(B w_+ + e_- b_+) + \xi_- \geq e_-, \ \ \xi_- \geq 0. \tag{3.165}$$

and

$$\min \frac{1}{2} c_4 \left(\| w_- \|^2 + b_-^2 \right) + \frac{1}{2} (B w_- + e_- b_-)^{\mathrm{T}} (B w_- + e_- b_-) + c_2 e_+^{\mathrm{T}} \xi_+ \tag{3.166}$$

$$s.t. \ (A w_- + e_+ b_-) + \xi_+ \geq e_+, \ \ \xi_+ \geq 0. \tag{3.167}$$

where c_i, $i = 1, 2, 3, 4$ are the penalty parameters and e_+ and e_- are vectors of ones of appropriate dimensions (Fig. 3.5).

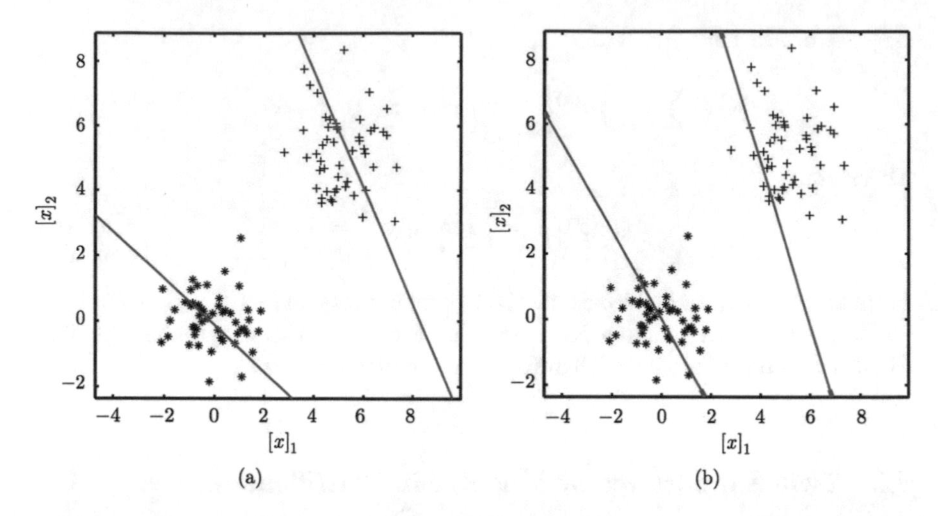

Fig. 3.5 Nonlinear TWSVM with the linear kernel is not equivalent to the linear TWSVM. The "+" and "$_*$" points are generated following two normal distributions respectively. (**a**) Two nonparallel lines obtained from linear TWSVM; (**b**) Two nonparallel lines obtained from nonlinear TWSVM with linear kernel

Comparing the problems (3.164)–(3.165) and (3.166)–(3.167) with (3.168)–(3.169) and (3.170)–(3.171) in linear TWSVM, we can see that the difference is only the introduction of the regularization terms $\|w_+\|^2 + b_+^2$ and $\|w_-\|^2 + b_-^2$, which leads to the following dual problems

$$\max\ e_-^{\mathrm{T}}\alpha - \frac{1}{2}\alpha^{\mathrm{T}}G\left(H^{\mathrm{T}}H + c_3I\right)^{-1}G^{\mathrm{T}}\alpha \tag{3.168}$$

$$s.t.\ \ 0 \leq \alpha \leq c_1e_-. \tag{3.169}$$

and

$$\max\ e_+^{\mathrm{T}}\gamma - \frac{1}{2}\gamma^{\mathrm{T}}H\left(G^{\mathrm{T}}G + c_4I\right)^{-1}H^{\mathrm{T}}\gamma \tag{3.170}$$

$$s.t.\ \ 0 \leq \gamma \leq c_2e_+. \tag{3.171}$$

The matrices $(H^{\mathrm{T}}H + c_3I)$ and $(G^{\mathrm{T}}G + c_4I)$ are nonsingular naturally, therefore their inverse matrices can be calculated without any extra assumption and need not be modified any more.

Nonlinear TBSVM

For the nonlinear case, two other regularization terms $\left(\|u_+\|^2 + b_+^2\right)$ and $\left(\|u_-\|^2 + b_-^2\right)$ are introduced into the problems respectively, and the primal problems turn to be

$$\min \frac{1}{2}c_3\left(\|u_+\|^2 + b_+^2\right) + \frac{1}{2}\left\|K\left(A, C^{\mathrm{T}}\right)u_+ + e_+b_+\right\|^2 + c_1e_-^{\mathrm{T}}\xi_- \tag{3.172}$$

$$s.t. \quad -\left(K\left(B, C^{\mathrm{T}}\right)u_+ + e_-b_+\right) + \xi_- \geq e_-, \;\; \xi_- \geq 0. \tag{3.173}$$

and

$$\min \frac{1}{2}c_4\left(\|u_-\|^2 + b_-^2\right) + \frac{1}{2}\left\|K\left(B, C^{\mathrm{T}}\right)u_- + e_-b_-\right\|^2 + c_2e_+^{\mathrm{T}}\xi_+ \tag{3.174}$$

$$s.t. \quad \left(K\left(A, C^{\mathrm{T}}\right)u_- + e_+b_-\right) + \xi_+ \geq e_+, \;\; \xi_+ \geq 0. \tag{3.175}$$

The corresponding dual problems are

$$\max e_-^{\mathrm{T}}\alpha - \frac{1}{2}\alpha^{\mathrm{T}}R\left(S^{\mathrm{T}}S + c_3I\right)^{-1}R^{\mathrm{T}}\alpha \tag{3.176}$$

$$s.t. \;\; 0 \leq \alpha \leq c_1. \tag{3.177}$$

and

$$\max e_+^{\mathrm{T}}\gamma\gamma - \frac{1}{2}\gamma^{\mathrm{T}}S\left(R^{\mathrm{T}}R + c_4I\right)^{-1}S^{\mathrm{T}}\gamma \tag{3.178}$$

$$s.t. \;\; 0 \leq \gamma \leq c_2. \tag{3.179}$$

Similar to the linear TBSVM, the matrices $(S^{\mathrm{T}}S + c_3I)$ and $(R^{\mathrm{T}}R + c_4I)$ are nonsingular, therefore their inverse matrices can be calculated without any extra assumption and need not be modified any more. Though TBSVM is claimed more rigorous and complete than TWSVM, it still suffers from the two drawbacks discussed earlier. First, it cannot avoid computing the inverse matrices; second, the nonlinear

TBSVM with the linear kernel is not equivalent to the linear TBSVM. A toy example in Fig. 3.6 illustrates this. We can further verify the second drawback of

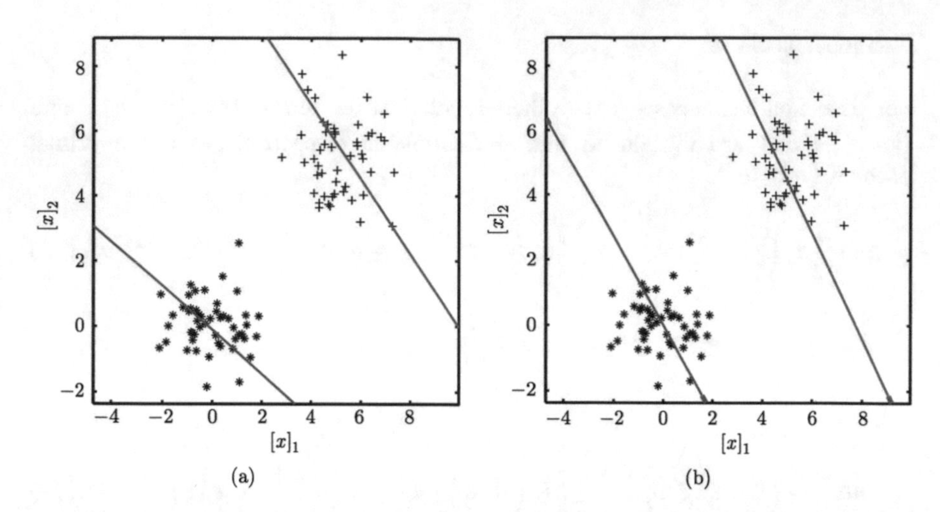

Fig. 3.6 Nonlinear TBSVM with the linear kernel is not equivalent to the linear TBSVM. The +" and $_*$" points are generated following two normal distributions respectively. (**a**) Two nonparallel lines obtained from linear TBSVM; (**b**) Two nonparallel lines obtained from nonlinear TBSVM with linear kernel

TBSVM and TWSVM from the experiments reported in [58] and [21]. For example, for the dataset "Australian", the reported best results of TWSVM and TBSVM are 85.80% and 85.94% for linear case respectively, while not reported in [58] and 75.8% in [21] for the radial basis function (RBF) kernel. However, as we all know, RBF kernel

$$K\left(x, x'\right) = \exp\left(-\frac{\left\|x - x'\right\|^2}{\sigma}\right) \tag{3.180}$$

performs approximately like linear kernel when the parameter σ is chosen large enough, they should get the similar best results with linear case after parameters tuning. Unfortunately, nonlinear TBSVM or TWSVM with optimal kernel parameters performs worse than linear TBSVM or TWSVM.

3.2.1.2 Improved TWSVM

In this section, we propose a different TWSVM based on TBSVM, termed as ITSVM, which inherits the essence of the SVMs and has the unexpected merits compared with TWSVM and TBSVM.

Linear ITSVM

For the linear case, the primal problems are the same as (3.164)–(3.165) and (3.166)–(3.167) of TBSVM

$$\min \frac{1}{2}c_3\left(\|w_+\|^2 + b_+^2\right) + \frac{1}{2}\eta_+^{\mathrm{T}}\eta + +c_1 e_-^{\mathrm{T}}\xi_- \tag{3.181}$$

$$s.t.\ \ Aw + +e_+ b + = \eta +, \tag{3.182}$$

$$-\left(Bw_+ + e_- b_+\right) + \xi_- \geq e_-, \xi_- \geq 0. \tag{3.183}$$

and

$$\min \frac{1}{2}c_4\left(\|w_-\|^2 + b_-^2\right) + \frac{1}{2}\eta_-^{\mathrm{T}}\eta_- + c_2 e_+^{\mathrm{T}}\xi_+ \tag{3.184}$$

$$s.t.\ \ Bw - +e_- b_- = \eta_-, \tag{3.185}$$

$$\left(Aw - +e_+ b_-\right) + \xi_+ \geq e_+, \xi_+ \geq 0. \tag{3.186}$$

where c_i, $i = 1, 2, 3, 4$ are the penalty parameters, $e+$ and e_- are vectors of ones of appropriate dimensions, $\xi+$ and ξ_- are slack vectors of appropriate dimension, $\eta+$ and η_- are vectors of appropriate dimension. It is worth noting that here we only add two variables $\eta+$ and η_- to (3.1) and (3.2) respectively. Therefore, we introduce a different Lagrangian corresponding to the problem (3.1) as

$$L\left(w_+, b_+, \eta+, \xi_-, \alpha, \beta, \lambda\right) = \frac{1}{2}c_3\left(\|w_+\|^2 + b_+^2\right) + \frac{1}{2}\eta_+^{\mathrm{T}}\eta + +c_1 e_-^{\mathrm{T}}\xi_-$$

$$+\lambda^{\mathrm{T}}\left(Aw + +e_+ b + -\eta_+\right) + \alpha^{\mathrm{T}}\left(Bw + +e_- b + -\xi_- + e_-\right) - \beta^{\mathrm{T}}\xi_-, \tag{3.187}$$

where $\alpha = (\alpha_1, \ldots, \alpha_q)^{\mathrm{T}}$, $\beta = (\beta_1, \ldots, \beta_q)^{\mathrm{T}}$ and $\lambda = (\lambda_1, \ldots, \lambda_p)^{\mathrm{T}}$ are vectors of Lagrange multipliers. That is, the Lagrangian (3.3) has extra variables $\eta+$ and λ compared with the TBSVM. The KarushKuhn-Tucker (KKT) necessary and sufficient optimality conditions of the problem (3.1) are given by

$$c_3 w_+ + A^{\mathrm{T}}\lambda + B^{\mathrm{T}}\alpha = 0, \tag{3.188}$$

$$c_3 b_+ + e_+^{\mathrm{T}}\lambda + e_-^{\mathrm{T}}\alpha = 0, \tag{3.189}$$

$$\lambda - \eta_{+} = 0, \tag{3.190}$$

$$c_1 e_{-} - \alpha - \beta = 0, \tag{3.191}$$

$$Aw_{+} + e + b_{+} = \eta +, \tag{3.192}$$

$$-(Bw_{+} + e_{-}b_{+}) + \xi_{-} \geq e_{-}, \xi_{-} \geq 0, \tag{3.193}$$

$$\alpha^{\mathrm{T}}(Bw + +e_{-}b + -\xi_{-} + e_{-}) = 0, \beta^{\mathrm{T}}\xi_{-} = 0, \tag{3.194}$$

$$\alpha \geq 0, \beta \geq 0. \tag{3.195}$$

Since $\beta \geq 0$, from (3.191) we have

$$0 \leq \alpha \leq c_1 e_{-}. \tag{3.196}$$

(3.188) and (3.189) imply that

$$w_{+} = -\frac{1}{c_3}\left(A^{\mathrm{T}}\lambda + B^{\mathrm{T}}\alpha\right), \tag{3.197}$$

$$b_{+} = -\frac{1}{c_3}\left(e_{+}^{\mathrm{T}}\lambda + e_{-}^{\mathrm{T}}\alpha\right). \tag{3.198}$$

Using (3.197), (3.198) and (3.190), we obtain the Wolfe dual of the problem (3.181)–(3.183) as follows:

$$\max\ \lambda, \alpha - \frac{1}{2}\left(\lambda^{\mathrm{T}}\alpha^{\mathrm{T}}\right)\hat{Q}\left(\lambda^{\mathrm{T}}\alpha^{\mathrm{T}}\right)^{\mathrm{T}} + c_3 e_{-}^{\mathrm{T}}\alpha \tag{3.199}$$

$$s.t.\ \ 0 \leq \alpha \leq c_1 e_{-}. \tag{3.200}$$

where

$$\hat{Q} = \begin{pmatrix} AA^{\mathrm{T}} + c_3 I & AB^{\mathrm{T}} \\ AB^{\mathrm{T}} & BB^{\mathrm{T}} \end{pmatrix} + E \tag{3.201}$$

and I is the $p \times p$ identity matrix, E is the $l \times l$ matrix with all entries equal to 1. Similarly, the dual of the problem (3.2) is obtained as

$$\max\ \theta, \gamma - \frac{1}{2}\left(\theta^{\mathrm{T}}\gamma^{\mathrm{T}}\right)\widetilde{Q}\left(\theta^{\mathrm{T}}\gamma^{\mathrm{T}}\right)^{\mathrm{T}} + c_4 e_+^{\mathrm{T}}\gamma \tag{3.202}$$

$$s.t.\ \ 0 \le \gamma \le c_2 e_+ \tag{3.203}$$

where

$$\widetilde{Q} = \begin{pmatrix} BB^{\mathrm{T}} + c_4 I & BA^{\mathrm{T}} \\ BA^{\mathrm{T}} & AA^{\mathrm{T}} \end{pmatrix} + E \tag{3.204}$$

and I is the $q \times q$ identity matrix, E is the $l \times l$ matrix with all entries equal to 1.

The pair of nonparallel hyperplanes is obtained from the solutions (λ^*, α^*) and (θ^*, γ^*) of (3.199)–(3.200) and (3.202)–(3.203) by

$$\left(w_+^* \cdot x\right) + b_+^* = 0, \tag{3.205}$$

where

$$w_+^* = -\left(A^{\mathrm{T}}\lambda^* + B^{\mathrm{T}}\alpha^*\right), b_+^* = -\left(e_+^{\mathrm{T}}\lambda^* + e_-^{\mathrm{T}}\alpha^*\right), \tag{3.206}$$

and

$$\left(w_-^* \cdot x\right) + b_-^* = 0, \tag{3.207}$$

where

$$w_-^* = -\left(B^{\mathrm{T}}\theta^* + A^{\mathrm{T}}\gamma^*\right), b_-^* = -\left(e_-^{\mathrm{T}}\theta^* + e_+^{\mathrm{T}}\gamma^*\right). \tag{3.208}$$

The linear ITSVM is equivalent to the linear TBSVM because of the same primal problems. However, it is just the difference between their Lagrangians leads to the final difference between their dual problems. Problems (3.199)–(3.200) and (3.202)–(3.203) are obvious QPPs and have nothing to do with the computation of inverse matrices compared with the problems (3.176)–(3.177) and (3.178)–(3.179). More importantly, they can be easily extended to the nonlinear case.

Nonlinear ITSVM

Totally different with the nonlinear TBSVM, we do not need to consider the kernel-generated surfaces and construct two other problems (3.172)–(3.173) and (3.174)–(3.175), since we can introduce the kernel function directly into the problems (3.199)–(3.200) and (3.202)–(3.203) just as the standard SVMs usually do.

Introducing the kernel function $K(x,x') = (\Phi(x) \cdot \Phi(x'))$ and the corresponding transformation

$$\mathrm{x} = \Phi(x) \tag{3.209}$$

where $\mathrm{x} \in \mathcal{H}$, $\mathcal{H}$ is the Hilbert space. So, the training set (2.1) becomes

$$\widetilde{T} = \left\{(\mathrm{x}_1, y_1), \ldots \left(\mathrm{x}_p, y_p\right), \left(\mathrm{x}_{p+1}, y_{p+1}\right), \ldots \left(\mathrm{x}_{p+q}, y_{p+q}\right)\right\} \tag{3.210}$$

and the corresponding two primal problems in the Hilbert space $\mathcal{H}$ are

$$\min \frac{1}{2}c_3\left(\|\mathrm{w}_+\|^2 + b_+^2\right) + \frac{1}{2}\eta_+^{\mathrm{T}}\eta + +c_1 e_-^{\mathrm{T}}\xi_- \tag{3.211}$$

$$s.t.\ \Phi(A)\mathrm{w} + +e_+ b+ = \eta+, \tag{3.212}$$

$$-\left(\Phi(B)\mathrm{w}_+ + e_- b_+\right) + \xi_- \geq e_-, \xi_- \geq 0. \tag{3.213}$$

and

$$\min \frac{1}{2}c_4\left(\|\mathrm{w}_-\|^2 + b_-^2\right) + \frac{1}{2}\eta_-^{\mathrm{T}}\eta_- + c_2 e_+^{\mathrm{T}}\xi + \tag{3.214}$$

$$s.t.\ \Phi(B)\mathrm{w}_- + e_- b_- = \eta_-, \tag{3.215}$$

$$\left(\Phi(A)\mathrm{w}_- + e_+ b_-\right) + \xi + \geq e_+, \xi + \geq 0. \tag{3.216}$$

Their dual problems can be obtained as

$$\max \lambda, \alpha - \frac{1}{2}\left(\lambda^{\mathrm{T}}\alpha^{\mathrm{T}}\right)\overline{Q}\left(\lambda^{\mathrm{T}}\alpha^{\mathrm{T}}\right)^{\mathrm{T}} + c_3 e_-^{\mathrm{T}}\alpha \tag{3.217}$$

$$s.t.\ 0 \leq \alpha \leq c_1 e_-. \tag{3.218}$$

where

$$\overline{Q} = \begin{pmatrix} K\left(A^{\mathrm{T}}|A^{\mathrm{T}}\right) + c_3 I & K\left(A^{\mathrm{T}}|B^{\mathrm{T}}\right) \\ K\left(A^{\mathrm{T}}|B^{\mathrm{T}}\right) & K\left(B^{\mathrm{T}}|B^{\mathrm{T}}\right) \end{pmatrix} + E \tag{3.219}$$

and

$$\max\ \theta, \gamma - \frac{1}{2}\left(\theta^{\mathrm{T}}\gamma^{\mathrm{T}}\right) Q^{=}\left(\theta^{\mathrm{T}}\gamma^{\mathrm{T}}\right)^{\mathrm{T}} + c_4 e_+^{\mathrm{T}}\gamma \tag{3.220}$$

$$s.t.\ \ 0 \leq \gamma \leq c_2 e_+. \tag{3.221}$$

where

$$\bar{\bar{Q}} = \begin{pmatrix} K(B^{\mathrm{T}}|B^{\mathrm{T}}) + c_4 I & K(B^{\mathrm{T}}|A^{\mathrm{T}}) \\ K(B^{\mathrm{T}}|A^{\mathrm{T}}) & K(A^{\mathrm{T}}|A^{\mathrm{T}}) \end{pmatrix} + E \tag{3.222}$$

respectively.

The pair of nonparallel hyperplanes in the Hilbert space $\mathcal{H}$ is obtained from the solutions (λ^*, α^*) and (θ^*, γ^*) of (3.217)–(3.218) and (3.220)–(3.221) by

$$K\left(x^{\mathrm{T}}, A^{\mathrm{T}}\right)\lambda^* + K\left(x^{\mathrm{T}}, B^{\mathrm{T}}\right)\alpha^* + b_+^* = 0 \tag{3.223}$$

where

$$b_+^* = e_+^{\mathrm{T}}\lambda^* + e_-^{\mathrm{T}}\alpha^* \tag{3.224}$$

and

$$K\left(x^{\mathrm{T}}, B^{\mathrm{T}}\right)\theta^* + K\left(x^{\mathrm{T}}, A^{\mathrm{T}}\right)\gamma^* + b_-^* = 0 \tag{3.225}$$

where

$$b_-^* = e_-^{\mathrm{T}}\theta^* + e_+^{\mathrm{T}}\gamma^* \tag{3.226}$$

Obviously the problems (3.217)–(3.218) and (3.220)–(3.221) are QPPs and have nothing to do with the computation of inverse matrices any more, and can degenerate to the problems (3.199)–(3.200) and (3.202)–(3.203) of linear ITSVM when the linear kernel is applied.

Algorithm 3.5 (ITSVM)

---

(1) Input the training set (2.1);
(2) Choose appropriate kernels (x, x'), and penalty parameters $c_i > 0$, $i = 1, 2, 3, 4$;
(3) Construct and solve the two convex QPPs (3.217)–(3.218) and (3.220)–(3.221) separately, get the solutions (λ^*, α^*) and (θ^*, γ^*);
(4) Construct two decision functions

$$f_+(x) = K\left(x^{\mathrm{T}}, A^{\mathrm{T}}\right)\lambda^* + K\left(x^{\mathrm{T}}, B^{\mathrm{T}}\right)\alpha^* + b_+^*, \tag{3.227}$$

and

$$f_{-}(x) = K\left(x^{\mathrm{T}}, B^{\mathrm{T}}\right)\theta^{*} + K\left(x^{\mathrm{T}}, A^{\mathrm{T}}\right)\gamma^{*} + b_{-}^{*}, \tag{3.228}$$

(5) For any unknown input x, assign it to the class $k(k = -, +)$ by

$$\arg\ \min \mid f_k(x) \mid \tag{3.229}$$

where $| \cdot |$ is the perpendicular distance of point x from the hyperspheres $f(x) = 0$.

3.2.1.3 Fast Solvers for ITSVM

SOR for ITSVM

The two convex QPPs (3.217)–(3.218) and (3.220)–(3.221) can be solved efficiently by the following successive overrelaxation (SOR) technique, see [21, 42]. Taking the problem (3.217)–(3.218) as the example, we rewrite it as

$$\min\ \frac{1}{2}\mu^{\mathrm{T}}Q\mu - \kappa^{\mathrm{T}}\mu \tag{3.230}$$

$$s.t.\ \ 0 \le \mu_i \le c_1,\ \ i = p+1, \ldots, p+q, \tag{3.231}$$

where $\mu = (\lambda^{\mathrm{T}}, \alpha^{\mathrm{T}})^{\mathrm{T}}$, $Q \in \mathbb{R}^{l \times l}$ is defined by (3.28) and positive semi-definite,

$$\kappa = \left(\underbrace{0, \ldots, 0}_{p}, \underbrace{c_3, \ldots, c_3}_{q}\right)^{\mathrm{T}} \tag{3.232}$$

Algorithm 3.6 (SOR for ITSVM)

Choose $\in(0, 2)$. Start with any $\mu^0 \in \mathbb{R}^m$. Having μ^i, compute μ^{i+1} as follows

$$\mu^{i+1} = \left(\mu^i - \omega D^{-1}\left(Q\mu^i + \kappa + L\left(\mu^{i+1} - \mu^i\right)\right)\right)_{\#} \tag{3.233}$$

until $\|\mu^{i+1} - \mu^i\|$ is less than some prescribed tolerance, where $(\cdot)_{\#}$ denotes the 2-norm projection on the feasible region of (5.1), that is,

$$((\mu)_{\#})_i = \begin{cases} 0 \ \ \text{if } \mu_i \le 0, \\ \mu_i \ \text{if } 0 < \mu_i < c_1, \\ c_1 \ \text{if } \mu_i \ge c_1, \end{cases} \quad \text{for } i = p+1, \ldots, p+q\ , \tag{3.234}$$

the nonzero elements of $L \in \mathbb{R}^{m \times m}$ constitute the strictly lower triangular part of the symmetric matrix Q, and the nonzero elements of $D \in \mathbb{R}^{m \times m}$ constitute the diagonal of Q.

Without residing in memory, SOR can process huge quantities of datasets, which has been proved to converge to a solution linearly, and therefore it is regarded as an excellent solver.

SMO for ITSVM

If we make a little change in the two primal problems (3.181)–(3.183) and (3.184)–(3.186), ITSVM can be solved efficiently by the SMO-type decomposition methods.

For the linear case, we take the terms $\|w_+\|^2$ and $\|w_-\|^2$ instead of $\left(\|w_+\|^2 + b_+^2\right)$ and $\left(\|w_-\|^2 + b_-^2\right)$ respectively, then change the two primal problems (3.181)–(3.183) and (3.184)–(3.186) to be

$$\min \frac{1}{2}c_3 \left\| w_+^2 \right\| + \frac{1}{2}\eta_+^{\mathrm{T}}\eta + +c_1 e_-^{\mathrm{T}}\xi_- \tag{3.235}$$

$$s.t.\ \ Aw + +e + b+ = \eta+, \tag{3.236}$$

$$-\left(Bw_+ + e_- b_+\right) + \xi_- \geq e_-, \xi_- \geq 0. \tag{3.237}$$

and

$$\min \frac{1}{2}c_4 \|w_-\|^2 + \frac{1}{2}\eta_-^{\mathrm{T}}\eta_- + c_2 e_+^{\mathrm{T}}\xi + \tag{3.238}$$

$$s.t.\ \ Bw - +e_- b_- = \eta_-, \tag{3.239}$$

$$(Aw - +e_+ b_-) + \xi + \geq e_+, \xi + \geq 0. \tag{3.240}$$

It is easy to obtain the Wolfe dual of the problem (5.4) as

$$\max \lambda, \alpha - \frac{1}{2}\left(\lambda^{\mathrm{T}}\alpha^{\mathrm{T}}\right)\hat{Q}\left(\lambda^{\mathrm{T}}\alpha^{\mathrm{T}}\right)^{\mathrm{T}} + c_3 e_-^{\mathrm{T}}\alpha \tag{3.241}$$

$$s.t.\ \ e_+^{\mathrm{T}}\lambda + e_-^{\mathrm{T}}\alpha = 0, \tag{3.242}$$

$$0 \leq \alpha \leq c_1 e_-. \tag{3.243}$$

where

$$\hat{Q} = \begin{pmatrix} AA^{\mathrm{T}} + c_3 I & AB^{\mathrm{T}} \\ AB^{\mathrm{T}} & BB^{\mathrm{T}} \end{pmatrix} \tag{3.244}$$

and the Wolfe dual of the problem (3.241)–(3.243) as

$$\max\ \theta, \gamma - \frac{1}{2}\left(\theta^{\mathrm{T}}\gamma^{\mathrm{T}}\right)\widetilde{Q}\left(\theta^{\mathrm{T}}\gamma^{\mathrm{T}}\right)^{\mathrm{T}} + c_4 e_+^{\mathrm{T}}\gamma \tag{3.245}$$

$$s.t.\ \ e_-^{\mathrm{T}}\theta + e_+^{\mathrm{T}}\gamma = 0, \tag{3.246}$$

$$0 \le \gamma \le c_2 e + . \tag{3.247}$$

where

$$\widetilde{Q} = \left(BA^{\mathrm{T}}BB^{\mathrm{T}} + c_4 I\ AA^{\mathrm{T}}BA^{\mathrm{T}}\right. \tag{3.248}$$

For the nonlinear case, we only need to introduce the kernel functions to the problems (3.241)–(3.243) and (3.245)–(3.247). We can see that problems (3.241)–(3.243) and (3.245)–(3.247) have the same formulation

$$\min\ \frac{1}{2}\mu^{\mathrm{T}}Q\mu + \kappa^{\mathrm{T}}\mu \tag{3.249}$$

$$s.t.\ \ L_i \le \mu \le U_i,\ \ i = 1, \ldots, p+q, \tag{3.250}$$

$$y^{\mathrm{T}}\mu = 0. \tag{3.251}$$

which has been proved to be efficiently solved by the SMO-type technique [59].

3.2.2 *Extending Twin Support Vector Machine Classifier for Multi-category Classification Problems*

3.2.2.1 One-Versus-All Twin Support Vector Machines

In this section, we propose a new *k*-category classifier, one-versus-all twin support vector machine classifier, which we will term as OVA-TWSVM. As mentioned earlier, TWSVM obtains two nonparallel hyperplanes by solving two comparatives smaller QPPs, one for each class. Based on this idea, we extend TWSVM to solve multicategory data classification problems.

Given a dataset containing m datapoints represented by $A \in R^{m \times n}$, each element is labeled by one of $k(k \geq 2)$ labels. Let matrix $A_i \in R^{m_i \times n}$ represent the datapoints of class $(i = 1, 2, \ldots, k)$. We define

$$A = \begin{bmatrix} A_1 \\ \vdots \\ A_k \end{bmatrix} \tag{3.252}$$

$$\widetilde{A}_i = \begin{bmatrix} A_1 \\ \vdots \\ A_{i-1} \\ A_{i+1} \\ \vdots \\ A_k \end{bmatrix} \tag{3.253}$$

$i \in \{1, 2, \ldots, k\}$ and $m = m_1 + m_2 + \ldots + m_k$. For class $(i = 1, 2, \ldots, k)$, we solve the following QPP Eqs. (3.254)–(3.256).

$$\min \frac{1}{2} \left\| A_i w^{(i)} + e_i b^{(i)} \right\|^2 + C_i \widetilde{e}_i' q \tag{3.254}$$

$$s.t. \quad -\left(\widetilde{A}_i w^{(i)} + \widetilde{e}_i b^{(i)} \right) + q \geq \widetilde{e}_i, \tag{3.255}$$

$$q \geq 0. \tag{3.256}$$

where $C_i(>0)$ is a penalty parameter, and q is a vector of error or slack variables associated with samples, and e_i and $\tilde{\imath}$ are vectors of ones of appropriate dimensions. In the above QPP Eqs. (3.254)–(3.256), the first term in the objective function is the sum of squared distance from the points of class i to the hyperplane. Therefore, minimizing it means to keep the data points of class i clustered around the hyperplane. The second term of the objective function minimizes the sum of error variables, thus trying to minimize misclassification due to points belonging to the other $k - 1$ classes. The constraints require the hyperplane to be at a distance of at least 1 from points of the other $k - 1$ classes.

Linear One-Versus-All Twin Support Vector Machines

The linear OVA-TWSVM classifier obtains k nonparallel hyperplanes by solving k QPPs, one for each class, around which the corresponding data points get clustered. We can classify points according to which hyperplane a given point is closest to.

The Lagrangian corresponding to the QPP Eqs. (3.254)–(3.256) is given by

$$L\left(w^{(i)}, b^{(i)}, q, \alpha, \beta\right) = \frac{1}{2}\left\| A_i w^{(i)} + e_i b^{(i)} \right\|^2 + C_i \tilde{e_i}' q$$

$$-\alpha' \left(-\left(\tilde{A}_i w^{(i)} + \tilde{e}_i b^{(i)}\right) + q - \tilde{e}_i\right) - \beta' q \tag{3.257}$$

where $\alpha = (\alpha_1, \alpha_2, \ldots, \alpha_s)'$, $\beta = (\beta_1, \beta_2, \ldots, \beta_s)'$, and $s = m - m_i$. Here α, β are vectors of Lagrange multipliers. The Karush-Kuhn-Tucker (K.K.T) necessary and sufficient optimality conditions [60] for Eq. (3.257) are given by

$$A_{\mathrm{i}}' \left(A_i w^{(i)} + e_i b^{(i)}\right) + \widetilde{A}_i' \alpha = 0 \tag{3.258}$$

$$e_{\mathrm{i}}' \left(A_{\mathrm{i}} w^{(\mathrm{i})} + e_{\mathrm{i}} b^{(\mathrm{i})}\right) + \widetilde{e_i}' \alpha = 0 \tag{3.259}$$

$$C_{\mathrm{i}}\ \widetilde{ei} - \alpha - \beta = 0 \tag{3.260}$$

$$-\left(\widetilde{A}_{\mathrm{i}} w^{(\mathrm{i})} + \widetilde{ei}\ b^{(i)}\right) + q \geq \widetilde{e_i}', q \geq 0 \tag{3.261}$$

$$\alpha' \left(-\left(\widetilde{A}_i w^{(i)} + \tilde{e}_i b^{(i)}\right) + q - \widetilde{ei}\right) = 0, \beta' q = 0 \tag{3.262}$$

$$\alpha \geq 0, \beta \geq 0 \tag{3.263}$$

Since $\beta \geq 0$, from Eq. (3.260) we get Eq. (3.264)

$$0 \leq \alpha \leq C_i \tag{3.264}$$

Next, combining Eqs. (3.258) and (3.259) leads to Eq. (3.265).

$$\left[A_{\mathrm{i}}' e_{\mathrm{i}}'\right]\left[A_i e_i\right]\left[w^{(i)}\ b(i)\right]' + \left[\widetilde{A}_i' \widetilde{e_i}'\right]\alpha = 0 \tag{3.265}$$

Then we define Eq. (3.266),

$$E = \left[A_i e_i\right], F = \left[\tilde{A}_i \tilde{e}_{\mathrm{i}}\right], u_i = \left[w^{(i)}\ b^{(i)}\right]' \tag{3.266}$$

with these notations, Eq. (3.265) can be rewritten as Eq. (3.267).

$$E'Eu_i + F'\alpha = 0, \text{i.e.,} \;\; u_i = -\left(E'E\right)^{-1} F'\alpha \tag{3.267}$$

Because $E'E$ is always positive semidefinite, we can introduce a regularization term εI, $\varepsilon > 0$, to take care of problems due to possible ill-conditioning of $E'E$. Here, I is an identity matrix of appropriate dimensions. Therefore, Eq. (3.267) can be modified to Eq. (3.268).

$$u_i = -\left(E'E + \varepsilon I\right)^{-1} F'\alpha \tag{3.268}$$

However, in the following, we shall continue to use Eq. (3.267) with the understanding that, if needed, Eq. (3.268) is to be used for the determination of u_i.

Using Eq. (3.257) and K.K.T. conditions, we can obtain the Wolfe dual of QPP Eqs. (3.254)–(3.256) as follows:

$$\max \; \widetilde{e_i}'\alpha - \frac{1}{2}\alpha' F\left(E'E\right)^{-1} F'\alpha \tag{3.269}$$

$$s.t. \;\; 0 \leq \alpha \leq C_i. \tag{3.270}$$

Once vector u_i is known from Eqs. (3.267) and (3.269)–(3.270), the separating plane Eq. (3.271) of class $i (i = 1, 2, \ldots, k)$

$$x'w^{(i)} + b^{(i)} = 0 \tag{3.271}$$

is obtained. A new data sample x is assigned to class i $(i = 1, 2, \ldots, k)$, depending on which of the k planes given by Eq. (3.271) it lies closest to, i.e.,

$$x'w^{(i)} + b^{(i)} = \min \mid x'w^{(l)} + b^{(l)} \mid \tag{3.272}$$

where $\mid \cdot \mid$ is the perpendicular distance from point x to the hyperplane $x'w^{(l)} + b^{(l)} = 0$, $l = 1, 2, \ldots, k$. According to TWSVM, we can define such patterns of the other $k - 1$ classes for which $0 \leq \alpha_j \leq C_i (j = 1, 2, \ldots, m - m_i)$ as support vectors with respect to class $i (i = 1, 2, \ldots, k)$ because they play an important role in determining the required hyperplane.

For clarity, our linear OVA-TWSVM is described in the following Algorithm 3.7.

Algorithm 3.7 Linear OVA-TWSVM

Given a dataset containing m data points represented by $A \in R^{m \times n}$, each element of which is labeled by one of $k (k \geq 2)$ labels. Let matrix $A_i \in R^{m_i \times n}$ represent the m_i data points of class $(i = 1, 2, \ldots, k)$, with $m = \sum_{i=1}^{k} m_i$. The linear OVA-TWSVM is described as following:

(i) Start with $i = 1$.
(ii) Iterate (iii), (iv)~ and (v) until $i = k$.
(iii) Define A and A_i in Eqs. (3.15) and (3.16), respectively.
(iv) Select the penalty parameter C_i. This parameter in our study is determined via 10-fold cross validation experiments.
(v) Define $E = [A_i, e_i]$, and $F=[\widetilde{A}_i\ \widetilde{e}_i]$ in Eq. (3.27). Solve QPPs Eqs. (3.269)–(3.270) and calculate u_i in Eq. (3.29) to get the augmented vector $u_i = [w^{(i)}, b^{(i)}]'$ in Eq. (3.267).
(vi) Calculate the perpendicular distances $\lvert x' w^{(i)} + b^{(i)} \rvert$ $(i = 1, 2, \ldots, k)$ for a new data point x.
(vii) Assign the new data point x to class l based on which of the distance $\lvert x' w^{(l)} + b^{(l)} \rvert$ is the minimum one.

Nonlinear One-Versus-All Twin Support Vector Machines

In this section, we extend our linear OVA-TWSVM to nonlinear OVA-TWSVM by considering the following k kernel generated surfaces Eq. (3.273).

$$K\left(x', A'\right) w^{(i)} + b^{(i)} = 0\ (i = 1, 2, \ldots, k) \tag{3.273}$$

where K is an appropriately chosen kernel. The primal two QPPs of nonlinear OVA-TWSVM can be modified to the QPPs as showed in Eqs. (3.274)–(3.276).

$$min\ \frac{1}{2}\left\| K\left(A_i, A'\right) w^{(i)} + e_i b^{(i)} \right\|^2 + C_i \widetilde{\text{ï}}' q \tag{3.274}$$

$$s.t. - \left(K\left(\widetilde{A}_i, A'\right) w^{(i)} + \widetilde{e_i}'\, b^{(i)}\right) + q \geq \widetilde{\text{ï}}, \tag{3.275}$$

$$q \geq 0,\ \ i = 1, 2, \ldots, k. \tag{3.276}$$

where $C_i \geq 0$ is a penalty parameter, q is a vector of error variables associated with samples, and $\tilde{\imath}$ and $\tilde{e}_i'$ are vectors of ones of appropriate dimensions.

The Lagrangian corresponding to the problem Eqs. (3.274)–(3.276) is given by the following Eq. (3.277),

$$L\left(w^{(i)}, b^{(i)}, q, \alpha, \beta\right) = \frac{1}{2}\left\| K\left(A_i, A'\right) w^{(i)} + e_i b^{(i)} \right\|^2$$

$$+C_i\ \widetilde{e_i}' q - \alpha' \left(-\left(K\left(\widetilde{A}_i, A'\right) w^{(i)} + \tilde{e}_i b^{(i)}\right) + q - \widetilde{ei}\right) - \beta' q \tag{3.277}$$

We can obtain the K.K. T conditions for Eq. (3.277) as the following Eqs. (3.278)–(3.283).

$$K\left(A_i', A'\right)'\left(K\left(A_i, A'\right)w^{(i)} + e_i b^{(i)}\right) + K\left(\widetilde{A}_i, A'\right)'\alpha = 0 \tag{3.278}$$

$$e_i'\left(K\left(A_i, A'\right)w^{(i)} + e_i b^{(i)}\right) + e_i'\alpha = 0 \tag{3.279}$$

$$C_i\tilde{e}_i - \alpha - \beta = 0 \tag{3.280}$$

$$-\left(K\left(\widetilde{A}_i, A'\right)w^{(i)} + \tilde{e}_i b^{(i)}\right) + q \geq \tilde{e}_i, q \geq 0 \tag{3.281}$$

$$\alpha'\left(-\left(K\left(\widetilde{A}_i, A'\right)w^{(i)} + \tilde{e}_i b^{(i)}\right) + q - \tilde{e}_i\right) = 0, \beta' q = 0 \tag{3.282}$$

$$\alpha \geq 0, \beta \geq 0 \tag{3.283}$$

Since $\beta \geq 0$, from Eq. (3.280) we have the Eq. (3.284).

$$0 \leq \alpha \leq C_i \tag{3.284}$$

Combining Eqs. (3.278) and (3.279), we get the Eq. (3.285).

$$\left[K\left(A_i, A'\right)' e_i'\right]\left[K\left(A_i, A'\right)e_i\right]\left[w^{(i)} b^{(i)}\right]' + \left[K\left(\widetilde{A}_i, A'\right)' e_i'\right]\alpha = 0 \tag{3.285}$$

Define

$$E = \left[K\left(A_i, A'\right)e_i\right], \quad F = \left[K\left(\widetilde{A}_i, A'\right)\tilde{e}_i\right], \quad u_i = \left[w^{(i)} b^{(i)}\right]' \tag{3.286}$$

Then, Eq. (3.285) can be modified as Eq. (3.287),

$$E'Eu_i + F'\alpha = 0, \ \ i.e., \ \ u_i = -\left(E'E\right)^{-1}F'\alpha \tag{3.287}$$

The Wolfe dual QPPs of Eqs. (3.274)–(3.276) is given as follows Eqs. (3.288)–(3.289),

$$\max \tilde{e_i}'\alpha - \frac{1}{2}\alpha' F\left(E'E\right)^{-1}F'\alpha \tag{3.288}$$

$$s.t. \ \ 0 \leq \alpha \leq C_i \tag{3.289}$$

Once the k QPPs Eqs. (3.288)–(3.289) are solved to obtain the k hyperplanes of Eq. (3.273), a new pattern x is assigned to class i ($i = 1, 2, \ldots, k$) in a similar way to the linear case.

Here, we will give an explicit statement of our nonlinear OVA-TWSVM algorithm.

Given a dataset containing m data points represented by $A \in R^{m \times n}$, each element is labeled by one of $k(k \geq 2)$ labels. Let matrix $A_i \in R^{m_i \times n}$ represent the m_i data points of class $i(i = 1, 2, \ldots, k)$ with $m = \sum_{i=1}^{k} m_i$, then our nonlincar OVA-TWSVM is described in the following Algorithm 3.8.

Algorithm 3.8 Nonlinear OVA-TWSVM

(i) Choose a kernel function K and start with $i = 1$.
(ii) Iterate (iii), (iv)~ and (v) until $i = k$.
(iii) Define A and $\widetilde{A}_i$ in Eqs. (3.15) and (3.16), respectively.
(iv) Select the penalty parameter C_i. This parameter is selected using 10-fold cross validation experiments in our study.
(v) Define $E = [K(A_i, A') \ e_i]$, $F = \left[K\left(\widetilde{A}_i, A'\right) \ e\tilde{\imath}\right]$, and $u_i = [w^{(i)}, b^{(i)}]'$ in Eq. (3.44). Solve QPPs Eq. (3.46) and calculate u_i in Eq. (3.45) to get the augmented vector $u_i = [w^{(i)}, b^{(i)}]'$ in Eq. (3.44).
(vi) Calculate the perpendicular distances $|K(x', A')w^{(i)} + b^{(i)}|$ $(i = 1, 2, \ldots, k)$ for a new data point x.
(vii) Assign the new data point x to class l based on the distance $|x' w^{(l)} + b^{(l)}|$ is the minimum distance among $|x' w^{(i)} + b^{(i)}|$, $i = 1, 2, \ldots, k$.

Complexity Analysis of One-Versus-All Twin Support Vector Machines

In the OVA-SVMs classifier for k-category data classification, it requires solving k Wolfe dual QPPs, one of which contains m parameters, so the complexity of the conventional one-from-rest classifier is no more than $k \times m^3$. However, OVA-TWSVM only solves k Wolfe duals of QPP Eqs. (3.269)–(3.270) for linear or Eqs. (3.288)–(3.289) for non-linear separable classification problems. Suppose that the size of each class is roughly m/k. Thus, each Wolfe dual QPP of Eqs. (3.269)–(3.270) or (3.288)–(3.289) contains of $\frac{m}{k} \times (k-1)$ parameters. The ratio of runtime of OVA-SVMs to OVA-TWSVM is approximately as:

$$\frac{k \times m^3}{k \times \left(\frac{m}{k} \times (k-1)\right)^3} = \left(\frac{k}{k-1}\right)^3 \quad (k \geq 3) \tag{3.290}$$

$$\frac{m^3}{2\left(\frac{m}{2}\right)^3} = 4 \quad (k = 2) \tag{3.291}$$

That is, our OVA-TWSVM classifier is approximately $\left(\frac{k}{k-1}\right)^3$ times faster than traditional OVA-SVMs classifier. It should be noted that this holds when k here is greater or equal to three. When k equals two, the OVA-SVMs will degenerate to classical SVMs and has the complexity of m^3, whilst OVA-TWSVM to TWSVM and has $2\times\ \left(\frac{m}{2}\right)^3$ complexity, so the proportion of runtime between them is $\frac{m^3}{2\times\left(\frac{m}{2}\right)^3}=3$.

The experimental study can *be found in [5].*

3.2.3 *Robust Twin Support Vector Machine for Pattern Classification*

3.2.3.1 Robust Twin Support Vector Machine ($\mathcal{R}$-TWSVM)

Linear $\mathcal{R}$-TWSVM

We firstly give the formal representation of robust classification learning problem. Given a training set

$$T=\{(\mathcal{X}_1,y_1),\ldots(\mathcal{X}_l,y_l)\},\tag{3.292}$$

where $y_i\in\mathcal{Y}=\{1,-1\}$, $i=1,\ \ldots\ l$, and input set $\mathcal{X}_i$ is a sphere within r_i radius of the x_i center:

$$\mathcal{X}_i=\{\overline{x}_i|\overline{x}_i=x_i+r_iu_i\},i=1,\ldots,l,\ \|u_i\|\leq 1,\tag{3.293}$$

$\overline{x}_i$ is the true value of the training data, $u_i\in\mathfrak{R}^n$, r_i is a given constant. The goal is to induce a real-valued function

$$y=\text{sgn}\left(g(x)\right)\tag{3.294}$$

to infer the label y corresponding to any example x in $\mathfrak{R}^n$ space. Generally, such problem is caused by measurement errors, where r_i reflects the measurement accuracy.

In order to obtain the optimization decision function of (3.294), by introducing $\frac{1}{2}\|w_+\|^2$, can be written as the following robust optimization problem:

$$\min_{w_+,b_+,\xi}\ \frac{1}{2}\|\ w_+\ \|^2+\frac{1}{2}\ \|\ [w_+\cdot x_1)+b_+,\ldots,\left(w_+\cdot x_{l_1}\right)\ +b_+]\ \|^2+c_1\sum_{i=l_1+1}^{1}\xi_i\tag{3.295}$$

$$\text{s.t.} - ((w_+ \cdot (x_i + r_i u_i)) + b_+) + \geq 1 - \xi_i, \forall \|u_i\| \leq 1 i = l_1 + 1, \ldots l. \quad (3.296)$$

$$\xi_i \geq 0, i = l_1 + 1, \ldots l \quad (3.297)$$

Since

$$\min \{y_i r_i (w \cdot u_i), \|u_i\| \leq 1\} = -r_i \|w_+\| \quad (3.298)$$

problem (3.295)–(3.297) can be converted to

$$\underset{w_+, b_+, \xi}{min} \frac{1}{2} \| w_+ \|^2 + \frac{1}{2} \| [w_+ \cdot x_1) + b_+, \ldots, (w_+ \cdot x_{l_1}) + b_+ \Big] \|^2 + c_1 \sum_{i=1_1+1}^{1} \xi_i \quad (3.299)$$

$$s.t. \quad - ((w_+ \cdot x_i) + b_+) - r_i \| w_+ \| \geq 1 - \xi_i, \quad (3.300)$$

$$i = l_1 + 1, \ldots, l \xi_i \geq 0, i = l_1 + 1, \ldots, l. \quad (3.301)$$

By introducing new variables t_1, t_2 and setting $\| w_+ \| \leq t_1$, $\| [(w_+ \cdot x_1) + b_+, \ldots (w_+ \cdot x_{l_1}) + b_+] \| \leq t_2$, The above problem becomes

$$\min \frac{1}{2} t_1^2 + \frac{1}{2} t_2^2 + c_{1_{i=l}} X_{1+1}^l \xi_i \quad (3.302)$$

$$s.t. \quad - ((w_+ \cdot x_i) + b_+) - r_i t_1 \geq 1 - \xi_i, \quad i = l_1 + 1, \ldots l, \quad (3.303)$$

$$\xi_i \geq 0, \quad i = l_1 + 1, \ldots, l, \quad (3.304)$$

$$\| w_+ \| \leq t_1, \quad (3.305)$$

$$\| [(w_+ \cdot x_1) + b_+, \ldots (w_+ \cdot x_{l_1}) + b_+] \| \leq t_2. \quad (3.306)$$

For replacing t_1^2, t_2^2 in the objective function (3.302)–(3.306), we introduce new variables u_1, u_2, v_1, v_2 and satisfy the linear constraints $u_i + v_i = 1$, $i = 1; 2$ and second order cone constraints $\sqrt{t_i^2 + v_i^2} \leq u_i$. Therefore, problem (3.302)–(3.306) can be reformulated as the following second order cone program (SOCP):

$$\min \frac{1}{2}(u_1 - v_1) + \frac{1}{2}(u_2 - v_2) + c_1 \sum\nolimits_{i=l_1+1}^{l} \xi_i \tag{3.307}$$

$$s.t. \quad -\left((w_+ \cdot x_i)) + b_+\right) - r_i t_1 \geq 1 - \xi_i, \ \ i = l_1 + 1, \ldots, l, \tag{3.308}$$

$$\xi_i \geq 0, \ \ i = l_1 + 1, \ldots, l, \tag{3.309}$$

$$u_1 + v_1 = 1, \tag{3.310}$$

$$u_2 + v_2 = 1, \tag{3.311}$$

$$\| w_+ \| \leq t_1, \tag{3.312}$$

$$\sqrt{t_1^2 + v_1^2} \leq u_1, \tag{3.313}$$

$$\sqrt{t_2^2 + v_2^2} \leq u_2, \tag{3.314}$$

$$\left\| \left[(w_+ \cdot x_1) + b_+, \ldots \left(w_+ \cdot x_{l_1}\right) + b_+ \right] \right\| \leq t_2. \tag{3.315}$$

where

$$\Theta_1 = \left[w_+^{\mathrm{T}}, b_+, \xi^{\mathrm{T}}, t_1, t_2, u_1, v_1, u_2, v_2 \right]^{\mathrm{T}}. \tag{3.316}$$

By the optimization theory [44], the dual problem of (19) can be expressed as

$$\max \ \beta_1 + \beta_2 + \sum\nolimits_{i=l_1+1}^{l} \alpha_i \tag{3.317}$$

$$s.t. \ \ \beta_1 + z_{u_1} = \frac{1}{2}, \ \ \beta_1 + z_{v_1} = -\frac{1}{2}, \tag{3.318}$$

$$\beta_2 + z_{u_2} = \frac{1}{2}, \ \ \beta_2 + z_{v_2} = -\frac{1}{2}, \tag{3.319}$$

$$\sum\nolimits_{i=l_1+1}^{l}\alpha_i-\sum\nolimits_{i=1}^{l_1}\lambda_i=0, \tag{3.320}$$

$$\left\|\sum\nolimits_{i=1}^{l_1}\lambda_i x_i-\sum\nolimits_{j=l_1+1}^{l}\alpha_j x_j\right\|\le\sum\nolimits_{i=l_1+1}^{l}r_i\alpha_i-\gamma_1, \tag{3.321}$$

$$\|\lambda\|\le-\gamma_2, \tag{3.322}$$

$$\sqrt{\gamma_1^2+z_{v_1}^2}\le z_{u_1}, \tag{3.323}$$

$$\sqrt{\gamma_2^2+z_{v_2}^2}\le z_{u_2}, \tag{3.324}$$

$$0\le\alpha_i\le c_1, i=l_1+1,\ldots,l. \tag{3.325}$$

where

$$\Theta_2=\left[\alpha^{\mathrm{T}},\beta_1,\beta_2,\gamma_1,\gamma_2,z_{u_1},z_{v_1},z_{u_2},z_{v_2},\lambda^{\mathrm{T}}\right]^{\mathrm{T}}. \tag{3.326}$$

Theorem 3.3 *Suppose that Θ_2^* is a solution of the dual problem* (20), *where* $\Theta_2^*=\left[\alpha^*\mathrm{T},\beta_1^*,\beta_2^*,\gamma_1^*,\gamma_2^*,z_{\mathrm{u}_1}^*,z_{v_1}^* z_{\mathrm{u}_2}^*,z_{v_2}^*,\lambda^{*\mathrm{T}}\right]^{\mathrm{T}}$. *If there exists* $0<\alpha_j^*<c_1$ *we will obtain the solution* (w^*,b^*) *of the primal problem* (3.295)–(3.297):

$$w_+^*=\frac{\gamma_1^*}{\left(\Sigma_{i=l_1+1}^{l}r_i\alpha_i^*-\gamma_1^*\right)}\left(\sum\nolimits_{l_1+1}^{l}\alpha_i^* x_1\cdot-\sum\nolimits_{i=1}^{l_1}\lambda_i^* x_1\right)' \tag{3.327}$$

$$b_+^*=-1+\gamma_1^*\mathrm{r}_j-\frac{\gamma_1^*}{\left(\Sigma_{i=l_1+1}^{l}r_i\cdot\alpha_j^*-\gamma_1^*\right)}\left(\sum\nolimits_{l_1+1}^{l}\alpha_{\mathrm{i}}^*\left(x_1\cdot x_j\right)-\sum\nolimits_{i=1}^{l_1}\lambda_{\mathrm{i}}^*\left(x_{\mathrm{i}}\cdot x_j\right)\right) \tag{3.328}$$

Similarly, the dual of problem can be written as

$$\underset{\Theta_3}{max}\,\beta_1+\beta_2+.\sum\nolimits_{i=l_1+1}^{l}\alpha_i \tag{3.329}$$

$$\text{s.t.}\beta_1+z_{\mathrm{u}_1}=\frac{1}{2},\beta_1+\mathrm{z}_{v_1)}=-\frac{1}{2}, \tag{3.330}$$

$$\beta_2+z_{\mathrm{u}_2}=\frac{1}{2},\beta_2+\mathrm{z}_{v_2}=-\frac{1}{2}, \tag{3.331}$$

$$-\sum_{=1}^{l_1}\alpha_i - \sum_{i=l_1+1}^{1}\lambda_i = 0, \tag{3.332}$$

$$\left\|\sum_{i=l_1+1}^{1}\lambda_i x_i + \sum_{j=1}^{l_1}\alpha_j x_j\right\| \le \sum_{i=1}^{l_1} r_i\alpha_i - \gamma_1, \tag{3.333}$$

$$\|\lambda\| \le -\gamma_2, \tag{3.334}$$

$$\sqrt{\rho_1 + z_{v_1}^2} \le z_{u_1}, \tag{3.335}$$

$$\sqrt{P_2 + z_{v_2}^2} \le z_{u_2}, \tag{3.336}$$

$$0 \le \alpha_i \le c_2, i = 1_t \mid .\ /1. \tag{3.337}$$

where

$$\Theta_3 = \left[\alpha^T, \beta_1, \beta_2, \gamma_1, \gamma_2 z_{u_1}, z_{U_1}, z_{u_2} z_{v_2}, \lambda^T\right]^T \tag{3.338}$$

The corresponding solution is

$$w_-^* = \frac{\gamma_1^*}{\left(\Sigma_{i=1}^{l_1} r_i\alpha_i^* - \gamma_1^*\right)}\left(-\sum_{1=1}^{l_1}\alpha_i^* x_i - \sum_{i=l_1+1}^{l}\lambda_i^* x_i\right) \tag{3.339}$$

$$b_-^* = -1 + \gamma_1^* r_j \frac{\gamma_1^*}{\left(\Sigma_{i=1}^{l1} r_i\alpha_i^* - \gamma_1^*\right)}\left(-\sum_{1=1}^{l_1}\alpha_i^*\left(x_i \cdot x_j\right) - \sum_{i=l_1+1}^{l}\lambda_i^* x_i x_j\right) \tag{3.340}$$

Once vectors w_+, b_+ and w_-, b_- are obtained from (3.317)–(3.325) and (3.329)–(3.337), the separating planes

$$w_+^T x + b_+ = 0, w_-^T x + b_- = 0 \tag{3.341}$$

are known. A new data point $x \in \Re^n$ is then assigned to the positive or negative class, depending on which of the two hyperplanes given by (26) it lies closest to, i.e.

$$f(x) = \text{argmin}\ \{d_+(x), d_-(x)\}, \tag{3.342}$$

where

$$d_+(x) = \left|w_+^{\mathrm{T}}x + b_+\right|, \quad d_-(x) = \left|w_-^{\mathrm{T}}x + b_-\right| \tag{3.343}$$

where $|\cdot|$ is the perpendicular distance of point x from the planes $w_+^{\mathrm{T}}x + b_+$ and $w_-^{\mathrm{T}}x + b_-$.

Nonlinear $\mathcal{R}$-TWSVM

The above discussion is restricted to the linear case. Here, we will analyze nonlinear $\mathcal{R}$-TWSVM by introducing kernel function

$K(x, x') = (\Phi(x) \cdot \Phi(x'))$, and the corresponding transformation:

$$\mathrm{x} = \Phi(x) \tag{3.344}$$

where $\mathrm{x} \in \mathcal{H}$, $\mathcal{H}$ is the Hilbert space. So, the training set (12) becomes

$$T = \left\{ (X_i, y_i), \ldots, (X_l, y_l) \right. \tag{3.345}$$

where

$\mathrm{X_i} = \{\Phi(\widetilde{\mathrm{x}}) | \widetilde{\mathrm{x}}$ is in the sphere of the radius r and the center $x_\mathrm{i}\}$. So, when $| \left|\overline{x}_\mathrm{i} - x_i\right| | \leq r_\mathrm{i}$ and choosing, we have

$$\|\Phi(\tilde{x}_i) - \Phi(x_\mathrm{i})\|^2 = (\Phi(\tilde{x}_i) - \Phi(x_\mathrm{i})) \cdot (\Phi(\tilde{x}_i) - \Phi(x_\mathrm{i}))$$

$$= K(\tilde{x}_i, \tilde{x}_i) - 2K(\tilde{x}_i, x_i) + K\left(x_i, x_i\right)$$

$$= 2 - 2\exp\left(-\|\tilde{x}_i - x_\mathrm{i}\|^2 / 2\sigma^2\right)$$

$$\leq \mathrm{r}_i^2, \tag{3.346}$$

where

$$r_\mathrm{i} = \sqrt{2 - 2\exp\left(-\|\tilde{x}_i - x_\mathrm{i}\|^2 / 2\sigma^2\right)} \tag{3.347}$$

Thus χ_i becomes a sphere of the center $\Phi(x_i)$ and the radius r_i

$$X_1 = \left\{ \widetilde{x} \middle\| |\widetilde{x} - \Phi(x_i)|| \le r_i \right\}. \tag{3.348}$$

For nonlinear case of $\mathcal{R}$-TWSVM, $\left\| \sum_{i=1}^{l_1} \lambda_i \Phi(x_i) - \sum_{j=l_1+1}^{l} \alpha_j \Phi(x_j) \right\|^2$ can be expressed as

$$\begin{aligned} &\sum_{1=1}^{l_1} \sum_{j=1}^{l_1} \lambda_i \lambda_j K(x_i \cdot x_j) - 2 \sum_{i=1}^{l_1} \sum_{j=l_1+1}^{l} \lambda_1 \cdot \alpha_j K(x_i \cdot x_j) \\ &\quad + \sum_{i=l_1+1}^{l} \sum_{j=l_1+1}^{l} \alpha_i \alpha_j K(x_i x_j) \end{aligned}. \tag{3.349}$$

Similarly, $\left\| \sum_{i=l_1+1}^{l} \lambda_i x_1 + \sum_{j=1}^{l_1} \alpha_j x_j \right\|^2$ can be expressed as

$$\begin{aligned} &\sum_{1=1}^{l_1} \sum_{j=1}^{l_1} \lambda_i \lambda_j K(x_i \cdot x_j) + 2 \sum_{i=l_1+1}^{l} \sum_{j=1}^{l_1} \lambda_i \cdot \alpha_j K(x_i \cdot x_j) \\ &\quad + \sum_{i=1}^{l_1} \sum_{j=1}^{l_1} \alpha_i \alpha_j K(x_i \cdot x_j) \end{aligned} \tag{3.350}$$

So we can easily obtain the nonlinear $\mathcal{R}$-TWSVM only by taking $K(x, x')$ instead of (xx') of the optimization problem (3.317)–(3.325) and (3.329)–(3.337).

3.2.4 *Structural Twin Support Vector Machine for Classification*

3.2.4.1 Structural Twin Support Vector Machine (S-TWSVM)

Extracting Structural Information Within Classes

Following the strategy of the SLMM and SRSVM, S-TWSVM also has two steps. The first step is to extract the structural information within classes by some clustering method; the second step is the model learning. In order to compare the main difference of the second step between S-TWSVM and the other two methods, here we also adopt the same clustering method: Ward's linkage clustering (WIL) [61–64], which is one of the hierarchical clustering analysis. A main advantage of WIL is that clusters derived from this method are compact and spherical, which provides a meaningful basis for the computation of covariance matrices [64]. Concretely, if S and T are two clusters with means μ_S and μ_T, the Ward's linkage

$W(S, T)$ between clusters S and T is computed as [64]

$$W\left(S,T\right)=\frac{\mid S\mid\cdot\mid T\mid\cdot\|\mu_{s}-\mu_{\mathrm{T}}\|}{\mid S\mid+\mid T\mid} \tag{3.351}$$

Initially, each sample is considered as a cluster. The Wards linkage of two samples x_i and x_j is $W(x_i, x_j) = \|x_{\mathrm{i}} - x_j\|^2/2$. When two clusters are being merged to a new cluster $A^{'}$, the linkage $W(A^{'}, C)$ can be conveniently derived from $W(A, C)$, $W(B, C)$ and $W(A, B)$ by [64]

$$W\left(A',C\right)=\frac{\left(|A|+|C|\right)W\left(A,C\right)+\left(|B|+|C|\right)W\left(B,C\right)-\mid C\mid W\left(A,B\right)}{\mid A\mid+\mid B\mid+\mid C\mid} \tag{3.352}$$

During the hierarchical clustering, the Ward's lineage between clusters to be merged increases as the number of clusters decreases [64]. A relation curve between the merge distance and the number of clusters are able to be drawn to represent this process. The optimal number of clusters is determined by finding the knee point [65]. Furthermore, the WIL can also be extended to the kernel space. More details of WIL are able to be found in [64].

We obtain two groups of P and N clusters in class P and N by the first step, i. e. , $P = P_1 \cup \cdots P_1 \cup \cdots P_{c_p}, N = N_1 \cup \cdots N_j \cup \cdots N_{\mathrm{c_N}}$. Suppose that data points belong to positive class are denoted by $A \in R^{\mathrm{m}_1 \times n}$, where each row $A_i \in R^n$ represents a data point. Similarly, $B \in R^{m_2 \times n}$ represents all of the data points belong to negative class, where $\mathrm{m}_1 + \mathrm{m}_2 = l$. For the linear case, the S-TWSVM determines two nonparallel hyperplanes:

$$f_{+}(x)=w_{+}^{\mathrm{T}}x+b_{+}=0 \text{ and } f_{-}(x)=w^{\mathrm{T}}x+b=0, \tag{3.353}$$

where $w_+, w_- \in R^{\mathrm{n}}$, $b_+, b_- \in R$. Here, each hyperplane is closer to one of the two classes and is at least one distance from the other, at the same time, minimizes the compactness within the class by the structural information obtained by clustering technology. A new data point is assigned to positive class or negative class depending upon its proximity to the two nonparallel hyperplanes. By introducing the data distributions of the clusters in different classes into the object functions of TBSVM, (Notice we only consider one class' structural information for each model. In other words, each model only considers these structural information of which the hyperplane is closer to the class.) the S-TWSVM model can be formulated as

$$\min\frac{1}{2}\|Aw_{+}+e_{+}b_{+}\|_{2}^{2}+c_{1}e_{-}^{\mathrm{T}}\xi+\frac{1}{2}c_{2}\left(\|w_{+}\|_{2}^{2}+b_{+}^{2}\right)+\frac{1}{2}C_{3}w_{+}^{\mathrm{T}}\Sigma_{+}w_{+}$$

$$s.t.\quad -\left(Bw_{+}+eb_{+}\right)+\xi\geq e,\ \ \xi\geq 0. \tag{3.354}$$

and

$$\min \frac{1}{2}\| Bw + e\_b\_ \|_2^2 + \frac{1}{2}c_4 e_+^{\mathrm{T}}\eta + \frac{1}{2}c_5\left(\|w_-\|_2^2 + b_-{}^2\right) + \frac{1}{2}c_6 w^{\mathrm{T}}\Sigma_- w_-$$

$$\text{st.}\ (Aw_- + e_+b_-) + \eta \geq e_+,\quad \eta \geq 0. \tag{3.355}$$

where $c_1, \ldots, c_6 \geq 0$ are the pre-specified penalty factors, e_+, e_- are vectors of ones of appropriate dimensions, ξ_1 is the slack variables, $\Sigma_+ = \Sigma_{P_1} + \cdots + \Sigma_{\mathrm{P}_{c_P}}$, $\Sigma = \Sigma_{N_1} + \cdots + \Sigma_{N_{c_N}}$, Σ_{P_i} and Σ_{N_j} are respectively the covariance matrices corresponding to the ith andjth clusters in the two classes, $i = 1, \ldots, \mathrm{C}_p, j = 1, \ldots, \mathrm{C}_N$.

The Wolfe dual of the problem is as follow:

$$\begin{aligned} &\max\ e_-^{\mathrm{T}}\alpha - \tfrac{1}{2}\alpha^{\mathrm{T}}G\left(H^{\mathrm{T}}\mathrm{H} + c_2\mathrm{I} + c_3J\right)^{-1}G^{\mathrm{T}}\alpha \\ &\text{s.t.} 0 \leq \alpha \leq c_1 e_- \end{aligned} \tag{3.356}$$

where

$$\mathrm{H} = [A\ e_+],\quad J = \begin{bmatrix} \Sigma_+ & 0 \\ 0 & 0 \end{bmatrix},\quad G = [Be_-\ ], \tag{3.357}$$

and the augmented vector $\theta_+ = \left[W_+^{\mathrm{T}} b_+^{\mathrm{T}}\right]^{\mathrm{T}}$ is given by

$$\theta_+ = -\left(H^T H + c_2 I + c_3 J\right)^{-1}\left(G^T\alpha\right). \tag{3.358}$$

I is an identity matrix of appropriate dimensions. According to matrix theory [66], it is very easy to prove that $H^{\mathrm{T}}H + c_2 I + c_3 J$ is a positive definite matrix.

Similarly, the dual of (3.355) is

$$\max\ e_+^T\beta - \frac{1}{2}\beta^{\mathrm{T}}P\left(\mathrm{Q}^{\mathrm{T}}\mathrm{Q} + c_5\mathrm{I} + c_6F\right)^{-1}P^{\mathrm{T}}\beta$$

$$s.t.\ \ 0 \leq \beta \leq c_4 e_+, \tag{3.359}$$

where

$$P = [A\ e_-],\ F = \begin{bmatrix} \Sigma & 0 \\ 0 & 0 \end{bmatrix}\ \ \mathrm{Q} = [B\ e_+], \tag{3.360}$$

and the augmented vector $\theta_- = [w_- b_-]^{\mathrm{T}}$ given by

$$\Theta_- = -\left(\left(\mathrm{Q}^{\mathrm{T}}\mathrm{Q} + c_5\mathrm{I} + c_6 F\right)^{-1} P^{\mathrm{T}}\beta, \tag{3.361}$$

where $\mathrm{Q}^{\mathrm{T}}\mathrm{Q} + c_5\mathrm{I} + c_6 F$ is a positive definite matric. Once vectors ϑ_+ and ϑ_- are obtained from (3.358) and (3.361), the separating planes

$$w_+^{\mathrm{T}} x + b_+ = 0, w^{\mathrm{T}} x + b_- = 0 \tag{3.362}$$

are known. A new data point $x \in R^{\mathrm{n}}$ is then assigned to the positive or negative class, depending on which of the two hyperplanes given by (3.362) it lies closest to, i.e.

$$f(x) = \arg\ \min\ \{d_+(\mathrm{x}), d_-(\mathrm{x})\} \tag{3.363}$$

where

$$d_+(x) = \left|w_+^{\mathrm{T}} x + b_+\right|, d_-(\mathrm{x}) = \left|w^{\mathrm{T}} x + b_-\right|, \tag{3.364}$$

where $|\cdot|$ is the perpendicular distance of point x from the planes $w_+^{\mathrm{T}} x + b_+$ or $w^{\mathrm{T}} x + b_-$.

Nonlinear S-TWSVM

Now we extend the linear *S*-TWSVM to the nonlinear case.

Similar to linear case, the decision function is written as $f_+(x) = (w_+ \ \Phi(x)) + b_+$ and $f_-(x) = (w_- \cdot \Phi(x)) + b_-$, where $\Phi(\cdot)$ is a nonlinear mapping from *a* low dimensional space to a higher dimensional Hilbert space $\mathcal{H}$. According to Hilbert space theory [67], w_+ and w_- can be expressed as $w_+ = \sum_{i=1}^{m_1+m_2} (\lambda_+)_{\mathrm{i}} \Phi(x_{\mathrm{i}}) = \Phi(M)\lambda_+$ and $w_- = \sum_{i=1}^{m_1+m_2} (\lambda_-)_{\mathrm{i}} \Phi(x_{\mathrm{i}}) = \Phi(M)\lambda_-$, respectively. So the following kernel-generated hyperplane:

$$K\left(x^{\mathrm{T}}, M^{\mathrm{T}}\right)\lambda_+ + b_+ = 0, \tag{3.365}$$

$$K\left(x^{\mathrm{T}}, M^{\mathrm{T}}\right)\lambda_- + b_- = 0. \tag{3.366}$$

where K is a chosen kernel function: $K(x_j \cdot x_j) = (\Phi(x_\mathrm{i}\cdot)\ \Phi(x_j))$, $M = [A^\mathrm{T} B^\mathrm{T}]$. The nonlinear optimization problem can be expressed as

$$\min \frac{1}{2}\left\| K\left(A, M^\mathrm{T}\right)\lambda_+ + e_+ b_+ \right\|^2 + c_1 e_-^\mathrm{T}\xi + \frac{1}{2}c_2\left(\|\lambda_+\|^2 + b_+^2\right)$$
$$+\frac{1}{2}c_3\lambda_+^\mathrm{T}\Phi(M)^\mathrm{T}\Sigma_+^\Phi\Phi(M)\lambda_+,$$

$$s.t.\ -\left(K\left(B, M^\mathrm{T}\right)\lambda_+ + e_- b_+\right) + \xi \geq e,\ \ \xi \geq 0, \tag{3.367}$$

and

$$\min \frac{1}{2}\left\| K, B\, M^\mathrm{T}\right)\lambda_- + e_- b_- \Big\|^2 + c_4 e_+^\mathrm{T}\eta + \frac{1}{2}c_5\|\lambda_-\|^2 + b_-^2\Big)$$

$$+\frac{1}{2}c_6\lambda^\mathrm{T}\Phi(M)^\mathrm{T}\Sigma^\varnothing\Phi\,(\mathrm{M})\,\lambda_-\mathrm{st.}\left(K\left(A, M^\mathrm{T}\right)\lambda_- + e_+ b_-\right) + \eta \geq e_+, \eta \geq 0, \tag{3.368}$$

where $\Sigma_+^\Phi = \Sigma_{P_1}^\Phi + \cdots + \Sigma_{P_{c_p}}^\Phi$, $\Sigma^\Phi = \Sigma_{N_1}^\Phi + \cdots + \Sigma_{N C_N}^\Phi$, Σ_{P_i} and Σ_{N_j} are respectively the covariance matrices corresponding to the i-th and j th clusters in the two classes by the kernel Ward's linkage clustering [63, 64], $i = 1, \ldots, \mathrm{C_p}, j = 1, \ldots, \mathrm{C}_N$.

The Wolfe dual of the problem (3.367) is formulated as follow:

$$\max\ e_-^\mathrm{T}\alpha - \frac{1}{2}\left(\alpha^\mathrm{T} G_\Phi\right)\left(H_\Phi^\mathrm{T}\mathrm{H}_\Phi + c_2\mathrm{I} + c_3\mathrm{J}_\Phi\right)^{-1}\left(C_\Phi^\mathrm{T}\alpha\right)$$

$$s.t.\ 0 \leq \alpha \leq c_1 e_- \tag{3.369}$$

where
$H_\Phi = [K(A, M^\mathrm{T})e_+]$, $\mathrm{C}_\Phi = [K(B, M^\mathrm{T})e_-]$

$$J_\Phi = \begin{bmatrix} \Phi(M)^\mathrm{T}\sum_+^\Phi\Phi(M) & 0 \\ 0 & 0 \end{bmatrix} \tag{3.370}$$

and the augmented vector $\rho_+ = [\lambda_+ b_+]^\mathrm{T}$

$$\rho_+ = -\left(\mathrm{H}_\Phi^\mathrm{T}\mathrm{H}_\Phi + c_2 I + c_3 J_\Phi\right)^{-1}\left(\mathrm{C}_\Phi^\mathrm{T}\alpha\right). \tag{3.371}$$

In a similar manner, the dual of (3.368) is

$$\max\ e_+^{\mathrm{T}}\beta - \frac{1}{2}\left(\beta^{\mathrm{T}}\mathrm{P}_\Phi\right)\left(\mathrm{Q}_\Phi^{\mathrm{T}}\mathrm{Q}_\Phi + c_5 I + c_6\mathrm{F}_\Phi\right)^{-1}\left(P_\Phi^{\mathrm{T}}\beta\right)$$

$$\text{s.t.}0 \le \beta \le c_4 e_+, \tag{3.372}$$

where

$$P_\Phi = \left[K\left(A, M^{\mathrm{T}}\right)e_-\right], \mathrm{Q}_\Phi = \left[K\left(B, M^{\mathrm{T}}\right)e_+\right]$$

$$F_\Phi = \begin{bmatrix} \Phi(M)^{\mathrm{T}}\Sigma^\Phi\Phi(M) & 0 \\ 0 & 0 \end{bmatrix}, \tag{3.373}$$

and the augmented vector $\rho_- = [\lambda_- b_-]^{\mathrm{T}}$, which is given by

$$\rho_{-=-}\left(\mathrm{Q}_\Phi^{\mathrm{T}}\mathrm{Q}_\Phi + c_5\mathrm{I} + c_6\mathrm{F}_\Phi^{-1}\right)\left(\mathrm{P}_\Phi^{\mathrm{T}}\alpha\right) \tag{3.374}$$

Once vectors ρ_+ and ρ_- are obtained from (3.371) and (3.374), a new data point $x \in R^n$ is then assigned to the positive or negative class, depending on a manner similar to the linear case (Fig. 3.7).

Now we consider how to compute the kernel matrix J_Φ. Suppose T_P is a matrix corresponding to the cluster P_{i}, $T_{\mathrm{P}} \in R^{P_1 \times \mathrm{n}}$, in which the kth row is $x_{\mathrm{k}}^{\mathrm{T}} O_{P_i}$ is a mean matrix of cluster P_1, $\mathrm{O}_{\mathrm{P}} \in R^{\mathrm{P} \times \cap}$ Each row of O_P is the same, i.e.

$$\mu_{p_{\mathrm{i}}} = \frac{1}{P_{\mathrm{i}}}\sum\nolimits_{\mathrm{x_k} \in \mathrm{P_i}} .x_k. \tag{3.375}$$

The related covariance matrix for cluster P_j can be expressed as

$$\Sigma_{p_{\mathrm{i}}}^\Phi = \frac{1}{P_{\mathrm{i}}}\left(\Phi\left(T_{\mathrm{P_i}}\right) - \Phi\left(0_{\mathrm{P_i}}\right)\right)^{\mathrm{T}}\left(\Phi\left(T_{p_1}\right) - \Phi\left(\mathrm{O}_{p_{\mathrm{i}}}\right)\right). \tag{3.376}$$

So, we obtain

$$\Phi(M)^T\Sigma_+^\Phi\Phi(M) = \left(\frac{1}{\sqrt{P_i}}\left(\Phi\left(T_{P_i}\right) - \Phi\left(O_{P_i}\right)\right)\Phi(M)\right)^T$$

$$\left(\frac{1}{\sqrt{P_i}}\left(\Phi\left(T_{P_i}\right) - \Phi\left(O_{P_i}\right)\right)\Phi(M)\right)$$

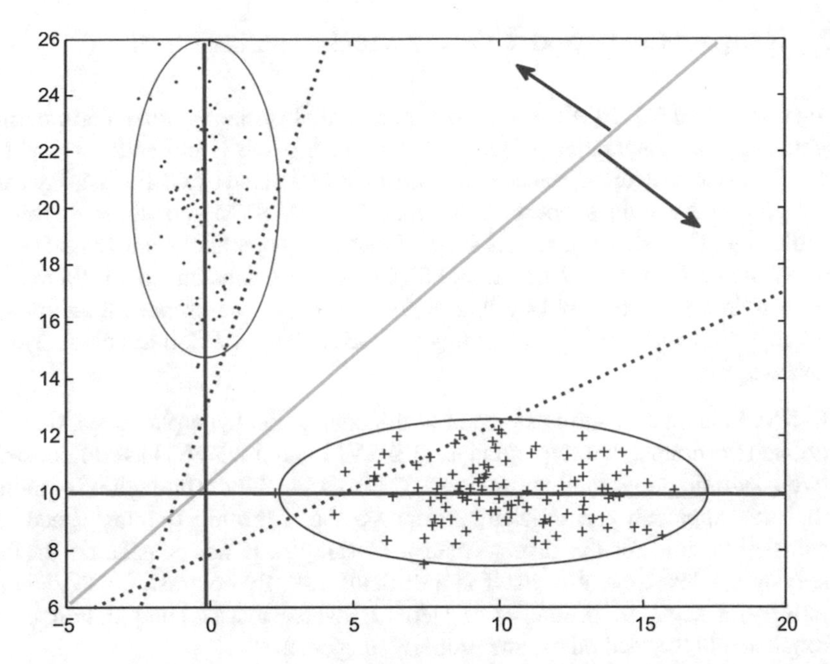

Fig. 3.7 The geometric interpretation of existing the structural information confliction between positive class and negative class. The red and blue solid line denotes the classifier of *S*-TWSVM. The red and blue dotted line denotes the classifier of Snuctural TWSVM of each model consider two class s structural information (for simplify we called it *SS-TWSVM*). Obviously, S-TWSVM is able to better predict the data distribution tendency than *SS*-TWSVM. The cyan line denotes the classifier based on one hyperplane such as SLMM or SRSVM and the red and blue arrows denotes the tendency of the two-class structural information hoping the classifier to rotate. In the case the classifier is almost the same as the that traditional SVM and these structural information does not play a role and make the classifier change. *S*-TWSVM is obviously superior to SLMM SRSVM and *SS*-TWSVM

$$= \Big(\frac{1}{\sqrt{P_i}}(K\left(T_{P_i}, M\right) - K\left(O_{P_i}, M\right)\Big)^{\mathrm{T}}$$

$$\frac{1}{\sqrt{P_i}}\left(K\left(T_{P_i}, M\right) - K\left(O_{P_i}, M\right)\right) \tag{3.377}$$

Similarly, $\Phi(M)^{\mathrm{T}}\Sigma_{\mathrm{M}}+^{\Phi}\Phi(M)$ of F_{Φ} are computed as

$$\Phi(M)^{\mathrm{T}}\Sigma_{-}^{\Phi}\Phi(M) = \Big(\frac{1}{\sqrt{\mathrm{P}_j}}(K\left(T_{N_{\mathrm{i}}}, M\right) - K\left(\mathrm{O}_{N_{\mathrm{i}}}.M\right)\Big)^{\mathrm{T}}$$

$$\Big(\frac{1}{\sqrt{P_1}}\left(\mathrm{K}\left(T_{N_{\mathrm{i}}}, M\right) - K\left(\mathrm{O}_{N_{\mathrm{i}}}, M\right)\right) \tag{3.378}$$

where $T_{N_{\mathrm{i}}}$ is a matrix of cluster N_{i}, $\mathrm{O}_{N_{\mathrm{i}}}$ is a mean matrix of cluster N_{i}.

3.3 Nonparallel Support Vector Machine Classifiers

Mangasarian and Wild [38] proposed a nonparallel plane classifier that attempts to generate two nonparallel planes such that each plane is closer to one of two classes and is at least one distance from the other. Motivated by GEPSVM, Jayadeva et al. proposed a twin support vector machine (TWSVM) classifier for binary classification. Experimental results showed that the nonparallel plane classifier can improve the performance of traditional SVMs. Other extensions to TWSVM have also been described. Inspired by this previous success, we propose a nonparallel SVM algorithm with universum learning that we call U-NSVM. It has the following innovative points:

- U-NSVM is a very useful extension of the nonparallel hyperplanes classifier. To obtain two nonparallel hyperplanes, GEPSVM and TWSVM have to construct two quadratic programming problems (QPPs) separately. Although it is claimed that this approach can efficiently improve the algorithm training speed, the calculation time for the inverse matrix of samples is not considered. In fact, solving the inverse matrix itself is a difficult task. By contrast, U-NSVM uses universum samples, so the extra step of inverse matrix computation is not required and the method has the property of sparseness.
- As the U-NSVM classifier combines two nonparallel hyperplanes, compared to U-SVM it has better algorithm flexibility and can yield a more reasonable classifier in most cases. In addition, U-NSVM includes fewer parameters and is therefore easier to implement. In practice, the U-SVM algorithm uses an ε-insensitive loss function to divide the universum data, while our algorithm does not include a corresponding parameter. Experiments confirm that our method is superior to a traditional SVM and U-SVM.

3.3.1 A Nonparallel Support Vector Machine for a Classification Problem with Universum Learning

3.3.1.1 Nonparallel SVM for Classification with a Universum: U-NSVM

Linear U-NSVM

U-SVM requires that the hyperplane satisfies the maximum margin principle for labeled data and that all the universum data are as close as possible to it. Following previous success, we achieve U-SVM using two nonparallel planes. We first construct two nonparallel hyperplanes to divide the training set into three parts, with the universum data sandwiched between the two hyperplanes. This can be achieved by maximizing the two margins associated with the two closest neighboring classes (labeled data and universum data). Then a new data point x can be predicted as belonging to the positive or negative class, depending on the distances between x and the two hyperplanes. It is not hard to imagine that the perpendicular bisector (decision function) of the two hyperplanes is as close as possible to the central area of the universum data distribution and far away from the labeled data. According to this idea, the primal optimization of U-NSVM can be expressed as

$$
\begin{gathered}
min\ \frac{1}{2}\left(\|w_1\|_2^2+\|w_2\|_2^2\right)+C\left(\sum_{1=1}^{l_1}\xi_1^1+\sum_{i=1}^{l_2}\left(\xi_i^2+\xi_i^{*2}\right)+\sum_{i=1}^{l_3}\xi_j^{*3}\right)\\
s.t.\ \left(w_1\cdot x_i^1\right)-b_1\le -1+\xi_i^1,\ \ i=1,\dots,l_1\\
\left(w_2\cdot x_i^2\right)-b_2\le -1+\xi_i^2,\ \ i=1,\dots,l_2\\
\left(w_1\cdot x_1^3\right)-b_1\ge 1-\xi_i^{*2},\ \ i=1,\dots,l_2\\
\left(w_2\cdot x_i^3\right)-b_2\ge 1-\xi_i^{*3},\ \ i=1,\dots,l_3\\
\xi_{\mathrm{i}}^1\ge 0,\ \ i=1,\dots,l_1\\
\xi_i^2\ge 0,\ \ \xi_i^{*2}\ge 0,\ \ i=1,\dots,l_2\\
\xi_j^{*3}\ge 0,\ \ i=1,\dots,1_3,
\end{gathered}
\tag{3.379}
$$

where $w=(w_1,w_2)$, $b=(b_1,b_2)$, and $\xi^{(*)}=(\xi_1^1,\dots,\xi_{l_1}^1,\xi_1^2,\dots,\xi_{l_2}^2,\xi_1^{*2},\dots,\xi_{l_2}^{*2},\xi_1^{*3},\dots,\xi_{l_3}^{*3})$.

By introducing the Lagrange function

$$
\begin{aligned}
L(\Theta)=&\frac{1}{2}\left(\|w_1\|_2^2+\|w_2\|_2^2\right)+C\left(\sum_{1=1}^{l_1}\xi_i^1+\sum_{1=1}^{l_2}\left(\xi_i^2+\xi_i^{*2}\right)\right.\\
&\left.+\sum_{i=1}^{l_3}\xi_1^{*3}\right)+\sum_{k=1}^{2}\sum_{\mathrm{i}=1}^{l_{\mathrm{k}}}\alpha_{\mathrm{i}}^{\mathrm{k}}\left(\left(w_k\cdot x_{\mathrm{i}}^{\mathrm{k}}\right)-b_{\mathrm{k}}+1-\xi_{\mathrm{i}}^k\right)\\
&-\sum_{k=2}^{3}\sum_{1=1}^{l_k}\alpha_1^{*k}\left(\left(w_{k-1}\,x_i^k\right)-b_{k-1}-1+\xi_i^{*k}\right)-\sum_{k=1}^{2}\sum_{i=1}^{l_k}\eta_i^k\xi_i^k-\sum_{k=2}^{3}\sum_{i=1}^{l_k}\eta_i^{*k}\xi_i^{*k},
\end{aligned}
\tag{3.380}
$$

where

$$\Theta = \left(w, b, \xi^{(*)}, \alpha^{(*)}, \eta^{(*)}\right), \tag{3.381}$$

$$\alpha^{(*)} = \left(\alpha_1^1, \ldots, \alpha_{l_1}^1, \alpha_1^{*2}, \ldots, \alpha_{l_2}^{*2}, \alpha_1^2, \ldots, \alpha_{l_2}^2, \alpha_1^{*3}, \ldots, \alpha_{l_3}^{*3}\right), \tag{3.382}$$

$$\eta^{(*)} = \left(\eta_1^1, \ldots, \eta_{l_1}^1, \eta_1^{*2}, \ldots, \eta_{l2}^{*2}, \eta_1^2, \ldots, \eta_{l_2}^2, \eta_1^{*3}, \ldots, \eta_{l3}^{*3}\right). \tag{3.383}$$

are the Lagrange multipliers, the dual problem can be formulated as

$$\begin{aligned} &\max L(\Theta) \\ &s.t.\ \nabla_{w,b.\xi^{(*)}} L(\Theta) = 0, \\ &\alpha^{(*)}, \eta^{(*)} \geq 0. \end{aligned} \tag{3.384}$$

From Eq. (3.384) we obtain

$$\nabla_{w_k} L = w_k + \sum\nolimits_{i=1}^{l_k} \alpha_i^k x_i^k - \sum\nolimits_{\iota=1}^{l_{k+1}} \alpha_1^{\dot{*}k+1} x_i^{k+1} = 0, k = 1, 2, \tag{3.385}$$

$$\nabla_{b_k} L = -\sum\nolimits_{i=1}^{l_k} \alpha_1^{\dot{k}} + \sum\nolimits_{j=1}^{lk+1} \alpha_j^{*k+1} = 0, k = 1, 2, \tag{3.386}$$

$$\nabla_{\xi_i^k} . L = C - \alpha_i^k - \eta_i^k = 0, k = 1, 2, i = 1, \ldots, l_k, \tag{3.387}$$

$$\nabla_{\xi_{\mathrm{i}}} L = \mathrm{C} - \alpha_i^{*k} - \eta_{\mathrm{i}*\mathrm{k}} = 0, k = 2; 3, i = 1, \ldots, l_k. \tag{3.388}$$

Substituting the above equations into, the dual problem can be expressed as

$$\begin{aligned} &\max -\frac{1}{2}\sum\nolimits_{k=1}^{2} \left(\sum_{j=1}^{l_k}\sum_{j=1}^{l_k} \alpha_i^k \alpha_j^k \left(x_i^k\ X_j^k\right) - 2\sum_{1=1}^{l_k}\sum_{j=1}^{l_{k+1}} \alpha_i^k \alpha_j^{*k+1} \left(x_i^k x_j^{\dot{k}+1}\right)\right. \\ &\left. + \sum_{j=1}^{1_{k+1}}\sum_{j=1}^{1_{k+1}} \alpha_1^{*k+1} \alpha_j^{*k+1} \left(x_i^{k+1}\ x_j^{\dot{k}+1}\right)\right) + \sum_{k=1}^{2}\sum_{i=1}^{l_k} \alpha_i^k + \sum_{k=2}^{3}\sum_{i=1}^{l_k} \alpha_i^{*k} \\ &s.t \sum\nolimits_{\mathrm{i}=1}^{\mathrm{l_k}} \alpha_{\mathrm{i}}^k = \sum\nolimits_{i=1}^{\mathrm{l_{k+1}}} \alpha_i^{*k+1}, \mathrm{k} = 1, 2, \\ &0 \leq \alpha_i^k \leq C, k = 1, 2; = 1, \ldots, l_k, \\ &0 \leq \alpha_{\mathrm{i}}^{\dot{*}k} < \mathrm{C},\ \ k = 2, 3; i = 1, \ldots, 1_k. \end{aligned} \tag{3.389}$$

Theorem 3.4 *Optimization problem is a convex quadratic* program.

Proof Concisely, it can be reformulated as

$$\max -\frac{1}{2}\hat{\lambda}^{\mathrm{T}}\hat{\Lambda}\hat{\lambda} + \hat{\kappa}^{\mathrm{T}}\hat{\lambda}$$

$$s.t.\ \ \hat{\Omega}\hat{\lambda} = 0, \tag{3.390}$$

$$0 \le \hat{\lambda} \le \hat{\mathrm{C}},$$

where

$$\hat{\lambda} = \left(\alpha_1^1, \ldots, \alpha_{1_1}^1, \alpha_1^{*2}, \ldots, \alpha_{1_2}^{*2}, \alpha_1^2, \ldots, \alpha_{1_2}^2, \alpha_1^{*3}, \ldots, \alpha_{13}^{*3}\right), \tag{3.391}$$

$$\hat{\kappa} = e, e \in \Re^{\Sigma_{\mathrm{k}=1}^{2}(\mathrm{l_k}+l_{k+1})}, \tag{3.392}$$

$$\hat{\mathrm{C}} = (C, \mathrm{C}, \mathrm{C})\, e, e \in \Re^{\Sigma_{\mathrm{k}=1}^{2}(\mathrm{l_k}+l_{k+1})}, \tag{3.393}$$

$$\hat{\Omega} = \begin{pmatrix} e_1^{\mathrm{T}} & -e_2^{\mathrm{T}} & & \\ & & e_2^{\mathrm{T}} & -e_3^{\mathrm{T}} \end{pmatrix}, e_k \in R^{l_k}, k = 1, 2, \hat{\Omega}_1 \in R^{2\times\Sigma_{k=1}^{2}(l_k+l_{\mathrm{k}+1})}, \tag{3.394}$$

$$\hat{\Lambda} = \begin{pmatrix} Q^1 & \\ & Q^2 \end{pmatrix}, \quad \hat{\Lambda} \in R^{(\Sigma_{k=1}^{2}(l_k+l_{k+1}))\times(\Sigma_{k=1}^{2}(l_k+l_{k+1}))}, \tag{3.395}$$

and

$$Q^k = \begin{pmatrix} Q_1^{\prime\langle} & -Q_2^{\prime\langle} \\ -Q_3^k & \mathrm{Q}_4^k \end{pmatrix}, \quad \mathrm{Q}^k \in R^{(l_k+l_{\mathrm{k}+1})\times(l_k+l_{\mathrm{k}+1})}, \tag{3.396}$$

$$Q_1^k = \begin{pmatrix} \left(x_1^k \cdot x_1^k\right) & \cdots & \left(x_1^k \cdot x_{l_k}^k\right) \\ \vdots & \ddots & \vdots \\ \left(x_{l_k}^k \cdot x_1^k\right) & \cdots & \left(x_{l_k}^k \cdot x_{l_k}^k\right) \end{pmatrix}, \quad Q_1^k \in \Re^{l_k \times l_k}, k = 1, 2 \tag{3.397}$$

$$Q_2^k = Q_3^k = \begin{pmatrix} \left(x_1^k \cdot x_1^{k+1}\right) & \cdots & \left(x_1^k \cdot x_{l_{k+1}}^{k+1}\right) \\ \vdots & \ddots & \vdots \\ \left(x_{l_k}^k \cdot x_1^{k+1}\right) & \cdots & \left(x_{l_k}^k \cdot x_{l_{k+1}}^{k+1}\right) \end{pmatrix}, \quad Q_2^k \in \Re^{l_k \times l_{k+1}}, k = 1, 2 \tag{3.398}$$

$$Q_4^k = \begin{pmatrix} \left(x_1^{k+1} \cdot x_1^{k+1}\right) & \cdots & \left(x_1^{k+1} \cdot x_{l_{k+1}}^{k+1}\right) \\ \vdots & \ddots & \vdots \\ \left(x_{l_{k+1}}^{k+1} \cdot x_1^{k+1}\right) & \cdots & \left(x_{l_{k+1}}^{k+1} \cdot x_{l_{k+1}}^{k+1}\right) \end{pmatrix}, \quad Q_4^k \in \Re_{k+1}^{l_{k+1} \times l_{k+1}}, k = 1, 2 \tag{3.399}$$

It is easy to see that $\hat{\Lambda}$ is a positive semi-definite matrix. According to convex programming theory, we can obtain that is a convex quadratic program.

Theorem 3.5 *Suppose that* $\hat{\lambda} = \left(\alpha_1^1, \ldots, \alpha_{l_1}^1, \alpha_1^{*2}, \ldots, \alpha_{l_2}^{*2}, \alpha_1^2, \ldots, \alpha_{l_2}^2, \alpha_1^{*3}, \ldots, \alpha_{l_3}^{*3}\right)^{\mathrm{T}}$ *is a solution of the dual problem*

If there exist components of $\hat{\lambda}$ *with values in the interval* $\left(0, \hat{C}\right)$, *then the solution* (w_1, b_1), (w_2, b_2) *of* (12) *can be obtained in the following way.*

Let

$$w_k = \sum\nolimits_{i=1}^{l_{k+1}} \alpha_i^{*k+1} x_i^{k+1} - \sum\nolimits_{i=1}^{l_k} \alpha_i^k x_i^k, k = 1, 2. \tag{3.400}$$

Choose a component of α^k, $\alpha_j^k \in (0, C)$, $k = 1, 2$ *and compute*

$$b_k = 1 + \sum\nolimits_{i=1}^{l_{k+1}} \alpha_i^{*k+1} \left(x_i^{k+1} \cdot x_j^k\right) - \sum\nolimits_{i=1}^{l_k} \alpha_i^k \left(x_i^k \cdot x_j^k\right), \tag{3.401}$$

or choose a component of α^{*k+1}, $\alpha^{*k+1} \in (0, C)$, $k = 1, 2$ *and compute*

$$b_k = -1 + \sum\nolimits_{i=1}^{l_{k+1}} \alpha_i^{*k+1} \left(x_i^{k+1} \cdot x_j^{k+1}\right) - \sum\nolimits_{i=1}^{l_k} \alpha_i^k \left(x_i^k \cdot x_j^{k+1}\right). \tag{3.402}$$

Proof First, we show that for w^* given by (3.400), there exists $\overline{b} = \left(\overline{b}^1, \overline{b}^2\right)$ such that $\left(-w, \overline{b}\right)$ is the solution to (3.415). In fact, Theorem 3.5 shows that (3.415) can be rewritten as (3.390). It is easy to see that (3.390) satisfies the Slater condition. Accordingly, if $\alpha^{(*)}$ is a solution to (3.390), there exists a multiplier $\overline{b}, \overline{s}$, and $\overline{\xi}$

such that

$$0 \leq \hat{\lambda} - \leq \hat{C}, \hat{\Omega}\hat{\lambda} = 0-, \tag{3.403}$$

$$-\hat{\Lambda}\hat{\lambda} + \hat{\kappa} + \overline{b}_1 \hat{\Omega}_1^{\mathrm{T}}. + \overline{b}_2 \hat{\Omega}_2^{\mathrm{T}}. - \overline{s} + \overline{\xi} = 0-, \tag{3.404}$$

$$\overline{s} \geq 0, \overline{\xi} \geq 0, \overline{\xi}^{\mathrm{T}} \left(\hat{\lambda} - -\hat{C} \right) = 0, \overline{s}^{\mathrm{T}} \hat{\lambda} - = 0. \tag{3.405}$$

According to (3.404), we have

$$-\hat{\Lambda}\hat{\lambda} - +\hat{\kappa} + \overline{b}_1 \hat{\Omega}_1^{\mathrm{T}}. + \overline{b}_2 \hat{\Omega}_2^{\mathrm{T}}. + \overline{\xi} \geq 0. \tag{3.406}$$

From (30), this is equivalent to

$$\left(\overline{w}_1 \cdot x_i^1 \right) - \overline{b}_1 \leq -1 + \overline{\xi}_i^1, i = 1, \ldots, l_1, \tag{3.407}$$

$$\left(\overline{w}_2 \cdot x_i^2 \right) - \overline{b}_2 \leq -1 + \overline{\xi}_i^2, i = 1, \ldots, l_2, \tag{3.408}$$

$$\left(\overline{w}_1 \cdot x_i^2 \right) - \overline{b}_1 \geq 1 - \overline{\xi}_i^{*2}, i = 1, \ldots, l_2, \tag{3.409}$$

$$\left(\overline{w}_2 \cdot x_i^3 \right) - \overline{b}_2 \geq 1 - \overline{\xi}_i^{*3}, i = 1, \ldots, l_3, \tag{3.410}$$

which implies that $(\overline{w}, \overline{b})$ is a feasible solution to the primal problem Furthermore, we have

$$\begin{aligned}
&\frac{1}{2} \left(\|\overline{w}_1\|_2^2 + \|w_2 - \|_2^2 \right) + C \sum_{i=1}^{l_1} \overline{\xi}_i^1 + C \sum_{i=1}^{l_2} \left(\overline{\xi}_i^2 + \xi_i^{*2} \right) + \mathrm{C} \sum_{i=1}^{l_3} \overline{\xi}_i^{*3} \\
&= \frac{1}{2} \hat{\lambda}^{\mathrm{T}} \hat{\Lambda} \hat{\lambda} + C \sum_{i=1}^{l_1} \overline{\xi}_i^1 + \mathrm{C} \sum_{i=1}^{l_2} \left(\overline{\xi}_i^2 + \xi_i^{*2} \right) + \mathrm{C} \sum_{i=1}^{l_3} \overline{\xi}_i^{*3} + - - \hat{\lambda}^{\mathrm{T}} \\
&\times \left(-\hat{\Lambda}\hat{\lambda} + \hat{\kappa} + \overline{b}_1 \hat{\Omega}_1^{\mathrm{T}}. + \overline{b}_2 \hat{\Omega}_2^{\mathrm{T}}. - \overline{s} + \overline{\xi} \right) - -,, \\
&= -\frac{1}{2} \hat{\lambda}^{\mathrm{T}} \hat{\Lambda} \hat{\lambda} - - + \hat{\kappa}^{\mathrm{T}} \hat{\lambda} - .
\end{aligned} \tag{3.411}$$

This shows that the value of the objective function for the primal problem at $(\overline{w}, \overline{b})$ is equal to the optimum value of its dual problem. Thus, $(\overline{w}, \overline{b})$ is the optimal solution to the primal problem.

If there exists a feasible solution $(\overline{w}, \overline{b})$ of the primal problem, we know that $\hat{\lambda}-$ is nonzero by (30). According to convex duality theory, $(\overline{w}, \overline{b})$ obtained from (3.415) is the unique solution to the primal problem (12). In fact, note that $\hat{\lambda} \neq 0$ implies $s_j^* = 0$ from (35). This implies that the j-th entry of $-\hat{\Lambda}\hat{\lambda} - +\hat{\kappa} + \overline{b}_1\hat{\Omega}_1^{\mathrm{T}}. +\overline{b}_2\hat{\Omega}_2^{\mathrm{T}}. +\overline{\xi}$ is zero. Solving the equation w.r. t.$\overline{b}$ leads to the expressions (3.401) and (3.402).

Nonlinear U-NSVM

Now we extend the linear U-NSVM to the nonlinear case by introducing Gaussian kernel function

$$K\left(x, x'\right) = \Phi(x)\Phi\left(x'\right) \tag{3.412}$$

and the corresponding transformation

$$\mathrm{x} = \Phi(x), \tag{3.413}$$

where x ∈ H, H represents Hilbert space. Thus, the training set becomes

$$\begin{aligned}\widetilde{T}\bigcup\widetilde{U} = &\left\{\left(\Phi\left(x_1^1\right), 1\right), \ldots, \left(\Phi\left(x_{l_1}^1\right), 1\right), \left(\Phi\left(x_1^3\right), -1\right), \left(\Phi\left(x_{l_3}^3\right), -1\right)\right\} \\ &\bigcup\left\{\Phi\left(x_1^2\right), \ldots, \Phi\left(x_{l_2}^2\right)\right\}\end{aligned} \tag{3.414}$$

The nonlinear optimization problem to be solved is

$$\max -\frac{1}{2}\sum_{k=1}^{2}\left(\sum_{i=1}^{l_k}\sum_{j=1}^{l_k}\alpha_i^k\alpha_j^k K\left(x_i^k \cdot x_j^k\right) - 2\sum_{i=1}^{l_k}\sum_{j=1}^{l_{k+1}}\alpha_i^k\alpha_j^{*k+1} K\left(x_i^k \cdot x_j^{k+1}\right)\right.$$

$$\left.+\sum\nolimits_{i=1}^{l_{k+1}}\sum\nolimits_{j=1}^{l_{k+1}}\alpha_i^{*k+1}\alpha_j^{*k+1} K\left(x_i^{k+1} \cdot x_j^{k+1}\right)\right) + \sum\nolimits_{k=1}^{2}\sum\nolimits_{i=1}^{l_k}\alpha_i^k + \sum\nolimits_{k=2}^{3}\sum\nolimits_{i=1}^{l_k}\alpha_i^{*k}$$

$$s.t. \sum\nolimits_{i=1}^{l_k}\alpha_i^k = \sum\nolimits_{i=1}^{l_{k+1}}\alpha_i^{*k+1}, k = 1, 2,$$

$$0 \le \alpha_i^k \le C, k = 1, 2; i = 1, \ldots, l_k,$$

$$0 \le \alpha_i^{*k} \le C, k = 2, 3; i = 1, \ldots, l_k. \tag{3.415}$$

The corresponding theorems in the nonlinear case are similar to Theorems 3.4 and 3.5. In fact, we only need to take $K(x, x^{'})$ instead of $(x, x^{'})$. Now we establish U-NSVM as follows.

(U-NSVM)
(1) Input the training set (3.414);
(2) Choose appropriate kernels $K(x, x^{'})$, appropriate parameters and $C > 0$;
(3) Construct and solve optimization problem (3.415) to obtain the solutions

$$\hat{\lambda} = \left(\alpha_1^1, \ldots, \alpha_{l_1}^1, \alpha_1^{*2}, \ldots, \alpha_{l_2}^{*2}, \alpha_1^2, \ldots, \alpha_{l_2}^2, \ \alpha_1^{*3}, \ldots, \alpha_{l_3}^{*3}\right) \tag{3.416}$$

(4) Construct the decision functions

$$f_1(x) = \sum\nolimits_{i=1}^{l_2} \alpha_i^{*2} K\left(x_i^2 \cdot x\right) - \sum\nolimits_{i=1}^{l_1} \alpha_i^1 K\left(x_i^1 \cdot x\right) - b_1, \tag{3.417}$$

$$f_2(x) = \sum\nolimits_{i=1}^{l_3} \alpha_i^{*3} K\left(x_i^3 \cdot x\right) - \sum\nolimits_{i=1}^{l_2} \alpha_i^2 K\left(x_i^2 \cdot x\right) - b_2, \tag{3.418}$$

where $b.$, $b+$ are computed according to Theorem 3.1 and Theorem 3.2 for the kernel cases;
(5) For any new input x, assign it to class $k(k = 1, 2)$ according to

$$\arg\ \min \frac{|\ f_k(x)\ |}{\|\Delta_k\|}, \tag{3.419}$$

where

$$\Delta_1 = \hat{\lambda}_1^{\mathrm{T}} Q^1 \hat{\lambda}_1, \Delta_2 = \hat{\lambda}_2^{\mathrm{T}} Q^2 \hat{\lambda}_2, \tag{3.420}$$

and

$$\hat{\lambda}_1 = \left(\alpha_1^1, \ldots, \alpha_{l_1}^1, \alpha_1^{*2}, \ldots, \alpha_{l_2}^{*2}\right), \tag{3.421}$$

$$\hat{\lambda}_2 = \left(\alpha_1^2, \ldots, \alpha_{l_2}^2, \alpha_1^{*3}, \ldots, \alpha_{l_3}^{*3}\right). \tag{3.422}$$

3.3.2 A Divide-and-Combine Method for Large Scale Nonparallel Support Vector Machines

3.3.2.1 NPSVM

Consider the binary classification problem with the training set

$$T = \left\{(x_1, +1), \ldots, (x_p, +1), \left(x_{p+1}, -]\right), \ldots, (x_{\rho+q}, -1)\right\}, \tag{3.423}$$

where $x_i \in R^n$, $i = 1, \ldots, p + q$, Let$A = (x_1, \ldots, x_p)^T \in R^{p \times n}$, $B = (x_{p+1}, \ldots, x_{\rho+q})^T \in R^{q \times n}$, and $n = p + q$. NPSVM seeks two nonparallel hyperplanes

$$(w_+ \cdot x) + b_+ = 0 \text{ and } (w_- \cdot x) + b_- = 0 \tag{3.424}$$

by solving two convex quadratic programming problems (QPPs):

$$\min \frac{1}{2}\|w_+\|^2 + C_1 \sum\nolimits_{1=1}^{p} (\eta_i + \eta_i^*) + C_2 \sum\nolimits_{j=p+1}^{\rho+q} \xi_j,$$

$$s.t. \quad (w_+ \cdot x_i) + b_+ \leq \varepsilon + \eta_i, i = 1, \ldots, p,$$

$$-(w_+ \cdot x_i) - b_+ \leq \varepsilon + \eta_j^*, i = 1, \ldots, p,$$

$$\left(w_+ \cdot x_j\right) + b_+ \leq -1 + \xi_j,$$

$$j = p + 1, \ldots, p + q,$$

$$\eta_i, \eta_i^* \geq 0, i = 1, \ldots, p,$$

$$\xi_j \geq 0, j = p + 1, \ldots, p + q, \tag{3.425}$$

and

$$min \frac{1}{2}\|w_-\|^2 + C_3 \sum\nolimits_{i=p+1}^{p+q} (\eta_i + \eta_i^*) + C_4 \sum\nolimits_{j=1}^{p} \xi_j,$$

$$s.t. \quad (w_- \cdot x_i) + b_- \leq \varepsilon + \eta_i,$$

$$i = p + 1, \ldots, p + q,$$

$$
\begin{aligned}
&-(w_-\cdot x_\mathrm{i}) - b_- \leq \varepsilon + \eta_j^*,\\
&i = p+1, \ldots, \mathrm{p}+q,\\
&\left(w_-\cdot x_j\right) + b_- \geq 1 - \xi_j,\ j = 1, \ldots, p,\\
&\eta_i, \eta_i^* \geq 0, i = p+1, \ldots, p+q,\\
&\xi_i \geq 0, j = 1, \ldots, p,
\end{aligned} \tag{3.426}
$$

where x_i, $i = 1, \ldots,$ p are positive inputs, and x_i, $i = \mathrm{p}+1, \ldots, \mathrm{p}+q$ are negative inputs, $\mathrm{C}_i \geq 0$, $i = 1, \ldots, 4$ are penalty parameters, $\xi_+ = (\xi_1, \ldots, \xi_\mathrm{p})^\mathrm{T}$, $\xi_- = (\xi_{\mathrm{p}+1}, \ldots, \xi_{p+\mathrm{q}})^\mathrm{T}$, $\eta_+^{(*)} = (\eta_+^\mathrm{T}, \eta_+^{*\mathrm{T}})^\mathrm{T} = \left(\eta_1, \ldots, \eta_\rho, \eta_1^*, \ldots, \eta_\mathrm{p}^*\right)^\mathrm{T}$, $\eta_-^{(*)} = (\eta_-^\mathrm{T}, \eta_-^{*\mathrm{T}})^\mathrm{T} = \left(\eta_{p+1}, \ldots, \eta_{\mathrm{p+q}}, \eta_{\rho+1}^*, \ldots, \eta_{p+\mathrm{q}}^*\right)^\mathrm{T}$, are slack variables.

In order to get the solutions of problems (3.425) and (3.426), we need to solve their dual problems:

$$
\begin{aligned}
&\min \frac{1}{2}\theta^\mathrm{T} \Lambda\theta + \kappa^\mathrm{T}\theta,\\
&\mathrm{st.}e^\mathrm{T}\theta = 0,\\
&0 \leq \theta \leq \overline{\mathrm{C}},
\end{aligned} \tag{3.427}
$$

where

$$
\begin{aligned}
&\theta = \left(\alpha_+^{*\mathrm{T}}, \alpha_+^\mathrm{T}, \beta_-^\mathrm{T}\right)^\mathrm{T}, \kappa = \left(\varepsilon e_+^\mathrm{T}, \varepsilon e_+^\mathrm{T}, -e_-^\mathrm{T}\right)^\mathrm{T},\\
&e = \left(-e_+^\mathrm{T}, e_+^\mathrm{T}, -e_-^\mathrm{T}\right)^\mathrm{T},\ \overline{\mathrm{C}} = \left(\mathrm{C}_1 e_+^\mathrm{T}, \mathrm{C}_1 e_+^\mathrm{T}, \mathrm{C}_2 e_-^\mathrm{T}\right)^\mathrm{T},\\
&\Lambda = \begin{pmatrix} H_1 & \mathrm{H}_2 \\ H_2^\mathrm{T} & H_3 \end{pmatrix}, H_1 = \begin{pmatrix} K(A, A)^\mathrm{T} & -K(A, A)^\mathrm{T} \\ -K(A, A)^\mathrm{T} & K(A, A)^\mathrm{T} \end{pmatrix},\\
&\mathrm{H}_2 = \begin{pmatrix} K(A, B)^\mathrm{T} \\ -K(A, B)^\mathrm{T} \end{pmatrix}, \mathrm{H}_3 = K(B, B)^\mathrm{T},
\end{aligned} \tag{3.428}
$$

and

$$\min \frac{1}{2}\gamma^{\mathrm{T}}\Lambda \mathrm{y}+\kappa^{\mathrm{T}}\mathrm{y},$$
$$s.t.\ e^{\mathrm{T}}\gamma=0, \tag{3.429}$$
$$0\leq\gamma\leq\overline{\mathrm{C}},$$

where

$$y=\left(\alpha_{-}^{*\mathrm{T}},\alpha_{-}^{\mathrm{T}},\beta_{+}^{\mathrm{T}}\right)^{\mathrm{T}},\kappa=\left(\varepsilon e_{-}^{\mathrm{T}},\varepsilon e_{-}^{\mathrm{T}},-e_{+}^{\mathrm{T}}\right)^{\mathrm{T}},$$
$$e=\left(-e_{-}^{\mathrm{T}},e_{-}^{\mathrm{T}},-e_{+}^{\mathrm{T}}\right)^{\mathrm{T}},\ \ \overline{\mathrm{C}}=\left(\mathrm{C}_3e_{-}^{\mathrm{T}},\mathrm{C}_3e_{-}^{\mathrm{T}},\mathrm{C}_4e_{+}^{\mathrm{T}}\right)^{\mathrm{T}},$$
$$\Lambda=\begin{pmatrix}\mathrm{Q}_1 & \mathrm{Q}_2\\ \mathrm{Q}_2^{\mathrm{T}} & \mathrm{Q}_3\end{pmatrix},\mathrm{Q}_1=\begin{pmatrix}\mathrm{B}K(B,)^{\mathrm{T}} & B-K(B,)^{\mathrm{T}}\\ -K(B,\mathcal{B})^{\mathrm{T}} & K(B,B)^{\mathrm{T}}\end{pmatrix},$$
$$\mathrm{Q}_2=\begin{pmatrix}K(B,A)^{\mathrm{T}}\\ -K(B,A)^{\mathrm{T}}\end{pmatrix},\mathrm{Q}_3=K(A,A)^{\mathrm{T}}, \tag{3.430}$$

then construct the decision functions

$$f_{+}(x)=\sum\nolimits_{1=1}^{p}\left(\alpha_{\mathrm{i}}^{*}-\alpha_{\mathrm{i}}\right)K\left(x_1,x\right)-\sum\nolimits_{j=p+1}^{p+\mathrm{q}}\beta_jK\left(x_j,x\right)+b_{+}, \tag{3.431}$$

and

$$f_{-}(x)=.\sum\nolimits_{i=p+1}^{p+q}\left(\alpha_{i^*}-\alpha_1\right)K\left(x_i,x\right)+\sum\nolimits_{j=1}^{p}\beta_jK\left(x_{j^{\cdot}},x\right)+b_{-}, \tag{3.432}$$

separately, a new point $x\in R^{\mathrm{n}}$ is therefore predicted to the class $k(k=-,+)$ by

$$\arg\ \min\ \frac{|\ f_k(x)\ |}{\|\Delta_k\|}, \tag{3.433}$$

where

$$\Delta_{+}=\theta^{\mathrm{T}}\Lambda\theta,\Delta_{-}=\gamma^{\mathrm{T}}\Lambda\gamma. \tag{3.434}$$

3.3.2.2 A Divide-and-Combine NPSVM Solver with a Single Level

In this section we present DCNPSVM with a single level. As a first step, we divide the full samples into smaller subsets $\{v_1, \ldots, v_k\}$, and then solve the respective subproblems of (3.355) and (3.358) independently.

$$\min \frac{1}{2}\theta_{(c)}^{\mathrm{T}} \Lambda_{(c)}\theta_{(c)} + \kappa^{\mathrm{T}}\theta_{(c)},$$
$$\mathrm{st.} e^{\mathrm{T}}\theta_{(c)} = 0, \tag{3.435}$$
$$0 \leq \theta_{(c)} \leq \overline{\mathrm{C}},$$

and

$$\min \frac{1}{2}\mathrm{y}_{(c)}^{\mathrm{T}} \Lambda_{(c)}\mathrm{y}_{(c)} + \kappa^{\mathrm{T}}\mathrm{y}_{(c)},$$
$$s.t.\ e^{\mathrm{T}}\gamma_{(c)} = 0, \tag{3.436}$$
$$0 \leq \gamma_{(c)} \leq \overline{\mathrm{C}},$$

where $=1, \ldots, k$, $\theta_{(c)}$ and $\gamma_{(c)}$ denotes sub-vector obtained by the c-th subsets. $\Lambda_{(c)}$ is the sub-matrix of A with row and column indexed by subsets.

The training time complexity for the two convex QPPs (5) and (8), an SMO-type decomposition method [68] implemented in LIBSVM has the complexity

$$\#\text{iterations} \times O(1.5n). \tag{3.437}$$

Chang and Chih-Jen [69] also pointed out that there is no theoretical result yet on LIBSVM's number of iterations, however, empirically, it is known that the number of iterations may be higher than linear to the number of training data. Supposed that the number of iterations is the number of training data and removing all coefficients, the time complexity for solving two QPPs is $\mathrm{O}(n^2)$. By dividing the whole samples into k subproblems with almost equal sizes and solved by (15) and (16), the time complexity for solving the subproblems can be reduced to $\mathrm{O}\left(\frac{n^2}{\mathrm{k}}\right)$. Besides, DCNPSVM enhances flexibility and generalization of DCSVM, so it can build models more accurately than existing scale-up SVM models. Therefore, DCNPSVM can reduce time complexity and enhance accuracy for classification.

With the solutions obtained from all the subproblems, we combine them to form initial solvers $\theta = [\theta_{(1)}, \ldots, \theta_{(k)}]$ and $\gamma = [\gamma_{(1)}, \ldots, \gamma_{(k)}]$, where $\theta_{(c)}$ and $\gamma_{(c)}$ are the optimal solutions for the cth subproblem for the whole problem. Subsequently, θ and γ are used initial solvers for the whole problem. Here we give a toy experiment which applies NPSVM and DCNPSVM using RBF kernel on iris with two main

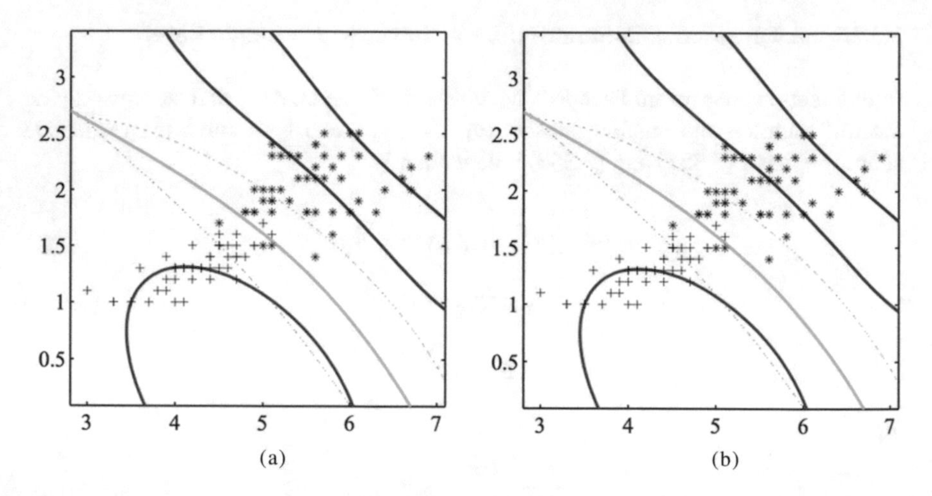

Fig. 3.8 Hyperplanes comparison between (**a**) NPSVM and (**b**) DCNPSVM on iris data set

features in R^2. Figure 3.8 shows the hyperplanes obtained from the two methods. The hyperplanes are almost the same and their solutions are only different in thousandth.

The results above are expected due to the following reasons: (1) θ and γ are close to the optimal solutions for the whole problem $\theta*$ and $\gamma*$; (2) the objective function is close to the optimal objective function; (3) the set of support vectors of the subproblems are close to the set of support vectors of the whole problem. Now we establish theoretical bounds on the difference between solutions, objective functions glued by subproblems and the whole problems. Without loss of generality, we take the second QPP as an example.

Lemma 3.4 *$\overline{\gamma}$ is the* optimal *solution of* (8) *with kernel matrix $\Lambda(x_{\mathrm{i}}, x_j)$ replaced by*

$$\overline{\Lambda}_{\mathrm{i}j} = I\left(\pi\left(x_{\mathrm{i}}\right), \pi\left(x_{\mathrm{j}}\right)\right)\Lambda_{ij}, \tag{3.438}$$

where $\pi(x_i)$ is the cluster that x_{i} belongs to; *$I(a, b) = 1$ iff $a = b$ and* I (a,*b*) =0 *otherwise.*

Proof By clustering, the matrix Λ in (3.429) becomes $\overline{\Lambda}$ as follows.

$$\overline{\Lambda}_{ij} = \begin{cases} \Lambda_{\mathrm{i}j}, \text{if } \pi\left(x_{\mathrm{i}}\right) = \pi\left(x_j\right) \\ 0, \text{if } \pi\left(x_{\mathrm{i}}\right) \neq \pi\left(x_j\right). \end{cases} \tag{3.439}$$

The quadratic term in becomes

$$\gamma^{\mathrm{T}}\Lambda\gamma = -\sum\nolimits_{c=1}^{\mathrm{k}} \gamma_{(\mathrm{c})}^{\mathrm{T}}\Lambda_{(\mathrm{c})}\gamma_{(c)}. \tag{3.440}$$

Meanwhile, other terms in (3.429) are changed. The subproblems are independent and the composite of their optimal solutions is the optimal solution of the whole problem in when Λ is replaced by $\overline{\Lambda}$.

Theorem 3.6 *Given data points* $x_1, \ldots, x_n$ *and a partition indicator* $\{\pi(x_1), \ldots, \pi(x_n)\}$,

$$0 \leq f(\overline{\gamma}) - f(\gamma^*) \leq (1/2)\, C^2 D(\pi) \tag{3.441}$$

Where $f(\gamma)$ *is the objective function in* (3.429), $\overline{\gamma}$ *is as in Lemma 3.4,* γ^* *is the global optimal of* (3.429) *and* $D(\pi) = \sum_{i,j:\pi(x_i)\neq\pi(x_j)} \mid \Lambda(x_i, x_j) \mid$, $C = \max\{C_3, C_4\}$. *Furthermore,* $\|\gamma^* - \overline{\gamma}\|_2^2 \leq C^2 D(\pi) / \mid \sigma_n \mid$, *where* σ_n *is the smallest eigenvalue of the kernel matrix.*

Proof We use $\overline{f}(\gamma)$ to denote the objective function of (3.429) with kernel $\overline{\Lambda}$. By Lemma 3.4, $\overline{\gamma}$ is the minimizer of (3.429) with kernel Λ replaced by Λ, thus $\overline{f}(\overline{\gamma}) \leq \overline{f}(\gamma^*)$.

$$\overline{f}(\gamma^*) = f(y^*) - \frac{1}{2}\sum\nolimits_{i,j:\pi(x_i)\neq\pi(x_j)} \gamma_i^* \gamma_j^* \Lambda_{ij}. \tag{3.442}$$

Similarly, we have

$$\overline{f}(\overline{\gamma}) = f(\overline{\gamma}) - \frac{1}{2}\sum\nolimits_{i,j:\pi(x_i)\neq\pi(x_j)} \overline{\gamma}_i \overline{\gamma}_j \Lambda_{ij}. \tag{3.443}$$

Combining with $\overline{f}(\overline{\gamma}) \leq \overline{f}(\gamma^*)$ we have

$$f(\overline{\gamma}) \leq \overline{f}(\gamma^*) + \frac{1}{2} \sum_{i,j:\pi(x_i)\neq\pi(x_j)} \overline{\gamma}_i \overline{\gamma}_j \Lambda_{ij}$$

$$= f(\gamma^*) + \frac{1}{2}\sum\nolimits_{i,j:\pi(x_i)\neq\pi(x_j)} \left(\overline{\gamma}_i \overline{\gamma}_j - \gamma_i^* y_j^*\right) \Lambda_{ij}$$

$$\leq f(\gamma^*) + \frac{1}{2} C^2 D(\pi), \tag{3.444}$$

where $C \in \max\{C_3, C_4\}$. Also, since γ^* is the optimal solution of (8) and $\overline{\gamma}$ is a feasible solution, $(\gamma^*) < f(\overline{\gamma})$, thus proving the first part ofthe theorem.

Let σ_n be the smallest singular value of the positive definite kernel matrix Λ. Suppose we write $\overline{\gamma} = \gamma + \Delta\gamma$,

$$f(\overline{\gamma}) = f(\gamma^*) + (\gamma^*)^T \Lambda \Delta\gamma + \frac{1}{2}(\Delta\gamma)^T \Lambda \Delta\gamma + \kappa^T \Lambda\gamma. \tag{3.445}$$

If $i \in \{1, \ldots, 2q\}$, the optimality condition for is

$$\nabla_{\mathrm{i}} f\left(\gamma^{*}\right)\begin{cases} = 0 \text{ if } 0 < \gamma_j^* < c_3, \\ \geq 0 \text{ if } \gamma_{\mathrm{i}}^* = 0, \\ \leq 0 \text{ if } \gamma_j^* = C_3, \end{cases} \tag{3.446}$$

where $\nabla_{\mathrm{i}} f(\gamma_*) = \Lambda\gamma_* + \kappa$. Since $\overline{\gamma}$ is a feasible solution, it is easy to see that $(\Delta\gamma)_i > 0$ if $\gamma_{\mathrm{i}}^* = 0$, and $(\Delta\gamma^*)_{\mathrm{i}} < 0$ if $\gamma_{\mathrm{i}}^* = C_3$.

If $\mathrm{i} \in \{2q+1, \ldots, 2q+\mathrm{p}\}$, the optimality condition for (3.429) is

$$\nabla_i f\left(y^{*}\right)\begin{cases} = 0 \text{ if } 0 < \gamma_1^* < C_4, \\ \geq 0 \text{ if } \gamma_i^* = 0, \\ \leq 0 \text{ if } \gamma_{\mathrm{i}}^* = C_4, \end{cases} \tag{3.447}$$

where $\nabla_{\mathrm{i}} f(\gamma_*) = \Lambda\gamma_* + \kappa$. Since $\overline{\gamma}$ is a feasible solution, it is easy to see that $(\Delta\gamma)_i > 0$ if $\gamma_{\mathrm{i}}^* = 0$, and $(\Delta\gamma)_i \leq 0$ if $\gamma_{\mathrm{i}}^* = C_4$.

Thus,

$$(\Delta\gamma)^T\left(\Lambda\gamma^* + \kappa\right) = \sum\nolimits_{1=1}^{2\mathrm{q}} (\Delta\gamma)_{\mathrm{i}}\left(\left(\Lambda\gamma^*\right)_{\mathrm{i}} + \varepsilon\right)$$

$$+\sum\nolimits_{\mathrm{i}=2\mathrm{q}+1}^{2\mathrm{q}+\rho} (\Delta\gamma)_{\mathrm{i}} \cdot \left(\left(\Lambda\gamma^*\right)_{\mathrm{i}} - 1\right). \tag{3.448}$$

Combining with (3.445), we have $f(\overline{\gamma}) > f(\gamma^*) + \frac{1}{2}\Delta\gamma^T\Lambda\Delta\gamma > f(\gamma^*) + \frac{1}{2}\sigma_{\mathrm{n}}\|\Delta\gamma\|_2^2$. Since we already know that $(\overline{\gamma}) \leq f(\gamma^*) + \frac{1}{2}\mathrm{C}^2 D(\pi)$, this implies $\|\gamma^* - \overline{\gamma}\|_2^2 \leq \mathrm{C}^2 D(\pi) / \mid \sigma_n \mid$.

In Theorem 3.6, in order to make $f(\overline{\gamma})$ close to $f(\gamma_*)$, and $\overline{\gamma}$ close to γ_*, we want to find a partition with small $D(\pi)$ and faster training speed. Kernel kmeans algorithm can minimize the off-diagonal values of the kernel matrix. In addition, each partition should include "positive" samples and "negative" samples aiming at solving by NPSVM. Therefore, after partitioning with kernel k-means algorithm, we combine partitions including no "positive" samples with including no "negative" samples. This can further make the upper bound tight. In order to show the efficiency of the upper bound, we show an experiment result on a subset of the Covtype data set. The number of clusters is $k = 4,16,64,128$ and for each cluster, we compute the upper bound $\mathrm{C}^2 D(\pi)/2$ and $(\overline{\gamma}) - f(\gamma^*)$. The results present that the upper bound is quite close to the difference in objectives and therefore our this strategy can lead to good approximates to global NPSVM problem.

Another important conclusion is that the support vectors from subproblems are very close to the support vectors of the whole problems. We define the set of support vectors from subproblems as $\overline{S}$ and the set of support vectors of the whole problems as S. Theorem 3.7 shows that if x_{i} is not a support vector of the subproblem, and then x_{i} will not be a support vector of the whole problem.

Theorem 3.7 *For any* $i \in \{1, \ldots, \mathrm{q}\}$, *if* $\overline{\gamma}_i = 0$, $\gamma_{i+\mathrm{q}}^{-} = 0$ *and*

$$\nabla_i \overline{f}\,(\overline{\gamma}) > CD\,(\pi)\left(-\varepsilon + \sqrt{n}\Lambda_{\max}/\sqrt{\sigma_n D\,(\pi)}\right), \tag{3.449}$$

$$\nabla_{\mathrm{i+q}} \overline{f}\,(\overline{\gamma}) > CD\,(\pi)\left(-\varepsilon + \sqrt{\mathrm{n}}\Lambda_{\max}/\sqrt{\sigma_{\mathrm{n}} D\,(\pi)}\right), \tag{3.450}$$

Or for any $i \in \{2\mathrm{q} + 1, \ldots, 2q + p\}$, *if* $\overline{\gamma}_1 = 0$ *and*

$$\nabla_i \overline{f}\,(\overline{\gamma}) > CD\,(\pi)\left(1 + \sqrt{\mathrm{n}}\Lambda_{\max}/\sqrt{\sigma_{\mathrm{n}} D\,(\pi)}\right), \tag{3.451}$$

where $\Lambda_{\max} = \max_j \Lambda(x_i, x_j)$, *then* x_i *will not be a support vector of the whole problem.*

Proof Let $\Delta\Lambda = \Lambda - \overline{\Lambda}$ and $\Delta\gamma = \gamma^* - \overline{\gamma}$. From the optimality condition for (8) (see (3.446) and (3.447)), we know that $\gamma_i^* = 0$ if $(\Lambda\gamma^*)_{\mathrm{i}} > -\varepsilon$ which $i \in \{\mathrm{i}, \ldots, 2\mathrm{q}\}$, and $\gamma_{\mathrm{i}}^* = 0$ if $(\Lambda\gamma^*)_{\mathrm{i}} > \mathrm{i}$ which $i \in \{2\mathrm{q} + \mathrm{i}, \ldots, 2\mathrm{q} + p\}$. Since $\gamma^* = \left(\overline{\Lambda} + \Delta\right)(\overline{\gamma} + \Delta\gamma)$, we see that

$$
\begin{aligned}
\left(\Lambda\gamma^*\right)_{\mathrm{i}} &= \left(\overline{\Lambda}\overline{\gamma}\right)_{\mathrm{i}} + (\Delta\Lambda\overline{\gamma})_{\mathrm{i}} + (\Lambda\Delta\gamma)_{\mathrm{i}} \\
&= \left(\overline{\Lambda}\overline{\gamma}\right)_{\mathrm{i}} \cdot + \sum_{i,j:\pi(x_i)\neq\pi(x_j)} K\left(x_{\mathrm{i}}, x_{\mathrm{j}}\right)\overline{\gamma}_j \\
&\quad + \sum_j K\left(x_{\mathrm{i}}, x_j\right)(\Delta\gamma)_j \\
&\geq \left(\overline{\Lambda}\overline{\gamma}\right)_{\mathrm{i}} - CD\,(\pi) - K_{\max}\|\Delta\gamma\|_1 \\
&\geq \left(\overline{\Lambda}\overline{\gamma}\right)_i - CD\,(\pi) - \sqrt{n}K_{\max}\mathrm{C}\sqrt{D\,(\pi)}/\sqrt{\sigma_{\mathrm{n}}} \\
&= \left(\overline{\Lambda}\overline{\gamma}\right)_{\mathrm{i}} - CD\,(\pi)\left(\mathrm{i} + \frac{\sqrt{\mathrm{n}}K_{\max}}{\sqrt{\sigma_{\mathrm{n}} D\,(\pi)}}\right)
\end{aligned}
\tag{3.452}
$$

The condition stated in the theorem implies for any $i \in \{1, \ldots, q\}$, $\left(\overline{\Lambda}\overline{\gamma}\right)_{\mathrm{i}} > -\varepsilon + CD\,(\pi)\left(1 + \frac{\sqrt{n}\mathrm{K}_{\max}}{\sqrt{\sigma_{\mathrm{n}} D(\pi)}}\right)$, and $\nabla_{\mathrm{i+q}} \overline{f}\,(\overline{\gamma}) > CD\,(\pi)\left(-\varepsilon + \sqrt{n}\Lambda_{\max}/\sqrt{\sigma_{\mathrm{n}} D\,(\pi)}\right)$

which implied $(\Lambda)^{i^*}\big)_{\mathrm{i}} + \varepsilon > 0$, and $(\Lambda\gamma^*)_{\mathrm{i}+\mathrm{q}} + \varepsilon > 0$, so from the optimality condition (3.446)

$\gamma_{\mathrm{i}}^* = 0$ and $\gamma_{i+\mathrm{q}}^* = 0$. And for any $\in\{2\mathrm{q} + \mathrm{i}, \ldots, 2\mathrm{q} + p\}$, $\left(\overline{\Lambda\gamma}\right)_{\mathrm{i}} > 1 + CD(\pi)\left(\mathrm{i} + \frac{\sqrt{\mathrm{n}}K_{\max}}{\sqrt{\sigma_{\mathrm{n}}D(\pi)}}\right)$, which implied $(\Lambda\gamma^*)_{\mathrm{i}} - 1 > 0$, so from the optimality condition (3.447) $\gamma_{\mathrm{i}}^* = 0$.

In order to illustrate the efficiency of Theorem 3.7, we also show an experiment result on a subset of the Covtype data set. Figure 3.8 demonstrate DCNPSVM can find support vectors effectively and efficiently.

After getting solutions from subproblems, we carry on the combination step. With the solutions obtained from all the subproblems, we combine them to form an initial solver $=[\gamma_{(\mathrm{i})}, \ldots, \gamma_{(\mathrm{k})}]$, where $\gamma_{(c)}$ is the optimal solution for the cth subproblem for the whole problem. Subsequently, γ is used as initial solver to solve the whole problem by global coordinate descent which can avoid unnecessary computing if γ_i never changes from zero to nonzeros so this algorithm can converge quickly.

3.3.2.3 Divide and Combine NPSVM with Multiple Levels

In divide-and-combine NPSVM with a single level, there is a trade-off in selecting the number of clusters k. On the one hand, when k is small, the solutions and objective function obtained by the subproblems are close to the optimal solutions and objective function according to Theorem 3.6, however the training time is very long. On the other hand, if we increase k, although the training time reduces, the difference between solutions and objective function is large. Therefore, we adopt a multiple structure used in [70] to avoid this situation.

Now we introduce the main idea of divide-and-combine NPSVM with multiple levels. At the lth level, we divide the whole samples into k^{l} sub-samples $\left\{v_{\mathrm{i}}^{(\mathrm{l})}, \ldots, v_{k^{\mathrm{l}}}^{(\mathrm{l})}\right\}$, and solve each sub-sample with NPSVM to get $\overline{\gamma}^{(\mathrm{l})}$ independently. In a higher level, we choose a bigger k^{l} aiming to get rough solutions in a short time. Subsequently, we use the solutions from the higher level $\overline{\gamma}^{(\mathrm{l}+1)}$ to initial the solver at the lth level and therefore each level requires few iterations.

In this section, we also adopt an "adaptive clustering" method used in [70] to achieve fast kernel k-means algorithm. The time complexity of kernel k-means algorithm is $\mathrm{O}(\mathrm{n}^2 d)$, where d is the feature of the samples. This algorithm takes too much time for large scale problems so a two-step kernel k-means approach is considered. First of all, the two-step kernel k-means approach run kernel k-means on m ($\mathrm{m} \ll n$) random samples to construct cluster centers. Then, other samples are partitioned based on the distance each sample to cluster centers and decide which cluster they belong to. The time complexity of this approach is $0(\mathrm{mn}d)$ and therefore m cannot be too large. However, when we solve problems with large scale data sets, smaller m cannot represent the whole data sets efficiently and the performance of two-step kernel k-means may not be good. Therefore, we adopt an "adaptive clustering" method for clustering. The main idea is to utilize the sparsity of γ in NPSVM efficiently and perform two-step kernel k-means algorithm on the set of the support vectors. At the lth level, we suppose the current set of support vectors defined by S and the set of support vector of the final solution defined by $S*$. Then we can define the sum of off-diagonal elements on $S^* \cup \overline{S}$ as

$D_{S^*\cup\overline{S}}(\pi) = \sum_{i,j\in S^*\cup\overline{S} \text{ and } \pi(x_i)\neq\pi(x_j)} \mid \mathrm{K}\left(\mathrm{x}_i, \mathrm{x}_j\right) \mid$. Therefore, we can refine the bound in as the following theorem.

Theorem 3.8 *Given data points* $x_1, \ \ldots, \ x_n$ *and a partition* $\{v_1, \ldots, v_k\}$ *with indicators* π,

$$0 < f\left(\overline{\gamma}\right) - f\left(\gamma^*\right) \leq (1/2)\,\mathrm{C}^2\mathrm{D}_{S^*\cup\overline{S}}(\pi)\,. \tag{3.453}$$

$$\textit{Furthermore, } \left\|\gamma^* - \overline{\gamma}\right\|_2^2 \leq \mathrm{C}^2 D_{S^*\cup\overline{S}}(\pi) \,/\mid \sigma_\mathrm{n}\mid.$$

Proof Similar to the proof in Theorem 3.6, we use $\overline{f}(\gamma)$ to denote the objective function of (3.429) with kernel K. Combine (24) with the fact that $\gamma_\mathrm{i}^* = 0, \forall i \notin S*$ and $\overline{\gamma}_\mathrm{i}, \forall i \notin S$, we have

$$\overline{f}\left(\gamma^*\right) \leq f\left(\gamma^*\right) - \frac{1}{2}\sum\nolimits_{i,j\in S^* and \ \pi(x_\mathrm{i})\neq\pi(x_\mathrm{j})}\left(\overline{\gamma}_\mathrm{i}\cdot\overline{\gamma}_\mathrm{j} - \gamma_\mathrm{i}^*\gamma_\mathrm{j}^*\right)\Lambda_\mathrm{ij}$$

$$\leq f\left(\gamma^*\right) + \frac{1}{2}C^2 D\left(\{x_\mathrm{i}\cdot\}_{i\in S^*\cup\overline{S}}, \pi\right), \tag{3.454}$$

where $\mathrm{C} \in \max\{C_3, C_4\}$. The second part ofthe proofis exactly the same as the second part of Theorem 3.6.

In practice, at the $l-$ 1-th level, we do not know $\overline{S}$ and $S*$ before solving the problems. However, we give a good guess of support vectors at lth level based on both Theorem 3.7 and experiments as shown, so we can use the support vectors at lth level to run two-step kernel k-means for finding the clusters at the l–1-th level. We first run LIBSVM for NPSVM to obtain the final set of support vectors $S*$ and then run DCNPSVM with this multiple structure.

We use an "early prediction" framework used in [70] to predict a new sample too. From Lemma 3.4, $\overline{\gamma}$ is the optional to the second NPSVM dual problem (8) on the whole data sets with the approximated kernel $\overline{\Lambda}$ defined in so the same kernel function in the testing phase is used, which leads to the prediction for the second

QPP in NPSVM

$$\sum_{c=1}^{k}\sum_{\mathrm{i}\in v_c}\left(\sum_{1}(\alpha_i^*-\alpha_i)\overline{\Lambda}(\mathrm{x_i},x)+\sum_{\mathrm{i}}\beta_\mathrm{i}\overline{\Lambda}(\mathrm{x_i},\mathrm{x})+b_-\right)$$

$$=\sum_{\mathrm{i}\in v_{\pi(\mathrm{x})}}\left(\sum_{\mathrm{i}}(\alpha_i^*-\alpha_i)\right)\left(\Lambda(\mathrm{x_i},x)+\sum_{\mathrm{i}}\beta_\mathrm{i}\cdot\Lambda(x_\mathrm{i},x)+b_-\right)\quad(3.455)$$

The testing phase for "early prediction" is that (1) find the cluster x belongs to; (2) use the model obtained by data within that cluster to compute the two decision values. and (3) x belongs to the label which the smaller decision absolute value belongs to. This approach can reduce the testing time from O(| S| d) to O(| S| d/k) where S is the set of support vectors.

Algorithm 3.9 DCNPSVM

Input: Training data sets $\{x_i, y_i\}$, $i = 1\cdots n$, parameters C_1, C_2, C_3, C_4, and kernel function parameter.

Output: The NPSVM dual solutions θ and.

for $l = l^{\max}, \cdots, 1$ **do**

Set number of clusters in the current level $k_l = l^l$;

if $l = l^{\max}$ **then**

Sample m points $\{x_{i_1}, \cdots, x_{i_m}\}$ from the whole training set;

else

Sample m points $\{x_{i_1}, \cdots, x_{i_m}\}$ from $\overline{S}$;

end

Run kernel k-means on $\{x_{i_1}, \cdots, x_{i_m}\}$ to get cluster centers and obtain partition $\{v_1, \cdots, v_{k^l}\}$ for all samples;

for $c = 1, \ldots, k^l$ **do**

Calculate the number of positive points n_{c+}^l and the number of negative points n_{c-}^l, and then set parameters $\varepsilon_{c_+}^l$ and ε_{c-}^l.

Obtain $\overline{\theta}_{v_c}^{(l)}$ and $\overline{\gamma}_{v_c}^{(l)}$ by solving NPSVM for the data in the c-th cluster v_c with $\overline{\theta}_{v_c}^{(l+1)}$ and $\overline{y}_{v_c}^{(l+1)}$ as the initial points.

end

end

Solve NPSVM on the whole data using $\theta^{(1)}$ and $\gamma^{(1)}$ as the initial points.

3.3.3 Nonparallel Support Vector Machines for Pattern Classification

In this section, we propose a novel nonparallel SVM, termed NPSVM for binary classification. NPSVM has the following incomparable advantages.

1. The semi-sparseness is promoted to the whole sparseness.
2. The regularization term is added naturally due to the introduction of ε-insensitive loss function, and two primal problems are constructed implementing the SRM principle.
3. The dual problems of these two primal problems have the same advantages as that of the standard SVMs, i.e., only the inner products appear so that the kernel trick can be applied directly.
4. The dual problems have the same formulation with that of standard SVMs and can certainly be solved efficiently by SMO, we do not need to compute the inverses of the large matrices as TWSVMs usually do.
5. The initial TWSVM or improved TBSVM are the special cases of our models. Our NPSVM degenerates to the initial TWSVM or TBSVM when the parameters of which are appropriately chosen, therefore, our models are certainly superior to them theoretically.

3.3.3.1 NPSVM

Now we propose our nonparallel SVM, termed as NPSVM, which has several unexpected and incomparable advantages compared with the existing TWSVMs.

Linear NPSVM

We seek the two nonparallel hyperplanes by solving two convex QPPs

$$
\begin{aligned}
\min \tfrac{1}{2}\|w_+\|^2 + C_1\textstyle\sum_{j=1}^{p}\left(\eta_i + \eta_i^*\right) + C_2.\textstyle\sum_{j=p+1}^{p+q}\xi_j \\
s.t. \left(w_+ x_j\right) + b_+ \le \varepsilon + \eta_j, i = 1, \cdots p \\
-\left(w_+ \cdot x_j\right) - b_+ \le \varepsilon + \eta_j^*, i = 1, \cdots \\
\left(w_+ \cdot x_j\right) + b_+ \le -1 + \xi_j \\
j = p+1, \cdots p+q \\
\eta_i, \eta_i^* \ge 0, i = 1, \cdots p \\
\xi_j \ge 0, j = p+1, \cdots p+q
\end{aligned}
\tag{3.456}
$$

and

$$\min \frac{1}{2}\|w_-\|^2 + C_3\sum_{i=p+1}^{p+q}\left(\eta_i + \eta_j^*\right) + C_4\sum_{j=1}^{p}\xi_j$$

$$\text{s.t.} \left(w_- \cdot x_j\right) + b_- \le \varepsilon + \eta_j$$

$$i = p+1, \cdots p+q$$

$$-\left(w_- \cdot x_j\right) - b_- \le \varepsilon + \eta_j^*$$

$$i = p + 1, \cdots p + q$$

$$\left(w_{-} \cdot x_j\right) + b_{-} \geq 1 - \xi_j, j = 1, \cdots p$$

$$\eta_i, \eta_i^* \geq 0, i = p + 1, \cdots p + q$$

$$\xi_j \geq 0, j = 1, \cdots p \tag{3.457}$$

where x_j, $i = 1, \ldots, p$ are positive inputs, and x_i, $i = p + 1, \ldots, p + q$ are negative inputs, $C_i \geq 0$, $i = 1, \ldots, 4$ are penalty parameters, $\xi_+ = (\xi_1, \ldots, \xi_p)^{\mathrm{T}}$, $\xi_- = \left(\xi_{p+1}, \cdots, \xi_{p+q}\right)^{\mathrm{T}}, \eta_+^{(*)} = \left(\eta_+^{\mathrm{T}}, \eta_+^{*\mathrm{T}}\right)^{\mathrm{T}} = \left(\eta_1, \cdots, \eta_p, \eta_1^*, \cdots, \eta_l^*\right)^{\mathrm{T}}, \eta_-^{(*)} = \left(\eta_-^{\mathrm{T}}, \eta_-^{*\mathrm{T}}\right)^{\mathrm{T}} = \left(\eta_{p+1}, \cdots \eta_{p+q}, \eta_{p+1}^*, \ldots, \eta_{p+q}^*\right)^{\mathrm{T}}$, are slack variables.

Now, we discuss the primal problem (3.456) geometrically in $\mathcal{R}^2$. First, we hope that the positive class locate as much as possible in the ε-band between the hyperplanes $(w_+ . x) + b_+ = \varepsilon$ and $(w_+ \cdot x) + b_+ = -\varepsilon$ (red thin solid lines), the errors $\eta_i + \eta_i^*$, $i = 1, \ldots, p$ are measured by the ε-insensitive loss function. Second, we hope to maximize the margin between the hyperplanes $(w_+ \cdot x) + b_+ = \varepsilon$ and $(w_+ \cdot x) + b_+ = -\varepsilon$, which can be expressed by $\frac{2\varepsilon}{\|w\|}$. Third, similar with the TWSVM, we also need to push the negative class from the hyperplane $(w_+ \cdot x) + b_+ = -1$ (red thin dotted line) as far as possible, the errors ξ_i, $i = p + 1, \ldots, p + q$ are measured by the soft margin loss function.

Based on the above three considerations, problem (3.456) is established and the structural risk minimization principle is implemented naturally. Problem (3.457) is established similarly. When the parameter s is set to be zero, and the penalty parameters are chosen to be $C_i = \frac{c_i}{2}$, i = 1,3 and $C_i = c_i$, $i = 2, 4$, problems (3.456) and (3.457) of NPSVM degenerate to problems except that the L_1-loss $\mid \eta_i + \eta_i^* \mid$ is taken instead of the L_2-loss $(w_\pm \cdot x_i) + b_\pm)^2$, and an additional term $\frac{1}{2}b^2$. Furthermore, if the parameter ε is set to be zero, and C_i, $i = 1, \ldots, 4$ are chosen large enough and satisfying $\frac{C_2}{C_1} = 2d_1$, $\frac{C_4}{C_3} = 2d_2$, problems (3.456) and (3.457) degenerate to problems except that the L_1-loss is taken instead of the L_2-loss.

In order to get the solutions of problems (3.456) and (3.457), we need to derive their dual problems. The Lagrangian of the problem (3.456) is given by

$$L\left(w_+, b_+, \eta_+^{(*)}, \xi_-, \alpha_+^{(*)}, \gamma_+^{(*)}, \beta_-, \lambda_-\right)$$

$$= \frac{1}{2}\|w_+\|^2 + C_1 \sum_{j=1}^{P}\left(\eta_j + \eta_j^*\right) + C_2 \sum_{/j=,+1}^{\rho+q} \xi_j$$

$$+ \sum_{j=1}^{p} \alpha_i \left(\left(w_+ \cdot x_j\right) + b_+ - \eta_i - \varepsilon\right)$$

$$+\sum_{j=1}^{\rho}\alpha_i^*\left(-\left(w_+\cdot x_i\right)-b_+-\eta_i^*-\varepsilon\right)$$

$$+\sum_{j=p+1}^{p+q}\beta_j\left(\left(w_+\cdot x_j\right)+b_++1-\xi_j\right)$$

$$-\sum\nolimits_{j=1}^{p}\gamma_i\eta_i-\sum\nolimits_{i=1}^{p}\gamma_i^*\eta_i^*-\sum\nolimits_{j=p+1}^{p+q}\lambda_j\xi_j \tag{3.458}$$

where $\alpha_+^{(*)}=\left(\alpha_+^{\mathrm{T}},\alpha_+^{*\mathrm{T}}\right)^{\mathrm{T}}=\left(\alpha_1,\cdots,\alpha_p,\alpha_1^*,\cdots,\alpha_p^*\right)^{\mathrm{T}}, \gamma_+^{(*)}=\left(\gamma_+^{\mathrm{T}},\gamma_+^{*\mathrm{T}}\right)^{\mathrm{T}}=\left(\gamma_1,\cdots,\gamma_\rho,\gamma_1^*,\cdots,\gamma_p^*\right)^{\mathrm{T}}, \beta_-=(\beta_{p+1},\cdots,\beta_{p+q})^{\mathrm{T}}, \lambda_-=(\lambda_{p+1},\cdots,\beta_{p+q})^{\mathrm{T}}$ are the Lagrange multiplier vectors. The Karush–Kuhn–Tucker (KKT) conditions [71] for $w_+, b_+, \eta_+^{(*)}, \xi_-$ and $\alpha_+^{(*)}, \gamma_+^{(*)}, \beta_-, \lambda_-$ are given by

$$\nabla_{w_+}L=w_++\sum\nolimits_{i=1}^{p}\alpha_ix_l\cdot-\sum\nolimits_{j=1}^{p}\alpha_i^*x_i+.\sum\nolimits_{j=p+1}^{p+q}\beta_jx_j=0 \tag{3.459}$$

$$\nabla_{b_1}L=\sum\nolimits_{i=1}^{p}\alpha_i-\sum\nolimits_{i=1}^{P}\alpha_i^*+\sum\nolimits_{j=p+1}^{p+q}\beta_j=0 \tag{3.460}$$

$$\nabla_{\eta_+}L=C_1e_+-\alpha_+-\gamma_+=0 \tag{3.461}$$

$$\nabla_{\eta_+^*}L=C_1e_+-\alpha_+^*-\gamma_+^*=0 \tag{3.462}$$

$$\nabla_{\xi}L=C_2e_--\beta_--\lambda_-=0 \tag{3.463}$$

$$(w_+\cdot x_i)+b_+\le\varepsilon+\eta_j, i=1,\cdots,p \tag{3.464}$$

$$-\left(w_+\cdot x_j\right)-b_+\le\varepsilon+\eta_i^\star, i=1,\cdots,p \tag{3.465}$$

$$\left(w_+\cdot x_j\right)+b_+\le-1+\xi_j, j=p+1,\cdots,p+q \tag{3.466}$$

$$\eta_i,\eta_i^*\ge0, i=1,\cdots,p \tag{3.467}$$

$$\xi_j\ge0, j=p+1,\cdots,p+q \tag{3.468}$$

where $e_+ = (1, \cdots, 1)^{\mathrm{T}} \in \mathcal{R}^{p}$, $e_- = (1, \cdots, 1)^{\mathrm{T}} \in \mathcal{R}^{q}$. Since $\gamma_+, \gamma_+^* \geq 0$, $\lambda_- \geq 0$, from (3.461), (3.462) and (3.463) we have

$$0 \leq \alpha_+, \alpha_+^* \leq C_1 e_+, \tag{3.469}$$

$$0 \leq \beta_- \leq C_2 e_-. \tag{3.470}$$

And from (3.459), we have

$$w_+ = \sum_{i=1}^{p} \left(\alpha_i^* - \alpha_i\right) x_i - \sum_{j=p+1}^{p+q} \beta_j x_j. \tag{3.471}$$

Then putting (3.471) into the Lagrangian (3.458) and using (3.459)–(3.468), we obtain the dual problem of the problem (3.456)

$$\min \ \frac{1}{2} \sum_{i=1}^{p} \sum_{j=1}^{p} \left(\alpha_i^* - \alpha_i\right) \left(\alpha_j^* - \alpha_j\right) \left(x_i \cdot x_j\right) - \sum_{i=1}^{p} \sum_{j=p+1}^{p+q} \left(\alpha_i^* - \alpha_i\right) \beta_j \left(x_i \cdot x_j\right)$$

$$+ \frac{1}{2} \sum_{i=p+1}^{p+q} \sum_{J=p+1}^{p+q} \beta_i \beta_j \left(x_i \cdot x_j\right) + \varepsilon \sum_{i=1}^{p} \left(\alpha_i^* + \alpha_i\right) - \sum_{i=p+1}^{p+q} \beta_i,$$

$$\text{s.t.} \sum_{i=1}^{p} \left(\alpha_i - \alpha_i^*\right) + \sum_{j=p+1}^{p+q} \beta_j = 0,$$

$$0 \leq \alpha_+, \alpha_+^* \leq C_1 e_+$$

$$0 \leq \beta_- \leq C_2 e_-. \tag{3.472}$$

Concisely, this problem can be further formulated as

$$\min \ \frac{1}{2} \left(\alpha_+^* - \alpha_+\right)^{\mathrm{T}} A A^{\mathrm{T}} \left(\alpha_+^* - \alpha_+\right) - \left(\alpha_+^* - \alpha_+\right)^{\mathrm{T}} A B^{\mathrm{T}} \beta_- + \frac{1}{2} \beta_-^{\mathrm{T}} B B^{\mathrm{T}} \beta_-$$

$$+ \varepsilon e_+^{\mathrm{T}} \left(\alpha^* + \alpha\right) - e_-^{\mathrm{T}} \beta_-$$

$$s.t. \ e_+^{\mathrm{T}} \left(\alpha_+ - \alpha_+^*\right) + e_-^{\mathrm{T}} \beta_- = 0,$$

$$0 \leq \alpha_+, \alpha_+^* \leq C_1 e_+,$$

$$0 \leq \beta_- \leq C_2 e_-. \tag{3.473}$$

where $A = \left(x_1, \cdots, x_p\right)^{\mathrm{T}} \in \mathcal{R}^{p\times n}$, $B = \left(x_{p+1}, \cdots, x_{p+q}\right) \in \mathcal{R}^{q\times n}$. Furthermore, let

$$\widetilde{\pi} = \left(\alpha_{+}^{*\mathrm{T}}, \alpha_{+}^{\mathrm{T}}, \beta_{-}^{\mathrm{T}}\right)^{\mathrm{T}} \tag{3.474}$$

$$\widetilde{k} = \left(\varepsilon e_{+}^{\mathrm{T}}, \varepsilon e_{+}^{\mathrm{T}}, -e_{-}^{\mathrm{T}}\right)^{\mathrm{T}} \tag{3.475}$$

$$\widetilde{e} = \left(-e_{+}^{\mathrm{T}}, e_{+}^{\mathrm{T}}, e_{-}^{\mathrm{T}}\right)^{\mathrm{T}} \tag{3.476}$$

$$\widetilde{C} = \left(C_1 e_{+}^{\mathrm{T}}, C_1 e_{+}^{\mathrm{T}}, C_2 e_{-}^{\mathrm{T}}\right)^{\mathrm{T}} \tag{3.477}$$

and

$$\overline{\Lambda} = \begin{pmatrix} H_1 & -H_2 \\ -H_2^{\mathrm{T}} & H_3 \end{pmatrix}, \tag{3.478}$$

$$H_1 = \begin{pmatrix} AA^{\mathrm{T}} & -AA^{\mathrm{T}} \\ -AA^{\mathrm{T}} & AA^{\mathrm{T}} \end{pmatrix}, \tag{3.479}$$

$$H_2 = \begin{pmatrix} AB^{\mathrm{T}} \\ -AB^{\mathrm{T}} \end{pmatrix} \quad H_3 = BB^{\mathrm{T}} \tag{3.480}$$

then problem (3.473) is reformulated as

$$\begin{aligned} \min \quad & \frac{1}{2}\widetilde{\pi}^{\mathrm{T}} \Lambda \widetilde{\pi} + \widetilde{k}^{\mathrm{T}}\widetilde{\pi} - \\ s.t. \quad & \widetilde{e}^{\mathrm{T}}\widetilde{\pi} = 0 \\ & 0 \le \widetilde{\pi} \le \widetilde{C}. \end{aligned} \tag{3.481}$$

1. Obviously, problem (3.481) is a convex QPP and exactly the same elegant formulation as problem, the well-known SMO can be applied directly with a minor modification. For (3.481), by applying the KKT conditions, we can get the following conclusions without proof, which is similar with the conclusions in [67] and [18].

Theorem 1 Suppose that $\widetilde{\pi} = \left(\alpha_{+}^{*\mathrm{T}}, \alpha_{+}^{\mathrm{T}}, \beta_{-}^{\mathrm{T}}\right)^{\mathrm{T}}$ is a solution of the problem (3.481), then for $=1, \ldots, p$, each pair of α_i and $\alpha_{\mathrm{i}}^{\star}$ can not be both simultaneously nonzero, i.e.,$\alpha_i \alpha_i^* = 0$, $i = 1, \ldots, p$.

Theorem 2 Suppose that $\widetilde{\pi} = \left(\alpha_{+}^{*\mathrm{T}}, \alpha_{+}^{\mathrm{T}}, \beta_{-}^{\mathrm{T}}\right)^{\mathrm{T}}$ is a solution of the problem (3.481), if there exist components of $\widetilde{\pi}$ of which value is in the interval $\left(0, \overline{C}\right)$, then the solution (w_{+}, b_{+}) of the problem (3.456) can be obtained in the following way. Let

$$w_{+} = \sum\nolimits_{i=1}^{p} \left(\alpha_i^{*} - \alpha_i\right) x_i - \sum\nolimits_{j=p+1}^{p+q} \beta_{\mathrm{j}} x_j, \tag{3.482}$$

and choose a component of α_{+}, $\alpha_{+j} \in (0, C_1)$, compute

$$b_{+} = -\left(w_{+} \cdot x_j\right) + \varepsilon \tag{3.483}$$

or choose a component of α_{+}^{*}, $\alpha_{+k}^{*} \in (0, C_1)$, compute

$$b_{+} = -\left(w_{+} \cdot x_k\right) - \varepsilon \tag{3.484}$$

or choose a component of β_{-}, $\beta_{-m} \in (0, C_2)$, compute

$$b_{+} = -\left(w_{+} \cdot x_m\right) - 1. \tag{3.485}$$

In the same way, the dual of the problem (11) is obtained

$$\min \ \frac{1}{2} \sum_{i=p+1}^{p+\mathrm{q}} \sum_{j=p+1}^{p+q} \left(\alpha_i^{*} - \alpha_i\right)\left(\alpha_j^{*} - \alpha_j\right)\left(x_j \cdot x_j\right)$$

$$+ \sum_{Reject=p+1}^{p+\mathrm{q}} \sum_{j=1}^{\rho} \left(\alpha_j^{*} - \alpha_j\right) \beta_j \left(x_i x_j\right)$$

$$+\frac{1}{2} \sum_{j=1}^{p} \sum_{j=1}^{p} \beta_i \beta_j \left(x_i \cdot x_j\right) + \varepsilon \sum_{i=p+1}^{p+q} \left(\alpha_i^{*} + \alpha_i\right) - \sum_{i=1}^{\rho} \beta_i$$

$$\text{s.t.} \sum_{\mathrm{i}=p+1}^{p+\mathrm{q}} \left(\alpha_i - \alpha_i^{*}\right) - \sum_{i=1}^{p} \beta_i = 0$$

$$0 \le \alpha_i, \alpha_i^{*} \le C_3, i = p+1, \cdots, p+q$$

$$0 \le \beta_{\mathrm{i}} \le C_4, i = 1, \cdots, p \tag{3.486}$$

where $\alpha_{-}^{(*)}$, β_{+} are the Lagrange multiplier vectors. It can also be rewritten as

$$\min \ \frac{1}{2}\left(\alpha_{-}^{*} - \alpha_{-}\right)^{\mathrm{T}} B B^{\mathrm{T}} \left(\alpha_{-}^{*} - \alpha_{-}\right)$$

$$+\left(\alpha_{-}^{*} - \alpha_{-}\right)^{\mathrm{T}} B A^{\mathrm{T}} \beta_{+} + \frac{1}{2} \beta_{+}^{\mathrm{T}} A A^{\mathrm{T}} \beta_{+}$$

$$+\varepsilon e_-^{\mathrm{T}}\left(\alpha^*+\alpha\right)-e_+^{\mathrm{T}}\beta_+ \tag{3.487}$$

$$\text{s.t. } e_-^{\mathrm{T}}\left(\alpha_- - \alpha_-^*\right)-e_+^{\mathrm{T}}\beta_+=0,$$

$$0\le \alpha_-,\alpha_-^*\le C_3e_-,$$

$$0\le \beta_+\le C_4e_+.$$

Concisely, it is reformulated as

$$\min\ \frac{1}{2}\hat{\pi}^{\mathrm{T}}\hat{\Lambda}\hat{\pi}+\hat{\kappa}^{\mathrm{T}}\hat{\pi}$$

$$s.t.\ \hat{e}^{\mathrm{T}}\hat{\pi}=0, \tag{3.488}$$

$$0\le\hat{\pi}\le\hat{C}.$$

where

$$\hat{\pi}=\left(\alpha_-^{*\mathrm{T}},\alpha_-^{\mathrm{T}},\beta_+^{\mathrm{T}}\right)^{\mathrm{T}} \tag{3.489}$$

$$\overline{\kappa}=\left(\varepsilon e_-^{\mathrm{T}}\varepsilon e_-^{\mathrm{T}},-e_+^{\mathrm{T}}\right)^{\mathrm{T}} \tag{3.490}$$

$$\hat{e}=\left(-e_-^{\mathrm{T}},e_-^{\mathrm{T}},-e_+^{\mathrm{T}}\right)^{\mathrm{T}} \tag{3.491}$$

$$\hat{C}=\left(C_3e_-^{\mathrm{T}},C_3e_-^{\mathrm{T}},C_4e_+^{\mathrm{T}}\right)^{\mathrm{T}} \tag{3.492}$$

and

$$\hat{\Lambda}=\begin{pmatrix} Q_1 & Q_2\\ Q_2^{\mathrm{T}} & Q_3\end{pmatrix} \tag{3.493}$$

$$Q_1=\begin{pmatrix} BB^{\mathrm{T}} & -BB^{\mathrm{T}}\\ -BB^{\mathrm{T}} & BB^{\mathrm{T}}\end{pmatrix} \tag{3.494}$$

$$Q_2 = \begin{pmatrix} BA^{\mathrm{T}} \\ -BA^{\mathrm{T}} \end{pmatrix} \tag{3.495}$$

$$Q_3 = AA^{\mathrm{T}} \tag{3.496}$$

For (3.476), we have the following conclusions corresponding to problem (3.481).

Theorem 3 Suppose that $\hat{\pi} = \left(\alpha_-^{*\mathrm{T}}, \alpha_-^{\mathrm{T}}, \beta_+^{\mathrm{T}}\right)^{\mathrm{T}}$ is a solution of the problem (3.488), then for $=p+1, \cdots, p+q$, each pair of α_i and α_i^* cannot be both simultaneously nonzero, i.e., $\alpha_i \alpha_i^* = 0$, $i = p+1, \cdots, p+q$.

Theorem 4 Suppose that $\hat{\pi} = \left(\alpha_-^{*\mathrm{T}}, \alpha_-^{\mathrm{T}}, \beta_+^{\mathrm{T}}\right)^{\mathrm{T}}$ is a solution of the problem (3.488), if there exist components of $\hat{\pi}$ of which value is in the interval $\left(0, \hat{C}\right)$, then the solution (w_-, b_-) of the problem (11) can be obtained in the following way. Let

$$w_- = \sum\nolimits_{j=p+1}^{p+q} \left(\alpha_i^* - \alpha_i\right) x_i + \sum\nolimits_{j=1}^{p} \beta_j x_j \tag{3.497}$$

and choose a component of α_+, $\alpha_{+j} \in (0, C_3)$, compute

$$b_- = -\left(w_- \cdot x_j\right) + \varepsilon, \tag{3.498}$$

or choose a component of α_+^*, $\alpha_{+k}^* \in (0, C_3)$, compute

$$b_- = -\left(w_- \cdot x_k\right) - \varepsilon, \tag{3.499}$$

or choose a component of β_-, $\beta_{-m} \in (0, C_4)$, compute

$$b_- = -\left(w_- \cdot x_m\right) + 1. \tag{3.500}$$

From Theorems 2 and 4, we can see that the inherent semi sparseness in the existing TWSVMs is improved to the whole sparseness in our linear NPSVM, because of the introduction of ε-insensitive loss function instead of the quadratic loss function for each class itself.

Once the solutions (w_+, b_+) and (w_-, b_-) of the problems (10) and (11) are obtained, a new point $x \in \mathcal{R}^n$ is predicted to the class by

$$Class = \arg \ \min \mid (w_k \cdot x) + b_k \mid \tag{3.501}$$

where $| \cdot |$ is the perpendicular distance of point x from the planes $(w_k \cdot x) + b_k = 0$, $k = -, +$.

Nonlinear NPSVM

Now, we extend the linear NPSVM to the nonlinear case. Totally different with all the existing TWSVMs, we do not need to consider the extra kernel-generated surfaces since only inner products appear in the dual problems, so the kernel functions are applied directly in the problems and the linear NPSVM is easily extended to the nonlinear classifiers.

In detail introducing the kernel function $K(x,x^{'}) = (\Phi(x)\Phi(x^{'}))$ and the corresponding transformation

$$\mathrm{x} = \Phi(x) \tag{3.502}$$

where $\mathrm{x} \in \mathcal{H}$, $\mathcal{H}$ is the Hilbert space, we can construct the corresponding problems (10) and (11) in $\mathcal{H}$, the only difference is that the weight vectors w_+ and w_- in $\mathcal{R}^n$ change to be w_+ and w_-, respectively. Two dual problems to be solved are

$$\begin{aligned} \min\ & \tfrac{1}{2}(\alpha_+^* - \alpha_+)^{\mathrm{T}} K, AA^{\mathrm{T}}\Big) (\alpha_+^* - \alpha_+) \\ -(\alpha_+^* - \alpha_+)^{\mathrm{T}} & K, AB^{\mathrm{T}}\Big)\beta_- + \tfrac{1}{2}\beta_-^{\mathrm{T}} K\left(B, B^{\mathrm{T}}\right)\beta_- + \varepsilon e_+^{\mathrm{T}}(\alpha^* + \alpha) - e_-^{\mathrm{T}}\beta_- \\ s.t.\ & e_+^{\mathrm{T}}(\alpha_+ - \alpha_+^*) + e_-^{\mathrm{T}}\beta_- = 0 \\ & 0 \le \alpha_+, \alpha_+^* \le C_1 e_+ \\ & 0 \le \beta \ \ \le C_2 e_- \end{aligned} \tag{3.503}$$

and

$$\begin{aligned} \min & \frac{1}{2}(\alpha_-^* - \alpha_-)^{\mathrm{T}} K\left(B, B^{\mathrm{T}}\right)^{\mathrm{T}} (\alpha_-^* - \alpha_-) \\ +(\alpha_-^* - \alpha_-)^{\mathrm{T}} & K\left(B, A^{\mathrm{T}}\right)\beta_+ + \frac{1}{2}\beta_+^{\mathrm{T}} K\left(A, A^{\mathrm{T}}\right)\beta_+ + \varepsilon e_-^{\mathrm{T}}(\alpha^* + \alpha) - e_+^{\mathrm{T}}\beta_+ \\ s.t.\ & e_-^{\mathrm{T}}(\alpha_- - \alpha_-^*) - e_+^{\mathrm{T}}\beta_+ = 0 \\ & 0 \le \alpha_-, \alpha_-^* \le C_3 e_- \\ & 0 \le \beta_+ \le C_4 e_+ \end{aligned} \tag{3.504}$$

respectively.

Corresponding Theorems are similar to Theorems 1–4 and we only need to take $K(x,x^{'})$ instead of $(x \cdot x^{'})$.

Now we establish the NPSVM as follows.

Algorithm 3.10 (NPSVM)

(1) Input the training set (8).

(2) Choose appropriate kernels (x, x'), appropriate parameters $\varepsilon > 0$, C_1, C_2 for problem (3.476), and C_3, $C_4 > 0$ for problem (3.487).

(3) Construct and solve the two convex QPPs separately, get the solutions $\alpha^{(*)}$ $\left(\alpha_1, \cdots, \alpha_{p+q}, \alpha_1^*, \cdots, \alpha_{p+q}^*\right)^{\mathrm{T}}$ and $\beta = (\beta_1, \cdots, \beta_{p+q})^{\mathrm{T}}$

(4) Construct the decision functions

$$f_+(x) = \sum_{i=1}^{p} \left(\alpha_i^* - \alpha_i\right) K\left(x_{\mathrm{i}}, x\right) - \sum_{j=p+1}^{p+q} \beta_j K\left(x_j, x\right) + b_+ \tag{3.505}$$

and

$$f_-(x) = \sum_{j=p+1}^{p+q} \left(\alpha_{\mathrm{i}}^\star - \alpha_i\right) K\left(x_i, x\right) + \sum_{j=1}^{p} \beta_j K\left(x_j, x\right) + b_- \tag{3.506}$$

separately, where b_-, b_+ are computed by Theorems 2 and 4 for the kernel cases;

(5) For any new input x, assign it to the class $k(k = -, +)$ by

$$argmin \frac{\mid f_k(x) \mid}{\|\Delta_k\|} \tag{3.507}$$

where

$$\Delta_+ = \widetilde{\pi}^{\mathrm{T}} \Lambda \widetilde{\pi} -, \Delta_- = \hat{\pi}^{\mathrm{T}} \hat{\Lambda} \hat{\pi}. \tag{3.508}$$

Advantages of NPSVM

As NPSVM degenerates to TBSVM and TWSVM when parameters are chosen appropriately, it is theoretically superior to them. Furthermore, it is more flexible and has better generalization ability than typical SVMs since it pursues two nonparallel surfaces for discrimination. Although NPSVM has an additional parameter s, which leads to two larger optimal problems than TBSVM (about three times), it still has the following advantages.

1. Although TWSVM and TBSVM solve smaller QPPs in which successive overrelaxation (SOR) technique or coordinate descent method can be applied [21, 72]; they have to compute the inverse matrices before training which is in practice intractable or even impossible for a large dataset. More detailed, suppose the size of the training set is one, and the size of negative training set is roughly equal to the size of positive set, i. e. $p \approx q \approx 0.5l$, the computational complexity

of TWSVM or TBSVM solved by SOR is estimated as

$$O\left(l^3\right) + \#iteration \times O(0.5l) \tag{3.509}$$

where $O(l^3)$ is the complexity of computing $l \times l$ inverse matrix, and *#iteration* $\times O(0.5l)$ is of SOR for $0.5l$ sized problem (*#iteration* is the number of the iterations, experiments in [42] has shown that *#iteration* is almost linear scaling with the size one). While NPSVM does not require the inverse matrices and can be solved efficiently by the SMO-type technique, [69] has proved that for the two convex QPPs, an SMO-type decomposition method [68] implemented in LIBSVM has the complexity

$$\#iterations \times O(1.51) \tag{3.510}$$

if most columns of the kernel matrix are cached throughout iterations ([69] also pointed out that there is no theoretical result yet on LIBSVM's number of iterations. Empirically, it is known that the number of iterations may be higher than linear to the number of training data). Comparing (3.509) and (3.510), obviously NPSVM is faster than TWSVMs.

2. Although TBSVM improved TWSVM by introducing the regularization terms $\left(\|w_+\|^2 + b_+^2\right)$ (for example, in problem (8), another regularization term, $\|w_+\|^2$, can be found in [72] and [73] to make the SRM principle implemented, it can only be explained for the linear case that $\frac{1}{\sqrt{\|w_+\|^2+b_+^2}}$ is the margin of two parallel hyperplanes $(w_+ \cdot x) + b_+ = 0$ (the proximal hyperplane) and $(w_+ \cdot x) + b_+ = -1$ (the bounding hyperplane) in $\mathcal{R}^{n+1}$ space. However, for the nonlinear case, it is not a real kernel method like the standard SVMs usually do, it considers the kernel-generated surfaces, and apply the regularization terms, for example, $\left(\|u_+\|^2 + b_+^2\right)$ [21]. This term cannot be explained clearly, since it is only an approximation of the term $\left(\|\mathrm{w}_+\|^2 + b_+^2\right)$ in Hilbert space. NPSVM introduces the regularization terms $\|w_+\|^2$ [for example, in (10)] for linear case and $\|\mathrm{w}_\pm\|^2$ for nonlinear case naturally and reasonably, since $\frac{2}{\|w\|}$ is the margin of two parallel hyperplanes $(w \cdot x) + b = \mathrm{s}$ and $(w \cdot x) + b = -\varepsilon$ in $\mathcal{R}^n$ space, while $\frac{2}{\|\mathrm{w}\|}$ is the margin of two parallel hyperplanes $(\mathrm{w} \cdot \mathrm{x}) + b = \varepsilon$ and $(\mathrm{w} \cdot \mathrm{x}) + b = -\varepsilon$ in Hilbert space.
3. For the nonlinear case, TWSVMs have to consider the kernel-generated surfaces instead of the hyperplanes in the Hilbert space, they are still parametric methods. NPSVM constructs two primal problems for both cases via using different kernels, which is the marrow of the standard SVMs.

3.3.4 A Multi-instance Learning Algorithm Based on Nonparallel Classifier

3.3.4.1 MI-NSVM

Multi-instance Learning Problem

One of the drawbacks of applying the supervised learning model is that it is not always possible for a teacher to provide labeled examples for training. MIL provides a new way of modeling the teacher's weakness [74]. MIL considers a particular form of weak supervision in which training class labels are associated with sets of patterns, or bags, instead of individual patterns. A negative bag only consists of negative instances, whereas a positive bag comprises both positive and negative instances. The goal of MIL is to find a separating hyperplane which can decide the label of any new instance or bag.

In the following, we give the formal description of MIL problem. Given a training set

$$\{\mathbf{B}_1^+, \ldots, \mathbf{B}_{m^+}^+, \mathbf{B}_1^-, \ldots, \mathbf{B}_{m^-}^-\} \tag{3.511}$$

where a bag $\mathrm{B}_i^+ = \left\{x_{i1}, \ldots, x_{im_i^+}\right\}; x_{ij} \in \mathfrak{R}^n; j = 1, \ldots, m_i^+; i = 1, \ldots, m^+; \mathrm{B}_i^- = \left\{x_{i1}, \ldots, x_{im_i^-}\right\}; x_{ij} \in \mathfrak{R}^n; j = 1, \ldots, m^-; i = 1, \ldots, m^-; \mathrm{B}^+$ means that the positive bag B^+ contains at least one positive instance x_{ij}; B^- means that all instance x_{ij} of the negative bag are negative. The goal is to induce a real-valued function

$$f(x) = sgn\,(bfg(x)) \tag{3.512}$$

such that the label of any instance x in $\mathfrak{R}^n$ space can be predicted. Obviously, for a new bag $l3 = \{\widetilde{x}_1, \ldots, \tilde{x}_m\}$, its label $\mathcal{Y}$ can be decided by

$$\mathcal{Y} = sgn\,(\max\ f\,(\tilde{x}_i)) \tag{3.513}$$

Now we rewrite the training set (1) as

$$Train = \left\{\mathrm{B}_1^+, \ldots, \mathrm{B}_{m^+}^+; \mathrm{B}_{m+1}^-+, \ldots, \mathrm{B}_{m^++m-;}^-\right\} \tag{3.514}$$

$$= \left\{\mathrm{B}_{1;}^+ \ldots; \mathrm{B}_{m;}^+ + x_{z+1;} \ldots; x_{z+f}\right\} \tag{3.515}$$

where z is the number of the instances in all positive bags and f the number of the instances in negative bags. The set consisting of subscripts of B_i is expressed as:

$$s \leftarrow (i) = \{i | x_i \in \mathrm{B}_i\} \tag{3.516}$$

Linear MI-NSVM

For the usual MIL methods based on SVMs, the "witness" instance of each positive bag is always obtained by selecting the farthest from the hyperplane constructed by SVMs. For a separable multi-instance classification problem, if a positive bag can be correctly classified, it should satisfy the following constraint:

$$j \in_{\sim}, (i) \max \left(w \cdot x_j\right) + b > 1 \tag{3.517}$$

Mangasarian and Wild [40, 75] show that it is equivalent to the fact that there exist convex combination coefficients set $\left\{\lambda_j^i, j \in s \leftarrow (i); i = 1, \ldots, m^+\right\}$, such that

$$\left(w \cdot \sum\nolimits_{j \in \mathcal{S}(i)} \lambda_j^i x_j\right) + b \geqslant 1 \tag{3.518}$$

$$\lambda_j^i \geqslant 0, \sum\nolimits_{j \in \mathcal{F}(i)} \lambda_j^i = 1 \tag{3.519}$$

In the first step of MI-NSVM, our goal is to construct a hyperplane, which is closer to the negative instances and is at least one distance from the positive instances. According to the conclusion of [40, 75] above. The first model can be expressed as

$$\begin{aligned} &\min_{w_-, b_-, \eta} \ \frac{1}{2} \| Bw_- + e_- b_- \|^2 + c_1 e_+^{\top} \eta \\ \text{s.t. } &\left(w_- \cdot \sum\nolimits_{j \in 3(i)} \lambda_j^i x_j\right) + b_- \geqslant 1 - \eta_i, i = 1, \ldots, m^+ \\ &\lambda_j^i \geqslant 0, i = 1, \ldots, m^+ \\ &\sum\nolimits_{j \in \Im(i)} \lambda_j^i = 1, i = 1, \ldots, m^+ \end{aligned} \tag{3.520}$$

where $c_1 \geq 0$ is the pre-specified penalty factors, e_+ is the vector of ones of appropriate dimensions, $B = (x_{z+1}; \ldots; x_{z+f})^{\mathrm{T}}$; $\eta = (\eta_1; \ldots; \eta_{m^+})$. By solving the optimization problem, we can obtain the first hyperplane about MI-NSVM, and then, we may estimate a score for each instance of positive bags according to the distance between them and the optimal hyperplane $\left(w_-^* \cdot x\right) + b_-^* = 0$ (bigger the distance is, bigger the score of the corresponding instance is).

In the second step, we first pick up the "most positive" instance of each positive bag:

$$s_i = \text{argmax} \left\{\text{score}\left(x_j\right)\right\}; j \in s \leftarrow (i); i \in 1, \ldots, m^+ \tag{3.521}$$

where score (x) denotes the score of x computed by the first step, and s_i denotes the index of the "most positive" instance of the i-th bag. Then, we construct the second hyperplane which is closer to the "the most positive" instances and is at least one distance from the negative instances. The corresponding model is as follows

$$\begin{aligned} &\min \frac{1}{2}\|Aw_+ + e_+b_+\|^2 + c_2 e_-^{\mathrm{T}}\xi \\ &\text{s.t.} \ -(Bw_+ + e_-b_+) + \xi \geq e_-; \\ &\xi \geq 0. \end{aligned} \tag{3.522}$$

where $c_2 \geq 0$ is the pre-specified penalty factors, e_- is the vector of ones of appropriate dimensions, $A = \left(x_{1s_1}; \ldots; x_{m^+s_{m^+}}\right)^{\mathrm{T}}$; $B = \left(x_{z+1}; \ldots; x_{z+f}\right)^{\mathrm{T}}$. By solving the two optimization problems (10) and (12), we can obtain the following two nonparallel hyperplanes:

$$f_+(x) = (w_+ \cdot x) + b_+ = 0 \tag{3.523}$$

$$f_-(x) = \frac{(w_+ \cdot x) + b_+}{\|w_+\|} + \frac{(w_- \cdot x) + b_-}{\|w_-\|} = 0 \tag{3.524}$$

A new point $x \in \Re^n$ is then assigned to the positive or negative class, depending on which of the two hyperplanes given by it lies closest to, i.e.

$$f(x) = \arg\ \min \{d_+(x); d_-(x)\} \tag{3.525}$$

where

$$d_+(x) = \left|w_+^{\mathrm{T}}x + b_+\right|; d_-(x) = \frac{(w_+ \cdot x) + b_+}{\|w_+\|} + \frac{(w_- \cdot x) + b_-}{\|w_-\|} = 0 \tag{3.526}$$

where $|\cdot|$ is the perpendicular distance of point x from the planes $w_+^{\mathrm{T}}x + b_+$ and $\frac{(w_+ \cdot x)+b_+}{\|w_+\|} + \frac{(w_- \cdot x)+b_-}{\|w_-\|}$.

Nonlinear MI-NSVM

The above discussion is restricted in the linear case. Here, we will analyze nonlinear MI-NSVM by introducing the Radial Basis Function (RBF)

$$K\left(x;x^{\mathrm{T}}\right) = \exp\left(-\left\|x - x^{\mathrm{T}}\right\|^{2} = 2\sigma^{2}\right); \tag{3.527}$$

where σ is a real parameter, and the corresponding transformation:

$$x = \Phi(x) \tag{3.528}$$

where $x \in \mathcal{H}$; $\mathcal{H}$ is a Hilbert space.

Consider the following kernel-generated hyperplanes:

$$K\left(x_{;}^{\mathrm{T}}C^{\mathrm{T}}\right)k_{+} + b_{+} = 0_{;} \tag{3.529}$$

$$K\left(x_{;}^{\mathrm{T}}C^{\mathrm{T}}\right)k_{-} + b_{-} = 0_{;} \tag{3.530}$$

where

$$C^{\top} = \left[\overline{A}B\right]^{\top}, \overline{A} = \left(\frac{1}{m_{1}^{+}}\sum\nolimits_{j\in 3(1)} x_{j}, \ldots, \frac{1}{m_{m^{+}}^{+}}\sum\nolimits_{j\in 3(m^{+})} x_{j}\right)+^{\top} \tag{3.531}$$

and K is the chosen kernel function. The first nonlinear optimization problem can be expressed as

$$\begin{aligned} &\underset{k_{-},b_{-},\eta}{min}\ \tfrac{1}{2}\|\ K\left(B, C^{\top}\right)k_{-} + e_{-}b_{-}\ \|^{2} + c_{1}e^{\top}\eta \\ &\text{s.t.}K\left(\sum\nolimits_{V\in\mathfrak{I}(i)}\lambda_{j}^{i}x_{j}, C^{\top}\right)k_{-} + b_{-} \geqslant 1 - \eta_{i}, i = 1, \ldots, m^{+} \\ &\qquad \lambda_{j}^{i} \geqslant 0, i = 1, \ldots, m^{+} \\ &\qquad \sum\nolimits_{j\in\mathfrak{I}(i)}\lambda_{j}^{i} = 1, i = 1, \ldots, m^{+} \end{aligned} \tag{3.532}$$

Correspondingly, the second optimization problem can be written as

$$\begin{aligned} &\underset{k_{+},b_{+},\xi}{min}\ \tfrac{1}{2}\|\ K\left(\overline{A}, C^{\top}\right)k_{+} + e_{+}b_{+}\ \|^{2} + c_{2}e_{-}^{\top}\xi \\ &\text{s.t.} - \left(K\left(B, C^{\top}\right)k_{+} + e_{-}b_{+}\right) + \xi \geqslant e_{-}, \xi \geqslant 0 \end{aligned} \tag{3.533}$$

How to Solve MI-NSVM

We firstly discuss to how to solve the linear MI-NSVM.

Consider the optimization problem. This a typical non-convex optimization problem. Mangasarian et al. solve a similar optimization problem via successive linear programming [40, 75]. Therefore, we can also apply the same technique. However, this section mainly focuses on the construction of the model. In order to simplify, we give an approximate iterative solution via solving successive QPP and LPP. Firstly, we fix λ, and solve a quadratic programming with respect to w_-; b_-; η and then fix w_-; b_-, and solve a quadratic programming with respect to λ;η.

1. For fixed $\hat{\lambda}^i_{j;} i = 1, \ldots, m^+; j \in s^{\infty}(i)$, we can obtain

$$\hat{x}_i = \sum\nolimits_{j \in \mathfrak{I}(i)} \hat{\lambda}^i_j x_j, i = 1, \ldots, m^+ \tag{3.534}$$

so the problem can be written as

$$\min \frac{1}{2} \|Bw_- + e_- b_-\|^2 + c_1 e_+^{\mathrm{T}} \eta$$

$$\text{s.t.} \quad \left(\hat{A} w_- + e_+ b_-\right) + \eta \geq e_+; \eta \geq 0 \tag{3.535}$$

where $\hat{A} = \left(\hat{x}_1; \ldots; \hat{X}_m +\right)^{\mathrm{T}}$. The problem (3.535) is a standard quadratic programming problem and its dual problem can be formulated as

$$\max e_+^{\mathrm{T}} \beta - \frac{1}{2} \beta^{\mathrm{T}} P \left(Q^{\mathrm{T}} Q\right)^{-1} P^{\mathrm{T}} \beta$$

$$\text{s.t.} \quad 0 \leq \beta \leq c_1 e_+; \tag{3.536}$$

where $P = \left[\hat{A} e_+\right]$ and $Q = [Be_-]$; $\beta \in \mathfrak{R}^{m^+}$ are Lagrangian multipliers.

We can compute: $\hat{\beta} = \left(\hat{\beta}_1; \ldots; \hat{\beta}_m +\right)^{\mathrm{T}}$ by solving the problem of (3.536), and $\left(\hat{w}_-; \hat{b}_-\right)$ can be obtained by computing

$$\hat{\mathrm{v}}_1 = \left[\hat{w}_-^{\mathrm{T}} \hat{b}_-\right]^{\mathrm{T}} = -\left(Q^{\mathrm{T}} Q\right)^{-1} P^{\mathrm{T}} \hat{\beta}; \tag{3.537}$$

$\left(\hat{w}_-; \hat{b}_-\right)$ is the updating $\wedge$value of $(w_-; b_-)$.

2. For fixed $\hat{w}_-$ and b_-, the optimization problem can be substituted by the LPP as follows

$$\begin{aligned} &\min_{\lambda,\eta} e_+^\top \eta \\ \text{s.t.} &\left(\hat{w}_- \cdot \sum_{j\in \mathcal{S}(i)} \lambda_j^i x_j\right) + \hat{b}_- \geqslant 1 - \eta_i, i = 1, \ldots, m^+ \\ &\sum_{j\in \mathfrak{I}(i)} \lambda_j^i = 1, i = 1, \ldots, m^+ \\ &\lambda_j^i \geqslant 0, i = 1, \ldots, m^+ \end{aligned} \tag{3.538}$$

Now consider the optimization problem (3.522). The dual of (3.522) can be written as

$$\begin{aligned} &\max_{\alpha} e_-^\top \alpha - \tfrac{1}{2}\alpha^\top G\left(H^\top H\right)^{-1} G^\top \alpha \\ &\text{s.t.} 0 \leqslant \alpha \leqslant c_2 e_- \end{aligned} \tag{3.539}$$

The second hyperplane can be obtained by

$$\mathrm{v}_2 = \left[w_+^\mathrm{T} b_+\right]^\mathrm{T} = -\left(H^\mathrm{T} H\right)^{-1} G^\mathrm{T} \alpha; \tag{3.540}$$

where $G = [Be_-]$; $H = [Ae_+]$.

Thus, we are able to establish the following Algorithm 3.11 based on the discussion above.

Algorithm 3.11 Linear MI-NSVM

Initialize: Given a training set;

Choose appropriate penalty parameters c_1,$c_2 > 0$;

Setting initial values for ($k = 1$), where $\left\{\lambda_j^i(1) j \in s \leftarrow (i); i = 1; \ldots; m^+\right\}$;

Process 1: 1. For fixed $\lambda(k) = \left\{\lambda_j^i(k)\right\}$, the goal is to compute $w(k)$:

1.1. Compute $\left\{\hat{x}_1; \ldots; \hat{x}_m+\right\}$ by

1.2. Solve the QPP (3.536), obtaining the solution $\hat{\beta} = \left(\hat{\beta}_1; \ldots; \hat{\beta}_m+\right)^\mathrm{T}$;

1.3. Compute $\hat{w}_-$; $\hat{b}_-$ from

1.4. Set $w(k) = \hat{w}$; $b(k) = \hat{b}$.

2. For fixed $w_-(k)$;$b_-(k)$, the goal is to compute $\lambda(k + 1)$:

2.1. Solve quadratic programming (3.520) with the variables λ;η; obtaining the solution $\hat{\lambda}$.

2.2. Set $\lambda\ (k + 1) = \hat{\lambda}$;

3. If $\|\lambda(k + 1) - \lambda(k)\| < \varepsilon$, goto 4; Otherwise, goto the step 1, setting $k = k + 1$.

4. Obtain the first optimal hyperplane $(w_-^* \cdot x) + b_-^* = 0$ and construct the "most positive" training set according to

Process 2: Construct and solve the optimization problem (3.539), get the second hyperplane: $(w_+^* \cdot x) + b_+^* = 0$.

Output: For any new input x, assign it to the class $k(k = -; +)$ by (3.526)

Next, we explain to how to solve the nonlinear MI-NSVM. This process is similar to that of linear MI-NSVM. In order to simplify, we only discuss the different part with linear MI-NSVM. For fixed $\hat{\lambda}^i_{j;} i = 1, \ldots, m^+; j \in s \leftarrow (i)$, the optimization problem (3.535) and its dual problem are replaced by

$$\begin{aligned} &\min_{k_-, b_-, \eta} \frac{1}{2} \| K\left(B, C^\top\right) k_- + e_- b_- \|^2 + c_1 e_+^\top \eta \\ &\text{s.t.} -\left(K\left(\hat{A}, C^\top\right) k_- + e_+ b_-\right) + \eta \geqslant e_+, \eta \geqslant 0 \end{aligned} \tag{3.541}$$

and

$$\begin{aligned} &\max\ e_+^{\mathrm{T}} \beta\beta - \frac{1}{2} \beta^{\mathrm{T}} S \left(R^{\mathrm{T}} R\right)^{-1} S^{\mathrm{T}} \beta \\ &\text{s.t.}\ = 0 \leq \beta \leq c_1 e_+; \end{aligned} \tag{3.542}$$

where $S = \left[K\left(\hat{A}; C^{\mathrm{T}}\right) e_+\right]$; $R = [K(B; C^{\mathrm{T}}) e_-]$. For fixed $\hat{k}_-$ and $\hat{b}_-$, the optimization (26) can be substituted by

$$\begin{aligned} &\min_{\lambda, \eta} c_1 e_+^\top \eta \\ &\text{s.t.} K\left(\sum_{V \in 3(i)} \lambda_j^i x_j, C^\top\right) \hat{k}_- + \hat{b}_- \geqslant 1 - \eta_i, i = 1, \ldots, m^+ \\ &\lambda_j^i \geqslant 0, i = 1, \ldots, m^+ \\ &\sum_{j \in \mathcal{Y}(i)} \lambda_j^i = 1, i = 1, \ldots, m^+ \end{aligned} \tag{3.543}$$

Correspondingly, the optimization problem (3.535) and its dual problem are replaced by

$$\begin{aligned} &\max\ e_-^{\mathrm{T}} \alpha - \frac{1}{2} \alpha^{\mathrm{T}} L \left(M^{\mathrm{T}} M\right)^{-1} L^{\mathrm{T}} \alpha \\ &\text{s.t.}\ 0 \leq \alpha \leq c_2 e_-; \end{aligned} \tag{3.544}$$

Where $L = \left[K\left(B, C^{\top}\right) e_{-}\right], M = \left[K\left(\hat{A}, C^{\top}\right) e_{+}\right]$. The formula (3.523)–(3.524) is replaced by

$$f_{+}(x) = K\left(x^{\top}, C^{\top}\right) k_{+} + b_{+} = 0 \tag{3.545}$$

$$f_{-}(x) = \frac{K\left(x^{\top}, C^{\top}\right) k_{+} + b_{+}}{\sqrt{k_{+}^{\top} K\left(C, C^{\top}\right) k_{+}^{\top}}} + \frac{K\left(x^{\top}, C^{\top}\right) k_{-} + b_{-}}{\sqrt{k_{-}^{\top} K\left(C, C^{\top}\right) k_{-}^{\top}}} = 0 \tag{3.546}$$

The detailed algorithm's procedure is similar to Algorithm 3.1. In the following, we compare the computational complexity of MI-NSVM and MI-SVM1. See optimization problems (3.520) or (3.532) the number of variables is only approximative half of MI-SVM. We know that MI-NSVM and MI-SVM can be solved by successive QPPs and LPPs. Here we rough estimates the computational complexity of QPP and LPP is $\mathcal{O}(n)^3$ (n denotes the number of variables). So, the computational complexity of MI-SVM can be expressed as $2k\mathcal{O}(n)^3$ (Suppose MI-SVM exists n variables, k is iterations of the successive QPP and LPP method), and the computational complexity of MI-NSVM as$(2k + 1)\,\mathcal{O}\left(\frac{n}{2}\right)^3$. So, these sizes of the optimization problem of MI-SVM are about four times than that of MI-NSVM.

3.4 Laplacian Support Vector Machine Classifiers

3.4.1 Successive Overrelaxation for Laplacian Support Vector Machine

In this section, we propose a novel fast Laplacian SVM classification (FLapSVM), which is deduced by the traditional SVM progress, and can overcome two drawbacks mentioned above effectively. Finally, FLapSVM can be solved efficiently by the successive overrelaxation (SOR) technique, which converges linearly to a solution and can process very large data sets that need not reside in memory, which make it more suitable for large scale problems.

3.4.1.1 Background

In this section, we give a brief outline of LapSVM.

SSL Framework

Regularization [76] is a key technology for obtaining smooth decision functions and avoiding overfitting of the training data, which is widely used in machine learning [77, 78]. Recently, the regularization framework has been recently extended in the SSL field by [77] as follows.

Given a set of labeled data

$$T = \{(x_1, y_1), \ldots, (x_l, y_l)\} \in \left(R^n \times \mathcal{Y}\right)^l \tag{3.547}$$

where $x_i \in R^n$, $y_i \in \mathcal{Y} = \{1, -1\}$, $i = 1, \ldots, l$, and a set of unlabeled data

$$(xl + 1, \ldots, xl + u) \tag{3.548}$$

where $x_i \in R^n$. For a kernel function $(\cdot, \cdot)$, which associates a reproducing kernel Hilbert space $\mathcal{H}_k$, the decision function can be obtained by minimizing

$$f^* = \arg\ \min \sum\nolimits_{i=1}^{l} V\ (x_i, y_i, f) + \gamma\mathcal{H} \| f \|_{\mathcal{H}}^2 + \gamma\mathcal{M} \| f \|_{\mathcal{M}}^2 \tag{3.549}$$

where f is an unknown decision function, V represents some loss function on the labeled data, $\gamma_{\mathcal{H}}$ is the weight of $\| f \|_{\mathcal{H}}^2$ and controls the complexity of f in the reproducing kernel Hilbert space. $\gamma\mathcal{M}$ is the weight of $\| f \|_{\mathcal{M}}^2$ and controls the complexity of the function in the intrinsic geometry of marginal distribution, $\| f \|_{\mathcal{M}}^2$ is able to penalize f along the Riemann manifold $\mathcal{M}$.

LapSVM

The same as traditional SVM, LapSVM also uses the hinge function

$$V\ \ (x_i, y_i, f) = \max\ \{0, 1 - y_i f\ (x_i)\} \tag{3.550}$$

as its loss function.

To solve nonlinear classification problem, according to reproducing theorem [67], weights w can be expressed as

$$\mathrm{w} = \sum\nolimits_{i=1}^{l+u} \alpha_i \Phi\ (x_i) = \Phi\alpha \tag{3.551}$$

where $\Phi = \{\varphi_1, \ldots, \varphi_{l+u}\}^{\mathrm{T}}$, $\alpha = \{\alpha 1, \ldots, \alpha l + u\}^{\mathrm{T}}$, φ is the feature map from the data space into the feature space.

Next, Belkin et al. [77] assume that the probability distribution of data has the geometric structure of a Riemannian manifold $\mathcal{M}$. The labels of two points that are close in the intrinsic geometry of $P_{\mathcal{X}}$ should be the same or similar, and use the intrinsic regularizer $\|f\|_{\mathcal{M}}^2$ to describe the constraint above

$$\|f\|_{\mathcal{M}}^2 = \sum_{i=1}^{l+u}\sum_{i=1}^{l+u} \mathrm{w}_{ij}\left(f(x_i) - f(x_j)\right)^2 = f^{\mathrm{T}} L f \tag{3.552}$$

where L is the graph Laplacian defined as $L = D - W$, where D is a diagonal matrix with its i th diagonal $D_{ii} = \sum_{d=1}^{l+u} W_{ij}$, and the edge weight matrix W can be determined by k nearest neighbor or graph kernels [77]. In practice, choosing exponential weights for the adjacency matrix leads to convergence of the graph Laplacian to the Laplace-Beltrami operator on the manifold [79].

Let

$$\|f\|_{\mathcal{H}}^2 = \|\mathrm{w}\|^2 = (\Phi\alpha)^{\mathrm{T}}(\Phi\alpha) = \alpha^{\mathrm{T}} K^{-}\alpha \tag{3.553}$$

and

$$f(x) = \sum_{i=1}^{l+u} \alpha_i K(x_i, x) \tag{3.554}$$

and then introducing the slack variables $\xi = \{\xi_i, \ldots, \xi_l\}$, the primal optimization problem can be written as

$$\begin{aligned} &\min_{a,\xi} \sum_{i=1}^{l} \xi_i + \gamma_{\mathcal{H}} \alpha^{\top} \overline{K} \alpha + \gamma_{\mathcal{M}} \alpha^{\top} \overline{K} L \overline{K} \alpha \\ &\text{s.t.}\, y_i \left(\sum_{j=1}^{l+u} \alpha_i K(x_i, x_j) + B \right) \geq 1 - \xi_i, i = 1, \ldots, l \\ &\quad \xi_i \geq 0, i = 1, \ldots, l \end{aligned} \tag{3.555}$$

Correspondingly, the dual problem of (3.555) can be expressed as

$$\min \frac{1}{2}\beta^{\mathrm{T}} Q \beta - \sum_{i=1}^{l} \beta_i$$

$$s.t. \sum_{i=1}^{l} \beta_i y_i = 0$$

$$0 \leq \beta_i \leq 1, i = 1, \ldots, l \tag{3.556}$$

where

$$Q = Y J_{\mathcal{L}} \overline{K} \left(2\gamma \mathcal{H} + 2\gamma \mathcal{M} \overline{K} L \right)^{-1} J_{\mathcal{L}}^{\mathrm{T}} Y \tag{3.557}$$

$Y = \text{diag}(y_1, \ldots, y_l)$, $J_{\mathcal{L}} = [I, 0]$ is a matrix of $l \times (l + u)$, where I is a $l \times l$ identity matrix. To obtain the decision function, α is computed as

$$\alpha = \left(2\gamma\mathcal{H}^I + 2\gamma\mathcal{M}\overline{K}L\right)^{-1} J_{\mathcal{L}}^{\mathrm{T}} Y\beta \tag{3.558}$$

As we can see that, LapSVM needs to solve the inverse matrix of $\left(2\gamma\mathcal{H}^I + 2\gamma\mathcal{M}\overline{K}L\right)^{-1}$ and burdens the computations related to the variable switching (3.558), which is in practice intractable or even impossible for large-scale data.

3.4.1.2 FLAPSVM

In this section, we describe our new algorithm: FLapSVM.

Linear Case

Unlike LapSVM, we first propose our algorithm from linear case. Suppose $\| f \|_{\mathcal{H}}^2 = \|w\|^2$, $\| f \|_{\mathcal{M}}^2 = f^{\mathrm{T}} L f$, where

$$f(x) = (w \cdot x) + b. \tag{3.559}$$

Introducing b^2 into the object function to ensure the solution uniqueness, the primal problem is expressed as

$$\min \frac{1}{2}\left(\|w\|^2 + b^2\right) + Ce_1^{\mathrm{T}}\xi + \frac{\gamma}{2}\eta^{\mathrm{T}} L\eta \tag{3.560}$$

$$s.t. \; Mw + e_2 b = \eta \tag{3.561}$$

$$Y\ (Aw + eb) \geq e1 - \xi \tag{3.562}$$

$$\xi \geq 0 \tag{3.563}$$

where $\xi = \{\xi_1, \ldots, \xi_l\}$, $\eta = \{\eta_1, \ldots, \eta_l\}$, $Y = \text{diag}(y_1, \ldots, y_l)$ is a diagonal matrix, C and γ are the penalty parameters, e_1, e_2 are the vectors of one of appropriate dimensions, $M = [A^{\mathrm{T}}\ U^{\mathrm{T}}]^{\mathrm{T}}$, where $A \in R^{l \times n}$ denotes the training data with labels, $U \in R^{u \times n}$ denotes the unlabeled data.

Note, we do not put $Mw + e_2 b = \eta$ into the object function (3.560). Our key idea is to make a very simple, but very fundamental change in the formulation, namely take (3.561) as an equality constraint to infer its dual problem. Therefore, the Lagrangian corresponding to the problem (3.560)–(3.563) is given by

$$L(\Theta) = \frac{1}{2}\left(\|w\|^2 + b^2\right) + Ce_1^{\mathrm{T}}\xi + \frac{\gamma}{2}\eta^{\mathrm{T}}L\eta - \beta^{\mathrm{T}}\xi$$

$$+\delta^{\mathrm{T}}(Mw + e_2 b - \eta) - \alpha^{\mathrm{T}}(Y(Aw + e_1 b) - e_1 + \xi) \tag{3.564}$$

where $\alpha = (\alpha 1, \ldots, \alpha l)^{\mathrm{T}}$, $\beta = (\beta_1, \ldots, \beta_l)^{\mathrm{T}}$, and $\delta = (\delta_1, \ldots, \delta_{l+u})^{\mathrm{T}}$ are vectors of Lagrange multipliers, $\xi = (\xi_1, \ldots, \xi_l)^{\mathrm{T}}$, $\eta = (\eta_1, \ldots, \eta_{l+u})^{\mathrm{T}}$, $\Theta = \{w, \xi, b, \eta, \alpha, \beta, \delta\}$. Therefore, the dual problem can be formulated as

$$\begin{aligned} &\max_{\Theta} \quad L(\Theta) \\ &\text{s.t.} \quad \nabla_{w,b,\xi,\eta} L(\Theta) = 0 \\ &\qquad \alpha, \beta \quad \geq 0 \end{aligned} \tag{3.565}$$

From (3.565), we get

$$\nabla_w L = w + M^{\mathrm{T}}\delta - A^{\mathrm{T}}Y^{\mathrm{T}}\alpha = 0 \tag{3.566}$$

$$\nabla_b L = b + e_2^{\mathrm{T}}\delta - e_1^{\mathrm{T}}Y^{\mathrm{T}}\alpha = 0 \tag{3.567}$$

$$\nabla_\xi L = Ce1 - \alpha - \beta = 0 \tag{3.568}$$

$$\nabla_\eta L = \gamma L\eta - \delta = 0. \tag{3.569}$$

Equations (3.566), (3.567), and (3.569) imply

$$w = A^{\mathrm{T}}Y^{\mathrm{T}}\alpha - \gamma M^{\mathrm{T}}L\eta \tag{3.570}$$

$$b = e_1^{\mathrm{T}}Y^{\mathrm{T}}\alpha - \gamma e_2^{\mathrm{T}}L\eta. \tag{3.571}$$

Using (3.566)–(3.571), the Wolfe dual of the problem (3.565) can be expressed as

$$\min \frac{1}{2}\left(\alpha^{\mathrm{T}}\eta^{\mathrm{T}}\right) Q \left(\alpha^{\mathrm{T}}\eta^{\mathrm{T}}\right)^{\mathrm{T}} - e_1^{\mathrm{T}}\alpha$$

$$s.t. \; 0 \leq \alpha_i \leq C, i = 1, \ldots, l \tag{3.572}$$

where Q is expressed as

$$\begin{pmatrix} Y\left(AA^{\mathrm{T}}+e_1e_1^{\mathrm{T}}\right)^{\mathrm{T}}Y^{\mathrm{T}}-\gamma Y\left(AM^{\mathrm{T}}+e_2e_2^{\mathrm{T}}\right)L^{\mathrm{T}} \\ -\gamma\left(Y\left(AM^{\mathrm{T}}+e_1e_2^{\mathrm{T}}\right)L^{\mathrm{T}}\gamma^2L\left(MM^{\mathrm{T}}+e_2e_2^{\mathrm{T}}\right)L^{\mathrm{T}}+\gamma L^{\mathrm{T}}\right. \end{pmatrix} \tag{3.573}$$

which is a positive definite matrix of $(2l+u)\times(2l+u)$.

The optimization problem (3.572) does not need to both solve a corresponding inverse matrix.

Nonlinear Case

Different with LapSVM, only inner products appear in the dual problems (26), so the kernel functions can be applied directly into the problem (26) and the linear FLapSVM is easily extended to the nonlinear classifiers. In detail, the dual problem of FLapSVM in nonlinear case can be easily written as

$$\min \frac{1}{2}\left(\alpha^{\mathrm{T}}\eta^{\mathrm{T}}\right)Q^{-}\left(\alpha^{\mathrm{T}}\eta^{\mathrm{T}}\right)^{\mathrm{T}}-e_1^{\mathrm{T}}\alpha$$

$$s.t.\ 0\le\alpha_i\le C, i=1,\ldots,l \tag{3.574}$$

where $\overline{Q}$ is

$$\begin{pmatrix} YK\left(A,A^{\top}\right)^{\top}Y^{\top} & -\gamma YK\left(A,M^{\top}\right)L^{\top} \\ -\gamma YK\left(A,M^{\top}\right)L^{\top} & \gamma^2LK\left(M,M^{\top}\right)L^{\top}+\gamma L^{\top} \end{pmatrix}$$

$$+\begin{pmatrix} Yele_1^{\mathrm{T}}Y^{\mathrm{T}} & -\gamma Yele_2^{\mathrm{T}}L^{\mathrm{T}} \\ -\gamma Yele_2^{\mathrm{T}}L^{\mathrm{T}} & \gamma^{2Le}2e_2^{\mathrm{T}}L^{\mathrm{T}} \end{pmatrix} \tag{3.575}$$

K is the kernel matrix formed by kernel functions

$$K\left(x_i,x_j\right)=\left(\varphi\left(x_i\right),\varphi\left(x_j\right)\right) \tag{3.576}$$

where β is a feature map from the data space into the feature space. The decision function is

$$f(x)=K\left(x,A^{\mathrm{T}}\right)Y^{\mathrm{T}}\alpha-\gamma K\left(x,M^{\mathrm{T}}L\right)\eta+b \tag{3.577}$$

x denotes a new input and b is expressed in (3.571).

3.4.1.3 Implementation Issues

SOR Technique

The optimization problem (3.572) and (3.574) are the standard QPPs. Compared with the traditional SVM, (3.572) or (3.574) has more concise constraint conditions. In fact, our algorithm can be solved efficiently by SOR technique [42]. Take the QPP (3.574) as an example, it can be rewritten as

$$\min \frac{1}{2}\lambda^{\mathrm{T}} Q^{-} \lambda - \kappa^{\mathrm{T}} \lambda$$

$$s.t.\ 0 \le \lambda_i \le C, i = 1, \ldots, l \tag{3.578}$$

where $\lambda = \left(\alpha^{\mathrm{T}} \eta^{\mathrm{T}}\right)^{\mathrm{T}}$, $\overline{Q}$ is defined by (29) and $\kappa = \Bigg(\underbrace{1, \ldots, 1}_{l}, \underbrace{0, \ldots, 0}_{l+u}\Bigg)^{\mathrm{T}}$

In practice, λ_j^{i+1} is computed by $\left(\lambda_1^{i+1}, \ldots, \lambda_{j-1}^{i+1}, \lambda_j^{i+1}, \ldots, \lambda_l^{i+1}\right)$, and the latest computed components of λ are used in the computation of λ_j^{i+1}. The strictly lower triangular matrix G in (3.579) can be seen as a substitution operator,

---

Algorithm 3.12 (SOR for FLapSVM)

---

(1) Input the training set (1) and the unlabeled data (2);

(2) Choose appropriate kernels $K(x, x')$, and appropriate parameters $C, \gamma > 0$;

(3) Choose $\omega \in (0, 2)$. Start with any $\lambda^0 \in R^{2l+u}$. Having λ^i, compute λ^{i+1} as follows:

$$\lambda^{i+1} = \left(\lambda^i - \omega E^{-1}\left(Q^{-}\lambda^i + \kappa + G\left(\lambda^{i+1} - \lambda^i\right)\right)\right)_{\#} \tag{3.579}$$

until $\|\lambda^{i+1} - \lambda^i\|$ is than some prescribed tolerance, where $(\cdot)\#$ is the two-norm projection on the feasible region of (32), that is

$$((\lambda)_{\#})_i = \begin{Bmatrix} 0 \ \text{ if } \lambda_i \le 0 \\ \lambda_i \text{ if } 0 < \lambda_i < C \\ C \ \text{ if } \lambda_i \ge C \end{Bmatrix}, i = 1, \ldots, l \tag{3.580}$$

and

$$((\lambda)_1)_i = \lambda_i, i = l + 1, \ldots, 2l + u \tag{3.581}$$

where the nonzero elements of $G \in R^{(2l+u) \times (2l+u)}$ constitute the strictly lower triangular part of the symmetric matrix $\overline{Q}$, and the nonzero elements of $E \in R^{(2l+u) \times (2l+u)}$ constitute the diagonal of $\overline{Q}$.

---

i.e., using $\left(\lambda_1^{i+1}, \ldots, \lambda_{j-1}^{i+1}\right)$ in the place of $\left(\lambda_1^{i}, \ldots, \lambda_{j-1}^{i}\right)$.. The iterative formula can be written as

$$\lambda_j^{i+1} = \left(\lambda_j^i - \omega E_{jj}^{-1}\left(\sum\nolimits_{k=1}^{j-1} Q_{jk}^{-}\lambda_k^{i+1} + \sum\nolimits_{k=j}^{2l+u} Q_{jk}^{-}\lambda_k^i - 1\right)\right)_{\#} \tag{3.582}$$

From (3.582), we can find that only one variable needs to be updated in each iteration of Algorithm 3.1.

Mangasarian and Musicant [42] pointed out that SOR can process very large data sets, and does not need to reside in memory. The algorithm converges linearly to a solution, and has shown that on smaller problems, SOR is faster than SVM^{light} and comparable or faster than Sequential Minimal Optimization (SMO) [53]. Remarkably Hsieh et al. [43]. applied the SOR method to the famous dual coordinate descent method for large-scale linear SVM. To our knowledge, we did not find any SOR version that can deal with semi-supervised classification problem. Thus, it is an important contribution in this section that FLapSVM can be effectively solved by SOR technology.

Random Scheduling of Subproblem Technology

In Algorithm 3.12, we have known that FLapSVM only handles the one-point subproblems in each iteration. Chang et al. [43, 80] showed that solving subproblems in an arbitrary order may give the faster convergence. This motivates us to resort the subproblems randomly after all elements of μ^i are updated. Concretely, at the kth outer iteration,1 we sort randomly $\{i = 1, \ldots, l\}$ to $\{\varpi(1), \ldots, \varpi(l)\}$, and handle sub-problems in the order of $\lambda_{\varpi(1)}, \ldots, \lambda_{\varpi(l)}$. Each outer iteration generates vectors $\lambda^{i,j} \in R^l, j = 1, \ldots, l+1$, such that $\lambda^{i,1} = \lambda^i$, $\lambda^{i,l+1} = \lambda^{i+1}$, and

$$\lambda^{i,j} = \left(\lambda_{\varpi(1)}^{i+1}, \ldots, \lambda_{\varpi(j-1)}^{i+1}, \lambda_{\varpi(j)}^{i}, \ldots, \lambda_{\varpi(l)}^{i}\right)^{\mathrm{T}} \tag{3.583}$$

where $j = 2, \ldots, l$.

Finally, update $\lambda^{i,j}$ to $\lambda^{i,j+1}$ by

$$\lambda_{\omega(j)}^{i+1} = \left(\lambda_{\omega(j)}^{i} - \omega E_{\omega(j,j)}^{-1}\Big(\sum_{\omega(k)=\omega(1)}^{\omega(j-1)} \overline{Q}_{\omega(j,k)}\lambda_{\omega(k)}^{i+1}\right.$$

$$\left. + \sum\nolimits_{\omega(k)=\omega(j)}^{\omega(2l+u)} \overline{Q}_{\omega(j,k)}\lambda_{\omega(k)}^{i} - 1\Big)\right)_{\#} \tag{3.584}$$

where $\omega(j,j)$ and $\omega(j,k)$ mean $\omega(j)\omega(j)$ and $\omega(j)\omega(k)$, respectively.

In addition, due to the existing high amount of unlabeled data in the SSL framework, we also use the stability of the decision $y(x) =$ sign $(f(x))$, $x \in u$ as the early stopping criteria (SC). The detail progress can be found in [81].

3.4.1.4 Complexity Analysis

For the QPP problem (3.578), despite eliminating the computation of the corresponding inverse matrix, the model has more $l + u$ variables than the traditional LapSVM. Is the computational complexity of FLapSVM really smaller than that of the LapSVM? We know, at present, solving the inverse matrix effectively is still a difficult problem. First, we suppose that the computational cost of $K\left(2\gamma_{\mathcal{H}} I + 2\gamma \mathcal{M}^{KL}\right)^{-1}$ of $(l + u) \times (l + u)$ in the problem of (3.556) is $\mathcal{O}(l + u)^3$, so the computational cost of LapSVM is at least $\mathcal{O}(l + u)^3$. Next, let us analysis the computational complexity of FLapSVM. Each SOR iteration requires to compute the $2l + u$ product, leading to a complexity of $\mathcal{O}\,(2l + u)$ to update the current λ. So, the computational cost of FLapSVM is approximately (the number of iterations). $\mathcal{O}\,(2l + u)$. All experiments indicate that the average value of the number of iterations is about two third of $2l + u$ (the average computational complexity of FLapSVM is roughly $(m^{1.8} \sim m^{2.2})$, where $m = l + u$ in all experiments), which is quicker than LapSVM and outperforms PlapSVM in the most cases.

3.4.2 *Laplacian Twin Support Vector Machine for Semi-supervised Classification*

3.4.2.1 Laplacian Twin Support Vector Machine for Semi-supervised Classification (Called Lap-TSVM)

Semi-supervised Learning Framework

Regularization [76] is a key technology for obtaining smooth decision functions and thus avoiding overfitting to the training data, which is widely used in machine learning [77, 78, 82, 83]. Recently, the regularization framework has been recently extended in the SSL field as follows [77, 84].

Given a set of labeled data and a set of unlabeled data

$$(x_{l+1}, \ldots x_{l+u}) \tag{3.585}$$

where $x_{l+i} \in R^n$, $i = 1, \ldots, u$. Suppose the labeled data are generated according to the distribution P on $X \times R$, whereas unlabeled examples are drawn according to the marginal distribution P_X of P. Labels of samples can be obtained from the conditional probability distribution $(y|x)$. The manifold regularization approach exploits the geometry information of the marginal distribution P_X. An important premise of this kind of approach is to assume that the probability distribution of data has the geometric structure of a Riemannian manifold M. The labels of two points that are close in the intrinsic geometry of P_X should be the same or similar. Belkin et al. [77] applied the intrinsic regularizer $\|f\|_M^2$ to describe the constraint above,

$$\|f\|_M^2 = \sum_{i=1}^{l+u} \sum_{i=1}^{l+u} \left(f(x_i) - f(x_j)\right)^2 = f^{\mathrm{T}} L f \tag{3.586}$$

where L is the graph Laplacian. In practice, choosing exponential weights for the adjacency matrix leads to convergence of the graph Laplacian to the Laplace-Beltrami operator on the manifold [84]. For a kernel function $(\cdot, \cdot)$, which is associated with a reproducing kernel Hilbert space H_k, the decision function can be obtained by minimizing

$$f^* = \arg \min \sum_{i=1}^{l} V(x_i, y_i, f) + \gamma_H \left\| f\|_H^2 + \gamma_M \right\| f \Big\|_M^2 \tag{3.587}$$

where f is an unknown decision function, V represents some loss function on the labeled data, and γ_H is the weight of $\|f\|_H^2$ and controls the complexity of f in the Reproducing Kernel Hilbert Space. γ_M is the weight of $\|f\|_M^2$ and controls the complexity of the function in the intrinsic geometry of marginal distribution, and $\|f\|_M^2$ is able to penalize f along the Riemann manifold M. More detailed discussion can be found in [77].

Linear Lap-TSVM

Similar to the TSVM, we use the square loss function and hinge loss function for Lap-TSVM. $V_{\pm}(x_i, y_i, f_{\pm})$ can be expressed as

$$V_+(x_i, y_i, f_+) = ((A_i, \cdot \cdot w_+) + b_+)^2 + \max(0, 1 - f_+(B_i, \cdot)), \tag{3.588}$$

$$V_-(x_i, y_i, f_-) = ((B_i, \cdot \cdot w_-) + b_-)^2 + \max(0, 1 - f_-(A_i, \cdot)). \tag{3.589}$$

A_i, or B_i, denotes the i-th row of A or B.

Correspondingly, the decision functions are written as

$$f_+(x) = (w_+ \cdot x) + b_+, \tag{3.590}$$

$$f_-(x) = (w_- \cdot x) + b_-. \tag{3.591}$$

The regularization terms $\| f_+ \|_H^2$ and $\| f_- \|_H^2$ can be expressed by

$$\| f_+ \|_H^2 = \frac{1}{2}\left(\| w_+ \|_2^2 + b_+^2\right), \tag{3.592}$$

$$\| f_- \|_H^2 = \frac{1}{2}\left(\| w_- \|_2^2 + b_-^2\right). \tag{3.593}$$

For manifold regularization, a data adjacency graph $W_{(l+u)\times(l+u)}$ is defined by nodes $W_{i,j}$, which represents the similarity of every pair of input samples. The weight matrix W may be defined by k nearest neighbor or graph kernels as follows [77]:

$$W_{ij} = \begin{cases} \exp\left(-\| x_i - x_j \|_2^2 / 2\sigma^2\right), & \text{if } x_i, x_j \text{ are neighbor;} \\ 0, & \text{Otherwise,} \end{cases} \tag{3.594}$$

where $\| x_i - x_j \|_2^2$ denotes the Euclidean norm in R^n. So, the manifold regularization is defined by

$$\| f_+ \|_M^2 = \frac{1}{(l+u)^2} \sum\nolimits_{i,j=1}^{l+u} W_{i,j}\left(f_+(x_i) - f_+(x_j)\right)^2 = f_+^{\mathrm{T}} L f_+, \tag{3.595}$$

$$\| f_- \|_M^2 = \frac{1}{(l+u)^2} \sum\nolimits_{i,j=1}^{l+u} W_{i,j}\phi_-(x_i) - f_-(x_j)\,)^2 = f_-^{\mathrm{T}} L f_-, \tag{3.596}$$

where $L = D - W$ is the graph Laplacian, D is a diagonal matrix with its i-th diagonal $D_{ii} = \sum_{j=1}^{l+u} W_{ij}$, $f_+ = [f_+(x_1), \ldots, f_+(x_{l+u})]^{\mathrm{T}} = Mw + eb_+$, $f_- = [f_-(x_1), \ldots, f_-(x_{l+u})]^{\mathrm{T}} = Mw- + eb -$, $M \in R^{(l+u)\times n}$ includes all of labeled data and unlabeled data, and e is an appropriate ones vector. When (3.595) or (3.596) is used as a penalty item of Eq. (3.586), we can understand them by these means: if the neighbor of x_i, x_j has the higher similarity (W_{ij} is larger), the difference of $f_\pm(x_i)$, $f_\pm(x_j)$ will obtain a big punishment. More intuitively, the smaller $|f_\pm(x_i) - f_\pm(x_j)|$ is, the smoother $f_\pm(x)$ in the data adjacency graph is.

Substituting (3.589)–(3.597) into (3.588), the primal problems of Linear Lap-TSVM can be written as

$$\begin{aligned} \min_{w_+, b_+, \xi} \ & \frac{1}{2}\| Aw_+ + e_+ b_+ \|_2^2 + c_1 e_-^{\top}\xi + c_2\left(\| w_+ \|_2^2 + b_+\right) \\ & + c_3\left(w_+^{\top} M^{\top} + e^{\top} b_+\right) L\left(Mw_+ + eb_+\right) \\ & \text{s.t.} - (Bw_+ + eb_+) + \xi \geq e_-, \xi \geq 0 \end{aligned} \tag{3.597}$$

and

$$\begin{aligned}\min_{w-,b-\pi}\ & \tfrac{1}{2}\| Bw_- + e_- b_- \|_2^2 + c_1 e_+^\top \eta + c_2 \left(\| w_- \|_2^2 + b_-\right) \\ & + c_3 \left(w_-^\top M^\top + e^\top b_-\right) L \left(M w_- + e b_-\right) \\ & \text{s.t. } (A w_- + e_+ b_-) + \eta \geq e_+, \eta \geq 0\end{aligned} \tag{3.598}$$

The Lagrangian corresponding to the problem (26) is given by

$$\begin{aligned} L(\Theta) = & \frac{1}{2}(Aw_+ + e_+ b_+)^{\mathrm{T}} (Aw_+ + e_+ b_+) + c_1 e_-^{\mathrm{T}} \xi \\ & + \frac{1}{2} c_2 \left(\|w_+\|_2^2 + b_+^2\right) + \frac{1}{2} c_3 \left(w_+^{\mathrm{T}} M^{\mathrm{T}} + e^{\mathrm{T}} b_+\right) \\ & \times L(Mw + +eb_+) - \alpha^{\mathrm{T}} \Big(- (Bw + +e_- b_+) \\ & + \xi - e_- \Big) - \beta^{\mathrm{T}} \xi \end{aligned} \tag{3.599}$$

where $\Theta = \{w_+, b_+, \xi, \alpha, \beta\}$, $\alpha = \left(\alpha_1, \ldots, \alpha_{m_1}\right)^{\mathrm{T}}$, $\beta = \left(\beta_1, \ldots, \beta_{m_1}\right)^{\mathrm{T}}$ are the Lagrange multipliers. The dual problem can be formulated as

$$\begin{aligned} & \max\ L(\Theta) \\ & s.t.\ \nabla_{w+, b_+, \xi} L(\Theta) = 0, \\ & \alpha, \beta \geq 0. \end{aligned} \tag{3.600}$$

From Eq. (3.600), we get

$$\nabla_{w+} L = A^{\mathrm{T}} (Aw + +e_+ b_+) + c_2 w + c_3 M^{\mathrm{T}} L (Mw + +eb_+) + B^{\mathrm{T}} \alpha = 0, \tag{3.601}$$

$$\nabla_{b_+} L = e_+^{\mathrm{T}} (Aw + +e_+ b_+) + c_2 b_+ + c_3 e^{\mathrm{T}} L (Mw + +eb_+) + e_-^{\mathrm{T}} \alpha = 0, \tag{3.602}$$

$$\nabla_{\xi} L = c_1 e_- - \alpha - \beta = 0. \tag{3.603}$$

Since $\beta \geq 0$, (3.603) turns out to be

$$0 \leq \alpha \leq c_1 e_-. \tag{3.604}$$

Next, combining (3.601) and (3.602) leads to

$$\begin{bmatrix} A^{\top} \\ e_{+}^{\top} \end{bmatrix} [Ae_{+}] \begin{bmatrix} w_{+} \\ b_{+} \end{bmatrix} + c_2 \begin{bmatrix} w_{+} \\ b_{+} \end{bmatrix}$$

$$+c_3 \begin{bmatrix} M^{\top} \\ e^{\top} \end{bmatrix} L \left[M\ e_{+} \right] \begin{bmatrix} w_{+} \\ b_{+} \end{bmatrix} + \begin{bmatrix} B^{\top} \\ e_{-}^{\top} \end{bmatrix} \alpha = 0 \tag{3.605}$$

Let

$$H = [Ae_{+}]\,,\ J = [Me]\,,\ G = [Be_{-}] \tag{3.606}$$

and the augmented vector $\vartheta + = \left[w_{+}^{\mathrm{T}} b_{+}^{\mathrm{T}}\right]^{\mathrm{T}}$. Equation (3.34) can be rewritten as:

$$\left(H^{\mathrm{T}}H + c_2 I + c_3 J^{\mathrm{T}} L J\right) \vartheta + + G^{\mathrm{T}} \alpha = 0,$$

$$i.e.,\ \ + = -\left(H^{\mathrm{T}}H + c_2 I + c_3 J^{\mathrm{T}} L J\right)^{-1} \left(G^{\mathrm{T}} \alpha\right), \tag{3.607}$$

where I is an identity matrix of appropriate dimensions. According to matrix theory [66], it can be easily proved that $H^{\mathrm{T}}H + c_2 I + c_3 J^{\mathrm{T}} LJ$ is a positive definite matrix.

Substituting the above equations into problem, we obtain the Wolfe dual of the problem as follows:

$$\max\ e_{-}^{\mathrm{T}} \alpha - \frac{1}{2} \alpha^{\mathrm{T}} G \left(H^{\mathrm{T}}H + c_2 I + c_3 J^{\mathrm{T}} L J\right)^{-1} G^{\mathrm{T}} \alpha$$

$$s.t.\ \ 0 \le \alpha \le c_1 e_{-}. \tag{3.608}$$

Similarly, the dual of (27) is

$$\max\ \beta e_{+}^{\mathrm{T}} \beta - \frac{1}{2} \beta^{\mathrm{T}} P \left(Q^{\mathrm{T}} Q + c_2 I + c_3 F^{\mathrm{T}} L F\right)^{-1} P^{\mathrm{T}} \beta$$

$$s.t. 0 \le \beta \le c_2 e+, \tag{3.609}$$

where

$$Q = [Ae_{-}]\,,\quad F = [Me]\ \ P = [Be_{+}] \tag{3.610}$$

and the augmented vector $\vartheta_- = [w_- b_-]^{\mathrm{T}}$ is given by

$$\vartheta_- = -\left(Q^{\mathrm{T}}Q + c_2 I + c_3 F^{\mathrm{T}} L F\right)^{-1} P^{\mathrm{T}} \beta, \tag{3.611}$$

where $Q^{\mathrm{T}}Q + c_2 I + c_3 F^{\mathrm{T}} L F$ is a positive definite matrix. Once vectors $\vartheta+$ and ϑ_- are obtained from (3.607) and (3.611), the separating planes

$$w_+^{\mathrm{T}} x + b_+ = 0,\ w_-^{\mathrm{T}} x + b_- = 0 \tag{3.612}$$

are known. A new data point $x \in R^n$ is then assigned to the positive or negative class, depending on which of the two hyperplanes it lies closest to, i.e.

$$f(x) = \underset{+.-}{\mathrm{argmin}} d_{\pm}(x), \tag{3.613}$$

where

$$d_{\pm}(x) = \left| w_{\pm}^{\mathrm{T}} x + b_{\pm} \right|, \tag{3.614}$$

where $|\cdot|$ is the perpendicular distance of point x from the planes $w_{\pm}^{\mathrm{T}} x + b_{\pm}$.

Nonlinear Lap-TSVM

Now we extend the linear Lap-TSVM to the nonlinear case.

The same as in the linear case, the cost function of the errors $V_+\ (x_i, y_i, f_+)$ and $V_-(x_i, y_i, f_-)$ can be expressed as (3.588) and (3.589). The decision function can be written as $f_{\pm}(x) = (w\pm\ \cdot \Phi(x)) + b_{\pm}$, where $\Phi(\cdot)$ is a nonlinear mapping from a low dimensional space to a higher dimensional Hilbert space H. According to Hilbert space theory [67], $w\pm$ can be expressed as $w\pm = \sum_{i=1}^{l+u} \lambda_{\pm i} \Phi\ (x_i) = \Phi(M)\lambda\pm$. So the following kernel-generated hyperplanes are:

$$K\left(x^{\mathrm{T}}, M^{\mathrm{T}}\right)\lambda_+ + b_+ = 0, \tag{3.615}$$

$$K\left(x^{\mathrm{T}}, M^{\mathrm{T}}\right)\lambda_- + b_- = 0, \tag{3.616}$$

where K is a chosen kernel function: $K(x_i \cdot x_j) = (\Phi(x_i) \cdot \Phi(x_j))$. By means of the kernel matrix K and relevant coefficients $\lambda\pm$, the regularization term $\| f_+ \|_{\mathrm{H}}^2$ and

$\| f_- \|_{\mathrm{H}}^2$ can be expressed as

$$\| f_+ \|_{\mathrm{H}}^2 = \frac{1}{2}\left(\lambda_+^{\mathrm{T}} K \lambda_+ + b_+^2\right), \tag{3.617}$$

$$\| f_- \|_{\mathrm{H}}^2 = \frac{1}{2}\left(\lambda_-^{\mathrm{T}} K \lambda_- + b_-^2\right). \tag{3.618}$$

For manifold regularization, on the basis of $f_\pm = [f_\pm(x_1), \ldots, f_\pm(x_{l+u})]^{\mathrm{T}} = K\lambda \pm + eb_\pm$, $\| f_+ \|_{\mathrm{M}}^2$ and $\| f_- \|_{\mathrm{M}}^2$ can be written as

$$\| f_+ \|_{\mathrm{M}}^2 = f_+^{\mathrm{T}} L f_+ = \left(\lambda_+^{\mathrm{T}} K + e^{\mathrm{T}} b_+\right) L \left(K\lambda + + eb_+\right), \tag{3.619}$$

$$\| f_- \|_{\mathrm{M}}^2 = f_-^{\mathrm{T}} L f_- = \left(\lambda_-^{\mathrm{T}} K + e^{\mathrm{T}} b_-\right) L \left(K\lambda_- + eb_-\right). \tag{3.620}$$

So, the nonlinear optimization problems can be expressed as

$$\begin{array}{c} \min\limits_{\lambda+.b+} \frac{1}{2} \| \; K\left(A, M^\top\right)\lambda_+ + e_+ b_+ \; \|^2 + c_1 e_-^\top \xi \\ + \frac{1}{2} c_2 \left(\lambda_+^\top K \lambda_+ + b_+^2\right) + c_3 \frac{1}{2}\left(\lambda_+^\top K + e^\top b_+\right) L \left(K\lambda_+ + eb_+\right) \\ \text{s.t.} - \left(K\left(B, M^\top\right)\lambda_+ + e_- b_+\right) + \xi \geq e_-, \xi \geq 0 \end{array} \tag{3.621}$$

and

$$\begin{array}{c} \min\limits_{\lambda-,b-,\eta} \frac{1}{2} \| \; K\left(B, M^\top\right)\lambda_- + e_- b_- \; \|^2 + c_1 e_+^\top \eta \\ + \frac{1}{2} c_2 \left(\lambda_-^\top K \lambda_- + b_-^2\right) + c_3 \frac{1}{2}\left(\lambda_-^\top K + e^\top b_-\right) L \left(K\lambda_- + eb_-\right) \\ \text{s.t.} \left(K\left(A, M^\top\right)\lambda_- + e_+ b_-\right) + \eta \geq e_+, \eta \geq 0 \end{array} \tag{3.622}$$

Define the Lagrangian corresponding to the problem (3.621) as follows

$$\begin{aligned} L(\Theta) = \frac{1}{2}\left\| K\left(A, M^{\mathrm{T}}\right)\lambda_+ + e_+ b_+ \right\|^2 \\ + c_1 e_-^{\mathrm{T}} \xi + \frac{1}{2} c_2 \left(\lambda_+^{\mathrm{T}} K \lambda_+ + b_+^2\right) \\ + c_3 \frac{1}{2}\left(\lambda_+^{\mathrm{T}} K + e^{\mathrm{T}} b_+\right) L \left(K\lambda_+ + eb_+\right) \\ -\alpha^{\mathrm{T}}\left(-\left(K\left(B, M^{\mathrm{T}}\right)\lambda + + e_- b_+\right) + \xi - e_-\right) - \beta^{\mathrm{T}} \xi, \end{aligned} \tag{3.623}$$

where $\Theta = \{\lambda_+, b_+, \xi, \alpha, \beta\}$.

The dual problem can be formulated as

$$\max\ L(\Theta)$$

$$s.t.\ \nabla_{\lambda_+, b_+, \xi} L(\Theta) = 0,$$

$$\alpha, \beta \geq 0. \tag{3.624}$$

From Eq. (3.624), we get

$$\nabla_{\lambda_+} L = K\left(A, M^{\mathrm{T}}\right)^{\mathrm{T}} \left(K\left(A, M^{\mathrm{T}}\right)\lambda + +e_+ b_+\right) + c_2 K\lambda$$

$$+c_3 K L\left(K\lambda + +eb_+\right) + K\left(B, M^{\mathrm{T}}\right)^{\mathrm{T}} \alpha = 0, \tag{3.625}$$

$$\nabla_{b_+} L = e_+^{\mathrm{T}}\left(K\left(A, M^{\mathrm{T}}\right)\lambda + +e_+ b_+\right) + c_2 b_+ + c_3 e^{\mathrm{T}} L\left(K\lambda + +eb_+\right) + e_-^{\mathrm{T}}\alpha = 0, \tag{3.626}$$

$$\nabla_{\xi} L = c_1 e_- - \alpha - \beta = 0. \tag{3.627}$$

Combining (3.625) and (3.626) leads to

$$\begin{bmatrix} K\left(A, M^{\top}\right)^{\top} \\ e_+^{\top} \end{bmatrix} \left[K\left(A, M^{\top}\right) e_+\right] \begin{bmatrix} \lambda_+ \\ b_+ \end{bmatrix} + c_2 \begin{bmatrix} K & 0 \\ 0 & 1 \end{bmatrix} \begin{bmatrix} \lambda_+ \\ b_+ \end{bmatrix} + c_3 \begin{bmatrix} K \\ e^{\top} \end{bmatrix} L\left[Ke\right] \begin{bmatrix} \lambda_+ \\ b_+ \end{bmatrix} + \begin{bmatrix} K\left(B, M^{\top}\right)^{\top} \\ e_-^{\top} \end{bmatrix} \alpha = 0 \tag{3.628}$$

Let

$$\begin{gathered} H_{\Phi} = \left[K\left(A, M^{\top}\right) e_+\right], O_{\phi} = \begin{bmatrix} K & 0 \\ 0 & 1 \end{bmatrix} \\ J_{\phi} = \left[Ke\right], G_{\phi} = \left[K\left(B, M^{\top}\right) e_-\right] \end{gathered} \tag{3.629}$$

and the augmented vector $\rho+ = [\lambda + b_+]^{\mathrm{T}}$, Eq. (3.628) can be rewritten as:

$$\begin{aligned} & H_\phi^{\top} H_\phi \rho_+ + c_2 O_\phi \rho_+ + c_3 J_\phi^{\top} L_\phi \rho_+ + G_\phi^{\top} \alpha_+ = 0 \\ & \text{i.e., } \rho_+ = -\left(H_\phi^{\top} H_\phi + c_2 O_\phi + c_3 J_\phi^{\top} L_\phi\right)^{-1} \left(G_\phi^{\top} \alpha\right) \end{aligned} \tag{3.630}$$

So, the Wolfe dual of the problem (3.621) is formulated as follows:

$$\max\ e_-^{\mathrm{T}} \alpha - \frac{1}{2}\left(\alpha^{\mathrm{T}} G_\Phi\right)\left(H_\Phi^{\mathrm{T}} H_\Phi + c_2 0_\Phi + c_3 J_\Phi^{\mathrm{T}} L J_\Phi\right)^{-1}\left(G_\Phi^{\mathrm{T}} \alpha\right)$$

$$s.t.\ 0 \le \alpha \le c_1 e_-. \tag{3.631}$$

In a similar manner, the dual of (3.622) is

$$\max\ \beta e_+^{\mathrm{T}} \beta - \frac{1}{2}\left(\beta^{\mathrm{T}} P_\Phi\right)\left(\mathrm{Q}_\Phi^{\mathrm{T}} Q_\Phi + c_2 U_\Phi + c_3 F_\Phi^{\mathrm{T}} L F_\Phi\right)^{-1}\left(P_\Phi^{\mathrm{T}} \beta\right)$$

$$s.t. 0 \le \beta \le c_2 e+, \tag{3.632}$$

where

$$\begin{aligned} Q_\phi &= \left[K\left(A, M^{\top}\right) e_-\right], U_\phi = \begin{bmatrix} K & 0 \\ 0 & 1 \end{bmatrix} \\ F_\phi &= [K e], P_\phi = \left[K\left(B, M^{\top}\right) e_+\right] \end{aligned} \tag{3.633}$$

and the augmented vector $\rho_- = [\lambda_- b_-]^{\mathrm{T}}$, which is given by

$$\rho_- = -\left(Q_\Phi^{\mathrm{T}} Q_\Phi + c_2 U_\Phi + c_3 F_\Phi^{\mathrm{T}} L F_\Phi\right)^{-1}\left(P_\Phi^{\mathrm{T}} \alpha\right). \tag{3.634}$$

Once vectors $\rho+$ and ρ_- are obtained from (3.630) and (3.634), a new data point $x \in \mathcal{T}^n$ is then assigned to the positive or negative class, depending on a manner similar to the linear case.

3.5 Loss Function of Support Vector Machine Classification

3.5.1 Ramp Loss Least Squares Support Vector Machine

In this section, by introducing a non-convex and non-differentiable loss function instead of the quadratic loss function to LSSVM, a robust and sparse LSSVM is constructed and named RLSSVM. Compared with the original LSSVM, RLSSVM can explicitly incorporate noise and outlier suppression in the training process, has less support vectors and the increased sparsity leads to its better scaling properties. Similar to RSVM, RLSSVM is non-convex and the CCCP procedure is applied to solve a sequence of convex QPPs. Experimental results on benchmark datasets confirm the effectiveness of the proposed algorithm.

3.5.1.1 Background

In this section, we briefly introduce the Hinge loss SVM, Ramp Loss SVM and LSSVM.

Hinge Loss SVM

Consider the binary classification problem with the training set

$$T = \{(x_1, y_1), \ldots, (x_l, y_l)\} \tag{3.635}$$

where $x_i \in \mathcal{R}^n$, $y_i \in \mathcal{Y} = \{1, -1\}$, $i = 1, \ldots, l$, the standard SVM relies on the classical Hinge loss function (see Fig. 3.9b)

$$H_s(z) = \max\ (0, s - z) \tag{3.636}$$

where the subscript s indicates the position of the Hinge point, to penalize examples classified with an insufficient margin and results (6) in the following primal problem

$$\min \frac{1}{2}\|\mathrm{w}\|^2 + C\sum\nolimits_{i=1}^{l} H_1\left(yf\left(x_i\right)\right), \tag{3.637}$$

where $f(x)$ is the decision function with the form of $f(x) = (\mathrm{w} \cdot \Phi(x)) + b$, and $\Phi(\cdot)$ is the chosen feature map, often implicitly defined by a Mercer kernel $K(x,x^{'}) = (\Phi(x) \cdot \Phi(x^{'}))$ [18]. For the choice of the kernel function $(x,x^{'})$, one has several possibilities: $K(x,x^{'}) = (x \cdot x^{'})$ (linear kernel); $K(x,x^{'}) = ((x \cdot x^{'}) + 1)^d$ (polynomial kernel of degree d); $KK(x,x^{'}) = \exp\left(-\|x - x^{'}\|^2/\sigma^2\right)$ (RBF kernel); $K(x,x^{'}) = \tanh\left(\kappa(x \cdot x^{'}) + \theta\right)$ (Sigmoid kernel), etc.

Due to the application of the Hinge loss, standard SVM has the sensitivity to outlier observations since they will normally have the largest hinge loss, thus the decision hyperplane is inappropriately drawn toward outlier samples so that its generalization performance is degraded [85]. Another property of the Hinge Loss function is that the number of Support Vectors (SVs) scales linearly with the number of examples [86], and since the SVM training and recognition times grow quickly with the number of SVs, it is obviously that SVMs cannot deal with very large datasets.

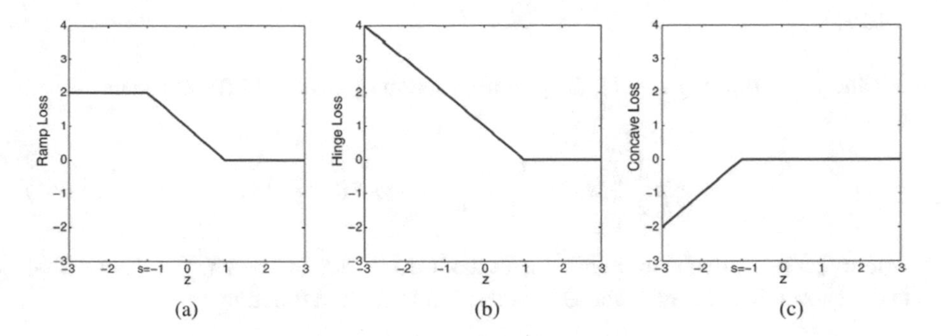

Fig. 3.9 The Ramp Loss function (**a**) can be decomposed into the sum of the convex Hinge Loss (**b**) and a concave loss (**c**)

Ramp Loss SVM

In order to increase the robustness of SVM and avoid converting the outliers into SVs, the Ramp Loss function [87] (see Fig. 3.9a), also known as the Robust Hinge Loss

$$R_s(z) = \begin{cases} 0, & z > 1 \\ 1 - z, & s \leqslant z \leqslant 1 \\ 1 - s, & z < s \end{cases} \tag{3.638}$$

was introduced to replace the Hinge loss function, by making the loss function flat for scores z smaller than a predefined value $s < 1$. $R_s(z)$ can be decomposed into the sum of the convex Hinge Loss and a concave loss (see Fig. 3.9c),

$$R_s(z) = H_1(z) - H_s(z), \tag{3.639}$$

therefore, the primal problem of the Ramp Loss SVM (RSVM) can be formulated as

$$\begin{aligned} &\min_{w,b} \frac{1}{2} \mid w\|^2 + C\sum\nolimits_{i=1}^{l} R_s\left(y_i f\left(x_i\right)\right) \\ = &\underbrace{\frac{1}{2} \parallel \mathrm{w}\|^2 + C\sum\nolimits_{i=1}^{l} H_1\left(y_i f\left(x_i\right)\right)}_{\text{convex}} \underbrace{-C\sum\nolimits_{i=1}^{l} H_s\left(y_i f\left(x_i\right)\right)}_{\text{concave}} \end{aligned} \tag{3.640}$$

which can be solved by the CCCP Procedure [88].

LSSVM

For the given training set (1), the primal problem of standard LSSVM to be solved is

$$\min_{\mathrm{w},b,} \frac{1}{2} \| \mathrm{w}\|^2 + \frac{\mathrm{C}}{2}\sum\nolimits_{i=1}^{l} \mathrm{Q}\left(y_i f\left(x_i\right) - 1\right), \tag{3.641}$$

where $f(x)$ is the decision function with the form of $f(x) = (\mathrm{w} \cdot \Phi(x)) + b$, and $\Phi(\cdot)$ is the chosen feature map, and Q(z) is the quadratic loss function

$$\mathrm{Q}(z) = z^2 \tag{3.642}$$

The geometric interpretation of the above problem with $x \in R^2$ is shown in Fig. 3.10, where minimizing (1/2) $\|\mathrm{w}\|^2$ realizes the maximum margin between the positive proximal straight line and negative proximal straight line

$$(\mathrm{w}\cdot \Phi(x)) + b = 1 \text{ and } (\mathrm{w}\cdot \Phi(x)) + b = -1 \tag{3.643}$$

while minimizing $\sum_{i=1}^{l} \mathrm{Q}\left(y_1 f\left(x_1\right) - 1\right)$ implies making the straight lines (9) to be proximal to all positive inputs and negative inputs respectively.

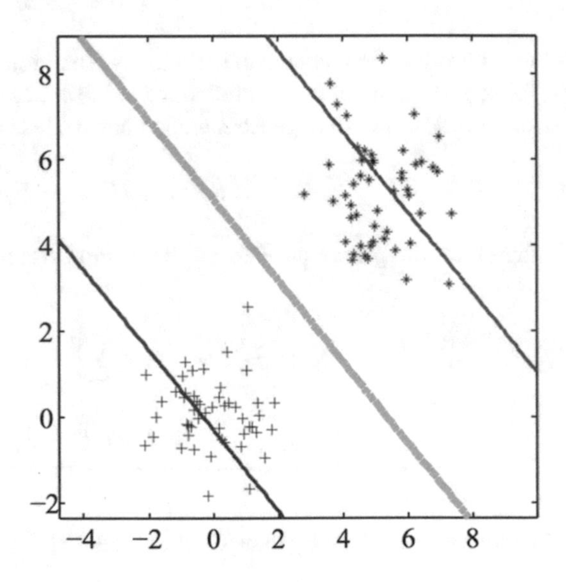

Fig. 3.10 Geometric interpretation of LSSVM: positive points represented by +s, negative points represented by $_*$'s, positive proximal line $(wx) + b = 1$ (down left line), negative proximal line $(wx) + b = -1$ (top right line), separating line $(wx) + b = 0$ (middle line)

Its dual problem is also a convex QPP

$$\&\min_{\alpha} \frac{1}{2} \sum_{i=1}^{1} \sum_{j=1}^{1} \alpha_i \alpha_j y_i y_j \left(K\left(x_i, x_j\right) + \frac{\delta_{ij}}{C} \right) - \sum_{i=1}^{1} \alpha_i$$

$$s.t. \ \& \sum_{i=1}^{l} \alpha_i y_i = 0 \tag{3.644}$$

where

$$\delta_{ij} = \begin{cases} 1, i = j; \\ 0, i \neq j. \end{cases} \tag{3.645}$$

The solution of the above problem is given by the following set of linear equations

$$\begin{bmatrix} 0 & -Y^{\top} \\ Y & \Omega + C^{-1} I \end{bmatrix} \begin{bmatrix} b \\ \alpha \end{bmatrix} = \begin{bmatrix} 0 \\ e \end{bmatrix}, \tag{3.646}$$

where $Y = (y_1, \ldots, y_l)^{\mathrm{T}}$, $\Omega = \left(\Omega_{ij}\right)_{1 \times 1} = \left(y_i y_j K\left(x_i, x_j\right)\right)_{1 \times 1}$, I is the identity matrix and $e = (1, 1)^{\mathrm{T}} \in R^1$, therefore the decision function is

$$f(x) = \operatorname{sgn}(g(x)) = \operatorname{sgn}\left(\sum_{i=1}^{1} \alpha_i y_i K\left(x_i, x\right) + b \right) \tag{3.647}$$

The support values α_1 are proportional to the errors at the data points since

$$\alpha_1 = C\eta_j, \ \ i = 1, \ldots, l. \tag{3.648}$$

Clearly, points located close to the two hyperplanes $(\mathrm{w} \cdot \Phi(x)) + b = \pm 1$ have the smallest support values, one could rather speak of the support value spectrum in the least squares case than the support vector in standard hinge loss SVM.

3.5.1.2 Ramp Loss ISSVM

In this section, we propose the Ramp Loss LSSVM, termed as RLSSVM, into which the Ramp Loss function is applied and is sparser and more robust than LSSVM.

Primal problem

As the points located close to the two hyperplanes $(\mathrm{w} \cdot \Phi(x)) + b \neq 1$ have the smallest support values, they contribute less to the decision function (3.647) at the same time, for the points located far away from the two hyperplanes, especially for the outliers, they tend to have large support values, and we want to eliminate the effects of such points. Therefore, following the idea of ε-insensitive loss function incorporated in ε-support vector regression (SVR), the following optimization problem is constructed

$$\min \frac{1}{2}\|\mathrm{w}\|^2 + \frac{C}{2}\sum\nolimits_{(=1}^{1} R_{\varepsilon.\mathrm{f}}\left(y_{\mathrm{i}} f\left(x_{\mathrm{i}}\right) - 1\right), \tag{3.649}$$

where $R_{\varepsilon I}(z)$ is our proposed ε-insensitive Ramp Loss function (see Fig. 3.11a),

$$R_{\varepsilon,\mathrm{f}}(z) = \begin{cases} (t-\varepsilon)^2, \mid Z \mid > t \\ (|z|-\varepsilon)^2, 8\xi \mid z \mid \leq t \\ 0, \mid z \mid < s \end{cases} \tag{3.650}$$

which makes the ε-insensitive quadratic loss function (see Fig. 3.11b)

$$l_{\varepsilon}(z) = \left(\max\{0, \mid z \mid -\varepsilon \mathrm{I}\right)^2 \tag{3.651}$$

is flat for scores $|z|$ larger than a predefined value $t > \varepsilon$. Obviously, that $R_{\varepsilon,\,t}(z)$ can be decomposed into the sum of the convex ε-insensitive quadratic loss and a concave loss (see Fig. 3.11c),

$$R_{\varepsilon,t} = I_{\varepsilon}(z) - I_f(z), \tag{3.652}$$

therefore, the problem can be reformulated as

$$\underbrace{\min_{w,b} \frac{1}{2} \| \mathrm{w}\|^2 + \frac{C}{2}\sum\nolimits_{i=1}^{l} I_{\varepsilon}\left(y_{\mathrm{i}} f\left(x_{\mathrm{i}}\right) - 1\right)}_{\text{convex}} \underbrace{- \frac{C}{2}\sum\nolimits_{i=1}^{l} I_t\left(y_{\mathrm{i}} f\left(x_{\mathrm{i}}\right) - 1\right)}_{\text{concave}} \tag{3.653}$$

Now we discuss the primal problem (3.653) geometrically in $\mathcal{R}^2$ (see Fig. 3.12). On the one hand, we hope that the positive points locate as much as possible in the ε-band, between the bounded hyperplanes $(\mathrm{w}\Phi(x)) + b = 1 + \varepsilon$ and $(\mathrm{w}\Phi(x)) + b = 1 - \varepsilon$, the negative class locates as much as possible in the ε-band between the hyperplanes $(\mathrm{w} \cdot \Phi(x)) + b \neq -1 + \varepsilon$ and $(\mathrm{w}\Phi(x)) + b = -1 - \varepsilon$, while the errors of the points clipped in the ε-band and the t-band are measured

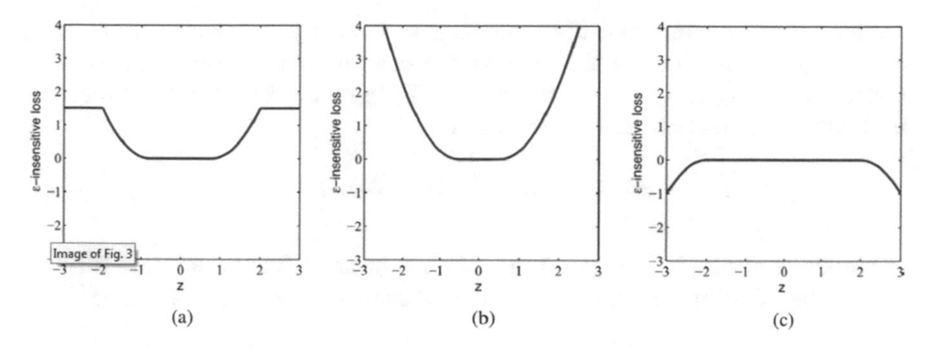

Fig. 3.11 The Ramp ε-insensitive loss function (**a**) can be decomposed into the sum of the convex ε-insensitive quadratic loss (**b**) and a concave loss (**c**)

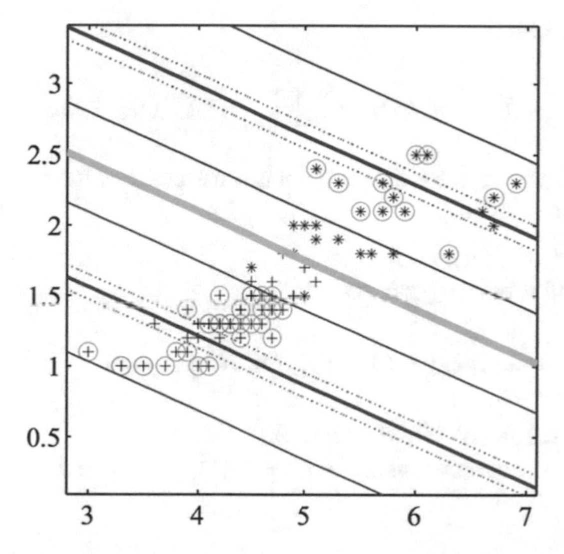

Fig. 3.12 Geometric interpretation of RLSSVM: positive proximal line $(w \cdot \Phi(x)) + b = 1$ (down left thick line), negative proximal line $(w \cdot \Phi(x)) + b = -1$ (top right thick line), positive ε-bounded lines $(w \cdot \Phi(x)) + b = 1 \pm \varepsilon$ (down left dotted lines), positive t-bounded lines $(w \cdot \Phi(x)) + b = 1 \pm t$ (down left thin lines), negative ε-bounded lines $(w \cdot \Phi(x)) + b = -1 \pm \varepsilon$ (top right dotted lines), negative t-bounded lines $(w \cdot \Phi(x)) + b = -1 \pm t$ (top right thin lines), separating line $(w \cdot \Phi(x)) + b = 0$ (middle line)

as $(| y_j((w\Phi(x)) + b) - 1| -\varepsilon)^2$, and the points out of the r-band are measured as $(f - \varepsilon)^2$, on the other hand, we still hope to maximize the margin between the two proximal hyperplanes $(w\Phi(x)) + b = 1$ and $(w\Phi(x)) + b = -1$. Based on the above two considerations, problem (3.653) is formulated and the structural risk minimization principle is implemented.

Obviously, the problem (3.653) can be solved by the CCCP Procedure, which is closely related to the "Difference of Convex" (DC) methods, which were successfully applied to a lot of different and various non-differentiable non-convex optimization problems especially in the large-scale setting [89]. For such a problem as (3.653) with an objective function written as a sum of a convex part $u(x)$ and a concave part $v(x)$, i.e., $u(x) + v(x)$, the CCCP algorithm is an iterative procedure that solves a sequence of convex programs

$$x^{t+1} = arg\ \min \left\{ u(x) + x^{\mathrm{T}} \nabla v\left(x^{t}\right) \right\}. \tag{3.654}$$

Collobert et al. [87] proposed the CCCP procedure for the RSVM. We now present the CCCP framework for (19). Let the convex part of the problem be

$$P_{vex}\left(\mathrm{w}, b\right) = \frac{1}{2}\|\mathrm{w}\|^{2} + \frac{C}{2}\sum\nolimits_{i=1}^{l} I_{\varepsilon}\left(y_{i} f\left(x_{i}\right) - 1\right), \tag{3.655}$$

and the concave part be

$$P_{cav}\left(\mathrm{w}, b\right) = -\frac{C}{2}\sum\nolimits_{i=1}^{l} I_{t}\left(y_{i} f\left(x_{i}\right) - 1\right). \tag{3.656}$$

The CCCP framework for the problem is constructed as follows:
CCCP for the problem

1. Initialize (w^{0}, b^{0}), set $k = 0$;
2. Construct and solve the problem

$$\min\ P_{vex}\left(\mathrm{w}, b\right) + P'_{cav}\left(\mathrm{w}^{k}, b^{k}\right) \cdot \left(\mathrm{w}, b\right), \tag{3.657}$$

get the solution $(\mathrm{w}^{k+1}, b^{k+1})$; w, b
3. If (w^{k}, b^{k}) dose not convergence, set $k = k + 1$, go to step 2.

Dual Problem

The convex optimization problem (3.657) that constitutes the core of the CCCP algorithm is easily reformulated into a dual one using the standard LSSVM technique. Rewrite the problem (3.657) as

$$\begin{aligned} &\min \frac{1}{2}\|\mathrm{w}\|^{2} + \frac{C}{2}\sum\nolimits_{i=1}^{l}\left(\xi_{i}^{2} + \xi_{i}^{*2}\right) + P'_{cav}\left(\mathrm{w}^{k}, b^{k}\right) \cdot \left(\mathrm{w}, b\right) \\ &s.t.\ \ \left(\mathrm{w} \cdot \Phi\left(x_{i}\right)\right) + b - y_{i} \leq \varepsilon + \xi_{i}, i = 1, \ldots, l, \\ &\qquad y_{i} - \left(\mathrm{w} \cdot \Phi\left(x_{i}\right)\right) - b \leq \varepsilon + \xi_{i*}, i = 1, \cdots, l. \end{aligned} \tag{3.658}$$

where $\varepsilon \geq 0$ is a prior parameter. Note that $P_{cav}(\mathrm{w}, b)$ is nondifferentiable at some points, it can be shown that the CCCP remains valid when using any super-derivative of the concave function. For simplification purposes, we introduce the notations

$$\theta_i = -\frac{C}{2} y_i \frac{\partial I_t \left(y_i f\left(x_i\right) - 1\right)}{\partial \left(y f\left(x_i\right) - 1\right)}$$

$$= \begin{cases} -C\left(\left|f\left(x_i\right) - y_i\right| - t\right), & \text{if} y_i f\left(x_i\right) - 1 > t \\ C\left(\left|f\left(x_i\right) - y_i\right| - t\right), & \text{if} y_i f\left(x_i\right) - 1 < -t \\ 0, & \text{otherwise} \end{cases} \tag{3.659}$$

for $i = 1, \ldots, l$. Therefore, the problem (3.658) can be rewritten as

$$\min \frac{1}{2}\|\mathrm{w}\|^2 + \frac{C}{2}\sum\nolimits_{i=1}^{l}\left(\xi_i^2 + \xi_i^{*2}\right) + \sum\nolimits_{i=1}^{l}\theta_i\left(\left(\mathrm{w}\cdot \Phi\left(x_i\right)\right) + b\right)$$

$$s.t.\;\; \left(\mathrm{w}\cdot \Phi\left(x_i\right)\right) + b - y_i \le \varepsilon + \xi_i, i = 1, \cdots, l, \tag{3.660}$$

$$y_i - \left(\mathrm{w}\cdot \Phi\left(x_i\right)\right) - b \le \varepsilon + \xi_i^*, i = 1, \cdots, l.$$

In order to find the solution of problem (3.660) in $\mathcal{H}$, we need to derive its dual problem. By introducing the Lagrangian

$$L\left(\mathrm{w}, b, \xi, \xi^*, \alpha, \alpha^*\right)$$

$$= \frac{1}{2}\|\mathrm{w}\|^2 + \frac{C}{2}\sum_{i=1}^{l}\left(\xi_i^2 + \xi_i^{*2}\right) + \sum_{i=1}^{l}\theta_i\left(\left(\mathrm{w}\cdot \Phi\left(x_i\right)\right) + b\right)$$

$$+\sum_{i=1}^{l}\alpha_i\left(\left(\mathrm{w}\cdot \Phi\left(x_i\right)\right) + b - y_i - \varepsilon - \xi_i\right)$$

$$+\sum\nolimits_{i=1}^{l}\alpha_i^*\left(y_i - \left(\mathrm{w}\cdot \Phi\left(x_i\right)\right) - b - \varepsilon - \xi_i^*\right), \tag{3.661}$$

where α, α^* are the Lagrange multiplier vectors, the dual problem is obtained as

$$\frac{1}{2}\sum\nolimits_{i=1}^{l}\sum\nolimits_{j=1}^{l}\left(\alpha_i^* - \alpha_i - \theta_i\right)\left(\alpha_j^* - \alpha_j - \theta_j\right)K\left(x_i, x_j\right)$$

$$+\frac{1}{2C}\sum\nolimits_{i=1}^{l}\left(\alpha_i^2 + \alpha_i^{*2}\right) + \varepsilon\sum\nolimits_{i=1}^{l}\left(\alpha_i^* + \alpha_i\right) - \sum\nolimits_{i=1}^{l}y_i\left(\alpha_i^* - \alpha_i\right), \tag{3.662}$$

$$s.t.\;\; \sum\nolimits_{i=1}^{l}\left(\alpha_i - \alpha_i^* - \theta_i\right) = 0, \alpha_i, \alpha_i^* \ge 0, i = 1, \ldots, l.$$

For the problem (3.662), it is easy to prove the following theorem.

Theorem 3.9 *If* (α, α^*) *is the solution of the problem* (3.662), *then* $\alpha_i^* \alpha_i = 0$ *for* $i = 1, \ldots, l$.

If we let

$$\theta_i = \overline{\theta}_i + \widetilde{\theta}_i, \tag{3.663}$$

where

$$\overline{\theta}_i = \begin{cases} C\left(|f(x_i) - y_i| - t\right), & \text{if } y_i f(x_i) - 1 < -t \\ 0, & \text{otherwise} \end{cases} \tag{3.664}$$

and

$$\widetilde{\theta}_i = \begin{cases} -C\left(|f(x_i) - y_i| - t\right), & \text{if } y_i f(x_i) - 1 > t \\ 0, & \text{otherwise} \end{cases} \tag{3.665}$$

for $i = 1, \ldots, l$, furthermore, let

$$\overline{\alpha}_i = \alpha_i^* - \overline{\theta}_i, \quad i = 1, \cdots, l, \tag{3.666}$$

then

$$\begin{aligned} \alpha_i^{*2} + \alpha_i^2 &= \left(\overline{\alpha}_i + \overline{\theta}_i\right)^2 + \left(\widetilde{\alpha}_i - \widetilde{\theta}_i\right)^2 \\ &= \overline{\alpha}_i^2 + \widetilde{\alpha}_i^2 + 2\overline{\alpha}_i\overline{\theta}_i - 2\widetilde{\alpha}_i\widetilde{\theta}_i + \overline{\theta}_i^2 + \widetilde{\theta}_i^2 \\ &= (\overline{\alpha}_i - \widetilde{\alpha}_i)^2 + 2\overline{\alpha}_i\widetilde{\alpha}_i + 2\overline{\alpha}_i\overline{\theta}_i - 2\widetilde{\alpha}_i\widetilde{\theta}_i + \overline{\theta}_i^2 + \widetilde{\theta}_i^2 \\ &= (\overline{\alpha}_i - \widetilde{\alpha}_i)^2 + 2\theta_i\left(\overline{\alpha}_i - \widetilde{\alpha}_i\right) + \Delta\left(\overline{\theta}_i, \widetilde{\theta}_i\right) \end{aligned} \tag{3.667}$$

where $\Delta\left(\overline{\theta}_i, \widetilde{\theta}_i\right)$ is the constant decided by $\overline{\theta}_i$, $\widetilde{\theta}_i$. Therefore, the dual problem (3.662) equals to the following problem

$$\begin{aligned} &\min_{\overline{\alpha}, \widetilde{\alpha}} \frac{1}{2}\sum_{i=1}^{l}\sum_{j=1}^{l}\left(\overline{\alpha}_i - \widetilde{\alpha}_i\right)\left(\overline{\alpha}_j - \widetilde{\alpha}_j\right)\hat{K}\left(x_i, x_j\right) \\ &+ \varepsilon\sum_{i=1}^{l}\left(\overline{\alpha}_i + \widetilde{\alpha}_i\right) + \sum_{i=1}^{l}\left(\frac{\theta_i}{C} - y_i\right)\left(\overline{\alpha}_i - \widetilde{\alpha}_i\right) \\ &\text{s.t.} \sum_{i=1}^{l}\left(\overline{\alpha}_i - \widetilde{\alpha}_i^*\right) = 0, \overline{\alpha}_i \geqslant -\overline{\theta}_i, \widetilde{\alpha}_i \geqslant \widetilde{\theta}_i, i = 1, \cdots, l \end{aligned} \tag{3.668}$$

where $\hat{K}\left(x_i, x_j\right) = \left(K\left(x_i, x_j\right) + \frac{t_{ij}}{C}\right)$, $i, j = 1, \ldots, l$.

Now, we can construct the RLSSVM based on the CCCP procedure.

RLSSVM

(1) Input the training set (3.635)

(2) Choose the appropriate t, $\varepsilon > 0$ for the ramp loss (3.649); Choose appropriate penalty parameter $C > 0$ and kernel function (x, x'); Initialize $\overline{\theta}^0, \widetilde{\theta}^0$, set $k - 1$,

(3) Construct and solve the QPP (34) in the kth iteration step

$$\min_{\overline{\alpha}, \widetilde{\alpha}} \frac{1}{2} \sum_{i=1}^{l} \sum_{j=1}^{l} (\overline{\alpha}_i - \widetilde{\alpha}_i)(\overline{\alpha}_j - \widetilde{\alpha}_j) \hat{K}(x_i, x_j) + \varepsilon \sum_{i=1}^{l} (\overline{\alpha}_i + \widetilde{\alpha}_i)$$
$$+ \sum_{i=1}^{l} \left(\frac{\theta_i^k}{C} - y_i \right) (\overline{\alpha}_i - \widetilde{\alpha}_i)$$
$$\text{s.t.} \sum_{i=1}^{l} (\overline{\alpha}_i - \widetilde{\alpha}_i^*) = 0, \overline{\alpha}_i \geqslant -\overline{\theta}_i^k, \widetilde{\alpha}_i \geqslant \widetilde{\theta}_i^k, i = 1, \ldots, l \tag{3.669}$$

get the solutions $(\overline{\alpha}^k, \widetilde{\alpha}^k)$, compute b^k based on the KKT conditions; Construct the decision functions

$$f_k(x) = \sum_{i=1}^{l} \left(\overline{\alpha}_i^k - \widetilde{\alpha}_i^k \right) K(x_i, x) + b^k. \tag{3.670}$$

(4) Compute $\overline{\theta}^k, \widetilde{\theta}^k$ based on the equations

(5) If $\left(\overline{\theta}^k, \widetilde{\theta}^k\right) = \left(\overline{\theta}^{k-1}, \widetilde{\theta}^{k-1}\right)$, then we get the final decision function, go to step (6); else set $k = k + 1$, go to step (3);

(6) A new point $x \in \mathcal{R}^n$ is predicted to the *Class* by

$$Class = \text{sgn}\left(\sum_{i=1}^{l} \left(\overline{\alpha}_i^k - \widetilde{\alpha}_i^k \right) K(x_i, x) + b^k \right). \tag{3.671}$$

3.5.2 *Ramp Loss Nonparallel Support Vector Machine for Pattern Classification*

3.5.2.1 Ramp Loss NPSVM

In this section, we propose the Ramp loss NPSVM, termed as RNPSVM, into which the Ramp loss function is applied and is sparser and more robust than NPSVM.

Linear RNPSVM

Primal Problems

We seek the two nonparallel hyperplanes $f^{+}(x) = (w_{+} \cdot x) + b_{+} = 0$ and $f^{-}(x) = (w_{-} \cdot x) + b_{-} = 0$ by solving two problems

$$\min_{w_{+} b_{+}} \frac{1}{2}\| w_{+} \|^{2} + C_{1}\sum_{i=1}^{p} R_{\varepsilon,t}\left(f^{+}(x_{i})\right) + C_{2}\sum_{j=p+1}^{p+q} R_{s}\left(-f^{+}(x_{j})\right) \tag{3.672}$$

and

$$\min_{w_{-},b_{-}} \frac{1}{2}\| w_{-} \|^{2} + C_{3}\sum_{i=p+1}^{p+q} R_{\varepsilon,t}\left(f^{-}(x_{i})\right) + C_{4}\sum_{j=1}^{p} R_{s}\left(f^{-}(x_{j})\right) \tag{3.673}$$

where $C_i \geq 0$; $i = 1, \ldots, 4$ are penalty parameters, and $R_{\varepsilon;t}(z)$ is our proposed ε-insensitive Ramp loss function,

$$R_{\varepsilon;t}(z) = \begin{cases} t - \varepsilon; & |z| > t \\ |z| - \varepsilon; & \varepsilon \leq |z| \leq t \\ 0; & |z| < \varepsilon \end{cases} \tag{3.674}$$

which makes the ε-insensitive loss function flat for scores z larger than a predefined value $t > \varepsilon$. It is obviously that $R_{\varepsilon;t}(z)$ can be decomposed into the sum of the convex ε-insensitive loss and a concave loss,

$$R_{\varepsilon;t} = I_{\varepsilon}(z) - I_{t}(z) \tag{3.675}$$

Therefore, the problems (3.672) and (3.673) of the Ramp loss NPSVM can be reformulated as

$$\min_{w_{+},b_{+}} \frac{1}{2}\| w_{+} \|^{2} + C_{1}\sum_{i=1}^{p} R_{c,t}\left(f^{+}(x_{i})\right) + C_{2}\sum_{j=p+1}^{p+q} R_{s}\left(-f^{+}(x_{j})\right)$$

$$= \underbrace{\frac{1}{2}\|w_{+}\|^{2} + C_{1}\sum_{i=1}^{p} I_{\varepsilon}\left(f^{+}(x_{i})\right) + C_{2}\sum_{j=p+1}^{p+q} H_{1}\left(-f^{+}(x_{j})\right)}_{\text{convex}}$$

$$\underbrace{-C_{1}\sum_{i=1}^{p} I_{t}\left(f^{+}(x_{i})\right) - C_{2}\sum_{j=p+1}^{p+q} H_{s}\left(-f^{+}(x_{j})\right)}_{\text{concave}} \tag{3.676}$$

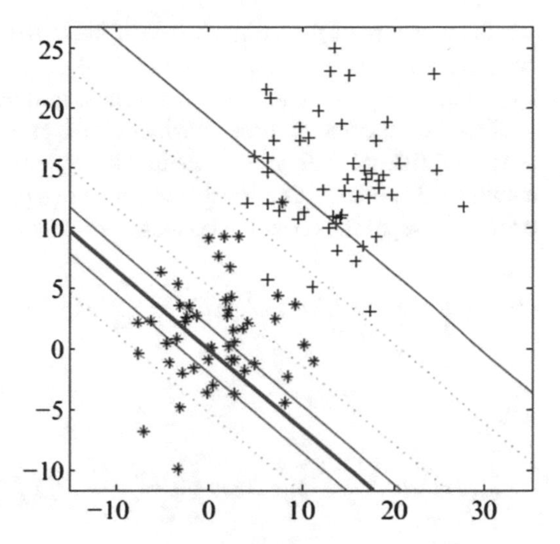

Fig. 3.13 Geometrical illustration of the primal problem (3.676) in R^2

and

$$\min_{w_-,b_-} \frac{1}{2}\| w_- \|^2 + C_3 \sum_{i=p+1}^{p+q} R_{\varepsilon,t}\left(f^-(x_i)\right) + C_4 \sum_{j=1}^{p} R_s\left(f^-(x_j)\right)$$

$$= \underbrace{\frac{1}{2}\|w_-\|^2 + C_3 \sum_{i=p+1}^{p+q} I_\varepsilon\left(f^\dagger(x_i)\right) + C_4 \sum_{j=1}^{p} H_1\left(f^-(x_j)\right)}_{\text{convex}}$$

$$\underbrace{-C_3 \sum\nolimits_{i=p+1}^{p+q} I_t\left(f^-(x_i)\right) - C_4 \sum\nolimits_{j=1}^{p} H_s\left(f^-(x_j)\right)}_{\text{concave}} \tag{3.677}$$

Now we discuss the primal problem (21) geometrically in $\mathcal{R}^2$ (see Fig. 3.13). First, the positive points (marked by $_*$,) are separated into three parts by the ε-band and the t-band, where the ε-band is between the hyperplanes $(w_+ \cdot x) + b_+ = \varepsilon$ and $(w_+ \cdot x) + b_+ = -\varepsilon$ (red thin solid lines), and the t-band is between the hyperplanes $(w_+ \cdot x) + b_+ = t$ and $(w_+ \cdot x) + b_+ = -t$ (red thin dotted lines). We hope that the positive points locate as much as possible in the ε-band, while the errors of the points clipped in the ε-band and the t-band are measured as $|(w_+ \cdot x_i) + b_+| - \varepsilon$, and the points out of the t-band are measured as $t - \varepsilon$; Second, we hope to maximize the margin between the hyperplanes $(w_+ \cdot x) + b_+ = \varepsilon$ and $(w_+ \cdot x) + b_+ = -\varepsilon$, which can be expressed by $\frac{2\varepsilon}{\|w\|}$; Third, the negative points (marked by $+'$) are separated into three parts by the hyperplane $(w_+ \cdot x) + b_+ = -1$ (blue thin solid line) and $(w_+ \cdot x) + b_+ = -s$ (blue thin dotted line), we need to push the negative class from the hyperplane $(w_+ \cdot x) + b_+ = -1$ (blue thin solid line) as far as possible, while the errors of the points between the hyperplanes $(w_+ \cdot x) + b_+ = -1$ and $(w_+ \cdot x) + b_+ = -s$ are measured as $1 + (w_+ \cdot x) + b_+$, and the points left to the hyperplane $(w_+ \cdot x) + b_+ = -s$ are measured by $1 - s$.

Based on the above three considerations, problem (3.676) is established and the structural risk minimization principle is implemented naturally. Problem (3.677) is established similarly. Obviously the above two problems can be solved by the CCCP procedure separately. Let the convex parts of the two problems be

$$P_{\text{vex}}\left(w_{+;}b_{+}\right) = \frac{1}{2}\|w_+\|^2 + C_1 \sum\nolimits_{i=1}^{p} I_\varepsilon\left(f^+(x_i)\right) + C_2 \sum\nolimits_{j=p+1}^{p+q} H_1\left(-f^+(x_j)\right) \tag{3.678}$$

and

$$N_{\text{vex}}\left(w_{-;}b_{-}\right) = \frac{1}{2}\|w_-\|^2 + C_3 \sum\nolimits_{i=p+1}^{p+q} I_\varepsilon\left(f^-(x_i)\right) + C_4 \sum\nolimits_{j=1}^{p} H_1\left(f^-(x_j)\right) \tag{3.679}$$

and the concave parts be

$$P_{cav}\left(w_{+;}b_{+}\right) = -C_1 \sum\nolimits_{i=1}^{p} I_t\left(f^+(x_i)\right) - C_2 \sum\nolimits_{j=p+1}^{p+q} H_s\left(-f^+(x_j)\right) \tag{3.680}$$

and

$$N_{cav}\left(w_{-;}b_{-}\right) = -C_3 \sum\nolimits_{i=p+1}^{p+q} I_t\left(f^-(x_i)\right) - C_4 \sum\nolimits_{j=1}^{p} H_s\left(f^-(x_j)\right) \tag{3.681}$$

separately. The CCCP framework for the two problems is constructed as follows:

Algorithm 3.13 (CCCP for the Problem (3.676)

---

(1) Initialize $\left(w^0_{+;} b^0_{+}\right)$, set $k = 0$;

(2) Construct and solve the problem

$$min\ P_{\text{vex}}\left(w_{+;}b_{+}\right) + P'_{cav}\left(w^k_{+};\, b^k_{+}\right) \cdot \left(w_{+;}b_{+}\right); \tag{3.682}$$

get the solution $\left(w^{k+1}_{+};\, b^{k+1}_{+}\right)$;

(3) If $\left(w^k_{+;} b^k_{+}\right)$ not convergence, set $k = k + 1$, go to step (2).

Algorithm 3.14 (CCCP for the problem (3.677)

(1) Initialize $\left(w_{-;}^{0}b_{-}^{0}\right)$, set $k = 0$;

(2) Construct and solve the problem

$$min\ N_{vex}\left(w_{-;}b_{-}\right) + N_{cav}'\left(w_{-}^{k};b_{-}^{k}\right)\cdot\left(w_{-;}b_{-}\right)_{;} \tag{3.683}$$

get the solution $\left(w_{-}^{k+1};b_{-}^{k+1}\right)$;

(3) If $\left(w_{-;}^{k}b_{-}^{k}\right)$ not convergence, set $k = k + 1$, go to step (2).

Dual Problems

The convex optimization problem (3.682) that constitutes the core of the CCCP algorithm is easily reformulated into dual variables using the standard NPSVM technique. Rewrite the problem (3.682) as

$$\min \frac{1}{2}\|w_+\|^2 + C_1\sum\nolimits_{i=1}^{p}\left(\eta_i + \eta_i^*\right) + C_2\sum\nolimits_{j=p+1}^{p+q}\xi_j + P_{cav}'\left(w_+^t;b_+^t\right)\cdot\left(w_{+;}b_+\right)$$

$$s.t.\ \ (w_+\cdot x_i) + b_+ \leq \varepsilon + \eta_i;\ i = 1,\ldots,p,$$

$$-(w_+\cdot x_i) - b_+ \leq \varepsilon + \eta_i^*;\ i = 1,\ldots,p,$$

$$\left(w_+\cdot x_j\right) + b_+ \leq -1 + \xi_j;\ \ j = p+1,\ldots,p+q,$$

$$\eta_i,\eta_i^* \geq 0;\ i = 1,\ldots,p,$$

$$\xi_j \geq 0;\ j = p+1,\ldots,p+q. \tag{3.684}$$

note that $P_{cav}(w_{+;}b_+)$ is non-differentiable at some points, it can be shown that the CCCP remains valid when using any super-derivative of the concave function. For simplification purposes, we introduce the notations

$$\delta_j = -C_2 y_j\frac{\partial H_s\left(yf^+\left(x_j\right)\right)}{\partial f^+\left(x_j\right)} = \begin{cases} C_2, & \text{if } y_j f^+\left(x_j\right) < s \\ 0, & \text{otherwise}\end{cases} \tag{3.685}$$

for $j = p + 1, \ldots, p + q$, and

$$\theta_i = -C_1 \frac{\partial I_t\left(f^+(x_i)\right)}{\partial f^+(x_i)} = \begin{cases} -C_1, & \text{if } f^+(x_i) > t \\ C_1, & \text{if } f^+(x_i) < -t \\ 0, & \text{otherwise} \end{cases} \tag{3.686}$$

for $i = 1, \ldots, p$. Therefore, the problem (29) turns to be

$$\min \frac{1}{2}\|w_+\|^2 + C_1 \sum\nolimits_{i=1}^{p} \left(\eta_i + \eta_i^*\right) + C_2 \sum\nolimits_{j=p+1}^{p+q} \xi_j$$

$$+ \sum_{\text{i}=1}^{p} \theta_i \left(\left(w_+ \cdot x_j\right) + b_+\right) + \sum_{j=p+1}^{p+q} \delta_j y_j \left(\left(w_+ \cdot x_j\right) + b_+\right)$$

$$s.t.\ \ (w_+ \cdot x_i) + b_+ \le \varepsilon + \eta_i; i = 1, \ldots, p$$

$$-(w_+ \cdot x_i) - b_+ \le \varepsilon + \eta_i^*; i = 1, \ldots, p$$

$$\left(w_+ \cdot x_j\right) + b_+ \le -1 + \xi_j; j = p + 1, \ldots, p + q$$

$$\eta_i, \eta_i^* \ge 0; i = 1, \ldots, p$$

$$\xi_j \ge 0; j = p + 1, \ldots, p + q \tag{3.687}$$

In order to get the solutions of problem (3.687), we need to derive its dual problem. Its Lagrangian is given by

$$L\left(w_{+;} b_{+;} \eta_{+;}^{(*)} \xi_{-;} \alpha_{+;}^{(*)} \gamma_{+;}^{(*)} \beta_{-;} \lambda_{-}\right) = \frac{1}{2}\|w_+\|^2 + C_1 \sum\nolimits_{\text{i}=1}^{p} \left(\eta_i + \eta_i^*\right)$$

$$+C_2 \sum_{j=p+1}^{p+q} \xi_i + \sum_{i=1}^{p} \theta_i \left(\left(w_+ \cdot x_i\right) + b_+\right) + \sum_{j=p+1}^{p+q} \delta_j y_j \left(\left(w_+ \cdot x_j\right) + b_+\right)$$

$$+\sum_{i=1}^{p} \alpha_i \left(\left(w_+ \cdot x_i\right) + b_+ - \eta_i - \varepsilon\right) + \sum_{i=1}^{p} \alpha_{\text{i}}^* \left(-\left(w_+ \cdot x_i\right) - b_+ - \eta_i^* - \varepsilon\right)$$

$$+\sum\nolimits_{j=p+1}^{p+q} \beta_j \left(\left(w_+ \cdot x_j\right) + b_+ + 1 - \xi_j\right) - \sum\nolimits_{i=1}^{p} \gamma_i \eta_i - \sum\nolimits_{i=1}^{p} \gamma_i^* \eta_i^* - \sum\nolimits_{j=p+1}^{p+q} \lambda_j \xi_{j;} \tag{3.688}$$

where $\alpha_+^{(*)} = \left(\alpha_+^{\mathrm{T}}; \alpha_+^{*\mathrm{T}}\right)^{\mathrm{T}} = \left(\alpha_1, \ldots, \alpha_p; \alpha_1^*, \ldots, \alpha_p^*\right)^{\mathrm{T}}$, $\gamma_+^{(*)} = \left(\gamma_+^{\mathrm{T}}; \gamma_+^{*\mathrm{T}}\right)^{\mathrm{T}} = \left(\gamma_1, \cdots, \gamma_p; \gamma_1^*, \ldots, \gamma_p^*\right)^{\mathrm{T}}$, $\beta_- = (\beta_{p+1}; \ldots; \beta_{p+q})^{\mathrm{T}}$, $\lambda_- = (\lambda_{p+1}; \ldots; \beta_{p+q})^{\mathrm{T}}$ are the Lagrange multiplier vectors. The Karush−Kuhn−Tucker (KKT) conditions [71] for $w_+, b_+, \eta_+^{(*)}, \xi_-$ and $\alpha_+^{(*)}, \gamma_+^{(*)}, \beta_-, \lambda_-$ are given by

$$\begin{aligned}\nabla_{w_+} L = w_+ + \sum\nolimits_{i=1}^{p} \alpha_i x_i - \sum\nolimits_{i=1}^{p} \alpha_i^* x_i - \sum\nolimits_{j=p+1}^{p+q} \beta_j x_j + \sum\nolimits_{i=1}^{p} \theta_i x_i \\ + \sum\nolimits_{i=p+1}^{p+q} \delta_j x_j = 0;\end{aligned} \tag{3.689}$$

$$\nabla_{b_+} L = \sum\nolimits_{i=1}^{p} \alpha_i - \sum\nolimits_{i=1}^{p} \alpha_i^* + \sum\nolimits_{j=p+1}^{p+q} \beta_j + \sum\nolimits_{i=1}^{p} \theta_i - \sum\nolimits_{i=p+1}^{p+q} \delta_j = 0 \tag{3.690}$$

$$\nabla_{\eta_+} L = C_1 e_+ - \alpha_+ - \gamma_+ = 0 \tag{3.691}$$

$$\nabla_{\eta_+^{Reject}} L = C_1 e_+ - \alpha_+^* - \gamma_+^* = 0 \tag{3.692}$$

$$\nabla_{\xi_-} L = C_2 e_- - \beta_- - \lambda_- = 0 \tag{3.693}$$

where $e_+ = (1, \ldots, 1)^{\mathrm{T}} \in \mathcal{R}^p$, $e_- = (1, \ldots, 1)^{\mathrm{T}} \in \mathcal{R}^q$. Since $\gamma_+, \gamma_+^* \geq 0, \lambda_- \geq 0$, from (3.691)–(3.693) we have

$$0 \leq \alpha_+, \alpha_+^* \leq C_1 e_+; \tag{3.694}$$

$$0 \leq \beta_- \leq C_2 e_- \tag{3.695}$$

and from (3.695), we have

$$w_+ = \sum\nolimits_{i=1}^{p} \left(\alpha_i^* - \alpha_i - \theta_i\right) x_i - \sum\nolimits_{j=p+1}^{p+q} \left(\beta_j - \delta_j\right) x_j \tag{3.696}$$

Then putting (3.696) into the Lagrangian (3.692) and using, we obtain the dual problem

$$\min \frac{1}{2}\sum_{i=1}^{p}\sum_{j=1}^{p} \left(\alpha_i^* - \alpha_i - \theta_i\right)\left(\alpha_j^* - \alpha_j - \theta_j\right)\left(x_i \cdot x_j\right)$$

$$-\sum_{i=1}^{p}\sum_{j=p+1}^{p+q} \left(\alpha_i^* - \alpha_i - \theta_i\right)\left(\beta_j - \delta_j\right)\left(x_i \cdot x_j\right)$$

$$+\frac{1}{2}\sum_{i=p+1}^{p+q}\sum_{j=p+1}^{p+q} \left(\beta_i - \delta_i\right)\left(\beta_j - \delta_j\right)\left(x_i \cdot x_j\right)$$

$$+\varepsilon\sum_{i=1}^{p}\left(\alpha_i^* + \alpha_i\right) - \sum_{i=p+1}^{p+q}\beta_i$$

$$s.t. \sum_{i=1}^{p}\left(\alpha_i^* - \alpha_i - \theta_i\right) - \sum_{j=p+1}^{p+q}\left(\beta_j - \delta_j\right) = 0,$$

$$0 \le \alpha_+; \alpha_+^* \le C_1 e_+,$$

$$0 \le \beta_- \le C_2 e_- \tag{3.697}$$

If we let

$$\theta_i = \overline{\theta}_i + \widetilde{\theta}_i \tag{3.698}$$

where

$$\overline{\theta}_i = \begin{cases} C_1, \text{ if } f^+(x_i) < -t \\ 0, \quad \text{otherwise} \end{cases} \tag{3.699}$$

and

$$\widetilde{\theta}_i = \begin{cases} -C_1; \text{ if } f^+(x_i) > t \\ 0; \quad \text{otherwise} \end{cases} \tag{3.700}$$

for $i = 1, \ldots, p$, furthermore, let

$$\overline{\alpha}_i = \alpha_i^* - \overline{\theta}_i; \widetilde{\alpha}_i = \alpha_i + \theta_i; Reject\ i = 1, \ldots, p \tag{3.701}$$

and

$$\overline{\beta}_j = \beta_j - \delta_j;\ j = p+1, \ldots, p+q \tag{3.702}$$

the dual problem (3.697) equals to the following problem

$$\min_{\alpha_+, \overline{x}_+, \hat{\beta}_-} \frac{1}{2}\sum_{i=1}^{p}\sum_{j=1}^{p} (\overline{\alpha}_i - \widetilde{\alpha}_i)(\overline{\alpha}_j - \widetilde{\alpha}_j)(x_i \cdot x_j) - \sum_{i=1}^{p}\sum_{j=p+1}^{p+q} (\overline{\alpha}_i - \widetilde{\alpha}_i)\overline{\beta}_j (x_i \cdot x_j)$$

$$+\frac{1}{2}\sum_{i=p+1 j=p+1}^{p+q}\sum_{i=p+1}^{p+q} \overline{\beta}_i\overline{\beta}_j (x_i \cdot x_j) + \varepsilon\sum_{i=1}^{p} (\overline{\alpha}_i + \widetilde{\alpha}_i) - \sum_{i=p+1}^{p+q} \overline{\beta}_i$$

$$s.t. \sum_{i=1}^{p} (\overline{\alpha}_i - \widetilde{\alpha}_i) - \sum_{j=p+1}^{p+4} \overline{\beta}_j = 0$$

$$-\overline{\theta}_+ \leqslant \overline{\alpha}_+ \leqslant C_1 e_+ - \overline{\theta}_+$$

$$\widetilde{\theta}_+ \leqslant \widetilde{\alpha}_+ \leqslant C_1 e_+ + \widetilde{\theta}_+$$

$$-\delta_- \leqslant \overline{\beta}_- \leqslant C_2 e_- - \delta_- \tag{3.703}$$

where $\overline{\alpha}_+ = (\overline{\alpha}_1; \ldots; \overline{\alpha}_p)^{\mathrm{T}}, \widetilde{\alpha}_+ = (\widetilde{\alpha}_1; \ldots; \widetilde{\alpha}_p)^{\mathrm{T}}, \overline{\beta}_- = (\overline{\beta}_{p+1}; \ldots; \overline{\beta}_{p+q})^{\mathrm{T}}$, and $\overline{\theta}_+ = (\overline{\theta}_1; \ldots; \overline{\theta}_p)^{\mathrm{T}}, \widetilde{\theta}_+ = (\widetilde{\theta}_1; \ldots; \widetilde{\theta}_p)^{\mathrm{T}}, \delta_- = (\delta_{p+1}; \ldots; \delta_{p+q})^{\mathrm{T}}$.

Concisely, this problem can be further formulated as

$$\min_{\pi} \quad \frac{1}{2}\widetilde{\pi}^{\top}\widetilde{\Lambda}\widetilde{\pi} + \widetilde{\kappa}^{\top}\widetilde{\pi}$$

$$s.t.\ \widetilde{e}^{\top}\widetilde{\pi} = 0, \widetilde{V} \leqslant \widetilde{\pi} \leqslant \widetilde{U} \tag{3.704}$$

where

$$\widetilde{\pi} = \left(\overline{\alpha}_+^{\mathrm{T}}; \widetilde{\alpha}_+^{\mathrm{T}}; \overline{\beta}_-^{\mathrm{T}}\right)^{\mathrm{T}}, \tag{3.705}$$

$$\widetilde{\kappa} = \left(\varepsilon e_+^{\mathrm{T}}; \varepsilon e_+^{\mathrm{T}}; -e_-^{\mathrm{T}}\right)^{\mathrm{T}}, \tag{3.706}$$

$$\widetilde{e} = \left(-e_+^{\mathrm{T}}; e_+^{\mathrm{T}}; e_-^{\mathrm{T}}\right)^{\mathrm{T}}, \tag{3.707}$$

$$\widetilde{U} = \left((C_1e_+ - \overline{\theta}_+)^{\mathrm{T}}_{;}(C_1e_+ + \widetilde{\theta}_+)^{\mathrm{T}}_{;}(C_2e_- - \delta_-)^{\mathrm{T}}\right)^{\mathrm{T}}, \tag{3.708}$$

$$\widetilde{V} = \left(-\overline{\theta}^{\mathrm{T}}_+; \widetilde{\theta}^{\mathrm{T}}_+; -\delta^{\mathrm{T}}_-\right)^{\mathrm{T}}, \tag{3.709}$$

and

$$\widetilde{\Lambda} = \begin{pmatrix} H_1 & -H_2 \\ -H_2^{\mathrm{T}} & H_3 \end{pmatrix}; H_1 = \begin{pmatrix} AA^{\mathrm{T}} & -AA^{\mathrm{T}} \\ -AA^{\mathrm{T}} & AA^{\mathrm{T}} \end{pmatrix}; H_2 = \begin{pmatrix} AB^{\mathrm{T}} \\ -AB^{\mathrm{T}} \end{pmatrix}; H_3 = BB^{\mathrm{T}} \tag{3.710}$$

where $A = \left(x_1; \ldots; x_p\right)^{\mathrm{T}} \in \mathcal{R}^{p\times n}$, $B = \left(x_{p+1}; \ldots; x_{p+q}\right) \in \mathcal{R}^{q\times n}$.

Obviously, the dual problem (3.708) is a convex QPP. In the same way, the dual of the problem (3.687) is obtained

$$\min \frac{1}{2}\hat{\pi}^{\mathrm{T}}\hat{\Lambda}\hat{\pi} + \hat{\kappa}^{\mathrm{T}}\hat{\pi}$$

$$s.t.\ \hat{e}\hat{\pi} = 0,$$

$$\hat{V} \leq \hat{\pi} \leq \hat{U}. \tag{3.711}$$

where

$$\hat{\pi} = \left(\overline{\alpha}^{\mathrm{T}}_{-;}\widetilde{\alpha}^{\mathrm{T}}_{-;}\overline{\beta}^{\mathrm{T}}_+\right)^{\mathrm{T}} \tag{3.712}$$

$$\hat{\kappa} = \left(\varepsilon e^{\mathrm{T}}_{-;}\varepsilon e^{\mathrm{T}}_{-;} - e^{\mathrm{T}}_+\right)^{\mathrm{T}} \tag{3.713}$$

$$\hat{e} = \left(-e^{\mathrm{T}}_{-;}e^{\mathrm{T}}_{-;} - e^{\mathrm{T}}_+\right)^{\mathrm{T}} \tag{3.714}$$

$$\hat{U} = \left((C_3e_- - \overline{\theta}_-)^{\mathrm{T}}_{;}(C_3e_- + \widetilde{\theta}_-)^{\mathrm{T}}_{;}(C_4e_+ - \delta_+)^{\mathrm{T}}\right)^{\mathrm{T}} \tag{3.715}$$

$$\hat{V} = \left(-\overline{\theta}^{\mathrm{T}}_{-;}\widetilde{\theta}^{\mathrm{T}}_{-;} - \overline{\delta}^{\mathrm{T}}_+\right)^{\mathrm{T}} \tag{3.716}$$

$$\overline{\alpha}_- = \left(\overline{\alpha}_{p+1}; \ldots; \overline{\alpha}_{p+q}\right)^{\mathrm{T}} \tag{3.717}$$

$$\widetilde{\alpha}_{-} = \left(\alpha_{p+1; } Reject \ldots; \widetilde{\alpha}_{p+q}\right)^{\mathrm{T}} \tag{3.718}$$

$$\overline{\beta}_{+} = \left(\overline{\beta}_{1;} \ldots; \overline{\beta}_{p}\right)^{\mathrm{T}} \tag{3.719}$$

$$\delta_{+} = \left(\delta_{1;} \ldots; \delta_{p}\right)^{\mathrm{T}} \tag{3.720}$$

$$\overline{\theta}_{-} = \left(\overline{\theta}_{p+1}; \ldots; \overline{\theta}_{p+q}\right)^{\mathrm{T}} \tag{3.721}$$

$$\widetilde{\theta}_{-} = \left(\widetilde{\theta}_{p+1;} \ldots; \widetilde{\theta}_{p+q}\right)^{\mathrm{T}} \tag{3.722}$$

$$\delta_j = -C_4 y_j \frac{\partial H_s(y_j f^{-}(x_j))}{\partial f^{-}(x_j)} = \begin{cases} C_4, & \text{if } y_j f^{-}\left(x_j\right) < s \\ 0, & \text{otherwise} \end{cases}, j = 1, \ldots, p \tag{3.723}$$

$$\overline{\theta}_i = \begin{cases} C_3, & \text{if } f^{-}\left(x_i\right) < -t \\ 0, & \text{otherwise} \end{cases}, i = p+1, \ldots, p+q \tag{3.724}$$

and

$$\widetilde{\theta}_i = \begin{cases} -C_3, & \text{if } f^{-}\left(x_i\right) < -t \\ 0, & \text{otherwise} \end{cases}, i = p+1, \ldots, p+q \tag{3.725}$$

and

$$\hat{\Lambda} = \begin{pmatrix} Q_1 & Q_2 \\ Q_2^{\mathrm{T}} & Q_3 \end{pmatrix}; Q_1 = \begin{pmatrix} BB^{\mathrm{T}} & -BB^{\mathrm{T}} \\ -BB^{\mathrm{T}} & BB^{\mathrm{T}} \end{pmatrix}$$

$$Q_2 = \begin{pmatrix} BA^{\mathrm{T}} \\ -BA^{\mathrm{T}} \end{pmatrix}; Q_3 = AA^{\mathrm{T}} \tag{3.726}$$

Now, we can construct the linear RNPSVM based on the CCCP procedure as follows:

Algorithm 3.15 (Linear RNPSVM)

(1) Input the training set

(2) Choose the appropriate $s < 1$ for the ramp loss and $t > \varepsilon$ for the ramp loss; Choose appropriate parameters $\varepsilon > 0; C_i > 0; i = 1; 2; 3; 4$; Initialize δ_-^0; $\overline{\theta}_+^0$; $\widetilde{\theta}_+^0$, and δ_+^0; $\overline{\theta}_-^0$; $\widetilde{\theta}_-^0$, set $k = 1$;

(3) Construct and solve the QPPs and in the k-th iterative step

$$\min \frac{1}{2}\mathrm{T}\widetilde{\Lambda}\widetilde{\pi} + \widetilde{\kappa}^{\mathrm{T}}\widetilde{\pi} \tag{3.727}$$

$$s.t. \ \widetilde{e}^{\mathrm{T}}\widetilde{\pi} = 0,$$

$$\widetilde{V}^k \leq \widetilde{\pi} \leq \widetilde{U}^k.$$

where

$$\widetilde{U}^k = \left(\left(C_1 e_+ - \overline{\theta}_+^k\right)^{\mathrm{T}}, \left(C_1 e_+ + \widetilde{\theta}_+^k\right)^{\mathrm{T}}, \left(C_2 e_- - \delta_-^k\right)^{\mathrm{T}}\right)^{\mathrm{T}} \tag{3.728}$$

$$\widetilde{V}^k = \left(-\overline{\theta}_{+;}^{k\mathrm{T}}\widetilde{\theta}_{+;}^{\prime\langle\mathrm{T}} - \delta_-^{k\mathrm{T}}\right)^{\mathrm{T}} \tag{3.729}$$

and

$$\min \frac{1}{2}\hat{\pi}^{\mathrm{T}}\hat{\Lambda}\hat{\pi} + \hat{\kappa}^{\mathrm{T}}\hat{\pi} \tag{3.730}$$

$$s.t. \ \hat{e}^T\hat{\pi} = 0,$$

$$\hat{V}^k \leq \hat{\pi} \leq \hat{U}^k.$$

where

$$\hat{U}^{\mathrm{k}} = \left(\left(C_3 e_- - \overline{\theta}_-^k\right)^{\mathrm{T}}_{;}\left(C_3 e_- + \widetilde{\theta}_-^k\right)^{\mathrm{T}}_{;}\left(C_4 e_+ - \delta_+^k\right)^{\mathrm{T}}\right)^{\mathrm{T}} \tag{3.731}$$

$$\hat{V}^k = \left(-\overline{\theta}_{-;}^{k\mathrm{T}}\widetilde{\theta}_{-;}^{1\langle\mathrm{T}} - \overline{\delta}_+^{k\mathrm{T}}\right)^{\mathrm{T}} \tag{3.732}$$

get the solutions $\left(\overline{\alpha}_+^k; \widetilde{\alpha}_{+;}^k \overline{\beta}^k\right)$ and $\left(\overline{\alpha}^k; \widetilde{\alpha}^k; \overline{\beta}_+^k\right)$, compute

$$w_+^k = \sum_{i=1}^{p} \left(\overline{\alpha}_i^k - \widetilde{\alpha}_i^k\right) x_i - \sum_{j=p+1}^{p+q} \overline{\beta}_j x_j \tag{3.733}$$

and

$$w_-^k = \sum_{i=p+1}^{p+q} \left(\overline{\alpha}_i^k - \widetilde{\alpha}_i^k\right) x_i + \sum_{j=1}^{p} \overline{\beta}_j x_j, \tag{3.734}$$

compute b_+^k, b_-^k based on the KKT conditions; Construct the decision functions

$$f_k^+(x) = \left(w_+^k \cdot x\right) + b_+^k \tag{3.735}$$

$$f_k^-(x) = \left(w_-^k \cdot x\right) + b_-^k \tag{3.736}$$

(4) Compute δ_-^k, $\overline{\theta}_+^k$, $\widetilde{\theta}_+^k$, and δ_+^k, $\overline{\theta}_-^k$, $\widetilde{\theta}_-^k$, based on the Equations.
(5) If $\left(\delta_-^k, \overline{\theta}_+^k, \widetilde{\theta}_+^k\right) = \left(\delta_-^{k-1}, \overline{\theta}_+^{k-1}, \widetilde{\theta}_+^{\prime(-1)}\right)$ and $\left(\delta_+^{le}, \overline{\theta}_-^1(, \widetilde{\theta}_-^k\right) = \left(\delta_+^{k-1}, \overline{\theta}_-^{k-1}, \widetilde{\theta}_-^{k-1}\right)$, then we get the solution the solutions $(w_+, b_+) = \left(w_+^k, b_+^k\right)$ and $(w_-, b_-) = \left(w_-^k, b_-^k\right)$, go to step (6); else set $k = \mathrm{k} + 1$, go to step (2);
(6) A new point $x \in \mathcal{R}^{\mathrm{n}}$ is predicted to the *Class* by

$$Class = \arg\ \min \left|(w_{\mathrm{m}} \cdot x) + b_{\mathrm{m}}\right|, \tag{3.737}$$

where $|\cdot|$ is the perpendicular distance of point x from the planes $(w_m \cdot x) + b_{\mathrm{m}} = 0$, $m = -, +$.

Nonlinear RNPSVM

Now we extend the linear RNPSVM to the nonlinear case. Introduce the kernel function $K(x, x') = (\Phi(x) \cdot \Phi(x'))$ and the corresponding transformation

$$x = \Phi(x) \tag{3.738}$$

where $x \in \mathcal{H}$, $\mathcal{H}$ is the Hilbert space, we can construct the corresponding problems (3.676) and (3.677) in $\mathcal{H}$ and apply the CCCP procedure to solve them. The resulting algorithm is different with Algorithm 3.4 only that

1. The inner products A^{T}, AB^{T}, BA^{T}, BB^{T} in the dual problems are instead of the appropriate kernel function $K(A, A^{\mathrm{T}})$, $K(A, B^{\mathrm{T}})$, $K(B, A^{\mathrm{T}})$ and $K(B, B^{\mathrm{T}})$.

2. The decision functions constructed in step (3) turn to be

$$f_k^+(x) = \sum_{i=1}^{p} \left(\overline{\alpha}_i^k - \widetilde{\alpha}_i^k\right) K\left(x_i, x\right) - \sum_{j=p+1}^{p+q} \beta_j^k K\left(x_j, x\right) + b_+^k, \tag{3.739}$$

$$f_k^-(x) = \sum_{i=p+1}^{p+q} \left(\overline{\alpha}_i^k - \widetilde{\alpha}_i^k\right) K\left(x_i, x\right) + \sum_{j=1}^{p} \beta_j'^k K\left(x_j, x\right) + b_-^k \tag{3.740}$$

3. For the new point $x \in \mathcal{R}^n$ in step (6), it is predicted to the *Class* by

$$Class = \arg\min \left\lfloor r_m(x) \right| . \tag{3.741}$$

Discussion

Complexity

Obviously, problems (3.732) and (3.736) solved in each iteration are convex QPPs and have the same formulation as the dual problem solved in the standard SVM, therefore the well-known SMO-type decomposition method [68] implemented in LIBSVM can be applied directly with minor modifications, of which the computational complexity of SMO for such problem (3.732) and (3.736) is about

$$\#iteration \times O(1.5l) \tag{3.742}$$

(*#iteration* is the number of the iterations in SMO, empirically it may be higher than linear to the number of training data, $l = p + q$ is the size of the training set) if most columns of the kernel matrix are cached throughout iterations. Therefore, the complexity of Algorithm 3.14 is

$$2 * k * \#\text{iteration} \times O(1.5l) \tag{3.743}$$

where k is the number of the iterations in CCCP.

Initialize

Setting the initial $\delta_-^0, \overline{\theta}_+^0, \widetilde{\theta}_+^0$, and $\delta_+^0, \overline{\theta}_-^0, \widetilde{\theta}_-^0$, to zero makes the first convex optimization identical to the standard NPSVM optimization. Useless support vectors are eliminated during the following iterations. However, we can also initialize them according to the outputs $f_0^+(x)$ and $f_0^-(x)$ of standard NPSVM on a subset of examples

$$\delta_j^0 = \begin{cases} C_2, \text{ if } y f_0^+(x_i) < s \\ 0, \quad \text{otherwise} \end{cases}, \quad j = p+1, \ldots, p+q \tag{3.744}$$

$$\overline{\theta}_i^0 = \begin{cases} C_1, \text{ if } f_0^+(x_i) < -r \\ 0, \quad \text{otherwise} \end{cases}, \quad i = 1, \ldots, p, \tag{3.745}$$

$$\widetilde{\theta}_i^0 = \begin{cases} -C_1, \text{ if } f_0^+(x_i) > f \\ 0, \qquad \text{otherwise} \end{cases}, \quad i = 1, \ldots, p, \tag{3.746}$$

$$\delta_j^0 = \begin{cases} C_4, \text{ if } y f_0^-(x_j) < s \\ 0, \quad \text{otherwise} \end{cases}, \quad j = 1, \ldots, p, \tag{3.747}$$

$$\overline{\theta}_i^0 = \begin{cases} C_3, \text{ if } f_0^-(x_i) < -\mathrm{f} \\ 0, \quad \text{otherwise} \end{cases}, \quad i = p+1, \ldots, p+q, \tag{3.748}$$

and

$$\widetilde{\theta}_i^0 = \begin{cases} -C_3, \text{ if } f_0^-(x_i) > r \\ 0, \qquad \text{otherwise} \end{cases}, \quad i = p+1, \ldots, p+q \tag{3.749}$$

In practice, this procedure is robust, and its overall training time can be significantly smaller than the standard NPSVM training time [87].

Sparsity

Due to the application of the ramp loss, RNPSVM becomes sparser than the standard NPSVM. We know that

$$w_+ = \sum\nolimits_{i=1}^{p} (\overline{\alpha}_i - \widetilde{\alpha}_i) x_i - \sum\nolimits_{j=p+1}^{p+q} \overline{\beta}_j x_j, \tag{3.750}$$

so the support vectors are the corresponding positive points with $\overline{\alpha}_i - \widetilde{\alpha}_i \neq 0$ and negative points with $\overline{\beta}_i \neq 0$. It is easy to show that

1. Each pair of α_i^* and α_i cannot be both simultaneously nonzero, i.e. $\alpha_i^* \alpha_i = 0$, $i = 1, \ldots, p$.
2. If $\alpha_i^* = C_1$, then $\overline{\alpha}_i = 0$, if $\alpha_i = C_1$, $\widetilde{\alpha}_i = 0$, for $i = 1, \ldots, p$.
3. If $\alpha_i^* = 0$, then $\overline{\alpha}_i = 0$, if $\alpha_i = 0$, then $\overline{\alpha}_i = 0$, for $i = 1, \ldots, p$.
4. If $\beta_j = 0$, then $\overline{\beta}_j = 0$, for $j = p+1, \ldots, p+q$.
5. If $\beta_j = C_2$, then $\overline{\beta}_j = 0$, for $j = p+1, \ldots, p+q$.

Therefore, the bounded support vectors (with $\alpha_i^* = C_1$ or $\alpha_i = C_1$ or $\beta_j = C_2$) of the standard NPSVM are not support vectors any more in RNPSVM, similar results can be derived for the problem (3.711). So RNPSVM is sparser than NPSVM.

Selection of s and t

If $s \to -\infty$ and $t \to \infty$, then $R_s \to H_1$ and $R_{\varepsilon f} \to I_\varepsilon$, in other words, if s takes large negative values and r takes large positive values, the Ramp loss used in RNPSVM cannot help to remove outliers, and the RNPSVM will degenerate to the standard NPSVM. For t, it will be better chosen in $(\varepsilon;1)$. Take Fig. 3.12 as the example, on one hand, we need to put the negative class from the hyperplane $(w_+ \cdot x) + b_+ = -1$ (blue thin solid line) as far as possible, on the other hand, if we take $t = 1$, the positive points outside $(w_+ \cdot x) + b_+ = -t = -1$ will be taken as outliers, so $t = 1$ is the largest value we suggest to choose. For s, $s = -2$ is the largest value we suggest, since almost all the points are right-up to $(w_+ \cdot x) + b_+ = -s = 2$, the negative points left to the hyperplane $(w_+ \cdot x) + b_+ = 2$ treated as outliers are few.

3.5.3 A New Classification Model Using Privileged Information and Its Application

3.5.3.1 Fast Twin Support Vector Machine Using Privileged Information (FTSVMPI)

Learning Model Using Privileged Information (LUPI)

Different with standard binary classification problem, LUPI is given a training set as follows:

$$T = \left(x_1, x_1^*, y_1\right), \ldots, \left(x_l, x_l^*, y_l\right) \tag{3.751}$$

where $x_i \in R^n, x_l^* \in R^m, y_i \in \{-1, 1\}, i = 1, \ldots, l$, and the privileged information x_i^* is only included in the training input $\left(x_i, x_i^*\right)$, while not in any testing input x. In order to find a real valued function $g(x)$ in R^n, such that the value of y for any x can be predicted by the decision function

$$f(x) = sgn\,(g(x)) \tag{3.752}$$

In order to explain the basic idea of LUPI, we first introduce the definition of oracle function [90].

Definition 3.7 *(Oracle function).* Given a traditional classification problem with the training set

$$T = \{(x_1, y_1), \ldots, (x_l, y_l)\} \tag{3.753}$$

Suppose there exists the best but unknown linear hyperplane:

$$(w_0 \cdot x) + b_0 = 0 \tag{3.754}$$

The oracle function $\xi(x)$ of the input x is defined as follows:

$$\xi^0 = \xi(x) = [1 - y((w_0 \cdot x) + b_0)]_+ \tag{3.755}$$

where

$$[\eta]_+ = \begin{cases} \eta, & if \eta \geq 0; \\ 0, & otherwise \end{cases} \tag{3.756}$$

If we could know the value of the oracle function on each training input x_i such as we know the triplets $\left(x_i, \xi_i^0, y_i\right)$ with $\xi_i^0 = \xi(x), i = 1, \ldots, l$, we can accelerate its learning rate. However, in fact, a teacher does not know either the values of slacks or the oracle function. Instead, Vapnik et al. use a so-called correcting function to approximate the oracle function. In the linear case,

$$\phi\left(x^*\right) = \left(w^* \cdot x^*\right) + b^* \tag{3.757}$$

Replacing $\xi_i(i = 1, \ldots, l)$ by $\phi(x_i*)$ in the primal problem of SVM, we can get the following primal problem:

$$\begin{aligned} &\min_{w, w^*, b, b^*} \frac{1}{2}\|w\|^2 + C\sum_{i=1}^{l}\left[(w^* \cdot x_i{}^*) + b^*\right], \\ &\text{s.t. } y_i\left[(w \cdot x_i) + b\right] \geq 1 - \left[(w^* \cdot x_i{}^*) + b^*\right], \\ &(w^* \cdot x_i{}^*) + b^* \geq 0, i = 1, \ldots, l. \end{aligned} \tag{3.758}$$

The corresponding dual problem is as follows:

$$\begin{aligned} &\max_{\alpha, \beta} \sum_{j=1}^{l}\alpha_j - \frac{1}{2}\sum_{i=1}^{l}\sum_{j=1}^{l} y_i y_j \alpha_i \alpha_j \left(x_i \cdot x_j\right), \\ &s.t. \sum_{i=1}^{l}\alpha_i y_i = 0, \\ &\sum_{i=1}^{l}(\alpha_i + \beta_i - C) = 0, \\ &\sum_{i=1}^{l}(\alpha_i + \beta_i - C) \cdot x_i = 0, \\ &\alpha_i \geq 0, \beta_i \geq 0, i = 1, \ldots, l \end{aligned} \tag{3.759}$$

For the nonlinear case, introducing two transformations: $x = \Phi(x) : R^n \rightarrow \mathsf{H}$ and $x^* = \Phi^*(x^*) : R^m \rightarrow \mathsf{H}^*$, the primal problem is constructed as follows:

$$\begin{aligned} &\min_{w,w^*,b,b^*} \frac{1}{2}\|w\|^2 + C\sum_{i=1}^{l}\left[(w^* \cdot x_i{}^*) + b^*\right], \\ &\text{s.t. } y_i\left[(w \cdot \Phi(x_i)) + b\right] \geq 1 - \left[(w^* \cdot \Phi^*(x_i{}^*)) + b^*\right], \\ &(w^* \cdot \Phi^*(x_i{}^*)) + b^* \geq 0, i = 1, \ldots, l. \end{aligned} \tag{3.760}$$

Similarly, we can give its dual programming:

$$\begin{aligned} &\min_{\alpha,\beta} \frac{1}{2}\sum_{i=1}^{l}\sum_{j=1}^{l} y_i y_j \alpha_i \alpha_j K\left(x_i \cdot x_j\right) - \sum_{j=1}^{l}\alpha_j, \\ &s.t. \quad \sum_{i=1}^{l}\alpha_i y_i = 0, \\ &\sum_{i=1}^{l}(\alpha_i + \beta_i - C) = 0, \\ &\sum_{i=1}^{l}(\alpha_i + \beta_i - C) K^*\left(x^*{}_i \cdot x^*{}_j\right) = 0, \\ &\alpha_i \geq 0, \beta_i \geq 0, i = 1, \ldots, l \end{aligned} \tag{3.761}$$

FTSVMPI

Let us reconsider the above classification problem with l_1 positive points and l_2 negative points. Suppose that the positive training points and their additional information (privileged information) are denoted by $A \in R^{l_1 \times n}$ and $A^* \in R^{l_1 \times m}$, where each row of $A \in R^n$ and $A^* \in R^m$ represents a training point and an additional information. Similarly, $B \in R^{l_2 \times n}$ and $B^* \in R^{l_2 \times m}$ represent all the data points, and its additional information that belongs to the negative class.

Linear Case

Similar to [12, 21, 58, 91, 92] in order to improve the training speed of LUPI, we first use two small models to contrast the classifier. Replacing slack variables by $\phi\left(A_i^*\right)$ and $\phi\left(B_i^*\right)$ in the primal problem of TWSVM (Twin support vector machine) [58] and using two linear correcting functions to approximate the related oracle functions:

$$\phi\left(A_i^*\right) = \left(w_* \cdot A_i^*\right) + b^* \tag{3.762}$$

and

$$\phi\left(B_i^*\right) = \left(w_+^* \cdot B_i^*\right) + b_+^* \tag{3.763}$$

where $w_+^* \in R^{m_1}, w_-^* \in R^{m_2}, b_+^*, b^* \in R$ and is a dot product operation. The corresponding model can be formulated as

$$\begin{aligned} &\min_{w_+, w_+^*, b_+, b_+^*} \tfrac{1}{2}\|Aw_+ + e_+b_+\|_2^2 + c_1 e\left(B^*w_+^* + eb_+^*\right), \\ &s.t. \ -(Bw_+ + e_-b_+) \geq e - \left(B^*w_+^* + eb_+^*\right) \\ &B^*w_+^* + eb_+^* \geq 0 \end{aligned} \tag{3.764}$$

and

$$\min_{w_-, w_-^* b_-^*, b_-} \frac{1}{2}\|Bw_- + e_-b_-\|_2^2 + c_2 e_+^{\mathrm{T}}\left(A^*w_-^* + e_+b_-^*\right)$$

$$s.t. \ (Aw_- + e_+b_-) \geq e_+ - \left(A^*w_-^* + e_+b_-^*\right), \tag{3.765}$$

$$A^*w_-^* + e_+b_-^* \geq 0,$$

where $c_1, c_2 \geq 0$ are the pre-specified penalty factors, e_+, e_- are vectors of ones of appropriate dimensions.

Next, we use 1-norm distance to replace the square of the 2-norm of model [92] and [67]. Specifically, $\|Aw_+ + e_+b_+\|_2^2$ is replaced by $\|Aw_+ + e_+b_+\|_1$, which can be easily converted to a linear term $e_+^{\mathrm{T}} a$ with the corresponding constraint $-\alpha \leq Aw_+ + e_+b_+ \leq \alpha$, where $\alpha = \{\alpha_1, \ldots, \alpha_{l_1}\}$. So the optimization problem is replaced by

$$\min_{w_+, w_+^* b_+^*, b_+} \frac{1}{2} e_+^{\mathrm{T}}\alpha + c_1 e_-^{\mathrm{T}}\left(B^*w_+^* + e_-b_+^*\right),$$

$$s.t. \ -(Bw_+ + e_-b_+) \geq e_- - \left(B^*w_+^* + e_-b_+^*\right), \tag{3.766}$$

$$-\alpha \leq Aw_+ + e_+b_+ \leq \alpha,$$

$$\mathrm{B}^*w_1^* + e_-b_1^* \geq 0$$

Similarly, the optimization problem can be converted to

$$\min_{w_-, w_-^* b_-^*, b_-, \beta} \frac{1}{2} e_-^{\mathrm{T}}\beta + c_2 e_+^{\mathrm{T}}\left(A^*w_-^* + e_+b_-^*\right)$$

$$s.t. \ (Aw_- + e_+b_-) \geq e_+ - \left(A^*w_-^* + e_+b_-^*\right),$$

$$-\beta \le Bw_- + e_-b_- \le \beta, \tag{3.767}$$

$$A^* w^*_- + e_+ b^*_- \ge 0$$

Finally, we get two nonparallel hyperplanes

$$f_+(x) = w_+^{\mathrm{T}} x + b_1 = 0 \text{ and } f(x) = w_-^{\mathrm{T}} x + b = 0, \tag{3.768}$$

where w_+, $w_- \in R^{\mathrm{n}}$, $b_+ b_- \in R$. A new data point $x \in R^{\mathrm{n}}$ is then assigned to the positive or negative class, depending on which of the two hyperplanes it lies closer to, i.e.

$$f(x) = \arg\ \min \{d_+(\mathrm{x}), d_-(x)\} \tag{3.769}$$

where

$$\mathrm{d}_+(x) = \left| w_+^{\mathrm{T}} x + b_+ \right|, d_-(x) = \left| w_-^{\mathrm{T}} x + b_- \right|, \tag{3.770}$$

and $|\cdot|$ is the perpendicular distance of point x from the planes $w_+^{\mathrm{T}} x + b_+ = 0$ or $w_-^{\mathrm{T}} x + b_- = 0$.

Non-linear Case

Now we extend the linear FTSVMPI to the non-linear case. Similar to the linear case, two hyperplanes $f_+(x) = (w_+ \cdot \Phi(x)) + b_+ = 0$ and $f_-(\mathrm{x}) = (w_- \cdot \Phi(x)) + b_- = 0$ are considered, where $\Phi(\cdot)$ is a non-linear mapping from a low dimensional space to a higher dimensional Hilbeit space $\mathcal{H}$. According to Hilbert space theory [67], w_+ and w_- can be expressed as $w_+ = \sum_{i=1}^{l1+l2} + (\lambda_+)_{\mathrm{i}} . \Phi\left(x_{\mathrm{i}}\right) = \Phi(M)\lambda_+$ and $w_- = \sum_{i=1}^{l1+l2} (\lambda_-)_{\mathrm{i}} \Phi\left(x_{\mathrm{i}}\right) = \Phi(M)\lambda_-$ respectively. Similarly $w_+^* = \sum_{i=1}^{l1+l2} \left(\lambda_+^*\right)_1 \cdot \Phi\left(x_1^*\right) = \Phi\left(M^*\right) \lambda_+^*$ and $w_-^* = \sum_{i=1}^{l1+l2} + l_2 \left(\lambda_-^*\right)_{\mathrm{I}} \Phi\left(x_1^*\right) = \Phi\left(M^*\right) \lambda_-^*$. So the two hyperplanes turn to be the following kernel-generated formulations:

$$\mathrm{K}\left(\mathrm{x}^{\mathrm{T}} M^{\mathrm{T}}\right) \lambda_+ + b_+ = 0, \tag{3.771}$$

$$\mathrm{K}\left(x^{\mathrm{T}} M^{\mathrm{T}}\right) \lambda_- + b_- = 0, \tag{3.772}$$

where K is a kernel function: $K(x_{\mathrm{i}}, x_j) = (\Phi(x_1) \cdot \Phi(x_j))$, $(\cdot)$ denotes dot product operation, $M^{\mathrm{T}} = [A^{\mathrm{T}} B^{\mathrm{T}}]_{\mathrm{n} \times 1}$, $\lambda_+, \lambda_- \in R^l$, and $b_+, b_- \in R$.

Correspondingly, the correcting function can be written as

$$K^{*}\left(x^{*\mathrm{T}}M^{*\mathrm{T}}\right)\lambda_{+}^{*}+b_{+}^{*}=0, \tag{3.773}$$

$$K^{*}\left(x^{*\mathrm{T}}M^{*\mathrm{T}}\right)\lambda_{-}^{*}+b_{-}^{*}=0, \tag{3.774}$$

where K^{*} is a kernel function: $K^{*}\left(x_{j}^{*}x_{j}^{*}\right)=\left(\Phi\left(x_{\mathrm{i}}^{*}\right)\cdot\Phi\left(x_{l}^{*}\right)\right)$, $M^{*}=[A^{*T}B^{*T}]_{n\times l}$, λ_{+}^{*}, $\lambda_{-}^{*}\in R^{l}$, and $b_{+}^{*}, b_{-}^{*}\in R$. Therefore, the optimization problems for the nonlinear case are constructed as

$$\begin{aligned}
&\min_{\lambda_{+}^{*},b_{+}^{*},\lambda_{+},b_{+},w_{+}^{*},\varphi}\frac{1}{2}e_{+}^{\mathrm{T}}\Phi+c_{1}e_{-}^{\mathrm{T}}\left(K^{*}\left(B^{*},M^{*\mathrm{T}}\right)\lambda_{+}^{*}+e_{-}b_{+}^{*},\right.\\
&s.t.\ -\left(\left(B,M^{T}\right)\lambda_{+}+e_{-}b_{+}\right)\geq\\
&e_{-}-\left(K\left(B^{*}M^{*T}\right)\lambda_{+}^{*}+e_{-}b_{+}^{*}\right),\\
&-\Phi\leq K\left(A,M^{\mathrm{T}}\right)\lambda_{+}+e_{+}b_{+}\leq\Phi,\\
&\mathrm{K}^{*}\left(B^{*}M^{*\mathrm{T}}\right)\lambda_{+}^{\mathrm{r}}+e_{-}b_{+}^{*}>0,
\end{aligned} \tag{3.775}$$

and

$$\begin{aligned}
&\min_{\lambda_{-}^{*},b_{-}^{*},\lambda_{-},b_{-},w_{-}^{*},\varphi}\frac{1}{2}e_{-}^{\mathrm{T}}\Phi+c_{2}e_{+}^{\mathrm{T}}\left(K^{*}\left(BA^{*},M^{*\mathrm{T}}\right)\lambda_{-}^{*}+e_{+}b_{-}^{*},\right.\\
&s.t.\ -\left(K\left(AM^{T}\right)\lambda_{+}+e_{-}b_{+}\right)\geq\\
&e_{+}-\left(K\left(A^{*}M^{*T}\right)\lambda_{-}^{*}+e_{+}b_{-}^{*}\right),\\
&-\Phi\leq K\left(BM^{\mathrm{T}}\right)\lambda_{-}+e_{-}b_{-}\leq\Phi,\\
&\mathrm{K}^{*}\left(A^{*}M^{*\mathrm{T}}\right)\lambda_{+}^{*}+e_{+}b_{-}^{*}\geq 0,
\end{aligned} \tag{3.776}$$

Notice that problems (3.775) and (3.776) are two standard Linear Programming (LP).

Discussion

Since LUPI-SVM model is more than two times slower than the standard SVM and usually needs to solve a more difficult optimization problem than the standard SVM, we improve the LUPI model by the following two ways: reducing the model size and using L-l norm 1 regularization term method. For the first way, not only does FTSVMPI accelerate the training speed but also inherits the virtue of TWSVM [58] which uses two nonparallel hyperplanes to construct a decision function, and has a better generalized capability than traditional LUCPI (Jayadeva et al. used a famous XOR datasets to fully confirm this viewpoint [58]. Furthermore, unlike TWSVM, our model avoids solving the inverse matrix whose computational complexity is more than $o(l^3)$ and further reduce the model's training time. For the second way, FTSVMPI can obtain advantages as follows: (1) our model can help to perform feature ranking and selection in the learning process. In the result, the final classification rule found by our FTSVMPI might be more interpretable. (2) Since the computational cost of solving LP is much cheaper than solving (l^{pp} with the same scale, our model is usually much faster and cheaper thani the training of LUPI. In fact, some recent works have adapted different strategies and methods to improve the speed and quality of SVM [93–95]. For example, Luo et al. proposed a manifold regularized multitask SVM learning algorithm to improve the quality of classification [96], Zhou and Tao et al. proposed a fast gradient method for SVM [95]. We are very interested in how to adding privileged information into these improved algorithms for SVM in the future work.

References

1. Shi, Y., Tian, Y., Kou, G., Peng, Y., Li, J.: Optimization Based Data Mining: Theory and Applications. Springer Science & Business Media, New York (2011)
2. Tian, Y., Shi, Y., Liu, X.: Recent advances on support vector machines research. Technol. Econ. Dev. Econ. **18**(1), 5–33 (2012)
3. Niu, L., Zhou, R., Zhao, X., Shi, Y.: Two new decomposition algorithms for training bound-constrained support vector machines. Found. Comput. Decis. Sci. **40**(1), 67–86 (2015)
4. Tian, Y., Ju, X., Qi, Z., Shi, Y.: Improved twin support vector machine. Sci. China Math. **57**(2), 417–432 (2014)
5. Xie, J., Hone, K., Xie, W., Gao, X., Shi, Y., Liu, X.: Extending twin support vector machine classifier for multi-category classification problems. Intell. Data Anal. **17**(4), 649–664 (2013)
6. Qi, Z., Tian, Y., Shi, Y.: Robust twin support vector machine for pattern classification. Pattern Recogn. **46**(1), 305–316 (2013)
7. Qi, Z., Tian, Y., Shi, Y.: Structural twin support vector machine for classification. Knowl. Based Syst. **43**, 74–81 (2013)
8. Qi, Z., Tian, Y., Shi, Y.: A nonparallel support vector machine for a classification problem with universum learning. J. Computat. Appl. Math. **263**, 288–298 (2014)
9. Tian, Y., Ju, X., Shi, Y.: A divide-and-combine method for large scale nonparallel support vector machines. Neural Netw. **75**, 12–21 (2016)
10. Qi, Z., Tian, Y., Shi, Y.: A new classification model using privileged information and its application. Neurocomputing. **129**, 146–152 (2014)

11. Qi, Z., Tian, Y., Shi, Y.: Successive overrelaxation for laplacian support vector machine. IEEE Trans. Neural Netw. Learn. Syst. **26**(4), 674–683 (2014)
12. Qi, Z., Tian, Y., Shi, Y.: Laplacian twin support vector machine for semi-supervised classification. Neural Netw. **35**, 46–53 (2012)
13. Liu, D., Shi, Y., Tian, Y., Huang, X.: Ramp loss least squares support vector machine. J. Computat. Sci. **14**, 61–68 (2016)
14. Liu, D., Shi, Y., Tian, Y.: Ramp loss nonparallel support vector machine for pattern classification. Knowl. Based Syst. **85**, 224–233 (2015)
15. Deng, N., Tian, Y.: Support Vector Machines: A New Method in Data Mining. Science Press, Beijing, China (2004)
16. Deng, N., Tian, Y.: Support Vector Machines-Theory, Algorithms and Development. Science Press, Beijing, China (2009)
17. Deng, N., Tian, Y., Zhang, C.: Support Vector Machines: Optimization Based Theory, Algorithms, and Extensions. CRC Press, Boca Raton, FL (2012)
18. Vapnik, V., Vapnik, V.: Statistical Learning Theory, pp. 156–160. Springer, Berlin (1998)
19. Zhang, C., Tian, Y., Deng, N.: The new interpretation of support vector machines on statistical learning theory. Sci China Ser A Math. **53**(1), 151–164 (2010)
20. Suykens, J.A., Van Gestel, T., De Brabanter, J.: Least squares support vector machines. World Sci. (2002)
21. Shao, Y.H., Zhang, C.H., Wang, X.B., Deng, N.Y.: Improvements on twin support vector machines. IEEE Trans. Neural Netw. **22**(6), 962–968 (2011)
22. Ataman, K., Street, W.N.: Optimizing area under the roc curve using ranking svms. In: Proceedings of International Conference on Knowledge Discovery in Data Mining (2005)
23. Brefeld, U., Scheffer, T.: Auc maximizing support vector learning. In: Proceedings of the ICML 2005 Workshop on ROC Analysis in Machine Learning (2005)
24. Goswami, A., Jin, R., Agrawal, G.: Fast and exact out-of-core k-means clustering. In: Fourth IEEE International Conference on Data Mining (ICDM'04), pp. 83–90. IEEE, New York (2004)
25. Lin, C.F., Wang, S.D.: Fuzzy support vector machines. IEEE Trans. Neural Netw. **13**(2), 464–471 (2002)
26. Akbani, R., Kwek, S., Japkowicz, N.: Applying support vector machines to imbalanced datasets. In: European Conference on Machine Learning, pp. 39–50. Springer, New York (2004)
27. Herbrich, R., Graepel, T., Obermayer, K.: Support vector learning for ordinal regression. In: 1999 Ninth International Conference on Artificial Neural Networks ICANN 99. IEEE, New York (1999)
28. Yang, Z.: Support vector ordinal regression and multi-class problems. Ph.D. thesis, China Agricultural University (2007)
29. Yang, Z., Deng, N., Tian, Y.: A multi-class classification algorithm based on ordinal regression machine. In: International Conference on Computational Intelligence for Modelling, Control and Automation and International Conference on Intelligent Agents, Web Technologies and Internet Commerce (CIMCA-IAWTIC'06), vol. 2, pp. 810–815. IEEE, New York (2005)
30. Joachims, T.: Svmlight: support vector machine. SVM-Light Support Vector Machine. http://svmlight.joachims.org/, University of Dortmund **19**(4) (1999)
31. Xu, L., Schuurmans, D.: Unsupervised and semi-supervised multi-class support vector machines. AAAI. **40**, 50 (2005)
32. Zhao, K., Tian, Y.J., Deng, N.Y.: Unsupervised and semi-supervised two-class support vector machines. In: Sixth IEEE International Conference on Data Mining-Workshops (ICDMW'06), pp. 813–817. IEEE, New York (2006)
33. Zhao, K., Tian, Y.J., Deng, N.Y.: Unsupervised and semi-supervised Lagrangian support vector machines. In: International Conference on Computational Science, pp. 882–889. Springer, New York (2007)
34. Angulo, C., Català, A.: K-svcr. A multi-class support vector machine. In: European Conference on Machine Learning, pp. 31–38. Springer, New York (2000)

35. Gao, T.: U-support vector machine and its applications. Master's thesis, China Agricultural University (2008)
36. Goldfarb, D., Iyengar, G.: Robust convex quadratically constrained programs. Math. Program. **97**(3), 495–515 (2003)
37. Fung, G., Mangasarian, O.L., Shavlik, J.W.: Knowledge-based support vector machine classifiers. In: NIPS, pp. 521–528. Citeseer (2002)
38. Mangasarian, O.L., Wild, E.W.: Multisurface proximal support vector machine classification via generalized eigenvalues. IEEE Trans. Pattern Anal. Mach. Intell. **28**(1), 69–74 (2005)
39. Vapnik, V., Vashist, A.: A new learning paradigm: learning using privileged information. Neural Netw. **22**(5–6), 544–557 (2009)
40. Mangasarian, O.L., Wild, E.W.: Nonlinear knowledge-based classification. IEEE Trans. Neural Netw. **19**(10), 1826–1832 (2008)
41. Frie, T.T., Cristianini, N., Campbell, C.: The kernel-adatron algorithm: a fast and simple learning procedure for support vector machines. In: Machine Learning: Proceedings of the Fifteenth International Conference (ICML'9m8), pp. 188–196. Citeseer (1998)
42. Mangasarian, O.L., Musicant, D.R.: Successive overrelaxation for support vector machines. IEEE Trans. Neural Netw. **10**(5), 1032–1037 (1999)
43. Hsieh, C.J., Chang, K.W., Lin, C.J., Keerthi, S.S., Sundararajan, S.: A dual coordinate descent method for large-scale linear svm. In: Proceedings of the 25th International Conference on Machine Learning, pp. 408–415 (2008)
44. Joachims, T.: Training linear svms in linear time. In: Proceedings of the 12th ACM SIGKDD International Conference on Knowledge Discovery and Data Mining, pp. 217–226 (2006)
45. Joachims, T., Finley, T., Yu, C.N.J.: Cutting-plane training of structural svms. Mach. Learn. **77**(1), 27–59 (2009)
46. Joachims, T., Yu, C.N.J.: Sparse kernel svms via cutting-plane training. Mach. Learn. **76**(2), 179–193 (2009)
47. Bottou, L., Chapelle, O., DeCoste, D., Weston, J.: Trading convexity for scalability. In: Proceedings of the 23th International Conference on Machine Learning. Google Scholar Digital Library (2006)
48. Shalev-Shwartz, S., Singer, Y., Srebro, N., Cotter, A.: Pegasos: primal estimated sub-gradient solver for svm. Math. Program. **127**(1), 3–30 (2011)
49. Yuan, G.X., Ho, C.H., Lin, C.J.: Recent advances of large-scale linear classification. Proc. IEEE. **100**(9), 2584–2603 (2012)
50. Hsu, C.W., Lin, C.J.: A simple decomposition method for support vector machines. Mach. Learn. **46**(1), 291–314 (2002)
51. Joachims, T.: Making large-scale svm learning practical. Technical report (1998)
52. Osuna, E., Freund, R., Girosit, F.: Training support vector machines: an application to face detection. In: Proceedings of IEEE Computer Society Conference on Computer Vision and Pattern Recognition, pp. 130–136. IEEE, New York (1997)
53. Platt, J.: Sequential minimal optimization: a fast algorithm for training support vector machines (1998)
54. Saunders, C., Stitson, M.O., Weston, J., Bottou, L., Smola, A., et al.: Support vector machine-reference manual (1998)
55. Zanni, L., Serafini, T., Zanghirati, G., Bennett, K.P., Parrado-Hernández, E.: Parallel software for training large scale support vector machines on multiprocessor systems. J. Mach. Learn. Res. **7**(54), 1467–1492 (2006)
56. Sun, W., Yuan, Y.X.: Optimization Theory and Methods: Nonlinear Programming, vol. 1. Springer Science & Business Media, New York (2006)
57. Mercer, J.: Functions of positive and negative type and their connection with the theory of integral equations. Philos. Trans. R Soc. **83**, 4–415 (1909)
58. Khemchandani, R., Chandra, S., et al.: Twin support vector machines for pattern classification. IEEE Trans. Pattern Anal. Mach. Intell. **29**(5), 905–910 (2007)
59. Chen, P.H., Fan, R.E., Lin, C.J.: A study on smo-type decomposition methods for support vector machines. IEEE Trans. Neural Netw. **17**(4), 893–908 (2006)

60. Burges, C.J.: A tutorial on support vector machines for pattern recognition. Data Min. Knowl. Disc. **2**(2), 121–167 (1998)
61. Ward Jr., J.H.: Hierarchical grouping to optimize an objective function. J. Am. Stat. Assoc. **58**(301), 236–244 (1963)
62. Xue, H., Chen, S., Yang, Q.: Structural support vector machine. In: International Symposium on Neural Networks, pp. 501–511. Springer, New York (2008)
63. Xue, H., Chen, S., Yang, Q.: Structural regularized support vector machine: a framework for structural large margin classifier. IEEE Trans. Neural Netw. **22**(4), 573–587 (2011)
64. Yeung, D.S., Wang, D., Ng, W.W., Tsang, E.C., Wang, X.: Structured large margin machines: sensitive to data distributions. Mach. Learn. **68**(2), 171–200 (2007)
65. Salvador, S., Chan, P.: Determining the number of clusters/segments in hierarchical clustering/segmentation algorithms. In: 16th IEEE International Conference on Tools with Artificial Intelligence, pp. 576–584. IEEE, New York (2004)
66. Gantmacher, F.R.: Matrix Theory. Chelsea, New York (1990)
67. Schölkopf, B., Smola, A.J., Bach, F., et al.: Learning with Kernels: Support Vector Machines, Regularization, Optimization, and Beyond. MIT Press, Cambridge, MA (2002)
68. Fan, R.E., Chen, P.H., Lin, C.J., Joachims, T.: Working set selection using second order information for training support vector machines. J. Mach. Learn. Res. **6**(12), 1889–1918 (2005)
69. Chang, C.C., Lin, C.J.: Libsvm: a library for support vector machines. ACM Trans. Intell. Syst. Technol. **2**(3), 1–27 (2011)
70. Hsieh, C.J., Si, S., Dhillon, I.: A divide-and-conquer solver for kernel support vector machines. In: International Conference on Machine Learning, pp. 566–574. PMLR, New York (2014)
71. Mangasarian, O.L.: Nonlinear Programming. SIAM, Philadelphia, PA (1994)
72. Shao, Y.H., Deng, N.Y.: A coordinate descent margin based-twin support vector machine for classification. Neural Netw. **25**, 114–121 (2012)
73. Peng, X.: Tpmsvm: a novel twin parametric-margin support vector machine for pattern recognition. Pattern Recogn. **44**(10–11), 2678–2692 (2011)
74. Maron, O., Lozano-Pérez, T.: A framework for multiple-instance learning. In: Advances in Neural Information Processing Systems, pp. 570–576 (1998)
75. Mangasarian, O.L., Wild, E.W.: Multiple instance classification via successive linear programming. J. Optim. Theory Appl. **137**(3), 555–568 (2008)
76. Tikhonov, A.N.: Regularization of incorrectly posed problems. Soviet Math. Doklady. **4**(6), 1624–1627 (1963)
77. Belkin, M., Niyogi, P., Sindhwani, V.: Manifold regularization: a geometric framework for learning from labeled and unlabeled examples. J. Mach. Learn. Res. **7**(85), 2399–2434 (2006)
78. Evgeniou, T., Pontil, M., Poggio, T.: Regularization networks and support vector machines. Adv. Comput. Math. **13**(1), 1–50 (2000)
79. Belkin, M., Niyogi, P.: Towards a theoretical foundation for laplacian-based manifold methods. J. Comput. Syst. Sci. **74**(8), 1289–1308 (2008)
80. Chang, K.W., Hsieh, C.J., Lin, C.J.: Coordinate descent method for large-scale l_2-loss linear support vector machines. J. Mach. Learn. Res. **9**(7) (2008)
81. Chapelle, O.: Training a support vector machine in the primal. Neural Comput. **19**(5), 1155–1178 (2007)
82. Cucker, F., Zhou, D.X.: Learning Theory: An Approximation Theory Viewpoint, vol. 24. Cambridge University Press, Cambridge (2007)
83. Gnecco, G., Sanguineti, M.: Regularization techniques and suboptimal solutions to optimization problems in learning from data. Neural Comput. **22**(3), 793–829 (2010)
84. Melacci, S., Belkin, M.: Laplacian support vector machines trained in the primal. J. Mach. Learn. Res. **12**(3), 1149–1184 (2011)
85. Wang, L., Jia, H., Li, J.: Training robust support vector machine with smooth ramp loss in the primal space. Neurocomputing. **71**(13–15), 3020–3025 (2008)
86. Steinwart, I.: Sparseness of support vector machines. J. Mach. Learn. Res. **4**(Nov), 1071–1105 (2003)

87. Weston, J., Collobert, R., Sinz, F., Bottou, L., Vapnik, V.: Inference with the universum. In: Proceedings of the 23rd International Conference on Machine Learning, pp. 1009–1016 (2006)
88. Yuille, A.L., Rangarajan, A.: The concave-convex procedure. Neural Comput. **15**(4), 915–936 (2003)
89. Tao, P.D., et al.: The dc (difference of convex functions) programming and dca revisited with dc models of real world nonconvex optimization problems. Ann. Oper. Res. **133**(1–4), 23–46 (2005)
90. Vapnik, V.: Estimation of Dependences Based on Empirical Data. Springer Science & Business Media, New York (2006)
91. Qi, Z., Tian, Y., Shi, Y.: Twin support vector machine with universum data. Neural Netw. **36**, 112–119 (2012)
92. Tian, Y., Qi, Z., Ju, X., Shi, Y., Liu, X.: Nonparallel support vector machines for pattern classification. IEEE Trans. Cybernetics. **44**(7), 1067–1079 (2013)
93. Guan, N., Tao, D., Luo, Z., Shawe-Taylor, J.: Mahnmf: Manhattan non-negative matrix factorization. arXiv preprint arXiv:1207.3438 (2012)
94. Tao, D., Tang, X., Li, X., Wu, X.: Asymmetric bagging and random subspace for support vector machines-based relevance feedback in image retrieval. IEEE Trans. Pattern Anal. Mach. Intell. **28**(7), 1088–1099 (2006)
95. Zhou, T., Tao, D., Wu, X.: Nesvm: a fast gradient method for support vector machines. In: 2010 IEEE International Conference on Data Mining, pp. 679–688. IEEE, New York (2010)
96. Luo, Y., Tao, D., Geng, B., Xu, C., Maybank, S.J.: Manifold regularized multitask learning for semi-supervised multilabel image classification. IEEE Trans. Image Process. **22**(2), 523–536 (2013)

Part II
Functional Analysis

Chapter 4
Feature Selection

In big data analytics, irrelevant and redundant features may not only deteriorate the performances of classifiers, but also slow down the prediction process. Although there is the availability of many classification models for prediction, it is a challenge to choose a set of important features that can lead to a satisfactory classifier. This chapter outlines some achievements of feature selection research in the last decade. Section 4.1 has three subsections. The first is an integrated scheme for feature selection and classifier evaluation in the context of prediction [1]. The second is about two-stage hybrid feature selection algorithms [2]. The third one is the feature selection with attributes clustering by maximal information coefficient [3]. Section 4.2 presents two regularizations for feature selections. They are feature selection with MCP^2 regularization [4] and feature selection with $\ell_{2,1-2}$ regularization [5]. Finally, Sect. 4.3 describes two distance-based feature selections. They are the spatial distance join based feature selection [1] and a domain driven two-phase feature selection method based on bhattacharyya distance and kernel distance measurements [6].

4.1 Systematic Methods for Feature Selection

4.1.1 An Integrated Feature Selection and Classification Scheme

This section presents the research scheme and the major components of the scheme, including feature selection methods, MCDM methods, and classification algorithms. Based on the findings of [7], this study designs the research scheme with careful consideration of these three factors. First, multiple datasets, representing different sizes and domains, are selected for the experimental study. Second, five accuracy indicators are used to evaluate classifiers. Third, tenfold cross-validation

Y. Shi, *Advances in Big Data Analytics*,
https://doi.org/10.1007/978-981-16-3607-3_4

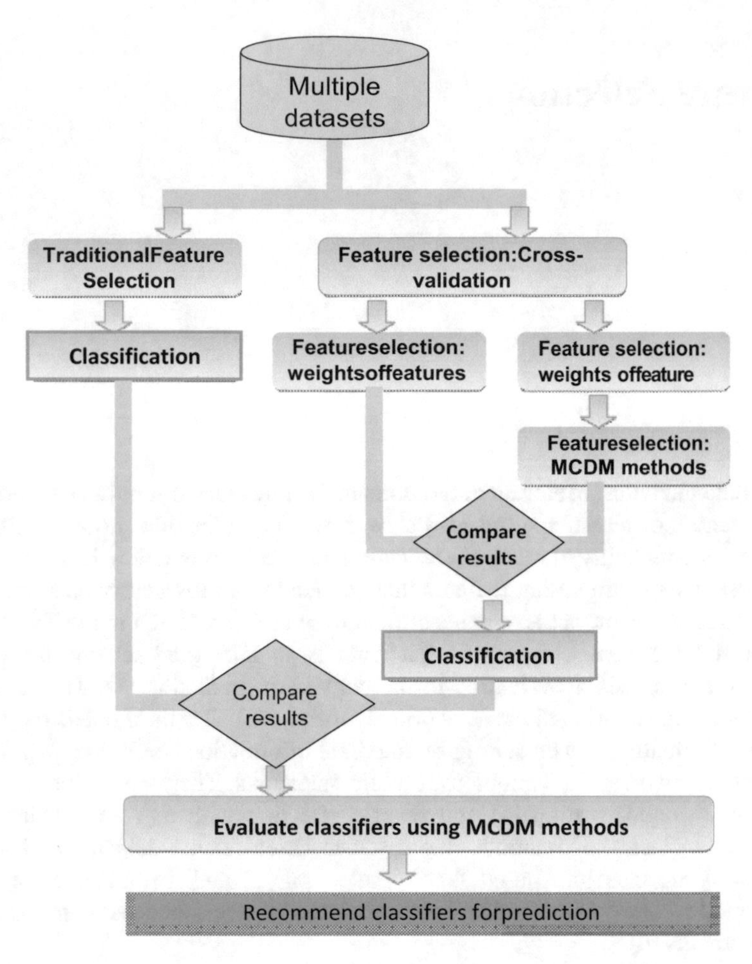

Fig. 4.1 Research scheme

technique is applied to the sample datasets to select features. The research scheme is summarized in Fig. 4.1.

The datasets are handled by two different approaches. The first approach applies traditional feature selection and classification algorithms to the datasets to get prediction results. In the second approach, feature selection and classification are conducted in four steps. First, feature selection is conducted using traditional techniques. Features are then ranked using the proposed feature selection method. The third step employs MCDM methods to evaluate feature selection techniques and choose the better performed techniques. In the last step, the selected features are used in the classification. The classification results of the first and second approaches are compared to examine whether the proposed feature selection and MCDM methods can improve the prediction accuracy. The performances of classifiers are also evaluated using MCDM methods and a recommendation of classifiers for prediction is made based on their accuracy and reliability.

4.1.1.1 Proposed Feature Selection Methods

The proposed feature selection approach makes use of both types of techniques. Weka is used throughout this study to implement feature selection and classification tasks [8]. Four types of feature selection evaluators are provided by Weka: CfsSubsetEval, ConsistencySubsetEval, ClassifierSubsetEval, and WrapperSubsetEval. CfsSubsetEval selects attributes that are highly correlated with the class label and have low inter-correlation. The second method projects training data onto attribute set to measure the level of consistency in the class values. The goal is to find the smallest subset with the highest consistency. The third and fourth methods belong to wrapper approach and they both use a classifier to evaluate attributes. The difference is that ClassifierSubsetEval measures attribute sets on the training data and WrapperSubsetEval uses cross-validation.

The proposed feature selection approach is based on the results of traditional feature selection methods. The importance of feature a_i is measured by a weight W_a, which is calculated as:

$$W_{a_i} = \frac{count_{j=1}^{n}\left(b_{ij}\right)}{n} \times \frac{\sum_{j=1}^{n}\left(b_{ij}\right)^2}{\sum_{j=1}^{n} b_{ij}}, i = 1, 2, \cdots, n \quad (4.1)$$

Where n is the number of feature selection techniques, m is the number of features, and b_j is the numeric value generated by each feature selection technique. A feature with a high weight indicates that it is chosen by many feature selection methods and the variations of values generated by different feature selection techniques for this feature are low. Therefore, features can be sorted according to their weights.

This study chooses WrapperSubsetEval, CfsSubsetEval, and ConsistencySubsetEval, as feature subset evaluators. Wrapper-SubsetEval uses nine classifiers as base learners for estimating the accuracy of subsets. The classifiers are described in a later section. Each classifier represents one feature selection method. Thus, there are total eleven feature selection methods. Some of them are reviewed as below.

4.1.1.2 MCDM Methods

Multiple criteria decision making (MCDM) aims at solving decision problems with multiple objectives and often conflictive constraints [9–11]. Algorithm evaluation or selection normally needs to examine more than one criterion and can be modeled as MCDM problems. Two types of algorithm evaluation tasks are considered: the evaluation of feature selection techniques and classification algorithms. In order to do so, the following provides an overview of five MCDM methods, and explains how they are used in the experimental study to evaluate algorithms.

Data Envelopment Analysis (DEA)

The original DEA model presented by [12] is called "CCR ratio model", which uses the ratio of outputs to inputs to measure the efficiency of DMUs. Assume that there are n DMUs with m inputs to produce s outputs. $\boldsymbol{x_{ij}}$ and $\boldsymbol{y_{rj}}$ represent the amount of input $\boldsymbol{i}$ and output r for DMU, respectively. Then the ratio-form of DEA can be represented as:

$$\begin{aligned} \max\ h_0(u, v) &= \textstyle\sum_r u_r y_{ro} / \sum_i v_i x_{io} \\ \text{subject to } \textstyle\sum_r u_r y_{rj} / \sum_i v_i x_{ij} &\le 1, \text{ for } j = 1, 2 \cdots, n, u_r, v_i \ge 0, \text{for all } i \end{aligned} \tag{4.2}$$

where the $\boldsymbol{u_r}'$ and the $\boldsymbol{v_i}'$ are the variables and the $\boldsymbol{y_{ro}}'$ and $\boldsymbol{x_{io}}'$ are the observed output and input values of the DMU to be evaluated (i.e., DMUo), respectively [13].

The equivalent linear programming problem using the Charnes-Cooper Transformation is

$$\begin{aligned} \max\ z &= \textstyle\sum_{r=1}^{s} \mu_r y_{ro} \\ \text{subject to } \textstyle\sum_{i=1}^{m} v_i x_{i0} &= 1, \mu_r, v_i \ge 0 \\ \textstyle\sum_{i=1}^{m} v_i x_{i0} &= 1, \mu_r, v_i \ge 0 \end{aligned} \tag{4.3}$$

Comparing with the CCR model, a constraint $\sum_{j=1}^{n} \lambda_j = 1$is added in the BCC model. These models can be solved using the simplex method for each DMUs. DMUs with value of 1 are efficient and others are inefficient.

Nakhaeizadeh and Schnabl [14] proposed to use DEA approach in data mining algorithms selection. They argued that in order to make an objective evaluation of data mining algorithms that all the available positive and negative properties of algorithms are important and DEA models can take both aspects into consideration. Positive and negative properties of data mining algorithms can be considered as output and input components in DEA, respectively. For example, the overall accuracy rate of a classification algorithm is an output component and the computation time of an algorithm is an input component. Using existing DEA models, it is possible to give a comprehensive evaluation of feature selection and classification algorithms.

ELimination and Choice Expressing REality (ELECTRE)

ELECTRE stands for ELimination Et Choix Traduisant la REalite (ELimination and Choice Expressing the REality) and was first proposed by Roy [15] to choose the best alternative from a collection of alternatives. ELECTRE III is chosen in this section because it is appropriate for the sorting problem. The procedure can be summarized as follows:

Step 1: define a concordance and discordance index set for each pair of alternatives A_j and A_k, and $j, k = 1, 2 \cdots m$; $i \neq k$.
Step 2: add all the indices of an alternative to get its global concordance index C_{ki}
Step 3: define an outranking credibility degree $\sigma_s(A_i, A_k)$ by combining the discordance indices and the global concordance index.
Step 4: define two outranking relations using descending and ascending distillation. Descending distillation selects the best alternative first and the worst alternative last. Ascending distillation selects the worst alternative first and the best alternative last.
Step 5: alternatives are ranked based on ascending and descending distillation processes.

Preference Ranking Organisation Method for Enrichment of Evaluations (PROMETHEE)

The Promethee methods use pairwise comparisons and outranking relationships to choose the best alternatives. Since the purpose of this section is to build a ranking of classification algorithms, PROMETHEE II is selected. The PROMETHEE II procedure as:

Step 1: define aggregated preference indices. Let $a, b \in A$, and let:

$$\begin{cases} \pi(a, b) = \sum_{j=1}^{k} P_j(a, b)\, w_j, \\ \pi(b, a) = \sum_{j=1}^{k} P_j(b, a)\, w_j \end{cases} \tag{4.4}$$

where A is a finite set of possible alternatives k represents the number of evaluation criteria and w_j is the weight of each criterion. Arbitrary numbers for the weights can be assigned by the DM. The weights are then normalized to ensure that $\sum_{j=1}^{k} w_j = 1$. $\pi(a, b)$ indicates how a is preferred to b and $\pi(b, a)$ indicates how b is preferred to a. $P_j(a, b)$ and $P_j(b, a)$ are the preference functions for alternatives a and b.
Step 2: calculate $\pi(a, b)$ and $\pi(b, a)$ for each pair of alternatives of A
Step 3: define the positive and the negative outranking flow as follows
The positive outranking flow:

$$\phi^{+}(a) = \frac{1}{n-1} \sum_{x \in A} \pi(a, x), \tag{4.5}$$

The negative outranking flow

$$\phi^{-}(a) = \frac{1}{n-1} \sum_{x \in A} \pi(x, a), \tag{4.6}$$

Step 4: compute the net outranking flow for each alternative as follows:

$$\phi^{-}(a) = \phi^{+}(a) - \phi^{-}(a), \tag{4.7}$$

When $\phi(a) > 0$, a is more outranking all the alternatives on all the evaluation criteria. When $\phi(a) < 0$, a is more outranked.

Technique for Order Preference by Similarity to Ideal Solution (TOPSIS)

The Technique for order preference by similarity to ideal solution (TOPSIS) method is proposed to rank alternatives over multiple criteria. It finds the best alternatives by minimizing the distance to the ideal solution and maximizing the distance to the nadir or negative-ideal solution. The following TOPSIS procedure adopted from Opricovic and Tzeng [11] is used:

Step 1: calculate the normalized decision matrix. The normalized value r_{ij} is calculated as

$$r_{ij} = x_{ij} / \sqrt{\sum_{j=1}^{J} x_{ij}^2}, i = 1, 2, \cdots, J; i = 1, 2, \cdots, n. \tag{4.8}$$

where J and n denote the number of alternatives and the number of criteria, respectively. For alternative A_j, the performance measure of the i-th criterion C_i is represented by x_{ij}.

Step 2: develop a set of weights w_i for each criterion and calculate the weighted normalized decision matrix. The weighted normalized value v_{ij} is calculated as:

$$v_{ij} = w_i r_{ij}, j = 1, 2, \cdots, J; i = 1, 2, \cdots, n.$$

where w_i is the weight of the ith criterion, and $\sum_{i=1}^{n} w_i = 1$.

Step 3: find the ideal alternative solution S^+, which is calculated as:

$$S^+ = \left\{v_1^+, \cdots, v_n^+\right\} = \left\{\left(\max_j | i \in I'\right), \left(\min_j v_{ij} | i \in I''\right)\right\}$$

where I' is associated with benefit criteria and I'' is associated with cost criteria.

Step 4: find the negative-ideal alternative solution S^-, which is calculated as:

$$S^- = \left\{v_1^-, \cdots, v_n^-\right\} = \left\{\left(\min_j | i \in I'\right), \left(\max_j v_{ij} | i \in I''\right)\right\}$$

Step 5: Calculate the separation measures, using the n-dimensional Euclidean distance. The separation of each alternative from the ideal solution is calculated as:

$$\boldsymbol{D_j^+ = \sqrt{\sum_{i=1}^{n} \left(v_{ij} - v_i^+\right)^2}, j = 1, \cdots, J.}$$

The separation of each alternative from the negative-ideal solution is calculated as:

$$D_j^- = \sqrt{\sum_{i=1}^{n} \left(v_{ij} - v_i^-\right)^2}, j = 1, \cdots, J.$$

Step 6: Calculate a ratio R_j^+ that measures the relative closeness to the ideal solution and is calculated as:

$$R_j^+ = D^- / \left(D_j^+ + D_j^-\right)$$

Step 7: Rank alternatives by maximizing the ratio R^+

VlseKriterijumska Optimizacija I Kompromisno Resenje (VIKOR)

VIKOR was proposed by Opricovic [10] for multicriteria optimization of complex systems. This section uses the following VIKOR algorithm provided by Opricovic and Tzeng [11] in the experiment:

Step 1: Determine the best f_i^* and the worst f_i^- values of all criterion functions, $i = 1, 2 \cdots n$ and $j = 1, 2 \cdots J$.

$$f_i^* = \begin{cases} \max\limits_j f_{ij}, & \text{for benefit criteria} \\ \min\limits_j f_{ij}, & \text{for cost criteria} \end{cases} \tag{4.9}$$

$$f_i^- = \begin{cases} \min\limits_j f_{ij}, & \text{for benefit criteria} \\ \max\limits_j f_{ij}, & \text{for cost criteria} \end{cases} \tag{4.10}$$

where J is the number of alternatives, n is the number of criteria, and f_{ij} is the rating of i-th criterion function for alternative a_i.

Step 2: Compute the values S_j and $R_j, j = 1, 2 \cdots J$, by the relations

$$S_j = \sum_{i=1}^{n} w_i \left(f_i^* - f_{ij}\right) / \left(f_i^* - f_i^-\right) \tag{4.11}$$

$$R_j = \max\left[w_i \left(f_i^* - f_{ij}\right) / \left(f_i^* - f_i^-\right)\right] \tag{4.12}$$

where w_i is the weight of ith criteria, S_j and R_j are used to formulate ranking measure.

Step 3: Compute the values $Q_j, j = 1, 2 \cdots J$ by the relations

$$Q_j = v\left(S_j - S^*\right) / \left(S^- - S^*\right) + (1 - v)\left(\left(R_j - R^*\right) / \left(R^- - R^*\right)\right) \tag{4.13}$$

$$S^* = \min_j S_j, \quad S^- = \max_j S_j \tag{4.14}$$

$$R^* = \min_j R_j, \quad R^- = \max_j R_j \tag{4.15}$$

where the solution obtained by S^* is with a maximum group utility, the solution obtained by R^* is with a minimum individual regret of the opponent, and v is the weight of the strategy of the most of criteria. The value of v is set to 0.5 in the experiment.

Step 4: Rank the alternatives in decreasing order. There are three ranking lists: S, R and Q.

Step 5: Propose the alternative a', which is ranked the best by Q, as a compromise solution if the following two conditions are satisfied:

$$(a)\, Q\left(a''\right) - Q\left(a'\right) \geq 1/(J - 1)$$

Alternative a' is ranked the best by S or/and R. If only the condition (b) is not satisfied, alternatives a' and a'' are proposed as compromise solutions, where a'' is ranked the second by Q. If the condition (a) is not satisfied, alternatives a', a'', $\cdots$, a^M are proposed as compromise solutions, where a^M is ranked the Mth by Q and is determined by the relation $Q(a^M) - Q(a') < 1/(J - 1)$ for maximum M.

4.1.1.3 Classification Algorithms

The experimental study selects nine classifiers. The same set of classifiers is also used as base learners by feature subset evaluator WrapperSubsetEval. These classifiers belong to six categories of classification methods: trees, functions, Bayesian classifiers, lazy classifiers, rules, and miscellaneous classifiers. All of them are implemented in WEKA [8, 16, 17]. C4.5 decision tree is selected to represent the trees category. It constructs decision trees in a top-down recursive divide-and-conquer manner. The functions category includes linear logistic regression, radial basis function (RBF) network, and sequential minimal optimization (SMO).

Bayesian classifiers category includes naive Bayes. IB1, a basic nearest-neighbor instance-based learner provided by WEKA, represents lazy classifiers. An unknown instance is assigned to the same class as the training instance that is the closest to it measured by Euclidean distance. For the rules category, decision table and Repeated Incremental Pruning to Produce Error Reduction (RIPPER) rule induction were selected. Decision table builds a decision table majority classifier by selecting the right feature subsets. Instances not covered by a decision table can be determined by the nearest-neighbor method. RIPPER is an optimized version of incremental reduced error pruning (IREP). In addition, fuzzy lattice reasoning (FLR), which induces rules using fuzzy lattices, is chosen to represent the miscellaneous category.

4.1.1.4 Performance Measures

Five commonly used performance measures in classification are precision, true positive rate, false positive rate, F-measure, and the area under receiver operating characteristic (AUC) [18]. The following paragraphs briefly describe these measures.

- True Positive (TP): TP is the number of correctly classified fault-prone modules. TP rate measures how well a classifier can recognize fault-prone modules. It is also called sensitivity measure.

$$\text{True Positive rate/Sensitivity} = \frac{TP}{FP + TN}$$

- False Positive (FP): FP is the number of non-fault-prone modules that is misclassified as fault-prone class. FP rate measures the percentage of non-fault-prone modules that were incorrectly classified.

$$\text{False Positive rate} = \frac{FP}{FP + TN}$$

- True Negative (TN): TN is the number of correctly classified non-fault-prone modules. TN rate measures how well a classifier can recognize non-fault-prone modules. It is also called specificity measure.

$$\text{True Negative rate} = \frac{TN}{TN + FP}$$

- False Negative (FN): FN is the number of fault-prone modules that is misclassified as non-fault-prone class. FN rate measures the percentage of fault-prone modules that were incorrectly classified.

$$\text{False Negative rate} = \frac{FN}{FN + TP}$$

- Precision: This is the number of classified fault-prone modules that actually are fault-prone modules.

$$\text{Precision} = \frac{TP}{TP + FP}$$

- Recall: This is the percentage of fault-prone modules that are correctly classified.

$$\text{Recall} = \frac{TP}{TP + FN}$$

- F-measure: It is the harmonic mean of precision and recall. F-measure has been widely used in information retrieval.

$$\text{F} - \text{measure} = \frac{2 \times Precision \times Recall}{Precision + Recall}$$

- AUC: ROC stands for Receiver Operating Characteristic, which shows the tradeoff between TP rate and FP rate. AUC represents the accuracy of a classifier. The larger the area, the better the classifier.

4.1.1.5 Experimental Design

The experiment was carried out according to the following process:

Input: datasets
Output: Ranking of classifiers

Step 1: Feature selection: apply 11 feature selection techniques to each dataset using WEKA 3.7 and calculate feature weights.

Step 2: Evaluate feature selection techniques using DEA, ELECTRE, PROMETHEE, TOPSIS, and VIKOR.

Step 3: Select the highly ranked feature selection techniques and use these techniques to re-calculate feature weights.

Step 4: Train and test the classification models on a randomly sampled partitions (i.e., tenfold cross-validation) of each dataset with features selected by traditional feature selected technique. All methods are implemented using WEKA 3.7.

Step 5: Train and test the classification models on a randomly sampled partitions (i.e., tenfold cross-validation) of each dataset with the features selected by Step 3. Compare these results with the results of Step 4.

Step 6: Evaluate classification algorithms using DEA, ELECTRE, PROMETHEE, TOPSIS, and VIKOR. All the MCDM methods are implemented using MATLAB.

Step 7: Generate four separate tables of the final ranking of classifiers provided by each MCDM method. END

The data analysis of the above experimental study can be found in Peng et al. [1].

4.1.2 Two-Stage Hybrid Feature Selection Algorithms

Feature selection plays an important role in building a classification system [19–21]. It can not only reduce the dimensionality of data, but also reduce the computational cost and gain a good classification performance.

The general feature selection algorithms comprise two categories: the filter and wrapper methods [22, 23] The filter methods identify a feature subset from original feature set via a given evaluation criterion that is independent of learning algorithms. While the wrappers choose those features with high prediction performance estimated by a specific learning algorithm. The filters are efficient because of its independence of learning algorithms, while wrappers can obtain higher classification accuracy with the deficiency in generalization and computational cost. So there are more and more experts focus on studying the hybrid feature selection methods in recent decades for the hybrid feature selection methods can combine the advantages of filters and wrappers to uncover the classifiers with excellent performance.

This subsection presents several two-stage hybrid feature selection algorithms. These algorithms take two steps to construct the stable and efficient classifiers. In the first step, the generalized F-score is adopted to rank features, and our extending SFS and SFFS and SBFS are used to select the necessary features to comprise the selected feature subset whist the performance of the temporary SVM evaluated with our modified accuracy is used to guide the feature selection procedure.

Figure 4.2 illustrates the main idea of our hybrid feature selection algorithms. Where, the Generalized F-score is used to guide the application of filters, while the extended SFS/SFFS/SBFS with SVM combined our modified accuracy criterion are employed as wrappers. We rank features in descending order. The extended SFS and SFFS and SBFS is adopted to select the important or necessary features one by one by constructing many temporary SVM classifiers, whilst SVM with our new accuracy criterion is as a classification tool to direct the feature selection procedure.

Here we respectively introduce the generalized F-score and the definition of our new accuracy and our proposed three hybrid feature selection algorithms in the following subsections.

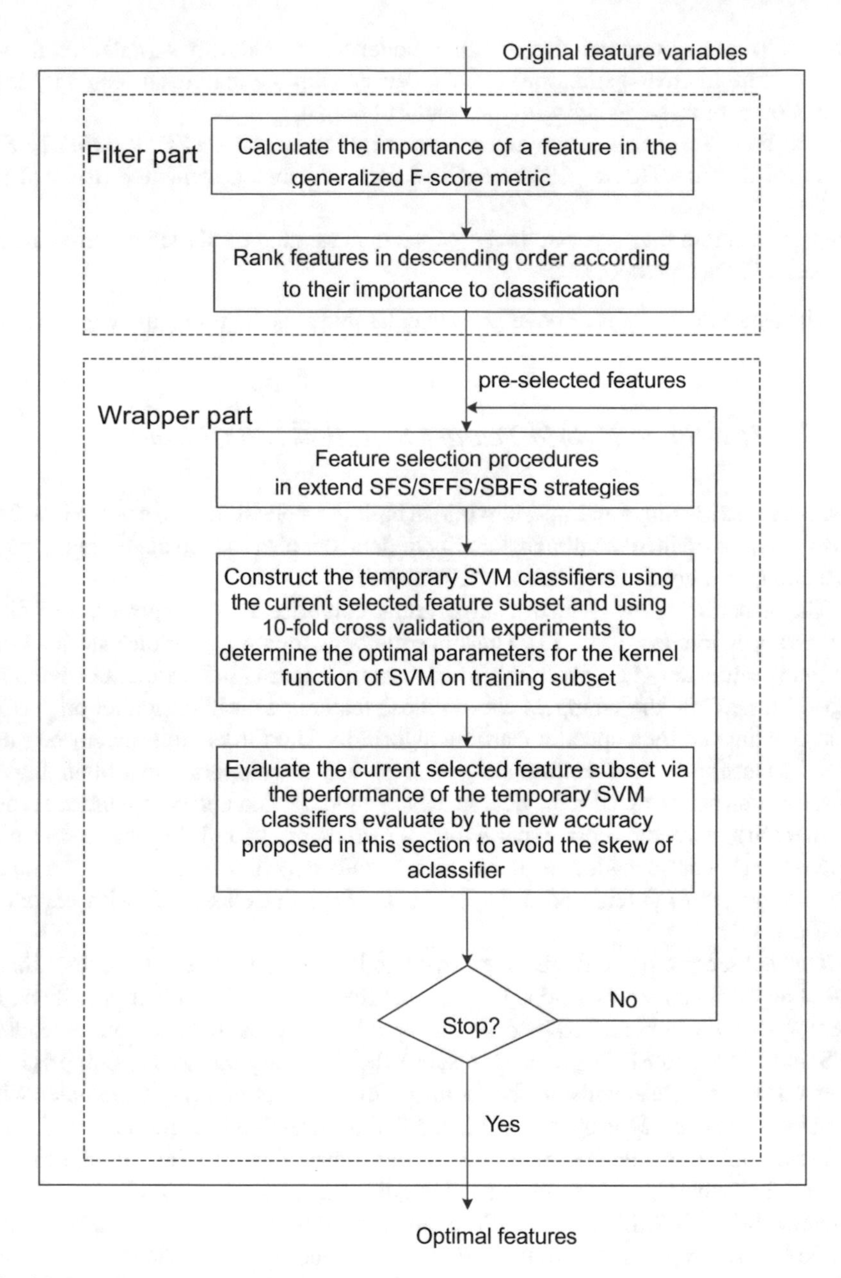

Fig. 4.2 New hybrid feature selection algorithms

4.1.2.1 Generalized F-Score

The original F-score is to measure the discrimination of one feature between two sets of real numbers [24]. We generalized it in [6] to measure the discrimination of one feature between more than two sets of real numbers, so that it can value the importance of a feature to the classification in a multi-category classification problem. Here is the definition of the generalized F-score. Given training vectors x_k, $k = 1, 2, \cdots$, and the number of subsets $l (l \geq 2)$, if the size of the jth subset is $n_j, j = 1, 2, \cdots, l$, then the F-score of the ith feature is F_i.

$$F_i = \frac{\sum_{j=1}^{l} \left(\overline{x}_i^{(j)} - \overline{x}_i \right)^2}{\sum_{j=1}^{l} \frac{1}{n_j - 1} \sum_{k=1}^{n_j} \left(x_{k,i}^{(j)} - \overline{x}_i^{(j)} \right)^2} \tag{4.16}$$

where $\overline{x}_i$ and $\overline{x}_i^{(j)}$ are the average of the ith feature on the whole dataset and on the jth subset respectively, and $x_{k,i}^{(j)}$ is the ith feature of the kth instance in the jth subset. The numerator of the right-hand side of Eq. (4.16) indicates the discrimination of the ith feature between each subset, and the denominator is the one within each subset. Thus, the larger the F_i is, the more discriminative the ith feature is.

4.1.2.2 The New Classification Accuracy Measure

The accuracy of a classifier is often measured in the following Eq. (4.17).

$$accuracy = \frac{N_r}{N} \tag{4.17}$$

where N_r is the number of samples which are classified correctly, and N is the total number of samples which are to be classified.

This accuracy does not consider the performance of a classifier on each class, which may lead the skew of a classifier on some classes. For example, there is a cancer diagnostic problem with 90 normal people and 10 cancer patients. Now we have got a classifier that can recognize all normal people and zero cancer patients. Although the accuracy of the classifier is 90%, it is not a good one. Thus we define a new accuracy in the Eq. (4.18).

$$new_{accuracy} = \frac{1}{l} \sum_{c=1}^{l} \frac{N_r^c}{N^c} \tag{4.18}$$

where l is the number of classes which are to be considered in a classification problem, and the N_r^c is the number of samples which are correctly classified in the cth class, and Nc is the total number of samples which are to be classified in the

cth class. This new accuracy does consider the performance of a classifier on each class that is considering in the classification problem, so that the new accuracy can overcome the skew of a classifier when it is used to evaluate the performance of a classifier to guild the feature selection procedure.

4.1.2.3 Several Hybrid Feature Selection Algorithms

Here are the related issues and our proposed hybrid feature selection algorithms which will comprise our two-stage hybrid feature selection algorithms for diagnosing erythemato-squamous diseases.

The traditional and still popular feature search strategies include sequential forward search (SFS) [25] and sequential backward search (SBS) [26] and sequential forward floating search (SFFS) and sequential backward floating search (SBFS) [27].

Here the aforementioned traditional SFS, SFFS, and SBFS strategies are extended as the followings. Firstly, the features are ranked according to their F-score values, here the generalized F-score is used, then the features are dealt with one by one. In the extended SFS, features are selected according to their rank order, not as the traditional SFS which selects the feature that must be the best one when combined with the selected ones. And in the extended SFFS, we first trying to add a feature according to its rank order, then in the floating procedure we test the feature to be indeed added or not according to the new accuracy of the temporary classifier goes up or not, if the new accuracy of the temporary classifier goes up, then the related feature will be added to the selected feature subset, otherwise it will not be added. Similarly, in the extended SBFS, the procedure starts with all feature included, then at the following steps, the current lowest rank feature is tested deleting or not, if the accuracy of the temporary classifier without the feature becomes worse evaluated in our new accuracy, then the feature will not be deleted, otherwise it will be deleted. These procedures continue until all features are tested. The extended SFS and SFFS and SBFS are respectively faster than the traditional SFS and SFFS and SBFS in determining one feature to be selected or not in feature selection procedures.

Here are the three hybrid feature selection algorithms, named new GFSFS, new GFSFFS and new GFSBFS, respectively. The generalized F-score plays the role of filters, and our extending SFS and SFFS and SBFS, respectively, with SVM and our new accuracy act as wrappers. Using the three new hybrid feature selection algorithms, new GFSFS, new GFSFFS and new GFSBFS, the necessary features are selected and the redundant ones are eliminated, so that a sound predictor to diagnose erythemato-squamous diseases is constructed. The detail procedures of our new GFSFS, new GFSFFS and new GFSBFS are, respectively, described as followings.

The new GFSFS uses extending SFS strategy to uncover the important features in building a classifier according to their F-score values, and uses SVM as a classification tool. The new accuracy is adopted to judge the performance of the temporary SVM classifiers to guide the feature selection procedure. The feature subset is composed of features with which the classifier on training subset has got the best diagnosing result. The pseudo code of our new GFSFS is here.

Step 1: Determine the training and testing subsets of exemplars; Initialize the selected-feature-subset empty, and selecting-feature-subset with all features;
Step 2: Computing the F-score value for each feature by using the Eq. (4.16) on the training subset, and sort features in descending order according to their F-score values;
Step 3: Add the top feature in the selecting-feature subset to the selected-feature-subset, and deleted it from the selecting-feature-subset as well;
Step 4: Train the training subset with features in selected-feature-subset to construct the temporary optimal SVM classifier, the optimal parameters of SVM are determined via the aforementioned grid search technique and tenfold cross validation experiments on the training subset;
Step 5: Classify exemplars in the test subset and record the accuracy;
Step 6: go to step 3, until the selecting-feature-subset becomes empty;

Although the new GFSFS can get a comparable good performance in diagnosing erythemato-squamous diseases, it may suffer the weakness of feature subset "nesting" that is the nature of SFS. That is, one feature will not be discarded once it has been selected and added to the selected-feature-subset.

The coming hybrid feature selection algorithm, new GFSFFS, will overcome this disadvantage of the new GFSFS by considering the correlation between features, so once the new accuracy of a temporary classifier on training subset doesn't go up, then the selected feature will only be deleted from the selecting-feature-subset but will not be added to the selected-feature-subset. The details of the new GFSFFS are as the followings.

Step 1: Calculate the F-score value for each feature via the generalized F-score defined in Eq. (4.16) on the training subset of this fold, and rank features in descending order according to their F-score values; Initialize the selected-feature-subset empty and the selecting-feature-subset with all features;
Step 2: Delete the top feature from the selecting-feature-subset and add it to the selected-feature-subset;
Step 3: Train the training subset to build the optimal predictor model where the optimal parameter for the kernel function of SVM is determined in the aforementioned grid search technique and tenfold cross-validation experiments on the training subset;
Step 4: If the new accuracy defined in Eq. (4.18) of the training subset is not improved, then the feature that has just been added will be eliminated from the selected-feature-subset;
Step 5: Go to Step 2 till all features in the selecting-feature-subset have been processed.

The features in the selected-feature-subset comprise the best feature subset of this fold, and the last SVM classifier is the optimal diagnostic model we are looking for on this fold. To get a further self-contained demonstration of our new accuracy, we propose new GFSBFS hybrid feature selection algorithm and its procedure is here.

Step 1: Compute the F-score for each feature via the generalized F-score in Eq. (4.16) on this fold training subset, and rank features in descending order according to their F-score values; Initialize the selected-feature-subset with all features, and the visited tag for each feature unvisited;
Step 2: Train the training subset with features in the selected-feature-subset to build the optimal predictor model where the optimal parameter for the kernel function of SVM is determined by the aforementioned grid search technique and tenfold cross validation experiments on the training subset; Record the accuracy of the model on training subset in the new accuracy defined in Eq. (4.18);
Step 3: Trying to delete the last unvisited feature in selected-feature-subset, and let the visited tag of it be visited;
Step 4: Train the training subset with features in the selected-feature-subset to build the optimal predictor model as step 2, and record the new accuracy of the model on training subset;
Step 5: If the new accuracy of training subset does not go up, keep the feature that it is trying to delete back to the selected-feature-subset, otherwise deleted it;
Step 6: Go to Step 3, until all features in the selected-feature-subset have been visited.

At last those features left in the selected-feature-subset are the necessary ones to build the optimal diagnostic model for this fold.

Because of the variation in the results of tenfold cross validation experiments, we propose the two-stage hybrid feature selection algorithms. We do tenfold cross validation experiments of our new GFSFS, new GFSFFS, and new GFSBFS in the first stage. Then we merge the 10 selected feature subsets of the tenfold cross validation experiments as a new full feature set on which to carry out the following feature selection procedure of the second stage of our two-stage hybrid feature selection algorithms. In the second stage we repeat our new hybrid feature selection algorithms which are described in above subsection on the one partition which has got the best performance during the first stage, that is, the partition of the corresponding fold that has got the optimal accuracy in the tenfold cross validation experiments in the pre-stage. In our experiment we choose the tenth fold, i.e., the last partition in the tenfold cross validation experiments, to finish our two-stage hybrid feature selection algorithms.

The experimental study of this two-stage hybrid feature selection algorithms can be found in [2].

4.1.3 Feature Selection with Attributes Clustering by Maximal Information Coefficient

Attribute clustering methods make features cluster together rather than instances. In this case, instances distance metric is replaced by feature similarity measure. Since clustering belongs to unsupervised learning, to obtain discriminating features it is better to put some supervision during the process of clustering [20]. Pereira et al. provided the idea of distributional clustering for feature selection in [28] which is based on the information bottleneck theory [29]. This kind of method try to find a suitable T and minimize the objective function $I(X;T) - \beta I(T;Y)$, where $I(X;T)$ and $I(T;Y)$ are mutual information between X; Y and T; Y. Baker and McCallum used this idea to generate a feature clustering method for document classification [30]. Similar with information bottleneck theory, information-theoretic framework was introduced by Dhillon for identifying feature clusters [31]. In [32], Wai-Ho Au et al. presented an attribute clustering method which had capability of group genes expression base on their interdependence. Recently, a self-constructing algorithm based on fuzzy similarity for feature clustering was introduced [33].

However, obstacles are still on the way of these methods. First of all, finding the optimal subset feature space with best classification or regression performance have been proved to be NP-hard problem [34]. Furthermore, feature selection procedure is usually a separate process which cannot benefit from result of the data exploration in advance. There are various kinds of data exploration strategies.

4.1.3.1 Maximal Information Coefficient

Relationship coefficient between features are usually used for measuring attribute similarity. For instance, Combarro et al. propose a way of choosing relevant features by linear measures for text categorization application in [35]. In [35], Combarro et al. propose a way of choosing relevant features by linear measures for text categorization application. Person coefficient is one of the most famous relationship metrics, because it is easy to calculate and has a naive explanation. However, only linear relationship can be captured well using this metric when other kinds of dependence work badly such as functional sin or cubic. Pearson coefficient could only capture the association limited to linear function well. That's mean a various of important relationships such as a superposition of functions cannot be scored properly.

Recently, Reshef et al. propose a new relationship measure called maximal information coefficient (MIC). With innovative idea, they show that MIC could capture a wide range of associations both functional and not. Furthermore, the value of MIC is roughly equal to the coefficient of determination R2 in statistics [36]. Next, we briefly introduce some concepts related with MIC in [36].

Given a finite set D whose elements are two dimensions data points, we consider one of the dimensions as x-values and the other as y-values. Suppose x-values is divided into x bins and y-values into y bins, this type of partition is called x-by-y grid G. Let $D|_G$ represent the distribution of D divided by one of x-by-y grids as G. $I^*(D,x,y) = \max I(D|G)$, where $I(D|_G)$ is the mutual information of $D|_G$. There are infinite number of x-by-y grids, as a result, there are infinite number of $I(D|_G)$ either. Choose the maximum one of them and present it as $I^*(D,x,y)$, thus a matrix named characteristic matrix is constructed as $M(D)_{x,y} = \frac{I^*\left(D,x,y\right)}{\log\min\left\{x,y\right\}}$. Furthermore, MIC can be obtained $MIC(D) = \max_{xy<B(n)}\{M(D)_{x,y}\}$, where $B(n)$ is the upper bound of the grid size need to be considered. The elements of characteristic matrix $I^*(D,x,y)$ is chosen from a infinite amount of $I(D|_G)$, thus authors of MIC develop an approximation algorithm and program for generating characteristic matrix and the estimators such as MIC.[1] With these sophisticated utilities, data exploration by MIC can be easily done before other complex data mining task.

4.1.3.2 Affinity Propagation Clustering

Clustering data points through a measure similarity is a crucial step in many scientific analysis and application systems [37]. Brendan et al. develop a modern clustering method named "affinity propagation" (AP) which constructs clusters by messages exchanged between data points. Given the similarities of each two distinct data points as input, AP algorithm considers all the instance as potential centroids at the beginning. And then, algorithm merges small cluster into bigger ones step by step. Being different with some of the typical clustering algorithms as k-means, each instances are regarded as one node in a network. Messages was Transactionsmitted between nodes, so each data point reconsidered their situation through new information and properly modified the cluster they belong to. This procedure went on until a good set of clusters and centroids produced.

In this process, there are mainly two categories of message exchanged between data points. One of them is sent from point i to point j which formulated as $r(i,j)$. It illustrates the strength point i choosing point j as its centroid. The other sort information is from point j to point i as $a(i,j)$. It shows the confidence that one point j recommends itself as the centroid of another point i. And the author of AP take $r(i, j) \leftarrow s(i, j) - \max_{j', j' \neq j}\left\{a\left(i, j'\right) + s\left(i, j'\right)\right\}$ and $a(i, j) \leftarrow \min\left\{0, r(k, k) + \sum_{i's.t.i' \neq i,j} \max\left\{0, r\left(i', k\right)\right\}\right\}$ to update current situation. Update is needed only for the pairs of points whose similarities are already known. This trait makes the algorithm much faster than other methods. To identify the centroid of point i, point j that maximizes $r(i,j) + a(i,j)$ should be considered during each iteration. AP clustering method requires a similarity matrix s as input, and the element of the matrix $s(i,j)$ provides the distance from point i to point j.

[1] http://www.sciencemag.org/cgi/content/full/334/6062/1518/DC1.http://exploredata.net

In addition, the diagonal values of the matrix is not assign 1 as usual. These values are called "preference" which show how point i is likely to be chosen as a centroid. That's to say, the larger $s(i,j)$ is, the more probability that point i play a role of a centroid. Obviously, $s(i,j)$ are key parameters which control the number of final clusters by AP method. After all, the algorithm can be terminated when the values exchanged are under some threshold or the clusters keep stable for some iterations.

4.1.3.3 Attributes Clustering by Maximal Information Coefficient

At present data exploration is an indispensable step for discovering valuable knowledge in large amount of data. MINE tool [36] has been recognized as one of the usual data exploration procedures. This exploration tool could detect novel association between a pair of variables and has been widely used in practice. Relationship information is contained in the results generated by MINE tool. However, this kind of information hasn't been made the best of. Methods that take the full advantage of MINE exploration result should be initiated.

For this purpose, we propose a new unsupervised learning method called MICAP for feature selection task. It needs no supervised information and directly selects the key attributes of a dataset. The proposed algorithm makes features with high dependence cluster together and only keep the center feature of each cluster left. The algorithm follows a simple idea that takes the MICs as the relationship metric for each pair of features, and cluster them through affinity propagation clustering method. It's combines the MIC and affinity propagation clustering method. That's why our algorithm is called MICAP.

First step, a MINE data exploration procedure is executed. As a result, each pair of features except label attribute has been explored by MINE tool. For most scientific research this kind of data exploration it's necessary because it provides general relationships among features. Next, based on the data exploration result, a maximal information coefficient matrix has be constructed. According to descending order list of the elements of the matrix, a preference value could be obtained by setting the quantile for the list. And then affinity propagation clustering method is applied with MIC matrix as the similarity matrix. After all, the centroid of each cluster is chosen as the selected subset for original feature space.

According to the general steps of affinity propagation clustering algorithm [37], a key parameter is preference which controls the number of the features left in the final result. Due to the property of the affinity propagation, the number of features selected need not be given in advance. When preference is large, more features will be reserved. When preference is small, fewer attributes will be kept. The detail of our algorithm is illustrated in Algorithm 4.1. The discussion of data analysis of Algorithm 4.1 can be found in [3].

Algorithm 4.1 MICAP Feature Selection Algorithm

Input: Training dataset $\mathcal{D} = \{\Omega, \mathrm{C}\}$, $\Omega = (\mathrm{f}_1, \mathrm{f}_2, \cdots, \mathrm{f}_\mathrm{n})$; The quantile q for MIC values listed in descending order.

Output: Selected features $\mathcal{S}$, $\mathcal{S} \subseteq \otimes$.

1: **Begin** Set $\mathcal{S} = \emptyset$
2: **for** all $\mathrm{f}_\mathrm{i}, \mathrm{f}_\mathrm{i} \in \mathcal{D}, \mathrm{i} \neq \mathrm{j}$ **do**
3: Calculate their MIC values and Set $\mathrm{M(i, j)} = \mathrm{MIC}(\mathrm{f}_\mathrm{i}, \mathrm{f}_\mathrm{j})$;
4: **end for**
5: Sort distinct values of M(i, j) elements in descending order as MICList;
6: Choose the q quantile value of MICList as preference;
7: Set All M(i, i) = preference
8: $\gamma = \{\mathrm{F}_1, \mathrm{F}_2, \cdots, \mathrm{F}_\mathrm{l}\} = \mathrm{APClustering}(\Omega, \mathrm{M})$
9: **for** all $\mathrm{F}_\mathrm{k} \in \gamma$ **do**
10: Set $\mathcal{S} = \mathcal{S} \cup \mathrm{centroid}(\mathrm{F}_\mathrm{k})$
11: **end for**
12: **End**

4.2 Regularizations for Feature Selections

4.2.1 Supervised Feature Selection with $\ell_{2,1-2}$ Regularization

4.2.1.1 Feature Selection with Sparse Learning

In the supervised learning scenario, let $\mathrm{X} = [\mathrm{x}_1, \cdots, \mathrm{x}_n]$ be a data matrix with n samples of d features. Suppose there are c classes, by one-hot encoding, the class labels can be represented as a $n \times c$ matrix Y, where $\mathrm{Y}_{ij} = 1$, if x_i belongs to the j-th class and $\mathrm{Y}_{ij} = 0$ otherwise. Feature selection with sparse learning is to find a Transformed matrix $\mathrm{W} \in \Re^{d \times c}$ to evaluate the correlation between labels and features, and select the discriminative features based on the weight of W. Mathematically, this task can be described as the following structured model:

$$\min_{\mathrm{W}} L_{\mathrm{X,Y}}(\mathrm{W}) + \alpha \mathcal{R}(\mathrm{W}), \tag{4.19}$$

where $\mathrm{L}_{\mathrm{X,Y}}(\cdot)$ is the loss which measures the model fidelity, $\mathcal{R}(\cdot)$ is the regularizer which is the preference of selecting features across all the classes with jointly sparsity, and $\alpha > 0$ is a hyper-parameter that controls the trade-off between these two terms.

The joint minimization of the fidelity term and the sparsity regularization enable Transformed matrix W to evaluate the correlation between labels and features, which is particularly suitable for feature selection. More specifically, the norm of the i-th row in W shrinks to zero or a number close to zero if the i-th feature is less discriminative to the labels.

As mentioned above, regularization term $\mathcal{R}(\mathrm{W})$ in (4.19) is for feature selection. A variety of sparsity regularizations and the corresponding models have been proposed. From the sparsity perspective, $\ell_{2,0}$ norm might be the most desirable. However, it will result in the NP-hard combinatory model and is very difficult to solve. Therefore, several proxies of $\ell_{2,0}$ norm have been studied. The most common one is the $\ell_{2,1}$ norm, which has been used as the sparse regularization in many works [38–41]. The $\ell_{2,p}$ norm is the only existing nonconvex regularizer in matrix-based feature selection. Although experiments have empirically demonstrated that $\ell_{2,p}$ outperforms $\ell_{2,1}$ [42]. The $\ell_{2,p}$ regularizer is more difficult to compute. Firstly, there is a-priori unknown hyper-parameter p, which controls the effect of feature selection. As stated in [42], the smaller p is, the better performance can be obtained. Secondly, $\ell_{2,p}$ is non-Lipschitz continuous in mathematics. Although [42] provides a proximal gradient algorithm together with rank-one update to solve it, when $p \to 0$, $\ell_{2,p}$ is still difficult to solve. Considering the good performance of $\ell_{2,p}$ rising from the nonconvexity and the difficulty of $\ell_{2,p}$ lying the non-Lipschitz continuity, we propose a hyper-parameter free regularizer on matrix, which is nonconvex and Lipschitz continuous.

4.2.1.2 ConCave-Convex Procedure

Consider the following structured optimization problem, whose objective function can be decomposed into two convex functions:

$$\min_{x} h(\mathrm{x}) := f(\mathrm{x}) - g(\mathrm{x}), \tag{4.20}$$

where $f(\cdot): \mathfrak{R}^n \to \mathfrak{R}$and $g(\cdot): \mathfrak{R}^n \to \mathfrak{R}$ are two convex functions. Problem (4.20) is nonconvex unless the function $g(\cdot)$ is affine, and it is difficult to solve in general.

ConCave-Convex Procedure (CCCP) tackles the nonconvex problem (4.20) by solving a series of convex subproblems. The idea of CCCP involves linearizing of the second term $g(x)$ in the objective function at the current solution x^k, and advancing to a new one x^{k+1} by solving a subproblem. In detail, CCCP solves problem (4.20) with the following iterations:

$$\begin{cases} \mathrm{y}^k \in \partial g\left(\mathrm{x}^k\right) \\ \mathrm{x}^{k+1} = \arg\min_{\mathrm{x}} f(\mathrm{x}) - \left(g\left(\mathrm{x}^k\right) + \langle \mathrm{y}^k, \mathrm{x} - \mathrm{x}^k \rangle\right). \end{cases} \tag{4.21}$$

Recalling the definition of sub-gradient, for any $x \in \mathfrak{R}^n$,

$$g(\mathrm{x}) \geq g\left(\mathrm{x}^k\right) + \left\langle \mathrm{y}^k, \mathrm{x} - \mathrm{x}^k \right\rangle. \tag{4.22}$$

In particular $g(\mathrm{x}^{k+1}) \geq g(\mathrm{x}^k) + \langle \mathrm{y}^k, \mathrm{x}^{k+1} - \mathrm{x}^k \rangle$. Note that x^{k+1} minimizes $f(\mathrm{x}) - (g(\mathrm{x}^k) + \langle \mathrm{y}^k, \mathrm{x} - \mathrm{x}^k \rangle)$, on can have,

$$\begin{aligned} h\left(\mathrm{x}^k\right) = f\left(\mathrm{x}^k\right) - g\left(\mathrm{x}^k\right) \geq f\left(\mathrm{x}^{k+1}\right) - \Big(g\left(\mathrm{x}^k\right) + \\ \langle \mathrm{y}^k, \mathrm{x}^{k+1} - \mathrm{x}^k \rangle\Big) \geq f\left(\mathrm{x}^{k+1}\right) - g\left(\mathrm{x}^{k+1}\right) \geq h\left(\mathrm{x}^{k+1}\right). \end{aligned} \tag{4.23}$$

Hence, the CCCP algorithm produces a monotonically decreasing sequence $\left\{h\left(\mathrm{x}^k\right)\right\}_{k=0}^{\infty}$ of objective function values. Moreover, when $h(\cdot)$ is bounded from below, the sequence $\left\{h\left(\mathrm{x}^k\right)\right\}_{k=0}^{\infty}$ is convergent. One reasonable stopping criterion is that the improvement in the objective value is less than a given threshold ϵ, i.e., $h(\mathbf{x}^k) - h(\mathbf{x}^{k+1}) \leq \epsilon$

Although the sequence of the objective function values $\left\{h\left(\mathrm{x}^k\right)\right\}_{k=0}^{\infty}$ is convergent, the sequence of iterative points $\left\{\mathrm{x}^k\right\}_{k=0}^{\infty}$ generated by CCCP is not always convergent. Therefore, researchers investigated the convergence of iteration points, which is usually called strong global convergence in optimization theory. Existing standard strong convergence analysis of CCCP is conducted on the assumption that the component functions are differentiable, strong convex or the non-smooth part is convex piecewise-linear [43–45]. None of them is suitable for the $\ell_{2,1-2}$ function.

In addition, the convergent proof of ℓ_{1-2} in [46] strongly relies on the specific formulation of the model and could not be extended to $\ell_{2,1-2}$. Therefore, how to analyze the strong global convergence of our CCCP algorithm for $\ell_{2,1-2}$ would be an interesting and challenging work.

4.2.1.3 Supervised Feature Selection with the $\ell_{2,1-2}$ Regularization

$\ell_{2,1-2}$ Function for Matrix

Here, the bold uppercase characters are used to denote matrices, and bold lowercase characters to denote vectors. The ℓ_p and ℓ_0 norms of vector $\mathrm{w} = [w_1, w_2, \cdots, w_n] \in \Re^n$ are defined as

$$\|\mathrm{w}\|_p = \left(\sum_{i=1}^{n} |w_i|^p\right)^{1/p} \quad \text{and} \quad \|\mathrm{w}\|_0 = \sum_{w_i \neq 0} |w_i|^0 \tag{4.24}$$

respectively. Correspondingly, the $\ell_{2,p}$ and $\ell_{2,0}$ norms of matrix $\mathrm{W} = \left[\mathrm{w}_1^{\mathrm{T}}, \mathrm{w}_2^{\mathrm{T}}, \cdots, \mathrm{w}_m^{\mathrm{T}}\right]^{\mathrm{T}} \in \Re^{m\times n}$ are defined as:

$$\|\mathrm{W}\|_{2,p} = \left(\sum_{i=1}^{m} \|\mathrm{w}_i\|^p\right)^{1/p} = \left(\sum_{i=1}^{m}\left(\sum_{j=1}^{n} \mathrm{W}_{ij}^2\right)^{p/2}\right)^{1/p} \tag{4.25}$$

where W_{ij} is the entry of W at the i-th row and the j-th column, and

$$\|\mathrm{W}\|_{2,0} = \sum_{\mathrm{w}_i \neq 0} \|\mathrm{w}_i\|_2^0 \tag{4.26}$$

respectively. Specifically, when $p = 1$, $\ell_{2,p}$ reduces into the $\ell_{2,1}$ norm and when $p = 2$, $\ell_{2,p}$ is the Frobenius norm denoted by $\|\cdot\|_F$.

The Euclidean inner product between two matrices with the same scale is defined as, where $\mathrm{Tr}(\cdot)$ is the trace operator. Obviously, $\|\mathrm{W}\|_F^2 = \langle \mathrm{W}, \mathrm{W}\rangle = \mathrm{Tr}\left(\mathrm{W}^T\mathrm{W}\right)$. Given $\mathrm{W} \in \Re^{m\times n}$, the sub-gradient of $\|\mathrm{W}\|_\mathrm{F}$ is:

$$\partial\|\mathrm{W}\|_F = \begin{cases} \left\{\frac{\mathrm{W}}{\|\mathrm{W}\|_F}\right\}, & \text{if } \mathrm{W} \neq 0; \\ \left\{\mathrm{M} \in \Re^{m\times n} : \|\mathrm{M}\|_F \leq 1\right\}, & \text{otherwise.} \end{cases} \tag{4.27}$$

and the sub-gradient of $\|\mathrm{W}\|_{2,1}$ is:

$$\partial\|\mathrm{W}\|_{2,1} = \left\{\left[\psi(\mathrm{w}_1)^{\mathrm{T}}, \psi(\mathrm{w}_2)^{\mathrm{T}}, \cdots, \psi(\mathrm{w}_m)^{\mathrm{T}}\right]^{\mathrm{T}}\right\},$$

where $\psi(\mathrm{w}_i)$ is defined as:

$$\psi(\mathrm{w}_i) = \begin{cases} \frac{\mathrm{w}_i}{\|\mathrm{w}_i\|_2}, & \text{if } \mathrm{w}_i \neq 0; \\ \hat{\mathrm{w}}_i \in \{\mathrm{w} \in \Re^n : \|\mathrm{w}\|_2 \leq 1\}, & \text{otherwise.} \end{cases} \tag{4.28}$$

Inspired by the good performance of ℓ_{1-2} function for vectors in [46–49], we propose a sparse metric $\ell_{2,1-2}$ for matrices as follows

$$\|\mathrm{W}\|_{2,1-2} = \|\mathrm{W}\|_{2,1} - \|\mathrm{W}\|_{2,2} = \|\mathrm{W}\|_{2,1} - \|\mathrm{W}\|_F. \tag{4.29}$$

The following proposition shows that for a given matrix, $\ell_{2,1-2}$ is less than or equal to 1, if and only if, the matrix has one nonzero row at most.

Proposition 4.1 *For any matrix* $\mathrm{W} \in \Re^{m\times n}$,

$$\|\mathrm{W}\|_{2,1-2} = 0 \text{ if and only if } \|\mathrm{W}\|_{2,0} \leq 1 \tag{4.30}$$

Proof To prove the proposition, for any $\mathrm{W} = \left[\mathrm{w}_1^{\mathrm{T}}, \mathrm{w}_2^{\mathrm{T}}, \cdots, \mathrm{w}_m^{\mathrm{T}}\right]^{\mathrm{T}} \in \Re^{m \times n}$, we first show that

$$\left(s - \sqrt{s}\right)\min_{i \in \Delta}\|\mathrm{w}_i\|_2 \leq \|\mathrm{W}\|_{2,1-2} \tag{4.31}$$

where $s = \|\mathrm{W}\|_{2,0}$ and $\Delta = \{i : \mathrm{w}_i \neq 0, i = 1, 2, \cdots, m\}$. Without loss of generality, suppose $\|\mathrm{w}_1\|_2 \geq \|\mathrm{w}_2\|_2 \geq \cdots \geq \|\mathrm{w}_m\|_2$ and $q = \left\lfloor \sqrt{s} \right\rfloor$, we have

$$\begin{aligned}
\|\mathbf{W}\|_F^2 &= \textstyle\sum_{i=1}^q \|\mathbf{w}_i\|_2^2 + \sum_{i=q+1}^s \|\mathbf{w}_i\|_2^2 \\
&\leq \textstyle\sum_{i=1}^q \|\mathbf{w}_i\|_2^2 + (s-q)\left\|\mathbf{w}_{q+1}\right\|_2^2 \\
&\leq \textstyle\sum_{i=1}^q \|\mathbf{w}_i\|_2^2 + \sum_{i=1}^q \sum_{j=1, j\neq i}^q \|\mathbf{w}_i\|_2 \left\|\mathbf{w}_j\right\|_2 \\
&\quad + 2\left(\sqrt{s} - q\right)\left\|\mathbf{w}_{q+1}\right\|_2 \textstyle\sum_{i=1}^q \|\mathbf{w}_i\|_2 + \left(\sqrt{s} - q\right)^2 \left\|\mathbf{w}_{q+1}\right\|_2^2 \\
&\leq \left(\textstyle\sum_{i=1}^q \|\mathbf{w}_i\|_2 + \left(\sqrt{s} - q\right)\left\|\mathbf{w}_{q+1}\right\|_2\right)^2
\end{aligned} \tag{4.32}$$

where the second inequality follows

$$\begin{aligned}
&\textstyle\sum_{i=1}^q \sum_{j=1, j\neq i}^q \|\mathrm{w}_i\|_2 \left\|\mathrm{w}_j\right\|_2 + 2\left(\sqrt{s} - q\right)\left\|\mathrm{w}_{q+1}\right\|_2 \sum_{i=1}^q \|\mathrm{w}_i\|_2 \\
&\qquad + \left(\sqrt{s} - q\right)^2 \left\|\mathrm{w}_{q+1}\right\|_2^2 \\
&\quad \geq \textstyle\sum_{i=1}^q \sum_{j=1, j\neq i}^q \left\|\mathrm{w}_{q+1}\right\|_2^2 + 2q\left(\sqrt{s} - q\right)\left\|\mathrm{w}_{q+1}\right\|_2^2 \\
&\qquad + \left(\sqrt{s} - q\right)^2 \left\|\mathrm{w}_{q+1}\right\|_2^2 = (s-q)\left\|\mathrm{w}_{q+1}\right\|_2^2
\end{aligned} \tag{4.33}$$

Therefore, it can be seen

$$\begin{aligned}
\|\mathrm{W}\|_{2,1-2} &= \|\mathrm{W}\|_{2,1} - \|\mathrm{W}\|_F \\
&\geq \|\mathrm{W}\|_{2,1} - \left(\textstyle\sum_{i=1}^q \|\mathrm{w}_i\|_2 + \left(\sqrt{s} - q\right)\left\|\mathrm{w}_{q+1}\right\|_2\right) \\
&= \textstyle\sum_{i=q+1}^s \|\mathrm{w}_i\|_2 + \left(q - \sqrt{s}\right)\left\|\mathrm{w}_{q+1}\right\|_2 \\
&\geq (s-q)\|\mathrm{w}_s\|_2 + \left(q - \sqrt{s}\right)\|\mathrm{w}_s\|_2 \\
&= \left(s - \sqrt{s}\right)\|\mathrm{w}_s\|_2 = \left(s - \sqrt{s}\right)\min_{i \in \Delta}\|\mathrm{w}_i\|_2
\end{aligned} \tag{4.34}$$

If $\|\mathrm{W}\|_{2,0} \leq 1$, we can easily see that $\|\mathrm{W}\|_{2,1-2} = 0$ holds. Conversely, if $\|\mathrm{W}\|_{2,1-2} = 0$, from (4.30), we have $\left(s - \sqrt{s}\right)\min_{i \in \Delta}\|\mathbf{w}_i\|_2 \leq 0$, then $\left(s - \sqrt{s}\right) \leq 0$, consequently $s \leq 1$, i.e., $\|\mathrm{W}\|_{2,0} \leq 1$.

Adopting the $\ell_{2,1-2}$ function in (4.29) as regularizer term $\mathcal{R}(\mathrm{W})$, the feature selection model in (4.19) can be written as:

$$\min_{\mathrm{W}} L_{\mathrm{X,Y}}(\mathrm{W}) + \alpha\|\mathrm{W}\|_{2,1-2}. \tag{4.35}$$

Denote by W^* the optimal solution to the model (4.35), we select the top-ranking features according to the value of $\left\|\mathrm{w}_i^*\right\|_2 \left(i = 1, 2 \cdots, d\right)$. Several popular loss functions can be used in (4.35), such as the Frobenius norm loss

$$L_{X,Y}(W) = \left\| Y - X^T W \right\|_F^2, \tag{4.36}$$

the Logistic loss

$$L_{X,Y}(W) = -\frac{1}{n} \sum_{i=1}^{n} \sum_{j=1}^{c} \delta(y_i, j) \log \frac{\exp^{x_i^T w^j}}{\sum_{l=1}^{c} \exp^{x_i^T w^l}}, \tag{4.37}$$

the $\ell_{2,1}$ norm loss

$$L_{X,Y}(W) = \left\| Y - X^T W \right\|_{2,1} \tag{4.38}$$

and the multi-class hinge loss

$$L_{X,Y}(W) = \sum_{i=1}^{n} \left(1 - x_i^T w^{y_i} + \max_{j \neq y_i} x_i^T w^j \right)_+, \tag{4.39}$$

where y_i is the label of x_i, $\delta(y_i, j)$ is the delta function that equals 1 if $y_i = j$ and equals 0 otherwise, w^j is the j-th column in W and the function $z_+ = \max(z, 0)$.

To make $\ell_{2,1-2}$ more applicable, instead of concentrating on some specific formulation of $L_{X,Y}(\cdot)$, we consider a general assumption for loss function, which is stated as follow:

Assumption 4.1
1. The loss function $L_{X,Y}(\cdot)$ is convex.
2. Suppose there is a linear transformation W with $Y = X^T W$, then $L_{X,Y}(W) = 0$.
3. For any matrix $W^0 \in \Re^{d \times c}$, the set $\{W : L_{X,Y}(W) < L_{X,Y}(W^0)\} \cap \{W : \|W\|_{2,1-2} = 0\}$ is bounded.

The following Lemma demonstrates that the loss functions in (4.36), (4.37), (4.38) and (4.39) satisfy Assumption 4.1.

Lemma 4.1 *Suppose that each row in* **X** *is nonzero. Then the Frobenius norm loss* (4.36)*, the Logistic loss* (4.37)*, the* $\ell_{2,1}$ *norm loss* (4.38) *and the multi-class hinge loss* (4.39) *all satisfy Assumption 4.1.*

Proof The verification of the first and second items in Assumption 4.1 for these loss functions is intuitive due to their convexity and special formulations. Therefore, we only check the third item in Assumption 4.1.

1. Because the Frobenius norm loss and the Logistic loss are strong convex, (3) in Assumption 4.1 is obviously satisfied.
2. To show that the $\ell_{2,1}$ norm loss satisfies (3) in Assumption 4.1, it suffices to prove that for any fixed nonzero matrix W satisfying $\|W\|_{2,1-2} = 0$,

$\|\mathrm{Y} - r\mathrm{X}^T\mathrm{W}\|_{2,1} \to \infty$ as $r \to \infty$. For any $\mathrm{W} \neq 0$, if $\|\mathrm{W}\|_{2,1-2} = 0$, from Proposition 4.1, we know that $\|\mathrm{W}\|_{2,0} = 1$. Let the j-th row be nonzero. Suppose for all $x_i (i = 1, 2, \cdots, n)$, $x_i^T \mathrm{W} = 0$ then the j-th row of X is zero, which is a contradiction to the fact that each row in $\mathbf{X}$ is nonzero. Therefore there exists at least an x_i such that $\mathrm{x}_i^T \mathrm{W} \neq 0$. Assume $\mathrm{x}_k^\mathrm{T} \mathrm{W} \neq 0$. Then when $r \to \infty$, we have

$$\begin{aligned} L_{\mathrm{X,Y}}(r\mathrm{W}) = \left\|\mathrm{Y} - r\mathrm{X}^\mathrm{T}\mathrm{W}\right\|_{2,1} = \textstyle\sum_{i=1}^n \left\|\mathrm{y}_i - r\mathrm{x}_i^\mathrm{T}\mathrm{W}\right\|_2 \geq \left\|\mathrm{y}_k - r\mathrm{x}_k^\mathrm{T}\mathrm{W}\right\|_2 \\ \geq \mid \|\mathrm{y}_k\|_2 - r\left\|\mathrm{x}_k^\mathrm{T}\mathrm{W}\right\|_2 \mid \to \infty, \end{aligned} \tag{4.40}$$

which implies that the $\ell_{2,1}$ norm loss satisfies (3).

3. Similarly, to show that the multi-class hinge loss satisfies (3) in Assumption 4.1, it suffices to prove that, for any fixed nonzero matrix W satisfying $\|\mathrm{W}\|_{2,1-2} = 0$, $\sum_{i=1}^n \left(1 - r\mathrm{x}_i^\mathrm{T}\mathrm{w}^{y_i} + \max_{j \neq y_i} r\mathrm{x}_i^\mathrm{T}\mathrm{w}^j\right)_+ \to \infty$as $r \to \infty$. Given $\mathrm{W} \neq 0$, which satisfies $\|\mathrm{W}\|_{2,1-2} = 0$, there exists at least an x_k with the corresponding label y_k such that $\max_{j \neq y_k} \mathrm{x}_j^T \mathrm{w}^{y_j} > \mathrm{x}_k^T \mathrm{w}^{y_k}$. Then, as $r \to \infty$, we have

$$\begin{aligned} L_{\mathrm{X,Y}}(r\mathrm{W}) = \textstyle\sum_{i=1}^n \left(1 - r\mathrm{x}_i^\mathrm{T}\mathrm{w}^{y_i} + r\max_{j \neq y_i} \mathrm{x}_i^\mathrm{T}\mathrm{w}^j\right)_+ \\ \geq r\left(\max_{j \neq y_k} \mathrm{x}_j^\mathrm{T}\mathrm{w}^{y_j} - \mathrm{x}_k^\mathrm{T}\mathrm{w}^{y_k}\right) + 1 \to \infty \end{aligned} \tag{4.41}$$

From (4.41), we know the multi-class hinge loss satisfies (3). Combing (1), (2) and (3), we know that the four loss functions satisfy Assumption 4.1.

Lets' discuss how to solve the nonconvex model in (4.34). Denote the objective function in (4.34) as $F(\mathrm{W})$ for short. Note that $F(\mathrm{W})$ can be naturally split into two convex functions

$$F(\mathrm{W}) = \left(L_{\mathrm{X,Y}}(\mathrm{W}) + \alpha\|\mathrm{W}\|_{2,1}\right) - \alpha\|\mathrm{W}\|_F, \tag{4.42}$$

An iterative algorithm in the framework of ConCave-Convex Procedure (CCCP) is proposed [45]. In detail, according to (4.27), the linearized convex subproblem of (4.42) is:

$$\min_{\mathrm{W}} \left(L_{\mathrm{X,Y}}(\mathrm{W}) + \alpha\|\mathrm{W}\|_{2,1}\right) - \alpha\left\langle \mathrm{W}, \mathrm{A}^k \right\rangle \tag{4.43}$$

where

$$\mathrm{A}^k = \begin{cases} \left\|\mathrm{W}^k\right\|_F^{-1}\mathrm{W}^k, & \mathrm{W}^k \neq 0; \\ 0, & \mathrm{W}^k = 0. \end{cases} \tag{4.44}$$

and the superscript k is the iteration index.

The summary of the proposed CCCP framework is shown in Algorithm 4.2. Noticing that at the first iteration of Algorithm 4.2, since $W^0 = 0$, the linearized convex subproblem in (4.43) becomes the classic feature selection model with the $\ell_{2,1}$ norm regularization. From this point of view, the proposed method could be considered as an improvement from the convex feature selection model with the $\ell_{2,1}$ norm.

Algorithm 4.2 Supervised Feature Selection with $\ell_{2,1\text{-}2}$

Input: data matrix $\mathbf{X}$, label matrix $\mathbf{Y}$ and regularization parameter α

1: Initialize $k = 0$ and $\mathbf{W}^0 = \mathbf{0}$
2: **repeat**
3: $\quad\mathbf{W}^{k+1} := \arg\min_{\mathbf{W}} \mathbf{L}_{\mathbf{X},\mathbf{Y}}(\mathbf{W}) + \alpha\|\mathbf{W}\|_{2,1} - \alpha\langle\mathbf{W}, \mathbf{A}^k\rangle$
4: $\quad k := k + 1$
5: **until** CCCP stopping criterion is satisfied

Output: The optimal solution $\mathbf{W}^*$

As mentioned in related work, existing convergence results of CCCP cannot be applied to Algorithm 4.2 directly. In [46], the least square loss is used and the convergent proof of CCCP for ℓ_{1-2} strongly depends on its specific formulation. However, in this section, the loss function in the model is in a general form. We could not prove the strong convergence of $\ell_{2,1-2}$ following the way of ℓ_{1-2} in [46]. To prove the convergence of Algorithm 4.2, we use Zangwill's theory [50], which is a powerful and general framework to deal with the convergence issues of the iterative algorithm. To this end, we first construct a majorization function as follow:

$$G(\mathrm{W}, \mathrm{Z}) = L_{\mathrm{X},\mathrm{Y}}(\mathrm{W}) + \alpha\|\mathrm{W}\|_{2,1} - \alpha\left(\|\mathrm{Z}\|_F + \langle \mathrm{A}(\mathrm{Z}), \mathrm{W} - \mathrm{Z}\rangle\right) \tag{4.45}$$

where $\mathrm{A(Z)} \in \partial\|\mathrm{Z}\|_F$. Let $S(\cdot) : \Re^{d\times c} \to \mathcal{X}$ be a point-to-set mapping, where $\mathcal{X}$ is the power set of $\Re^{d\times c}$. Denote

$$S(\mathrm{Z}) := \arg\min_{\mathrm{W}} G(\mathrm{W}, \mathrm{Z}). \tag{4.46}$$

Denote the set of the stationary points of $F(\cdot)$ by S. Then the following lemma is given.

Lemma 4.2 *Let $\mathrm{Z} \neq 0$. $\mathrm{Z} \in S(\mathrm{Z})$ if and only if Z is a stationary point of $F(\cdot)$.*

Proof If $0 \neq \mathrm{Z} \in S(\mathrm{Z})$, substituting $A(\mathrm{Z}) = \|\mathrm{Z}\|_F^{-1}\mathrm{Z}$ in Eq. (4.45), we have

$$G(\mathrm{W}, \mathrm{Z}) = L_{\mathrm{X,Y}}(\mathrm{W}) + \alpha\|\mathrm{W}\|_{2,1} - \alpha\left\langle \|\mathrm{Z}\|_F^{-1}\mathrm{Z}, \mathrm{W}\right\rangle \tag{4.47}$$

Since $G(\mathrm{W}, \mathrm{Z})$ is convex in W, from Proposition 4.4.6 of [51] (pp. 194), we have

$$\partial G_{\mathrm{W}}(\mathrm{W}, \mathrm{Z})\,|_{\mathrm{W=Z}} = \partial\left(L_{\mathrm{X,Y}}(\mathrm{W}) + \alpha\|\mathrm{W}\|_{2,1}\right)|_{\mathrm{W=Z}} - \|\mathrm{Z}\|_F^{-1}\mathrm{Z} \tag{4.48}$$

Then the optimality of $G(\cdot, \mathrm{Z})$ at Z gives

$$0 \in \partial\left(L_{\mathrm{X,Y}}(\mathrm{W}) + \alpha\|\mathrm{W}\|_{2,1}\right)|_{\mathrm{W=Z}} - \|\mathrm{Z}\|_F^{-1}\mathrm{Z} \tag{4.49}$$

which implies that Z is just a stationary point of $F(\cdot)$. Conversely, if $Z \neq 0$ is a stationary point of $F(\cdot)$, then $0 \in \partial F(\mathrm{Z})$. Noticing that $\|\cdot\|_F$ is smooth near $Z \neq 0$, from [52] (pp. 304), we have

$$\partial F(\mathrm{Z}) = \partial\left(L_{\mathrm{X,Y}}(\mathrm{W}) + \alpha\|\mathrm{W}\|_{2,1}\right)|_{\mathrm{W=Z}} - \|\mathrm{Z}\|_F^{-1}\mathrm{Z} \tag{4.50}$$

Therefore

$$0 \in \partial\left(L_{\mathrm{X,Y}}(\mathrm{W}) + \alpha\|\mathrm{W}\|_{2,1}\right)|_{\mathrm{W=Z}} - \|\mathrm{Z}\|_F^{-1}\mathrm{Z} \tag{4.51}$$

which together with the fact that $G(\cdot, \mathrm{Z})$ is convex, implies that $\mathrm{Z} \in S(\mathrm{Z})$.

Based on Lemma 4.2, the following theorem is given for the global convergence of Algorithm 4.2.

Theorem 4.1 *Assume that* $\{W^k\}_{k=0}^{\infty}$ *is the sequence of iterations generated by Algorithm 4.2, the following properties hold.*

1. $\|W^{k+1} - W^k\|_F \to 0$ *as* $k \to \infty$.
2. *Any nonzero limit point of the sequence* $\{W^k\}_{k=0}^{\infty}$ *is the stationary point of problem* (4.35).

Proof If $\mathrm{W}^1 = \mathrm{W}^0 = 0$, we stop the algorithm and produce the optimal solution $\mathrm{W}^* = 0$. Otherwise we assume that $\mathrm{W}^k \neq 0$, for $k = 1, 2, \cdots$. For convenience, denote $L_1 = \{\mathrm{W} : F(\mathrm{W}) \leq F(\mathrm{W}^1)\}$. We first show L_1 is compact.

1. If $\|\mathrm{W}\|_{2,0} = 1$, according to Proposition 4.1, $\|\mathrm{W}\|_{2,1-2} = 0$. From Assumption 4.1, the set $\{\mathrm{W} : L_{\mathrm{X,Y}}(\mathrm{W}) \leq L_{\mathrm{X,Y}}(\mathrm{W}^1)\}$ is bounded.
2. If $\|\mathrm{W}\|_{2,0} > 1$, according to Proposition 6.1, $\|\mathrm{W}\|_{2,1-2} > 0$. When $r \to \infty$, we have

$$F(r\mathrm{W}) = L_{\mathrm{X,Y}}(r\mathrm{W}) + \alpha r\|\mathrm{W}\|_{2,1-2} \to \infty \tag{4.52}$$

Combing (1) and (2), we know that the level set L_1 is bounded. Due to the continuity of $F(\cdot)$, we also know that the level set is closed. In all, the level set is compact.

1. For any $Z \in L_1$ and $Z \notin S$, assume $W \in S(Z)$. Then

$$F\left(W\right) = G\left(W, W\right) \leq G\left(W, Z\right) \leq G\left(Z, Z\right) = F\left(Z\right). \tag{4.53}$$

If $F(W) = F(Z)$, according to (4.53), we have $G(W, Z) = G(Z, Z)$, which means $Z \in S(Z)$. The assumption $F(W^1) < F(0)$ follows $Z \neq 0$, and from Lemma 4.2, Z is a stationary point of F, which is contradiction to $Z \notin S$. Namely, $S(\cdot)$ is strictly monotonic on L.
2. For any $Z \in L_1$, since $G(W, Z)$ is continuous in Z, $p(Z) = \min_W G(W, Z)$ is also continuous. Based on Theorem 1.17 of [52] (pp. 16), $S(\cdot)$ is upper semi-continuous on L_1.
3. For any $Z \in L_1$ and any $W \in S(Z)$, it follows that

$$F\left(W\right) \leq G\left(W, Z\right) \leq G\left(Z, Z\right) = F\left(Z\right) \leq F\left(W^1\right). \tag{4.54}$$

We know $W \in L_1$, and therefore $S(Z) \subset L_1$. Moreover, L_1 is compact. This means that $S(\cdot)$ is uniformly compact on L_1.

In summary, the point-to-set mapping $S(\cdot)$ on compact set L_1 is strictly monotonic, upper semi-continuous and uniformly compact. We can check that the point-to-set mapping satisfies the condition in Theorem 3.1 [53]. Therefore, the desired result in theorem follows.

The following discusses how to solve the linearized subproblem (4.43). Generally speaking, problem (4.43) can be efficiently solved by the Alternating Direction Methods of Multipliers (ADMM), which is a versatile algorithm [54]. As discussed in last subsection, several loss functions can be used in our model (4.43) with proven convergence. Since the algorithm strongly relies on the specific formulation of the loss function, we choose the $\ell_{2,1}$ norm loss, which is very popular and can efficiently tackle outliers and noise in data points [40], to present the algorithm. To be more specific, when the $\ell_{2,1}$ norm is used as the loss function, the linearized subproblem (4.43) becomes

$$\min_{W} \left\| Y - X^T W \right\|_{2,1} + \alpha \|W\|_{2,1} - \alpha \left\langle W, A^k \right\rangle. \tag{4.55}$$

To solve problem (4.55), we introduce two auxiliary variables $\mathrm{U} = \mathrm{W}$ and $\mathrm{V} = \mathrm{Y} - \mathrm{X}^{\mathrm{T}}\mathrm{W}$, and transforming (4.55) into the following equivalent form

$$\begin{aligned} &\min_{\mathrm{W,U,V}} \|\mathrm{V}\|_{2,1} + \alpha\|\mathrm{U}\|_{2,1} - \alpha\left\langle \mathrm{W}, \mathrm{A}^k \right\rangle \\ &\mathrm{V} = \mathrm{Y} - \mathrm{X}^{\mathrm{T}}\mathrm{W}, \mathrm{U} = \mathrm{W}, \end{aligned} \tag{4.56}$$

which can be solved by the following ADMM problem

$$\begin{aligned} \min_{\mathrm{W,U,V}} \|\mathrm{V}\|_{2,1} + \alpha\|\mathrm{U}\|_{2,1} - \alpha\left\langle \mathrm{W}, \mathrm{A}^k\right\rangle + \left\langle \Sigma, \mathrm{W} - \mathrm{U}\right\rangle + \left\langle \Lambda, \mathrm{Y} - \mathrm{X}^{\mathrm{T}}\mathrm{W} - \mathrm{V}\right\rangle \\ + \frac{\rho}{2}\left(\|\mathrm{W} - \mathrm{U}\|_F^2 + \left\|\mathrm{Y} - \mathrm{X}^{\mathrm{T}}\mathrm{W} - \mathrm{V}\right\|_F^2\right) \end{aligned} \tag{4.57}$$

where Σ and Λ are two Lagrangian multipliers corresponding to two equality constraints, respectively, and $\rho > 0$ is a penalty parameter that determines the penalty for infeasibility of the equality constraints. In the following part, we give the details of solving each ADMM subproblem.

Updating W: Fixing variables U and V, problem (4.57) is reduced to

$$\min_{\mathbf{W}} - \alpha\left\langle \mathbf{W}, \mathbf{A}^k\right\rangle + \left\langle \boldsymbol{\Sigma}, \mathbf{W} - \mathbf{U}\right\rangle + \left\langle \boldsymbol{\Lambda}, \mathbf{Y} - \mathbf{X}^{\mathrm{T}}\mathbf{W} - \mathbf{V}\right\rangle + \frac{\rho}{2}\|\mathbf{W} - \mathbf{U}\|_F^2 + \frac{\rho}{2}\left\|\mathbf{Y} - \mathbf{X}^{\mathrm{T}}\mathbf{W} - \mathbf{V}\right\|_F^2$$

Since the above problem is a differentiable unconstrained convex problem, for any optimal solution, the gradient of objective function must be zero. By setting the gradient of objective function to zero, we get

$$\rho\left(\mathrm{XX}^{\mathrm{T}} + \mathrm{I}_d\right)\mathrm{W} = \mathrm{B}^k \tag{4.58}$$

where I_d is a $d \times d$ identity matrix and

$$\mathrm{B}^k = \alpha\mathrm{A}^k - \Sigma + \rho\mathrm{U} + \mathrm{X}\left(\Lambda + \rho\mathrm{Y} - \rho\mathrm{V}\right) \tag{4.59}$$

The following proposition is used to update U and V.

Proposition 4.2 Given a positive scalar λ *and vectors* $\mathbf{a}, \mathbf{b} \in \Re^n$*, the optimal solution of*

$$\min_{\mathrm{w}\in\Re^n} f(\mathrm{w}) = \frac{1}{2}\|\mathrm{w} - \mathrm{a}\|_2^2 + \lambda\|\mathrm{w}\|_2 - \mathrm{w}^T\mathrm{b} \tag{4.60}$$

is

$$\mathrm{w}^* = \begin{cases} \left(1 - \frac{\lambda}{\|\mathrm{a}+\mathrm{b}\|_2}\right)(\mathrm{a} + \mathrm{b}), & \text{if } \|\mathrm{a} + \mathrm{b}\|_2 > \lambda; \\ 0, & \text{otherwise.} \end{cases} \tag{4.61}$$

Proof Since problem (4.60) is an unconstrained convex problem, we know that its optimal solution w^* will satisfy the optimal condition that is $0 \in \partial f(w^*)$. Note that when $w \neq 0$,

$$\partial f(\mathrm{w}) = \nabla f(\mathrm{w}) = \mathrm{w} - \mathrm{a} + \lambda \frac{\mathrm{w}}{\|\mathrm{w}\|_2} - \mathrm{b}, \tag{4.62}$$

we know that if $w^* \neq 0$,

$$\lambda \frac{\mathrm{w}^*}{\|\mathrm{w}^*\|_2} + \mathrm{w}^* - \mathrm{a} - \mathrm{b} = 0, \textit{i.e.}, \mathrm{a} + \mathrm{b} - \mathrm{w}^* = \lambda \frac{\mathrm{w}^*}{\|\mathrm{w}^*\|_2}, \tag{4.63}$$

from which we obtain the proposition.

Updating U: Problem (4.57) with respect to U becomes

$$\min_{\mathrm{U}} \alpha \|\mathrm{U}\|_{2,1} + \langle \Sigma, \mathrm{W} - \mathrm{U} \rangle + \frac{\rho}{2} \|\mathrm{W} - \mathrm{U}\|_F^2. \tag{4.64}$$

Expanding the objective function in problem (4.64) and removing the terms that are irrelevant of U, we arrive at

$$\min_{\mathrm{U}} \frac{1}{2} \left\| \mathrm{U} - \left(\mathrm{W} + \frac{\Sigma}{\rho} \right) \right\|_F^2 + \frac{\alpha}{\rho} \|\mathrm{U}\|_{2,1}. \tag{4.65}$$

Let $M = W + \Sigma/\rho$, we have

$$\min_{\mathrm{U}} \frac{1}{2} \|\mathrm{U} - \mathrm{M}\|_F^2 + \frac{\alpha}{\rho} \|\mathrm{U}\|_{2,1}. \tag{4.66}$$

Furthermore, problem (4.64) can be rewritten as:

$$\min_{\mathrm{u}_i} \sum_{i=1}^{d} \left(\frac{1}{2} \|\mathrm{u}_i - \mathrm{m}_i\|_2^2 + \frac{\alpha}{\rho} \|\mathrm{u}_i\|_2 \right), \tag{4.67}$$

where u_i and m_i are the i-th row of U and M, respectively. Obviously, problem (4.67) is equivalent to solving d independent subproblems simultaneously. Based on Proposition 4.2, the optimal solution is, for $i = 1, 2, \cdots, d$,

$$\mathrm{u}_i = \begin{cases} \left(1 - \frac{\alpha}{\rho \|\mathrm{m}_i\|_2}\right) \mathrm{m}_i, & \text{if } \alpha < \rho \|\mathrm{m}_i\|_2 \\ 0, & \text{otherwise.} \end{cases} \tag{4.68}$$

Updating V: Similar to updating U, the subproblem with respect to V is equivalent to

$$\min_{V} \frac{1}{2}\|V - N\|_F^2 + \frac{1}{\rho}\|V\|_{2,1} \tag{4.69}$$

where $N = Y - X^T W + \Lambda/\rho$. Then, the solution of problem (4.69) is $V = \left[v_1^T, v_2^T, \cdots, v_n^T\right]^T$, where for $i = 1, 2, \cdots, n$,

$$v_i = \begin{cases} \left(1 - \frac{1}{\rho\|n_i\|_2}\right)n_i, & \text{if } 1 < \rho\|n_i\|_2; \\ 0, & \text{otherwise.} \end{cases} \tag{4.70}$$

Updating Σ and Λ: After updating the variables, we also need to adjust the Lagrangian multipliers. The specific rule is

$$\begin{aligned} \Sigma \quad &:= \Sigma + \rho\left(W - U\right) \\ \Lambda &:= \Lambda + \rho\left(Y - X^T W - V\right) \end{aligned} \tag{4.71}$$

With the above updating rules, we summarize the process of solving problem (4.55) in Algorithm 4.3.

Algorithm 4.3 ADMM for CCCP Subproblem (4.54)

Input: data matrix **X**, label matrix **Y**, regularization parameter α and penalty parameter ρ

1: Initialize $k = 0$ and Σ^0, $\Lambda^0 = \mathbf{0}$ and $\mathbf{U}^0$, $\mathbf{V}^0 = \mathbf{0}$
2: **repeat**
3: Calculate $\mathbf{B}^k$ by Eq. (4.58)
4: Update $\mathbf{W}^{k+1}$ by Eq. (4.57)
5: Calculate $\mathbf{M}^k = \mathbf{W}^{k+1} + \rho^{-1}\Sigma^k$
6: **for** $i = 1, 2, \ldots, d$ **do**
7: **if** $\alpha < \rho\left\|\mathbf{m}_i^k\right\|_2$ **then**
8: $\mathbf{u}_i^{k+1} = \left(1 - \frac{\alpha}{\rho\left\|\mathbf{m}_i^k\right\|_2}\right)\mathbf{m}_i^k$
9: **else**
10: $\mathbf{u}_i^{k+1} = \mathbf{0}$
11: **end if**
12: **end for**
13: Calculate $\mathbf{N}^k = \mathbf{Y} - \mathbf{X}^T\mathbf{W}^{k+1} + \rho^{-1}\Lambda^k$
14: **for** $i = 1, 2, \ldots, n$ **do**
15: **if** $1 < \rho\left\|\mathbf{n}_i^k\right\|_2$ **then**
16: $\mathbf{v}_i^{k+1} = \left(1 - \frac{1}{\rho\left\|\mathbf{n}_i^k\right\|_2}\right)\mathbf{n}_i^k$

17: **else**
18: $\mathbf{v}_i^{k+1} = \mathbf{0}$
19: **end if**
20: **end for**
21: Update Σ^{k+1} and Λ^{k+1} by Eq. (4.70)
22: $k := k + 1$
23: **until** ADMM stopping criterion is satisfied
Output: The optimal solution $\mathbf{W}^*$

The data analysis of implementing the above method can be found in [4].

4.2.2 Feature Selection with $\ell_{2,1-2}$ Regularization

In unsupervised learning, we don't have label information that can guide to select the most discriminative and relevant features. One commonly used strategy is to seek cluster indicators and simultaneously perform the supervised feature selection within the unified framework [55, 56].

Here we use spectral clustering to obtain the cluster indicators of data points. Suppose that n data points are grouped into c clusters. Denote $F = [f_1, f_2, \cdots, f_n]^T \in \{0, 1\}^{n \times c}$, where $f_i \in \{0, 1\}^{c \times 1}$ is the cluster indicator vector for x_i. That is, $F_{ij} = 1$ if x_i is assigned to the j-th cluster, and $F_{ij} = 0$ otherwise. The scaled cluster indicator matrix Y is defined as $Y = \left[y_1, y_2, \cdots, y_n\right]^T = F\left(F^TF\right)^{-\frac{1}{2}}$. Then $Y^TY = \left(F^TF\right)^{-\frac{1}{2}}\left(F^TF\right)\left(F^TF\right)^{-\frac{1}{2}} = I_c$ where I_c is the $c \times c$ identity matrix. Following the work in [55], we can get the scaled cluster indictors by exploiting the local geometrical structure of original data as follows

$$\min_{Y} \mathrm{Tr}\left(Y^TLY\right), \quad s.t. \; Y^TY = I_c, Y \geq 0 \tag{4.72}$$

where L is the normalized Laplace matrix and defined as $L = D^{-1/2}(D - S)D^{-1/2}$ is the affinity matrix of data points and D is a diagonal matrix with $D_{ii} = \sum_{j=1}^{n} S_{ij}$. Problem (4.72) can be called nonnegative spectral clustering [55]. Combing (4.35) and (4.72), we propose unsupervised feature selection model in a general form as follows:

$$\begin{gathered}\min_{W,Y} L_{X,Y}(W) + \alpha\|W\|_{2,1-2} + \beta\mathrm{Tr}\left(Y^TLY\right) \\ s.t. \; Y^TY = I_c, Y \geq 0\end{gathered} \tag{4.73}$$

where $\alpha > 0$ and $\beta > 0$ are two balance hyper-parameters. By solving model (4.73), the optimal solution W^* can be obtained. Then the top-ranking features can be selected according to the values of $\|w_i^*\|_2\left(i = 1, 2, \cdots, d\right)$.

4.2.2.1 Algorithm

The problem (4.73) is a non-convex optimization with orthogonal and nonnegative constraints. In general, it is difficult to solve directly. Based on the work in [54], we use the Alternating Direction Method of Multiplier (ADMM) to effectively solve our unsupervised model (4.73). Similar to the supervised case, to give the detailed algorithm of (4.73), we choose the $\ell_{2,1}$ norm as the loss function. Then (4.73) becomes

$$\begin{gathered}\min_{\mathrm{W,Y}}\left\|\mathrm{Y}-\mathrm{X}^{\mathrm{T}}\mathrm{W}\right\|_{2,1}+\alpha\|\mathrm{W}\|_{2,1-2}+\beta\mathrm{Tr}\left(\mathrm{Y}^{\mathrm{T}}\mathrm{LY}\right)\\ s.t.\mathrm{Y}^{\mathrm{T}}\mathrm{Y}=\mathrm{I}_c,\ \mathrm{Y}\geq 0\end{gathered} \tag{4.74}$$

By introducing auxiliary variables $\mathrm{U}=\mathrm{W}$, $\mathrm{V}=\mathrm{Y}-\mathrm{X}^{\mathrm{T}}\mathrm{W}$ and $\mathrm{H}=\mathrm{Y}$, the optimization problem (4.74) can be rewritten as the following equivalent form

$$\begin{gathered}\min_{\mathrm{W,Y,U,V,H}}\|\mathrm{V}\|_{2,1}+\alpha\left(\|\mathrm{U}\|_{2,1}-\|\mathrm{W}\|_F\right)+\beta\mathrm{Tr}\left(\mathrm{Y}^{\mathrm{T}}\mathrm{L}\mathbf{H}\right)\\ s.t.\mathrm{Y}^{\mathrm{T}}\mathrm{Y}=\mathrm{I}_c,\ \mathrm{U}=\mathrm{W},\ \mathrm{V}=\mathrm{Y}-\mathrm{X}^{\mathrm{T}}\mathrm{W},\ \mathrm{H}=\mathrm{Y},\ \mathrm{H}\geq 0,\end{gathered} \tag{4.75}$$

which can be Transactionsformed into the following form

$$\begin{gathered}\min_{\mathrm{W,Y,U,V,H}}\|\mathrm{V}\|_{2,1}+\alpha\left(\|\mathrm{U}\|_{2,1}-\|\mathrm{W}\|_F\right)+\beta\mathrm{Tr}\left(\mathrm{Y}^{\mathrm{T}}\mathrm{LH}\right)+\left\langle\Sigma,\mathrm{W}-\mathrm{U}\right\rangle\\ +\langle\Gamma,\mathrm{Y}-\mathrm{H}\rangle+\left\langle\Lambda,\mathrm{Y}-\mathrm{X}^{\mathrm{T}}\mathrm{W}-\mathrm{V}\right\rangle+\frac{\rho}{2}\|\mathrm{W}-\mathrm{U}\|_F^2\\ +\frac{\rho}{2}\left(\|\mathrm{Y}-\mathrm{H}\|_F^2+\left\|\mathrm{Y}-\mathrm{X}^{\mathrm{T}}\mathrm{W}-\mathrm{V}\right\|_F^2\right)\\ \text{s.t.}\quad \mathrm{Y}^{\mathrm{T}}\mathrm{Y}=\mathrm{I}_c,\ \mathrm{H}\geq 0\end{gathered} \tag{4.76}$$

where Σ, Γ and Λ are the Lagrangian multipliers corresponding to the three equality constraints, respectively, and $\rho > 0$ is a penalty parameter that determines the penalty for infeasibility of the three equality constraints. According to ADMM, we solve a manageable subproblem with respect to a given variable while fixing the other variables. In our algorithm, except the subproblem with respect to W, we can get the closed form solutions for all subproblems. Noticing that the subproblem of U and V are the same as the supervised case, we here only present the details of solving W, Y, and H.

Updating W: To update W, we fix all the variables except W and remove irrelevant terms that are irrelevant of W. Problem (4.76) becomes:

$$\min_{\mathrm{W}}-\alpha\|\mathrm{W}\|_F+\langle\Sigma,\mathrm{W}-\mathrm{U}\rangle+\left\langle\Lambda,\mathrm{Y}-\mathrm{X}^{\mathrm{T}}\mathrm{W}-\mathrm{V}\right\rangle+\frac{\rho}{2}\left(\|\mathrm{W}-\mathrm{U}\|_F^2+\left\|\mathrm{Y}-\mathrm{X}^{\mathrm{T}}\mathrm{W}-\mathrm{V}\right\|_F^2\right)$$

Similar to the proposed supervised model, we solve the above problem with CCCP, in which the linearized subproblem is

$$\begin{aligned}\min_{\mathrm{W}} - \alpha \left\langle \mathrm{W}, \mathrm{A}^k \right\rangle + \left\langle \Sigma, \mathrm{W} - \mathrm{U} \right\rangle + \left\langle \Lambda, \mathrm{Y} - \mathrm{X}^{\mathrm{T}}\mathrm{W} - \mathrm{V} \right\rangle \\ + \tfrac{\rho}{2} \left(\|\mathrm{W} - \mathrm{U}\|_F^2 + \left\|\mathrm{Y} - \mathrm{X}^{\mathrm{T}}\mathrm{W} - \mathrm{V}\right\|_F^2 \right)\end{aligned} \tag{4.77}$$

where A^k is defined as (4.27).

Updating Y: Problem (4.76) with respect to Y turns to be,

$$\begin{aligned}\min_{\mathrm{Y}} \ \beta \mathrm{Tr}\left(\mathrm{Y}^{\mathrm{T}}\mathrm{L}\mathrm{H}\right) + \langle \Gamma, \mathrm{Y} - \mathrm{H} \rangle + \left\langle \Lambda, \mathrm{Y} - \mathrm{X}^{\mathrm{T}}\mathrm{W} - \mathrm{V} \right\rangle \\ + \tfrac{\rho}{2} \left(\|\mathrm{Y} - \mathrm{H}\|_F^2 + \left\|\mathrm{Y} - \mathrm{X}^{\mathrm{T}}\mathrm{W} - \mathrm{V}\right\|_F^2 \right) \\ s.t. \quad \mathrm{Y}^{\mathrm{T}}\mathrm{Y} = \mathrm{I}_c,\end{aligned} \tag{4.78}$$

which is equivalent to

$$\min_{\mathrm{Y}} \|\mathrm{Y} - \mathrm{J}\|_F^2 \ \text{ s.t. } \ \mathrm{Y}^{\mathrm{T}}\mathrm{Y} = \mathrm{I}_c \tag{4.79}$$

where $\mathrm{J} = (\rho \mathrm{X}^T\mathrm{W} + \rho \mathrm{V} - \Gamma - \beta \mathrm{L}\mathrm{H} - \Lambda + \rho \mathrm{H})/2\rho$. Using the constraint $\mathrm{Y}^{\mathrm{T}}\mathrm{Y} = \mathrm{I}_c$, we rewrite problem (4.79) as

$$\max_{\mathrm{Y}} \mathrm{Tr}\left(\mathrm{Y}^{\mathrm{T}}\mathrm{J}\right) \ \text{ s.t. } \ \mathrm{Y}^{\mathrm{T}}\mathrm{Y} = \mathrm{I}_c. \tag{4.80}$$

According to Lemma 3.2 in [57], we present the optimal solution of problem (4.80) as follow:

$$\mathrm{Y} = \mathrm{J}_1\mathrm{J}_2^{\mathrm{T}}, \tag{4.81}$$

where J_1 is the first c columns of the left singular values of singular value decomposition (SVD) of J, and J_2 is the right singular values.

Updating H: Problem (4.76) yields the following problem

$$\min_{\mathrm{H} \geq 0} \mathrm{Tr}\left(\mathrm{Y}^{\mathrm{T}}\mathrm{L}\mathrm{H}\right) + \langle \Gamma, \mathrm{Y} - \mathrm{H} \rangle + \frac{\rho}{2} \|\mathrm{Y} - \mathrm{H}\|_F^2. \tag{4.82}$$

We can rewrite the equivalent form of (4.82) as follow

$$\min_{\mathrm{H}} \|\mathrm{H} - \mathrm{Q}\|_F^2 \ \text{ s.t. } \ \mathrm{H} \geq 0, \tag{4.83}$$

where $Q = \mathrm{Y} + \Gamma/\rho - \mathrm{L}^T\mathrm{Y}/\rho$. Furthermore, problem (4.83) can be decomposed into element-wise

$$\min_{\mathrm{H}_{ij}} \sum_{i,j} \left(\mathrm{H}_{ij} - \mathrm{Q}_{ij}\right)^2 \text{ s.t. } \mathrm{H}_{ij} \geq 0. \tag{4.84}$$

It is straightforward to see that the solution is

$$\mathrm{H}_{ij} = \max\left(\mathrm{Q}_{ij}, 0\right), i = 1, \cdots, n \text{ and } j = 1, \cdots, c \tag{4.85}$$

Updating Γ: The Lagrangian multiplier Γ is updated as

$$\Gamma := \Gamma + \rho\left(\mathrm{Y} - \mathrm{H}\right) \tag{4.86}$$

Finally, we present the algorithm for unsupervised feature selection model with $\ell_{2,1-2}$ regularization in Algorithm 4.4.

Algorithm 4.4 Unsupervised Feature Selection with $\ell_{2,1-2}$

Input: data matrix **X**, hyper-parameters α and β, penalty parameter ρ

1: Initialize $\mathbf{Y}^0$ by K-means, construct the normalized graph Laplacian matrix **L** and set k = 0
2: **repeat**
3: **repeat**
4: Update $\mathbf{W}^{\mathrm{k}}$ by Eq. (4.57)
5: **until** CCCP stopping criterion is satisfied
6: Update $\mathbf{U}^{\mathrm{k}}$ by Eq. (4.67)
7: Update $\mathbf{V}^{\mathrm{k}}$ by Eq. (4.69)
8: Update $\mathbf{H}^{\mathrm{k}}$ by Eq. (4.84)
9: Update $\mathbf{Y}^{\mathrm{k}}$ by Eq. (4.80)
10: Update Σ^{k}, Γ^{k} and Λ^{k} by Eqs. (4.70) and (4.85)
11: $\mathrm{k} := \mathrm{k} + 1$
12: **until** ADMM stopping criterion is satisfied

Output: The optimal solution $\mathbf{W}^*$

4.2.3 *Feature Selection with MCP² Regularization*

4.2.3.1 Sparse Regularization for Vectors

This section gives a brief review on sparse regularizer for vectors. The sparsest solution can be obtained by ℓ_0, which is defined as the number of non-zero entries, i.e., $\|\mathrm{w}\|_0 = \#\{i : w_i \neq 0\}$. Because the resulting optimization is NP hard, several relaxation of ℓ_0 have been proposed and studied, including the ℓ_1 norm, ℓ_p, the capped ℓ_1 norm [58], ℓ_{1-2} [47], the SCAD [59] and the MCP [60].

The ℓ_1 norm and $\ell_p (0 < p < 1)$ of vector $\mathrm{w} = [w_1, w_2, \cdots, w_n] \in \Re^n$ is defined as

$$\|\mathrm{w}\|_1 = \sum_{i=1}^{n} | w_i | \text{ and } \|\mathrm{w}\|_p = \left(\sum_{i=1}^{n} |w_i|^p \right)^{1/p}, \tag{4.87}$$

respectively. The ℓ_{1-2} of w is defined as

$$\|\mathrm{w}\|_{1-2} = \|\mathrm{w}\|_1 - \|\mathrm{w}\|_2, \tag{4.88}$$

which approaches the x-axis and y-axis closer as the values get smaller. The capped ℓ_1 norm of $w \in \Re^n$ is defined as $\sum_{i=1}^{n} P_\lambda\left(w_i\right)$, where $P_\lambda(w_i)$ is defined as

$$P_\lambda (w_i) = \lambda \min \{|w_i|, a\} = \begin{cases} \lambda \mid w_i \mid, \; if \mid w_i \mid < a \\ \lambda a, \quad if \mid w_i \mid \geq a \end{cases} \tag{4.89}$$

where $a > 0$. The SCAD regularization for $w \in \Re^n$ can be written as $\sum_{i=1}^{n} P_\lambda\left(w_i\right)$, where the SCAD function is defined as

$$P_\lambda (w_i) = \begin{cases} \lambda \mid w_i \mid & if \; 0 \leq \mid w_i \mid \leq \lambda \\ \frac{-|w_i|^2 + 2a\lambda|w_i| - \lambda^2}{2\left(a-1\right)} & if \; \lambda < \mid w_i \mid \leq a\lambda \\ \frac{\left(a+1\right)\lambda^2}{2} & if \mid w_i \mid > a\lambda \end{cases} \tag{4.90}$$

where $a > 0$.

The MCP regularization for $w \in \Re^n$ can be written as $\sum_{i=1}^{d} P_\lambda\left(w_i\right)$, where the MCP function is defined as,

$$P_\lambda (w_i) = \lambda \int_0^{|w_i|} \left(1 - \frac{x}{a\lambda}\right)_+ dx, \tag{4.91}$$

where $a > 0$ and $(\cdot)_+ :=$ max $\{0, \cdot\}$. Then, we have

$$P_\lambda (w_i) = \begin{cases} \lambda \mid w_i \mid - \frac{|w_i|^2}{2a}, \; if 0 \leq \mid w_i \mid \leq a\lambda \\ \frac{1}{2}\lambda^2 a, \quad if \mid w_i \mid > a\lambda \end{cases} \tag{4.92}$$

4.2.3.2 Sparse Regularization for Matrices

Due to the complexity of the matrix formulation, there are relatively fewer investigations dedicated to sparse regularization on matrices. Several commonly used ones include $\ell_{2,0}$, $\ell_{2,1}$, $\ell_{2,p}(0 < p < 1)$ and $\ell_{2,1-2}$.

The $\ell_{2,0}$ of matrix $W = \left[w_1^T, w_2^T, \cdots, w_m^T\right]^T \in \mathfrak{R}^{m\times n}$ is defined as

$$\|W\|_{2,0} = \sum_{w_i \neq 0} \|w_i\|_2^0 \tag{4.93}$$

The $\ell_{2,1}$ is defined as the ℓ_1-norm of the vector containing of the ℓ_2-norm of the matrix rows,

$$\|W\|_{2,1} = \sum_{i=1}^{m} \|w_i\|_2 = \sum_{i=1}^{m} \sqrt{\sum_{j=1}^{n} W_{ij}^2} \tag{4.94}$$

Similarly, the $\ell_{2,p}(0 < p < 1)$ is defined as,

$$W\|_{2,p} = \left(\sum_{i=1}^{m} \|w_i\|_2^p\right)^{1/p} = \left(\sum_{i=1}^{m}\left(\sum_{j=1}^{n} W_{ij}^2\right)^{p/2}\right)^{1/p} \tag{4.95}$$

The $\ell_{2,1-2}$ is defined as $\|W\|_{2,1-2} = \|W\|_{2,1} - \|W\|_F$.

4.2.3.3 The Proposed Model

Here, a novel sparse regularization on matrices first is proposed and then apply it to feature selection. For $W \in \mathfrak{R}^{m\times n}$, we define MCP^2 as follow,

$$MCP^2 = \sum_{i=1}^{m} P_\lambda\left(\|w_i\|_2\right) \tag{4.96}$$

where w_i is the i-th row of W and $P_\lambda(\cdot)$ is defined as in Eq. (4.92). According to the properties of MCP, MCP^2 is non-convex and Lipschitz continuous.

Let $X = [x_1, \cdots, x_n] \in \mathfrak{R}^{d\times n}$ be a data matrix with n samples of d features. Suppose that n samples are sampled from c classes, by one-hot encoding, the class labels can be represented as matrix $Y \in \mathfrak{R}^{n\times c}$, where $Y_{ij} = 1$, if x_i belongs to the j-th class and $Y_{ij} = 0$ otherwise. Using the new proposed MCP^2 in Eq. (4.96) as the sparsity regularization term, the following supervised feature selection model can be obtained:

$$\min_{W} \frac{1}{2}\left\|Y - X^T W\right\|_F^2 + \sum_{i=1}^{d} P_\lambda\left(\|w_i\|_2\right), \tag{4.97}$$

where the first term is the loss function which measures the model fidelity, the second term is the regularizer which is the preference of selecting features across all the classes with joint sparsity.

The joint minimization of the fidelity term and the sparsity regularization enable the transformed matrix $W \in \Re^{d\times c}$ to evaluate the correlation between labels and features, which is particularly suitable for feature selection. More specifically, the i-th row in W shrinks to zero or a number close to zero if the i-th feature is less discriminative to the labels. In the rest part of this section, we denote the objective function in (4.97) as $F(W)$ for short.

In practice, we can perform feature selection with the optimal solution W^* to model (4.97). In detail, we rank all features in descending order according to $\left\|w_i^*\right\|_2$, for $i = 1, 2, \cdots, d$, and select those features with the highest rankings.

The following theorem shows that the new proposed sparse regularization can indeed give a sparse solution.

Theorem 4.2 (Sparsity): *Let* $W^* \neq 0$ *be a solution of model (4.97). Assume* $F(W^*) \le F(W^0)$. *If* $\lambda \ge \max_i \|x_i\|_2 \sqrt{2F\left(W^0\right)}$, *then*

$$\|W^*\|_{2,0} \le \frac{F\left(W^0\right)}{P_\lambda\left(a\lambda - a\sqrt{2F\left(W^0\right)}\max_{1\le i\le d}\|x_i\|_2\right)} \tag{4.98}$$

Proof Since $W^* \neq 0$ is a solution of (4.97), for any $i = 1, 2, \cdots, d$, there exists $\hat{w}_i^* \in \partial\left\|w_i^*\right\|_2$ such that

$$x_i\left(X^T W^* - Y\right) + P_\lambda'\left(\left\|w_i^*\right\|_2\right)\hat{w}_i^* = 0, \tag{4.99}$$

where x_i is the i-th row of X, that is the i-th feature and

$$\partial\left\|w_i^*\right\|_2 = \begin{cases} \left\{\frac{w_i^*}{\|w_i^*\|_2}\right\} & if\ w_i^* \neq 0 \\ \{e : \|e\|_2 \le 1\} & if\ w_i^* = 0 \end{cases} \tag{4.100}$$

Based on the property of norm, we have

$$\begin{aligned}\left\|\mathrm{x}_i\left(\mathrm{X}^{\mathrm{T}}\mathrm{W}^*-\mathrm{Y}\right)\right\|_2^2 &\le \|\mathrm{x}_i\|_2^2\left\|\mathrm{X}^{\mathrm{T}}\mathrm{W}^*-\mathrm{Y}\right\|_F^2 \\ &\le \|\mathrm{x}_i\|_2^2\left(\left\|\mathrm{X}^{\mathrm{T}}\mathrm{W}^*-\mathrm{Y}\right\|_F^2+2\textstyle\sum_{i=1}^d P_\lambda\left(\left\|\mathrm{w}_i^*\right\|_2\right)\right) \\ &= 2\|\mathrm{x}_i\|_2^2 F\left(\mathrm{W}^*\right) \\ &\le 2\|\mathrm{x}_i\|_2^2 F\left(\mathrm{W}^0\right)\end{aligned} \tag{4.101}$$

When $w_i^* \neq 0$, combing (4.100) and (4.101), we get

$$\|\mathrm{x}_i\|_2\sqrt{2F\left(\mathrm{W}^0\right)} \ge P'_\lambda\left(\left\|\mathrm{w}_i^*\right\|_2\right) = \lambda - \frac{\left\|\mathrm{w}_i^*\right\|_2}{a}, \tag{4.102}$$

which is equivalent to

$$\left\|\mathrm{w}_i^*\right\|_2 \ge a\left(\lambda - \|\mathrm{x}_i\|_2\sqrt{2F\left(\mathrm{W}^0\right)}\right) > 0. \tag{4.103}$$

Moreover,

$$\begin{aligned}F\left(\mathrm{W}^0\right) \ge F\left(\mathrm{W}^*\right) &\ge \textstyle\sum_{i=1}^d P_\lambda\left(\left\|\mathrm{w}_i^*\right\|_2\right) \\ &= \textstyle\sum_{\mathrm{w}_i\neq 0} P_\lambda\left(\left\|\mathrm{w}_i^*\right\|_2\right) \\ &\ge \textstyle\sum_{\mathrm{w}_i^*\neq 0} P_\lambda\left(a\lambda - a\|\mathrm{x}_i\|_2\sqrt{2F\left(\mathrm{W}^0\right)}\right) \\ &\ge \textstyle\sum_{\mathrm{w}_i^*\neq 0} P_\lambda\left(a\lambda - a\sqrt{2F\left(\mathrm{W}^0\right)}\max_{1\le j\le d}\|\mathrm{x}_i\|_2\right) \\ &= \|\mathrm{W}^*\|_{2,0} P_\lambda\left(a\lambda - a\sqrt{2F\left(\mathrm{W}^0\right)}\max_{1\le j\le d}\|\mathrm{x}_i\|_2\right)\end{aligned} \tag{4.104}$$

which completes the proof.

4.2.3.4 The Optimization Algorithm

This section presents the optimization algorithm for the proposed feature selection model in (4.97). As previously stated, the new proposed sparse regularization is non-convex, which leads to the proposed model be also non-convex. Because non-convex problems are generally more challenging to be minimized, we need to design a specific algorithm to solve model (4.97) efficiently.

The ConCave-Convex Procedure (CCCP) is one of the ways to deal with a variety of non-convex problems, which has been widely and successfully applied to sparse optimization problems [61, 62]. The CCCP is an iterative approach, whose idea is that it approximates the convex part by its tangent and minimizes the resulting convex function at each iteration. We notice that the objective of 4.97 has a naturally decomposition, which can be written into the difference of two convex function

$$F(\mathrm{W}) = \frac{1}{2}\left\|\mathrm{Y} - \mathrm{X}^{\mathrm{T}}\mathrm{W}\right\|_F^2 - \left(\sum_{i=1}^{d} -P_\lambda\left(\|\mathrm{w}_i\|_2\right)\right) \tag{4.105}$$

This decomposition allows us to apply the ConCave-Convex Procedure (CCCP). In details, each iteration of our algorithm needs to solve a convex program defined by linearizing the convex part $\sum_{i=1}^{d} -P_\lambda\left(\|\mathrm{w}_i\|_2\right)$. More concretely, the linearized convex subproblem is

$$\frac{1}{2}\left\|\mathrm{Y} - \mathrm{X}^{\mathrm{T}}\mathrm{W}\right\|_F^2 - \left(\sum_{i=1}^{d} -\left\langle \mathrm{w}_i, P_\lambda'\left(\left\|\mathrm{w}_i^k\right\|_2\right)\hat{\mathrm{w}_i^k}\right\rangle\right) \tag{4.106}$$

where w_i^k is the i-th row of W^k, $\langle\cdot\rangle$ is the inner product between two vectors of the same scale, $P_{\lambda'}\left(\|\mathrm{w}_i^k\|_2\right)$ is the derivative of $P_\lambda(\cdot)$ at $\|\mathrm{w}_i^k\|_2$and given by

$$P_\lambda'\left(\left\|\mathrm{w}_i^k\right\|_2\right) = \begin{cases} \lambda - \frac{\|\mathrm{w}_i^k\|_2}{a}, & if 0 \leq \|\mathrm{w}_i^k\|_2 \leq a\lambda \\ 0, & if \|\mathrm{w}_i^k\|_2 > a\lambda \end{cases} \tag{4.107}$$

and $\hat{\mathrm{w}_i^k} \in \partial\left\|\mathrm{w}_i^k\right\|_2$, where

$$\partial\left\|\mathrm{w}_i^k\right\|_2 = \begin{cases} \left\{\frac{\mathrm{w}_i^k}{\|\mathrm{w}_i^k\|_2}\right\} & if\ \mathrm{w}_i^k \neq 0 \\ \{e : \|e\|_2 \leq 1\} & if\ \mathrm{w}_i^k = 0 \end{cases} \tag{4.108}$$

For simplicity of our algorithm, we specify $\hat{w_i^k}$ as

$$\hat{\mathrm{w}_i^k} = \begin{cases} \frac{\mathrm{w}_i^k}{\|\mathrm{w}_i^k\|_2}, & if \mathrm{w}_i^k \neq 0 \\ 0, & if \mathrm{w}_i^k = 0 \end{cases} \tag{4.109}$$

In the following part of this subsection, we focus on how to minimize the following CCCP subproblem,

$$\min_{\mathrm{W}} \frac{1}{2}\left\|\mathrm{Y} - \mathrm{X}^{\mathrm{T}}\mathrm{W}\right\|_F^2 - \sum_{i=1}^{d} -\left\langle \mathrm{w}_i, P'\left(\left\|\mathrm{w}_i^k\right\|_2\right)\hat{\mathrm{w}_i^k}\right\rangle \tag{4.110}$$

As we can see, CCCP subproblem in (4.110) is a convex quadratic function. The solution must satisfy the first order necessary condition, i.e., the derivative of objective function with respect to W must be zero. The derivative of the first part of objective function is $X(X^TW - Y)$. And we can easily derive the second part with respect to w_i as $P'\left(\|w_i^k\|_2\right)\hat{w}_i^k$. Then we can get the derivative of the whole objective function as $X(X^TW - Y) + M^k$, where $M^k \in \Re^{d\times c}$ and the entry of i-th row in M^k is $P'\left(\|w_i^k\|_2\right)\hat{w}_i^k$. By setting the derivative of objective function in (4.110) with respect to W to be zero, we have

$$X\left(X^TW - Y\right) + M^k = 0 \tag{4.111}$$

Furthermore, we can easily solve the above linear system of equations

$$W = \left(XX^T + \epsilon I_d\right)^{-1}\left(XY - M^k\right) \tag{4.112}$$

where $I_d \in \Re^{d\times d}$ is the identity matrix and $\epsilon > 0$ is a sufficiently small constant.

We summarize our algorithm in Algorithm 4.4.

Algorithm 4.5 CCCP for Solving Model (4.92)

Input: data matrix **X** and label matrix **Y**

1: Initialize $k = 0$
2: **repeat**
3: **for** $i = 1, 2, \ldots, d$ **do**
4: **if** $\mathbf{w}_i^k \neq 0$ **then**
5: $\hat{\mathbf{w}}_i^k = \frac{\mathbf{w}_i^k}{\|\mathbf{w}_i^k\|_2}$
6: **else**
7: $\hat{\mathbf{w}}_i^k = 0$
8: **end if**
9: **end for**
10: Update $\mathbf{W}^k$ by (4.107)
11: $k := k + 1$
12: **until** Convergence criteria is satisfied

Output: $\mathbf{W}^*$

The related experimental study of the above CCCP can be found in [5].

4.2.3.5 Computational Complexity

The computational complexity of the proposed method described in Algorithm 4.4 can be analyzed. Recall that d is the number of features, n is the number of instances and c is the number of classes. At each step, we need to update M^k, which requires $\mathcal{O}\left(dc\right)$ operations. However, since the computation of the inverse matrix of $XX^T + \epsilon I_d$ and the matrix multiplication of X and Y are only related to input data, we can calculate them before we go to the loop. And the cost for them are $\mathcal{O}\left(d^3 + \frac{d^2n^2}{2}\right)$ operations and $\mathcal{O}\left(dcn^2\right)$ operations, respectively. Suppose that CCCP terminates after N steps. Then, the overall cost for Algorithm 4.5 is $\mathcal{O}\left(d^3 + \frac{d^2n^2}{2} + dcn^2 + N\left(d^3c + 2dc\right)\right)$.

4.3 Distance-Based Feature Selections

4.3.1 Spatial Distance Join Based Feature Selection

4.3.1.1 Fundamental Concepts

The Correlation Fractal Dimension (CFD) that underlies various remarkable unsupervised feature selection methods [63, 64] can be considered as a special case of Spatial Distance Join (SDJ) [65]. The SDJ extends the concept of CFD and provides a general description of both feature relevance and feature redundancy in the same framework. In order to investigate feature selection problems in the SDJ framework, some fundamental concepts of SDJ are introduced below along with the establishments of SDJ based feature relevance and redundancy measures.

Spatial Distance Join

In multi-dimensional and spatial databases, Spatial Distance Join (SDJ) is an important query, which, for example, calculates 'the number of ATM machines that locates within 2 miles from restaurants. Formally, given a distance metric L, a radius r, and two datasets A and B, SDJ search for $\{(a, b)|\, a \in A \text{ and } b \in B, L(a, b) \leq r\}$. In other words, a SDJ query enumerates all pairs of points (respectively from A and B) that are within distance r. Clearly, the total number of such point pairs is a function of r, namely pair count function $PC(r)$ (for simplicity datasets A and B are not explicitly referred). It has been shown that the pair-count usually follows the power law $PC(r) = Kr^{\phi}$ [65], in which K is a constant and φ is the Pair-Count Exponent (PCE). Based on the power law, the PCE can be derived by

$$\phi = \frac{\partial ln\left(PC(r)\right)}{\partial ln\left(r\right)} \tag{4.113}$$

In the above definition CFD arises as a special case of PCE when datasets A and B are identical. For such a case ($A = B$) the SDJ is called self-spatial join and is otherwise called cross-spatial join between different datasets ($A \neq B$).

In the context of feature selection, CFD has been proven a good feature redundancy measure [63, 64, 66–68], while, as elaborated later in the present work, the PCE between two classes of a dataset is capable in quantification of feature relevance. In parallel to the term CFD, the PCE defined between two classes of a dataset is hereinafter referred to as Relevance Fractal Dimension (RFD). From the SDJ view of point, both CFD and RFD are PCE of either self-spatial join or cross-spatial join, which therefore allows investigation of both feature relevance and feature redundancy in a general framework. Moreover, PCE is shown invariant to affine transformations (i.e., translation, rotation, and uniform scaling), sampling, and the distance norm (L) used [65]. Such invariance properties are desirable by feature selection methods.

CFD and Feature Redundancy

As introduced above, CFD is the PCE of a special SDJ for two identical datasets (i.e., self-spatial join), which essentially measures self-similarity of a dataset. A dataset that presents self-similarity for a meaningful range of distances is called a fractal dataset [69], which usually represents a spatial object of dimensionality lower than its address space (in which the dataset resides) [63]. In other words, the intrinsic dimension of a fractal dataset (i.e., the dimensionality of the spatial object represented by it) could be much smaller than its embedding dimension (i.e., the number of features). As a measure of self-similarity, CFD provides an estimate for the intrinsic dimension of a dataset [70]. Moreover, CFD has an important property that has been adopted to identify redundant features. In a dataset A, a feature x of no contribution to the CFD of the dataset (i.e., $CFD(A/\{x\}) = CFD(A)$) implies the full restriction of the feature x by others. In other words, x is a redundant feature that can be reproduced by others. Using CFD as a feature redundancy measure, various feature selection methods were developed to identify the smallest subset S of A ($S \subset A$) such that $CFD(S) = CFD(A)$ [63, 64]. A particular advantage of CFD based feature selection method over other feature redundancy removal methods lies in that, it can handle complex feature correlations (not limited to a specific family of mapping functions) and can easily deal with feature group correlations [63].

RFD and Feature Relevance

In order to investigate feature relevance within the SDJ framework, a cross-spatial join composed by two classes of a dataset is considered. In the following analysis, P^+ and P^- denote the positive and negative point sets of a binary-class dataset (normalized into [0, 1]), respectively. In addition, the infinity norm L1 is used (i.e., $\|\cdot\| = \|\cdot\|_\infty$) without loss of generality. According to the power law, the pair counts for a radius $r (r \leq 1)$ is given by $PC(r) = Kr^\phi$. It can be shown that the constant $K = s^+s^-$ (by letting $r = 1$), in which s^+ and s^- are the respective sizes of P^+ and P^-. Therefore, a larger PCE (i.e., ϕ) indicates that there are less pairs of positive and negative points that are within distance r to each other. In other words, positive and negative data points are less mixed (i.e., low class impurity), which in turn suggests that the binary-class dataset is more separable. Similarly, a feature subset of larger PCE is more relevant to class label and the data points are more separable on the feature subset. For this reason, Relevance Fractal Dimension (RFD) is used in the present work to refer to the PCE between two different classes.

As shown further by the following property (Property 1), a relationship can be established for different PCEs defined on a binary-class dataset. In the property, PC^* denotes the pair counts of all data points (irrespective of class), $PC^{+/+}$ denotes pair counts of positive class, $PC^{-/-}$ denotes the pair counts of negative class, and $PC^{+/-}$ denotes the pair counts across positive and negative class, along with their corresponding PCEs denoted by ϕ_*, $\phi^{+/+}$, $\phi^{-/-}$, and $\phi^{+/-}$, respectively.

Theorem 4.3 *For a given radius* $r \leq 1, PC^*(r) = PC^{+/+}(r) + PC^{-/-}(r) + 2PC^{+/-}(r)$ *and* $\min\{\phi^{+/+}, \phi^{-/-}, \phi^{+/-}\} \leq \phi_* \leq \max\{\phi^{+/+}, \phi^{-/-}, \phi^{+/-}\}$.

Proof For a binary-class dataset, the point pairs within distance r (i.e., $PC_*(r)$) are comprised by (1) self-point pairs of positive class $L\left(p_i^+, p_i^+\right) = 0$; (2) cross-point pair of positive class $L\left(p_i^+, p_j^+\right) \leq r, i \neq j$; (3) self-point pairs of negative class $L\left(p_i^-, p_i^-\right) = 0$; (4) cross-point pairs of negative class $L\left(p_i^-, p_j^-\right) \leq r$; (5) cross-point pairs of positive class to negative class $L\left(p_i^+, p_j^-\right) \leq r$ and (6) cross-point pairs of negative class to positive class $L\left(p_i^-, p_j^+\right) \leq r$. Among the above six types of point pairs, $PC^{+/+}(r)$= pair count of type 1 + pair count of type 2, $PC^{-/-}(r)$= pair count of type 3 + pair count of type 4, and $PC^{+/-}(r)$= pair count of type 5 $PC^{-/+}(r)$= pair count of type 6. Accordingly, as the sum of all point pairs of different types, $PC_*(r)$ can be collectively expressed by $PC_*(r) = PC^{+/+}(r) + PC^{-/-}(r) + 2PC^{+/-}(r)$. Based on the power law ($PC(r) = Kr^{\phi}$), the above expression can be rewritten as $K_* r^{\phi_*} = K^{+/+} r^{\phi+/+} + K^{-/-} r^{\phi-/-} + 2K^{+/-}$. It can be further shown that $K_* = K^{+/+} + K^{-/-} + 2K^{+/-}$ by letting $r = 1$. Finally, with the denotations of $\phi_{min} = \min\{\phi^{+/+}, \phi^{-/-}, \phi^{+/-}\}$ and $\phi_{max} = \max\{\phi^{+/+}, \phi^{-/-}, \phi^{+/-}\}$, $K_* r^{\phi_*} \leq K^{+/+} r^{\phi min} + K^{-/-} r^{\phi min} + 2K^{+/-} r^{\phi min} = K_* r^{\phi min}$, can be derived from theequation about K's. In a similar way, it can be also shown that $\phi_* \leq \phi_{max}$.

In the property, it should be noted that, ϕ_* is essentially the CFD of the entire dataset, $\phi^{+/+}$ is the CFD of the positive data point set (P^+), $\phi^{-/-}$ is the CFD of the negative data point set (P^-), while $\phi^{+/-}$ is actually the RFD of the entire dataset (i.e., the PCE between the two different classes). The pair-count relationship shows that the increase of $\phi^{+/-}$ leads to decreased $\phi^{+/+}$ and/or $\phi^{-/-}$. In other words, as RFD increases the positive data points and/or the negative data points become less spread, which is consistent with that a dataset of larger RFD is more separable (i.e., of lower class impurity). It is noted that the above analysis involves only binary-class datasets. In order to address multi-class problems with RFD, one-versus-one and one-versus-the-rest styles can be adopted. For one-versus-one style, RFDs between any two classes are calculated and the averaged RFD is used to measure feature relevance, while one-versus-the-rest style quantifies feature relevance by the average of RFDs between one class and all the rest classes.

Box Occupancy Counter for PCE Calculation

The Box Occupancy Counter (BOC) is a widely used approach [63, 64, 69] for the calculation of PCE. The BOC approach is based on the derivative definition of PCE given by Eq. (4.1) with PC(r) replaced by an estimate. In order to estimate PC(r), the address space of datasets A and B is divided into a hyper-cubic grid of cell-size r. As a result, all the points in the same grid cell are within distance r (measured by L_∞ distance) to each other. Given the points from dataset A and B counted respectively by $C_{A,\,i}$ and $C_{B,\,i}$ for the i-th grid cell (i.e., the box occupancies of the i-th grid cell for datasets A and B, respectively), the number of r-distant point pairs can be expressed as $C_{A,\,i}C_{B,\,i}$. Adding up the contribution of each grid, the total number of the r-distant point pairs is given by the Box-Occupancy-Product-Sum $BOPS(r) = \sum_i C_{A,i}C_{B,i}$, which provides an estimate for $PC(r)$. With $PC(r)$ in Eq. (4.1) replaced by BOPS(r), PCE is calculated by

$$\phi \approx \frac{\partial ln\left[BOP(r)\right]}{\partial ln\left(r\right)} = \frac{\partial ln\left(\sum_i C_{A,i} \cdot C_{B,i}\right)}{\partial ln\left(r\right)} \tag{4.114}$$

In real world applications, datasets are usually normalized into a unitary hyper-cube ($\in[0, 1]$) followed by a series calculations for the BOPSs corresponding to $r = 2^{-1}, \cdots, 2^{-5}$. The PCE can then be obtained through a curve fitting step, which, as described by Eq. (4.114), is given by the slope of the linear curve determined by the points ($ln(r)$, ($ln(BOPS(r)$))).

4.3.1.2 Feature Selection in the SDJ Framework

The above elaboration shows that, both feature redundancy and relevance can be assessed together within the SDJ framework. In order to select relevant features a large RFD should be specified, while a large CFD is required for identifying non-redundant features. However, these two objectives may not always be consistent with each other and a feature subset of large RFD does not necessarily have large CFD. In the present work, following an approach that is widely used in multiple criteria mathematical programming-based data mining [43, 71], the two objectives are combined into a single one by assigning a weight factor ($\lambda \geq 0$) to CFD. In other words, a weighted sum of RFD and CFD (i.e., $RFD + \lambda CFD$) is used as a quality measure for a feature subset. Accordingly, for a given dataset A the feature selection problem is formulated in the SDJ framework as

$$\arg\max_{S \subset A} RFD(S) + \lambda CFD(S), \tag{4.115}$$

where S is a feature subset of A (of a given size) and λ reflects the preference assigned to feature redundancy.

In order to select nonredundant features, a large λ should be specified, while a small λ indicates that feature subsets of large RFD (i.e., relevant feature subsets) are preferred. When $\lambda = \infty$ the above formulation (Eq. 4.115) collapses into a CFD based unsupervised feature selection (which is equivalent to FDR without consideration of feature relevance).

To calculate the required RFD and CFD, a Divide-Count algorithm is designed providing an efficient implementation of the BOC approach for feature selection. In the algorithm the BOPSs for boxes of increasing size (from $r = 2^{-b}$ to $r = 2^{-1}$, in which b is a positive integer identifying the depth/level of the box refinement) are calculated with an integer array based indexing scheme. In such a scheme, for the bottom level grid ($r = 2^{-b}$), a point $p(= [x_1, \cdots, x_n])$ is indexed by an integer array $key_b = \left[k_1^b, \cdots, k_n^b\right] = \left[\lfloor x_1/r \rfloor, \cdots, \lfloor x_n/r \rfloor\right]$ (where $\lfloor a \rfloor$ identifies the largest previous integer of a, i.e., the largest integer less than a; for example, $\lfloor 2.1 \rfloor = \lfloor 2.9 \rfloor = 2$). With the above indexing scheme, data points in the same grid cell will have the same index, which enables a simple counting for the box occupancies of a grid cell (i.e., the number of points in the cell) by points from different classes. One can then simply use $C_{j,\, keyb}$ to identify/store box occupancy of class j for the cell indexed by key_b. Once the box occupancies of the bottom level grid (i.e., the b-th level) are obtained, those of the upper levels can be calculated without having to re-count the entire data points again. For example, the index of the $(b - 1)$-th level grid (key_{b-1}) can be readily obtained by $key_{b-1} = key_{b\backslash 2} = \left[k_1^b\backslash 2, \cdots, k_n^b\backslash 2\right]$(in which "\" identifies integer division; for example, $3\backslash 2 = 2\backslash 2 = 1$). Therefore, the box occupancies of the $(b - 1)$-th level grid (i.e., $C_{j,\, keyb-1}$) are essentially the sum of the occupancies of the same index. A detailed description of the Divide-Count algorithm is presented in Algorithm 4.6.

Algorithm 4.6 Divide-Count( ) // Calculate the Box Occupancies of Each Grid Level

---

Input: $\mathbf{A} \in \mathfrak{R}^{m \times n}$ // A dataset of m points and n features

Output: C//$C_{j,keyi}$ contains the box occupancy of class j for the grid cell indexed by key_i

1: Divide-Count algorithm:

2: **for** each $p(= [x_1, \cdots, x_n]) \in A$//Calculate the box occupancy of the bottom level($r = 2^{-b}$) **do**

3: $key_b = \left[k_1^b, \cdots, k_n^b\right] = \left[\lfloor x_1\backslash 2^{-b} \rfloor, \cdots, \lfloor x_n\backslash 2^{-b} \rfloor\right]$ //identifies the largest integer less than a

4: **if** key_b does not exist before **then**

5: $C_{j,keyb} = 0$ for each j//Create an entry for key_b and set it as 0 for each class

6: **end if**

7: $C_{j,keyb}$ + +//Here j is the class of point p

8: **end for**
9: **for** i = b, ⋯, 2//Calculate the box occupancies of upper grid levels **do**
10: **for** each key_i, i.e., for each key of the i-th grid level **do**
11: $\text{key}_{\text{i}-1} = \text{key}_\text{i}\backslash 2 = \left[\text{key}_1^\text{i}\backslash 2, \text{key}_\text{n}^\text{i}\backslash 2\right]$ //"\" identifies integer division
12: **if** $\text{key}_{\text{i}-1}$ does not exist before **then**
13: $\text{C}_{\text{j,keyi}-1} = 0$ for each j//Create an entry for $\text{key}_{\text{i}-1}$ and sot it as 0 for each class
14: **end if**
15: $\text{C}_{\text{j,keyi}-1} += \text{C}_{\text{j,keyi}}$ for each j (i.e., for each class)
16: **end for**
17: **end for**

The algorithm designed above has certain advantages over the existing Linear Box Occupancy Counter algorithm (LiBOC) for the calculation of BOPSs. The LiBOC algorithm adopts a topdown style facilitated by an E-dim tree data structure to calculate the BOPSs for boxes of decreasing size (from $r = 2^{-1}$ to $r = 2^{-b}$). Compared with the proposed integer array based indexing scheme, the E-dim tree is a complex data structure, which makes the LiBOC algorithm not amenable to implementation. Moreover, the LiBOC algorithm requires re-count of the entire data points at each grid level, while the Divide-Count algorithm involves only one complete count for the bottom level (box occupancies of an upper level are obtained by merging the box occupancies of the same updated indices instead of re-counting each data point). As a result, although both algorithms have the same time complexity order of O(mn) (m and n identify the number of data points and features in a dataset, respectively), the Divide-Count algorithm is more efficient. It is noted that the Divide-Count algorithm only keeps non-empty boxes, which therefore has a maximum space complexity (i.e., memory usage) of O(m) since there are m points in the dataset.

In addition, the Divide-Count algorithm offers good scalability to "big data" [72], which does not require storing the entire dataset in memory but maintains only a small set of indexed box occupancy obtained in only one complete data count. Particularly, for a large distributed dataset (pieces of data are stored at different local databases), its box occupancies can be obtained by assembling all the local box occupancies that are calculated by conducting the Divide-Count algorithm locally for each piece of data. The assembly of local box occupancies can be easily done by merging/adding box occupancies of the same index together (similar to the calculation of the box occupancies of $(b-1)$-th grid level from b-th level). In this way, the Divide-Count algorithm is ready for extension to parallel and distributed feature selection with the advantage of disseminating only indexed box occupancies (of significantly reduced size) instead of raw data over network.

Another appealing property of the integer array based indexing scheme lies in that, once the box occupancies of a complete feature set are calculated, the box occupancies for any feature subset can be obtained without having to re-count the entire data again. For example, if the s-th feature is removed from a set of n features, the new box occupancies of the remainder $n - 1$ features can be obtained by discarding the s-the entry from all the indexing arrays and then merging the occupancies of the same updated index $key'b = \left[k_1^b, \cdots, k_{s-1}^b, k_s^b, \cdots, k_n^b\right]$. In other words, the data points of the same index after the s-th entry removed will be projected into the same box in the lower dimensional space (without the s-th feature). The detailed process for this backward calculation of the box occupancies with a feature removed is described in Algorithm 4.7.

Algorithm 4.7 Rem-Count ( ) // Calculate the Box Occupancies After Exclusion of a Single Feature

---

Input: C // C_j, key_b contains the box occupancy of class j for the grid cell indexed by key_b, s // the index of the feature to be removed

Output: D // $D_{j,key'_b} = 0$ contains the box occupancy of class j for the cell indexed by key'_b (with the s-th feature removed)

1: Rem-Count algorithm:
2: **for** i = 1, . . . , b // For each grid level **do**
3:     **for** each key_i (i.e., for each key of the i-th grid level) **do**
4:         $key'_i = \left[k_1^i, \cdots, k_{s-1}^i, k_{s+1}^i, \cdots, k_n^i\right]$ // Delete the s-th entry from the indexing array key_i
5:         **if** if key'_i does not exist in D **then**
6:             $D_{j,key'_i} = 0$ for each j // Create an entry for key'_i and set it as 0 for each class
7:         **end if**
8:         $D_{j,key'_i} + = C_{j,keyi}$ for each j (i.e., for each class)
9:     **end for**
10: **end for**

Once the box occupancies of a feature subset are obtained, its RFD and CFD can be computed using the ln(r)-versus-ln(BOPS(r)) plot approach described in Algorithm 4.8.

Algorithm 4.8 Cal-PCEs( ) // Calculate the RFD and CFD of a Feature Subset

---

Input: C // The box occupancies of a feature subset ($C_{j,keyb}$ contains the box occupancy of class j for the grid cell indexed by keyb)

Output: RFD // The relevance fractal dimension of the feature subset; CFD // The correlation fractal dimension of the feature subset

1: Cal-PCEs algorithm:
2: **for** j = 1, · · ·, b // For each grid level **do**
3:     //Box-Occupancy-Product-Sum using one-versus-the-rest style for multiple classes
4:     $R\_BOPS_{ij} = \sum_{keyi}\left(C_{j,keyi}\right) \times \left(\sum_{k\neq j} C_{k,keyi}\right)$ for each j (i.e., for each class)

5: // Box-Occupancy-Product-Sum for all the data points irrespective of class labels

6: $C\_BOPS_i = \sum_{keyi}(\sum_k C_{k,keyi})^2$

7: **end for**

8: //Calculate RFD and CFD using ln(r)-versus-ln(BOPS(r)) plot

9: **for** each j (i.e., for each class) **do**

19: RFD_j = the slope of the linear curve fitted to the point

11: set $\{\ln(2^{-i}), \ln(R\_BOPS_{ij}) | i = 1, \cdots, b\}$

12: **end for**

13: RFD = the average of all the RFD_j's

14: CFD = the slope of the linear curve fitted to the points set $\{\ln(2^{-i}), \ln(C\_BOPS_i) | i = 1, \ldots, b\}$

It should be noted in the above algorithm that, the RFD is calculated using one-versus-the-rest style for a multi-class dataset (each time one class is considered as a dataset with all the other classes held together as a separate dataset). One can also use one-versus-one style, which however requires more calculations for the RFDs between any class pairs. Given the efficient calculations of RFD and CFD, the next step is the identification of the feature subset of the largest weighted sum RFD+ λCFD (i.e., the solution of the optimization problem described by Eq. (4.115)). In order to keep feature selection processes computationally tractable, heuristic instead of exhaustive search (e.g., forward/backward selection) are adopted by a wide range of feature selection methods (including those based on CFD), albeit suboptimal feature subsets may be selected as a result of the partial exploration of feature space. In the present work, a SDJ based backward feature selection method is designed, which takes advantage of the efficient calculation for the box occupancies (required by the calculations of RFD and CFD) of a feature subset from those of its parent set (see Algorithm 4.7). In such a SDJ based feature selection method (SDJ-FS), at each step, the feature which exclusion leads to the smallest decrease in RFD+ λCFD is removed until a desired number of features are selected. Details of the SDJ-FS method are given in Algorithm 4.9.

Algorithm 4.9 SDJ-FS ( ) // Feature Selection Using SDJ

Input: $A \in \Re^{m \times n}$ // A dataset of m points and n features, i.e., a feature set of (A_1, ..., A_n})

1: c // Number of features to be selected

2: λ // The weight factor for CFD ($\lambda \geq 0$)

Output: $S \in \Re^{m \times c}$ // The selected feature subset

3: SDJ-FS algorithm:

4: F = [1, $\cdots$, n] // Initialize the indices of selected features

5: Divide-Count (A) // Calculate the box occupancies for each class and each grid level

6: **while** n > c // n is the number of selected features (initially all the features are selected) **do**

7: // Calculate RFD and CFD for all the subsets of n − 1 features
8: [RFD_i, CFD_i] = Cal − PCEs(Rem − Count(C, i)) for i = 1, · · ·, n
9: j = the index of the largest $RFD_i + \lambda CFD_i$ (among all the i's)
10: C = Rem-Count(C, j) // Exclude the j-th feature and update the box occupancies
11: F[j] = [ ] // Remove the j-th entry from the indices of selected features
12: n = n − 1 Count down the number of selected features
13: **end while**
14: S = {$A_{F[1]}$, . . . , $A_{F[c]}$} // Return the selected features

One can finds the numerical analysis of the proposed method in [73].

4.3.2 Domain Driven Two-Phase Feature Selection Method Based on Bhattacharyya Distance and Kernel Distance Measurements

4.3.2.1 Preliminary Feature Selection Based on Bhattacharyya Distance Measurement

In gene expression profile, some genes' expression levels almost have the same distribution in both "abnormal" and "normal" samples. Those genes provide no useful information for identifying certain diseases with not significant difference of means or variance between the two classes. The first phase of our method is to filter those irrelevant genes so that a generated smaller candidate feature subsets would shrink the searching space later. The measurement "signal to noise ratio" [74] proposed by Golub evaluates the classification information of a certain gene. The larger the measurement "signal to noise ration", the more useful gene is for classification. The measurement is shown as Eq. (4.116).

$$d = \frac{\mu_1 - \mu_2}{\sigma_1 + \sigma_2} \tag{4.116}$$

where μ_1, μ_2 is the mean of the certain gene expression level in "abnormal" and "normal" samples respectively and accordingly, σ_1 and σ_2 are the variances.

Evidently, the measurement "signal to noise ratio" has its limitation. A gene with the same mean of expression level but distinct variances in "abnormal" and "normal" samples may be closely relevant to the disease. However, its zero "signal to noise ratio" fails to figure it out as informative gene. Bhattacharyya distance overcomes the shortcoming of "signal to noise ratio", by taking both mean and variance differences between two classes into consideration, shown as Eq. (4.117) with the same parameter definitions.

$$B = \frac{1}{4}\frac{(\mu_1 - \mu_2)^2}{\sigma_1^2 + \sigma_2^2} + \frac{1}{2}ln\frac{(\sigma_1^2 + \sigma_2^2)}{2\sigma_1\sigma_2} \tag{4.117}$$

In the above equation, variance difference of "abnormal" and "normal" samples also makes up a part of classification information. Therefore, Bhattacharyya distance is more reliable to select potential informative genes. And gene with larger Bhattacharyya distance is better to separate the two class samples. In our approach, all the genes whose Bhattacharyya distances surpass the specified threshold are considered as potential informative genes and then form the candidate feature set from where the "informative genes" can be further selected.

4.3.2.2 Second-Phase Feature Selection Based on Kernel Distance Measurement

As discussed in related work review part, distance measurements are the most popular and acceptable metrics of class separability. Generally, feature subset X is considered to be better than Y if sum of Euclidean distances between the samples from two classes is larger when using X for classification. However, in real classification tasks, small samples and non-linear separation are often the cases. Thus, this section extends the distance measurement to kernel space, i.e. using kernel distance to evaluate feature subset's ability of classification. The feature subset is the optimal if sum of kernel distances between the samples from two classes is the largest when using it to classify the overall samples.

In kernel-based learning algorithm, original feature space R^d turns to higher dimensional kernel space K by non-linear mapping $\psi(\cdot)$. The inner product of two points in kernel space can be represented in the form of kernel function: $k(x_i, x_j) = \langle\psi(x_i), \psi(x_j)\rangle$, where x_i and x_j are the original points.

Using a selected feature subset F for classification, the sum of distance between every two points respectively from two classes in kernel space, i.e., kernel distance measurement can be indicated by Eq. (4.118).

$$D(F) = \sum_{i=1}^{n_1}\sum_{j=1}^{n_2} \|\psi(x_i) - \psi(x_j)\| \tag{4.118}$$

Based on kernel function, kernel distance of random two points can be expressed by the following equation:

$$\|\psi(x_i) - \psi(x_j)\| = \sqrt{\|\psi(x_i) - \psi(x_j)\|^2} = \sqrt{2 - 2k(x_i, x_j)} \tag{4.119}$$

where kernel function is Gaussian radius basis function: $k(x_i, x_j) = \exp\left(-\frac{\|x_i - x_j\|^2}{2\sigma^2}\right)$, with $2\sigma^2$ as parameter to be determined.

We can easily learn from Eq. (4.118) that kernel distance measurement $D(F)$ increases with $2\sigma^2$ descending. Thus, the optimal parameter $2\sigma^2$ varies with respect to different datasets. To eliminate the influence of kernel parameter $2\sigma^2$, adaptive method is applied in our method to select the optimal parameter $2\sigma^2$. Given a certain dataset, in each round, we first randomly select a pair of two feature subsets, denoted as F_a, F_b. The parameter $2\sigma^2$ that makes biggest disparity of kernel distance measurements of the two distinct subsets F_a and F_b is selected as the optimal solution by searching a given parameter space by a specified step length. After several rounds, the final kernel parameter $2\sigma^2$ is determined as the mean value of the previous series of parameters. And then, the kernel distance measurement with the definitive parameter $2\sigma^2$ can be used for selecting informative genes.

Floating sequential search method is applied in our approach to further select informative genes using kernel distance measurement to evaluate feature subset's ability for classification. Suppose the candidate informative gene set after preliminary feature selection is $S.F_{i\_max}$ is the feature subset with maximum kernel distance among all the feature subsets containing i genes. User can specify parameter n to select the optimal feature subset containing n genes.

The algorithm is described as follows:

Step 1: Initialize $F_{2\_max} = \{g_1, g_2\}$, where g_1 and g_2 are the genes with the largest kernel distance measurement

Step 2: If $i = n$, then break; else $S = S - F_{2\_max}$, search for $g \in S$ to maximize the kernel distance of F_{i+1_max}, satisfying $F_{i+1\_max} = \{F_{i\_max}, g\}$

Step 3: Search all the feature subsets of F_{i+1_max} containing $i$genes, and find out the one with the maximum kernel distance F'_{i_max}.

Step 4: If kernel distance measurement $D\left(F_{i\_max}\right) \geq F'_{i\_max}$, then $i++$, go to **Step 2**.

Step 5: If $D\left(F_{i\_max}\right) < F'_{i\_max}$, then $F_{i\_max} = F'_{i\_max}$. If $i = 2$, go to (2); else $i--$, go to **Step 3**.

Based on the floating sequential search method, optima feature subsets of different dimensions can be selected when different n is specified. Afterwards, the global optimal feature subset is then picked with the best classification performance verified by SVMs. For every candidate feature subset, tenfold validation on dataset for classification performance evaluation is applied to avoid overfitting, and best C is determined when the average classification accuracy reaches maximum. Therefore, the ideal feature subset is then decided after verification of SVMs.

The related data analysis of the above algorithm can be found on [6].

References

1. Peng, Y., Kou, G., Ergu, D., Wu, W., Shi, Y.: An integrated feature selection and classification scheme. Stud. Inf. Contr. **21**(3), 241–248 (2012)
2. Xie, J., Lei, J., Xie, W., Shi, Y., Liu, X.: Two-stage hybrid feature selection algorithms for diagnosing erythemato-squamous diseases. Health Inf. Sci. Syst. **1**(1), 1–14 (2013)
3. Zhao, X., Deng, W., Shi, Y.: Feature selection with attributes clustering by maximal information coefficient. Proc. Comput. Sci. **17**, 70–79 (2013)
4. Shi, Y., Miao, J., Niu, L.: Feature selection with MCP2 regularization. Neural Comput. Applic. **31**(10), 6699–6709 (2019)
5. Shi, Y., Miao, J., Wang, Z., Zhang, P., Niu, L.: Feature selection with $\ell 2,1-2$ regularization. IEEE Trans. Neural Netw. Learn. Syst. **29**(10), 4967–4982 (2018). https://doi.org/10.1109/TNNLS.2017.2785403
6. Chen, Y., Zhang, L., Li, J., Shi, Y.: Domain driven two-phase feature selection method based on Bhattacharyya distance and Kernel distance measurements. In: 2011 IEEE/WIC/ACM International Conferences on Web Intelligence and Intelligent Agent Technology, vol. 3, pp. 217–220. IEEE, New York (2011)
7. Myrtveit, I., Stensrud, E., Shepperd, M.: Reliability and validity in comparative studies of software prediction models. IEEE Trans. Softw. Eng. **31**(5), 380–391 (2005)
8. Witten, I.H., Frank, E.: Data mining: practical machine learning tools and techniques with java implementations. ACM SIGMOD Rec. **31**(1), 76–77 (2002)
9. Brans, J.P., De Smet, Y.: Promethee methods. In: Multiple Criteria Decision Analysis, pp. 187–219. Springer, New York (2016)
10. Opricovic, S.: Multicriteria Optimization of Civil Engineering Systems. Faculty of Civil Engineering, Belgrade 2(1), 5–21 (1998)
11. Opricovic, S., Tzeng, G.H.: Compromise solution by mcdm methods: a comparative analysis of vikor and topsis. Eur. J. Oper. Res. **156**(2), 445–455 (2004)
12. Charnes, A., Cooper, W.W., Rhodes, E.: Measuring the efficiency of decision making units. Eur. J. Oper. Res. **2**(6), 429–444 (1978)
13. Cooper, W.W., Seiford, L.M., Zhu, J.: Data envelopment analysis: history, models, and interpretations. In: Handbook on Data Envelopment Analysis, pp. 1–39. Springer, New York (2004)
14. Nakhaeizadeh, G., Schnabl, A.: Development of multi-criteria metrics for evaluation of data mining algorithms. In: KDD, pp. 37–42 (1997)
15. Roy, B.: Classement et choix en présence de points de vue multiples. RAIRO Oper. Res. Recherche Opérationnelle. **2**(V1), 57–75 (1968)
16. Kou, G., Lou, C.: Multiple factor hierarchical clustering algorithm for large scale web page and search engine clickstream data. Ann. Oper. Res. **197**(1), 123–134 (2012)
17. Kou, G., Lu, Y., Peng, Y., Shi, Y.: Evaluation of classification algorithms using mcdm and rank correlation. Int. J. Inf. Technol. Decis. Mak. **11**(01), 197–225 (2012)
18. Peng, Y., Kou, G., Wang, G., Wu, W., Shi, Y.: Ensemble of software defect predictors: an ahp-based evaluation method. Int. J. Inf. Technol. Decis. Mak. **10**(01), 187–206 (2011)
19. Fu, K.S., Min, P.J., Li, T.J.: Feature selection in pattern recognition. IEEE Trans. Syst. Sci. Cybernet. **6**(1), 33–39 (1970)
20. Guyon, I., Elisseeff, A.: An introduction to variable and feature selection. J. Mach. Learn. Res. **3**(Mar), 1157–1182 (2003)
21. Hua, J., Tembe, W.D., Dougherty, E.R.: Performance of feature-selection methods in the classification of high-dimension data. Pattern Recogn. **42**(3), 409–424 (2009)
22. Blum, A.L., Langley, P.: Selection of relevant features and examples in machine learning. Artif. Intell. **97**(1–2), 245–271 (1997)
23. Kohavi, R., John, G.H.: Wrappers for feature subset selection. Artif. Intell. **97**(1–2), 273–324 (1997)
24. Xie, J., Xie, W., Wang, C., Gao, X.: A novel hybrid feature selection method based on ifsffs and svm for the diagnosis of erythemato-squamous diseases. In: Proceedings of the First Workshop on Applications of Pattern Analysis, pp. 142–151. PMLR (2010)

25. Whitney, A.W.: A direct method of nonparametric measurement selection. IEEE Trans. Comput. **100**(9), 1100–1103 (1971)
26. Marill, T., Green, D.: On the effectiveness of receptors in recognition systems. IEEE Trans. Inf. Theory. **9**(1), 11–17 (1963)
27. Pudil, P., Novovičová, J., Kittler, J.: Floating search methods in feature selection. Pattern Recogn. Lett. **15**(11), 1119–1125 (1994)
28. Pereira, F., Tishby, N., Lee, L.: Distributional clustering of English words. In: Proceedings of the 31st Annual Meeting on Association for Computational Linguistics, pp. 183–190 (1993)
29. Tishby, N., Pereira, F.C., Bialek, W.: The information bottleneck method. arXiv preprint physics/0004057 (2000)
30. Baker, L.D., McCallum, A.K.: Distributional clustering of words for text classification. In: Proceedings of the 21st Annual International ACM SIGIR Conference on Research and Development in Information Retrieval, pp. 96–103 (1998)
31. Dhillon, I.S., Mallela, S., Kumar, R.: A divisive information theoretic feature clustering algorithm for text classification. J. Mach. Learn. Res. **3**, 1265–1287 (2003)
32. Au, W.H., Chan, K.C., Wong, A.K., Wang, Y.: Attribute clustering for grouping, selection, and classification of gene expression data. IEEE/ACM Trans. Comput. Biol. Bioinform. **2**(2), 83–101 (2005)
33. Jiang, J.Y., Liou, R.J., Lee, S.J.: A fuzzy self-constructing feature clustering algorithm for text classification. IEEE Trans. Knowl. Data Eng. **23**(3), 335–349 (2010)
34. Amaldi, E., Kann, V.: On the approximability of minimizing nonzero variables or unsatisfied relations in linear systems. Theor. Comput. Sci. **209**(1–2), 237–260 (1998)
35. Combarro, E.F., Montanes, E., Diaz, I., Ranilla, J., Mones, R.: Introducing a family of linear measures for feature selection in text categorization. IEEE Trans. Knowl. Data Eng. **17**(9), 1223–1232 (2005)
36. Reshef, D.N., Reshef, Y.A., Finucane, H.K., Grossman, S.R., McVean, G., Turnbaugh, P.J., Lander, E.S., Mitzenmacher, M., Sabeti, P.C.: Detecting novel associations in large data sets. Science. **334**(6062), 1518–1524 (2011)
37. Frey, B.J., Dueck, D.: Clustering by passing messages between data points. Science. **315**(5814), 972–976 (2007)
38. Cai, X., Nie, F., Huang, H., Ding, C.: Multi-class l2, 1-norm support vector machine. In: 2011 IEEE 11th International Conference on Data Mining, pp. 91–100. IEEE, New York (2011)
39. Ma, Z., Nie, F., Yang, Y., Uijlings, J.R., Sebe, N.: Web image annotation via subspace-sparsity collaborated feature selection. IEEE Trans. Multimedia. **14**(4), 1021–1030 (2012)
40. Nie, F., Huang, H., Cai, X., Ding, C.: Efficient and robust feature selection via joint 2, 1-norms minimization. Adv. Neural Inf. Proces. Syst. **23**, 1813–1821 (2010)
41. Xiang, S., Nie, F., Meng, G., Pan, C., Zhang, C.: Discriminative least squares regression for multiclass classification and feature selection. IEEE Trans. Neural Netw. Learn. Syst. **23**(11), 1738–1754 (2012)
42. Zhang, M., Ding, C., Zhang, Y., Nie, F.: Feature selection at the discrete limit. In: Proceedings of the AAAI Conference on Artificial Intelligence, vol. 28, (2014)
43. Sriperumbudur, B.K., Lanckriet, G.R.: On the convergence of the concave-convex procedure. In: Nips, vol. 9, pp. 1759–1767. Citeseer (2009)
44. Yen, I.E., Peng, N., Wang, P.W., Lin, S.D.: On convergence rate of concave-convex procedure. In: Proceedings of the NIPS 2012 Optimization Work-shop, pp. 31–35 (2012)
45. Yuille, A.L., Rangarajan, A., Yuille, A.: The concave-convex procedure (cccp). Adv. Neural Inf. Proces. Syst. **2**, 1033–1040 (2002)
46. Yin, P., Lou, Y., He, Q., Xin, J.: Minimization of 1-2 for compressed sensing. SIAM J. Sci. Comput. **37**(1), A536–A563 (2015)
47. Esser, E., Lou, Y., Xin, J.: A method for finding structured sparse solutions to nonnegative least squares problems with applications. SIAM J. Imag. Sci. **6**(4), 2010–2046 (2013)
48. Lou, Y., Osher, S., Xin, J.: Computational aspects of constrained l 1-l 2 minimization for compressive sensing. In: Modelling, Computation and Optimization in Information Systems and Management Sciences, pp. 169–180. Springer, New York (2015)

49. Lou, Y., Yin, P., He, Q., Xin, J.: Computing sparse representation in a highly coherent dictionary based on difference of l 1 and l 2. J. Sci. Comput. **64**(1), 178–196 (2015)
50. Zangwill, W.I.: Nonlinear Programming: A Unified Approach, vol. 52. Prentice-Hall, Englewood Cliffs, NJ (1969)
51. Bertsekas, D.P.: Convex Optimization Theory. Athena Scientific, Belmont (2009)
52. Rockafellar, R.T., Wets, R.J.B.: Variational Analysis, vol. 317. Springer Science & Business Media, New York (2009)
53. Meyer, R.R.: Sufficient conditions for the convergence of monotonic mathematical programming algorithms. J. Comput. Syst. Sci. **12**(1), 108–121 (1976)
54. Boyd, S., Parikh, N., Chu, E.: Distributed Optimization and Statistical Learning via the Alternating Direction Method of Multipliers. Now Publishers Inc, Delft, Netherlands (2011)
55. Li, Z., Liu, J., Yang, Y., Zhou, X., Lu, H.: Clustering-guided sparse structural learning for unsupervised feature selection. IEEE Trans. Knowl. Data Eng. **26**(9), 2138–2150 (2013)
56. Yang, Y., Shen, H.T., Ma, Z., Huang, Z., Zhou, X.: 2, 1-norm regularized discriminative feature selection for unsupervised learning. In: IJCAI International Joint Conference on Artificial Intelligence (2011)
57. Huang, J., Nie, F., Huang, H., Ding, C.: Robust manifold nonnegative matrix factorization. ACM Trans. Knowl. Discov. Data. **8**(3), 1–21 (2014)
58. Jiang, W., Nie, F., Huang, H.: Robust dictionary learning with capped l1-norm. In: Twenty-Fourth International Joint Conference on Artificial Intelligence (2015)
59. Fan, J., Li, R.: Variable selection via nonconcave penalized likelihood and its oracle properties. J. Am. Stat. Assoc. **96**(456), 1348–1360 (2001)
60. Zhang, C.H., et al.: Nearly unbiased variable selection under minimax concave penalty. Ann. Stat. **38**(2), 894–942 (2010)
61. Collobert, R., Sinz, F., Weston, J., Bottou, L., Joachims, T.: Large scale transductive svms. J. Mach. Learn. Res. **7**(8) (2006)
62. Zhen, Y., Yeung, D.Y.: Co-regularized hashing for multimodal data. Adv. Neural Inf. Proces. Syst. **2**, 1376 (2012)
63. de Sousa, E.P., Traina, C., Traina, A.J., Wu, L., Faloutsos, C.: A fast and effective method to find correlations among attributes in databases. Data Min. Knowl. Disc. **14**(3), 367–407 (2007)
64. Traina Jr., C., Traina, A., Wu, L., Faloutsos, C.: Fast feature selection using fractal dimension. J. Inf. Data Manag. **1**(1), 3–3 (2010)
65. Faloutsos, C., Seeger, B., Traina, A., Traina Jr., C.: Spatial join selectivity using power laws. In: Proceedings of the 2000 ACM SIGMOD International Conference on Management of Data, pp. 177–188 (2000)
66. Lee, H.D., Monard, M.C., Wu, F.C.: A fractal dimension based filter algorithm to select features for supervised learning. In: Advances in Artificial Intelligence-IBERAMIA-SBIA 2006, pp. 278–288. Springer, New York (2006)
67. Ni, L.P., Ni, Z.W., Gao, Y.Z.: Stock trend prediction based on fractal feature selection and support vector machine. Expert Syst. Appl. **38**(5), 5569–5576 (2011)
68. Pham, D., Packianather, M., Garcia, M., Castellani, M.: Novel feature selection method using mutual information and fractal dimension. In: 2009 35th Annual Conference of IEEE Industrial Electronics, pp. 3393–3398. IEEE, New York (2009)
69. Schroeder, M.: Fractals, Chaos, Power Laws: Minutes from an Infinite Paradise. Courier Corporation, Chelmsford, MA (2009)
70. Belussi, A., Faloutsos, C.: Estimating the selectivity of spatial queries using the 'correlation' fractal dimension. Tech. rep. (1998)
71. Bishop, C.M.: Pattern Recognition and Machine Learning. Springer, New York (2006)
72. Madden, S.: From databases to big data. IEEE Internet Comput. **16**(3), 4–6 (2012)
73. Liu, R., Shi, Y.: Spatial distance join based feature selection. Eng. Appl. Artif. Intell. **26**(10), 2597–2607 (2013)
74. Golub, T.R., Slonim, D.K., Tamayo, P., Huard, C., Gaasenbeek, M., Mesirov, J.P., Coller, H., Loh, M.L., Downing, J.R., Caligiuri, M.A., et al.: Molecular classification of cancer: class discovery and class prediction by gene expression monitoring. Science. **286**(5439), 531–537 (1999)

Chapter 5
Data Stream Analysis

Data stream is a typical big data. Data stream can be founded in many real-life applications, such as wireless sensor networks, power consumption, information security and financial market. Data stream classification has drawn increasing attention from the data mining community in recent years. Data stream classification in such real-world applications is typically subject to three major challenges: concept drifting, large volumes, and partial labeling. As a result, training examples in data streams can be very diverse and it is very hard to learn accurate models with efficiency. This chapter provides two related research findings in the field. Section 5.1 describes a novel framework for application-driven classification of data streams [1]. The section first reviews the concepts of data stream, then categorizes diverse training examples into four types and assign learning priorities to them. Following the discussion, it derives four learning cases based on the proportion and priority of the different types of training examples. Finally, the respective support vector machine models are presented. Section 5.2 studies the problem of learning from concept drifting data streams with noise, where samples in a data stream may be mislabeled or contain erroneous values [2]. It has three subsections. The first one is about noisy description for data stream, the second one is the ensemble frameworks for mining data stream and the third one is the theoretical studies of the Aggregate Ensemble.

5.1 Application-Driven Classification of Data Streams

5.1.1 Data Streams in Big Data

Recent advances in computing technology and networking architectures have enabled generation and collection of the unprecedented amount of *data streams* of various kinds, such as network traffic data, wireless sensor readings, Web page

Y. Shi, *Advances in Big Data Analytics*,
https://doi.org/10.1007/978-981-16-3607-3_5

visits, online financial transactions and phone call record [3]. Consequently, data stream mining has emerged to be one of the most important research frontiers in data mining. Common stream mining tasks include classification [4, 5], clustering [6] and frequent pattern mining [7]. Among them, data stream classification has drawn particular attention due to its vast real-world applications.

Example 5.1 In wireless sensor networks, data stream classification has been used to monitor environment changes. For example, in the sensor data collected by the Intel Berkeley Research Lab [8], each sensor reading contains information (temperature, humidity, light and sensor voltage) collected from 54 sensors deployed in the lab. The whole stream contains consecutive information recorded over a 2-month period (1 reading per 1–3 min). By using the sensor ID as class label, the learning task is to correctly identify the sensor ID (1 out of 54 sensors) purely based on the sensor data and the corresponding recording time.

Example 5.2 In power consumption analysis, data stream classification has been used to measure power consumptions. For example, the power supply stream collected by an Italian electricity company [8] contains hourly power supply of the company recording the power from two sources: power supplied from main grid and power transformed from other grids. The stream contains 3-year power supply records from 1995 to 1998, and the learning task is to predict which hour (1 out of 24 h) the current power supply belongs to.

Example 5.3 In information security, data stream classification has been widely used to monitor Web traffic streams. For example, the KDDCUP'99 intrusion detection dataset [9] was provided by the MIT Lincoln Labs collecting 9 weeks of raw TCP dump data for a local area network. The learning task is to build a predictive model capable of distinguishing between normal connections and intrusive connections such as DOS (denial-of-service), R2L (unauthorized access from a remote machine), U2R (unauthorized access to local super user privileges), and Probing (surveillance and other probing) attacks.

In these applications, the essential goal is *to efficiently build classification models from data streams for accurate prediction*. Comparing to traditional stationary data, building prediction models from stream data faces three additional challenges:

- Concept drifting. In data streams, hidden patterns continuously change with time [29]. For example, in the wireless sensor stream, lighting during working hours is generally stronger than off-hours. Figure 5.1 illustrates the concept drifting problem, where the classification boundary (concept) continuously drifts from b_1 to b_2, and finally to b_3 down the streams.
- Large volumes. Stream data come rapidly and continuously in large volumes. For example, the wireless sensor stream contains 2,219,803 examples recorded over a 2-month period (1 reading per 1–3 min). It is impossible to maintain all historical stream records for in-depth analysis.
- Partial labeling. Due to large volumes of stream data, it is infeasible to label all stream examples for building classification models. Thus, data streams are

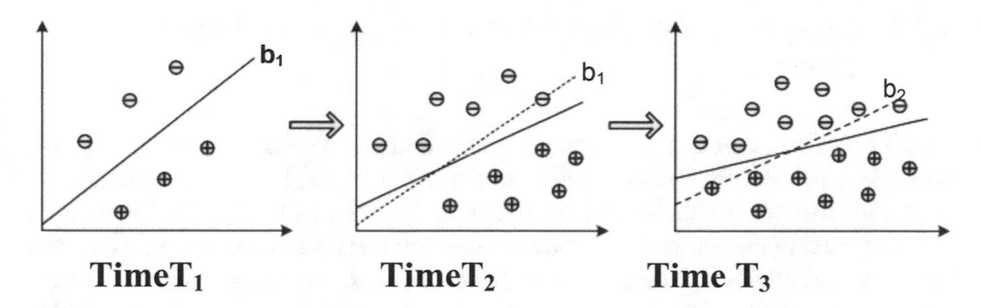

Fig. 5.1 An illustration of concept drifting in data streams. In the three consecutive time stamps T_1, T_2 and T_3, the classification boundary gradually drifts from b_1 to b_2 and finally to b_3

typically partially labeled and training data contain both labeled and unlabeled examples.

As a result, training examples in data streams are very diverse. To see why, let us assume data streams are buffered chunk by chunk. Examples in the most recent *up-to-date chunk* are training data, and examples in the *yet-to-come chunk* are testing data [8]. Due to concept drifting, training examples in the up-to-date chunk often exhibit two distributions: target domain and similar domain, where the former represents the distribution of the testing data, and the latter represents a distribution similar to the target domain [10]. Then, training examples can be categorized into four types: labeled and from the target domain (Type I), labeled and from a similar domain (Type II), unlabeled and from the target domain (Type III) and unlabeled and from a similar domain (Type IV).

In order to build accurate prediction models from such diverse training examples with efficiency, it is necessary to closely examine the characteristics, in particular, the proportion and learning priority, of the different types of examples in the training chunk.

- Proportion. The proportion of training examples from different types is determined by the concept drifting probability and labeling percentage (percentage of labeled examples). For example, when concept drifting is low and labeling percentage is high (low), the raining chunk will have a large portion of Type I (III) examples. When concept drifting is high and labeling percentage is high (low), the training chunk will have a large portion of Type II (IV) examples.
- Learning priority. Generally, examples from the target domains (Types I and III) are capable of capturing the genuine concept of the testing data and have a higher priority than examples from similar domains (Type II and IV). Besides, since Type I examples are labeled, they have a higher priority than Type III examples. Similarly, Type II examples have a higher priority than Type IV examples.

5.1.2 *Categorization of Training Examples and Learning Cases*

Consider a data stream S consisting of an infinite sequence of examples $\{x_i, y_i\}$, where $x_i \in \mathbb{R}^d$, d is the dimensionality and $y_i \in \{-1, +1\}$ indicates the class label of x_i. Note that y_i may not be always observed. Assume that the stream S arrives at a speed of n examples per second. The decision boundary (concept) underneath drifts with a probability of c, where $0 \leq c \leq 1$. Besides, assume that at each time stamp, a training chunk $D = \{x_1, \cdots, x_n\}$ is buffered and labeled by experts with a labeling rate of 1 per chunk where $0 < l < 1$.

5.1.2.1 Categorization of Training Examples

As discussed previously, due to concept drifting, not all examples in the up-to-date chunk share the same distribution with the testing data in the yet-to-come chunk. In other words, examples in the up-to-date chunk could be generated from some similar domain instead of the target domain. Besides, since it is impractical to label all examples in the up-to-date training chunk, the training chunk will contain both labeled and unlabeled examples. By combining these two factors, we categorize training examples in data streams into four types.

Definition 5.1 (Four types of training examples): In an up-to-date training chunk, there are four types of examples: labeled and from the target domain (Type I), labeled and from a similar domain (Type II), unlabeled and from the target domain (Type III) and unlabeled and from a similar domain (Type IV).

Figure 5.2 illustrates the four types of training examples, where blue solid circles denote the Type I examples, red solid circles denote the Type II examples, blue hollow circles denote the Type III examples, and red hollow circles denote the Type IV examples. Due to the temporal correlation of concepts [11], Type I and Type III examples are usually located at the tail of a training chunk and close to the yet-to-come chunk. Type II and Type IV examples are usually located at the head of a training chunk and relatively far away from the yet-to-come chunk.

Estimation of number of examples By estimating the number of examples of each type, we can gain insights into the training chunk and apply an appropriate learning model. Intuitively, the percentage of labeled examples depends on how fast labeling can be done by the experts, and the number of target domain examples depends on the concept drifting probability. By considering the two factors, the number of examples of each type can be estimated as follows.

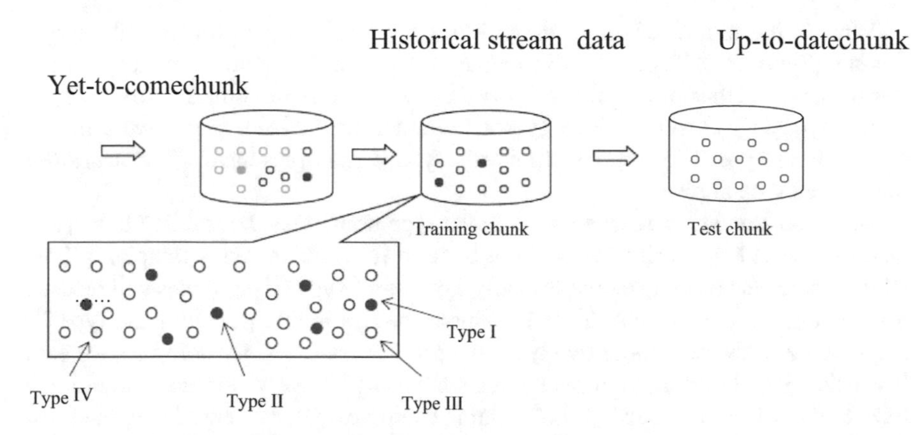

Fig. 5.2 An illustration of the four types of training examples in an up-to-date training chunk

Theorem 5.1 *Let L_1, L_2, L_3 and L_4 be the number of examples of Type I, Type II, Type III and Type IV respectively in the up-to-date chunk. Then,*

$$\begin{aligned} L_1 &\propto \gamma \cdot c^{-1} \cdot l \cdot n \\ L_2 &\propto \left(1 - \gamma \cdot c^{-1}\right) \cdot l \cdot n \\ L_3 &\propto \gamma \cdot c^{-1} \cdot (1 - l) \cdot n \\ L_4 &\propto \left(1 - \gamma \cdot c^{-1}\right) \cdot (1 - l) \cdot n \end{aligned} \tag{5.1}$$

where $\gamma > 0$ is a constant coefficient.

Proof Recall that stream S flows at a speed of n examples per second, the concept drifting probability is c, and the labeling rate is l. The number of target domain examples is inversely proportional to the concept drifting rate c with a coefficient of γ, so it can be easily estimated that $\gamma \cdot c^{-1} \cdot n$ examples in the up-to-date chunk have the same distribution as the testing data. The remaining $(1 - \gamma \cdot c^{-1}) \cdot n$ examples have a similar distribution to the testing examples. From the estimates the theorem follows immediately.

Learning priority Not all the four types of training examples have to be used in model construction. For example, consider a data stream where concept drifting is low and labeling rate is high, the training chunk will have a large portion of Type I examples. In this case, we are able to build a satisfactory model by training only on the Type I examples. We observe that the four types of training examples have the following learning priorities.

Remark 5.1 The learning priority of the four types of training examples are:

$$Type\ I > Type\ III > Type\ II > Type\ IV \tag{5.2}$$

What is the intuition behind Remark 5.1. Generally, examples from the target domain (Type I and Type III) are capable of capturing the genuine concept of the testing data, and thus have a high priority than examples from similar domains (Type II and Type IV). Besides, since Type I examples are labeled, they have a higher priority than Type III examples. Similarly, Type II examples have a higher priority than Type IV examples.

Based on Remark 5.1, when a particular type dominates the training examples, examples with lower priorities will not be used for training. For example, if Type III dominates the training examples, only Type I and Type III examples will be used for training. This is because Type I examples have a higher priority than Type III examples, and the remaining two types have lower priorities. By doing so, we gain in efficiency by building a simple model, comparing to a very complex model if we have to learn from all four types of training examples. On the other hand, the most informative examples are utilized in model construction and the learning accuracy is not sacrificed.

Learning cases Aiming at both accuracy and efficiency in learning prediction models, we categorize learning from data streams into the following four cases.

- Case 1: Type I dominates. When labeling rate is high and concept drifting probability is low, Type I dominates the training examples. In this case, we can train a satisfactory model by using only Type I examples.
- Case 2: Type III dominates. When both labeling rate and concept drifting probability are low, Type III dominates the training examples. According to the learning priority, it is necessary to combine both Type I and Type III examples for training.
- Case 3: Type II dominates. When both labeling rate and concept drifting probability are high, Type II dominates the training examples, and we will use Type I, Type II and Type III examples for training.
- Case 4: Type IV dominates. When labeling rate is low and concept drifting probability is high, Type IV dominates the training examples. This is the most difficult case because most examples are unlabeled and not from the target domain. According to the learning priority, we need to use all the four types of training examples for training.

These learning cases are further illustrated in Fig. 5.3.

5.1.3 Learning Models of Data Stream

We have introduced the four learning cases. In this section, we present their corresponding learning models.

Throughout the section, $T_1 = (x_1, y_1), \ldots, \left(x_{L_1}, y_{L_1}\right)$ denotes the set of Type I examples. $T_2 = \left\{ \left(x_{L_1+1}, y_{L_1+1}\right), \ldots, \left(x_L, y_L\right)\right\}$ denotes the set of Type II exam-

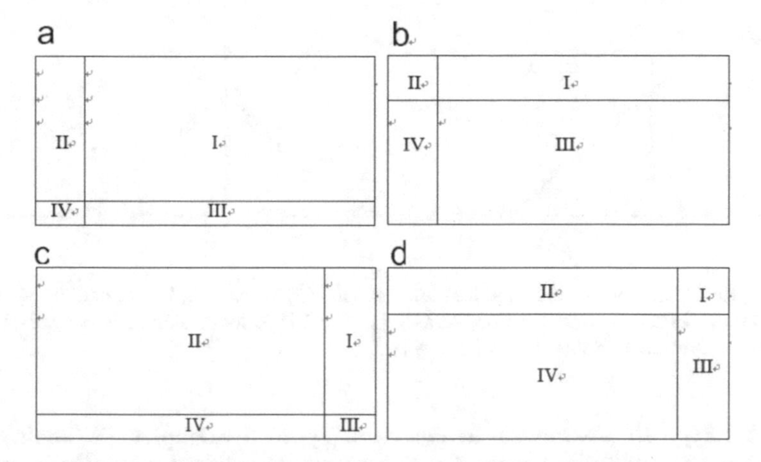

Fig. 5.3 The proportion of the four types of training examples with respect to different labeling rate l and concept drifting probability c. (**a**) l is high and c is low. Case 1, (**b**) both l and c are low. Case 2, (**c**) Both l and l are high. Case 3 and (**d**) l is low and c is high. Case 4

ples, where $L = L_1 + L_2$. $T_3 = \{x_{L+1}, \ldots, x_{L+U}\}$ denotes the set of Type III examples, where U is the set of unlabeled examples. $T_4 = \{x_{L+U+1}, \ldots, x_{L+U+N}\}$ denotes the set of Type IV examples, where N is the set of unlabeled examples.

Case 5.1 Type I Dominates In this case, Type I examples T_1 dominate the training chunk and has the highest learning priority. Thus, only T_1 will be used for training. Formally, to learn from $T_1 = \left\{ (x_1, y_1), \ldots, \left(x_{L_1}, y_{L_1}\right)\right\}$, a generic SVM model can be trained by maximizing the margin distance between classes while minimizing the error rates as,

$$\begin{array}{l} \min \frac{1}{2}\|w\|^2 + C\sum_{i=1}^{L_1} \xi_i \\ s.t. y_i\,(wx_i + b) \geq 1 - \xi_i \\ \xi_i \geq 0, \quad 1 \leq i \leq L_1 \end{array} \tag{5.3}$$

where w is the projection direction, b is the classification boundary, ξ_i is the error distance from x_i to b, and parameter C is the penalty for the examples inside the margin.

The SVM model given in Eq. (5.3) is a constrained convex optimization problem. To simplify the expression, the Hinge loss function [12] in Fig. 5.4 can be used to transform Eq. (5.3) into an unconstrained convex optimization problem as,

$$\min_{\theta} \frac{1}{2}\|w\|^2 + C\sum_{i=1}^{L_1} H\left(y_i f_\theta\,(x_i)\right) \tag{5.4}$$

where $\theta = (w, b)$ and $f_\theta(x) = (wx + b)$.

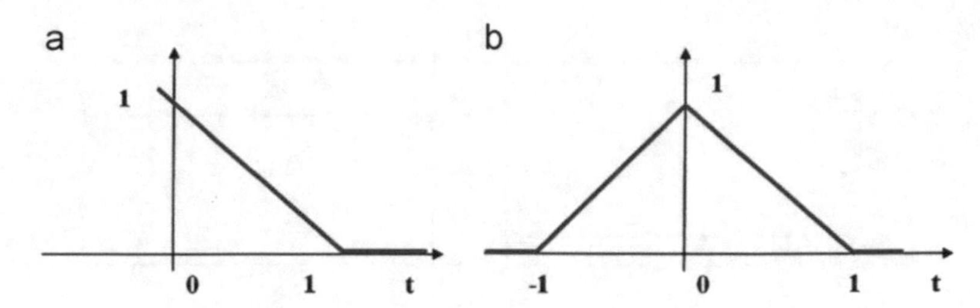

Fig. 5.4 An illustration of the Hinge loss function (**a**) $H(t) = \max(0, 1 - t)$, and the Symmetric Hinge loss function (**b**) $H(t) = \max(0, 1-|t|)$. The Hinge loss function is equivalent to the following optimization problem: $\min \xi \ s.\ t.\ \xi \geq 0, \xi \geq 1 - t$

Case 5.2 Type III Dominates In this case, Type III examples T_3 dominate the training chunk and Type I examples T_1 have a higher learning priority than Type III examples. Thus, both T_1 and T_3 will be used for training.

Learning from T_1 and T_3 is a semi-supervised learning problem [13]. Generally speaking, adding unlabeled T_3 examples into learning will further improve the performance for the following reasons: (1) Labeled examples in T are too few to build a satisfactory model. (2) T_3 contains a relatively large number of examples that come from the target domain, which can greatly help in differentiating the genuine classification boundaries.

Formally, in order to learn from both T_1 and T_3, semi-supervised SVM (TS^{3}VM) [14] can be used as the learning model. The logic behind TS^{3}VM is to find a classification boundary that achieves a maximum margin not only between labeled examples, but also unlabeled examples. That is, adding an extra term $C^* \sum_{i=L+1}^{L+U} H\left(| f_\theta (x_i) | \right)$ to penalize the misclassification of unlabeled examples located inside the margin as,

$$\min_{\theta} \frac{1}{2}\|w\|^2 + C \sum_{i=1}^{L_1} H\left(y_i f_\theta\left(x_i\right)\right) + C^* \sum_{i=L+1}^{L+U} H\left(|f_\theta\left(x_i\right)|\right) \tag{5.5}$$

Balance constraint A possible limitation of the TS^{3}VM model is that all unlabeled examples in T_3 may be classified into one class with a very large margin, leading to deteriorated performance. To address this issue, an additional balance constraint should be added to ensure that unlabeled examples in T_3 be assigned into both classes. In the case that we do not have any prior knowledge about the class ratio in T_3, a reasonable approach [12] is to estimate its class ratio from T_1 and T_2 as,

$$\frac{1}{U} \sum_{i=L+1}^{L+U} f_\theta\left(x_i\right) = \frac{1}{L_1} \sum_{i=1}^{L_1} y_i \tag{5.6}$$

where L denotes the number of labeled examples and U denotes the number of unlabeled examples.

By taking account of the balance constraint, we can derive a modified semi-supervised SVM model as,

$$\begin{aligned} &\min_{\theta} \tfrac{1}{2}\|w\|^2 + C\sum_{i=1}^{L_1} H\left(y_i f_\theta\left(x_i\right)\right) + C^* \sum_{i=L+1}^{L+U} H\left(\left|f_\theta\left(x_i\right)\right|\right) \\ &\qquad s.t. \tfrac{1}{U}\sum_{i=L+1}^{L+U} f_\theta\left(x_i\right) = \tfrac{1}{L_1}\sum_{i=1}^{L_1} y_i \end{aligned} \tag{5.7}$$

where $\theta = (w, b)$. Obviously, Eq. (5.7) is a standard TS^{3}VM model and can be easily solved by using off-the-shelf tools [15].

Case 5.3 Type II Dominates In this case, Type II examples T_2 dominate the training chunk, and Type I and Type III examples T_1 and T_3 have higher learning priorities than Type II examples. Thus, T_1, T_2 and T_3 will be used for training.

Accurately learning from these three types of examples is non-trivial. For this purpose, we design a novel transfer semi-supervised SVM model (TS^{3}VM for short). Intuitively, the TS^{3}VM model can be formulated by incorporating examples in T_1, T_2 and T_3 sequentially. Specifically, we can first formulate a generic SVM model by taking T_1 into consideration. Then, a transfer SVM model can be formulated by taking T_2 into consideration.

Finally, we can include T_2 and formulate the TS^{3}VM model.

Learning from T_1 has been discussed in Eq. (5.4), based on which T_2 can be incorporated by applying the transfer learning strategy. Practically, transfer learning can use labeled examples in T_2 to refine the classification boundary by transferring the knowledge from T_2 to T_1. An effective way of doing so is to consider the problem as a multi-task learning procedure [16]. A common two-task learning SVM model on $T_1$and T_2 can be formulated as,

$$\begin{aligned} &\min \tfrac{1}{2}\|w\|^2 + C_1\|v_1\|^2 + C_2\|v_2\|^2 + C\sum_{i=1}^{L} \xi_i \\ &s.t. \quad y_i\left(\left(w + v_1\right)x_i + b\right) \ge 1 - \xi_i, \quad 1 \le i \le L_1 \\ &\qquad y_i\left(\left(w + v_2\right)x_i + b\right) \ge 1 - \xi_i, \quad L_1 + 1 \le i \le L \\ &\qquad\qquad \xi_i \ge 0, \quad 1 \le i \le L \end{aligned} \tag{5.8}$$

where parameters C_1 and C_2 are the penalties on the two tasks, and v_1 and v_2 are the discrepancies between the global optimal decision boundary w and the local optimal decision boundary (i.e., $w + v_1$ for the task of learning from T_1 and $w + v_2$ for the task of learning from T_2).

In Eq. (5.8), parameters C_1 and C_2 control the preference between the two tasks. If $C_1 > C_2$, task 1 is preferred over task 2; otherwise, task 2 is preferred over task 1. By using the Hinge loss function, Eq. (5.8) can be transformed into an unconstrained form,

$$\min_{\theta} \frac{1}{2}\|w\|^2 + C_1\|V_1\|^2 + C_2\|V_2\|^2 + C\sum_{i=1}^{L} H\left(y_i f_\theta\left(x_i\right)\right) \tag{5.9}$$

where $\theta = (w, v_1, v_2, b), f_\theta(x) = (w + v_1)x + b$ for task 1 and $f_\theta(x) = (w + v_2)x + b$ for task 2.

In addition to T_1 and T_2, an additional semi-supervised learning method can be used to learn from the remaining T_3. As we discussed in Eq. (5.5), by adding an extra term $C^* \sum_{i=L+1}^{L+U} H\left(\left|f_\theta\left(x_i\right)\right|\right)$ to penalize the misclassification of unlabeled examples in T_3 located inside the margin decided by Eq. (5.9), as well as the balance constraint in Eq. (5.6), we can finally get the TS^{3}VM model as,

$$\begin{gathered} \min_{\theta} \tfrac{1}{2}\|w\|^2 + C_1\|v_1\|^2 + C_2\|v_2\|^2 \\ + C\sum_{i=1}^{L} H\left(y_i f_\theta\left(x_i\right)\right) + C^*\sum_{i=L+1}^{L+U} H\left(\left|f_\theta\left(x_i\right)\right|\right) \\ s.t. \tfrac{1}{U}\sum_{i=L+1}^{L+U} f_\theta\left(x_i\right) = \tfrac{1}{L}\sum_{i=1}^{L} y_i \end{gathered} \tag{5.10}$$

where $\theta = (w, v_1, v_2, b), f_\theta(x_i) = (w + v_1)x_i + b$ for $1 \le i \le L_1, f_\theta(x_i) = (w + v_2)x_i + b$ for $L_1 + 1 \le i \le L$, and $f_\theta(x_i) = wx_i + b$ for $L + 1 \le i \le L + U$.

5.1.3.1 Solution to the TS^{3}VM Objective Function

As shown in Eq. (5.10), optimizing the objective function of TS^{3}VM is a non-convex optimization problem, which is difficult to find global minima especially for large scale problems. We propose to solve this non-convex problem by using Concave-Convex Procedure (CCCP), which has been developed by the optimization community [6, 10, 26]. CCCP decomposes a non-convex function into the sum of a convex function and a concave function, and then approximates the concave part by using a linear function (a tangential approximation). By doing so, the whole optimization procedure can be carried out iteratively by solving a sequence of convex problems. Algorithm 5.1 describes the CCCP algorithm in detail.

Algorithm 5.1 CCCP Algorithm

Input: the objective function $J(\theta)$

1: Get the initial point θ_0 with a best guess
2: $J(\theta) = J_{vex}(\theta) + J_{cav}(\theta)$
3: **repeat**
4: $\quad \theta_{t+1} = \arg\min_\theta J_{vex}(\theta) + J'_{cav}(\theta_t) \cdot \theta$
5: **until** convergence of θ
6: **return** a local minima solution θ^*

From the CCCP perspective, we can observe that the first four terms of TS^{3}VM are convex functions, whereas the last Symmetric Hinge loss part $C^* \sum_{i=L+1}^{L+U} H\left(|f_\theta\left(x_i\right)|\right)$ makes it a non-convex model. Thus, we will decompose and analyze the last part by using the CCCP method. To simplify the notation, we denote $z_i = f_\theta(x_i)$, so the last part can be rewritten as $C^* \sum_{i=L+1}^{L+U} H\Big(\mid z_i \mid\Big)$. Considering a specific z_i (without loss of generality, we denote it as z here), the Symmetric Hinge loss on z can be denoted by $J(z)$ as,

$$J(z) = C^* H\Big(|z|\Big) \tag{5.11}$$

Equation (5.11) is a non-convex function, which can be split into a convex part and a concave part as,

$$J(z) = C^* H\left(|z|\right) = \underbrace{C^* \max\left(0, 1 - |z|\right) + C^* \mid z \mid}_{J_{vex}(t)} \underbrace{- C^* \mid z \mid}_{J_{cav}(t)} \tag{5.12}$$

According to Algorithm 5.1, the next iterative point can be calculated by the approximation of the concave part J_{cav} as,

$$\frac{\partial J_{cav}\Big(z\Big)}{\partial z} \cdot z = \begin{cases} C^* z, & z < 0 \\ -C^* z, & z \geq 0 \end{cases} \tag{5.13}$$

and then minimizing,

$$J(z) = C^* \cdot \max\left(0, 1 - |z|\right) + C^* \mid z \mid + \frac{\partial J_{cav}\Big(z\Big)}{\partial z} z \tag{5.14}$$

If $z < 0$ in the current iteration, then in the next iteration, the current effective loss can be denoted as

$$L\left(z, -1\right) = C^* \max\left(0, 1 - |z|\right) + C^* \mid z \mid + C^* z = \begin{cases} 2C^* z, & z \geq 1 \\ C^* \left(1 + z\right), & \mid z \mid < 1 \\ 0, & z \leq -1 \end{cases} \tag{5.15}$$

On the other hand, if $z > 0$, then in the next iteration, the current effective loss can be denoted as

$$L\left(z, +1\right) = C^* \max\left(0, 1 - |z|\right) + C^* \mid z \mid - C^* z = \begin{cases} 0, & z \geq 1 \\ C^* \left(1 - z\right), & \mid z \mid < 1 \\ -2C^* z, & z \leq -1 \end{cases} \tag{5.16}$$

By doing so, within each iteration, when taking all $z_i = f_\theta(x_i)$ into consideration, solving the TS^{3}VM model is equivalent to solving Eq. (5.17) under the balance constraint Eq. (5.6),

$$\begin{aligned} &\min_{\theta} \tfrac{1}{2}\|w\|^2 + +C_1\|v_1\|^2 + C_2\|v_2\|^2 \\ &+ C\sum_{i=1}^{L} H\left(y_i f_\theta\left(x_i\right)\right) + \sum_{i=L+1}^{L+U} L\left(f_\theta\left(x_i\right), y_i\right) \end{aligned} \tag{5.17}$$

where $y_i(L + 1 \leq i \leq L + U)$ is the class label of x_i that has been assigned in the previous iteration. If $y_i < 0$, Eq. (5.15) will be used to calculate the loss function; otherwise, Eq. (5.16) will be used to calculate the loss function.

The detailed description of solving TS^{3}VM is given in Algorithm 5.2.

Algorithm 5.2 TS^{3}VM Learning Model

Input: T_1, T_2 and T_3

Use T_1 and T_2 to build a transfer SVM model as shown in Eq. (6.8), and get the initial point $\theta_0 = (w_0, v_{10}, v_{20}, b_0)$

repeat

$y_i \leftarrow \text{sgn}(wx_i + b), \forall L + 1 \leq i \leq L + U$

$\theta \leftarrow$ Calculate Eq. (5.17) under the balance constraint Eq. (5.6)

until y_i remains unchanged, $\forall L + 1 \leq i \leq L + U$

return $f(x) = \text{sgn}(wx + b)$

Theorem 5.2 (Convergence of TS^{3}VM) *The TS^{3}VM learning model in Algorithm 5.2 converges after a limited number of iterations.*

Proof In Algorithm 5.2, in each iteration t, the objective function $J(\theta_t)$ is split into a convex part $J_{vex}(\theta_t)$ and a concave part $J_{cav}(\theta_t)$. Then, in the next iteration $t + 1$, the point θ_{t+1} is the minimal solution of the current objective function, and we have

$$J_{vex}\left(\theta_{t+1}\right) + J'_{cav}\left(\theta_t\right)\theta_{t+1} \leq J_{vex}\left(\theta_t\right) + J'_{cav}\left(\theta_t\right)\theta_t \tag{5.18}$$

Meanwhile, because the concavity of $J_{cav}(\theta)$, we have,

$$J_{cav}\left(\theta_{t+1}\right) \leq J_{cav}\left(\theta_t\right) + J'_{cav}\left(\theta_t\right)\left(\theta_{t+1} - \theta_t\right) \tag{5.19}$$

By adding both sides of Eq. (5.18) and Eq. (5.19), we have

$$\begin{aligned} &J_{vex}\left(\theta_{t+1}\right) + J_{cav}\left(\theta_{t+1}\right) + J'_{cav}\left(\theta_t\right)\theta_{t+1} \\ &\leq J_{vex}\left(\theta_t\right) + J'_{cav}\left(\theta_t\right)\theta_t + J_{cav}\left(\theta_t\right) + J'_{cav}\left(\theta_t\right)\left(\theta_{t+1} - \theta_t\right) \end{aligned} \tag{5.20}$$

Move the third item on the left-hand side of Eq. (5.20) to the right-hand side, we have

$$\begin{aligned} J_{vex}\left(\theta_{t+1}\right) + J_{cav}\left(\theta_{t+1}\right) \leq J_{vex}\left(\theta_{t}\right) + J'_{cav}\left(\theta_{t}\right)\theta_{t} \\ + J_{cav}\left(\theta_{t}\right) + J'_{cav}\left(\theta_{t}\right)\left(\theta_{t+1} - \theta_{t}\right) - J'_{cav}\left(\theta_{t+1}\right)\theta_{t+1} \end{aligned} \quad (5.21)$$

The right-hand side of the above inequation equals to $J_{vex}((\theta_t)) + J_{cav}(\theta_t)$. Therefore, the objective function will decrease after each iteration $J_{vex}(\theta_t + 1) \leq J(\theta_t)$.

Consequently, Algorithm 5.2 will converge after a limited number of iterations. In fact, as long as the initial point is carefully selected (i.e., using a multi-task SVM model built on T_1, and T_2 as the initial point), Algorithm 5.2 will converge very fast.

Case 5.4 Type IV Dominates This is the most complex learning case. In this case, Type IV examples T_4 dominate the training chunk and has the lowest learning priority. Thus, it is necessary to use all T_1, T_2, T_3 and T_4 for training.

To solve this learning problem, we design a novel Relational K-means-based Transfer Semi-Supervised learning model (RK-TS^{3}VM for short). The TS^{3}VM model, as discussed previously, is used to learn from T_1, T_2 and T_3. Now we discuss how to learn from T_4 using a Relational K-means model [17] (RK for short).

Learning from T_4 is more challenging than from other three types of training examples, mainly because examples in T_4 are unlabeled and have different distributions from the target domain. The aim of the RK model is to *transfer knowledge from T_4 to T_1, T_2 and T_3 by constructing some new features for the three types of examples using the relational information between T_1, T_2, T_3 and T_4.*

An example of RK learning is shown in Fig. 5.5, where T_4 examples are first clustered into k clusters, $G_1, \cdots, G_k$ based on a relational matrix built between T_1 and T_4. After that, k new features $f(x_i, G_\tau)$ $(\tau = 1 \cdots, k)$ are added to each example x_i in T_1 to construct a new data set T_1' by calculating the relationship between x_i and each cluster center. By doing so, the new data set T_1' will contain information transferred from T_4, which can help build a more accurate prediction model.

Given L_1 examples in T_1 and N examples in T_4, the purpose of the relational k-means clustering is to cluster instances in T_4 into k groups, by taking the relationships between instances in T_1 and T_4 into consideration. Let $W \in \mathbb{R}^{L_1 \times N}$ denote the similarity matrix between T_1 and T_4 with each $w_{i,j}$ indicating the

Information from T_1				Information from T_4			Class label for T_1
A_1	A_2	..	A_d	G_1	..	G_k	Y
1	2	..	5	$f(x_1, G_1)$	..	$f(x_1, G_k)$	1
..	..	..	..	..	..	..	..
3	7	..	1	$f(x_{L_1}, G_1)$	..	$f(x_{L_1}, G_k)$	2

Fig. 5.5 An illustration of the RK learning model

similarity (which can be calculated according to the Euclidian distance) between instance x_i in T_1 and instance x_j in T_4. For each cluster G_τ on W the average pairwise similarity for all examples in G_τ can be defined as

$$S_{G_\tau} = \frac{1}{|G_\tau|^2} \sum_{x \in G_\tau} \sum_{x' \in G_\tau} S\left(x, x'\right) \tag{5.22}$$

where $S(x, x^{'})$ denotes the similarity between two examples of x and $x^{'}$. On the other hand, the variance of the relationship values of all examples in G_τ can be calculated as

$$\delta_{G_\tau} = \frac{1}{|G_\tau|} \sum_{y_i \in G_\tau} \left(\beta_j - \beta_{G_\tau}\right)^T \left(\beta_j - \beta_{G_\tau}\right) \tag{5.23}$$

where β_{G_τ} denotes the average relationship vector of all instances in G_τ, and $\beta_i \in \mathbb{R}^{1 \times L_1}$ denotes the relationships of instance x_j with respect to all examples in T_1.

The objective of the relational k-means is to find k groups, G_τ, $\tau = 1, \cdots, k$, such that the sum of the similarities is maximized while the sum of variances is minimized as

$$J_e^{\prime} = \max \sum_{\tau=1}^{k} J_{G_\tau} = \max \sum_{\tau=1}^{k} \frac{S_{G_\tau}}{\delta_{G_\tau}} \tag{5.24}$$

Explicitly solving Eq. (5.24) is very difficult. Alternatively, we can use a recursive hill-climbing search process as an approximation solution. Assume that examples in T_4 are clustered into k clusters, $G_1, \cdots, G_k$. Moving an instance x from cluster G_i to cluster G_j changes only the cluster objective values J_{G_i}andJ_{G_j}. Therefore, in order to maximize Eq. (5.24), at each step t, we randomly select an example x from a cluster G_i, and move it to cluster G_j. Such a move is accepted only if the Inequity (5.25) achieves a higher value at step $t + 1$.

$$J_{G_i}(t) + J_{G_j}(t) < J_{G_i}(t+1) + J_{G_j}\left(t+1\right) \tag{5.25}$$

Based on the search process in Inequity (5.25), major steps of the relational k-means are listed in Algorithm 5.3.

Algorithm 5.3 has three tiers of loops. Within each tier, it needs to frequently recalculate $J_{G_c}\left(t\right)$ when the current examples are removed from its current group to another. Nevertheless, because $J_{G_c}\left(t\right)$, as shown in Eq. (5.24), contains information from both the similarity S_{G_i} and variance δ_{G_i} in the relationship matrix, frequently recalculating $J_{G_c}\left(t\right)$ will be time-consuming. To alleviate this problem, we introduce an addictive update method and a subtractive update method to recalculate $J_{G_c}\left(t\right)$.

Algorithm 5.3 Relational k-Means Clustering

Input: T_1, T_4, number of clusters k, and number of iterations T

1: W ← calculate similarity matrix between T_1 and T_4
2: $G_1, \ldots, G_k$ ← apply k-means to W
3: **for** t ← 1 to T **do**
4: x ← randomly select an example from T_4
5: G_i ← current cluster of example x
6: J_{G_i} (t) ← calculate G_i's objective value in Eq. (5.24)
7: J_{G_i} (t + 1) ← G_i's new value after excluding x
8: **for** j ← 1 to k, j ≠ i **do**
9: J_{G_j} (t) ← calculate G_j's objective value
10: J_{G_j} (t + 1) ← G_j's new value after including x
11: **if** inequity (6.25) is true **then**
12: $G_j \leftarrow G_j \cup x$; $G_i \leftarrow G_i \setminus x$
13: break
14: **end if**
15: **end for**
16: **end for**
17: $\mu_1, \ldots, \mu_k$ ← calculate cluster centers for $G_1, \ldots, G_k$
18: **return** $\mu_1, \ldots, \mu_k$

Consider an example x in T_4 that moves from group G_i to G_j. Before the move, β_{G_i} and β_{G_j} are the mean vectors, $\delta_{G_i}$and δ_{G_j} are the variance vectors. After the move, the new groups are G_i' and G_j'. Then the addictive update is given in the following theorem:

Theorem 5.3 (Additive Update) *When adding an example x into G_j, the mean vector of G_j, β_{G_i}, can be updated to β_{G_i}' as follows,*

$$\beta_{G_j}' = \frac{1}{|G_j'|}\sum_{x_i\in}\beta_l = \frac{1}{n_j+1}\left(n_j\cdot\beta_{G_j}+\beta_k\right) = \beta_{G_i} + \frac{\beta_k-\beta_{G_j}}{n_j+1}, \tag{5.26}$$

Meanwhile, the variance δ_{G_j} can be updated to δ_{G_j}' as follows,

$$\begin{aligned}\delta_{G_j}' &= \frac{1}{\left|G_j'\right|}\sum_{y_l\in G_j'}\left(\beta_l-\beta_{G_j}'\right)^T\left(\beta_l-\beta_{G_j}'\right)\\ &= \frac{n_j}{n_j+1}\delta_{G_j} + \frac{n_j}{\left(n_j+1\right)^2}\left(\beta_k-\beta_{G_j}\right)^T\left(\beta_k-\beta_{G_j}\right)\end{aligned} \tag{5.27}$$

where n_j is the number of examples in G_j.

Therefore, the updated mean and variance vectors of group G_j can be incrementally calculated, without recalculating Eq. (5.24). Similarly, for a group G_i, where an example x is removed, its mean and variance vectors can be updated using the following theorem.

Theorem 5.4 (Subtractive Update) *When an example x is removed from group G_i, the mean vector β_{G_i} can be updated to β'_{G_i} as follows,*

$$\beta'_{G_i} = \frac{1}{\mid G'_i \mid} \sum_{y_l \in G'_i} \beta_l = \frac{1}{n_i - 1} \left(n_i \cdot \beta_{G_i} - \beta_k \right) = \beta_{G_i} - \frac{\beta_k - \beta_{G_i}}{n_i - 1}, \tag{5.28}$$

Meanwhile, the variance δ_{G_i} can be updated to δ'_{G_i} as follows, where n_i is the number of examples in G_i.

$$\begin{aligned} \delta'_{G_j} &= \tfrac{1}{|G'_i|} \sum_{y_l \in G'_i} \left(\beta_l - \beta'_{G_i} \right)^T \left(\beta_l - \beta'_{G_i} \right) \\ &= \tfrac{n_i}{n_i - 1} \delta_{G_i} - \tfrac{n_i}{(n_i - 1)^2} \left(\beta_k - \beta_{G_i} \right)^T \left(\beta_k - \beta_{G_i} \right) \end{aligned} \tag{5.29}$$

where n_i is the number of examples in G_i.

Time complexity Now we analyze the time complexity of Algorithm 5.3. In Algorithm 5.3, when searching for a new group for each example in the relationship matrix, the updating operation, by using Theorems 5.3 and 5.4, can be executed within constant time $O(1)$. Besides, Algorithm 5.3 is a greedy algorithm. In each iteration, it uses a local optimization technique to cluster examples into groups that maximizes Eq. (5.24). There are three tiers of loops in the algorithm. The first tier aims to find the best group for each example x with the worst-case complexity of $O(k)$ (i.e., traversing all the k groups). The second tier aims to find the best groups for all examples in T_4, which has the worst-case complexity of $O(N)$ (i.e., searching over all the N examples). The last tier aims to make the algorithm converge to a stable solution. Obviously, the first two tiers dominate the time consumption of the whole algorithm, and thus the time complexity of Algorithm 5.3 is $O(k) \times O(N) = O(kN)$.

RK- TS^{3}VM learning model Algorithm 5.4 lists the detailed procedures of the RK- TS^{3}VM learning model, which is the combination of the TS^{3}VM and RK learning models. Given a training chunk D, Step 1 identifies the four types of examples T_1, T_2, T_3 and T_4. Step 2 constructs a group of k feature vectors, denoted by $\mu = \{\mu_1, \cdots, \mu_k\}$, by applying RK to T_1 and T_4. In Step 3 and Step 4, the k new features are appended to each example in T_1, T_2 and T_3 to form three new sets denoted by T'_1, $T'_2$and T'_3, respectively. Step 5 builds a TS^{3}VM model F from T'_1, T'_2, and T'_3. In Step 6, the feature vectors μ and F are combined to form the final prediction model. For any example x in the testing chunk, RK- TS^{3}VM first calculates k new features for x, then uses the TS^{3}VM model to predict a label for x.

Algorithm 5.4 RK-TS^{3}VM Learning Framework

Input: training chunk D, chunk size n, labeled rate l, concept drifting rate c, number of clusters k

Step 1: Identify the four types of data T_1, T_2, T_3, T_4 in D according to the labeled rate l and concept drifting probability c using Eq. (5.1)

Step 2: Using RK model on T_1 and T_4 to get k cluster centers denoted by $\mu = \{\mu_1, \ldots, \mu_k\}$

Step 3: for each instance x in T_1, T_2, and T_3, add k attributes using the inner produce between x and μ

Step 4: Get the new samples T'_1, T'_2, and T'_3 from **Step 3**

Step 5: Construct a TS^3 VM model using T'_1, T'_2, and T'_3, and get the model F

return μ and **F** together as the prediction model

The data analysis of implementing the above algorithms can be found in Zhang et al. [13].

5.2 Robust Ensemble Learning for Mining Noisy Data Streams

5.2.1 Noisy Description for Data Stream

Based on the characteristics of the stream data, existing work roughly describes data streams into the following two styles: stationary data streams [11, 18–20] and dynamic data streams [5, 21–23].

According to the stationary description, if data streams are divided into data chunks as shown in Fig. 5.6, then training data chunks (which include both historical data chunks and the up-to-date chunk) will have a similar or identical distribution to the yet-to-come data chunk. Thus, classifiers built from the training data chunks will have reasonably good performance in classifying data from the yet-to-come data chunk. The advantage of the stationary description is that we may directly apply traditional classification techniques to the data streams. For example, since the up-to-date data chunks have the same distribution as the yet-to-come data chunk, we can collect all historical classifiers to build a classifier ensemble. However, this stationary description takes no consideration of the concept drifting in stream data, so it can hardly, if not impossible, be used to describe most real-world data streams.

Noticing the limitations of the stationary description, a recent work [21] describes the data streams in a dynamic scenario where training chunks have different distributions *p(x,y)* (where *x* denotes the feature vector and *y* denotes the class label) from that of the yet-to-come data chunk, and classifiers built on the training set may perform only slightly better than random guessing or simply predicting all examples to belong to one single class. Comparing to the stationary description, the dynamic description emphasizes on the situation that training data

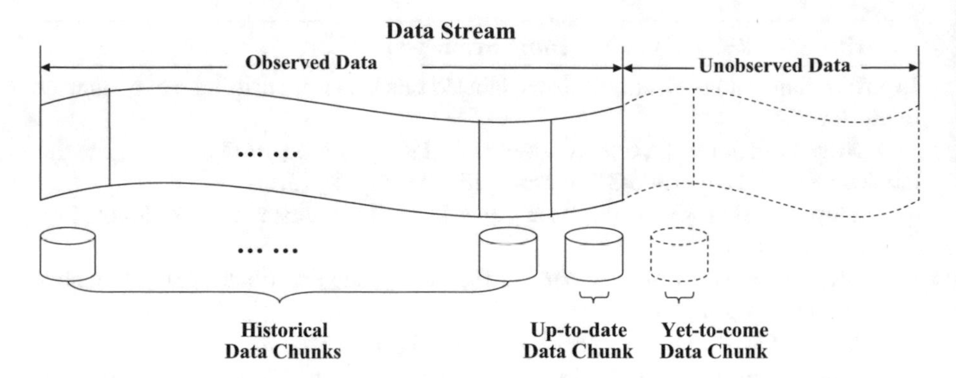

Fig. 5.6 An illustration of the "historical", "up-to-date" and "yet-to-come" data chunks. A data stream can be split into two parts: the observed data stream (which is denoted by the solid lines) and the unobserved data stream (which is denoted by the dotted lines). Assume the data stream is processed chunk-by-chunk. The observed data stream can be further categorized into two types: the latest data chunk is called the "up-to-date" chunk, while the remaining data chunks are called the "historical" data chunks. Besides, the "yet-to-come" data chunk is the first data chunk of the unobserved data streams

chunks do not necessarily have the same distribution as the yet-to-come data chunk. Under this description, building classifiers from the up-to-date data chunk to predict the yet-to-come data chunk is better than building classifiers from the aggregation of all historical chunks because the buffered chunks (probably outdated with respect to the newly arrived data chunk) will deteriorate the ensemble performance. In a narrow sense, this dynamic description is much looser than the stationary description, which makes it more applicable for mining concept drifting data streams. However, the disadvantage of the dynamic description is also obvious, in the sense that it doesn't discriminate concept drifting from data errors. If the up-to-date data chunk contains noisy samples, building classifiers on this noisy data chunk to predict the yet-to-come data chunk may cause more errors than using a classifier ensemble built on previously buffered data chunks. Consequently, although the dynamic description is more reasonable than the stationary description for data streams, in practice, it is still not capable of describing all the realistic data streams.

Consider a data stream management system whose buffer contains five consecutive data chunks as shown in Fig. 5.7. The stationary description can only cover the process from D_1 to D_2, where the distribution $p_1(x,y)$ remains unchanged. The dynamic description covers the process from D_2 to D_3, where the concept drifts from $p_1(x,y)$ to $p_2(x,y)$ without being interrupted by noisy data chunks. A more general situation, as depicted in the process from D_3 to D_5, is that the concept drifting ($p_2(x,y)$ evolves to $p_3(x,y)$) is mixed with noise (a noisy data chunk D_4 is observed). To explicitly describe this type of data streams, we define a noisy description of data streams as follows:

Noisy Description for Data streams Mining from real-world data streams may confront the challenges of concept drifting and data errors simultaneously.

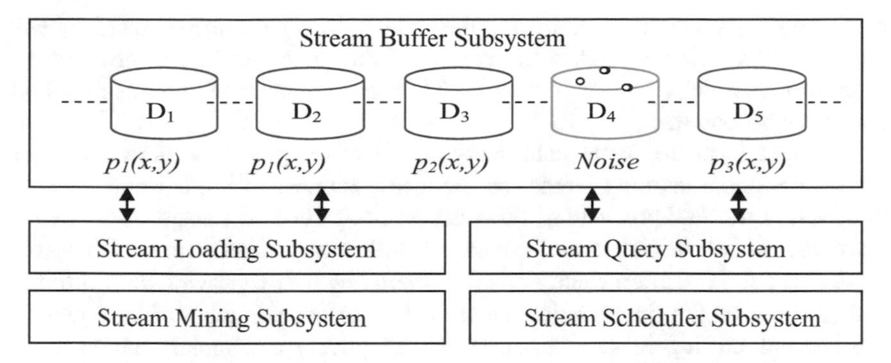

Fig. 5.7 A conceptual view of noisy data in data stream management system. The data stream management system can be separated into five parts: a stream buffer subsystem, a stream loading subsystem, a stream query subsystem, a stream mining subsystem, and a stream scheduler subsystem. In the stream buffer subsystem, there are five buffered data chunks, $D_1, D_2, \ldots, D_5$, of which D_4 is a noise data chunk. D_1 and D_2 share the same distribution $P_1(x,y)$. From D_2 to D_3, the underlying concept changes from $P_1(x,y)$ to $P_2(x,y)$. From D_3 to D_4 and finally to D_5, the concept changes from $P_2(x,y)$ to $P_3(x,y)$, meanwhile, a noisy chunk D_4 is observed between D_3 and D_5. The stationary description of data streams can only cover the process from D_1 to D_2, while the dynamic description of data streams only covers the process from D_2 to D_3. Our noisy description covers a much more common process from D_3 to D_5

The noisy description addresses both concept drifting and data errors in a data stream management system. It is much more general than the stationary and dynamic descriptions. It then can be adapted for generic data streams.

5.2.2 Ensemble Frameworks for Mining Data Stream

The nature of continuous volumes of the stream data raises the needs of designing effective classifiers with high accuracy in predicting future testing instances as well as good efficiency in handling massive volumes of training instances. In the past few years, many solutions have been proposed to build prediction models from data streams. An early solution is to build model by using online incremental methods [18, 19] which update a single model by incorporating newly arrived data. During the learning process, incremental methods continuously revise the model to discover new patterns in the most recent data chunk. For example, Domingos and Hulten [10] introduced an ultra fast decision tree learner VFDT which incrementally builds Hoeffding trees from the high-volume data streams. Similar approach was extended to CVFDT [19] which handles time changing and concept drifting streams. By doing so, most of the incremental methods violate the efficiency rule because updating a classifier according to the newly arrived data can be a time-consuming process. An alternative solution is to build a single and simple classifier on the up-to-date chunk without considering historical data chunks, i.e., discarding old classifiers and rebuilding a new classifier on the new data chunk. This build-then-discard method, unfortunately, may not work well because of the important loss incurred by the discarded classifiers. To overcome this challenge, a number of ensemble methods have been proposed.

Different from the incremental learning where the goal is to deliver a single model, ensemble learning intends to produce a number of models and relies on their voting for final predictions. Such design brings two advantages for ensemble learning to handle data streams: (1) because models are trained from a small portion of stream data, it can efficiently handle streams with fast growing data volumes; and (2) because the final predictions are the voting of a number of base models, the concept drifting in the stream can be adaptively and rapidly addressed by changing the weight value of each voting member. For example, Street and Kim [24] proposed a SEA algorithm, which combines decision tree models using majority-voting. Kolter and Maloof [25] proposed an ensemble method by using weighted online learners to handle drifting concepts. Wang et al. [11] proposed a weighted ensemble, in which they assign each classifier a weight reversely proportional to the classifier's accuracy on the most recent data chunk. Yang et al. [26] proposed proactive learning where concepts (models) learnt from previous chunks are used to foresee the best model to predict data in the current chunk. Zhu et al. [27] proposed an active learning framework to selectively label instances for concept drifting data streams. Gao et al. [21] proposed to build different base classifiers on a most recent data chunk to construct the classifier ensemble.

In summary, the above ensemble frameworks for stream data mining can be roughly categorized into the following two categories, according to their ways of forming the base classifiers: horizontal ensemble (including weighted ensemble) frameworks which build base classifiers using several buffered data chunks (as illustrated in Fig. 5.8a), and vertical ensemble framework which build base classifiers on the up-to-date data chunk using different algorithms (as illustrated in Fig. 5.8b).

5.2.2.1 Horizontal Ensemble and Weighted Ensemble Frameworks

Consider a data stream containing an infinite number of data chunks $\{Di\}_{i=1}^{+\infty}$. Due to the limitation of the storage space, the system buffer can only accommodate at most n consecutive chunks each of which contains a certain number of instances. Assume at the current time stamp we are observing the nth chunk D_n, and the buffered data chunks are denoted by $D_1, D_2, \ldots, D_n$. In order to predict data in a newly arrived chunk D_{n+1}, one can choose a learning algorithm L to build a base classifier f_i from each of the buffered data chunks D_i, say $f_i = \mathcal{L}(D_i)$, and then predict each instance x in D_{n+1} by combining the predictions of the base classifiers f_i $(i = 1, 2, \ldots, N)$ to form a classifier ensemble through the model averaging mechanism shown in Eq. (5.30) [11, 25, 27, 28]:

$$f_{HE}(x) = \frac{1}{N}\sum\nolimits_{i=1}^{N} f_i(x) \tag{5.30}$$

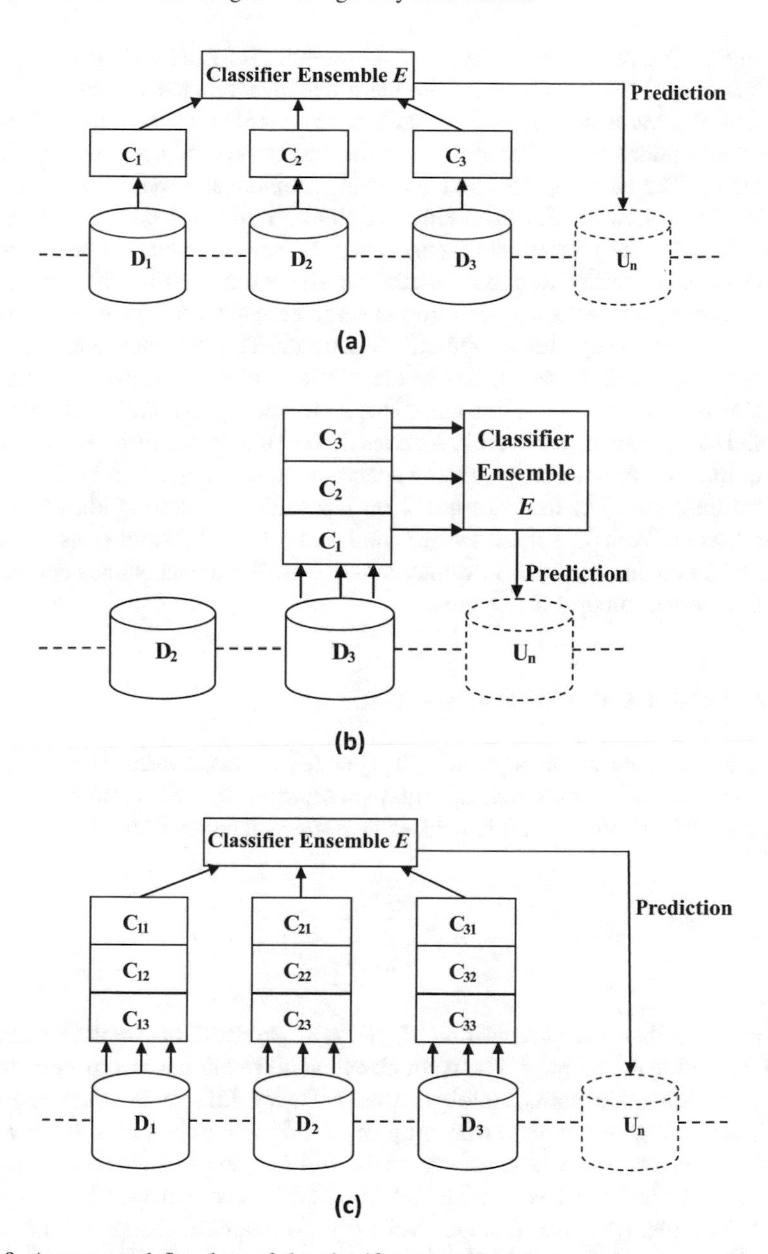

Fig. 5.8 A conceptual flowchart of the classifier ensemble framework for stream data mining where (**a**) shows the horizontal ensemble framework, which builds different classifiers on different data chunks; (**b**) shows the vertical ensemble framework, which builds different classifiers on the up-to-date data chunk with different learning algorithms; and (**c**) shows the aggregate ensemble framework, which builds classifiers on different data chunks using different learning algorithms

An alternative version of the horizontal ensemble is to add weight values to the base classifiers [11, 27]. Different from the model averaging, a weighted ensemble minimizes the variance error e_v of each base classifier on the up-to-date data chunk, then assigns each classifier a weight that is reversely proportional to the error rate e_v. The advantage of the horizontal ensemble and weighted ensemble is twofold: (1) they can reuse information of the buffered data chunks, which may be beneficial for the testing data chunk; and (2) they are robust to noisy streams because the final decisions are based on the classifiers trained from different chunks. Even if noisy data chunks may deteriorate some base classifiers, the ensemble can still maintain relatively stable prediction accuracy. The disadvantage of such an ensemble framework, however, lies in the fact that if the concepts of the stream continuously change, information contained in previously buffered classifiers may be invalid to the current data chunk. As a result, combining old-fashioned classifiers may not improve the overall prediction accuracy. In summary, both horizontal and weighted ensembles, in fact, are based on the stationary description of the data streams that buffered data chunks share similar or identical distributions to the yet-to-come data chunk, such that information in the buffered data chunks can be used to predict the yet-to-come data chunk.

5.2.2.2 Vertical Ensemble Framework

Assume we have *m* learning algorithms L_j ($j = 1,2,\ldots,m$), a vertical ensemble [17] builds base classifiers using each algorithm on the up-to-date data chunk D_n as $f_j = \mathcal{L}_| (D_n)$, and then combines all base classifiers through model averaging as given in Eq. (5.31),

$$f_{VE}^{n}(x) = \frac{1}{m}\sum_{i=1}^{m} f_{in}(x) \tag{5.31}$$

In the case that prior knowledge of the yet-to-come data chunk is unknown, model averaging on the most recent chunk can achieve minimum expectation error on the test set. In other words, building classifiers using different learning algorithms can decrease the expected bias error compared to any single classifiers. For example, assuming a data stream whose joint probability *p(x,y)* evolves continuously, if we only use a stable learner such as SVM, then SVM may perform better than an unstable classifier when *p(x)* changes while *p(y|x)* remains unchanged. On the other hand, if we only use an unstable learner such as decision trees, then decision trees may perform better than SVM when *p(x)* does not evolve much but *p(y|x)* changes dramatically. When we have no prior knowledge on whether the evolving of *p(x,y)* is triggered by *p(x)* or *p(y|x)*, it is difficult to determine whether a stable classifier or an unstable classifier is better, so combining these two types of classifiers is likely to be a better solution than simply choosing either of them. Although the vertical ensemble has a much looser condition (distribution *p(x,y)* may continuously change) than the stationary description (distribution *p(x,y)* remains unchanged), it also has a severe pitfall for realistic data streams. The vertical ensemble builds classifiers only on a single up-to-date data chunk, but as we have discussed before, a realistic data stream system may contain data errors. If the up-to-date data chunk is a noisy data chunk, the results may suffer from severe performance deterioration. Without realizing the noise problems, the vertical ensemble limits itself merely to the concept drifting scenarios, but not to the realistic data streams.

5.2.2.3 Aggregate Ensemble Framework

The disadvantages of the above two ensemble frameworks motivate the proposed Aggregate Ensemble framework (which is illustrated in Fig. 5.8c). We first use m learning algorithms L_i $(i = 1, 2, \ldots, m)$ to build classifiers on n buffered data chunks j $(j = 1, \ldots, n)$, and then train m-by-n base classifiers $f_{ij} = \mathcal{L}_i\left(D_j\right)$, where i denotes the ith algorithm, and j denotes the jth data chunk. Then we combine these base classifiers to form an aggregate ensemble through model averaging defined in Eq. (5.32), which indicates that the aggregate ensemble is a mixture of the horizontal ensemble and vertical ensemble, and its base classifiers constitute a *Classifier Matrix* (*CM*) in Eq. (5.33).

$$f_{AE} = \frac{1}{mn}\sum_{i=1}^{n}\sum_{j=1}^{m} f_{ij}(x) \tag{5.32}$$

$$CM = \begin{bmatrix} f_{11}\ f_{12} \cdots\cdots f_{1n} \\ f_{21}\ f_{22} \cdots\cdots f_{2n} \\ \cdots\cdots\cdots\cdots\cdots\cdots \\ f_{m1}\ f_{m2} \cdots\cdots f_{mn} \end{bmatrix}_{m*n} \tag{5.33}$$

In Eq. (5.33), each element f_{ij} in *CM* represents a base classifier built by using algorithm i on data chunk j. As we have mentioned in the vertical ensemble, classifiers on each column of *CM* (*i.e.*, classifiers built on the same data chunk using different learning algorithms) are used to reduce the expected classifier bias error on unknown test data. Classifiers on each row of *CM* (*i.e.*, classifiers built on different data chunks using the same learning algorithm) are used to reduce the impact of noisy data chunks. For example, when the up-to-date training chunk is a noisy chunk, combining classifiers built from the historical data chunks may alleviate the noisy impact. By building a classifier matrix *CM*, the aggregate ensemble is capable of solving a realistic data stream containing both concept drifting and data errors.

5.2.3 *Theoretical Studies of the Aggregate Ensemble*

5.2.3.1 Performance Study of AE Framework

In this section, we explore why and when AE performs better than HE and VE methods. As we have described in the above section, on each data chunk, the aggregate ensemble builds m classifiers by using m different learning algorithms. For a specific test instance x in the yet-to-come data chunk, the horizontal ensemble uses classifiers on a row in matrix CM to predict x, *i.e.*, if we choose learning algorithm i ($1 \leq i \leq m$), then the horizontal ensemble can be denoted by Eq. (5.34)

$$f_{HE}^{i}(x) = \frac{1}{n}\sum_{j=1}^{n} f_{ij}(x) \tag{5.34}$$

The vertical ensemble can be denoted by model averaging on the last column (column n) of the Matrix CM, which is given in Eq. (5.35)

$$f_{VE}^{n}(x) = \frac{1}{m}\sum_{i=1}^{m} f_{in}(x) \tag{5.35}$$

An aggregate ensemble combines all classifiers in CM as base classifiers, through the averaging rule defined by Eq. (5.35). Accordingly, the horizontal ensemble and vertical ensemble are, in fact, two special cases of the aggregate ensemble. Gao et al. [21] proved that in data stream scenario, the performance of a single classifier within a classifier ensemble is expected to be inferior to the performance of the entire classifier ensemble. The horizontal ensemble and vertical ensemble, as special cases of the aggregate ensemble, are not expected as good as the aggregate ensemble. For example, when combining each column in CM, one can have a variant of CM as $CM_c = [g_1, g_2, \ldots, g_n]$, where each $g_i = [f_{1i}, f_{2i}, \ldots, f_{mi}]^T$ is independent of each other and shares the same distribution, say $p(g)$. Then the mean squared error of the horizontal ensemble (with the ith learning algorithm) on a test instance x (with class label y) can be denoted by

$$MSE_{HE}^{i}(x) = E_{p(g)}(y - g_i(x))^2 = y^2 - 2y \cdot E_{p(g)}g_i(x) + E_{p(g)}g_i^2(x) \tag{5.36}$$

For the aggregate ensemble, the mean squared error on x can be calculated as

$$MSE_{AE}(x) = E_{p(g)}\big(y - E_{p(g)}g_i(x)\big)^2 = y^2 - 2y \cdot E_{p(g)}g_i(x) + E_{p(g)}^2 g_i(x) \tag{5.37}$$

Thus, the difference between Eqs. (5.35) and (5.36) is denoted by Eq. (5.38),

$$MSE_{AE}(x) - MSE^{i}_{HE}(x) = E^2_{p(g)} g_i(x) - E_{p(g)} g_i^2(x) \leq 0. \quad \left(\text{since } E^2(x) \leq E\left(x^2\right)\right) \tag{5.38}$$

Accordingly, we assert that the error rate of the aggregate ensemble is expected to be less or equal to the error rate of the horizontal ensemble. Similarly, if we regard *CM* as a column vector where each element is a combination of different rows in *CM*, we can show that the mean squared error of the aggregate ensemble is also expected to be less or equal to that of the vertical ensemble.

In the following we provide some intuitive explanations on why and when AE performs better than HE and VE by using two toy examples in Figs. 5.9 and 5.10. Note that our comparisons here are rather intuitive and qualitative, and rigorous numeric comparisons will be reported in the experimental results in the next section. As shown in Fig. 5.8, assume that AE is trained using three learning algorithms M_1, M_2, and M_3, where $HE(M_i)$ denotes an HE model trained using learning algorithm M_i. For each model, we list three results: (1) training accuracy at time A, (2) test accuracy at time A, and (3) test accuracy at time B which immediately follows A. We can observe that for concept drifting data streams, it is difficult to find a single "optimal" learning algorithm with the best performance across the whole stream. For example, model $HE(M_2)$ has the best prediction accuracy at time stamp A, but unfortunately, it has the worst performance at the next time stamp B. Model $HE(M_3)$ has the worst performance at time A, but it performs the best at time stamp B. On the other hand, AE can guarantee the most reliable performance by combining different learning algorithms. This is because in dynamic data stream

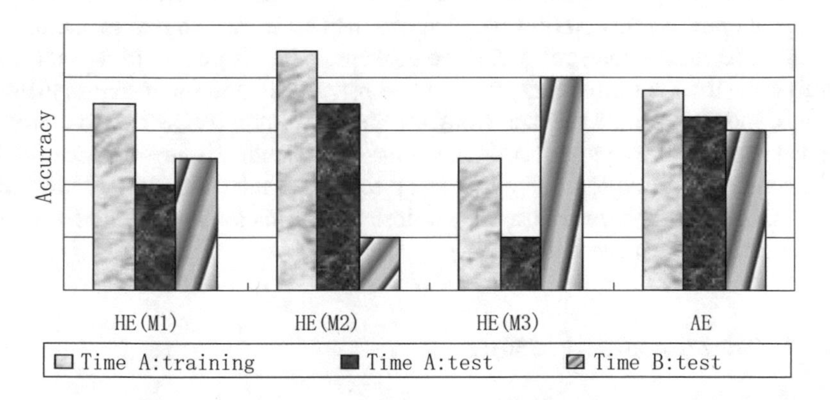

Fig. 5.9 A toy example for comparisons between AE and three HE ensemble methods trained with different learning algorithms (*i.e.*, algorithms M_1, M_2, and M_3). For each ensemble method, three results (bars) are listed for comparisons. The left bar denotes the training accuracy at time A, the bar in the middle denotes the test accuracy at time A, and the bar on the right denotes the test accuracy at time B which follows time stamp A. It is obvious that at time A, the higher the training accuracy, the better the prediction result. However, this result doesn't hold when the concept drifts at the next time stamp B

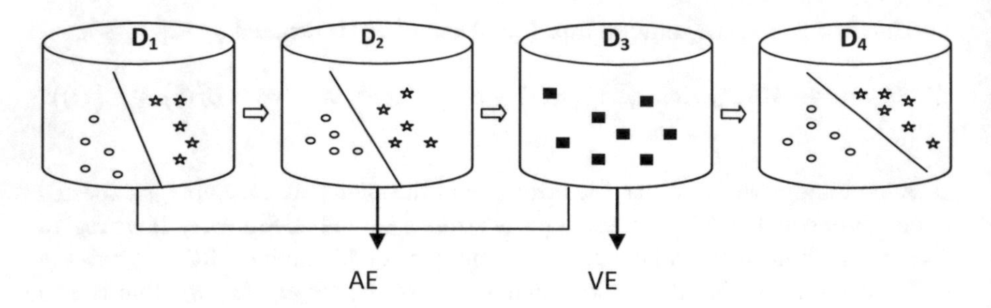

Fig. 5.10 A toy example for comparison between AE and VE. The concept (*i.e.*, the classification boundary) drifts marginally from chunk D_1 to D_2, and finally to D_4. Notice that the up-to-date chunk D_3 is a noisy chunk that carries useless or erroneous. Information when predicting the yet-to-come data chunk D_4

environments it is essentially difficult to know which learning algorithm performs the best at a particular time point. By integrating different learning algorithms as a unified model, we can expect AE to have the smallest variance error and thus have the best prediction accuracy.

AE performs better than VE when the concept drifts marginally and the up-to-date training chunk contains a significant amount of noisy samples. As illustrated in Fig. 5.10, assume that the concept drifts slightly along data chunks, and the up-to-date chunk D_3 is a noisy chunk. VE built on the up-to-date chunk D_3 will show deteriorated performance in predicting D_4. On the other hand, AE can largely avoid such a limitation by incorporating information from classifiers trained from the historical data chunks D_1 and D_2.

Although we have demonstrated that AE, on average, outperforms HE and VE, we are not claiming that AE always performs the best in data stream scenarios. For example, HE may outperform AE if the concept drifts marginally in data streams. In this case, the joint probability distribution $p(x,y)$ will stay stable across the data streams, and thus we can select a strong learning algorithm (*i.e.*, SVM) to construct HE and expect HE to outperform AE. On the other hand, VE may outperform AE if the concept drifts significantly and the up-to-date chunk contains very few noisy samples. In such a case, old-fashioned historical information in AE will deteriorate the learner performance even worse.

5.2.3.2 Time Complexity Analysis

In this section, we study the time complex of the AE framework and discuss whether it is a suitable model, from computational cost perspective, for mining noisy data streams. As discussed above, compared to its peers, AE combines much more base classifiers to build an ensemble predictor. This raises the concern on the efficiency of AE due to its additional cost for training extra base classifiers.

To study AE's time complexity, let's consider the following example. Assume the buffer of the system contains d data chunks, each of which contains N instances. Assume further that m learning algorithms are used to build models. Each time when a new data chunk arrives, we need to follow two steps to update an ensemble: (1) build new base classifier(s) on the new data chunk; and (2) update classifier ensemble by incorporating new base classifier(s). Without loss of generality, we assume that training a new base classifier needs $O(N \lg N)$ time on average, while updating the classifier ensemble to include one base classifier requires $O(\Gamma)$ time, where Γ is related to the dimensionality of attributes. Then the updating of the HE ensemble for each new data chunk needs to (1) build a new base classifier (which costs $O(N \lg N)$time), and then (2) combine the most recent d base classifiers (which costs $O(d\Gamma)$ time) together for prediction. The total time cost can be calculated by Eq. (5.39),

$$O(HE) = O\left(N \lg N\right) + O\left(d\Gamma\right). \tag{5.39}$$

Since training a base classifier dominates the total cost (*i.e.*, $O(\Gamma) < < O(N \lg N)$), and the number of data chunks d in the buffer is rather small. The time complexity of the HE ensemble can be simplified as in Eq. (5.40),

$$O(HE) = O\left(N \lg N\right) + O\left(d\Gamma\right) = O\left(N \lg N\right). \tag{5.40}$$

In comparison, VE builds m base classifiers for each new data chunk. Accordingly, its time complexity $O(VE)$ can be calculated by Eq. (5.41),

$$O(VE) = O(m) * \left(O\left(N \lg N\right) + O\left(\Gamma\right)\right) = O\left(mN \lg N\right) \tag{5.41}$$

For AE, it first builds m classifiers when a new data chunk arrives and combines all the d*m base classifiers to build an ensemble. Therefore, its time complexity can be calculated by Eq. (5.42),

$$O(AE) = O\left(mN \lg N\right) + O\left(dm\Gamma\right) = O\left(mN \lg N\right). \tag{5.42}$$

Combining Eqs. (5.40), (5.41), and (5.42), we have the following two conclusions: (1) AE is, asymptotically, as efficient as VE. Both of them have the same time complexity $O(mN \lg N)$; (2) AE requires more time complexity than HE because AE needs to train m base classifiers for each new data chunk. This limitation, in practice, can be alleviated by using a multi-core or multi-processor computing system, where base classifiers can be dispatched and trained on different computation units in parallel. The detailed data analysis of this section can be found in [4].

References

1. Zhang, P., Zhu, X., Shi, Y., Wu, X.: An aggregate ensemble for mining concept drifting data streams with noise. In: Pacific-Asia Conference on Knowledge Discovery and Data Mining, pp. 1021–1029. Springer, New York (2009)
2. Zhu, X.: Semi-supervised learning literature survey (2005)
3. Babcock, B., Babu, S., Datar, M., Motwani, R., Widom, J.: Models and issues in data stream systems. In: Proceedings of the Twenty-First ACM SIGMOD-SIGACT-SIGART Symposium on Principles of Database Systems, pp. 1–16 (2002)
4. Zhang, P., Zhu, X., Shi, Y., Guo, L., Wu, X.: Robust ensemble learning for mining noisy data streams. Decis. Support. Syst. **50**(2), 469–479 (2011)
5. Zhu, X., Zhang, P., Lin, X., Shi, Y.: Active learning from stream data using optimal weight classifier ensemble. IEEE Trans. Syst. Man Cybernetics Part B Cybernetics. **40**(6), 1607–1621 (2010)
6. Aggarwal, C.C., Philip, S.Y., Han, J., Wang, J.: A framework for clustering evolving data streams. In: Proceedings 2003 VLDB Conference, pp. 81–92. Elsevier, Amsterdam (2003)
7. Han, J., Cheng, H., Xin, D., Yan, X.: Frequent pattern mining: current status and future directions. Data Min. Knowl. Disc. **15**(1), 55–86 (2007)
8. Zhang, P., Li, J., Wang, P., Gao, B.J., Zhu, X., Guo, L.: Enabling fast prediction for ensemble models on data streams. In: Proceedings of the 17th ACM SIGKDD International Conference on Knowledge Discovery and Data Mining, pp. 177–185 (2011)
9. Asuncion, A., Newman, D.: Uci machine learning repository (2007)
10. Domingos, P., Hulten, G.: Mining high-speed data streams. In: Proceedings of the Sixth ACM SIGKDD International Conference on Knowledge Discovery and Data Mining, pp. 71–80 (2000)
11. Wang, H., Fan, W., Yu, P.S., Han, J.: Mining concept-drifting data streams using ensemble classifiers. In: Proceedings of the Ninth ACM SIGKDD International Conference on Knowledge Discovery and Data Mining, pp. 226–235 (2003)
12. Collobert, R., Sinz, F., Weston, J., Bottou, L., Joachims, T.: Large scale transductive svms. J. Mach. Learn. Res. **7**(8) (2006)
13. Zhang, P., Gao, B.J., Liu, P., Shi, Y., Guo, L.: A framework for application-driven classification of data streams. Neurocomputing. **92**, 170–182 (2012)
14. Chapelle, O., Sindhwani, V., Keerthi, S.S.: Optimization techniques for semi-supervised support vector machines. J. Mach. Learn. Res. **9**(2), 203–233 (2008)
15. Joachims, T., et al.: Transductive inference for text classification using support vector machines. ICML. **99**, 200–209 (1999)
16. Evgeniou, T., Pontil, M.: Regularized multi-task learning. In: Proceedings of the Tenth ACM SIGKDD International Conference on Knowledge Discovery and Data Mining, pp. 109–117 (2004)
17. Zhang, P., Zhu, X., Guo, L.: Mining data streams with labeled and unlabeled training examples. In: 2009 Ninth IEEE International Conference on Data Mining, pp. 627–636. IEEE, New York (2009)
18. Dietterich, T.G.: Ensemble methods in machine learning. In: International Workshop on Multiple Classifier Systems, pp. 1–15. Springer, New York (2000)
19. Hulten, G., Spencer, L., Domingos, P.: Mining time-changing data streams. In: Proceedings of the Seventh ACM SIGKDD International Conference on Knowledge Discovery and Data Mining, pp. 97–106 (2001)
20. Yuille, A.L., Rangarajan, A.: The concave-convex procedure (cccp). Adv. Neural Inf. Proces. Syst. **2**, 1033–1040 (2002)
21. Gao, J., Fan, W., Han, J.: On appropriate assumptions to mine data streams: analysis and practice. In: Seventh IEEE International Conference on Data Mining (ICDM 2007), pp. 143–152. IEEE, New York (2007)

22. Zhang, P., Zhu, X., Shi, Y.: Categorizing and mining concept drifting data streams. In: Proceedings of the 14th ACM SIGKDD International Conference on Knowledge Discovery and Data Mining, pp. 812–820 (2008)
23. Zhu, X.: Stream data mining repository. www.cse.fau.edu/xqzhu/S (2009)
24. Street, W.N., Kim, Y.: A streaming ensemble algorithm (sea) for large-scale classification. In: Proceedings of the Seventh ACM SIGKDD International Conference on Knowledge Discovery and Data Mining, pp. 377–382 (2001)
25. Kolter, J.Z., Maloof, M.A.: Using additive expert ensembles to cope with concept drift. In: Proceedings of the 22nd International Conference on Machine Learning, pp. 449–456 (2005)
26. Yang, Y., Wu, X., Zhu, X.: Combining proactive and reactive predictions for data streams. In: Proceedings of the Eleventh ACM SIGKDD International Conference on Knowledge Discovery in Data Mining, pp. 710–715 (2005)
27. Zhu, X., Jin, R.: Multiple information sources cooperative learning. In: Proceedings of the IJCAI, pp. 1369–1375 (2007)
28. Dai, W., Yang, Q., Xue, G., Yu, Y.: Boosting for transfer learning. In: Proceedings of the ICML, pp. 193–200 (2007)
29. Tsymbal, A.: The problem of concept drift: definitions and related work. Computer Science Department, Trinity College Dublin **106**(2), 58 (2004)

Chapter 6
Learning Analysis

It is event that most big data represented as non-structured or semi-structured forms, such as images, text and others. It is important to study how to use an abstract form to show data, either structured, non-structured or semi-structured or use label proportions to categorize the nature of data so that a data mining or data analytic algorithm can be performed smoothly. Leaning methods are very useful tools for understanding the data. Learning algorithms can be considered from different aspects, such as cognitive computing, mathematics, and machine learning.

This chapter deals with different learning techniques in the contexts of data science. Section 6.1 discusses the view of learning through the concept (the abstract of big data), which includes four subsections. Section 6.1.1 is about concept-cognitive learning model for Incremental concept learning [58]. Section 6.1.2 is a concurrent concept-cognitive learning model for classification [60]. Section 6.1.3 is a semi-supervised concept learning by concept-cognitive learning and conceptual clustering [42]. Section 6.1.4 is a fuzzybased concept learning method-exploiting data with fuzzy conceptual clustering [43]. Section 6.2 presents how to use the label proportion for learning that consists of another four subsections. Section 6.2.1 is a fast algorithm for learning from label proportions [84]. Section 6.2.2 is a learning from label proportions with generative adversarial networks [39]. Section 6.2.3 is a learning from label proportions on high-dimensional data [57]. Section 6.2.4 is a learning from label proportions with pinball loss [59]. Section 6.3 explores other enlarged learning models with two subsections. Section 6.3.1 is about classifying with adaptive hyper-spheres: an incremental classifier based on competitive learning [38]. Section 6.3.2 is a construction of robust representations for small data sets using broad learning system [66].

Y. Shi, *Advances in Big Data Analytics*,
https://doi.org/10.1007/978-981-16-3607-3_6

6.1 Concept of the View of Learning

6.1.1 Concept-Cognitive Learning Model for Incremental Concept Learning

Cognitive computing is viewed as an emerging computing paradigm of intelligent science that implements computational intelligence by trying to solve the problems of imprecision, uncertainty and partial truth in biological system [44, 68, 72]. As far as we know, it has been investigated by simulating human cognitive processes such as memory [33, 62] learning [14, 30, 36], thinking [69] and problem solving [70].

In this subsection, a novel CCLM is proposed based on a formal decision context. Moreover, to reduce its computational complexity, granular computing is included in our model. The main contributions are as follows:

(1) We describe a new model for incremental learning from the perspective of cognitive learning by the fusion of concept learning, granular computing, and formal decision context theory. More precisely, it is an attempt to construct a novel incremental algorithm by imitating human cognitive processes, and a new theory has been proposed for concept classification under a formal decision context.
(2) Beyond traditional CCL systems such as approximate CCL system [35, 36], three-way CCL system [37] and theoretical CCL system [68–70], CCLM has obtained incremental concept learning and generalization ability.
(3) Different from other classifiers, similar to the human learning processes, the previously acquired knowledge can be directly stored into concept lattice space in CCLM and it performs a good interpretation by concept hierarchies (e.g., Hasse diagram [16]).

6.1.1.1 Preliminaries

Now, we briefly review some basic notions related to (1) formal context, (2) formal decision context and (3) concept-cognitive learning.

A. Formal Context and Formal Decision Context

Definition 6.1 ([75]) A formal context is a triplet (G, M, I), where G is a set of objects, M is a set of attributes, and $I \subseteq G \times M$ is a binary relation between G and M. Here, gIm means that the object g has the attribute m.

Furthermore, the derivation operator $(\cdot)'$ is defined for $A \subseteq G$ and $B \subseteq M$ as follows:

$$A' = \{m \in M | gIm \text{ for all } g \in A\},$$
$$B' = \{g \in G | gIm \text{ for all } m \in B\}. \tag{6.1}$$

A' is the maximal set of the attributes that all the objects in A have in common and B' is the maximal set of the objects shared by all the attributes in B. A *concept* in the context (G, M, I) is defined to be an ordered pair (A, B) if $A' = B$ and $B' = A$, where the elements A and B of the concept (A, B) are called the extent and intent, respectively. The set of all concepts forms a complete lattice, called the concept lattice and denoted by $L(G, M, I)$.

Definition 6.2 ([74, 82]) A formal decision context is a quintuple (G, M, I, D, J), where (G, M, I) and (G, D, J) are two formal contexts. M and D are respectively called the conditional attribute set and the decision attribute set with $M \cap D = \emptyset$.

Definition 6.3 ([34]) Let (G, M, I, D, J) be a formal decision context and $E \subseteq M$. For any $(A, B) \in L(G, E, I_E)$ and $(Y, Z) \in L(G, D, J)$, if $A \subseteq Y$, and A, B, Y and Z are nonempty, then we say that (Y, Z) can be implied by (A, B), which is denoted by $(A, B) \to (Y, Z)$.

By Definitions 6.2 and 6.3, we obtain the relationship between the conditional attribute set and the decision attribute set.

B. Concept-Cognitive Learning

Let G be an object set and M be an attribute set. We denote the power sets of G and M by 2^G and 2^M, respectively. In addition, $\mathcal{F} : 2^G \to 2^M$ and $\mathcal{H} : 2^M \to 2^G$ are supposed to be two set-valued mappings, and they are rewritten as $\mathcal{F}$ and $\mathcal{H}$ for short.

Definition 6.4 ([36]) Set-valued mappings $\mathcal{F}$ and $\mathcal{H}$ are called cognitive operators if for any $A_1, A_2 \subseteq G$ and $B \subseteq M$, the following properties hold:

$$\text{(i)} \quad A_1 \subseteq A_2 \Rightarrow \mathcal{F}(A_2) \subseteq \mathcal{F}(A_1),$$
$$\text{(ii)} \quad \mathcal{F}(A_1 \cup A_2) \supseteq \mathcal{F}(A_1) \cap \mathcal{F}(A_2),$$
$$\text{(iii)} \quad \mathcal{H}(B) = \{g \in G | B \subseteq \mathcal{F}(\{g\})\}.$$

For convenience, hereinafter $\mathcal{F}(\{g\})$ is rewritten as $\mathcal{F}(g)$ for short when there is no confusion.

Definition 6.5 ([36]) Let $\mathcal{F}$ and $\mathcal{H}$ be cognitive operators. For $g \in G$ and $m \in M$, we say that $(\mathcal{H}\mathcal{F}(g), \mathcal{F}(g))$ and $(\mathcal{H}(m), \mathcal{F}\mathcal{H}(m))$ are granular concepts.

Definition 6.6 ([36]) Let G_{i-1}, G_i be object sets of $\{G_t\}\uparrow$ and M_{i-1}, M_i be attribute sets of $\{M_t\}\uparrow$, where $\{G_t\}\uparrow$ is a non-decreasing sequence of object sets $G_1, G_2, \ldots, G_n$ and $\{M_t\}\uparrow$ is a non-decreasing sequence of attribute sets $M_1, M_2, \ldots, M_m$. Denote $\Delta G_{i-1} = G_i - G_{i-1}$ and $\Delta M_{i-1} = M_i - M_{i-1}$. Suppose

1) $\mathcal{F}_{i-1} : 2^{G_{i-1}} \rightarrow 2^{M_{i-1}}$, $\mathcal{H}_{i-1} : 2^{M_{i-1}} \rightarrow 2^{G_{i-1}}$,

2) $\mathcal{F}_{\Delta G_{i-1}} : 2^{\Delta G_{i-1}} \rightarrow 2^{M_{i-1}}$, $\mathcal{H}_{\Delta G_{i-1}} : 2^{M_{i-1}} \rightarrow 2^{\Delta G_{i-1}}$,

3) $\mathcal{F}_{\Delta M_{i-1}} : 2^{G_i} \rightarrow 2^{\Delta M_{i-1}}$, $\mathcal{H}_{\Delta M_{i-1}} : 2^{\Delta M_{i-1}} \rightarrow 2^{G_i}$,

4) $\mathcal{F}_i : 2^{G_i} \rightarrow 2^{M_i}$, $\mathcal{H}_i : 2^{M_i} \rightarrow 2^{G_i}$

are four pairs of cognitive operators satisfying the following properties:

$$\mathcal{F}_i(g) = \begin{cases} \mathcal{F}_{i-1}(g) \cup \mathcal{F}_{\Delta M_{i-1}}(g), & \text{if } g \in G_{i-1}, \\ \mathcal{F}_{\Delta G_{i-1}}(g) \cup \mathcal{F}_{\Delta M_{i-1}}(g), & \text{otherwise}, \end{cases} \quad (6.2)$$

$$\mathcal{H}_i(m) = \begin{cases} \mathcal{H}_{i-1}(m) \cup \mathcal{H}_{\Delta G_{i-1}}(m), & \text{if } m \in M_{i-1}, \\ \mathcal{H}_{\Delta M_{i-1}}(m), & \text{otherwise}, \end{cases} \quad (6.3)$$

where $\mathcal{F}_{\Delta G_{i-1}}(g)$ and $\mathcal{H}_{\Delta G_{i-1}}(m)$ are set to be empty when $\Delta G_{i-1} = \emptyset$, and $\mathcal{F}_{\Delta M_{i-1}}(g)$ and $\mathcal{H}_{\Delta M_{i-1}}(m)$ are set to be empty when $\Delta M_{i-1} = \emptyset$. Then we say that $\mathcal{F}_i$ and $\mathcal{H}_i$ are extended cognitive operators of $\mathcal{F}_{i-1}$ and $\mathcal{H}_{i-1}$ with the newly input information ΔG_{i-1} and ΔM_{i-1}.

In other words, based on Definitions 6.4 and 6.5, the basic mechanism of concept-cognitive process is shown in Definition 6.6.

6.1.1.2 Theoretical Foundation

In this section, for adapting to dynamic learning and classification task, we show some new notions and properties for the proposed CCLM.

A. Initial Concept Generation

Definition 6.7 A regular formal decision context is a quintuple (G, M, I, D, J), where for any $z_1, z_2 \in D$, $\mathcal{H}(z_1) \cap \mathcal{H}(z_2) = \emptyset$. (G, M, I) and (G, D, J) are called the conditional formal context and the decision formal context, respectively.

Note that it means that each real-world object is associated with a single label.

Definition 6.8 Let (G, M, I, D, J) be a regular formal decision context, and $\mathcal{F}$ and $\mathcal{H}$ be cognitive operators. For $g \in G$ and $m \in M$, we say that $(\mathcal{HF}(g), \mathcal{F}(g))$

and $(\mathcal{H}(m), \mathcal{FH}(m))$ are conditional granular concepts. Similarly, for $y \in G$ and $z \in D$, $(\mathcal{HF}(y), \mathcal{F}(y))$ and $(\mathcal{H}(z), \mathcal{FH}(z))$ are decision granular concepts. For simplicity, we denote

$$\mathcal{G}^C = \{(\mathcal{HF}(g), \mathcal{F}(g))|g \in G\} \cup \{(\mathcal{H}(m), \mathcal{FH}(m))|m \in M\},$$
$$\mathcal{G}^D = \{(\mathcal{HF}(y), \mathcal{F}(y))|y \in G\} \cup \{(\mathcal{H}(z), \mathcal{FH}(z))|z \in D\},$$

where $\mathcal{G}^C$ and $\mathcal{G}^D$ are respectively called as condition-concept space and decision-concept space (or class-concept space) under cognitive operators $\mathcal{F}$ and $\mathcal{H}$.

Property 6.1 Let (G, M, I, D, J) be a regular formal decision context, and $\mathcal{F}$ and $\mathcal{H}$ be cognitive operators. Then for any $A, Y \subseteq G$, $B \subseteq M$ and $Z \subseteq D$, we have

$$\mathcal{F}(A) = \bigcap_{g \in A} \mathcal{F}(g), \mathcal{F}(Y) = \bigcap_{y \in Y} \mathcal{F}(y),$$
$$\mathcal{H}(B) = \bigcap_{m \in B} \mathcal{H}(m), \mathcal{H}(Z) = \bigcap_{z \in Z} \mathcal{H}(z). \tag{6.4}$$

Proof It is immediate from Definitions 6.4 and 6.7. □

Property 6.2 Let (G, M, I, D, J) be a regular formal decision context. For any $(A_G, B_G) \in \mathcal{G}^C$ and $(Y_G, Z_G) \in \mathcal{G}^D$, if $A_G \subseteq Y_G$, and A_G, B_G, Y_G and Z_G are nonempty, then we say that A_G is associated with the class of Z_G under the attribute set B_G. It means that the object g can be represented by a single label z when $A_G = \mathcal{HF}(g)$ and $Z_G = \mathcal{FH}(z)$.

Proof It is immediate from Definitions 6.3 and 6.8. □

Definition 6.9 Let (G, M, I, D, J) be a regular formal decision context and $D_1, D_2, \ldots, D_l$ be nonempty and finite class sets of D, where $D = D_1 \cup D_2 \cup \ldots \cup D_l$ and $D_r \cap D_j = \emptyset (1 \le r, j \le l, r \ne j)$. We call $G_i^D = G_i^{D_1} \cup G_i^{D_2} \cup \ldots \cup G_i^{D_l}$ is class-object set under the i-th cognitive state.

For brevity, we write G_i^D as G_i and the corresponding set of G_i is denoted by $\{G_i\} = \bigcup_{j=1}^{l} \{G_i^{D_j}\}$. Considering that the information will be updated by different classes, we initiate and learn concepts by different labels. For convenience, for any $D_j \subseteq D$, the subclass-object sets $G_1^{D_j}, G_2^{D_j}, \ldots, G_n^{D_j}$ with $G_1^{D_j} \subseteq G_2^{D_j} \subseteq \ldots \subseteq G_n^{D_j}$ are denoted by $\{G_t^{D_j}\}\uparrow$.

Property 6.3 Let (G, M, I, D, J) be a regular formal decision context, we have

$$\left\{G_t^D\right\}\uparrow = \{G_t^{D_1}\}\uparrow \cup \{G_t^{D_2}\}\uparrow \cup \ldots \cup \{G_t^{D_l}\}\uparrow . \tag{6.5}$$

Proof It is immediate from Definitions 6.6 and 6.9. □

From Definitions 6.7 and 6.8, and Property 6.1, the initial concepts can be constructed by condition-concept space and class-concept space in a regular formal decision context. Then, an object can be associated with a single label by the interaction between condition-concept space and class-concept space from Property 6.2. Property 6.3 means that a cognitive state can be decomposed into some cognitive sub-states by different categories in a regular formal decision context. Therefore, hereinafter we only discuss the situation under a cognitive sub-state D_j.

B. Concept-Cognitive Process

Considering the information on the object set G and the attribute set M will be updated as time goes by in the real world, we discuss that how the concept spaces are timely updated in a regular formal decision context.

Definition 6.10 Let (G, M, I, D, J) be a regular formal decision context, $G_{i-1}^{D_j}, G_i^{D_j}$ be two subclass-objects of $\{G_t^{D_j}\}\uparrow$ and M_{i-1}, M_i be attribute sets of $\{M_t\}\uparrow$. Denote $\Delta G_{i-1}^{D_j} = G_i^{D_j} - G_{i-1}^{D_j}$, $\Delta M_{i-1} = M_i - M_{i-1}$. Suppose

1) $\mathcal{F}_{D_j,i-1}^{M}: 2^{G_{i-1}^{D_j}} \to 2^{M_{i-1}}$, $\quad \mathcal{H}_{D_j,i-1}^{M}: 2^{M_{i-1}} \to 2^{G_{i-1}^{D_j}}$,

2) $\mathcal{F}_{D_j,i-1}^{D}: 2^{G_{i-1}^{D_j}} \to 2^{D}$, $\quad \mathcal{H}_{D_j,i-1}^{D}: 2^{D} \to 2^{G_{i-1}^{D_j}}$,

3) $\mathcal{F}_{D_j,\Delta G_{i-1}^{D_j}}^{M}: 2^{\Delta G_{i-1}^{D_j}} \to 2^{M_{i-1}}$, $\quad \mathcal{H}_{D_j,\Delta G_{i-1}^{D_j}}^{M}: 2^{M_{i-1}} \to 2^{\Delta G_{i-1}^{D_j}}$,

4) $\mathcal{F}_{D_j,\Delta G_{i-1}^{D_j}}^{D}: 2^{\Delta G_{i-1}^{D_j}} \to 2^{D}$, $\quad \mathcal{H}_{D_j,\Delta G_{i-1}^{D_j}}^{D}: 2^{D} \to 2^{\Delta G_{i-1}^{D_j}}$,

5) $\mathcal{F}_{D_j,\Delta M_{i-1}}^{M}: 2^{G_i^{D_j}} \to 2^{\Delta M_{i-1}}$, $\quad \mathcal{H}_{D_j,\Delta M_{i-1}}^{M}: 2^{\Delta M_{i-1}} \to 2^{G_i^{D_j}}$,

6) $\mathcal{F}_{D_j,i}^{M}: 2^{G_i^{D_j}} \to 2^{M_i}$, $\quad \mathcal{H}_{D_j,i}^{M}: 2^{M_i} \to 2^{G_i^{D_j}}$,

7) $\mathcal{F}_{D_j,i}^{D}: 2^{G_i^{D_j}} \to 2^{D}$, $\quad \mathcal{H}_{D_j,i}^{D}: 2^{D} \to 2^{G_i^{D_j}}$

are seven pairs of cognitive operators in a regular formal decision context satisfying the following properties:

$$\mathcal{F}_{D_j,i}^{M}(g) = \begin{cases} \mathcal{F}_{D_j,i-1}^{M}(g) \cup \mathcal{F}_{D_j,\Delta M_{i-1}}^{M}(g), & \text{if } g \in G_{i-1}^{D_j}, \\ \mathcal{F}_{D_j,\Delta G_{i-1}^{D_j}}^{M}(g) \cup \mathcal{F}_{D_j,\Delta M_{i-1}}^{M}(g), & \text{otherwise}, \end{cases} \tag{6.6}$$

$$\mathcal{H}_{D_j,i}^{M}(m) = \begin{cases} \mathcal{H}_{D_j,i-1}^{M}(m) \cup \mathcal{H}_{D_j,\Delta G_{i-1}^{D_j}}^{M}(m), & \text{if } m \in M_{i-1}, \\ \mathcal{H}_{D_j,\Delta M_{i-1}}^{M}(m), & \text{otherwise}, \end{cases} \tag{6.7}$$

$$\mathcal{F}^D_{D_j,i}(y) = \begin{cases} \mathcal{F}^D_{D_j,i-1}(y), & \text{if } y \in G^{D_j}_{i-1}, \\ \mathcal{F}^D_{D_j,\Delta G^{D_j}_{i-1}}(y), & \text{otherwise}, \end{cases} \tag{6.8}$$

$$\mathcal{H}^D_{D_j,i}(z) = \mathcal{H}^D_{D_j,i-1}(z) \cup \mathcal{H}^D_{D_j,\Delta G^{D_j}_{i-1}}(z), \text{ if } z \in D, \tag{6.9}$$

where $\mathcal{F}^M_{D_j,\Delta G^{D_j}_{i-1}}(g)$, $\mathcal{H}^M_{D_j,\Delta G^{D_j}_{i-1}}(m)$ and $\mathcal{H}^D_{D_j,\Delta G^{D_j}_{i-1}}(z)$ are set to be empty when $\Delta G^{D_j}_{i-1} = \emptyset$, and $\mathcal{F}^M_{D_j,\Delta M_{i-1}}(g)$ and $\mathcal{H}^M_{D_j,\Delta M_{i-1}}(m)$ are set to be empty when $\Delta M_{i-1} = \emptyset$.

Then we say that $\mathcal{F}^M_{D_j,i}, \mathcal{F}^D_{D_j,i}$ and $\mathcal{H}^M_{D_j,i}, \mathcal{H}^D_{D_j,i}$ are respectively extended cognitive operators of $\mathcal{F}^M_{D_j,i-1}, \mathcal{F}^D_{D_j,i-1}$ and $\mathcal{H}^M_{D_j,i-1}, \mathcal{H}^D_{D_j,i-1}$ with the newly input data $\Delta G^{D_j}_{i-1}$ and ΔM_{i-1}. For convenience, cognitive operators $\mathcal{F}^M_{D,i}$ and $\mathcal{H}^M_{D,i}$ denote the combination of $\mathcal{F}^M_{D_1,i}, \mathcal{F}^M_{D_2,i}, \ldots, \mathcal{F}^M_{D_l,i}$ and $\mathcal{H}^M_{D_1,i}, \mathcal{H}^M_{D_2,i}, \ldots, \mathcal{H}^M_{D_l,i}$, respectively. Similarly, we can define $\mathcal{F}^D_{D,i}$ and $\mathcal{H}^D_{D,i}$.

Meanwhile, for any $D_j \subseteq D$, $\mathcal{G}^C_{\mathcal{F}^M_{D_j,i-1}, \mathcal{H}^M_{D_j,i-1}}$ means subcondition-concept space under cognitive operators $\mathcal{F}^M_{D_j,i-1}$ and $\mathcal{H}^M_{D_j,i-1}$, and $\mathcal{G}^C_{\mathcal{F}^M_{D,i-1}, \mathcal{H}^M_{D,i-1}}$ is called as condition-concept space under cognitive operators $\mathcal{F}^M_{D,i-1}$ and $\mathcal{H}^M_{D,i-1}$. In a similar manner, we can define $\mathcal{G}^D_{\mathcal{F}^D_{D,i-1}, \mathcal{H}^D_{D,i-1}}$ and $\mathcal{G}^D_{\mathcal{F}^D_{D_j,i-1}, \mathcal{H}^D_{D_j,i-1}}$. In $\mathcal{G}^C_{\mathcal{F}^M_{D,i-1}, \mathcal{H}^M_{D,i-1}}$, we can obtain the k-th granular concept $\left(A^{D_j}_{G,k}, B^{D_j}_{G,k}\right)$ from $\mathcal{G}^C_{\mathcal{F}^M_{D,i-1}, \mathcal{H}^M_{D,i-1}}$ with a class set D_j. Moreover, for dynamic information $\Delta G^{D_j}_{i-1}$, we write $\mathcal{G}^C_{\mathcal{F}^M_{D_j,\Delta G^{D_j}_{i-1}}, \mathcal{H}^M_{D_j,\Delta G^{D_j}_{i-1}}}$ and $\mathcal{G}^D_{\mathcal{F}^D_{D_j,\Delta G^{D_j}_{i-1}}, \mathcal{H}^D_{D_j,\Delta G^{D_j}_{i-1}}}$ as $\mathcal{G}^C_{\Delta G^{D_j}_{i-1}}$ and $\mathcal{G}^D_{\Delta G^{D_j}_{i-1}}$ $\left(\text{Similarly, } \mathcal{G}^C_{\Delta M_{i-1}} \text{ and } \mathcal{G}^D_{\Delta M_{i-1}} \text{ for } \Delta M_{i-1}\right)$ under operators $\mathcal{F}^M_{D_j,\Delta G^{D_j}_{i-1}}, \mathcal{H}^M_{D_j,\Delta G^{D_j}_{i-1}}$ and $\mathcal{F}^D_{D_j,\Delta G^{D_j}_{i-1}}, \mathcal{H}^D_{D_j,\Delta G^{D_j}_{i-1}}$ $\left(\mathcal{F}^M_{D_j,\Delta M_{i-1}}, \mathcal{H}^M_{D_j,\Delta M_{i-1}} \text{ and } \mathcal{F}^D_{D_j,\Delta M_{i-1}}, \mathcal{H}^D_{D_j,\Delta M_{i-1}}\right)$.

In theory, although we can update concepts by objects and attributes simultaneously, we are extremely interested in the new object information because the attributes can be regarded as relatively stable under certain conditions.

Theorem 6.1 *Let $G^{D_j}_i$ be a subclass-object set under a set D_j and $(\mathcal{G}_{\mathcal{F}_{D_j,i-1},\mathcal{H}_{D_j,i-1}}, \mathcal{F}^M_{D_j,\Delta G^{D_j}_{i-1}}, \mathcal{F}^D_{D_j,\Delta G^{D_j}_{i-1}}, \mathcal{H}^M_{D_j,\Delta G^{D_j}_{i-1}}, \mathcal{H}^D_{D_j,\Delta G^{D_j}_{i-1}})$ be an object-oriented cognitive computing state, where $\mathcal{G}_{\mathcal{F}_{D_j,i-1},\mathcal{H}_{D_j,i-1}}$ is the concept space under cognitive operators $\mathcal{F}_{D_j,i-1}$ and $\mathcal{H}_{D_j,i-1}$. Then the following statements hold:*

1) *For any* $g \in G_i^{D_j}$, *if* $g \in G_{i-1}^{D_j}$, *then*

$$\left(\mathcal{H}^M_{D_j,i}\mathcal{F}^M_{D_j,i}(g), \mathcal{F}^M_{D_j,i}(g)\right) = \left(\mathcal{H}^M_{D_j,i-1}\mathcal{F}^M_{D_j,i-1}(g) \cup \mathcal{H}^M_{D_j,\Delta G^{D_j}_{i-1}}\mathcal{F}^M_{D_j,i-1}(g), \mathcal{F}^M_{D_j,i-1}(g)\right);$$

otherwise,

$$\left(\mathcal{H}^M_{D_j,i}\mathcal{F}^M_{D_j,i}(g), \mathcal{F}^M_{D_j,i}(g)\right) = \left(\mathcal{H}^M_{D_j,i-1}\mathcal{F}^M_{D_j,\Delta G^{D_j}_{i-1}}(g) \cup \mathcal{H}^M_{D_j,\Delta G^{D_j}_{i-1}}\mathcal{F}^M_{D_j,\Delta G^{D_j}_{i-1}}(g), \mathcal{F}^M_{D_j,\Delta G^{D_j}_{i-1}}(g)\right).$$

2) *For any* $m \in M_{i-1}$, *we have*

$$\left(\mathcal{H}^M_{D_j,i}(m), \mathcal{F}^M_{D_j,i}\mathcal{H}^M_{D_j,i}(m)\right) = \left(\mathcal{H}^M_{D_j,i-1}(m) \cup \mathcal{H}^M_{D_j,\Delta G^{D_j}_{i-1}}(m), \mathcal{F}^M_{D_j,i-1}\mathcal{H}^M_{D_j,i-1}(m) \cap \mathcal{F}^M_{D_j,\Delta G^{D_j}_{i-1}}\mathcal{H}^M_{D_j,\Delta G^{D_j}_{i-1}}(m)\right).$$

3) *For any* $y \in G_i^{D_j}$, *if* $y \in G_{i-1}^{D_j}$, *then*

$$\left(\mathcal{H}^D_{D_j,i}\mathcal{F}^D_{D_j,i}(y), \mathcal{F}^D_{D_j,i}(y)\right) = \left(\mathcal{H}^D_{D_j,i-1}\mathcal{F}^D_{D_j,i-1}(y) \cup \mathcal{H}^D_{D_j,\Delta G^{D_j}_{i-1}}\mathcal{F}^D_{D_j,i-1}(y), \mathcal{F}^D_{D_j,i-1}(y)\right);$$

otherwise,

$$\left(\mathcal{H}^D_{D_j,i}\mathcal{F}^D_{D_j,i}(y), \mathcal{F}^D_{D_j,i}(y)\right) = \left(\mathcal{H}^D_{D_j,i-1}\mathcal{F}^D_{D_j,\Delta G^{D_j}_{i-1}}(y) \cup \mathcal{H}^D_{D_j,\Delta G^{D_j}_{i-1}}\mathcal{F}^D_{D_j,\Delta G^{D_j}_{i-1}}(y), \mathcal{F}^D_{D_j,\Delta G^{D_j}_{i-1}}(y)\right).$$

4) *For any* $z \in D$, *we obtain*

$$\left(\mathcal{H}^D_{D_j,i}(z), \mathcal{F}^D_{D_j,i}\mathcal{H}^D_{D_j,i}(z)\right) = \left(\mathcal{H}^D_{D_j,i-1}(z) \cup \mathcal{H}^D_{D_j,\Delta G^{D_j}_{i-1}}(z), \mathcal{F}^D_{D_j,i-1}\mathcal{H}^D_{D_j,i-1}(z) \cap \mathcal{F}^D_{D_j,\Delta G^{D_j}_{i-1}}\mathcal{H}^D_{D_j,\Delta G^{D_j}_{i-1}}(z)\right).$$

Proof The proof of Theorem 6.1 can be found in the original paper [58]. □

From Theorem 6.1, we observe that the i-th concept space can be constructed under cognitive operators $\mathcal{F}_{D_j,i-1}$ and $\mathcal{H}_{D_j,i-1}$, and the concept space $\mathcal{G}_{\mathcal{F}_{D_j,i},\mathcal{H}_{D_j,i}}$

can be obtained by $\mathcal{G}_{\mathcal{F}_{D_j,i-1},\mathcal{H}_{D_j,i-1}}$ with the newly input data $\Delta G_{i-1}^{D_j}$. This means that we can obtain concepts based on the past concepts rather than reconstructing them from the beginning.

However, in the previous discussions, we still do not know which class-object set $G_{i-1}^{D_*}$ should be theoretically updated with $\Delta G_{i-1}^{D_j}$. In other words, although we obtain class-object set $G_{i-1}^{D_j}$ which will be actually updated by $\Delta G_{i-1}^{D_j}$, we are not sure if the updated class-object set in the model is in accordance with $G_{i-1}^{D_j}$. Thus, we will further discuss the relationship between $G_{i-1}^{D_j}$ and $G_{i-1}^{D_*}$.

Definition 6.11 Let $(\mathcal{HF}(g), \mathcal{F}(g))$ be a granular concept, for any $(A_{G,e}, B_{G,e}) \in \mathcal{G}^C$, where $e \in \{1, 2, \ldots, |\mathcal{G}^C|\}$. Then we can define concept-similarity degree (CS) as follows:

$$\theta_{CS} = CS(\mathcal{F}(g), B_{G,e}) = \frac{W_p \cdot M^T}{|\mathcal{F}(g) \cup B_{G,e}|}, \tag{6.10}$$

where M^T is the transpose of the vector M, and $W_p = (w_1, w_2, \ldots, w_m)$ is a cognitive weight vector that is associated with an attribute vector $M = (m_1, m_2, \ldots, m_m)$ consisting of (1) the elements from $\mathcal{F}(g) \cap B_{G,e}$ which are all set to be 1, and (2) the elements from $M - (\mathcal{F}(g) \cap B_{G,e})$ which are all set to be 0.

Let E be training times. For any $t \in E$, the cognitive weight vector of the t-th training is denoted by $W_{i,p}^t = (w_{i,1}^t, w_{i,2}^t, \ldots, w_{i,m}^t)$. Then we denote

$$\begin{bmatrix} W_{1,p}^t \\ \vdots \\ W_{n,p}^t \end{bmatrix} = \begin{bmatrix} w_{1,1}^t & \cdots & w_{1,m}^t \\ \vdots & \cdots & \vdots \\ w_{n,1}^t & \cdots & w_{n,m}^t \end{bmatrix}, \tag{6.11}$$

where $n = \left|\bigcup_{i=1}^{n}\{G_i\}\right|$. Our purpose is to obtain an optimal cognitive weight vector $W_{n,p}^t$ by computing concept-similarity degree vectors.

Definition 6.12 Let $\{G_{i-1}\}$ be a class-object set under G_{i-1}^D, $\Delta G_{i-1}^{D_*}$ be a new object set under D_*, and $G_{i-1}^{D_j}$ and $G_{i-1}^{D_r}$ be class-object sets under class sets D_j and D_r ($D_j \cap D_r = \emptyset$), respectively. For any granular concept $\left(A_{G,e}^{D_j}, B_{G,e}^{D_j}\right) \in \mathcal{G}^C_{\mathcal{F}^M_{D_j,i-1},\mathcal{H}^M_{D_j,i-1}}$ and a new granular concept $\left(\mathcal{H}^M_{D_*,\Delta G_{i-1}^{D_*}}\mathcal{F}^M_{D_*,\Delta G_{i-1}^{D_*}}(g), \mathcal{F}^M_{D_*,\Delta G_{i-1}^{D_*}}(g)\right)$, the degree of similarity between the concepts is defined as $CS_{i-1}^{D_j} = CS\left(B_{G,e}^{D_j}, \mathcal{F}^M_{D_*,\Delta G_{i-1}^{D_*}}(g)\right)$. Then, we denote

$$MCS^{D_j} = \max_{e=1}^{n}\left(CS_{i-1}^{D_j}\right) = \max_{e=1}^{n}\left(CS\left(B_{G,e}^{D_j}, \mathcal{F}^M_{D_*,\Delta G_{i-1}^{D_*}}(g)\right)\right),$$

where $n = \left|\mathcal{G}^{C}_{\mathcal{F}^{M}_{D_j,i-1},\mathcal{H}^{M}_{D_j,i-1}}\right|$. Then, we further denote

$$MMCS^{D_j} = \max_{j=1}^{l}\left(MCS^{D_j}\right). \tag{6.12}$$

From (6.12), we know that the subclass-object set $G_{i-1}^{D_j}$ should be updated in the class-object set G_{i-1}^{D}. Therefore, if $D_* = D_j$, it means that the theoretically updated subclass-object set $G_{i-1}^{D_*}$ is in accordance with the actually updated subclass-object set $G_{i-1}^{D_j}$. Otherwise, we should adjust cognitive weight vectors as follows.

$$\begin{aligned} w_i^t &\leftarrow w_i^t \pm \Delta w_i^t, \\ \Delta w_i^t &= activationFunction(\eta w_i^t), \end{aligned} \tag{6.13}$$

where the operator $+$ is adopted when the attributes are from $B_{G,e}^{D_j} \bigcap \mathcal{F}^{M}_{D_* \Delta G_{i-1}^{D_*}}(g)$ and the another operator $-$ is used for the elements from $B_{G,e}^{D_r} \bigcap \mathcal{F}^{M}_{D_* \Delta G_{i-1}^{D_*}}(g)$, and $activationFunction(\eta w_i^t) = \frac{exp(\eta w_i^t) - exp(-\eta w_i^t)}{exp(\eta w_i^t) + exp(-\eta w_i^t)}$ with the learning rate $\eta \in (0, 1)$.

6.1.1.3 Proposed Model

In this section, based on the above discussion, we put forward a CCLM with dynamic learning, which can perform a good performance in incremental learning and classification task.

A. Initial Concept Learning

We split raw data into training data $\overline{G}$ and testing data $\overline{\overline{G}}$. For the training data, let $\{\overline{G}\}$ be the set of the objects sets $\overline{G}_1, \overline{G}_2, \ldots, \overline{G}_n$ with $\overline{G}_i \cap \overline{G}_j = \emptyset (i \neq j)$, we denote

$$\{\overline{G}\} = \bigcup_{i=1}^{n}\{\overline{G}_i\}. \tag{6.14}$$

Here, $\overline{G}_1$ is an initial training data, and the rest of training data $\bigcup_{i=2}^{n}\{\overline{G}_i\}$ is used for concept cognition.

From Definitions 6.7 and 6.8, the initial concept learning consists of two parts: constructing condition-concept space and decision-concept space. The details are shown in Algorithm 6.1, and its time complexity is $O\big(|\{\overline{G}_1\}|(|M|+|D|+|\overline{G}_1^{D_j}|)\big)$.

Algorithm 6.1 Initial concept learning

1: **Input:** the initial training data set $\overline{G}_1$.
2: **Output:** the initial concept space.
3: **for** each $\overline{G}_1^{D_j} \in \{\overline{G}_1\}$ **do**
4: **for** each $m \in M$ **do**
5: $\mathcal{G}^C_{\mathcal{F}^M_{D,1},\mathcal{H}^M_{D,1}} \leftarrow (\mathcal{H}^M_{D_j,1}(m), \mathcal{F}^M_{D_j,1}\mathcal{H}^M_{D_j,1}(m))$
6: **end for**
7: **for** each $g \in \overline{G}_1^{D_j}$ **do**
8: $\mathcal{G}^C_{\mathcal{F}^M_{D,1},\mathcal{H}^M_{D,1}} \leftarrow (\mathcal{H}^M_{D_j,1}\mathcal{F}^M_{D_j,1}(g), \mathcal{F}^M_{D_j,1}(g))$
9: **end for**
10: **for** each $z \in D$ **do**
11: $\mathcal{G}^D_{\mathcal{F}^D_{D,1},\mathcal{H}^D_{D,1}} \leftarrow (\mathcal{H}^D_{D_j,1}(z), \mathcal{F}^D_{D_j,1}\mathcal{H}^D_{D_j,1}(z))$
12: **end for**
13: **for** each $y \in \overline{G}_1^{D_j}$ **do**
14: $\mathcal{G}^D_{\mathcal{F}^D_{D,1},\mathcal{H}^D_{D,1}} \leftarrow (\mathcal{H}^D_{D_j,1}\mathcal{F}^D_{D_j,1}(y), \mathcal{F}^D_{D_j,1}(y))$
15: **end for**
16: **end for**
17: **Return** $\mathcal{G}^C_{\mathcal{F}^M_{D,1},\mathcal{H}^M_{D,1}}$ and $\mathcal{G}^D_{\mathcal{F}^D_{D,1},\mathcal{H}^D_{D,1}}$

B. Concept-Cognitive Process

Let E, err_0, W, AW, IW be the training epochs, learning error rate, cognitive weight vector, active weight vector and inhibited weight vector, respectively. It should be pointed out that AW and IW are to enhance and weaken the corresponding attributes, respectively. Based on the theory in Sect. 6.1.1 Theoretical Foundation, the concept-cognitive process can be briefly represented as follows:

Firstly, construct a conditional granular concept $\big(A^{D_*}_{G,k}, B^{D_*}_{G,k}\big)$ and a decision granular concept $\big(Y^{D_*}_{G,k}, Z^{D_*}_{G,k}\big)$.

Secondly, for a new concept $\big(A^{D_*}_{G,k}, B^{D_*}_{G,k}\big)$, we compute its concept-similarity degree with each granular concept $\big(A^{D_j}_{G,e}, B^{D_j}_{G,e}\big)$ from $\mathcal{G}^C_{\mathcal{F}^M_{D,i-1},\mathcal{H}^M_{D,i-1}}$.

Thirdly, if the predicted label is not in accordance with the actual label, the weight vectors W, AW and IW will be updated.

Finally, for the first training, we will update the condition-concept space and decision-concept space by dynamic concepts $\big(A^{D_*}_{G,k}, B^{D_*}_{G,k}\big)$ and $\big(Y^{D_*}_{G,k}, Z^{D_*}_{G,k}\big)$, respectively. Using recursive approach, we can obtain a final cognitive weight vector

$W_{n,p}^{E}$ and a final concept space (i.e., the condition-concept space $\mathcal{G}_{\mathcal{F}_{D,n}^{M},\mathcal{H}_{D,n}^{M}}^{C}$ and decision-concept space $\mathcal{G}_{\mathcal{F}_{D,n}^{D},\mathcal{H}_{D,n}^{D}}^{D}$).

The details of concept-cognitive process are shown in Algorithm 6.2. Note that params[1], params[2] and params[3] are $\Big((A_{G,k}^{D_*}, B_{G,k}^{D_*}), \mathcal{G}_{\mathcal{F}_{D,i-1}^{M},\mathcal{H}_{D,i-1}^{M}}^{C}, \mathcal{G}_{\mathcal{F}_{D,i-1}^{D},\mathcal{H}_{D,i-1}^{D}}^{D}, W_{i-1,p}^{t-1}\Big), \Big(\eta, j, type, B_{G,k}^{D_{type}}, \theta L_{max}[|D|], W_{i-1,p}^{t-1}, AW_{i-1,p}^{t-1}, IW_{i-1,p}^{t-1}\Big)$ and $\Big(\mathcal{G}_{\mathcal{F}_{D,i-1}^{M},\mathcal{H}_{D,i-1}^{M}}^{C}, \mathcal{G}_{\mathcal{F}_{D,i-1}^{D},\mathcal{H}_{D,i-1}^{D}}^{D}, (A_{G,k}^{D_{type}}, B_{G,k}^{D_{type}}), (Y_{G,k}^{D_{type}}, Z_{G,k}^{D_{type}})\Big)$, respectively.

Now, we analyze the time complexity of Algorithm 6.2. Running Step 18 takes $O(1)$ because of updating objects one by one in CCLM. In Step 20, it will revoke Algorithm 6.3, and the running time is decided by two for loops. Thus, running Steps 18–26 takes $O\Big(|\mathcal{G}_{\mathcal{F}_{D,i-1}^{M},\mathcal{H}_{D,i-1}^{M}}^{C}||\mathcal{G}_{\mathcal{F}_{D,i-1}^{D},\mathcal{H}_{D,i-1}^{D}}^{D}|\Big)$, where $|\mathcal{G}_{\mathcal{F}_{D,i-1}^{D},\mathcal{H}_{D,i-1}^{D}}^{D}|$ is the number of $|D|$ and often very small. For Steps 27–32, it will call Algorithms 6.4 and 6.5. Therefore, the time complexity of Steps 27–32 is $O\Big(|D|\big((|activeSet| + |inhibitSet|) + (|M| + |D|)\big)\Big)$. To sum up, the time complexity of Algorithm 6.2 is $O\Big(P|\bigcup_{i=2}^{n}\{\overline{G}_i\}|\,(|\overline{G}_i^{D_*}| + |\mathcal{G}_{\mathcal{F}_{D,i-1}^{M},\mathcal{H}_{D,i-1}^{M}}^{C}||D| + Q)\Big)$ $(P = max\{E, E_{erro}\}, Q = (|M| + |D|)(|D| + 1) + |D|(|activeSet| + |inhibitSet|))$, where E is the number of training epochs and E_{erro} is the running times about $erro$.

C. Overall Procedure and Concept Prediction

Figure 6.1 shows the overall procedure of CCLM which includes three stages: initial concept generation, concept-cognitive process and concept prediction. Suppose there are still three classes to predict. The stage of initial concept generation is to generate concept space by mapping objects into concepts, and then the second stage will update the concept space by the concept-similarity degree with labeled data.

In the stage of concept prediction, for any test instance, concept-similarity degree is further used to compute similarity degree, and then the final prediction will be completed by the sum of the maximum class vector as shown in the right of Fig. 6.1. Note that, compared with the second stage, the concept space will not be updated in the third stage.

Based on the final concept space $\mathcal{G}_{\mathcal{F}_{D,n}^{M},\mathcal{H}_{D,n}^{M}}^{C}, \mathcal{G}_{\mathcal{F}_{D,n}^{D},\mathcal{H}_{D,n}^{D}}^{D}$, and the final weight vector $W_{n,p}^{E}$, we can make predictions in $\overline{\overline{G}}$. The details are described in Algorithm 6.6. Considering that running Step 6 will revoke the function of Algorithm 6.3, it is easy to verify that the time complexity of Algorithm 6.6 is $O\big(|\overline{\overline{G}}||\mathcal{G}_{\mathcal{F}_{D,i-1}^{M},\mathcal{H}_{D,i-1}^{M}}^{C}||D|\big)$.

Algorithm 6.2 Concept-cognitive process

1: **Input:** initial concept spaces $\mathcal{G}^C_{\mathcal{F}^M_{D,1},\mathcal{H}^M_{D,1}}$ and $\mathcal{G}^D_{\mathcal{F}^D_{D,1},\mathcal{H}^D_{D,1}}$.
2: **Output:** a final concept space and a final weight vector.
3: Initialize $W^1_{1,p}$, $AW^1_{1,p}$, $IW^1_{1,p}$, η, err_0, and E.
4: **while** $t \leq E||err_{min} \leq err_0$ **do** ▷ Initialize t=2.
5:     **for** each $\overline{G}_i^{D_*} \in \bigcup_{i=2}^{n}\{\overline{G}_i\}$ **do**
6:         **for** each $m \in M$ **do**
7:             $\mathcal{G}^C_{\overline{G}_i^{D_*}} \leftarrow \big(\mathcal{H}^M_{D_*,i}(m), \mathcal{F}^M_{D_*,i}\mathcal{H}^M_{D_*,i}(m)\big)$
8:         **end for**
9:         **for** each $g \in \overline{G}_i^{D_*}$ **do**
10:             $\mathcal{G}^C_{\overline{G}_i^{D_*}} \leftarrow \big(\mathcal{H}^M_{D_*,i}\mathcal{F}^M_{D_*,i}(g), \mathcal{F}^M_{D_*,i}(g)\big)$
11:         **end for**
12:         **for** each $z \in D$ **do**
13:             $\mathcal{G}^D_{\overline{G}_i^{D_*}} \leftarrow \big(\mathcal{H}^D_{D_*,i}(z), \mathcal{F}^D_{D_*,i}\mathcal{H}^D_{D_*,i}(z)\big)$
14:         **end for**
15:         **for** each $y \in \overline{G}_i^{D_*}$ **do**
16:             $\mathcal{G}^D_{\overline{G}_i^{D_*}} \leftarrow \big(\mathcal{H}^D_{D_*,i}\mathcal{F}^D_{D_*,i}(y), \mathcal{F}^D_{D_*,i}(y)\big)$
17:         **end for**
18:         **for** each$(A^{D_*}_{G,k}, B^{D_*}_{G,k}) \in \mathcal{G}^C_{\overline{G}_i^{D_*}}$ **do**
19:             Get a concept $(Y^{D_*}_{G,k}, Z^{D_*}_{G,k}) \in \mathcal{G}^D_{\overline{G}_i^{D_*}}$.
20:             Get concept-similarity degrees by Algorithm 6.3.
21:             $\theta_{max}[index] \leftarrow max\big(\theta_{max}[|D|]\big)$
22:             type ← indexType$\big(Y^{D_*}_{G,k}\big)$ ▷ Get a label.
23:             **if** index ≠ type **then**
24:                 errFunction$_i(A^{D_*}_{G,k})$=1 ▷ Misclassification.
25:             **end if**
26:         **end for**
27:         **for** j=0 to $|D|$ **do**
28:             **if** $\theta_{max}[j] \geq \theta_{max}[type]$ **then**
29:                 Update weight vector by Algorithm 6.4.
30:             **end if**
31:             Update concept space by Algorithm 6.5.
32:         **end for**
33:     **end for**
34:     $err_{min} \leftarrow \text{minimize}\left(\frac{\sum_{i=2}^{n} \text{errFunction}_i\left(A^{D_*}_{G,k}\right)}{\left|\bigcup_{i=2}^{n} \overline{G}_i\right|}\right)$
35:     $++t$
36: **end while**
37: **Return** $\mathcal{G}^C_{\mathcal{F}^M_{D,n},\mathcal{H}^M_{D,n}}$, $\mathcal{G}^D_{\mathcal{F}^D_{D,n},\mathcal{H}^D_{D,n}}$ and $W^E_{n,p}$.

Algorithm 6.3 Concept-similarity degree

1: **function** GETCONCEPTSIMILARITY(params[1])
2: Initialize $\theta_{max}[|D|], \theta L_{max}[|D|]$
3: **for** $\left(A_{G,e}^{D_j}, B_{G,e}^{D_j}\right) \in \mathcal{G}^C_{\mathcal{F}^M_{D,i-1}, \mathcal{H}^M_{D,i-1}}$ **do**
4: **for** $\left(Y_{G,q}^{D_r}, Z_{G,q}^{D_r}\right) \in \mathcal{G}^D_{\mathcal{F}^D_{D,i-1}, \mathcal{H}^D_{D,i-1}}$ **do**
5: **if** $A_{G,e}^{D_j} \subseteq Y_{G,q}^{D_r}$ **then**
6: $\theta^{t-1}_{CS_{i-1}} = CS_{i-1}\left(B_{G,k}^{D_*}, B_{G,e}^{D_j}\right)$
7: type $\leftarrow$ indexType($Y_{G,q}^{D_r}$)
8: **if** $\theta^{t-1}_{CS_{i-1}} \geq \theta_{max}[type]$ **then**
9: $\theta_{max}[type] = \theta^{t-1}_{CS_{i-1}}$
10: $\theta L_{max}[type] = B_{G,e}^{D_j}$
11: **end if**
12: **end if**
13: **end for**
14: **end for**
15: **return** $\theta_{max}[|D|], \theta L_{max}[|D|]$
16: **end function**

Algorithm 6.4 Adjust weight

1: **function** ADJUSTWEIGHT(params[2])
2: typeSet=$\theta L_*[type] \bigcap B_{G,k}^{type}$
3: jSet=$\theta L_*[j] \bigcap B_{G,k}^{type}$
4: activeSet=typeSet -jSet
5: inhibitSet=jSet - typeSet
6: **while** $m_1 \in$ activeSet **do**
7: indexA=indexAttribtue(m_1)
8: Update $AW^{t-1}_{i-1,p}$ by $aw^{t-1}_{i-1,indexA}++$.
9: Update $W^{t-1}_{i-1,p}$ by (6.13). ▷ Input $\eta aw^{t-1}_{i-1,indexA}$.
10: **end while**
11: **while** $m_2 \in$ inhibitSet **do**
12: indexB=indexAttribtue(m_2)
13: Update $IW^{t-1}_{i-1,p}$ by $iw^{t-1}_{i-1,indexB}++$.
14: Update $W^{t-1}_{i-1,p}$ by (6.13). ▷ Input $\eta iw^{t-1}_{i-1,indexB}$.
15: **end while**
16: **return** $W^{t-1}_{i-1,p}, AW^{t-1}_{i-1,p}, IW^{t-1}_{i-1,p}$
17: **end function**

Algorithm 6.5 Update concept space

1: **function** UPDATECONCEPTS(params[3])
2: **for** each $m \in M$ **do**
3: Update $\mathcal{G}^{C}_{\mathcal{F}^{M}_{D,i-1},\mathcal{H}^{M}_{D,i-1}}$ by Theorem 6.1.
4: **end for**
5: **for** each $z \in D$ **do**
6: Update $\mathcal{G}^{D}_{\mathcal{F}^{D}_{D,i-1},\mathcal{H}^{D}_{D,i-1}}$ by Theorem 6.1.
7: **end for**
8: **return** $\mathcal{G}^{C}_{\mathcal{F}^{M}_{D,i-1},\mathcal{H}^{M}_{D,i-1}}, \mathcal{G}^{D}_{\mathcal{F}^{D}_{D,i-1},\mathcal{H}^{D}_{D,i-1}}$
9: **end function**

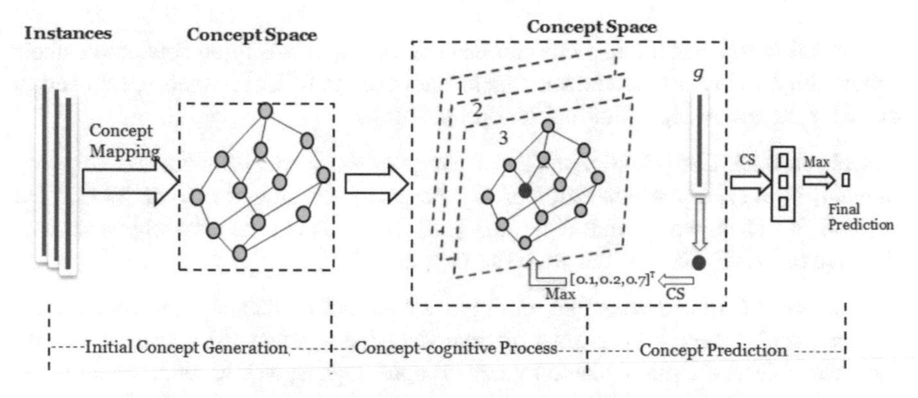

Fig. 6.1 Illustration of overall procedure for CCLM. Suppose there are three classes to predict, and the maximum class vector is obtained by concept-similarity degree

Algorithm 6.6 Concept prediction

1: **Input:** the testing data $\overline{\overline{G}}$, $W^{E}_{n,p}$, $\mathcal{G}^{C}_{\mathcal{F}^{M}_{D,n},\mathcal{H}^{M}_{D,n}}, \mathcal{G}^{D}_{\mathcal{F}^{D}_{D,n},\mathcal{H}^{D}_{D,n}}$.
2: **Output:** the class labels of test data.
3: **for** each $g_i \in \overline{\overline{G}}$ **do**
4: $\left(A^{D_*}_{G,k}, B^{D_*}_{G,k}\right) \leftarrow \left(\mathcal{H}^{M}_{D_*,i}\mathcal{F}^{M}_{D_*,i}(g_i), \mathcal{F}^{M}_{D_*,i}(g_i)\right)$
5: $\left(Y^{D_*}_{G,k}, Z^{D_*}_{G,k}\right) \leftarrow \left(\mathcal{H}^{D}_{D_*,i}\mathcal{F}^{D}_{D_*,i}(y_i), \mathcal{F}^{D}_{D_*,i}(y_i)\right)$
6: $\theta_{max}[|D|] \leftarrow$ getConceptSimilarity(params[1])
7: $\theta_{max}[index] \leftarrow \max(\theta_{max}[|D|])$
8: type $\leftarrow$ indexType$\left(Y^{D_*}_{G,k}\right)$
9: **if** index=type **then**
10: $correctNum+=1$
11: **else**
12: $incorrectNum+=1$
13: **end if**
14: **end for**

6.1.2 Concurrent Concept-Cognitive Learning Model for Classification

In this subsection, we discuss the design of a new theoretical framework for concurrent computing, which comprises three aspects: initial concurrent concept learning, the concurrent concept-cognitive process, and the concept generalization process.

6.1.2.1 Initial Concurrent Concept Learning in C3LM

In the real world, not all methods can be concurrent, as this often depends on their separability. In order to guarantee concurrency for the C3LM in theory, we need to consider the following definitions and propositions.

Definition 6.13 Let (G, M, I, D, J) be a regular formal decision context. Suppose that $D_1, D_2, \ldots, D_K$ is a partition of D by class labels, and let $G = G^{D_1} \cup G^{D_2} \cup \ldots \cup G^{D_K}$. Then, we say that G^{D_k} $(k \in \{1, 2, \ldots, K\})$ is a subclass-object set. For the sake of brevity, hereinafter we write G^{D_k} as G^k.

Definition 6.13 indicates that an object set G can be decomposed into several subclass-object sets in a regular formal decision context. Moreover, we only consider objects that are updated by newly input objects, as attributes can be taken as relatively stable in real life. Therefore, in the following, we discuss the scenario of a subclass-object G^k.

Let G^k be a subclass-object set, and M and D be attribute sets. The set-valued mappings $\mathcal{F}^k : 2^{G^k} \to 2^M, \mathcal{H}^k : 2^M \to 2^{G^k}$ and $\widetilde{\mathcal{F}}^k : 2^{G^k} \to 2^D, \widetilde{\mathcal{H}}^k : 2^D \to 2^{G^k}$ are respectively referred to as the conditional and decision cognitive operators with a subclass-object set G^k when no confusion exists.

Definition 6.14 Let $G_1^k, G_2^k, \ldots, G_n^k$ be a partition of an object set G^k. If the following cognitive operators:

$$\mathcal{F}_j^k : 2^{G_j^k} \to 2^M, \qquad \mathcal{H}_j^k : 2^M \to 2^{G_j^k}, j = 1, 2, \ldots, n,$$

$$\mathcal{F}^k : 2^{G^k} \to 2^M, \qquad \mathcal{H}^k : 2^M \to 2^{G^k}$$

satisfy $\mathcal{F}^k(g) = \mathcal{F}_j^k(g)$, where $g \in G_j^k$, we say that $\mathcal{HS}_{\mathcal{F}^k\mathcal{H}^k} = (\mathcal{F}_1^k, \ldots, \mathcal{F}_n^k; \mathcal{H}_1^k, \ldots, \mathcal{H}_n^k)$ is a conditional horizontal partition state.

Proposition 6.1 *Let* $\mathcal{HS}_{\mathcal{F}^k\mathcal{H}^k} = (\mathcal{F}_1^k, \ldots, \mathcal{F}_n^k; \mathcal{H}_1^k, \ldots, \mathcal{H}_n^k)$ *be a conditional horizontal partition state. For any* $g \in G_{j_1}^k$ $(j_1 \in \{1, 2, \ldots, n\})$, *if there exist objects* $g_1, g_2, \ldots, g_n \in G_{j_2}^k$ $(j_2 \in \{1, 2, \ldots, n\})$ *such that* $\mathcal{F}_{j_1}^k(g) \subseteq \mathcal{F}_{j_2}^k(g_i)$ $(i =$

$1, 2, \ldots, n$), *we have*

$$(\mathcal{H}^k\mathcal{F}^k(g), \mathcal{F}^k(g)) = \Big(\{g \cup (\bigcup_{i=1}^{n} g_i)\}, \mathcal{F}_{j_1}^k(g)\Big); \tag{6.15}$$

otherwise,

$$(\mathcal{H}^k\mathcal{F}^k(g), \mathcal{F}^k(g)) = (\{g\}, \mathcal{F}_{j_1}^k(g)). \tag{6.16}$$

Proof The proof of Proposition 6.1 can be found in the original paper [60]. □

In fact, from the perspective of objects, Definition 6.14 and Proposition 6.1 demonstrate that the separability holds for C3LM in the conditional formal context (G, M, I). Analogously, we can determine that the separability also holds for C3LM in the decision formal context (G, D, J) under the decision cognitive operators $\widetilde{\mathcal{F}}^k$ and $\widetilde{\mathcal{H}}^k$.

Definition 6.15 Let $M_1, M_2, \ldots, M_d$ be a partition of M. For any $G^k \subseteq G$, if the following cognitive operators:

$$\mathcal{F}_j^k : 2^{G^k} \to 2^{M_j}, \qquad \mathcal{H}_j^k : 2^{M_j} \to 2^{G^k}, \ j = 1, 2, \ldots, d,$$

$$\mathcal{F}^k : 2^{G^k} \to 2^{M}, \qquad \mathcal{H}^k : 2^{M} \to 2^{G^k}$$

satisfy $\mathcal{F}^k\mathcal{H}^k(m) = \bigcup_{j=1}^{d} \mathcal{F}_j^k\mathcal{H}^k(m)$ where $m \in M$, we say that $\mathcal{VS}_{\mathcal{F}^k\mathcal{H}^k} = (\mathcal{H}_1^k, \ldots, \mathcal{H}_d^k; \mathcal{F}_1^k, \ldots, \mathcal{F}_d^k)$ is a conditional vertical partition state.

Proposition 6.2 *Let* $\mathcal{VS}_{\mathcal{F}^k\mathcal{H}^k} = (\mathcal{H}_1^k, \ldots, \mathcal{H}_d^k; \mathcal{F}_1^k, \ldots, \mathcal{F}_d^k)$ *be a conditional vertical partition state. For any* $m \in M_{j_1}$ ($j_1 \in \{1, 2, \ldots, d\}$), *if there exist attributes* $m_1, m_2, \ldots, m_r \in M_{j_2}$ ($j_2 \in \{1, 2, \ldots, d\}$) *such that* $\mathcal{H}_{j_1}^k(m) \subseteq \mathcal{H}_{j_2}^k(m_i)$ ($i = 1, 2, \ldots, r$), *we have*

$$(\mathcal{H}^k(m), \mathcal{F}^k\mathcal{H}^k(m)) = \Big(\mathcal{H}^k(m), \{m \cup (\bigcup_{i=1}^{r} m_i)\}\Big); \tag{6.17}$$

otherwise,

$$(\mathcal{H}^k(m), \mathcal{F}^k\mathcal{H}^k(m)) = (\mathcal{H}^k(m), \{m\}). \tag{6.18}$$

Proof The proof of Proposition 6.2 can also be found in the original paper [60]. □

From Definition 6.15 and Proposition 6.2, we know that the separability holds for C3LM in the conditional formal context (G, M, I) from the attribute perspective. Similarly, under decision cognitive operators $\widetilde{\mathcal{F}}^k$ and $\widetilde{\mathcal{H}}^k$, there exists the same property for C3LM in the decision formal context (G, D, J).

Based on the above theory, we present an initial concurrent computing framework (see Fig. 6.2 for details) and its corresponding algorithm (see Algorithm 6.7)

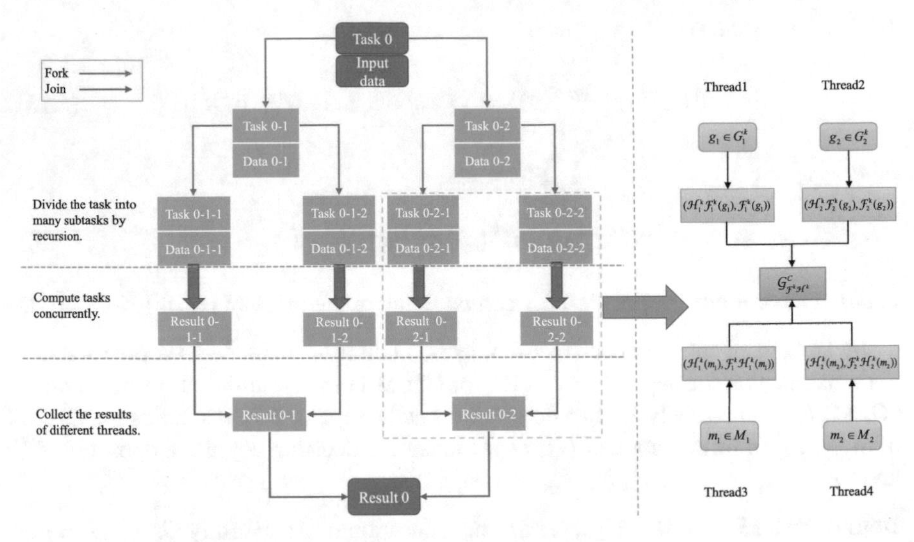

Fig. 6.2 Framework of constructing initial concepts in C3LM

Algorithm 6.7 Concurrent computation of initial concept space

1: **Input:** Initial training dataset G^k and chunk size.
2: **Output:** The initial concept spaces $\mathcal{G}^C_{\mathcal{F}^k,\mathcal{H}^k}$ and $\mathcal{G}^D_{\widetilde{\mathcal{F}}^k,\widetilde{\mathcal{H}}^k}$.
3: $n = \lceil |G^k|/\text{chunk-size} \rceil$, $d = \lceil |M|/\text{chunk-size} \rceil$, and $l = \lceil |D|/\text{chunk-size} \rceil$ are the numbers of threads for the objects, conditional attributes, and decision attributes, respectively.
4: **for** $G^k_j = G^k_1$ to G^k_n **do** in parallel
5: **for** each $g \in G^k_j (j \in \{1, 2, \ldots, n\})$ **do**
6: $\mathcal{G}^C_{\mathcal{F}^k,\mathcal{H}^k} \leftarrow \left(\mathcal{H}^k_j \mathcal{F}^k_j(g), \mathcal{F}^k_j(g)\right)$
7: **end for**
8: **end for**
9: **for** $M_j = M_1$ to M_d **do** in parallel
10: **for** each $m \in M_j (j \in \{1, 2, \ldots, d\})$ **do**
11: $\mathcal{G}^C_{\mathcal{F}^k,\mathcal{H}^k} \leftarrow \left(\mathcal{H}^k_j(m), \mathcal{F}^k_j \mathcal{H}^k_j(m)\right)$
12: **end for**
13: **end for**
14: **for** $G^k_j = G^k_1$ to G^k_n **do** in parallel
15: **for** each $y \in G^k_j (j \in \{1, 2, \ldots, n\})$ **do**
16: $\mathcal{G}^D_{\widetilde{\mathcal{F}}^k,\widetilde{\mathcal{H}}^k} \leftarrow \left(\widetilde{\mathcal{H}}^k_j \widetilde{\mathcal{F}}^k_j(y), \widetilde{\mathcal{F}}^k_j(y)\right)$
17: **end for**
18: **end for**
19: **for** $D_j = D_1$ to D_l **do** in parallel
20: **for** each $z \in D_j (j \in \{1, 2, \ldots, l\})$ **do**
21: $\mathcal{G}^D_{\widetilde{\mathcal{F}}^k,\widetilde{\mathcal{H}}^k} \leftarrow \left(\widetilde{\mathcal{H}}^k_j(z), \widetilde{\mathcal{F}}^k_j \widetilde{\mathcal{H}}^k_j(z)\right)$
22: **end for**
23: **end for**
24: **Return** $\mathcal{G}^C_{\mathcal{F}^k,\mathcal{H}^k}$ and $\mathcal{G}^D_{\widetilde{\mathcal{F}}^k,\widetilde{\mathcal{H}}^k}$

---

for constructing the initial concepts. The overall process in Fig. 6.2 can be described as follows: first, a task can be divided into many subtasks by the recursion method, based on Definitions 6.14 and 6.15, and Propositions 6.1 and 6.2. Second, according to Propositions 6.1 and 6.2, threads can concurrently calculate the concepts of each task. Finally, the results of different threads will be collected by Propositions 6.1 and 6.2. Moreover, the right of Fig. 6.2 illustrates that four threads calculate granular concepts based on the object and attribute sets. It should be pointed out that the proposed C3LM is based on the fork/join framework.[1]

Furthermore, it is easy to determine that the time complexity of Algorithm 6.7 is $O(\frac{1}{n}|G^k| + \frac{1}{d}|M| + \frac{1}{l}|D|)$. For an object set G, by means of Algorithm 6.7, we can obtain the conditional concept space $\mathcal{G}^C_{\mathcal{FH}}$ and decision concept space $\mathcal{G}^D_{\widetilde{\mathcal{F}}\widetilde{\mathcal{H}}}$.

6.1.2.2 Concurrent Concept-Cognitive Process in C3LM

In the real world, objects will be updated as time passes, which means that the obtained concept spaces need to be updated accordingly. For a person, learning is not simply a matter of acquiring a description, but involves taking something new and integrating it sufficiently with the existing thought processes [41]. The learning ability in humans is known as a gradual cognitive process. Therefore, in this subsection, we explore the concept-cognitive process under a concurrent environment.

As with the classical cognitive process [36], combining Definitions 6.6 and 6.13, we obtain the cognitive operators for C3LM with the newly input objects $\Delta G^k_{i-1} = G^k_i - G^k_{i-1}$, as follows:

$$\begin{aligned}
&\text{(i)} && \mathcal{F}^k_{i-1} : 2^{G^k_{i-1}} \to 2^M, && \mathcal{H}^k_{i-1} : 2^M \to 2^{G^k_{i-1}}, \\
&\text{(ii)} && \mathcal{F}^k_{\Delta G^k_{i-1}} : 2^{\Delta G^k_{i-1}} \to 2^M, && \mathcal{H}^k_{\Delta G^k_{i-1}} : 2^M \to 2^{\Delta G^k_{i-1}}, \\
&\text{(iii)} && \mathcal{F}^k_i : 2^{G^k_i} \to 2^M, && \mathcal{H}^k_i : 2^M \to 2^{G^k_i},
\end{aligned} \tag{6.19}$$

and

$$\begin{aligned}
&\text{(iv)} && \widetilde{\mathcal{F}}^k_{i-1} : 2^{G^k_{i-1}} \to 2^D, && \widetilde{\mathcal{H}}^k_{i-1} : 2^D \to 2^{G^k_{i-1}}, \\
&\text{(v)} && \widetilde{\mathcal{F}}^k_{\Delta G^k_{i-1}} : 2^{\Delta G^k_{i-1}} \to 2^D, && \widetilde{\mathcal{H}}^k_{\Delta G^k_{i-1}} : 2^D \to 2^{\Delta G^k_{i-1}}, \\
&\text{(vi)} && \widetilde{\mathcal{F}}^k_i : 2^{G^k_i} \to 2^D, && \widetilde{\mathcal{H}}^k_i : 2^D \to 2^{G^k_i}.
\end{aligned} \tag{6.20}$$

[1] https://docs.oracle.com/javase/tutorial/essential/concurrency/forkjoin.html.

Definition 6.16 Let $\Delta G_{i-1}^k = G_i^k - G_{i-1}^k$ be a singleton set with a new object, and $\mathcal{F}^k_{\Delta G_{i-1}^k}, \mathcal{H}^k_{\Delta G_{i-1}^k}$ and $\widetilde{\mathcal{F}}^k_{\Delta G_{i-1}^k}, \widetilde{\mathcal{H}}^k_{\Delta G_{i-1}^k}$ be cognitive operators. For any $g \in \Delta G_{i-1}^k$, if $(\mathcal{H}^k_{\Delta G_{i-1}^k} \mathcal{F}^k_{\Delta G_{i-1}^k}(g), \mathcal{F}^k_{\Delta G_{i-1}^k}(g)) = (\{g\}, \mathcal{F}^k_{\Delta G_{i-1}^k}(g))$ and $(\widetilde{\mathcal{H}}^k_{\Delta G_{i-1}^k} \widetilde{\mathcal{F}}^k_{\Delta G_{i-1}^k}(g), \widetilde{\mathcal{F}}^k_{\Delta G_{i-1}^k}(g)) = (\{g\}, \widetilde{\mathcal{F}}^k_{\Delta G_{i-1}^k}(g))$, $(\mathcal{H}^k_{\Delta G_{i-1}^k} \mathcal{F}^k_{\Delta G_{i-1}^k}(g), \mathcal{F}^k_{\Delta G_{i-1}^k}(g))$ and $(\widetilde{\mathcal{H}}^k_{\Delta G_{i-1}^k} \widetilde{\mathcal{F}}^k_{\Delta G_{i-1}^k}(g), \widetilde{\mathcal{F}}^k_{\Delta G_{i-1}^k}(g))$ are referred to as the newly formed conditional atomic concept and decision atomic concept, respectively, with a single object g.

In fact, we consider that the obtained concept spaces $\mathcal{G}^C_{\mathcal{F}^k\mathcal{H}^k}$ and $\mathcal{G}^D_{\widetilde{\mathcal{F}}^k\widetilde{\mathcal{H}}^k}$ are updated by a newly input object, rather than adding multiple objects simultaneously. For the sake of convenience, we denote the initial concept spaces obtained by Algorithm 6.7, namely $\mathcal{G}^C_{\mathcal{F}^k\mathcal{H}^k}, \mathcal{G}^D_{\widetilde{\mathcal{F}}^k\widetilde{\mathcal{H}}^k}$ and $\mathcal{G}^C_{\mathcal{F}\mathcal{H}}, \mathcal{G}^D_{\widetilde{\mathcal{F}}\widetilde{\mathcal{H}}}$, as $\mathcal{G}^C_{\mathcal{F}_0^k\mathcal{H}_0^k}, \mathcal{G}^D_{\widetilde{\mathcal{F}}_0^k\widetilde{\mathcal{H}}_0^k}$ and $\mathcal{G}^C_{\mathcal{F}_0\mathcal{H}_0}, \mathcal{G}^D_{\widetilde{\mathcal{F}}_0\widetilde{\mathcal{H}}_0}$, respectively. According to Eqs. (6.19) and (6.20), the cognitive operators $\mathcal{F}_i^k, \mathcal{H}_i^k$ and $\widetilde{\mathcal{F}}_i^k, \widetilde{\mathcal{H}}_i^k$ in the i-th period can be obtained by the cognitive operators $\mathcal{F}_{i-1}^k, \mathcal{H}_{i-1}^k$ and $\widetilde{\mathcal{F}}_{i-1}^k, \widetilde{\mathcal{H}}_{i-1}^k$ in the $(i-1)$-th period with incremental objects, respectively. Moreover, we denote their corresponding concept spaces by $\mathcal{G}^C_{\mathcal{F}_i^k\mathcal{H}_i^k}$ and $\mathcal{G}^D_{\widetilde{\mathcal{F}}_i^k\widetilde{\mathcal{H}}_i^k}$. Furthermore, the entire concept spaces in the i-th period are further denoted by $\mathcal{G}^C_{\mathcal{F}_i\mathcal{H}_i}$ and $\mathcal{G}^D_{\widetilde{\mathcal{F}}_i\widetilde{\mathcal{H}}_i}$.

Proposition 6.3 *Let* $(\mathcal{H}^k_{\Delta G_{i-1}^k} \mathcal{F}^k_{\Delta G_{i-1}^k}(g), \mathcal{F}^k_{\Delta G_{i-1}^k}(g))$ *and* $(\widetilde{\mathcal{H}}^k_{\Delta G_{i-1}^k} \widetilde{\mathcal{F}}^k_{\Delta G_{i-1}^k}(g), \widetilde{\mathcal{F}}^k_{\Delta G_{i-1}^k}(g))$ *be the newly formed conditional and decision atomic concepts, respectively. Then, the following statements hold:*

(i) For any granular concept $(A_{k,j}, B_{k,j}) \in \mathcal{G}^C_{\mathcal{F}_{i-1}^k\mathcal{H}_{i-1}^k}$ $(j \in \{1, 2, \ldots, |\mathcal{G}^C_{\mathcal{F}_{i-1}^k\mathcal{H}_{i-1}^k}|\})$, *if*

$$B_{k,j} \cap \mathcal{F}^k_{\Delta G_{i-1}^k}(g) \neq \emptyset, (A_{k,j}, B_{k,j}) = (A_{k,j} \cup \mathcal{H}^k_{\Delta G_{i-1}^k} \mathcal{F}^k_{\Delta G_{i-1}^k}(g), B_{k,j} \cap \mathcal{F}^k_{\Delta G_{i-1}^k}(g));$$

otherwise,

$$\mathcal{G}^C_{\mathcal{F}_i^k\mathcal{H}_i^k} = \mathcal{G}^C_{\mathcal{F}_{i-1}^k\mathcal{H}_{i-1}^k} \cup (\mathcal{H}^k_{\Delta G_{i-1}^k} \mathcal{F}^k_{\Delta G_{i-1}^k}(g), \mathcal{F}^k_{\Delta G_{i-1}^k}(g)).$$

(ii) For any granular concept $(Y_{k,j}, Z_{k,j}) \in \mathcal{G}^D_{\widetilde{\mathcal{F}}_{i-1}^k\widetilde{\mathcal{H}}_{i-1}^k}$ $(j \in \{1, 2, \ldots, |\mathcal{G}^D_{\widetilde{\mathcal{F}}_{i-1}^k\widetilde{\mathcal{H}}_{i-1}^k}|\})$, *if*

$$Z_{k,j} \cap \widetilde{\mathcal{F}}^k_{\Delta G_{i-1}^k}(g) \neq \emptyset, (Y_{k,j}, Z_{k,j}) = (Y_{k,j} \cup \widetilde{\mathcal{H}}^k_{\Delta G_{i-1}^k} \widetilde{\mathcal{F}}^k_{\Delta G_{i-1}^k}(g), Z_{k,j} \cap \widetilde{\mathcal{F}}^k_{\Delta G_{i-1}^k}(g));$$

otherwise,

$$\mathcal{G}^D_{\widetilde{\mathcal{F}}_i^k\widetilde{\mathcal{H}}_i^k} = \mathcal{G}^D_{\widetilde{\mathcal{F}}_{i-1}^k\widetilde{\mathcal{H}}_{i-1}^k} \cup (\widetilde{\mathcal{H}}^k_{\Delta G_{i-1}^k} \widetilde{\mathcal{F}}^k_{\Delta G_{i-1}^k}(g), \widetilde{\mathcal{F}}^k_{\Delta G_{i-1}^k}(g)).$$

Proof *The proof of Proposition 6.3 can be found in the original paper [60].* □

For any object g, according to Definitions 6.5 and 6.6, we can obtain a concept $(\{g\}, \mathcal{F}_{\Delta G_{i-1}}(g))$, as the concept spaces are updated by adding objects sequentially. The concept similarity (CS) degree [58] is used in this study to explore the interaction of attributes in the concept-cognitive process.

Definition 6.17 ([58]) Suppose that $(\{g\}, \mathcal{F}_{\Delta G_{i-1}}(g))$ is a new concept. For any $(A_{k,j}, B_{k,j}) \in \mathcal{G}^C_{\mathcal{F}^k_{i-1},\mathcal{H}^k_{i-1}}$ $(j \in \{1, 2, \ldots, |\mathcal{G}^C_{\mathcal{F}^k_{i-1},\mathcal{H}^k_{i-1}}|\})$, the CS degree can be defined as follows:

$$\theta_{k,j} = \frac{W \cdot M^T}{\left|\mathcal{F}_{\Delta G_{i-1}}(g) \bigcup B_{k,j}\right|}, \tag{6.21}$$

where $W = (w_1, w_2, \ldots, w_m)$ is a cognitive weight vector regarding a conditional attribute set M, and $M = (m_1, m_2, \ldots, m_m)$ is an attribute vector that contains (1) the value of attributes from $\mathcal{F}_{\Delta G_{i-1}}(g) \cap B_{k,j}$, which are set to 1, and (2) the elements from $M - (\mathcal{F}_{\Delta G_{i-1}}(g) \cap B_{k,j})$, which are all set to 0.

For any object, there always exists a unique class that is most similar to it by the sample separation axiom [79]. Thus, based on Definition 6.17, we can determine the maximum CS degree $\theta^*_{k,j^*} = \max_{j \in \{1,2,\ldots,|\mathcal{G}^C_{\mathcal{F}^k_{i-1},\mathcal{H}^k_{i-1}}|\}} \{\theta_{k,j}\}$ and its corresponding concept (A_{k,j^*}, B_{k,j^*}) in the concept space $\mathcal{G}^C_{\mathcal{F}^k_{i-1},\mathcal{H}^k_{i-1}}$. Moreover, for the entire concept space $\mathcal{G}^C_{\mathcal{F}_{i-1},\mathcal{H}_{i-1}}$, we can further determine the global maximum CS degree $\theta^*_{k^*,j^*} = \max_{k \in \{1,2,\ldots,K\}} \{\theta^*_{k,j^*}\}$ and its corresponding concept $(A_{k^*,j^*}, B_{k^*,j^*})$.

Definition 6.18 If $\theta^*_{k^*,j^*}$ is the global maximum CS degree in the entire concept space $\mathcal{G}^C_{\mathcal{F}_{i-1},\mathcal{H}_{i-1}}$, we say that a new concept $(\{g\}, \mathcal{F}_{\Delta G_{i-1}}(g))$ can be classified into the concept space $\mathcal{G}^C_{\mathcal{F}^{k^*}_{i-1},\mathcal{H}^{k^*}_{i-1}}$ by the optimal concept $(A_{k^*,j^*}, B_{k^*,j^*})$. Moreover, for any $(Y_k, Z_k) \in \mathcal{G}^D_{\widetilde{\mathcal{F}}_{i-1},\widetilde{\mathcal{H}}_{i-1}}$ $(k \in \{1, 2, \ldots, |\mathcal{G}^D_{\widetilde{\mathcal{F}}_{i-1},\widetilde{\mathcal{H}}_{i-1}}|\})$, if $A_{k^*,j^*} \subseteq Y_k$, we say that the object g is associated with a single label z, where $Z_k = \{z\}$ in a regular formal decision context.

From Definition 6.18, we can determine that an object g is associated with a class label z if and only if the real class label $\widetilde{\mathcal{F}}_{\Delta G_{i-1}}(g)$ is consistent with the predicted class label z. However, when the ground truth label is not the same as the predicted value, we adjust the cognitive weight as follows:

$$\begin{aligned} w_i &\leftarrow w_i \pm \Delta w_i, \\ \Delta w_i &= activationFunction(\eta w_i), \end{aligned} \tag{6.22}$$

where the operator $+$ is adopted when the attributes are from $\mathcal{F}_{\Delta G_{i-1}}(g) \cap B_{k^*,j^*}$, and the other operator $-$ is used for the elements from $\mathcal{F}_{\Delta G_{i-1}}(g) \cap B_{k,j^*}$. Moreover,

$activationFunction(\eta w_i) = \frac{e^{\eta w_i} - e^{-\eta w_i}}{e^{\eta w_i} + e^{-\eta w_i}}$, where $\eta \in (0, 1)$ is known as the learning rate.

In the following, a computational procedure for a concurrent concept-cognitive process (see Algorithm 6.8) is proposed based on the above discussion. The inputs of Algorithm 6.8 are the concept spaces obtained from the output results of Algorithm 6.7. In Algorithm 6.8, running steps 9 and 12 requires $O\big(|\mathcal{G}^C_{\mathcal{F}^k_{i-1}\mathcal{H}^k_{i-1}}|\big)$ and $O\big(|\mathcal{G}^D_{\widetilde{\mathcal{F}}^k_{i-1}\widetilde{\mathcal{H}}^k_{i-1}}|\big)$, respectively. In line 15, the runtime is $O\big(|\mathcal{G}^C_{\mathcal{F}^k_{i-1}\mathcal{H}^k_{i-1}}|\big)$. Hence, it is easy to determine that the time complexity of Algorithm 6.8 is $O\big(n(\frac{1}{m}|\mathcal{G}^C_{\mathcal{F}^k_{i-1}\mathcal{H}^k_{i-1}}| + \frac{1}{p}|\mathcal{G}^D_{\widetilde{\mathcal{F}}^k_{i-1}\widetilde{\mathcal{H}}^k_{i-1}}| + |\mathcal{G}^D_{\widetilde{\mathcal{F}}_{i-1},\widetilde{\mathcal{H}}_{i-1}}|)\big)$. Then, we can obtain the collections of all conditional and decision concepts in the final period, which are denoted by $\mathcal{G}^C_{\mathcal{F}_n,\mathcal{H}_n}$ and $\mathcal{G}^D_{\widetilde{\mathcal{F}}_n,\widetilde{\mathcal{H}}_n}$, respectively.

6.1.2.3 Concept Generalization Process in C3LM

Based on the final concept spaces obtained, we can achieve classification ability. This can be understood in terms of two aspects: (1) it can complete the static classification task when the final concept spaces are directly obtained from the initial concept learning, and (2) by combining the initial concept construction process with the CCL process, it is suitable for the dynamic classification task. However, both methods predict label information by means of the CS degree.

For a test instance g, let $\Delta G_{i-1} = \{g\}$, and we obtain a new concept $\big(\mathcal{H}_{\Delta G_{i-1}}\mathcal{F}_{\Delta G_{i-1}}(g), \mathcal{F}_{\Delta G_{i-1}}(g)\big) = \big(\{g\}, \mathcal{F}_{\Delta G_{i-1}}(g)\big)$ by Definitions 6.5 and 6.16. Furthermore, according to Definitions 6.17 and 6.18, a procedure is proposed for the concept generalization task (see Algorithm 6.9). It is easy to determine that the time complexity of Algorithm 6.9 is $O\big(|\overline{G}|(|\mathcal{G}^C_{\mathcal{F}_n,\mathcal{H}_n}| + |\mathcal{G}^D_{\widetilde{\mathcal{F}}_n,\widetilde{\mathcal{H}}_n}|)\big)$.

6.1.3 Semi-Supervised Concept Learning by Concept-Cognitive Learning and Conceptual Clustering

In this subsection, we will first introduce the initial concept spaces with labeled data, and then the concept-cognitive process with unlabeled data, followed by the concept recognition and theoretical analysis of S2CL. Finally, we present the whole procedure and computational cost of our methods.

6.1.3.1 Concept Space with Structural Information

Definition 6.19 Suppose G^k is a sub-object set which is associated with a label k, and a quintuple (G^k, M, I, D, J) is known as a regular sub-object formal decision

Algorithm 6.8 Concurrent concept-cognitive process

1: **Input:** Initial concept spaces $\mathcal{G}^C_{\mathcal{F}^k_0,\mathcal{H}^k_0}$, $\mathcal{G}^D_{\widetilde{\mathcal{F}}^k_0,\widetilde{\mathcal{H}}^k_0}$ and $\mathcal{G}^C_{\mathcal{F}_0,\mathcal{H}_0}$, $\mathcal{G}^D_{\widetilde{\mathcal{F}}_0,\widetilde{\mathcal{H}}_0}$, chunk size, and adding new object set ΔG^k.
2: **Output:** The final concept spaces $\mathcal{G}^C_{\mathcal{F}^k_n,\mathcal{H}^k_n}$ and $\mathcal{G}^D_{\widetilde{\mathcal{F}}^k_n,\widetilde{\mathcal{H}}^k_n}$.
3: Initialize $\Delta G^k = \{\Delta G^k_0, \Delta G^k_1, \ldots, \Delta G^k_{n-1}\} = \{\{g_0\}, \{g_1\}, \ldots, \{g_{n-1}\}\}$ and $W = (w_1, w_2, \ldots, w_m)$
4: **for** i=1 to n **do**
5: $m = \lceil |\mathcal{G}^C_{\mathcal{F}^k_{i-1}\mathcal{H}^k_{i-1}}|/\text{chunk-size} \rceil$ and $p = \lceil |\mathcal{G}^D_{\widetilde{\mathcal{F}}^k_{i-1}\widetilde{\mathcal{H}}^k_{i-1}}|/\text{chunk-size} \rceil$ are the numbers of threads for $\mathcal{G}^C_{\mathcal{F}^k_{i-1}\mathcal{H}^k_{i-1}}$ and $\mathcal{G}^D_{\widetilde{\mathcal{F}}^k_{i-1}\widetilde{\mathcal{H}}^k_{i-1}}$
6: get g_{i-1} from ΔG^k_{i-1}
7: construct new concepts $(\mathcal{H}^k_{\Delta G^k_{i-1}}\mathcal{F}^k_{\Delta G^k_{i-1}}(g_{i-1}), \mathcal{F}^k_{\Delta G^k_{i-1}}(g_{i-1}))$ and $(\widetilde{\mathcal{H}}^k_{\Delta G^k_{i-1}}\widetilde{\mathcal{F}}^k_{\Delta G^k_{i-1}}(g_{i-1}), \widetilde{\mathcal{F}}^k_{\Delta G^k_{i-1}}(g_{i-1}))$ by Definition 6.16
8: **for** $\mathcal{G}^C_{\mathcal{F}^k_{i-1}\mathcal{H}^k_{i-1},j} = \mathcal{G}^C_{\mathcal{F}^k_{i-1}\mathcal{H}^k_{i-1},1}$ to $\mathcal{G}^C_{\mathcal{F}^k_{i-1}\mathcal{H}^k_{i-1},m}$ **do** in parallel
9: get $\mathcal{G}^C_{\mathcal{F}^k_i\mathcal{H}^k_i}$ by updating $\mathcal{G}^C_{\mathcal{F}^k_{i-1}\mathcal{H}^k_{i-1}}$ based on Proposition 6.3
10: **end for**
11: **for** $\mathcal{G}^D_{\widetilde{\mathcal{F}}^k_{i-1}\widetilde{\mathcal{H}}^k_{i-1},j} = \mathcal{G}^D_{\widetilde{\mathcal{F}}^k_{i-1}\widetilde{\mathcal{H}}^k_{i-1},1}$ to $\mathcal{G}^D_{\widetilde{\mathcal{F}}^k_{i-1}\widetilde{\mathcal{H}}^k_{i-1},p}$ **do** in parallel
12: get $\mathcal{G}^D_{\widetilde{\mathcal{F}}^k_i\widetilde{\mathcal{H}}^k_i}$ by updating $\mathcal{G}^D_{\widetilde{\mathcal{F}}^k_{i-1}\widetilde{\mathcal{H}}^k_{i-1}}$ based on Proposition 6.3
13: **end for**
14: **for** $\mathcal{G}^C_{\mathcal{F}^k_{i-1}\mathcal{H}^k_{i-1},j} = \mathcal{G}^C_{\mathcal{F}^k_{i-1}\mathcal{H}^k_{i-1},1}$ to $\mathcal{G}^C_{\mathcal{F}^k_{i-1}\mathcal{H}^k_{i-1},m}$ **do** in parallel
15: compute the maximum CS degree θ^*_{k,j^*} and the corresponding concept (A_{k,j^*}, B_{k,j^*}) by Eq. (6.21)
16: **end for**
17: compute the global maximum CS degree $\theta^*_{k^*,j^*}$ and corresponding concept $(A_{k^*,j^*}, B_{k^*,j^*})$ in $\mathcal{G}^C_{\mathcal{F}_{i-1},\mathcal{H}_{i-1}}$
18: **for** each $(Y_k, Z_k) \in \mathcal{G}^D_{\widetilde{\mathcal{F}}_{i-1},\widetilde{\mathcal{H}}_{i-1}}$ $(k \in \{1, 2, \ldots, |\mathcal{G}^D_{\widetilde{\mathcal{F}}_{i-1},\widetilde{\mathcal{H}}_{i-1}}|\})$ **do**
19: **if** $A_{k^*,j^*} \subseteq Y_k$ **then**
20: get the predicted label z by Definition 6.18
21: **end if**
22: **if** $z \neq \widetilde{\mathcal{F}}^k_{\Delta G^k_{i-1}}(g_{i-1})$ **then**
23: update the cognitive weight vector W by Eq. (6.22)
24: **end if**
25: **end for**
26: **end for**
27: **Return** $\mathcal{G}^C_{\mathcal{F}^k_n,\mathcal{H}^k_n}$ and $\mathcal{G}^D_{\widetilde{\mathcal{F}}^k_n,\widetilde{\mathcal{H}}^k_n}$

Algorithm 6.9 Generalization process

1: **Input:** The final concept spaces $\mathcal{G}^C_{\mathcal{F}_n,\mathcal{H}_n}$ and $\mathcal{G}^D_{\widetilde{\mathcal{F}}_n,\widetilde{\mathcal{H}}_n}$, and test data $\overline{G}$.
2: **Output:** The class labels of test data.
3: **for** each $g \in \overline{G}$ **do**
4: construct a new concept $\big(\mathcal{H}_{\Delta G_{i-1}}\mathcal{F}_{\Delta G_{i-1}}(g), \mathcal{F}_{\Delta G_{i-1}}(g)\big)$
5: **for** each $(A_{k,j}, B_{k,j}) \in \mathcal{G}^C_{\mathcal{F}_n,\mathcal{H}_n}$ **do**
6: get $\theta^*_{k^*,j^*}$ by Eq. (6.21)
7: **end for**
8: **for** each $(Y_k, Z_k) \in \mathcal{G}^D_{\widetilde{\mathcal{F}}_n,\widetilde{\mathcal{H}}_n}$ $(k \in \{1, 2, \ldots, |\mathcal{G}^D_{\widetilde{\mathcal{F}}_n,\widetilde{\mathcal{H}}_n}|\})$ **do**
9: get the class label z by Definition 6.18
10: **end for**
11: **end for**
12: **Return** class labels

context. Then (G^k, M, I) and (G^k, D, J) are respectively called the conditional sub-object formal context and decision sub-object formal context.

Moreover, the set-valued mappings $\mathcal{F}^k : 2^{G^k} \to 2^M, \mathcal{H}^k : 2^M \to 2^{G^k}$, and $\widetilde{\mathcal{F}}^k : 2^{G^k} \to 2^D, \widetilde{\mathcal{H}}^k : 2^D \to 2^{G^k}$ are respectively called the conditional sub-object cognitive operators and decision sub-object cognitive operators with a sub-object set G^k.

Definition 6.20 Let (G^k, M, I) be a conditional sub-object formal context, and $\mathcal{F}^k$, $\mathcal{H}^k$ be the conditional sub-object cognitive operators. For any $x', x'' \in G^k$, if $\mathcal{H}^k\mathcal{F}^k(x') = \{x'\}$ and $\mathcal{H}^k\mathcal{F}^k(x'') \supset \{x''\}$, then the pairs $(\mathcal{H}^k\mathcal{F}^k(x'), \mathcal{F}^k(x'))$ and $(\mathcal{H}^k\mathcal{F}^k(x''), \mathcal{F}^k(x''))$ are referred to as object-oriented conditional granular concepts (or simply object-oriented conditional concepts). For convenience, we denote

$$\begin{aligned}\mathcal{OG}_{\mathcal{F}^k\mathcal{H}^k} =& \{(\mathcal{H}^k\mathcal{F}^k(x'), \mathcal{F}^k(x'))|x' \in G^k\}\cup\\ & \{(\mathcal{H}^k\mathcal{F}^k(x''), \mathcal{F}^k(x''))|x'' \in G^k\}.\end{aligned}$$

Simultaneously, for any $a', a'' \in M$, if $\mathcal{F}^k\mathcal{H}^k(a') = \{a'\}$ and $\mathcal{F}^k\mathcal{H}^k(a'') \supset \{a''\}$, then the pairs $(\mathcal{H}^k(a'), \mathcal{F}^k\mathcal{H}^k(a'))$ and $(\mathcal{H}^k(a''), \mathcal{F}^k\mathcal{H}^k(a''))$ are called attribute-oriented conditional granular concepts (or simply attribute-oriented conditional concepts). For brevity, we further denote

$$\begin{aligned}\mathcal{AG}_{\mathcal{F}^k\mathcal{H}^k} =& \{(\mathcal{H}^k(a'), \mathcal{F}^k\mathcal{H}^k(a'))|a' \in M\}\cup\\ & \{(\mathcal{H}^k(a''), \mathcal{F}^k\mathcal{H}^k(a''))|a'' \in M\}.\end{aligned}$$

Definition 6.21 Let (G^k, D, J) be a decision sub-object formal context and $\widetilde{\mathcal{F}}^k$, $\widetilde{\mathcal{H}}^k$ be the decision sub-object cognitive operators. For any $x', x'' \in G^k$, if $\widetilde{\mathcal{H}}^k\widetilde{\mathcal{F}}^k(x') = \{x'\}$ and $\widetilde{\mathcal{H}}^k\widetilde{\mathcal{F}}^k(x'') \supset \{x''\}$, then the pairs $(\widetilde{\mathcal{H}}^k\widetilde{\mathcal{F}}^k(x'), \widetilde{\mathcal{F}}^k(x'))$ and $(\widetilde{\mathcal{H}}^k\widetilde{\mathcal{F}}^k(x''), \widetilde{\mathcal{F}}^k(x''))$ are known as object-oriented decision granular concepts (or simply object-oriented decision concepts). For convenience, we denote

$$\begin{aligned} \mathcal{OG}_{\widetilde{\mathcal{F}}^k\widetilde{\mathcal{H}}^k} =& \{(\widetilde{\mathcal{H}}^k\widetilde{\mathcal{F}}^k(x'), \widetilde{\mathcal{F}}^k(x'))|x' \in G^k\} \cup \\ & \{(\widetilde{\mathcal{H}}^k\widetilde{\mathcal{F}}^k(x''), \widetilde{\mathcal{F}}^k(x''))|x'' \in G^k\}. \end{aligned}$$

Meanwhile, for any $k', k'' \in D$, if $\widetilde{\mathcal{F}}^k\widetilde{\mathcal{H}}^k(k') = \{k'\}$ and $\widetilde{\mathcal{F}}^k\widetilde{\mathcal{H}}^k(k'') \supset \{k''\}$, then the pairs $(\widetilde{\mathcal{H}}^k(k'), \widetilde{\mathcal{F}}^k\widetilde{\mathcal{H}}^k(k'))$ and $(\widetilde{\mathcal{H}}^k(k''), \widetilde{\mathcal{F}}^k\widetilde{\mathcal{H}}^k(k''))$ are called attribute-oriented decision granular concepts (or simply attribute-oriented decision concepts). For brevity, we further denote

$$\begin{aligned} \mathcal{AG}_{\widetilde{\mathcal{F}}^k\widetilde{\mathcal{H}}^k} =& \{(\widetilde{\mathcal{H}}^k(k'), \widetilde{\mathcal{F}}^k\widetilde{\mathcal{H}}^k(k'))|k' \in D\} \cup \\ & \{(\widetilde{\mathcal{H}}^k(k''), \widetilde{\mathcal{F}}^k\widetilde{\mathcal{H}}^k(k''))|k'' \in D\}. \end{aligned}$$

To facilitate the subsequent discussion, in a regular sub-object formal decision context, the conditional concept space and decision concept space are respectively denoted by

$$\begin{aligned} \mathcal{G}_{\mathcal{F}^k\mathcal{H}^k} &= \mathcal{OG}_{\mathcal{F}^k\mathcal{H}^k} \cup \mathcal{AG}_{\mathcal{F}^k\mathcal{H}^k} \\ &= \{(\mathcal{H}^k\mathcal{F}^k(x), \mathcal{F}^k(x))|x \in G^k\} \cup \\ &\quad \{(\mathcal{H}^k(a), \mathcal{F}^k\mathcal{H}^k(a))|a \in M\}, \text{ and} \\ \mathcal{G}_{\widetilde{\mathcal{F}}^k\widetilde{\mathcal{H}}^k} &= \mathcal{OG}_{\widetilde{\mathcal{F}}^k\widetilde{\mathcal{H}}^k} \cup \mathcal{AG}_{\widetilde{\mathcal{F}}^k\widetilde{\mathcal{H}}^k} \\ &= \{(\widetilde{\mathcal{H}}^k\widetilde{\mathcal{F}}^k(x), \widetilde{\mathcal{F}}^k(x))|x \in G^k\} \cup \\ &\quad \{(\widetilde{\mathcal{H}}^k(k'), \widetilde{\mathcal{F}}^k\widetilde{\mathcal{H}}^k(k'))|k' \in D\}. \end{aligned}$$

It means that the concept spaces of sub-object set G^k can be constructed by means of the object-oriented concepts and attribute-oriented concepts.

Theorem 6.2 *Let $\mathcal{G}_{\mathcal{F}^k\mathcal{H}^k}$ and $\mathcal{G}_{\widetilde{\mathcal{F}}^k\widetilde{\mathcal{H}}^k}$ be the conditional concept space and decision concept space, respectively. Then the following statements hold:*

(1) For any conditional concepts $(\mathcal{H}^k\mathcal{F}^k(x), \mathcal{F}^k(x))$ and $(\mathcal{H}^k(a), \mathcal{F}^k\mathcal{H}^k(a))$, if there exists a conditional concept $(\mathcal{H}^k\mathcal{F}^k(x_i), \mathcal{F}^k(x_i)) \in \mathcal{G}_{\mathcal{F}^k\mathcal{H}^k}$ such that $\mathcal{F}^k(x) \subseteq \mathcal{F}^k(x_i)$ $(i \in \{1, 2, \ldots, |G^k|\})$ and a conditional concept $(\mathcal{H}^k(a_j), \mathcal{F}^k\mathcal{H}^k(a_j))$ $\in \mathcal{G}_{\mathcal{F}^k\mathcal{H}^k}$ such that $\mathcal{H}^k(a) \subseteq \mathcal{H}^k(a_j)$ $(j \in \{1, 2, \ldots, |M|\})$, then we have

$$\begin{aligned} (\mathcal{H}^k\mathcal{F}^k(x), \mathcal{F}^k(x)) &= (\{x \cup \bigcup_{i\in\{1,2,\ldots,|G^k|\}} x_i\}, \mathcal{F}^k(x)), \\ (\mathcal{H}^k(a), \mathcal{F}^k\mathcal{H}^k(a)) &= (\mathcal{H}^k(a), \{a \cup \bigcup_{j\in\{1,2,\ldots,|M|\}} a_j\}); \end{aligned} \tag{6.23}$$

otherwise,

$$\begin{aligned}(\mathcal{H}^k\mathcal{F}^k(x), \mathcal{F}^k(x)) &= (\{x\}, \mathcal{F}^k(x)),\\ (\mathcal{H}^k(a), \mathcal{F}^k\mathcal{H}^k(a)) &= (\mathcal{H}^k(a), \{a\}).\end{aligned} \tag{6.24}$$

(2) *For any decision concepts* $(\widetilde{\mathcal{H}}^k\widetilde{\mathcal{F}}^k(x), \widetilde{\mathcal{F}}^k(x))$ *and* $(\widetilde{\mathcal{H}}^k(k'), \widetilde{\mathcal{F}}^k\widetilde{\mathcal{H}}^k(k'))$, *if there exists a decision concept* $(\widetilde{\mathcal{H}}^k\widetilde{\mathcal{F}}^k(x_i), \widetilde{\mathcal{F}}^k(x_i)) \in \mathcal{G}_{\widetilde{\mathcal{F}}^k\widetilde{\mathcal{H}}^k}$ *such that* $\widetilde{\mathcal{F}}^k(x) \subseteq \widetilde{\mathcal{F}}^k(x_i)$ $(i \in \{1, 2, \ldots, |G^k|\})$ *and a decision concept* $(\widetilde{\mathcal{H}}^k(k_j), \widetilde{\mathcal{F}}^k\widetilde{\mathcal{H}}^k(k_j)) \in \mathcal{G}_{\widetilde{\mathcal{F}}^k\widetilde{\mathcal{H}}^k}$ *such that* $\widetilde{\mathcal{H}}^k(k') \subseteq \widetilde{\mathcal{H}}^k(k_j)$ $(j \in \{1, 2, \ldots, |D|\})$, *then following statements hold:*

$$\begin{aligned}(\widetilde{\mathcal{H}}^k\widetilde{\mathcal{F}}^k(x), \widetilde{\mathcal{F}}^k(x)) &= (\{x \cup \bigcup_{i\in\{1,2,\ldots,|G^k|\}} x_i\}, \widetilde{\mathcal{F}}^k(x)),\\ (\widetilde{\mathcal{H}}^k(k'), \widetilde{\mathcal{F}}^k\widetilde{\mathcal{H}}^k(k')) &= (\widetilde{\mathcal{H}}^k(k'), \{k' \cup \bigcup_{j\in\{1,2,\ldots,|D|\}} k_j\});\end{aligned} \tag{6.25}$$

otherwise,

$$\begin{aligned}(\widetilde{\mathcal{H}}^k\widetilde{\mathcal{F}}^k(x), \widetilde{\mathcal{F}}^k(x)) &= (\{x\}, \widetilde{\mathcal{F}}^k(x)),\\ (\widetilde{\mathcal{H}}^k(k'), \widetilde{\mathcal{F}}^k\widetilde{\mathcal{H}}^k(k')) &= (\widetilde{\mathcal{H}}^k(k'), \{k'\}).\end{aligned} \tag{6.26}$$

Proof The proof of Theorem 6.2 can be found in the original paper [43]. □

Property 6.4 Let $\mathcal{G}^{\diamond}_{\mathcal{F}^k\mathcal{H}^k}$ and $\mathcal{G}^{\diamond}_{\widetilde{\mathcal{F}}^k\widetilde{\mathcal{H}}^k}$ be two concept spaces, $\mathcal{AG}_{\mathcal{F}^k\mathcal{H}^k}$ and $\mathcal{AG}_{\widetilde{\mathcal{F}}^k\widetilde{\mathcal{H}}^k}$ be the attribute-oriented conditional concept space and attribute-oriented decision concept space, respectively; meanwhile, initialize $\mathcal{G}^{\diamond}_{\mathcal{F}^k\mathcal{H}^k} = \mathcal{AG}_{\mathcal{F}^k\mathcal{H}^k}$ and $\mathcal{G}^{\diamond}_{\widetilde{\mathcal{F}}^k\widetilde{\mathcal{H}}^k} = \mathcal{AG}_{\widetilde{\mathcal{F}}^k\widetilde{\mathcal{H}}^k}$. Then we have

(1) For each $x \in G^k$, if there exists $(\mathcal{H}^k(a), \mathcal{F}^k\mathcal{H}^k(a)) \in \mathcal{AG}_{\mathcal{F}^k\mathcal{H}^k}$ such that $\mathcal{F}^k(x) = \mathcal{F}^k\mathcal{H}^k(a)$, then $(\mathcal{H}^k\mathcal{F}^k(x), \mathcal{F}^k(x)) = (\mathcal{H}^k(a), \mathcal{F}^k\mathcal{H}^k(a))$; otherwise,
$\mathcal{G}^{\diamond}_{\mathcal{F}^k\mathcal{H}^k} = \mathcal{G}^{\diamond}_{\mathcal{F}^k\mathcal{H}^k} \cup (\mathcal{H}^k\mathcal{F}^k(x), \mathcal{F}^k(x))$.
(2) For each $x \in G^k$, if there exists $(\widetilde{\mathcal{H}}^k(k'), \widetilde{\mathcal{F}}^k\widetilde{\mathcal{H}}^k(k')) \in \mathcal{AG}_{\widetilde{\mathcal{F}}^k\widetilde{\mathcal{H}}^k}$ such that $\widetilde{\mathcal{F}}^k(x) = \widetilde{\mathcal{F}}^k\widetilde{\mathcal{H}}^k(k')$, then $(\widetilde{\mathcal{H}}^k\widetilde{\mathcal{F}}^k(x), \widetilde{\mathcal{F}}^k(x)) = (\widetilde{\mathcal{H}}^k(k'), \widetilde{\mathcal{F}}^k\widetilde{\mathcal{H}}^k(k'))$; otherwise,
$\mathcal{G}^{\diamond}_{\widetilde{\mathcal{F}}^k\widetilde{\mathcal{H}}^k} = \mathcal{G}^{\diamond}_{\widetilde{\mathcal{F}}^k\widetilde{\mathcal{H}}^k} \cup (\widetilde{\mathcal{H}}^k\widetilde{\mathcal{F}}^k(x), \widetilde{\mathcal{F}}^k(x))$.

Proof The proof of Property 6.4 can be found in the original paper [43]. □

Property 6.4 means that we do not need to construct concepts $(\mathcal{H}^k\mathcal{F}^k(x), \mathcal{F}^k(x))$ and $(\widetilde{\mathcal{H}}^k\widetilde{\mathcal{F}}^k(x), \widetilde{\mathcal{F}}^k(x))$ like [58] when $\mathcal{F}^k(x) = \mathcal{F}^k\mathcal{H}^k(a)$ and $\widetilde{\mathcal{F}}^k(x) =$

$\widetilde{\mathcal{F}}^k\widetilde{\mathcal{H}}^k(k')$. Then, using this approach, we can finally obtain $\mathcal{G}_{\mathcal{F}^k\mathcal{H}^k} = \mathcal{G}^{\diamond}_{\mathcal{F}^k\mathcal{H}^k}$ and $\mathcal{G}_{\widetilde{\mathcal{F}}^k\widetilde{\mathcal{H}}^k} = \mathcal{G}^{\diamond}_{\widetilde{\mathcal{F}}^k\widetilde{\mathcal{H}}^k}$.

For convenience, we denote the labeled dataset S_L by G_0, and the initial concept spaces by $\mathcal{G}_{\mathcal{F}_0\mathcal{H}_0}$ and $\mathcal{G}_{\widetilde{\mathcal{F}}_0\widetilde{\mathcal{H}}_0}$. Note that, in the initial concept space period, if the object set G^k is replaced with G_0^k, then the corresponding cognitive operators $\mathcal{F}^k, \mathcal{H}^k$ and $\widetilde{\mathcal{F}}^k, \widetilde{\mathcal{H}}^k$ can be expressed as $\mathcal{F}_0^k, \mathcal{H}_0^k$ and $\widetilde{\mathcal{F}}_0^k, \widetilde{\mathcal{H}}_0^k$, respectively.

6.1.3.2 Cognitive Process with Unlabeled Data in Concept Learning

In the concept-cognitive process, suppose the obtained concept spaces will be updated by a newly added object instead of inputting multi-objects simultaneously. Then, for the unlabeled set S_U, we can denote S_U as $\Delta G = \{\Delta G_0, \Delta G_1, \ldots, \Delta G_{n-1}\}$ in which each learning step only consists of one object x (i.e., $\Delta G_i = \{x_i\}$). For brevity, in what follows, we write $\{x_i\}$ as x_i and then we have $\Delta G = \{x_0, x_1, \ldots, x_{n-1}\}$.

Different from [58], we assume that an object x is connected with a virtual label k^* due to no label information. Then we have the conditional sub-object cognitive operators and decision sub-object cognitive operators with the newly input data $\Delta G_{i-1}^{k^*} = G_i^{k^*} - G_{i-1}^{k^*}$ as follows:

$$\begin{aligned}
&\text{(i) } \mathcal{F}_{i-1}^{k^*} : 2^{G_{i-1}^{k^*}} \to 2^M, && \mathcal{H}_{i-1}^{k^*} : 2^M \to 2^{G_{i-1}^{k^*}},\\
&\text{(ii) } \mathcal{F}_{\Delta G_{i-1}^{k^*}}^{k^*} : 2^{\Delta G_{i-1}^{k^*}} \to 2^M, && \mathcal{H}_{\Delta G_{i-1}^{k^*}}^{k^*} : 2^M \to 2^{\Delta G_{i-1}^{k^*}},\\
&\text{(iii) } \mathcal{F}_i^{k^*} : 2^{G_i^{k^*}} \to 2^M, && \mathcal{H}_i^{k^*} : 2^M \to 2^{G_i^{k^*}},
\end{aligned} \tag{6.27}$$

and

$$\begin{aligned}
&\text{(i) } \widetilde{\mathcal{F}}_{i-1}^{k^*} : 2^{G_{i-1}^{k^*}} \to 2^D, && \widetilde{\mathcal{H}}_{i-1}^{k^*} : 2^D \to 2^{G_{i-1}^{k^*}},\\
&\text{(ii) } \widetilde{\mathcal{F}}_{\Delta G_{i-1}^{k^*}}^{k^*} : 2^{\Delta G_{i-1}^{k^*}} \to 2^D, && \widetilde{\mathcal{H}}_{\Delta G_{i-1}^{k^*}}^{k^*} : 2^D \to 2^{\Delta G_{i-1}^{k^*}},\\
&\text{(iii) } \widetilde{\mathcal{F}}_i^{k^*} : 2^{G_i^{k^*}} \to 2^D, && \widetilde{\mathcal{H}}_i^{k^*} : 2^D \to 2^{G_i^{k^*}}.
\end{aligned} \tag{6.28}$$

Theorem 6.3 *Let $\mathcal{AG}_{\mathcal{F}_{i-1}^{k^*}\mathcal{H}_{i-1}^{k^*}}$, $\mathcal{AG}_{\widetilde{\mathcal{F}}_{i-1}^{k^*}\widetilde{\mathcal{H}}_{i-1}^{k^*}}$ and $\mathcal{OG}_{\mathcal{F}_{i-1}^{k^*}\mathcal{H}_{i-1}^{k^*}}$, $\mathcal{OG}_{\widetilde{\mathcal{F}}_{i-1}^{k^*}\widetilde{\mathcal{H}}_{i-1}^{k^*}}$ be the attribute-oriented concept spaces and object-oriented concept spaces, respectively. Then we have*

*(1) For any $a' \in M$ and $(\mathcal{H}_{i-1}^{k^*}(a''), \mathcal{F}_{i-1}^{k^*}\mathcal{H}_{i-1}^{k^*}(a'')) \in \mathcal{AG}_{\mathcal{F}_{i-1}^{k^*}\mathcal{H}_{i-1}^{k^*}}$, if $\mathcal{F}_{i-1}^{k^*}\mathcal{H}_{i-1}^{k^*}(a'') \cap$
$\mathcal{F}_{\Delta G_{i-1}^{k^*}}^{k^*}\mathcal{H}_{\Delta G_{i-1}^{k^*}}^{k^*}(a') \neq \emptyset$, then*

$(\mathcal{H}_i^{k^*}(a''), \mathcal{F}_i^{k^*}\mathcal{H}_i^{k^*}(a'')) = (\mathcal{H}_{i-1}^{k^*}(a'') \cup \mathcal{H}_{\Delta G_{i-1}^{k^*}}^{k^*}(a'), \mathcal{F}_{i-1}^{k^*}\mathcal{H}_{i-1}^{k^*}(a'') \cap \mathcal{F}_{\Delta G_{i-1}^{k^*}}^{k^*}\mathcal{H}_{\Delta G_{i-1}^{k^*}}^{k^*}(a'))$;

otherwise,
$(\mathcal{H}_i^{k^*}(a'), \mathcal{F}_i^{k^*}\mathcal{H}_i^{k^*}(a')) = (\mathcal{H}_{\Delta G_{i-1}^{k^*}}^{k^*}(a'), \mathcal{F}_{\Delta G_{i-1}^{k^*}}^{k^*}\mathcal{H}_{\Delta G_{i-1}^{k^*}}^{k^*}(a'))$.

(2) For any $x' \in \Delta G^{k^*}$ *and* $(\mathcal{H}_{i-1}^{k^*}\mathcal{F}_{i-1}^{k^*}(x''), \mathcal{F}_{i-1}^{k^*}(x'')) \in \mathcal{OG}_{\mathcal{F}_{i-1}^{k^*}\mathcal{H}_{i-1}^{k^*}}$, *if* $\mathcal{F}_{i-1}^{k^*}(x'') \subseteq \mathcal{F}_{\Delta G_{i-1}^{k^*}}^{k^*}(x')$, *then* $(\mathcal{H}_i^{k^*}\mathcal{F}_i^{k^*}(x''), \mathcal{F}_i^{k^*}(x'')) = (\mathcal{H}_{i-1}^{k^*}\mathcal{F}_{i-1}^{k^*}(x'') \cup \mathcal{H}_{\Delta G_{i-1}^{k^*}}^{k^*}\mathcal{F}_{\Delta G_{i-1}^{k^*}}^{k^*}(x'), \mathcal{F}_{i-1}^{k^*}(x''))$;

if $\mathcal{F}_{\Delta G_{i-1}^{k^*}}^{k^*}(x') \subseteq \mathcal{F}_{i-1}^{k^*}(x'')$, *then*
$(\mathcal{H}_i^{k^*}\mathcal{F}_i^{k^*}(x''), \mathcal{F}_i^{k^*}(x'')) = (\mathcal{H}_{i-1}^{k^*}\mathcal{F}_{i-1}^{k^*}(x'') \cup \mathcal{H}_{\Delta G_{i-1}^{k^*}}^{k^*}\mathcal{F}_{\Delta G_{i-1}^{k^*}}^{k^*}(x'), \mathcal{F}_{\Delta G_{i-1}^{k^*}}^{k^*}(x'))$;

otherwise,
$(\mathcal{H}_i^{k^*}\mathcal{F}_i^{k^*}(x'), \mathcal{F}_i^{k^*}(x')) = (\mathcal{H}_{\Delta G_{i-1}^{k^*}}^{k^*}\mathcal{F}_{\Delta G_{i-1}^{k^*}}^{k^*}(x'), \mathcal{F}_{\Delta G_{i-1}^{k^*}}^{k^*}(x'))$.

(3) For any $k' \in D$ *and* $(\widetilde{\mathcal{H}}_{i-1}^{k^*}(k''), \widetilde{\mathcal{F}}_{i-1}^{k^*}\widetilde{\mathcal{H}}_{i-1}^{k^*}(k'')) \in \mathcal{AG}_{\widetilde{\mathcal{F}}_{i-1}^{k^*}\widetilde{\mathcal{H}}_{i-1}^{k^*}}$, *if* $\widetilde{\mathcal{F}}_{i-1}^{k^*}\widetilde{\mathcal{H}}_{i-1}^{k^*}(k'') \cap \widetilde{\mathcal{F}}_{\Delta G_{i-1}^{k^*}}^{k^*}\widetilde{\mathcal{H}}_{\Delta G_{i-1}^{k^*}}^{k^*}(k') \neq \emptyset$, *then*
$(\widetilde{\mathcal{H}}_i^{k^*}(k''), \widetilde{\mathcal{F}}_i^{k^*}\widetilde{\mathcal{H}}_i^{k^*}(k'')) = (\widetilde{\mathcal{H}}_{i-1}^{k^*}(k'') \cup \widetilde{\mathcal{H}}_{\Delta G_{i-1}^{k^*}}^{k^*}(k'), \widetilde{\mathcal{F}}_{i-1}^{k^*}\widetilde{\mathcal{H}}_{i-1}^{k^*}(k'') \cap \widetilde{\mathcal{F}}_{\Delta G_{i-1}^{k^*}}^{k^*}\widetilde{\mathcal{H}}_{\Delta G_{i-1}^{k^*}}^{k^*}(k'))$;

otherwise,
$(\widetilde{\mathcal{H}}_i^{k^*}(k'), \widetilde{\mathcal{F}}_i^{k^*}\widetilde{\mathcal{H}}_i^{k^*}(k')) = (\widetilde{\mathcal{H}}_{\Delta G_{i-1}^{k^*}}^{k^*}(k'), \widetilde{\mathcal{F}}_{\Delta G_{i-1}^{k^*}}^{k^*}\widetilde{\mathcal{H}}_{\Delta G_{i-1}^{k^*}}^{k^*}(k'))$.

(4) For any $x' \in \Delta G^{k^*}$ *and* $(\widetilde{\mathcal{H}}_{i-1}^{k^*}\widetilde{\mathcal{F}}_{i-1}^{k^*}(x''), \widetilde{\mathcal{F}}_{i-1}^{k^*}(x'')) \in \mathcal{OG}_{\widetilde{\mathcal{F}}_{i-1}^{k^*}\widetilde{\mathcal{H}}_{i-1}^{k^*}}$, *if* $\widetilde{\mathcal{F}}_{i-1}^{k^*}(x'') \subseteq \widetilde{\mathcal{F}}_{\Delta G_{i-1}^{k^*}}^{k^*}(x')$, *then* $(\widetilde{\mathcal{H}}_i^{k^*}\widetilde{\mathcal{F}}_i^{k^*}(x''), \widetilde{\mathcal{F}}_i^{k^*}(x'')) = (\widetilde{\mathcal{H}}_{i-1}^{k^*}\widetilde{\mathcal{F}}_{i-1}^{k^*}(x'') \cup \widetilde{\mathcal{H}}_{\Delta G_{i-1}^{k^*}}^{k^*}\widetilde{\mathcal{F}}_{\Delta G_{i-1}^{k^*}}^{k^*}(x'), \widetilde{\mathcal{F}}_{i-1}^{k^*}(x''))$;

if $\widetilde{\mathcal{F}}_{\Delta G_{i-1}^{k^*}}^{k^*}(x') \subseteq \widetilde{\mathcal{F}}_{i-1}^{k^*}(x'')$, *then*
$(\widetilde{\mathcal{H}}_i^{k^*}\widetilde{\mathcal{F}}_i^{k^*}(x''), \widetilde{\mathcal{F}}_i^{k^*}(x'')) = (\widetilde{\mathcal{H}}_{i-1}^{k^*}\widetilde{\mathcal{F}}_{i-1}^{k^*}(x'') \cup \widetilde{\mathcal{H}}_{\Delta G_{i-1}^{k^*}}^{k^*}\widetilde{\mathcal{F}}_{\Delta G_{i-1}^{k^*}}^{k^*}(x'), \widetilde{\mathcal{F}}_{\Delta G_{i-1}^{k^*}}^{k^*}(x'))$;

otherwise,
$(\widetilde{\mathcal{H}}_i^{k^*}\widetilde{\mathcal{F}}_i^{k^*}(x'), \widetilde{\mathcal{F}}_i^{k^*}(x')) = (\widetilde{\mathcal{H}}_{\Delta G_{i-1}^{k^*}}^{k^*}\widetilde{\mathcal{F}}_{\Delta G_{i-1}^{k^*}}^{k^*}(x'), \widetilde{\mathcal{F}}_{\Delta G_{i-1}^{k^*}}^{k^*}(x'))$.

Proof The proof of Theorem 6.3 can be found in the original paper [42]. □

Although Theorem 6.3 shows how to update the concept spaces when adding an instance, concept recognition is exceedingly difficult since we cannot directly recognize the real class label of each instance $\boldsymbol{x}$. Namely, unlike the initial concept spaces generation, there is still a mystery that which sub-concept space will be updated when inputting a new object without label information.

6.1.3.3 Concept Recognition

For any newly input object x, the concept $(\mathcal{H}^{k^*}_{\Delta G_i^{k^*}}\mathcal{F}^{k^*}_{\Delta G_i^{k^*}}(x), \mathcal{F}^{k^*}_{\Delta G_i^{k^*}}(x))$ can be rewritten as $(\{x\}, \mathcal{F}^{k^*}_{\Delta G_i^{k^*}}(x))$ due to $|\Delta G_i^{k^*}| = 1$. Meanwhile, to meet the demand of lots of unlabeled data, a new similarity metric for concept learning is proposed in this subsection. As a matter of fact, a good assessing similarity for concepts is a key success of S2CL.

Definition 6.22 Let $\mathcal{G}_{\mathcal{F}_{i-1},\mathcal{H}_{i-1}}$ be the concept space and $\mathcal{G}_{\mathcal{F}^{k^*}_{i-1},\mathcal{H}^{k^*}_{i-1}}$ be a sub-concept space with a virtual label k^* in the $(i-1)$-th state. For any concept $(X_j, B_j) \in \mathcal{G}_{\mathcal{F}^{k^*}_{i-1},\mathcal{H}^{k^*}_{i-1}}$, where $j \in \{1, 2, \ldots, |\mathcal{G}_{\mathcal{F}^{k^*}_{i-1},\mathcal{H}^{k^*}_{i-1}}|\}$, the global information w_{i-1,k^*} and the local information $z^{k^*}_{i-1,j}$ in the $(i-1)$-th state are, respectively, defined as

$$w_{i-1,k^*} = \frac{|\mathcal{G}_{\mathcal{F}^{k^*}_{i-1},\mathcal{H}^{k^*}_{i-1}}|}{|\mathcal{G}_{\mathcal{F}_{i-1},\mathcal{H}_{i-1}}|}, \tag{6.29}$$

$$z^{k^*}_{i-1,j} = \frac{|X_j|}{|\mathcal{G}_{\mathcal{F}^{k^*}_{i-1},\mathcal{H}^{k^*}_{i-1}}|}. \tag{6.30}$$

More generally, considering the entire concept space $\mathcal{G}_{\mathcal{F}_{i-1},\mathcal{H}_{i-1}}$ in the $(i-1)$-th state, we denote

$$\boldsymbol{w}_{i-1} = (w_{i-1,1}, w_{i-1,2}, \ldots, w_{i-1,l}), \tag{6.31}$$

$$\boldsymbol{z}_{i-1} = \begin{bmatrix} z^1_{i-1} \\ \vdots \\ z^l_{i-1} \end{bmatrix} = \begin{bmatrix} z^1_{i-1,1} & \cdots & z^1_{i-1,m_1} \\ \vdots & \ddots & \vdots \\ z^l_{i-1,1} & \cdots & z^l_{i-1,m_l} \end{bmatrix}, \tag{6.32}$$

where $m_{k^*} = \left|\mathcal{G}_{\mathcal{F}^{k^*}_{i-1},\mathcal{H}^{k^*}_{i-1}}\right|$ and $k^* \in \{1, 2, \ldots, K\}$.

Definition 6.23 Let $C = (\{x\}, \mathcal{F}_{\Delta G_i^{k^*}}(x))$ be a newly input concept. For any concept $(X_j, B_j) \in \mathcal{G}_{\mathcal{F}^{k^*}_{i-1},\mathcal{H}^{k^*}_{i-1}}$, where $j \in \{1, 2, \ldots, |\mathcal{G}_{\mathcal{F}^{k^*}_{i-1},\mathcal{H}^{k^*}_{i-1}}|\}$, the concept similarity (CS) can be defined as:

$$\theta^I_j = I\frac{|A^* \cap B_j|}{|A^* \cap B_j| + 2(\alpha|A^* - B_j| + (1-\alpha)|B_j - A^*|)}, \tag{6.33}$$

where $I = 1/(1 + w_{i-1,k^*} \times e^{-z^{k^*}_{i-1,j}})$, $A^* = \mathcal{F}_{\Delta G_i^{k^*}}(x)$ and $\alpha \in [0, 1]$.

For Eq. (6.33), I is set to be 1 when without considering the global and local information. In this case, Eq. (6.33) can further be formulated as

$$\theta_j = \frac{|A^* \cap B_j|}{|A^* \cap B_j| + 2(\alpha|A^* - B_j| + (1-\alpha)|B_j - A^*|)}. \tag{6.34}$$

In Eq. (6.34), $A^* - B_j$ represents the characteristics appearing in A^* but not in B_j, and it has the same meaning for $B_j - A^*$. Moreover, the parameters α and $(1-\alpha)$ can be, respectively, considered as the weight information added to $|A^* - B_j|$ and $|B_j - A^*|$, which express the importance of different features of $A^* - B_j$ and $B_j - A^*$ relative to the overall similarity degree. In fact, when $\alpha = 0.5$, Eq. (6.34) is degenerated into Jaccard similarity [27, 65].

According to sample separation axiom [79], for any instance, there always exists a unique class that is most similar to it. Hence, given an instance $\boldsymbol{x}$, the class vector can be generated as follows: each sub-concept space will first produce a set of CS degrees by computing the CS degree between the given concept and any concept from a sub-concept space. Then, the maximum CS degree $(\widehat{\theta}_j^I)$ of each sub-concept space will be obtained, namely, $\widehat{\theta}_j^I = \max\limits_{j\in J}\{\theta_j^I\}$, where $J = \{1, 2, \ldots, |\mathcal{G}_{\mathcal{F}_{i-1}^{k*}, \mathcal{H}_{i-1}^{k*}}|\}$. Finally, the estimated class distribution will form a maximum class vector $(\widehat{\theta}_1^I, \widehat{\theta}_2^I, \ldots, \widehat{\theta}_l^I)^{\mathrm{T}}$. In the same manner, we can obtain an average class vector $(\overline{\theta}_1^I, \overline{\theta}_2^I, \ldots, \overline{\theta}_l^I)^{\mathrm{T}}$.

Note that a SSL method, which is designed by combining the concept-cognitive process with the structural concept similarity θ_j, is referred to as a semi-supervised concept learning method, and it is abbreviated as S2CL for convenience. In the meanwhile, an extended version of S2CL is further proposed by taking full advantage of the global and local conceptual information (i.e., the structural concept similarity θ_j^I) within a concept space. For conciseness, we also write it as S2CL$^\alpha$ when no confusion exists.

6.1.3.4 Theoretical Analysis

Essentially, α mainly reflects the influences of different characteristics in sets $A^* - B_j$ and $B_j - A^*$ for the overall concept similarity measure. Hence, it is very important to discuss how to select an appropriate α on each dataset.

Let $\mathcal{Y} = \{1, 2, \ldots, l\}$ be the label space. The concept spaces with different α_r ($\alpha_r \in [0, 1]$) in the $(i-1)$-th period can be formulated as

$$\begin{bmatrix} \mathcal{G}^{\alpha_1}_{\mathcal{F}_{i-1},\mathcal{H}_{i-1}} \\ \vdots \\ \mathcal{G}^{\alpha_n}_{\mathcal{F}_{i-1},\mathcal{H}_{i-1}} \end{bmatrix} = \begin{bmatrix} \mathcal{G}^{\alpha_1,1}_{\mathcal{F}_{i-1},\mathcal{H}_{i-1}} & \cdots & \mathcal{G}^{\alpha_1,l}_{\mathcal{F}_{i-1},\mathcal{H}_{i-1}} \\ \vdots & \ddots & \vdots \\ \mathcal{G}^{\alpha_n,1}_{\mathcal{F}_{i-1},\mathcal{H}_{i-1}} & \cdots & \mathcal{G}^{\alpha_n,l}_{\mathcal{F}_{i-1},\mathcal{H}_{i-1}} \end{bmatrix}, \tag{6.35}$$

where $\sum\limits_{r=1}^{n} \alpha_r = 1$.

For an object x_i, we can obtain its corresponding concept $C_i = (\{x_i\}, B_i)$. Then, based on Definition 6.23, we denote

$$Sim(C_i, \mathcal{G}^{\alpha_r,k}_{\mathcal{F}_{i-1},\mathcal{H}_{i-1}}) = \{Sim(C_i, C^{\alpha_r}_j)\}^{m_k}_{j=1} = \{\theta^I_j\}^{m_k}_{j=1}, \tag{6.36}$$

where $C^{\alpha_r}_j \in \mathcal{G}^{\alpha_r,k}_{\mathcal{F}_{i-1},\mathcal{H}_{i-1}}$ $(k \in \mathcal{Y})$ and $m_k = |\mathcal{G}^{\alpha_r,k}_{\mathcal{F}_{i-1},\mathcal{H}_{i-1}}|$.

Combining Eqs. (6.35) with (6.36), the corresponding concept similarity in the $(i-1)$-th state can be described as

$$\begin{bmatrix} S(C_i, \mathcal{G}^{\alpha_1}_{i-1}) \\ \vdots \\ S(C_i, \mathcal{G}^{\alpha_n}_{i-1}) \end{bmatrix} = \begin{bmatrix} S(C_i, \mathcal{G}^{\alpha_1,1}_{i-1}) & \cdots & S(C_i, \mathcal{G}^{\alpha_1,l}_{i-1}) \\ \vdots & \ddots & \vdots \\ S(C_i, \mathcal{G}^{\alpha_n,1}_{i-1}) & \cdots & S(C_i, \mathcal{G}^{\alpha_n,l}_{i-1}) \end{bmatrix}, \tag{6.37}$$

where $S(C_i, \mathcal{G}^{\alpha_r}_{i-1}) = Sim(C_i, \mathcal{G}^{\alpha_r}_{\mathcal{F}_{i-1},\mathcal{H}_{i-1}})$ $(r \in \{1, 2, \ldots, n\})$ and $S(C_i, \mathcal{G}^{\alpha_r,k}_{i-1}) = Sim(C_i, \mathcal{G}^{\alpha_r,k}_{\mathcal{F}_{i-1},\mathcal{H}_{i-1}})$.

Furthermore, inspired by [79], the category similarity function between the given concept C_i and a class space $\mathcal{G}^{\alpha_r,k}_{\mathcal{F}_{i-1},\mathcal{H}_{i-1}}$ can be defined as

$$\phi_{Sim}(C_i, \mathcal{G}^{\alpha_r,k}_{\mathcal{T}_{i-1},\mathcal{H}_{i-1}}) = \frac{|N^{\alpha_r}_k(C_i)|}{K}, \tag{6.38}$$

where $N^{\alpha_r}_k(C_i) = \{C_j | C_j \in \mathcal{G}^{\alpha_r,k}_{\mathcal{F}_{i-1},\mathcal{H}_{i-1}} \wedge C_j \in N^{\alpha_r}_K(C_i)\}$, and $N^{\alpha_r}_K(C_i)$ is a set of near neighbor instances related to $\boldsymbol{x}_i$ under the parameter α_r.

According to top-K set similarity [77], if $\widehat{k} = \arg\max_{k \in \mathcal{Y}} \frac{|N^{\alpha_r}_k(C_i)|}{K}$, then the instance $\boldsymbol{x}_i$ is classified into the $\widehat{k}$-th class. Therefore, given the parameter K, the objective function can be formulated as

$$\begin{aligned} \widehat{\alpha}_r &= \arg\min_{\alpha_r \in [0,1]} \sum_{i=1}^{m} \left(\frac{|N^{\alpha_r}_k(C_i)|}{K} - y_i\right)^2 \\ \text{s.t.} & \sum_{r=1}^{n} \alpha_r = 1. \end{aligned} \tag{6.39}$$

In Eq. (6.39), our aim is to capture an optimal concept space with the concept structural information.

6.1.3.5 Framework and Computational Complexity Analysis

For brevity, we can consider that there are three classes to predict. Figure 6.3 illustrates the whole procedure of S2CL. From a dataset (that contains a small set of labeled data and a large amount of unlabeled data), we first obtain a corresponding regular formal decision context. Then, the initial concept spaces (that include a conditional concept space and its corresponding decision concept space) with concept structural information will be constructed based on the cognitive operators. Specifically speaking, the conditional concept space contains three sub-concept spaces, where each sub-concept space is composed of different concepts. As shown in the stage of initial concept spaces of Fig. 6.3 (see the left of Fig. 6.3 for details), there exist three sub-concept spaces corresponding to three classes in a conditional concept space, and each sub-concept space contains two different types of concepts, namely object-oriented conditional concepts (indicated by red shapes in Fig. 6.3) and attribute-oriented conditional concepts (denoted by black shapes in Fig. 6.3). Meanwhile, each sub-concept space is also associated with a decision concept in the corresponding decision concept space as shown in the first stage of Fig. 6.3. Thirdly, for any newly input unlabeled data, they are first used to form concepts, and then the concept-cognitive process is completed by concept recognition. Finally, given the parameter K, S2CL (or S2CL$^{\alpha}$) trys to learn an optimal concept space based on the concept recognition and concept-cognitive process under different parameters $\alpha_r (r = 1, 2, \dots, n)$. In other words, the objective of S2CL (or S2CL$^{\alpha}$) is to seek an appropriate concept space to represent the underlying data distributions by the concept-cognitive process.

In the prediction stage, given an instance, the final concept space can produce two estimates of class distribution (including a maximum class vector and an average class vector) by employing the CS degree θ_j (or θ_j^I). Then the final CS degree vector will be obtained by the sum of the two 3-dimensional class vectors, and the class with maximum value will be output as shown in Fig. 6.3.

Based on the above discussion, we are ready to propose the corresponding algorithm of S2CL (see Algorithm 6.10 for details). In Algorithm 6.10, Step 3 is to generate the initial concept spaces; then the concept recognition and concept-cognitive process are conducted by running Steps 4–8; at last, the final prediction will be completed by Steps 9–12. In Steps 9–12, if the prediction value $\widehat{k}$ is consistent with the ground truth label, then it means that the predicted value of S2CL is correct. Formally, the accuracy on a test dataset T can be descried as $acc = \frac{N}{|T|}$, where N denotes the number of correct predicted values. Simultaneously, it will be easy to obtain the algorithm of S2CL$^{\alpha}$ by means of replacing the structural concept similarity θ_j with θ_j^I in Step 6 of Algorithm 6.10.

The time complexity of S2CL is mainly composed of two parts, i.e. constructing the initial concept spaces and the concept-cognitive process with concept structural information. Let the time complexity of constructing a concept, computing the CS degree and updating the concept space be $O(t_1)$, $O(t_2)$ and $O(t_3)$, respectively. Then, it is easy to verify that the time complexity of Step 3 is $O(t_1|S_L|(|M|+|D|))$, and the complexity of accomplishing the concept-cognitive process by concept recognition is $O(|S_U|(t_1 + t_2 + t_3))$. Note that, CCL is an incremental learning process, as the proposed method is updated by inputting objects one by one. Therefore, S2CL can also be regarded as an incremental method for SSL in dynamic environments. For convenience, let E and C (that randomly selects instances from

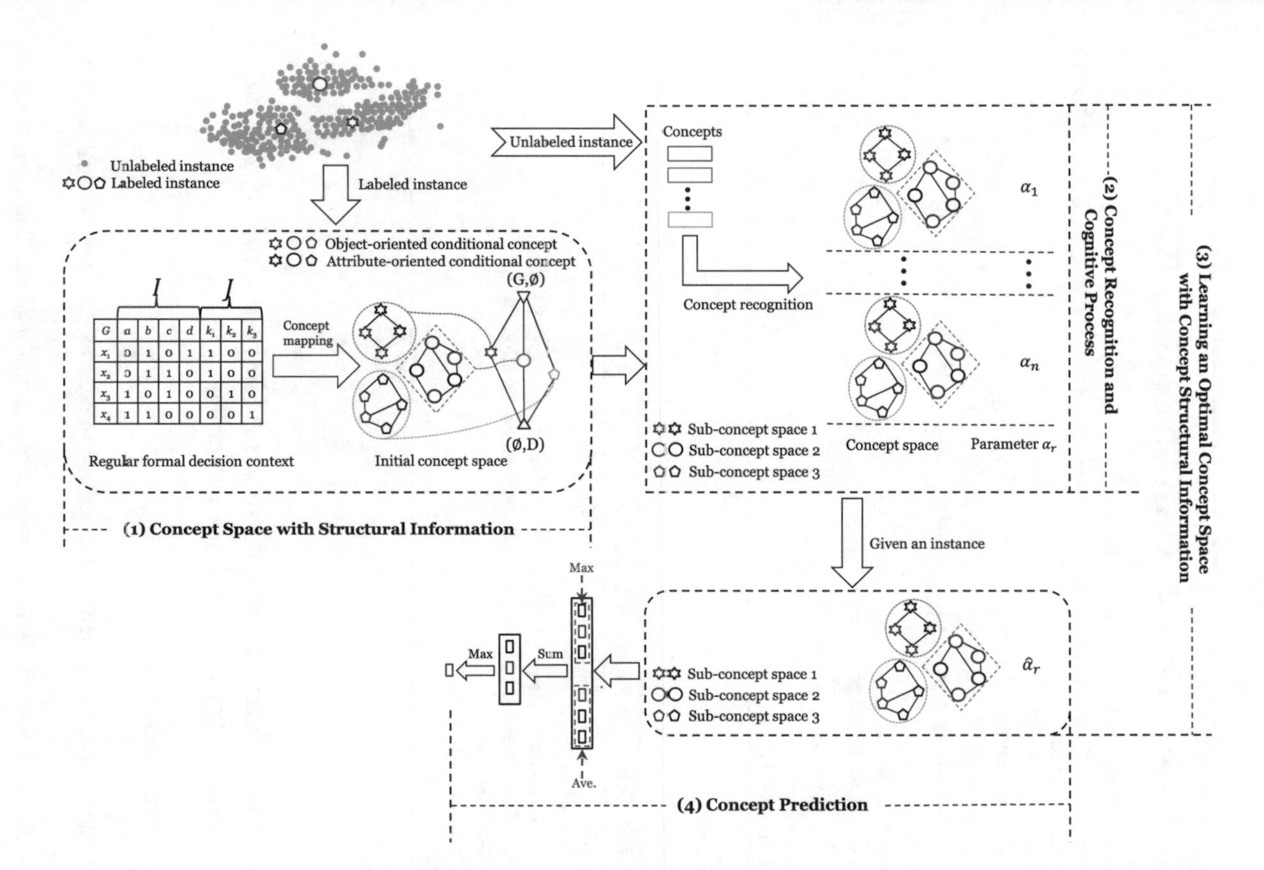

Fig. 6.3 Illustration of the framework of the proposed methods. Considering that there are three classes to predict, the concept spaces with concept structural information will be constructed, which include three different sub-concept spaces

S_U) be the incremental learning step and the sample size of each incremental learning step, respectively. Thus, the time complexity of incremental learning (see Algorithm 6.11 for details) is $O(E(|S_U|(t_1 + t_2 + t_3)) + |T|)$.

Algorithm 6.10 S2CL algorithm

1: **Input:** Labeled dataset S_L, unlabeled dataset S_U, test dataset T, and hyperparameters K and α_r.
2: **Output:** The class labels of the test dataset T.
3: Based on labeled dataset S_L, S2CL can construct two initial concept spaces $\mathcal{G}_{\mathcal{F}_0\mathcal{H}_0}$ and $\mathcal{G}_{\widetilde{\mathcal{F}}_0\widetilde{\mathcal{H}}_0}$ by Theorem 6.2.
4: **for** each $x_i \in S_U$ **do**
5: Get two concepts $(\{x_i\}, \mathcal{F}_{\Delta G_i^{k*}}(x_i))$ and $(\{x_i\}, \widetilde{\mathcal{F}}_{\Delta G_i^{k*}}(x_i))$.
6: Compute the CS degree by Eq. (6.33) (or Eq. (6.34)).
7: Update concept spaces $\mathcal{G}^{\alpha_r}_{\mathcal{F}_{i-1}\mathcal{H}_{i-1}}$ and $\mathcal{G}^{\alpha_r}_{\widetilde{\mathcal{F}}_{i-1}\widetilde{\mathcal{H}}_{i-1}}$ by Theorem 6.3.
8: **end for**
9: **for** each $x_j \in T$ **do**
10: Construct a concept $C_j = (\{x_j\}, B_j)$.
11: $\widehat{k} = \arg\max_{k \in \mathcal{Y}} \frac{|N_k^{\alpha_r}(C_j)|}{K}$.
12: **end for**
13: **return** class labels.

Algorithm 6.11 Incremental learning

1: **function** INCREMENTALLEARNINGMETHOD
2: **for** e=1 to E **do**
3: Conduct the same operation as Steps 4–8 of Algorithm 6.10.
4: Conduct the same operation as Steps 9–12 of Algorithm 6.10.
5: **end for**
6: **return** class labels.
7: **end function**

6.1.4 *Fuzzy-Based Concept Learning Method: Exploiting Data with Fuzzy Conceptual Clustering*

6.1.4.1 Preliminaries

In this subsection, we review some notions related to the fuzzy formal decision context.

In a classical formal decision context, the conditional attributes are discrete. However, in the real world, many tasks (e.g., classification, image segmentation, etc.) are described with numerical (or fuzzy) data, which means that classical formal decision contexts cannot cope with them directly. Therefore, a fuzzy formal decision context is proposed based on fuzzy sets [81].

Let G be a universe of discourse. A fuzzy set $\widetilde{X}$ on G can be defined as follows:

$$\widetilde{X} = \{< x, \mu_{\widetilde{X}}(x) > | x \in G\},$$

where $\mu_{\widetilde{X}} : G \to [0, 1]$, and $\mu_{\widetilde{X}}(x)$ is referred to as the membership degree to $\widetilde{X}$ of the object $x \in G$. And we denote by L^G the set of all fuzzy sets on G.

Definition 6.24 ([83]) A fuzzy formal context $(G, M, \widetilde{I})$ is a triple, where G is a set of objects, M is a set of attributes, and $\widetilde{I}$ is a fuzzy relation between G and M. Each relation $(x, a) \in \widetilde{I}$ has a membership degree $\mu_{\widetilde{I}}(x, a)$ in $[0, 1]$, and we denote by $\widetilde{I}(x, a) = \mu_{\widetilde{I}}(x, a)$ for the sake of convenience.

Definition 6.25 ([4, 78, 83]) Let $(G, M, \widetilde{I})$ be a fuzzy formal context. For $X \subseteq G$ and $\widetilde{B} \in L^M$, the operator $(\cdot)^*$ is defined as follows:

$$\begin{aligned} X^*(a) &= \bigwedge_{x \in X} \widetilde{I}(x, a), a \in M, \\ \widetilde{B}^* &= \{x \in G | \forall a \in M, \widetilde{B}(a) \leq \widetilde{I}(x, a)\}. \end{aligned} \tag{6.40}$$

Then, we say that a pair $(X, \widetilde{B})$ is a fuzzy concept of a fuzzy formal context $(G, M, \widetilde{I})$ if $X^* = \widetilde{B}$, $\widetilde{B}^* = X$, and X and $\widetilde{B}$ are respectively known as the extent and intent of the fuzzy concept $(X, \widetilde{B})$. For convenience, the set of all fuzzy concepts is denoted by $L(G, M, \widetilde{I})$. In [83], $L(G, M, \widetilde{I})$ is called a special crisp-fuzzy variable threshold concept lattice under the circumstance of the threshold being set to be 1. For $(X_1, \widetilde{B}_1), (X_2, \widetilde{B}_2) \in L(G, M, \widetilde{I})$, we define the order relation $(X_1, \widetilde{B}_1) \leq (X_2, \widetilde{B}_2)$ if and only if $X_1 \subseteq X_2$ (or $\widetilde{B}_2 \subseteq \widetilde{B}_1$). Then we say that $(X_1, \widetilde{B}_1)$ is a sub-concept of $(X_2, \widetilde{B}_2)$ and $(X_2, \widetilde{B}_2)$ is a super-concept of $(X_1, \widetilde{B}_1)$.

Definition 6.26 ([55]) Let $(G, M, \widetilde{I})$ and $(G, D, \widetilde{J})$ be two fuzzy formal contexts, where $\widetilde{I} : G \times M \to [0, 1]$ and $\widetilde{J} : G \times D \to [0, 1]$. Then $(G, M, \widetilde{I}, D, \widetilde{J})$ is referred to as a fuzzy formal decision context, where $M \cap D = \emptyset$, and M and D are the conditional and decision attribute sets, respectively.

Note that a quintuple $(G, M, I, D, \widetilde{J})$ is called a crisp-fuzzy formal decision context in [48], where (G, M, I) and $(G, D, \widetilde{J})$ are respectively a classical formal context and fuzzy formal context.

6.1.4.2 Fuzzy Concept Learning Method

In this subsection, we first show some new notions and properties for the proposed FCLM, which includes a regular fuzzy formal decision context, an object-oriented fuzzy conceptual clustering, and the related theoretical analysis. Based on them, we further present the detailed procedure of FCLM.

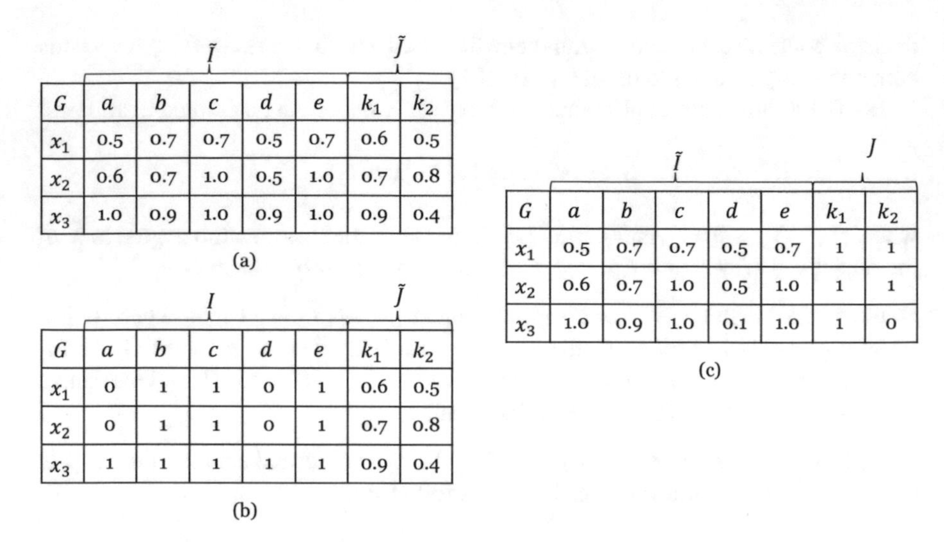

G	a	b	c	d	e	k_1	k_2
x_1	0.5	0.7	0.7	0.5	0.7	0.6	0.5
x_2	0.6	0.7	1.0	0.5	1.0	0.7	0.8
x_3	1.0	0.9	1.0	0.9	1.0	0.9	0.4

(a)

G	a	b	c	d	e	k_1	k_2
x_1	0	1	1	0	1	0.6	0.5
x_2	0	1	1	0	1	0.7	0.8
x_3	1	1	1	1	1	0.9	0.4

(b)

G	a	b	c	d	e	k_1	k_2
x_1	0.5	0.7	0.7	0.5	0.7	1	1
x_2	0.6	0.7	1.0	0.5	1.0	1	1
x_3	1.0	0.9	1.0	0.1	1.0	1	0

(c)

Fig. 6.4 Illustration of three different forms of fuzzy formal decision contexts

A. Regular Fuzzy Formal Decision Context

According to Definition 6.26, Fig. 6.4a and b represents two different fuzzy formal decision contexts $(G, M, \widetilde{I}, D, \widetilde{J})$ and $(G, M, I, D, \widetilde{J})$, respectively. More precisely, Fig. 6.4a expresses a fuzzy formal decision context in which M and D are both numerical; Fig. 6.4b denotes a fuzzy formal decision context, where M is discrete and D is numerical. However, in the real application, most original data are often presented in the form of Fig. 6.4c. It means that the decision attribute set D is described with discrete label information and the conditional attribute set M is constitutive of fuzzy data.

Definition 6.27 Let $(G, M, \widetilde{I})$ be a fuzzy formal context and (G, D, J) be a classical formal context. Then the quintuple $(G, M, \widetilde{I}, D, J)$ is known as a fuzzy-crisp formal decision context, where $\widetilde{I} : G \times M \to [0, 1]$ and $J : G \times D \to \{0, 1\}$.

Definition 6.28 Let $(G, M, \widetilde{I}, D, J)$ be a fuzzy-crisp formal decision context. For any $k_1, k_2 \in D$, if $\mathcal{H}^d(k_1) \cap \mathcal{H}^d(k_2) = \emptyset$, then we say that $(G, M, \widetilde{I}, D, J)$ is a regular fuzzy-crisp formal decision context.

Generally speaking, constructing a fuzzy concept lattice in a standard fuzzy context is sometimes quite complicated, as it is completed in exponential time complexity in the worst case. Hence, GrC should be introduced into the process of generating fuzzy concept lattices for greatly reducing the amount of calculation.

Let $(G, M, \widetilde{I})$ be a fuzzy formal context. $\widetilde{\mathcal{F}}^c : 2^G \to L^M$ and $\widetilde{\mathcal{H}}^c : L^M \to 2^G$ are supposed to be two mappings. Hence, $X^*(a)$ and $\widetilde{B}^*$ in Definition 4 can be rewritten as $\widetilde{\mathcal{F}}^c(X)(a)$ and $\widetilde{\mathcal{H}}^c(\widetilde{B})$, respectively. Especially, for an object set $\{x\}(x \in G)$, $\widetilde{\mathcal{F}}^c(\{x\})(a)$ is abbreviated as $\widetilde{\mathcal{F}}^c(x)$ for brevity.

Definition 6.29 Let $(G, M, \widetilde{I}, D, J)$ be a fuzzy-crisp formal decision context, and $\widetilde{\mathcal{F}}^c : 2^G \to L^M$,$\widetilde{\mathcal{H}}^c : L^M \to 2^G$ and $\mathcal{F}^d : 2^G \to 2^D$,$\mathcal{H}^d : 2^D \to 2^G$ be four mappings. For any $x \in G$, $(\widetilde{\mathcal{H}}^c\widetilde{\mathcal{F}}^c(x), \widetilde{\mathcal{F}}^c(x))$ and $(\mathcal{H}^d\mathcal{F}^d(x), \mathcal{F}^d(x))$ are called a fuzzy conditional granular concept and classical decision granular concept, respectively. The sets of all fuzzy conditional granular concepts and classical decision granular concepts are respectively represented as follows:

$$\mathcal{G}_{\widetilde{\mathcal{F}}^c\widetilde{\mathcal{H}}^c} = \{(\widetilde{\mathcal{H}}^c\widetilde{\mathcal{F}}^c(x), \widetilde{\mathcal{F}}^c(x))|x \in G\},$$

$$\mathcal{G}_{\mathcal{F}^d\mathcal{H}^d} = \{(\mathcal{H}^d\mathcal{F}^d(x), \mathcal{F}^d(x))|x \in G\},$$

where $\mathcal{G}_{\widetilde{\mathcal{F}}^c\widetilde{\mathcal{H}}^c}$ and $\mathcal{G}_{\mathcal{F}^d\mathcal{H}^d}$ are respectively referred to as the fuzzy conditional concept space and classical decision concept space.

It should be pointed out that fuzzy concept lattice has a good performance on classification but is very time-consuming. To the best of our knowledge, the reason is that fuzzy concept lattice may consist of many redundant elements. So, similar to classical concept lattice, it is better to replace fuzzy concept lattice with fuzzy concept space (only containing part of elements of fuzzy concept lattice) in achieving classification tasks with the purpose of improving learning efficiency.

Property 6.5 Let $(G, M, \widetilde{I}, D, J)$ be a fuzzy-crisp formal decision context. For any $(X_1, \widetilde{B}) \in \mathcal{G}_{\widetilde{\mathcal{F}}^c\widetilde{\mathcal{H}}^c}$ and $(X_2, K) \in \mathcal{G}_{\mathcal{F}^d\mathcal{H}^d}$, if $X_1 \subseteq X_2$, and X_1, $\widetilde{B}$, X_2 and K are nonempty, then the object set X_1 is connected with the decision attribute set K under the conditional attribute set $\widetilde{B}$.

Proof The proof is immediate from Definition 6.3 and Property 6.2. □

From Definition 6.29 and Property 6.5, we know that an object can also be connected with a label in a fuzzy-crisp formal decision context.

Based on the above discussion, the complete algorithm of constructing two concept spaces (including a fuzzy conditional concept space and classical decision concept space) is presented in Algorithm 6.12.

Algorithm 6.12 Constructing two concept spaces

1: **Input:** A dataset G.
2: **Output:** The fuzzy conditional concept space $\mathcal{G}_{\widetilde{\mathcal{F}}^c\widetilde{\mathcal{H}}^c}$ and classical decision concept space $\mathcal{G}_{\mathcal{F}^d\mathcal{H}^d}$.
3: **for** each $x \in G$ **do**
4: % Construct a conditional concept space.
5: Construct a fuzzy concept $(\widetilde{\mathcal{H}}^c\widetilde{\mathcal{F}}^c(x), \widetilde{\mathcal{F}}^c(x))$.
6: $\mathcal{G}_{\widetilde{\mathcal{F}}^c\widetilde{\mathcal{H}}^c} \leftarrow (\widetilde{\mathcal{H}}^c\widetilde{\mathcal{F}}^c(x), \widetilde{\mathcal{F}}^c(x))$.
7: % Construct a decision concept space.
8: Construct a classical concept $(\mathcal{H}^d\mathcal{F}^d(x), \mathcal{F}^d(x))$.
9: $\mathcal{G}_{\mathcal{F}^d\mathcal{H}^d} \leftarrow (\mathcal{H}^d\mathcal{F}^d(x), \mathcal{F}^d(x))$.
10: **end for**
11: **Return** $\mathcal{G}_{\widetilde{\mathcal{F}}^c\widetilde{\mathcal{H}}^c}$ and $\mathcal{G}_{\mathcal{F}^d\mathcal{H}^d}$.

B. Object-Oriented Fuzzy Conceptual Clustering

In order to generate fuzzy ontologies, a fuzzy conceptual clustering [50] was adopted in [67]. In fact, it was based on a crisp-crisp variable threshold concept lattice and implemented conceptual clustering via fuzzy sets intersection and union. However, to adapt to granular concepts based on a crisp-fuzzy variable threshold concept lattice, we need to consider the following notions.

Let $(G, M, \widetilde{I})$ be a fuzzy formal context. For any $(X, \widetilde{B}) \in \mathcal{G}_{\widetilde{\mathcal{F}}^c\widetilde{\mathcal{H}}^c}$, $|X|$ is called the object-oriented cardinality with reference to $(X, \widetilde{B})$.

Definition 6.30 Let $(X_j, \widetilde{B}_j)$ be a fuzzy granular concept and $(X_i, \widetilde{B}_i)$ be its sub-concept, then the object-oriented fuzzy concept similarity (object-oriented FCS) is defined as follows:

$$\theta^o = C^O(X_i, X_j) = \frac{|X_i \bigcap X_j|}{|X_i \bigcup X_j|}. \tag{6.41}$$

Definition 6.31 Let $(X_j, \widetilde{B}_j)$ and $(X_l, \widetilde{B}_l)$ be two fuzzy granular concepts, then the attribute-oriented fuzzy concept similarity (attribute-oriented FCS) is defined as follows:

$$\theta^a = C^A(\widetilde{B}_j, \widetilde{B}_l) = ||\widetilde{B}_j - \widetilde{B}_l||_2^2. \tag{6.42}$$

Definition 6.32 Let $\mathcal{G}^{S_\lambda}_{\widetilde{\mathcal{F}}^c\widetilde{\mathcal{H}}^c}$ be a sub-concept space of $\mathcal{G}_{\widetilde{\mathcal{F}}^c\widetilde{\mathcal{H}}^c}$. For any $(X_i, \widetilde{B}_i) \in \mathcal{G}^{S_\lambda}_{\widetilde{\mathcal{F}}^c\widetilde{\mathcal{H}}^c}$, we say that $\mathcal{G}^{S_\lambda}_{\widetilde{\mathcal{F}}^c\widetilde{\mathcal{H}}^c}$ is an object-oriented conceptual cluster of the concept space with an object-oriented FCS threshold λ if the following properties hold:

1. There exists a supremum concept $(X_p, \widetilde{B}_p) \in \mathcal{G}^{S_\lambda}_{\widetilde{\mathcal{F}}^c\widetilde{\mathcal{H}}^c}$ that is not similar to any of its super-concepts.
2. There exists at least one super-concept $(X_j, \widetilde{B}_j) \in \mathcal{G}^{S_\lambda}_{\widetilde{\mathcal{F}}^c\widetilde{\mathcal{H}}^c}$ such that $C^O(X_i, X_j) > \lambda$ when $X_i \neq X_p$.
3. Any fuzzy concept $(X_i, \widetilde{B}_i)$ only belongs to one object-oriented conceptual cluster $\mathcal{G}^{S_\lambda}_{\widetilde{\mathcal{F}}^c\widetilde{\mathcal{H}}^c}$.

Definition 6.33 Let $\mathcal{G}^{S_\lambda}_{\widetilde{\mathcal{F}}^c\widetilde{\mathcal{H}}^c}$ be an object-oriented conceptual cluster. For $(X_1, \widetilde{B}_1)$, $(X_2, \widetilde{B}_2), \ldots, (X_p, \widetilde{B}_p) \in \mathcal{G}^{S_\lambda}_{\widetilde{\mathcal{F}}^c\widetilde{\mathcal{H}}^c} (p = |\mathcal{G}^{S_\lambda}_{\widetilde{\mathcal{F}}^c\widetilde{\mathcal{H}}^c}|)$, let $X_{S_\lambda} = \bigcup_{i=1}^{p} X_i$ and $\widetilde{B}_{S_\lambda} = (\widetilde{B}_{S_\lambda}(a_1), \widetilde{B}_{S_\lambda}(a_2), \ldots, \widetilde{B}_{S_\lambda}(a_{|M|}))$, where $\widetilde{B}_{S_\lambda}(a_j) = \frac{1}{p}\sum_{i=1}^{p} \widetilde{B}_i(a_j)$ $(j \in \{1, 2, \ldots, |M|\})$. Then we say that the crisp-fuzzy pair $(X_{S_\lambda}, \widetilde{B}_{S_\lambda})$ is a pseudo concept induced by the object-oriented conceptual cluster $\mathcal{G}^{S_\lambda}_{\widetilde{\mathcal{F}}^c\widetilde{\mathcal{H}}^c}$.

In what follows, the pseudo concept $(X_{S_\lambda}, \widetilde{B}_{S_\lambda})$ is called the representation of the object-oriented conceptual cluster $\mathcal{G}^{S_\lambda}_{\widetilde{\mathcal{F}}^c\widetilde{\mathcal{H}}^c}$. Note that the process of generating a new pseudo concept is known as concept generation. Hereinafter, we do not distinguish pseudo concepts from fuzzy concepts since pseudo concepts are only intermediate variables in the subsequent fuzzy conceptual clustering. In other words, sometimes we also call pseudo concepts as fuzzy concepts when no confusion exists.

Statistically speaking, Definition 6.33 can completely characterize a new fuzzy concept. However, in cognitive science, concept cognition was often considered to be incremental due to individual cognitive limitations and incomplete cognitive environments. Inspired by this issue, the process of constructing a new fuzzy concept can be rephrased as follows.

Definition 6.34 Let $(X_p, \widetilde{B}_p)$ be the supremum concept of $\mathcal{G}^{S_\lambda}_{\widetilde{\mathcal{F}}^c\widetilde{\mathcal{H}}^c}$. For $(X_1, \widetilde{B}_1)$, $(X_2, \widetilde{B}_2), \ldots, (X_p, \widetilde{B}_p) \in \mathcal{G}^{S_\lambda}_{\widetilde{\mathcal{F}}^c\widetilde{\mathcal{H}}^c}$, each dimension of the intent of a new fuzzy concept $(X_{S_\lambda}, \widetilde{B}_{S_\lambda})$ can be rewritten as follows:

$$\begin{aligned}\widetilde{B}_{S_\lambda}(a_j) =& \frac{1}{2^{p-1}}(\widetilde{B}_1(a_j) + \widetilde{B}_2(a_j) + 2\widetilde{B}_3(a_j) + \\ & 4\widetilde{B}_4(a_j) +, \ldots, + 2^{p-2}\widetilde{B}_p(a_j)),\end{aligned} \tag{6.43}$$

where $j \in \{1, 2, \ldots, |M|\}$.

Theorem 6.4 *Let $\widetilde{B}_{S_\lambda}(a_j)$ be any dimension of the intent of a new fuzzy concept $(X_{S_\lambda}, \widetilde{B}_{S_\lambda})$. Then we have*

$$\frac{\widetilde{B}_p(a_j)}{2} \leq \widetilde{B}_{S_\lambda}(a_j) \leq 1. \tag{6.44}$$

Proof It is immediate from Definitions 6.24 and 6.34. □

For any $(X_i, \widetilde{B}_i), (X_j, \widetilde{B}_j) \in \mathcal{G}^{S_\lambda}_{\widetilde{\mathcal{F}}^c\widetilde{\mathcal{H}}^c}$, if $(X_j, \widetilde{B}_j)$ is a super-concept of $(X_i, \widetilde{B}_i)$, we say that $(X_j, \widetilde{B}_j)$ presents more strongly conceptual representation ability than $(X_i, \widetilde{B}_i)$. Equation (6.43) represents that the process of incremental cognition for concept formation by means of the hierarchical relations between sub-concepts and super-concepts, and the coefficient of each dimension will be heighten along with the increase of conceptual representation ability. Equation (6.44) denotes that the upremum concept has a great influence on the process of constructing new fuzzy concepts.

Definition 6.35 Let $\mathcal{G}^{S_{\lambda,1}}_{\widetilde{\mathcal{F}}^c\widetilde{\mathcal{H}}^c}, \mathcal{G}^{S_{\lambda,2}}_{\widetilde{\mathcal{F}}^c\widetilde{\mathcal{H}}^c}, \ldots, \mathcal{G}^{S_{\lambda,m}}_{\widetilde{\mathcal{F}}^c\widetilde{\mathcal{H}}^c}$ be a partition of $\mathcal{G}_{\widetilde{\mathcal{F}}^c\widetilde{\mathcal{H}}^c}$ with an object-oriented FCS threshold λ. Then a new concept space can be defined as follows:

$$\mathcal{G}^{S_{\lambda,*}}_{\widetilde{\mathcal{F}}^c\widetilde{\mathcal{H}}^c} = \bigcup_{i=1}^{m} \mathcal{G}^{S_{\lambda,i}}_{\widetilde{\mathcal{F}}^c\widetilde{\mathcal{H}}^c} = \bigcup_{i=1}^{m} (X_{S_{\lambda,i}}, \widetilde{B}_{S_{\lambda,i}}). \tag{6.45}$$

Theorem 6.5 *Let $\mathcal{G}^{S_{\lambda},*}_{\widetilde{\mathcal{F}}^c\widetilde{\mathcal{H}}^c}$ be a concept space with an object-oriented FCS threshold λ. We have*

$$1 \leq |\mathcal{G}^{S_{\lambda},*}_{\widetilde{\mathcal{F}}^c\widetilde{\mathcal{H}}^c}| \leq |\mathcal{G}_{\widetilde{\mathcal{F}}^c\widetilde{\mathcal{H}}^c}|. \tag{6.46}$$

Proof The proof of Theorem 6.5 can be found in the original paper [43]. □

Based on the above theory, the procedure of object-oriented fuzzy conceptual clustering is summarized in Algorithm 6.13.

Algorithm 6.13 Object-oriented fuzzy conceptual clustering method

---

1: **Input:** A fuzzy concept space $\mathcal{G}_{\widetilde{\mathcal{F}}^c\widetilde{\mathcal{H}}^c}$ and an object-oriented FCS threshold λ.
2: **Output:** A new fuzzy conceptual cluster space $\mathcal{G}^{S_{\lambda},*}_{\widetilde{\mathcal{F}}^c\widetilde{\mathcal{H}}^c}$.
3: $\mathcal{G}^{S_{\lambda},*}_{\widetilde{\mathcal{F}}^c\widetilde{\mathcal{H}}^c} = \emptyset$ and $\mathcal{G}^{S_{\lambda},i}_{\widetilde{\mathcal{F}}^c\widetilde{\mathcal{H}}^c} = \emptyset$.
4: $\mathcal{G}^{S_{\lambda},i}_{\widetilde{\mathcal{F}}^c\widetilde{\mathcal{H}}^c} \leftarrow (X_p, \widetilde{B}_p)$.
5: **for** each sub-concept $(X_j, \widetilde{B}_j) \in \mathcal{G}_{\widetilde{\mathcal{F}}^c\widetilde{\mathcal{H}}^c}$ of $(X_p, \widetilde{B}_p)$ **do**
6: Get θ^O by Definition 6.30.
7: **if** $\theta^O > \lambda$ **then**
8: $\mathcal{G}^{S_{\lambda},i}_{\widetilde{\mathcal{F}}^c\widetilde{\mathcal{H}}^c} \leftarrow (X_j, \widetilde{B}_j)$.
9: **end if**
10: **end for**
11: Construct a new fuzzy concept $(X_{S_{\lambda},i}, \widetilde{B}_{S_{\lambda},i})$ by Definition 6.34.
12: $\mathcal{G}^{S_{\lambda},*}_{\widetilde{\mathcal{F}}^c\widetilde{\mathcal{H}}^c} = \mathcal{G}^{S_{\lambda},*}_{\widetilde{\mathcal{F}}^c\widetilde{\mathcal{H}}^c} \bigcup (X_{S_{\lambda},i}, \widetilde{B}_{S_{\lambda},i})$.
13: **Return** $\mathcal{G}^{S_{\lambda},*}_{\widetilde{\mathcal{F}}^c\widetilde{\mathcal{H}}^c}$.

6.1.4.3 Theoretical Analysis

From Definition 6.35 and Theorem 6.5, we know that the object-oriented FCS threshold has a significant impact on the construction of a new concept space. Hence, it is very necessary to select an optimal (or approximate optimal) λ for each dataset.

Let $\lambda = \lambda(i)$ $(i \in \{1, 2, \ldots, n\})$, and $\lambda(i) \propto i$. For all the newly constructed concept spaces with different $\lambda(i)$, we denote

$$\begin{bmatrix} \mathcal{G}^{S_{\lambda(1)},*}_{\widetilde{\mathcal{F}}^c\widetilde{\mathcal{H}}^c} \\ \mathcal{G}^{S_{\lambda(2)},*}_{\widetilde{\mathcal{F}}^c\widetilde{\mathcal{H}}^c} \\ \vdots \\ \mathcal{G}^{S_{\lambda(n)},*}_{\widetilde{\mathcal{F}}^c\widetilde{\mathcal{H}}^c} \end{bmatrix} = \begin{bmatrix} \mathcal{G}^{S_{\lambda(1)},1}_{\widetilde{\mathcal{F}}^c\widetilde{\mathcal{H}}^c} & \mathcal{G}^{S_{\lambda(1)},2}_{\widetilde{\mathcal{F}}^c\widetilde{\mathcal{H}}^c} & \cdots & \mathcal{G}^{S_{\lambda(1)},m_1}_{\widetilde{\mathcal{F}}^c\widetilde{\mathcal{H}}^c} \\ \mathcal{G}^{S_{\lambda(2)},1}_{\widetilde{\mathcal{F}}^c\widetilde{\mathcal{H}}^c} & \mathcal{G}^{S_{\lambda(2)},2}_{\widetilde{\mathcal{F}}^c\widetilde{\mathcal{H}}^c} & \cdots & \mathcal{G}^{S_{\lambda(2)},m_2}_{\widetilde{\mathcal{F}}^c\widetilde{\mathcal{H}}^c} \\ \vdots & \vdots & \ddots & \vdots \\ \mathcal{G}^{S_{\lambda(n)},1}_{\widetilde{\mathcal{F}}^c\widetilde{\mathcal{H}}^c} & \mathcal{G}^{S_{\lambda(n)},2}_{\widetilde{\mathcal{F}}^c\widetilde{\mathcal{H}}^c} & \cdots & \mathcal{G}^{S_{\lambda(n)},m_n}_{\widetilde{\mathcal{F}}^c\widetilde{\mathcal{H}}^c} \end{bmatrix}, \tag{6.47}$$

where $m_i = |\mathcal{G}^{S_{\lambda(i)},*}_{\widetilde{\mathcal{F}}^c\widetilde{\mathcal{H}}^c}|$, and $\mathcal{G}^{S_{\lambda(i)},*}_{\widetilde{\mathcal{F}}^c\widetilde{\mathcal{H}}^c}$ is computed with $\lambda(i)$.

In Eq. (6.47), we say that $\mathcal{G}^{S_{\lambda(i)},j}_{\widetilde{\mathcal{F}}^c\widetilde{\mathcal{H}}^c}(j \in \{1, 2, \ldots, m_i\})$ is a conceptual subcluster of $\mathcal{G}^{S_{\lambda(i)},*}_{\widetilde{\mathcal{F}}^c\widetilde{\mathcal{H}}^c}$. Meanwhile, according to Definition 6.33, each object-oriented conceptual cluster can be represented as a new fuzzy concept. Hence, Eq. (6.47) can be rewritten as Eq. (6.48).

Note that there is only the fuzzy conditional concept space $\mathcal{G}_{\widetilde{\mathcal{F}}^c\widetilde{\mathcal{H}}^c}$ which will be influenced by the object-oriented FCS threshold $\lambda(i)$. The concept space $\mathcal{G}^{S_{\lambda(i)},*}_{\widetilde{\mathcal{F}}^c\widetilde{\mathcal{H}}^c}$ can be simplified by omitting the suffix $\widetilde{\mathcal{F}}^c\widetilde{\mathcal{H}}^c$ when no confusion exists, namely $\mathcal{G}^{S_{\lambda(i)},*}$.

Property 6.6 Let $\mathcal{G}^{S_{\lambda(i)},*}$ be a set of object-oriented conceptual clusters with the object-oriented FCS threshold $\lambda(i)$. Then we have

$$\begin{bmatrix} \mathcal{G}^{S_{\lambda(1)},*}_{\widetilde{\mathcal{F}}^c\widetilde{\mathcal{H}}^c} \\ \mathcal{G}^{S_{\lambda(2)},*}_{\widetilde{\mathcal{F}}^c\widetilde{\mathcal{H}}^c} \\ \vdots \\ \mathcal{G}^{S_{\lambda(n)},*}_{\widetilde{\mathcal{F}}^c\widetilde{\mathcal{H}}^c} \end{bmatrix} = \begin{bmatrix} (X_{S_{\lambda(1)},1}, \widetilde{B}_{S_{\lambda(1)},1}) & (X_{S_{\lambda(1)},2}, \widetilde{B}_{S_{\lambda(1)},2}) & \cdots & (X_{S_{\lambda(1)},m_1}, \widetilde{B}_{S_{\lambda(1)},m_1}) \\ (X_{S_{\lambda(2)},1}, \widetilde{B}_{S_{\lambda(2)},1}) & (X_{S_{\lambda(2)},2}, \widetilde{B}_{S_{\lambda(2)},2}) & \cdots & (X_{S_{\lambda(2)},m_2}, \widetilde{B}_{S_{\lambda(2)},m_2}) \\ \vdots & \vdots & \ddots & \vdots \\ (X_{S_{\lambda(n)},1}, \widetilde{B}_{S_{\lambda(n)},1}) & (X_{S_{\lambda(n)},2}, \widetilde{B}_{S_{\lambda(n)},2}) & \cdots & (X_{S_{\lambda(n)},m_n}, \widetilde{B}_{S_{\lambda(n)},m_n}) \end{bmatrix}. \tag{6.48}$$

$$|\mathcal{G}^{S_{\lambda(i)},*}| \propto \lambda(i). \tag{6.49}$$

Proof The proof can be derived by means of $\lambda(i) = \lambda$, and Definition 6.35. □

In the above discussion, we only consider the situation that there exists one concept cluster in FCLM. However, in the real-life world, studying the situation of multiple concept clusters with the label information is also highly desirable, as there are at least two concept clusters for classification tasks.

We denote by $G = \{\boldsymbol{x}_1, \boldsymbol{x}_2, \ldots, \boldsymbol{x}_m\}$ a set of instances and $\mathcal{K} = \{1, 2, \ldots, l\}$ the label space. There does exist a partition of the instances into l clusters $C_1, C_2, \ldots, C_l$ by means of the label information such that they can cover all the instances, and formally, $C_1 \cup C_2 \cup \cdots \cup C_l = G$, where $C_i \cap C_j = \emptyset$ $(\forall i \neq j)$. Meanwhile, we denote the corresponding fuzzy conceptual clusters by $\mathcal{G}_1^{S_{\lambda(i)},*}, \mathcal{G}_2^{S_{\lambda(i)},*}, \ldots, \mathcal{G}_l^{S_{\lambda(i)},*}$ with $\lambda(i)$. Moreover, the set of all fuzzy conceptual clusters with $\lambda(i)$ is denoted by $\mathcal{C}^{S_{\lambda(i)}}$, namely $\mathcal{C}^{S_{\lambda(i)}} = \{\mathcal{G}_1^{S_{\lambda(1)},*}, \mathcal{G}_2^{S_{\lambda(1)},*}, \cdots, \mathcal{G}_l^{S_{\lambda(1)},*}\}$. For different object-oriented FCS thresholds, we further denote

$$\begin{bmatrix} \mathcal{C}^{S_{\lambda(1)}} \\ \mathcal{C}^{S_{\lambda(2)}} \\ \vdots \\ \mathcal{C}^{S_{\lambda(n)}} \end{bmatrix} = \begin{bmatrix} \mathcal{G}_1^{S_{\lambda(1)},*} & \mathcal{G}_2^{S_{\lambda(1)},*} & \cdots & \mathcal{G}_l^{S_{\lambda(1)},*} \\ \mathcal{G}_1^{S_{\lambda(2)},*} & \mathcal{G}_2^{S_{\lambda(2)},*} & \cdots & \mathcal{G}_l^{S_{\lambda(2)},*} \\ \vdots & \vdots & \ddots & \vdots \\ \mathcal{G}_1^{S_{\lambda(n)},*} & \mathcal{G}_2^{S_{\lambda(n)},*} & \cdots & \mathcal{G}_l^{S_{\lambda(n)},*} \end{bmatrix}. \tag{6.50}$$

Our aim is to select an optimal $\lambda(i)$ in the interval [0,1] for each dataset. Let $(X_r, \widetilde{B}_r)$ $(r \in \{1, 2, \ldots, m\})$ be a fuzzy granular concept. Then, the objective function can be formulated as

$$\begin{aligned} E(\lambda(i), j) = & \min_{i \in \mathcal{I}, j \in \mathcal{J}, k'} \sum_{r=1}^{m} ||(X_r, \widetilde{B}_r) - \mathcal{G}_{k'}^{S_{\lambda(i),j}}||_2^2 - \\ & \max_{i \in \mathcal{I}} \min_{j \in \mathcal{J}} \sum_{k'' \in \overline{\mathcal{K}}} \sum_{r=1}^{m} ||(X_r, \widetilde{B}_r) - \mathcal{G}_{k''}^{S_{\lambda(i),j}}||_2^2 \\ & \text{s.t. } m_i \propto \lambda(i), 0 \leq \lambda(i) \leq 1, \end{aligned} \tag{6.51}$$

where $\mathcal{I} = \{1, 2, \ldots, n\}$, $\mathcal{J} = \{1, 2, \ldots, m_i\}$, $\overline{\mathcal{K}} = \mathcal{K} \setminus \{k'\}$, and k' represents the real class label of the fuzzy granular concept $(X_r, \widetilde{B}_r)$. Hence, in Eq. (6.51), the first item denotes that samples are classified into the ground truth conceptual subcluster, while the second item indicates the opposite situation.

Let $(X_{S_{\lambda(i),j}}^k, \widetilde{B}_{S_{\lambda(i),j}}^k)$ $(k \in \mathcal{K})$ be the representation of the conceptual subcluster $\mathcal{G}_k^{S_{\lambda(i),j}}$. For any fuzzy granular concept $(X_r, \widetilde{B}_r)$, it can be considered as an instance $\boldsymbol{x}_r$ with M-dimensional features. Therefore, according to Definition 6.31 and Eq. (6.48), the objective function can be reformulated as

$$\begin{aligned} E(\lambda(i), j) = & \min_{i \in \mathcal{I}, j \in \mathcal{J}, k'} \sum_{r=1}^{m} ||\widetilde{B}_r - \widetilde{B}_{S_{\lambda(i),j}}^{k'}||_2^2 - \\ & \max_{i \in \mathcal{I}} \min_{j \in \mathcal{J}} \sum_{k'' \in \overline{\mathcal{K}}} \sum_{r=1}^{m} ||\widetilde{B}_r - \widetilde{B}_{S_{\lambda(i),j}}^{k''}||_2^2 \\ & \text{s.t. } m_i \propto \lambda(i), 0 \leq \lambda(i) \leq 1. \end{aligned} \tag{6.52}$$

Based on Eq. (6.48) and Property 6.6, we know that the variable j is dependent on another variable $\lambda(i)$. Hence, we can optimize the objective function of our FCLM by means of updating $\lambda(i)$:

$$\begin{aligned} \widehat{\lambda}(i) = & \arg\min_{i \in \mathcal{I}, j \in \mathcal{J}} E(\lambda(i), j) \\ & \text{s.t. } m_i \propto \lambda(i), 0 \leq \lambda(i) \leq 1. \end{aligned} \tag{6.53}$$

In theory, we can obtain an optimal $\widehat{\lambda}(i)$ by solving Eq. (6.53) directly. Unfortunately, it is quite difficult to obtain analytical solutions due to lacking of a concrete functional expression between m_i and $\lambda(i)$. Hence, we select an approximate optimal $\widehat{\lambda}(i)$ by a method similar to grid search. The complete procedure for selecting an approximate optimal solution (see Algorithm 6.14 for details) is proposed based on the above discussion.

Algorithm 6.14 Select $\widehat{\lambda}(i)$ for FCLM

1: **Input:** Training set $\overline{G}$, validation set V, and step size ε.
2: **Output:** An approximate optimal $\widehat{\lambda}(i)$.
3: Construct a fuzzy conditional concept space $\mathcal{G}_{\widetilde{\mathcal{F}}^c\widetilde{\mathcal{H}}^c}$ and classical decision concept space $\mathcal{G}_{\mathcal{F}^d\mathcal{H}^d}$ by Algorithm 6.12.
4: **for** $\lambda(i) = 0$ to 1 **do**
5: Get $C^{S_{\lambda(i)}}$ by Algorithm 6.13.
6: **for** $x_r \in V$ **do**
7: Compute $E(\lambda(i), j)$ by Eq. (6.52).
8: **end for**
9: $\lambda(i)$=$\lambda(i)$+ε.
10: **end for**
11: Get $\widehat{\lambda}(i)$ by Eq. (6.53).
12: **Return** $\widehat{\lambda}(i)$.

6.2 Label Proportion for Learning

6.2.1 A Fast Algorithm for Multi-Class Learning from Label Proportions

Learning from label proportions (LLP) is a new kind of learning problem which has attracted wide interest in machine learning. Different from the well-known supervised learning, the training data of LLP is in form of bags and only the proportion of each class in each bag is available. In this subsection, we propose a fast algorithm called multi-class learning from label proportions by extreme learning machine (LLP-ELM), which takes advantage of extreme learning machine with fast learning speed to solve multi-class learning from label proportions.

6.2.1.1 Background

In this section, we give a brief introduction of the traditional extreme learning machine [21, 22]. Figure 6.5 shows the architecture of ELM. In detail, it is a single-hidden layer feed-forward networks with three parts: input neurons, hidden neurons and output neurons. In particular, $\mathbf{h}(\mathbf{x}) = [h_1(x), \ldots, h_L(x)]$ is nonlinear feature mapping of ELM with the form of $h_j(x) = g(\mathbf{w_j}.\mathbf{x} + b_j)$ and $\boldsymbol{\beta_j} = [\beta_{j1}, \ldots, \beta_{jc}]^T$, $j = 1, \ldots, L$ is the output weights between the jth hidden layer and the output nodes.

Given N samples $(x_i, t_i), i = 1, \ldots, N$, where $\mathbf{x_i} = [x_{i1}, \ldots, x_{id}]^T$ denotes the input feature vectors and $\mathbf{t_i} = [t_{i1}, \ldots, t_{ic}]^T$ is the corresponding label in a one-hot fashion. In particular, c and d respectively represent the total classes and feature

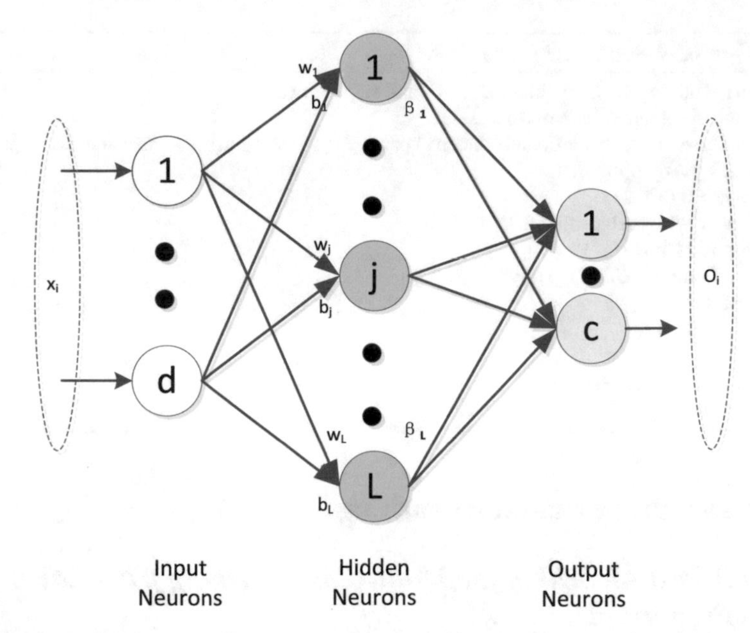

Fig. 6.5 The architecture of ELM. In detail, it is a single-hidden layer feed-forward network with three parts: input neurons, hidden neurons and output neurons. In particular, $\mathbf{h}(\mathbf{x}) = [h_1(x), \ldots, h_L(x)]$ is nonlinear feature mapping of ELM with the form of $h_j(x) = g(\mathbf{w_j}.\mathbf{x} + b_j)$ and $\boldsymbol{\beta}_j = [\beta_{j1}, \ldots, \beta_{jc}]^T, j = 1, \ldots, L$ is the output weights between the jth hidden layer and the output nodes

number. Consequently, a standard feed-forward neural network with L hidden nodes can be expressed as:

$$\sum_{j=1}^{L} \boldsymbol{\beta}_j g(\mathbf{w_j}.\mathbf{x_i} + b_j) = \mathbf{o_i}, i = 1, \ldots, N, \tag{6.54}$$

where $\mathbf{w_j} = [w_{j1}, w_{j2}, \ldots, w_{jL}]^T$ is the weight vector between the jth hidden neuron and the input neurons, and $\boldsymbol{\beta}_j = [\beta_{j1}, \beta_{j2}, \ldots, \beta_{jc}]^T, j = 1, \ldots, L$ is the weight vector connecting the output neuron and the jth hidden neurons. According to [21], the ELM can approximate those N samples to zero error with the equation $\sum_{i=1}^{N} \|\mathbf{o_i} - \mathbf{t_i}\| = 0$. Thus, the above equations can be expressed as:

$$\sum_{j=1}^{L} \boldsymbol{\beta}_j g(\mathbf{w_j}.\mathbf{x_i} + b_j) = \mathbf{t_i}, i = 1, \ldots, N. \tag{6.55}$$

In particular, we can use matrix to express the above N equations with form of:

$$\mathbf{H}\boldsymbol{\beta} = \mathbf{T}, \tag{6.56}$$

where **H** is the hidden layer output matrix of the single-hidden layer feed-forward network and **T** is output matrix. More specifically, **H** and **T** have the form of:

$$\mathbf{H} = \begin{bmatrix} \mathbf{h}(\mathbf{x_1}) \\ \vdots \\ \mathbf{h}(\mathbf{x_N}) \end{bmatrix} = \begin{bmatrix} h_1(\mathbf{x_1}) & \cdots & h_L(\mathbf{x_1}) \\ \vdots & \vdots & \vdots \\ h_1(\mathbf{x_N}) & \vdots & h_L(\mathbf{x_N}) \end{bmatrix} \tag{6.57}$$

and

$$\mathbf{T} = \begin{bmatrix} \mathbf{t_1^T} \\ \vdots \\ \mathbf{t_N^T} \end{bmatrix} = \begin{bmatrix} t_{11} & \cdots & t_{1c} \\ \vdots & \vdots & \vdots \\ t_{N1} & \vdots & t_{Nc} \end{bmatrix} \tag{6.58}$$

In practice, the hidden node parameters (**w**,b) of ELM are randomly generated and then fixed without iteratively tuning, which is different to the traditional BP neural networks [21]. As a result, training an ELM is equivalent to find the optimal solution to $\boldsymbol{\beta}$, which is in defined as:

$$\boldsymbol{\beta} = \begin{bmatrix} \boldsymbol{\beta}_\mathbf{1}^\mathbf{T} \\ \vdots \\ \boldsymbol{\beta}_\mathbf{L}^\mathbf{T} \end{bmatrix} = \begin{bmatrix} \beta_{11} & \cdots & \beta_{1c} \\ \vdots & \vdots & \vdots \\ \beta_{L1} & \vdots & \beta_{Lc} \end{bmatrix} \tag{6.59}$$

Furthermore, $\boldsymbol{\beta}$ can computed by the following expression:

$$\boldsymbol{\beta}^* = \mathbf{H}^\dagger \mathbf{T} \tag{6.60}$$

where $\mathbf{H}^\dagger$ is the Moore-Penrose generalized inverse of matrix **H**.

6.2.1.2 The LLP-ELM Algorithm

In this section, we propose a fast method for multi-class learning from label proportions algorithm called LLP-ELM, which employs extreme learning machine to solve multi-class LLP problem. In order to leverage extreme learning machine to LLP, we reshape the hidden layer output matrix **H** and the training data target matrix **T** to new forms, such that **H** is in bag level and **T** contains the proportion information instead of a label one.

A. Learning Setting

The LLP problem is described by a set of training data, which is divided into several bags. Furthermore, compared to the traditional supervised learning, we only know the proportions of different categories in each bag instead of the ground-truth labels. In this paper, we consider the situation that different bags are disjoint, and the nth bag of the training data can be denoted as $B_n, n = 1, \ldots, h$. Consequently, the total training data is in form of:

$$D = B_1 \cup B_2 \cup \ldots . \cup B_h \qquad (6.61)$$
$$B_i \cap B_j = \emptyset, \forall i \neq j.$$

where there are n bags and N is the number of total instances. Each bag consists of m_n instances with the constraint $\sum_{n=1}^{h} m_n = N$, and can be expressed as:

$$B_n = \{x_n^1, ..., x_n^{m_n}\}, n \in \{1, 2, \ldots, h\}. \qquad (6.62)$$

Meanwhile, $\mathbf{p_n}$ is the corresponding class proportion vector of B_n and c represents the total classes number. More specifically, $\mathbf{p_n}$ can be written as a vector form:

$$\mathbf{p_n} = \begin{bmatrix} p_{n1} \\ \vdots \\ p_{nc} \end{bmatrix}, \qquad (6.63)$$

where the mth element p_n^m is the proportion of the mth class in the nth bag with the constraint $\sum_{m=1}^{c} p_n^m = 1$. Furthermore, the total proportion information can be defined in form of matrix:

$$\mathbf{P} = \begin{bmatrix} \mathbf{p_1^T} \\ \vdots \\ \mathbf{p_h^T} \end{bmatrix} = \begin{bmatrix} p_{11} & \cdots & p_{1c} \\ \vdots & \vdots & \vdots \\ p_{h1} & \vdots & p_{hc} \end{bmatrix}. \qquad (6.64)$$

B. The LLP-ELM Framework

From the above learning setting of LLP, a classifier in instance level is the final objective. To this end, we modify the original equations in ELM to the new equations in bag level. Specifically, we add all the equations in each bag straightforward, and the final equations in nth bag can be expressed as follows:

$$\sum_{j=1}^{L} \sum_{k=1}^{m_n} \boldsymbol{\beta}_j g(\mathbf{w_j}.\mathbf{x_{nk}} + b_j) = \sum_{k=1}^{m_n} \mathbf{t_{nk}}, n = 1, \ldots, h \qquad (6.65)$$

where $\mathbf{t_{nk}}$ is the real label for the kth instances in nth bag. Obviously, the real label information in the right part is inaccessible to us, with only label proportions in each bag available. To this end, we derive the right part of the above equation as the following form:

$$\sum_{k=1}^{m_n} \mathbf{t_{nk}} = m_n * \mathbf{p_n}, n = 1, \ldots, h \tag{6.66}$$

where $\mathbf{p_n}$ is the label proportion of nth bag. Substituting the formula (6.66) to (6.65), we can naturally obtain the following equations:

$$\sum_{j=1}^{L} \boldsymbol{\beta}_j \sum_{k=1}^{m_n} g(\mathbf{w_j}.\mathbf{x_{nk}} + b_j) = m_n * \mathbf{p_j}, n = 1, \ldots, h, \tag{6.67}$$

In particular, similar to the method from ELM [21], we can write the above equations in the form of matrix computing as follows:

$$\mathbf{H_p}\boldsymbol{\beta} = \mathbf{P} \tag{6.68}$$

where $\mathbf{H_p}$ is the hidden layer output matrix in the bag level, and $\mathbf{P}$ is the training data target proportion matrix. More specifically, $\mathbf{H_p}$ and $\mathbf{P}$ are given in form of:

$$\mathbf{H_p} = \begin{bmatrix} \sum_{k=1}^{m_1} \mathbf{h}(\mathbf{x_{1k}}) \\ \vdots \\ \sum_{k=1}^{m_h} \mathbf{h}(\mathbf{x_{hk}}) \end{bmatrix} = \begin{bmatrix} \sum_{k=1}^{m_1} h_1(\mathbf{x_{1k}}) & \cdots & \sum_{k=1}^{m_1} h_L(\mathbf{x_{1k}}) \\ \vdots & \vdots & \vdots \\ \sum_{k=1}^{m_h} h_1(\mathbf{x_{hk}}) & \vdots & \sum_{k=1}^{m_h} h_L(\mathbf{x_{hk}}) \end{bmatrix}$$

and

$$\mathbf{P} = \begin{bmatrix} m_1 * \mathbf{p_1^T} \\ \vdots \\ m_h * \mathbf{p_h^T} \end{bmatrix} = \begin{bmatrix} m_1 * p_{11} & \cdots & m_1 * p_{1c} \\ \vdots & \vdots & \vdots \\ m_h * p_{h1} & \vdots & m_h * p_{hc} \end{bmatrix}$$

Meanwhile, the final solution $\boldsymbol{\beta}$ is the same with the original form in ELM with dimension $L \times c$. Again, the optimal solution to (6.68) is given by

$$\boldsymbol{\beta}^* = \mathbf{H_p^\dagger P} \tag{6.69}$$

where $\mathbf{H_p^\dagger}$ is the Moore-Penrose generalized inverse of matrix $\mathbf{H_p}$.

In order to obtain a better generalization performance of ELM, we also follow the method from [22] to study the regularized ELM. In detail, the final objective function of ELM is formulated as follows:

$$\min_{\beta \in R^{L\times c}} \frac{1}{2}\|\boldsymbol{\beta}\|^2 + \frac{C}{2}\sum_{i=1}^{N}\|\mathbf{e_i}\|^2$$

$$s.t.\ \ \mathbf{h}(\mathbf{x_i})\boldsymbol{\beta} = \mathbf{t_i^T} - \mathbf{e_i^T}, i = 1, \ldots, N, \tag{6.70}$$

in which the first term of the objective function is a regularization term and C is a parameter to make a trade-off between the first and second term.

We equivalently reformulate the problem (6.70) as follows by substituting the constraints to its objective function:

$$\min_{\beta \in R^{L\times c}} L_{ELM} = \frac{1}{2}\|\boldsymbol{\beta}\|^2 + \frac{C}{2}\|\mathbf{T} - \mathbf{H}\boldsymbol{\beta}\|^2 \tag{6.71}$$

Note that the second term of (6.71) can be replaced by $\frac{C}{2}\|\mathbf{P} - \mathbf{H_p}\boldsymbol{\beta}\|^2$, which is the matrix form in bag level. In other words, the final unconstrained optimization problem can be written as:

$$\min_{\beta \in R^{L\times c}} L_{ELM} = \frac{1}{2}\|\boldsymbol{\beta}\|^2 + \frac{C}{2}\|\mathbf{P} - \mathbf{H_p}\boldsymbol{\beta}\|^2 \tag{6.72}$$

In practice, the final objection is widely known as the ridge regression or regularized least squares.

C. How to Solve the LLP-ELM

We follow the strategy from [22] to solve (6.72), and the final purpose is to minimize the training error as well as the norm of the output weights. Obviously, the final objective function is a convex problem, which is always solved by way of gradient. More specifically, by setting the gradient of (6.72) to zero with respect to $\boldsymbol{\beta}$, we can obtain the following expression:

$$\boldsymbol{\beta} - C\mathbf{H_p^T}(\mathbf{P} - \mathbf{H_p}\boldsymbol{\beta}) = \mathbf{0}. \tag{6.73}$$

This yields

$$(\frac{\mathbf{I}}{\mathbf{C}} + \mathbf{H_p^T}\mathbf{H_p})\boldsymbol{\beta} = \mathbf{H_p^T}\mathbf{P}, \tag{6.74}$$

where **I** is an identity matrix with dimension L.

The above equation is very intuitive, and we can obtain the final optimization result by inverting a L×L matrix directly. However, it is less efficient to directly invert a L×L matrix when the number of bag is less than the number of hidden neurons(h < L). Therefore, there are two methods which are shown in **Remark 1** and **Remark 2**. In summary, in the case where the number of bags are plentiful than hidden neurons, we use **Remark 1** to compute the output weights, otherwise we use **Remark 2**.

Remark 1 The solution for formula (6.73) when h > L.

- $\mathbf{H_p}$ has more rows than columns, which means the number of bag is larger than the number of hidden neurons.
- By inverting a L×L matrix directly and multiplying both sides by $(\mathbf{H_p^T H_p} + \frac{\mathbf{I}}{\mathbf{C}})^{-1}$, we can obtain the following expression

$$\boldsymbol{\beta} = (\mathbf{H_p^T H_p} + \frac{\mathbf{I}}{\mathbf{C}})^{-1}\mathbf{H_p^T P}, \tag{6.75}$$

which is the optimal solution of (6.73).

Remark 2 The solution for formula (6.73) when h < L.

- Notice that $\mathbf{H_p}$ is full row rank and $\mathbf{H_p H_p^T}$ is invertible when h < L.
- Restrict $\boldsymbol{\beta}$ to be a linear combination of the row in $\mathbf{H_p} : \boldsymbol{\beta} = \mathbf{H_p^T}\boldsymbol{\alpha}$
- Substitute $\boldsymbol{\beta} = \mathbf{H_p^T}\boldsymbol{\alpha}$ into (6.73), and multiply by $(\mathbf{H_p H_p^T})^{-1}\mathbf{H_p}$.
- By the above step, we can obtain the following equation:

$$\boldsymbol{\alpha} - C(\mathbf{P} - \mathbf{H_p H_p^T}\boldsymbol{\alpha}) = \mathbf{0}. \tag{6.76}$$

- As a result, the final optimal solution of (6.73) is in form of

$$\boldsymbol{\beta} = \mathbf{H_p^T}\boldsymbol{\alpha} = \mathbf{H_p^T}(\mathbf{H_p H_p^T} + \frac{\mathbf{I}}{\mathbf{C}})^{-1}\mathbf{P} = \mathbf{0}. \tag{6.77}$$

The solution process of LLP-ELM model can be concluded to the following two steps:

- Compute training data target proportion matrix **P** and the hidden layer output matrix $\mathbf{H_p}$.
- Obtain the final optional solution of $\boldsymbol{\beta}$ according to **Remark 1** or **Remark 2**. The details of the process are shown in **Algorithm 6.15**.

D. Computational Complexity

From the **Remark 1** and **Remark 2**, we can observe that the main time cost of our method is to calculate the matrix inversion. Furthermore, the dimension of matrix is

Algorithm 6.15 LLP-ELM

1: **Input:** Training datasets in bags$\{B_n\}$; The corresponding proportion p_n of B_n; Activation function g(x) and the number of hidden nodes N.
2: **Output:** Classification model f(x,$\boldsymbol{\beta}$)
3: **Begin**
4: • Randomly initialize the value $\mathbf{w_j}$ and b_j for the jth node, $j = 1, \ldots, L$.
5: • Compute the training data target proportion matrix **P** by the proportion information of each bag.
6: • Compute the hidden layer output matrix in the bag level $\mathbf{H_p}$.
7: • Obtain the weight vector according to **Remark 1** or **Remark 2**.
8: **End**

minimum of the number of bags h and the hidden neurons L, which is determined by us. As we all know, the complexity of matrix inversion is proportional to the O^3, where O is the dimension of matrix, and is equal to Min(L,h) in this paper.

6.2.2 Learning from Label Proportions with Generative Adversarial Networks

6.2.2.1 Preliminaries

A. The Multi-Class LLP

Before further discussion, we formally describe multi-class LLP. For simplicity, we assume that all the bags are disjoint and let $\mathcal{B}_i = \{\mathbf{x}_i^1, \mathbf{x}_i^2, \cdots, \mathbf{x}_i^{N_i}\}, i = 1, 2, \cdots, n$ denote bags in training set. Then, training data is $\mathcal{D} = \mathcal{B}_1 \cup \mathcal{B}_2 \cup \cdots \cup \mathcal{B}_n, \mathcal{B}_i \cap \mathcal{B}_j = \emptyset, \forall i \neq j$, where the total number of bags is n.

In addition, $\mathbf{p}_i$ is a K-element vector where the kth element p_i^k is instance proportion in $\mathcal{B}_i$ belonging to the kth class with the constraint $\sum_{k=1}^K p_i^k = 1$ and K represents the total number of classes, i.e.,

$$p_i^k := \frac{|\{j \in [1 : N_i] | \mathbf{x}_i^j \in \mathcal{B}_i, y_i^{j*} = k\}|}{|\mathcal{B}_i|}. \tag{6.78}$$

Here, $[1 : N_i] = \{1, 2, \cdots, N_i\}$ and y_i^{j*} is the unaccessible ground-truth instance-level label of $\mathbf{x}_i^j$. In this way, we can denote the available training data as $\mathcal{L} = \{(\mathcal{B}_i, \mathbf{p}_i)\}_{i=1}^n$. The goal of LLP is to learn an instance-level classifier based on this kind of dataset.

B. Deep Discriminant Approach for LLP

In terms of deep learning, DLLP firstly leveraged CNNs to solve multi-class LLP problem [1]. Since CNNs can give a probabilistic interpretation for classification, it is straightforward to adapt cross-entropy loss into a bag-level version by averaging the probability outputs in every bag as the proportion estimation. To this end, inspired by [71], DLLP reshaped standard cross-entropy loss by substituting instance-level label with label proportion, in order to meet the proportion consistency.

In detail, suppose that $\tilde{\mathbf{p}}_i^j = p_\theta(\mathbf{y}|\mathbf{x}_i^j)$ is the vector-valued CNNs output for $\mathbf{x}_i^j$, where θ is the network parameter. Let $\oplus$ be element summation operator, then the bag-level label proportion in the ith bag is obtain by incorporating the element-wise posterior probability:

$$\overline{\mathbf{p}}_i = \frac{1}{N_i}\bigoplus_{j=1}^{N_i}\tilde{\mathbf{p}}_i^j = \frac{1}{N_i}\bigoplus_{j=1}^{N_i} p_\theta(\mathbf{y}|\mathbf{x}_i^j), \tag{6.79}$$

In order to smooth *max* function [5], $\tilde{\mathbf{p}}_i^j$ is in a vector-type softmax manner to produce the distribution for class probabilities. Taking *log* as element-wise logarithmic operator, objective of DLLP can be intuitively formulated using cross-entropy loss $L_{prop} = -\sum_{i=1}^n \mathbf{p}_i^\top log(\overline{\mathbf{p}}_i)$. It penalizes the difference between prior and posterior probabilities in bag-level, and commonly exists in GAN-based SSL [61].

C. Entropy Regularization for DLLP

Following the entropy regularization strategy [18], we can introduce an extra loss E_{in} with a trade-off hyperparameter λ to constrain instance-level output distribution in a low entropy accordingly:

$$L = L_{prop} + \lambda E_{in} = -\sum_{i=1}^{n}\mathbf{p}_i^\top log(\overline{\mathbf{p}}_i) - \lambda\sum_{i=1}^{n}\sum_{j=1}^{N_i}(\tilde{\mathbf{p}}_i^j)^\top log(\tilde{\mathbf{p}}_i^j). \tag{6.80}$$

This extension is similar to a KL divergence between two distributions. It takes advantage of DNN's output distribution to cater to the label proportions requirement, as well as minimizing output entropy as a regularization term to guarantee strong true-fake belief. This is believed to be linked with an inherent MAP estimation with certain prior distribution in network parameters.

6.2.2.2 Adversarial Learning for LLP

In this section, we propose LLP-GAN, which devotes GANs to harnessing LLP problem.

A. The Objective Function of Discriminator

We illustrate the LLP-GAN framework in Fig. 6.6. The generator is employed to generate images with input noise, which is labeled as fake. On the other hand, the discriminator yields class confidence maps for each class (including the fake one) by taking both fake and real data as the inputs. In particular, our discriminator is not only to identify whether it is a sample from the real data or not, but also to elaborately distinguish each real input's label assignment as a K classes classifier. This idea is fairly intuitive, and we conclude its loss as the L_{unsup} term.

Next, the main issue becomes how to exploit the proportional information to guide this unsupervised learning correctly. To this end, we replace the supervised information in semi-supervised GANs with label proportions, resulting in L_{sup}, same as L_{prop} in (6.80).

Definition 6.36 Suppose that $\mathcal{P}$ is a partition to divide the data space into n disjoint sections. Let $p_d^i(\mathbf{x}), i = 1, 2, \cdots, n$ be marginal distributions with respect to elements in $\mathcal{P}$ respectively. Accordingly, n bags in LLP training data spring from sampling upon $p_d^i(\mathbf{x}), i = 1, 2, \cdots, n$. In the meantime, let $p(\mathbf{x}, y)$ be the unknown holistic joint distribution.

We normalize the first K classes in $P_D(\cdot|\mathbf{x})$ into the instance-level posterior probability $\tilde{p}_D(\cdot|\mathbf{x})$ and compute $\overline{\mathbf{p}}$ based on (6.79). Then, the *ideal* optimization problem for the discriminator of LLP-GAN is:

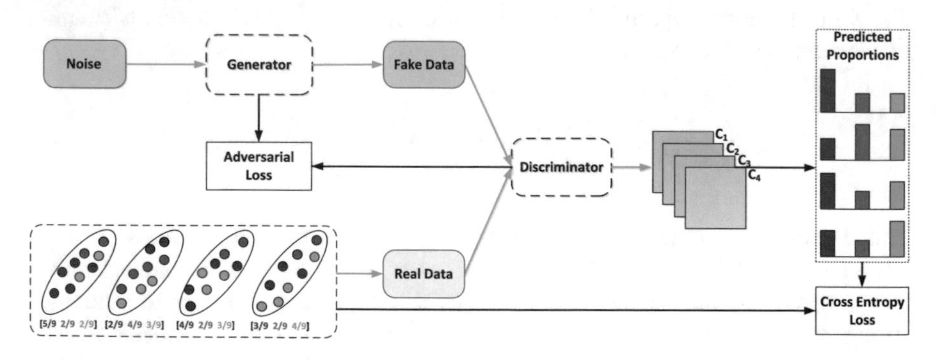

Fig. 6.6 An illustration of our LLP-GAN framework

$$\max_D V(G,D) = L_{unsup} + L_{sup} = L_{real} + L_{fake} - \lambda CE_{\mathcal{L}}(\mathbf{p},\overline{\mathbf{p}})$$

$$= \sum_{i=1}^{n} E_{\mathbf{x}\sim p_d^i}\Big[log P_D(y \leq K|\mathbf{x})\Big] + E_{\mathbf{x}\sim p_g}\Big[log P_D(K+1|\mathbf{x})\Big] + \lambda \sum_{i=1}^{n} \mathbf{p}_i^{\mathsf{T}} log(\overline{\mathbf{p}}_i). \tag{6.81}$$

Here, $p_g(\mathbf{x})$ represents the distribution of the synthesized data.

The normalized instance-level posterior probability $\tilde{p}_D(\cdot|\mathbf{x})$ is:

$$\tilde{p}_D(k|\mathbf{x}) = \frac{P_D(k|\mathbf{x})}{1 - P_D(K+1|\mathbf{x})}, k = 1, 2, \cdots, K. \tag{6.82}$$

Note that weight λ in (6.81) is added to balance between supervised and unsupervised terms, which is a slight revision of SSL with GANs [13, 54]. Intuitively, we reckon the proportional information is too weak to fulfill supervised learning pursuit. As a result, a relatively small weight should be preferable in the experiments. However, we fix $\lambda = 1$ in the following theoretical analysis on discriminator.

Aside from identifying the first two terms in (6.81) as that in semi-supervised GANs, the cross-entropy term harnesses the label proportions consistency. In order to justify the non-triviality of this loss, we first look at its lower bound. More important, it is easier to perform the gradient method on the lower bound, because it swaps the order of *log* and the summation operation. For brevity, the analysis will be done in a non-parametric setting, i.e. we assume that both D and G have infinite capacity.

Remark (The Lower Bound Approximation) Let $p_i(k) = p_i^k = \int p_i(y=k|\mathbf{x}) p_d^i(\mathbf{x}) d\mathbf{x}$ be the class k proportion in the ith bag. By applying Monte-Carlo sampling, we have:

$$-CE_{\mathcal{L}}(\mathbf{p},\overline{\mathbf{p}}) = \sum_{i=1}^{n}\sum_{k=1}^{K} p_i(k) log\Big[\frac{1}{N_i}\sum_{j=1}^{N_i} \tilde{p}_D(k|\mathbf{x}_i^j)\Big]$$

$$\backsimeq \sum_{i=1}^{n}\sum_{k=1}^{K} p_i(k) log\Big[\int p_d^i(\mathbf{x}) \tilde{p}_D(k|\mathbf{x}) d\mathbf{x}\Big] \geqslant \sum_{i=1}^{n}\sum_{k=1}^{K} p_i(k) E_{\mathbf{x}\sim p_d^i}\Big[log \tilde{p}_D(k|\mathbf{x})\Big]. \tag{6.83}$$

Similar to EM mechanism for mixture models, by approximating $-CE_{\mathcal{L}}(\mathbf{p},\overline{\mathbf{p}})$ with its lower bound, we can perform gradient ascend independently on every sample. Hence, SGD can be applied.

Property 6.7 The maximization on the lower bound in (6.83) induces an optimal discriminator D^* with a posterior distribution $\tilde{p}_{D^*}(y|\mathbf{x})$, which is consistent with the prior distribution $p_i(y)$ in each bag.

Proof Taking the aggregation with respect to one bag, for example, the ith bag, we have:

$$\begin{aligned}
E_{\mathbf{x}\sim p_d^i}[logp(\mathbf{x})] &= E_{\mathbf{x}\sim p_d^i}log\Big[\frac{p(\mathbf{x},y)}{\tilde{p}_D(y|\mathbf{x})}\frac{\tilde{p}_D(y|\mathbf{x})}{p(y|\mathbf{x})}\Big] \\
&= E_{\mathbf{x}\sim p_d^i}\int p_i(y)log\Big[\frac{p_i(y)p(\mathbf{x}|y)}{\tilde{p}_D(y|\mathbf{x})}\frac{\tilde{p}_D(y|\mathbf{x})}{p(y|\mathbf{x})}\Big]\mathrm{d}y \\
&= E_{\mathbf{x}\sim p_d^i}\int\Big[p(y_i)log\tilde{p}_D(y|\mathbf{x})+log\frac{p(\mathbf{x}|y)}{p(y|\mathbf{x})}\Big]\mathrm{d}y \\
&+ E_{\mathbf{x}\sim p_d^i}KL(p_i(y)\|\tilde{p}_D(y|\mathbf{x})) \\
&\geqslant \sum_{k=1}^{K}p_i(k)E_{\mathbf{x}\sim p_d^i}\Big[log\tilde{p}_D(k|\mathbf{x})\Big]+\sum_{k=1}^{K}p_i(k)E_{\mathbf{x}\sim p_d^i}\Big[log\frac{p(\mathbf{x}|k)}{p(k|\mathbf{x})}\Big]
\end{aligned} \tag{6.84}$$

Note that the last term in (6.84) is free of the discriminator, and the aggregation can be independently performed within every bag due to the disjoint assumption. Then, maximizing the lower bound in (6.83) is equivalent to minimizing the expectation of KL-divergence between $p_i(y)$ and $\tilde{p}_D(y|\mathbf{x})$. Because of the infinite capacity assumption on discriminator and the non-negativity of KL-divergence, we have:

$$D^* = \arg\min_D E_{\mathbf{x}\sim p_d^i}KL(p_i(y)\|\tilde{p}_D(y|\mathbf{x})) \Leftrightarrow \tilde{p}_{D^*}(y|\mathbf{x}) \overset{a.e.}{=} p_i(y), \mathbf{x}\sim p_d^i(\mathbf{x}). \tag{6.85}$$

That concludes the proof. □

Property 6.7 tells us that if there is only one bag, then $\tilde{p}_{D^*}(y|\mathbf{x}) \overset{a.e.}{=} p(y)$. However, there is normally more than one bag in LLP, the final classifier will somehow be a trade-off among all the prior proportions $p_i(y), i = 1, 2, \cdots, n$. Next, we will show how the adversarial learning on the discriminator helps to determine the formulation of this trade-off into a weighted aggregation.

B. Global Optimality

As shown in (6.83), in order to facilitate the gradient computation, we substitute cross entropy in (6.81) by its lower bound and denote this approximate objective function for discriminator by $\widetilde{V}(G, D)$.

Theorem 6.6 *For fixed G, the optimal discriminator D^* for $\widetilde{V}(G, D)$ satisfies:*

$$P_{D^*}(y = k|\mathbf{x}) = \frac{\sum_{i=1}^{n} p_i(k)p_d^i(\mathbf{x})}{\sum_{i=1}^{n} p_d^i(\mathbf{x}) + p_g(\mathbf{x})}, k = 1, 2, \cdots, K. \tag{6.86}$$

Proof According to (6.81) and (6.83) and given any generator G, we have:

$$\widetilde{V}(G, D) = \sum_{i=1}^{n} E_{\mathbf{x}\sim p_d^i}\Big[log(1 - P_D(K+1|\mathbf{x}))\Big] + E_{\mathbf{x}\sim p_g}\Big[log P_D(K+1|\mathbf{x})\Big] +$$

$$\sum_{i=1}^{n}\sum_{k=1}^{K} p_i(k) E_{\mathbf{x}\sim p_d^i}\Big[log\, \tilde{p}_D(k|\mathbf{x})\Big] = \int \Big\{ \sum_{i=1}^{n} p_d^i(\mathbf{x})\Big[log\big[\sum_{k=1}^{K} P_D(k|\mathbf{x})\big] + \tag{6.87}$$

$$\sum_{k=1}^{K} p_i(k) log \frac{P_D(k|\mathbf{x})}{1 - P_D(K+1|\mathbf{x})}\Big] + p_g(\mathbf{x}) log\Big[1 - \sum_{k=1}^{K} P_D(k|\mathbf{x})\Big]\Big\} d\mathbf{x}$$

By taking the derivative of the integrand, we find the maximum in [0, 1] as that in (6.86). □

Remark (Beyond the Incontinuity of p_g) According to [2], the problematic scenario is that the generator is a mapping from a low dimensional space to a high dimensional one, which results in the density of $p_g(\mathbf{x})$ infeasible. However, based on the definition of $\tilde{p}_D(y|\mathbf{x})$ in (6.82), we have:

$$\tilde{p}_{D^*}(y|\mathbf{x}) = \frac{\sum_{i=1}^{n} p_i(y) p_d^i(\mathbf{x})}{\sum_{i=1}^{n} p_d^i(\mathbf{x})} = \sum_{i=1}^{n} w_i(\mathbf{x}) p_i(y). \tag{6.88}$$

Hence, our final classifier does not depend on $p_g(\mathbf{x})$, and (6.88) explicitly expresses the weights of the aggregation.

Remark (Relationship to One-Side Label Smoothing) Notice that the optimal discriminator D^* is also related to the one-sided label smoothing mentioned in [54], which was inspirited by [64] and shown to reduce the vulnerability of neural networks to adversarial examples [73].

In our model, we only smooth labels of real data (multi-class classifier) in the discriminator by setting the targets as the holistic proportions (the prior) $p_i(y)$ in corresponding bags.

C. The Objective Function of Generator

Normally, for the generator, we should solve the following optimization problem with respect to p_g.

$$\min_G \widetilde{V}(G, D^*) = \min_G E_{\mathbf{x}\sim p_g} log P_{D^*}(K+1|\mathbf{x}). \tag{6.89}$$

If denoting $C(G) = \max_D \widetilde{V}(G, D) = \widetilde{V}(G, D^*)$, because $\widetilde{V}(G, D)$ is convex in p_g and the supremum of a set of convex function is still convex, we have the following conclusion.

Theorem 6.7 *The global minimum of $C(G)$ is achieved if and only if $p_g = \frac{1}{n}\sum_{i=1}^{n} p_d^i$.*

Proof Denote $p_d = \sum_{i=1}^{n} p_d^i$. Hence, according to Theorem 6.6, we can reformulate $C(G)$ as:

$$C(G) = \sum_{i=1}^{n} E_{\mathbf{x}\sim p_d^i}\Big[log \frac{p_d(\mathbf{x})}{p_d(\mathbf{x}) + p_g(\mathbf{x})}\Big] + E_{\mathbf{x}\sim p_g}\Big[log \frac{p_g(\mathbf{x})}{p_d(\mathbf{x}) + p_g(\mathbf{x})}\Big] + \sum_{i=1}^{n}\sum_{k=1}^{K} p_i(k) E_{\mathbf{x}\sim p_d^i}\Big[log\, \tilde{p}_D(k|\mathbf{x})\Big] = 2 \cdot JSD(p_d \| p_g) - 2log(2) - \sum_{i=1}^{n} E_{\mathbf{x}\sim p_d^i}\Big[CE(p_i(y), \tilde{p}_{D^*}(y|\mathbf{x}))\Big], \tag{6.90}$$

where $JSD(\cdot\|\cdot)$ and $CE(\cdot, \cdot)$ are the Jensen-Shannon divergence and cross entropy between two distributions, respectively. However, note that p_d is a summation of n independent distributions, so $\frac{1}{n}p_d$ is a well-defined probabilistic density. Then, we have:

$$C(G^*) = \min_G C(G) = nlog(n) - (n+1)log(n+1) - \sum_{i=1}^{n} E_{\mathbf{x}\sim p_d^i}\Big[CE(p_i(y), \tilde{p}_{D^*}(y|\mathbf{x}))\Big] \iff p_{g^*} \overset{\text{a.e.}}{=} \frac{1}{n} p_d. \tag{6.91}$$

That concludes the proof. □

Remark When there is only one bag, the first two terms in (6.91) will degenerate as $nlog(n) - (n+1)log(n+1) = -2log2$, which adheres to results in original GANs. On the other hand, the third term manifests the uncertainty on instance label, due to the concealment in the form of proportion.

Remark According to the analysis above, ideally, we can obtain the Nash equilibrium between the discriminator and the generator, i.e. the solution pair (G^*, D^*) satisfies:

$$\widetilde{V}(G^*, D^*) \geqslant \widetilde{V}(G^*, D), \forall D; \widetilde{V}(G^*, D^*) \leqslant \widetilde{V}(G, D^*), \forall G. \tag{6.92}$$

However, as shown in [13], a well-trained generator would lead to the inefficiency of supervised information. In other words, the discriminator would possess the same generalization ability as merely training it on L_{prop}. Hence, we apply feature matching (FM) to the generator, and obtain its alternative objective by matching the expected value of the features (statistics) on an intermediate layer of the discriminator [54]: $L(G) = \|E_{\mathbf{x}\sim\frac{1}{n}p_d} f(\mathbf{x}) - E_{\mathbf{x}\sim p_g} f(\mathbf{x})\|_2^2$. In fact, FM is similar to the perceptual loss for style transfer in a concurrent work [26] and the goal of this improvement is to impede the "perfect" generator resulting in unstable training and discriminator with low generalization.

D. LLP-GAN Algorithm

So far, we have clarified the objective functions of both discriminator and generator in LLP-GAN. In particular, note that we execute Monte-Carlo sampling for the expectations. When accomplishing the training stage in GAN manner, the discriminator can be put into effect as the final classifier.

The strict proof for algorithm convergence is similar to that in [17]. Because $\max_D \widetilde{V}(G, D)$ is convex in G and the subdifferential of $\max_D \widetilde{V}(G, D)$ contains that of $\widetilde{V}(G, D^*)$ in every step, the exact line search method gradient descent converges [7]. We present the LLP-GAN algorithm as follows.

Algorithm 6.16 LLP-GAN training algorithm

1: **Input:** The training set $\mathcal{L} = \{(\mathcal{B}_i, \mathbf{p}_i)\}_{i=1}^n$; L: number of total iterations; λ: weight parameter.
2: **Input:** The parameters of the final discriminator D.
3: Set m to the total number of training data points.
4: **for** i=1:L **do**
5: Draw m samples $\{\mathbf{z}^{(1)}, \mathbf{z}^{(2)}, \cdots, \mathbf{z}^{(m)}\}$ from a simple-to-sample noise prior $p(\mathbf{z})$ (e.g., $N(0, I)$).
6: Compute $\{G(\mathbf{z}^{(1)}), G(\mathbf{z}^{(2)}), \cdots, G(\mathbf{z}^{(m)})\}$ as sampling from $p_g(\mathbf{x})$.
7: Fix the generator G and perform gradient ascent on parameters of D in $\widetilde{V}(G, D)$ for one step.
8: Fix the discriminator D and perform gradient descent on parameters of G in $L(G)$ for one step.
9: **end for**
10: **Return** The parameters of the discriminator D in the last step.

6.2.3 *Learning from Label Proportions on High-Dimensional Data*

6.2.3.1 Background

In this subsection, the random forests which is used for our classification is presented.

Random forests are an ensemble learning method together with a bagging procedure for classification and other tasks, where each basic classifier is a decision tree and each tree depends on a collection of random variables. More specifically, during splitting of a randomized tree, each decision node randomly selects a set of features and then picks the best among them according to some quality measurement (*e.g.*, information gain or Gini index) [53]. Furthermore, as each tree in the forest is built and tested independently from other trees, the overall training and testing procedures can be performed in parallel [31].

We denote the mth tree of random forests as $f(x, \theta_m)$, where θ_m is a random vector representing the various stochastic elements of the tree. Meanwhile, let $p_m(k|x)$ represent the estimated density of class labels for the mth tree and M be total number of the trees in the forests. In practice, the final prediction results of random forests are given by probability towards different classes. As a result, the estimated probability for predicting class k in random forests can be defined as:

$$F_k(\mathbf{x}) = \frac{1}{M} \sum_{m=1}^{M} p_m(k|\mathbf{x}), k \in \gamma = \{1, 2, \ldots, K\}, \tag{6.93}$$

where K is the total number of classes. In particular, a decision can be made by simply taking the maximum over all individual probabilities of the trees for a class k with

$$C(\mathbf{x}) = \arg\max_{k \in \gamma} F_k(\mathbf{x}), \gamma = \{1, 2, \ldots, K\} \tag{6.94}$$

where the final result of $C(\mathbf{x})$ is the index of the corresponding class.

The classification margin measures the extent to which the average number of votes for the right class exceeds the average for any other class, which is introduced by Breiman [8], and is expressed as:

$$mg(\mathbf{x}, y) = F_y(\mathbf{x}) - \max_{k \neq y} F_k(\mathbf{x}). \tag{6.95}$$

Obviously, if the classification is correct, there should be $mg(\mathbf{x}, y) > 0$. In other words, the larger the margin is, the more confidence in the classification. The generalization error of random forests is in form of:

$$GE = E_{(X,Y)}(mg(\mathbf{x}, y) < 0), \tag{6.96}$$

where the expectation is measured over the entire distribution of (X,Y).

Random forests have shown its advantages in both classification [8] and clustering [45]. In particular, experiments have shown that high accuracy can be achieved by random forests when classifying high dimensional data [3]. Meanwhile, Caruana [9] presented an empirical evaluation on high dimensional data of different methods, and found that random forests perform consistently well across all dimensions compared with other methods. Additionally, it is easy for random forests to be parallelized, which makes them very easy for multi-core and GPU implementations. Sharp [56] have show that GPU can accelerate the random forests and have great advantage compared to CPU in processing speed, which is very useful for practical applications. Recently, random forests have been applied in video segmentation [49], object detection [15], image classification [6] and remote sensing [46] due to its advantages.

6.2.3.2 The LLP-RF Algorithm

In this subsection, we present a novel learning from label proportions algorithm called LLP-RF, which use random forests to solve high-dimensional LLP problem. In order to leverage random forests to LLP, the hidden class labels insides bags are defined as the optimization variables. Meanwhile, we formulate a robust loss function based on random forests and take the corresponding proportion information into LLP-RF by penalizing the difference between the ground-truth and estimated label proportion. A binary learning setting is considered in the following.

A. Learning Setting

Similar to the standard supervised learning, the problem is also described by a set of training data. But the training data of LLP is only provided in form of bags and the ground-truth labels of training data are not available. In this paper, we assume the bags are disjoint. Let $B_i, i = 1, \ldots, n$ denote the ith bag in the training set. As a result, the total training data can be expressed as:

$$D = B_1 \cup B_2 \cup \ldots . \cup B_n \qquad (6.97)$$
$$B_i \cap B_j = \emptyset, \forall i \neq j,$$

where the total number of training data is N. The ith bag consists of m_i instances and is in form of:

$$B_i = \{x_i^1, ..., x_i^{m_i}\}\{p_i\}, i \in \{1, 2, \ldots, n\}, \qquad (6.98)$$

where the associated p_i indicates the label proportion of the ith bag. As a result, the jth instance in the ith bag can be expressed as x_i^j.

The ground-truth labels of instances are modeled as $\mathbf{y} = (y_1, \ldots, y_N)^T$, where y_i is the unknown label of x_i. Furthermore, we can define the proportion of ith bag as:

$$p_i = \frac{|\{k|k \in B_i, y_k^* = 1\}|}{|B_i|}, \forall k \in \{1, 2, \ldots, N\}, \qquad (6.99)$$

in which $y_k^* \in \{1, -1\}$ is the unknown ground-truth label of x_k and $|B_i|$ denotes the bag size of ith bag. In practice, the above formulation is equivalent to the following:

$$p_i = \frac{\sum_{k \in B_i} y_k^*}{2|B_i|} + \frac{1}{2}, \forall k \in \{1, 2, \ldots, N\}. \qquad (6.100)$$

B. The LLP-RF Framework

The above LLP learning setting is very intuitive and the final objective is to train a classifier in the instance level. To this end, inspired by [32], we formulate a robust loss function based on random forests and take the corresponding proportion information into LLP-RF by penalizing the difference between the ground-truth and estimated label proportion. Therefore, the final objective function of LLP-RF is formulated as follows:

$$\begin{aligned} &\arg\min_{F(\cdot), y_i^j} \; C \sum_{i=1}^{n} \sum_{j=1}^{m_i} L[F_{y_i^j}(x_i^j)] + C_p \sum_{i=1}^{n} L_p[p_i(\mathbf{y}), p_i] \\ &s.t.\; \forall_{i=1}^{n}, \forall_{j=1}^{m_i} \quad y_i^j \in \{1, -1\}, \end{aligned} \tag{6.101}$$

where the hidden class labels $\mathbf{y}$ are defined as the optimization variables and the task is to simultaneously optimize the labels $\mathbf{y}$ and the model $F()$.

Specifically, $L()$ is a loss function which is defined over the entire set of instances and $L_p()$ is a loss function used to penalize the difference between the ground-truth label proportion and the estimated label proportion based on $\mathbf{y}$. Different weights can be added for the loss of bag proportions by changing the value of C_p.

Note that $F_k(x)$ is the confidence of classifier for the kth class, which is got from random forests.

Furthermore, our proposed framework permits choosing different loss functions for $L()$. In our paper, different loss function including hinge loss, logistic loss and entropy are tuned to obtain better classification results. In this paper, we consider $L_p()$ as the absolute loss:

$$L_p[p_i(\mathbf{y}), p_i] = |p_i(\mathbf{y}) - p_i|, \tag{6.102}$$

where p_i is the true label proportion of ith bag and $p_i(\mathbf{y})$ is the estimated label proportion of ith bag.

The above LLP-RF framework is fairly straightforward and intuitive. However, it leads to a non-convex integer programming problem because it needs to simultaneously optimize the labels y_i^j and trains a random forest. In practice, the problem is often NP-hard. Therefore, one key issue is how to solve the optimization problem efficiently. In this paper, a simple but efficient alternating optimization strategy based on annealing is employed to minimize the overall learning objective.

C. How to Solve the LLP-RF

The strategy to solve (6.101) is similar to the rule from [80]. There are two variables F and $\mathbf{y}$ in the optimization formula, where the unknown instance labels $\mathbf{y}$ can be seen as a bridge between supervised learning loss and label proportion loss.

Therefore, we solve the problem by alternating optimizing the two variables F and $\mathbf{y}$.

- We fix the $\mathbf{y}$. The optimization problem becomes a native random forests problem, which can be expressed as below:

$$\arg\min_{F(\cdot)} C \sum_{i=1}^{n} \sum_{j=1}^{m_i} L[F(x_i^j)]. \tag{6.103}$$

- Then, F is fixed. The problem can be transformed to the following:

$$\begin{aligned} \arg\min_{y_i^j} \ & C \sum_{i=1}^{n} \sum_{j=1}^{m_i} L[F_{y_i^j}(x_i^j)] + C_p \sum_{i=1}^{n} L_p[p_i(\mathbf{y}), p_i] \\ s.t. \ & \forall_{i=1}^{n}, \forall_{j=1}^{m_i} \quad y_i^j \in \{1, -1\}. \end{aligned} \tag{6.104}$$

The first term of the objective is defined over the entire instances. However, the proportion information p_i of the second term is provided in the bag level. In order to use the proportion information efficiently, the above formula can be written to the following:

$$\begin{aligned} \arg\min_{y_i^j} \ & \sum_{i=1}^{n} \left\{ C \sum_{j=1}^{m_i} L[F_{y_i^j}(x_i^j)] + C_p L_p[p_i(\mathbf{y}), p_i] \right\} \\ s.t. \ & \forall_{i=1}^{n}, \forall_{j=1}^{m_i} \quad y_i^j \in \{1, -1\}. \end{aligned} \tag{6.105}$$

As the bags are disjoint to each other, the contribution of each bag to the objective is independent. As a result, the objective can be optimized on each bag separately and the final result is equivalent to the summation of every bag. In particular, solving $\{y_i^j | j \in B_i\}$ yields the following optimization problem:

$$\begin{aligned} \arg \min_{\{y_i^j | j \in B_i\}} \ & C \sum_{j \in B_i} \ell[F_{y_i^j}(x_i^j)] + C_p L_p[p_i(\mathbf{y}), p_i] \\ s.t. \ & \forall j \in B_i, \quad y_i^j \in \{1, -1\}. \end{aligned} \tag{6.106}$$

Obviously, the original optimization problem has changed to solve the formula (6.106), whose solution can be found by the following optimization strategy.

Remark The steps for solving formula (6.106).

- Compute all the possible values of the second term in formula (6.106), where there are total $|B_i| + 1$ values. In practice, the kth value can be expressed as:

$$F_2(k) = |\frac{k-1}{|B_i|} - p_i|, k \in \{1, 2, \ldots, |B_i|, |B_i| + 1\}. \tag{6.107}$$

- Obtain all the values of first term $F_1(k)$ corresponding to the second term $F_2(k)$.
- Pick the smallest objective value from

$$C * F_1(k) + C_p * F_2(k), k \in \{1, 2, \ldots, |B_i|, |B_i| + 1\}, \tag{6.108}$$

  yielding the optimal solution of (6.106).

The above strategy is fairly intuitive and straightforward. The main focus is how to obtain the value of first term corresponding to the second term. In practice, there are total $|B_i| + 1$ values about the second term. For a fixed value of second term, steps can be taken as Proposition 6.4.

Proposition 6.4 *For a fixed $p_i(\mathbf{y}) = \theta$, we can find the solution of (6.106) by the iterative steps as below.*

- *Initialize $y_i^j = -1, \forall j \in \{1, 2, \ldots, |B_i|\}$, where $|B_i|$ is the number of instances in ith bag.*
- *Compute the value of $\ell[F_{-1}(x_i^j)]), j \in \{1, 2, \ldots, |B_i|\}$.*
- *Flip the sign of $y_i^j = 1, \forall j \in \{1, 2, \ldots, |B_i|\}$.*
- *Compute the value of $\ell[F_1(x_i^j)]), j \in \{1, 2, \ldots, |B_i|\}$.*
- *Let $\delta_i^j = C(\ell[F_1(x_i^j)] - \ell[F_{-1}(x_i^j)]), j \in \{1, 2, \ldots, |B_i|\}$ denote the reduction of the first term in (6.106) through flipping the sign of y_i^j.*
- *Sort $\delta_i^j, \forall j \in \{1, 2, \ldots, |B_i|\}$ in descending way. Then flip the signs of y_i^j of the top-R ($R = \theta|B_k|$) which have the highest reduction. For each bag, we only need to sort the $\delta_i^j, \forall j \in \{1, 2, \ldots, |B_i|\}$ once.*

Obviously, the minimum value of each bag and the corresponding $\mathbf{y}$ can be obtained using the above steps. In detail, the solution process of the LLP-RF model can be concluded to the following two alternative steps: solve random forests optimization problems and renovate the labels of $\mathbf{y}$ until the objective function value is no longer changing or the reduction of objective is smaller than a threshold. The details of the process are shown in **Algorithm** 6.17.

Furthermore, in order to avoid the local solutions, similar to T-SVM [10] and SVM [80], the novelly proposed LLP-RF algorithm also takes an additional annealing loop to gradually increase C. The annealing can be seen as a step to avoid the local optimal solution. In detail, the annealing loop is achieved by the following equation $C^* = \min\{(1 + \triangle)C^*, C\}$, where $\triangle$ is a step to control the increase of C. Throughout this work, we set $\triangle = 0.5$.

In practice, the different values of initializing **y** can lead to different results. In order to reduce the randomness, we should repeat the process several times and pick the smallest objective value as the final result.

Algorithm 6.17 LLP-RF

1: **Require:** Bags$\{B_i\}$;
2: The corresponding proportion p_i of B_i;
3: Randomly initialize $y_i^j \in \{1, -1\}, \forall_{i=1}^n, \forall_{j=1}^{m_i}$;
4: $C^* = 10^{-5}C$.
5: **while** $C^* < C$ **do**
6: $\quad C^* = \min\{(1+\triangle)C^*, C\}$.
7: $\quad$ **repeat**
8: $\quad\quad$ Fix **y** to solve **F**(Train the Random Forests: $trainRF(y_i^j)$).
9: $\quad\quad$ Fix **F** to solve **y**(using the strategy discussed in the above **Remark**).
10: $\quad\quad$ Update $y_i^j, \forall_{i=1}^n, \forall_{j=1}^{m_i}$.
11: $\quad$ **until** the decrease of the objective is smaller than a threshold or reach the setting iteration.
12: **end while**

6.2.4 Learning from Label Proportions with Pinball Loss

6.2.4.1 Preliminary

In this subsection, we introduce the basic formulation of learning from label proportions and give corresponding symbol description.

In learning from label proportions, although the proportion of each bag is given, the label of each instance is unknown. Suppose we are given a sample set $\{x_i, y_i^*\}_{i=1}^N$, where $x \in \mathcal{R}^n$ and $y_i^* \in \{1, 1\}$ denotes the unknown ground truth label of x_i. The sample set is grouped into K bags. In this subsection, we assume that the bags are disjoint.

The ground truth label proportion of the k-th bag S_k can be defined as

$$P_k := \frac{|\{i|i \in S_k, y_i^* = 1\}|}{|S_k|}.$$

The goal is to find a decision function $f(x) = sign(w^T\phi(x) + b)$ such that the label y for any instance x can be predicted, where $\phi(\cdot)$ is a map of the input data.

Assume the instance labels are explicitly modeled as $\{y_i\}_{i=1}^N$, where $y_i \in \{1, 1\}$. The modeled label proportion of the k-th bag can be defined as

$$P_k = \frac{|\{i|i \in S_k, y_i = 1\}|}{|S_k|}.$$

The Learning from label proportions model can be formulated as below:

$$\min_{y,w,b} \frac{1}{2}\|w\|^2 + C\sum_{i=1}^{N} L_\tau + C_2\sum_{k=1}^{K} |p_k(y) - P_k|, \quad \text{s.t. } y_i \in \{-1, 1\}, \tag{6.109}$$

in which $L_\tau(\cdot)$ is the supervised loss function. Notice the instance labels y is also a variable, which can be seen as a bridge between empirical loss and label proportion loss.

Here, we first discuss the noise generated in the framework of learning from label proportions, and introduce pinball loss to address this issue. Next, we give the learning from label proportions model with pinball loss. Also, the dual problem is given. Then, an alternating optimization method is applied to solve the proposed model. Finally, the complexity of our method is discussed.

6.2.4.2 Noise and Pinball Loss

Unlike traditional hinge loss, pinball loss pushes the surfaces that define the margin to quantile positions by penalizing also the correctly classified sampling points [24]. The distance between these two classes is easily affected by the noise on feature x. Also, improper initialization of label y causes noise as well. As a result, the classifier with hinge loss is sensitive to feature noise. The pinball loss is related to quantiles and has been well studied in regression (parametric methods [52] and nonparametric methods [12, 63]. And it is also used for binary classification recently [23].

The pinball loss is defined as follows:

$$L_\tau(u) = \begin{cases} u, & u \geq 0, \\ -\tau u, & u < 0. \end{cases} \tag{6.110}$$

Particularly, when $\tau = 0$, the pinball loss $L_\tau(u)$ reduces to the hinge loss. When a positive τ is used, minimizing the pinball loss results in the quantile value.

To intuitively show the properties of pinball loss, we are going to compare the classifiers based on the hinge loss and the pinball loss, respectively. Here, let's consider a two dimensional example: points are generated from two Gaussian distribution $N(\mu_1, \sigma)$ and $N(\mu_2, \sigma)$, where $\mu_1 = [0.5, -3]^T$, $\mu_2 = [0.5, 3]^T$ and $\sigma = [0.1, 0; 0, 2]$. As shown in Fig. 6.7, the solid lines indicate the classification hyperplane achieved by classifier based on the hinge loss and the dashed lines represent the hyperplane obtained by pinball loss. The data points are generated from the same distribution. However, the hinge loss classifier obtains the significantly different results while the pinball loss hyperplane achieve more stable results. It is mainly because that the hinge loss classifier measures the distance between two sets by the nearest points. But pinball loss takes the nearest τ (e.g. 35%) points

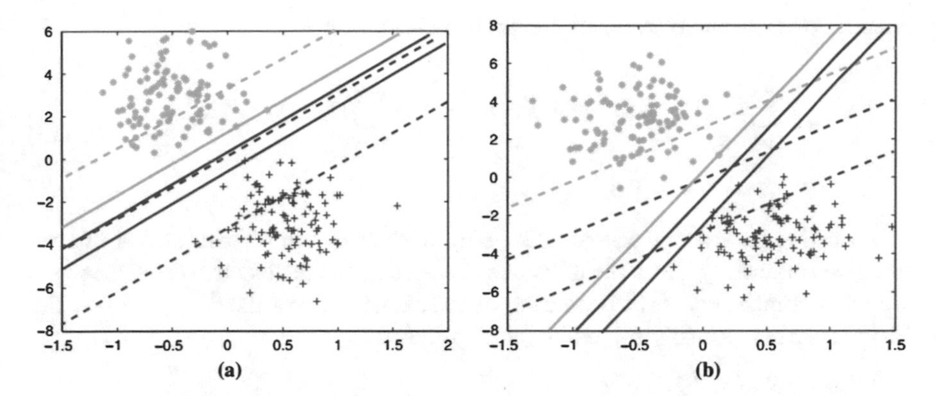

Fig. 6.7 Comparison between the classifiers based on hinge loss and pinball loss. As it is shown, the results of pinball loss classifier are more stable

to measure this distance, which makes its result less sensitive to noise around the boundary.

6.2.4.3 Learning from Label Proportions Model with Pinball Loss

With pinball loss, we can formulate the learning from label proportions model as below:

$$\begin{aligned}&\min_{y,w,b} \frac{1}{2}\|w\|^2 + C\sum_{i=1}^{N} L_\tau(1 - y_i(w^T\phi(x_i) + b)) + C_2\sum_{k=1}^{K} |p_k(y) - P_k|,\\&\quad\text{s.t. } y_i \in \{-1, 1\}.\end{aligned} \tag{6.111}$$

As the instance labels y is also a variable, one natural way for solving Eq. (6.111) is via alternating optimization.

Step 1 For a fixed y, the optimization of Eq. (6.109) w.r.t w and b becomes a classic SVM with pinball loss:

$$\min_{w,b} \frac{1}{2}\|w\|^2 + C\sum_{i=1}^{N} L_\tau(1 - y_i(w^T\phi(x_i) + b)). \tag{6.112}$$

Step 2 When w and b are fixed, the problem becomes:

$$\min_{y} \sum_{i=1}^{N} L_\tau(1 - y_i(w^T\phi(x_i) + b)) + \frac{C_2}{C} \sum_{k=1}^{K} |p_k(y) - P_k|, \\ \text{s.t. } y_i \in \{-1, 1\}. \tag{6.113}$$

By taking the strategy presented in [80], we show that the second step above can be solved efficiently. Since the influence of each bag on the objective is independent, we can optimize Eq. (6.113) on each bag separately. For a fixed $p_k(y) = \theta$, Eq. (6.113) can be optimally solved by the steps below.

- Initialize $y_i, i \in B_k$.
- Suppose the reduction of the first term in (6.113) is δ_i. Sort $\delta_i, i \in B_k$.
- Flip the signs of the top-R y_i which have the highest reduction δ_i, where $R = \theta|B_k|$.

  By conducting Step 1 and Step 2 alternately until the decrease of objective is smaller than a threshold (e.g. 10^{-4}), we can obtain the optimal solution.

6.2.4.4 Dual Problem

The problem in Eq. (6.112) can be transformed into:

$$\begin{aligned} &\min_{w,b} \frac{1}{2}\|w\|^2 + C\sum_{i=1}^{N} \xi_i, \\ &\text{s.t. } y_i(w^T\phi(x_i) + b) \geq 1 - \xi_i, i = 1, 2, \cdots, N, \\ &\quad y_i(w^T\phi(x_i) + b) \leq 1 + \frac{1}{\tau}\xi_i, i = 1, 2, \cdots, N. \end{aligned} \tag{6.114}$$

According to the Karush-Kuhn-Tucker (KKT) sufficient and necessary optimality conditions, the dual problem of Eq. (6.114) is obtained as follows,

$$\begin{aligned} &\max_{\alpha,\beta} -\frac{1}{2} \sum_{i=1}^{N}\sum_{j=1}^{N} (\alpha_i - \beta_i) y_i \phi(x_i)^T \phi(x_j) y_j (\alpha_j - \beta_j) + \sum_{i=1}^{N} (\alpha_i - \beta_i), \\ &\text{s.t. } \sum_{i=1}^{N} (\alpha_i - \beta_i) y_i = 0, \\ &\quad \alpha_i + \frac{1}{\tau}\beta_i = C, \ i = 1, 2, \cdots, N, \\ &\quad \alpha_i \geq 0, \ i = 1, 2, \cdots, N, \\ &\quad \beta_i \geq 0, \ i = 1, 2, \cdots, N. \end{aligned} \tag{6.115}$$

Introduce the variables γ_i, α_i and β_i. Let $\gamma_i = \alpha_i - \beta_i$. The dual problem Eq. (6.115) has the same solution set w.r.t. α as that to the following convex quadratic programming problem:

$$\begin{aligned} &\min_{\gamma,\beta} \frac{1}{2} \sum_{i=1}^{N} \sum_{j=1}^{N} \gamma_i y_i \phi(x_i)^T \phi(x_j) y_j \gamma_j - \sum_{i=1}^{N} \gamma_i, \\ &\text{s.t.} \sum_{i=1}^{N} \gamma_i y_i = 0, \\ &\quad -\tau C \leq \gamma_i \leq C,\ i = 1, 2, \cdots, N. \end{aligned} \tag{6.116}$$

Suppose $\gamma^* = (\gamma_1^*, \gamma_2^*, \ldots, \gamma_l^*)$ is the solution to problem Eq. (6.116). We can have

$$w^* = \sum_{i=1}^{N} \gamma_i^* y_i \phi(x_i), \text{ and}$$

$$b^* = y_j - \sum_{i=1}^{N} y_i \gamma_i^* \phi(x_i)^T \phi(x_j),$$

where $\forall j : -\tau C < \gamma_j^* < C$.

Then the obtained function can be represented as

$$f(x) = \sum_{i=1}^{N} y_i \gamma_i^* \phi(x_i)^T \phi(x_j) + b^*,$$

where $\forall j : -\tau C < \gamma_j^* < C$.

6.2.4.5 Overall Optimization Procedure

Based on the detailed explanation above, the overall optimization procedure is summarized in Algorithm 6.18.

By alternating between solving w^*, b^* and y, the objective is guaranteed to converge, for the reason that the objective function is lower bounded, and non-increasing. Empirically, the alternating optimization typically terminates fast within ten iterations.

In practice, the stopping criterion of the overall optimization procedure is that the objective function does not decrease any more (or if its decrease is smaller than a threshold).

Algorithm 6.18 Optimization procedure of learning from label proportions

1: **Input:** $\{x_i\}_{i=1}^N, \{P_k\}_{k=1}^K, C, C_2$ and $C^* = 10^{-5}C$.
2: **Output:** w^*, b^* and y.
3: Randomly initialize $y_i \in \{-1, 1\}$.
4: **while** $C^* < C$ **do**
5: $C^* = \min\{(1+\triangle)C^*, C\}$.
6: **while** not converged **do**
7: % Fix y.
8: $w^* = \sum_{i=1}^N \gamma_i y_i \phi(x_i)$.
9: $b^* = y_i - \sum_{i=1}^N y_i \gamma_i^* (x_i \cdot x_j)$.
10: % Fix w^* and b^*.
11: Solve y (Eq.(6.113) with $C \leftarrow C$).
12: **end while**
13: **end while**

6.2.4.6 Complexity

Step 1 takes the complexity of SVM with pinball loss. As described in the paper, the bags are disjoint, the influences of the bags are independent. In Step 2, for each bag S_k, sorting takes $O(|S_k|log(|S_k|))$, which is same with [80]. Overall, the complexity is $O(\sum_{k=1}^K |S_k|log(|S_k|))$. We know that $\sum_{k=1}^K |S_k| = N$ and denote $J = \max_{k=1,2,\ldots,K} |S_k|$. The complexity is $O(Nlog(J))$ time.

6.3 Other Enlarged Learning Models

6.3.1 Classifying with Adaptive Hyper-Spheres: An Incremental Classifier Based on Competitive Learning

6.3.1.1 Basic Theory

A. Basic Theory of Supervised Competitive Learning

We partially borrow the topological structure of CPN to introduce our model. CPNs are a combination of competitive networks and Grossberg's outstar networks [19]. The topological structure of CPN has three layers: input layer, hidden layer, and output layer (Fig. 6.8).

Suppose there are N elements in the input layer, M neurons in the hidden layer, and L neurons in the output layer. Let vector $V_i = (v_{i1}, \ldots, v_{iN})^T$ denote the weights of neuron i in the hidden layer connecting to each of the elements of the input layer. Then $V = (V_1, \ldots, V_M)$ denotes weight matrix of the instars. If the training in stage 1 can be viewed as a clustering process, then neuron i is cluster c_i and V_i is the centroid of cluster c_i.

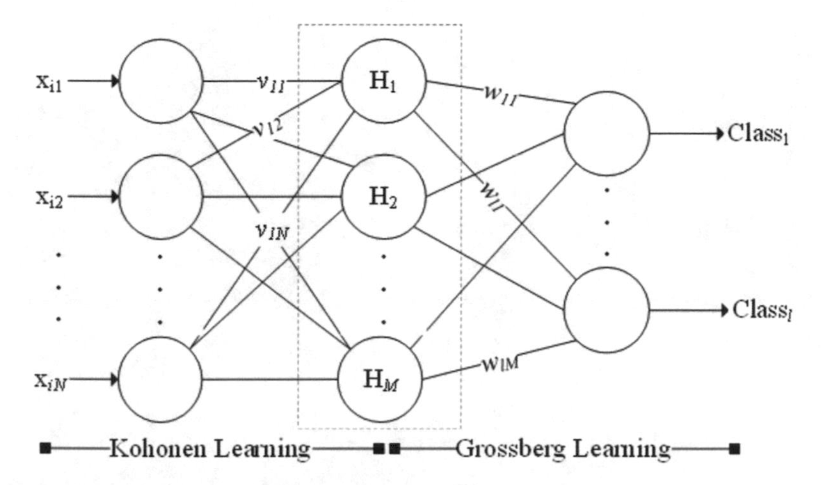

Fig. 6.8 Topological structure of CPN.

When an instance is coming, it will compute the proximity between the instance and each V_i in the weight matrix, i.e., the centroid of cluster c_i. Here, proximity can be measured by computing inner product $net_j = V_j^T x, (j = 1, 2, \ldots, m)$. It adopts a winner-takes-all strategy to determine which neuron's weights are to be adjusted. The winner is $net_{j*} = \max\{net_j\}$. In other words, the winner is c_{j*} whose centroid is the closest to the incoming instance. The winning neuron's weights would be adjusted as follows:

$$V_{j*}(t+1) = V_{j*}(t) + \alpha[x - V_{j*}(t)],$$

where α is the learning rate, indicating that the centroid of the winning cluster will move in the direction of x. As instances keep coming, the weights vector—i.e., the centroid of the hyper-spheres—tend to move toward the densest region of the space. This first stage of the CPN's training algorithm is a process of self-organizing clustering, although it is structured using a network.

The second part of the structure is a Grossberg learning [19]. We will redesign a different hidden layer and different connection from the hidden layer to the output layer.

B. Advantages and Disadvantages of the Original Model

To illustrate the advantage and disadvantage of original model, a set of two-dimensional artificial data were created and visualized in Fig. 6.9.

In Fig. 6.9a, instances can be grouped into six clusters. Setting the number of neurons in the hidden layer to six, the first training stage of the model in Fig. 6.8 can automatically find the centroids of the six clusters, which are represented by

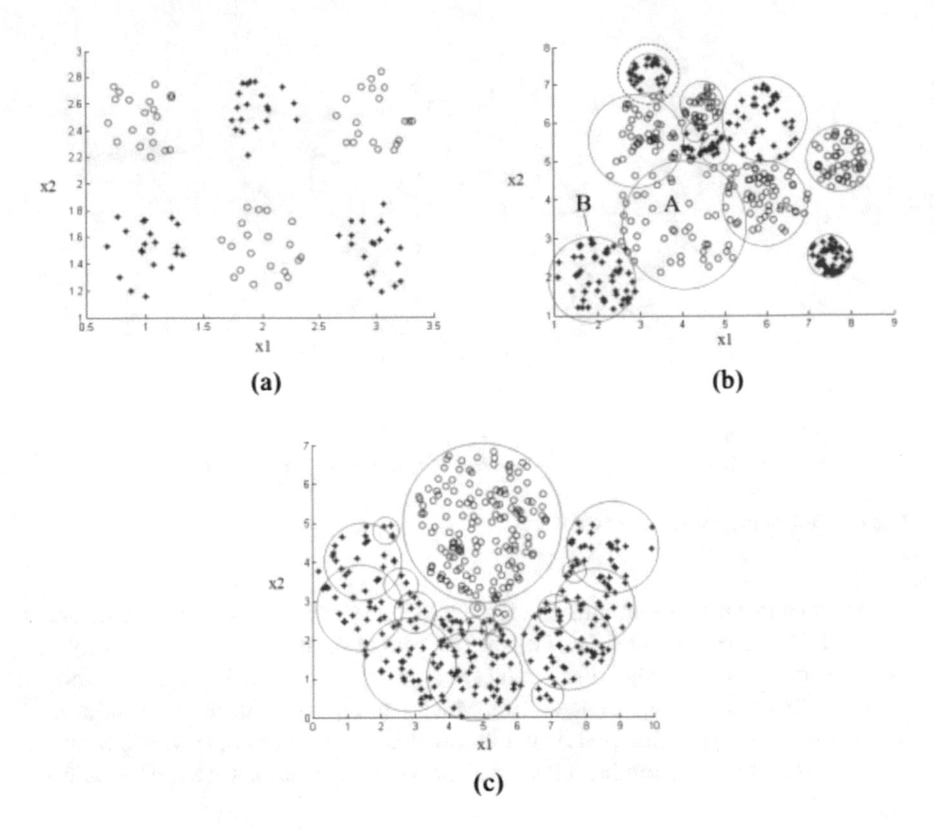

Fig. 6.9 Artificial datasets and the proposed clustering solutions

the weights of the six neurons. The second training stage can learn each cluster's connection to the right class. The distance from each instance in Fig. 6.9a to its cluster centroid is smaller than the distances to the centroids of other clusters. The dataset shown in Fig. 6.9 is ideal for CPN to classify.

Data distribution in Fig. 6.9a is simplified and idealistic. Data with distribution similar to Fig. 6.9b will cause two kinds of problems to the original model.

(1) First, the self-organized clustering process depends on the similarity measures between data points and hyper-sphere's centroid. Points closer to one cluster's centroid may belong to another cluster. Therefore, every cluster should have a definite scope or radius, and the scope should be as far away from others as possible.
(2) Second, the number of clusters in the hidden layer is fixed in the original model. However, it is difficult to estimate the number of clusters in advance. Given different numbers of neurons in the hidden layer, the accuracy varies dramatically. The training of the instar layer-i.e., the clustering process-is contingent on this fixed number.

C. Building of the DMZ

To solve the first aforementioned problem, we should have a general knowledge of the scope of the clusters. For example, points of cluster A (in Fig. 6.9b) near the border may be closer to the centroid of cluster B, so these points will be considered belong to cluster B in the original model. We must identify the decision border that separates clusters according to their labels. When two instances with conflicting labels fall into the same cluster, it gives us an opportunity to identify the border point that is somewhere between the two conflicting instances (as long as the instance is not an outlier). To maintain the maximum margin and for the sake of simplicity, the median point of two instances could be selected as a point in a zone called a Demilitarized Zone (DMZ), and clusters should be as far away from the DMZ as possible. As the number of conflicting instances increases, a general zone gradually forms as the DMZ. This mechanism can find borders of any shapes that are surrounded by many hyper-spheres.

To solve the second problem, the number of clusters should not be predetermined. The clusters should be formed dynamically and merged or split if necessary. The scope of the hyper-spheres, represented by the corresponding radii, should be adjusted on demand. As an example, consider the situation presented in Fig. 6.9b: with instances of conflicting labels found in the top cluster, the original cluster should tune its radius. After training, a new cluster would be formed beneath the top cluster containing instances of different labels from the ones in the top cluster. The radii of the two clusters should be tuned according to their distance to the borders.

One single hyper-sphere may not enclose an area whose shape is not hyper-spherical [51]. However, any shape could be enclosed as long as the number of the formed hyper-spheres is unlimited. Consider the clusters represented by the two-dimensional circles in Fig. 6.9c. All of the instances can be clustered no matter what the data distribution is and what the shape of the border is, as long as there are enough hyper-spheres of varying radii and are properly arranged.

D. Proposed Topological Structure

Given the solutions above, the structure of our improved model is as follows (Fig. 6.10):

The first difference is that our model has an adaptive dynamic hidden layer and the number of neurons in hidden layer is adaptive. The second difference is that each neuron H_i connects to only one particular neuron in the output layer, and w_{ij} is used to record the radius of neuron H_i .

E. Kernelization

It is challenging for competitive learning models to apply kernel methods because they cannot be denoted in inner-product forms. Some previous studies use approx-

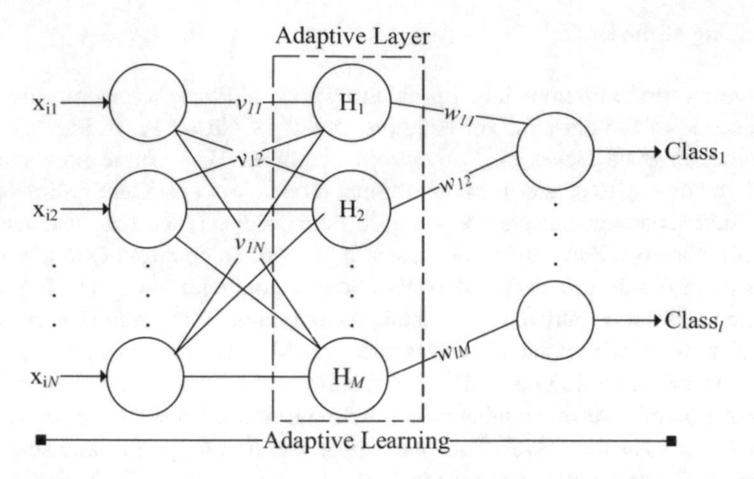

Fig. 6.10 Topological structure of the proposed model

imation methods for the kernelization of competitive learning [29, 76]. This paper uses Nyström method to kernelize the proposed model [28, 40].

Let the kernel matrix written in blocks form:

$$A = \begin{bmatrix} A_{11} & A_{12} \\ A_{21} & A_{22} \end{bmatrix},$$

Let $C = [A_{11}\ A_{12}]^T$, Nyström method uses A_{11} and C to approximate large matrix A. Suppose C is a uniform sampling of the columns, Nyström method generates a rank-k approximation of $A(k \le n)$ and is defined by:

$$A_k^{nys} = CA_{11}^{+}C^T = \begin{bmatrix} A_{11} & A_{21} \\ A_{21} & A_{21}A_{11}^{+}A_{21}^T \end{bmatrix} \approx A,$$

where A_{11}^{+} denotes the generalized pseudo inverse of A_{11}.

There exists an Eigen decomposition $A_{11}^{+} = V\Lambda^{-1}V^T$ such that each element $A_k^{nys}{}_{ij}$ in A_k^{nys} can be decomposed as:

$$\begin{aligned} A_k^{nys}{}_{ij} &= (C_i^T V \Lambda^{-1} V^T C_j) \\ &= (\Lambda^{-1/2} V^T C_i)^T (\Lambda^{-1/2} V^T C_i) \\ &= (\Lambda^{-1/2} V^T (\kappa(x_i, x_1), \ldots, \kappa(x_i, x_m)))^T \bullet (\Lambda^{-1/2} V^T (\kappa(x_j, x_1), \ldots, \kappa(x_j, x_m))), \end{aligned}$$

where $\kappa(x_i, x_j)$ is the base kernel function, $x_1, x_2, \ldots, x_m$ are representative data points and can be obtained by uniform sampling or clustering methods such as K-means and SOFM.

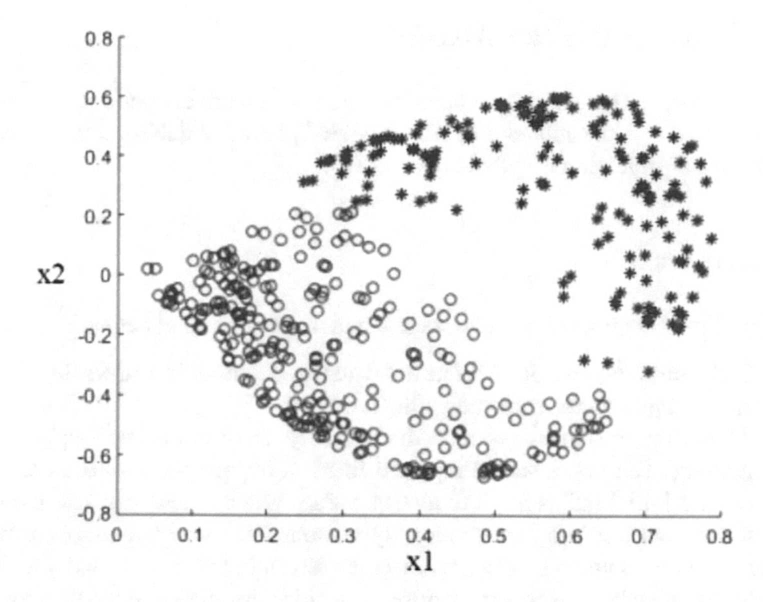

Fig. 6.11 Artificial dataset 3 after Nyström and SVD transformation

Let $\phi_m(x) = \Lambda^{-1/2}V^T(\kappa(x, x_1), \ldots, \kappa(x, x_m))^T$, such that ${A_k^{nys}}_{ij} = \phi_m(x_i)^T\phi_m(x_j) = \kappa(x_i, x_j)$.

With Nyström method, we can get an explicit approximation of the nonlinear projection $\phi_m(x)$, which is:

$$x \to \phi_m(x). \tag{6.117}$$

To justify why we use kernel methods for our model, we first used Nyström method to raise the dimension of dataset 3 to 403, then used Singular Value Decomposition (SVD) to reduce the dimension to 2 for the purpose of visualization. Figure 6.11 illustrates the transformed dataset 3 from Fig. 6.9c.

Compared with Fig. 6.9c, the data in Fig. 6.11 can be covered with less hyper-spheres, or each hyper-sphere can enclose more data points. Because the sampling points in Nyström methods can be obtained dynamically, the projection of Eq. (6.117) can be used for every single instance in competitive learning and can be applied directly to our incremental model.

Without loss of generality, we use $\phi_m(x)$ to denote a potential projection of x in the reminder of this paper. If it works in the original space, the projection of x is to itself.

6.3.1.2 Proposed Classifier: ADA-HS

The main characteristic of the proposed model is to adaptively build hyper-spheres. Therefore, we call the model Adaptive Hyper-Spheres (AdaHS), and the version after Nyström projection is called Nys-AdaHS.

A. Training Stages

Our algorithms are trained in three stages, which are described below.

Stage 1. Forming Hyper-Spheres and Adjusting Centroids and Radii

(1) Forming hyper-spheres and adjusting centroids

Given that instances are read dynamically, there is no hyper-sphere at the beginning. The first instance inputted forms a hyper-sphere whose centroid is itself and initial radius is set to a large value. When a new instance is inputted and does not fall into any existing hyper-spheres, a new hyper-sphere will be formed in the same way. If a new instance falls into one or more existing hyper-spheres, the winner is the one whose centroid is the closest to the new instance. The winning cluster's centroid is recalculated as:

$$c_i(t+1) = c_i(t) + \alpha[\phi(x) - c_i(t)],$$

where x is the new inputted instance, $c(t)$ is the original centroid of the hyper-sphere, $c(t+1)$ is the new centroid, and α is the learning rate.

When the number of instances that fall within a particular hyper-sphere grows, its centroid tends to move toward the densest zone.

In order to speed up the search of the winner, we build simple k-dimension trees for all hyper-spheres. With the knowledge of the radius, it is easy to figure out the upper and lower bounds of the selected k dimensions. In this way, it avoids extensive computation of all Euclidean distance of instance and hyper-sphere pairs.

(2) Building decision border zone: DMZ

The goal of this step is to find the DMZ's median points that approximate the shape of the DMZ.

We find the points using the following technique. The first time a labeled instance falls into a hyper-sphere, the hyper-sphere will be labeled using the label of this instance. If another instance with a conflicting label falls into the same hyper-sphere, it indicates that the hyper-sphere has entered the DMZ. We identify the nearest data point in the hyper-sphere to the newly inputted conflicting instance, and let p_i represent the median point as follows:

$$p_i = \frac{1}{2}(\phi(x_{conflicting}) + c_i),$$

where $\phi(x_{conflicting})$, $p_i \in c_i$ and p_i is recorded and used in the posterior clustering process.

(3) Adjusting the radii of hyper-spheres

Once a DMZ point is found in a hyper-sphere, the radius of the hyper-sphere should be updated such that it does not enter the DMZ. The new radius of hyper-sphere c_i should therefore be set as:

$$r_i = d(p_i, c_i) - d_{safe},$$

where d_{safe} represents a safe distance at which a hyper-sphere should be from the closest DMZ point. And the logics of this stage are outlined in Algorithm 6.19 below.

Algorithm 6.19 The forming of hyper-spheres and the adjusting of the centroids and radii

1: **Input:** x, the newly read instance.
2: **Output:** C: A set of hyper-spheres whose centroids and radii are tuned properly;
   DMZ: A set of points who shape the decision border approximately.
3: **Method:**
4: $c_t = Null, len = -1$.
5: **for** each c_t in C **do**
6:     **if** $\phi(x)$ falls into C **then**
7:         % Find the winner of the hyper-spheres.
8:         **if** $label(x) = label(c_i)$ **and** ($len = -1$ **or** $d_E(\phi(x), c_i) < len$) **then**
9:             $c_t = c_i$.% Store the present temporary nearest hyper-sphere
10:             $len = d_E(\phi(x), c_i)$.%Store the present temp nearest distance
11:         **else if** $label(x) \neq label(c_i)$ **then** %//Split the hyper-sphere
12:             $p_i = \frac{1}{2}(\phi(x_{conflicting}) + c_i), \phi(x_{conflicting}), p_i \in c_i$.
13:             Add p_i to DMZ.
14:             $r_l = d(p_i, c_i) - d_{safe}$.% Adjusting radii r_j of hyper-sphere c_j
15:             Mark c_i as "support hyper-sphere".
16:         **end if**
17:     **end if**
18: **end for**
19: **if** $c_t \neq Null$ **then**
20:     % Adjust the winning hyper-sphere's centroid
21:     $c_i(t+1) = c_i(t) + \alpha[\phi(x) - c_i(t)]$.
22: **else**
23:     Form a new hyper-sphere, and make $\phi(x)$ be the centroid.
24:     Let the label of the new hyper-sphere be $label(x)$.
25: **end if**

Stage 2. Merging Hyper-Spheres

Hyper-spheres may overlap with one another or even be contained in others. Therefore, after certain period of training, a merging operation should be performed. Suppose that we have two hyper-spheres, c_A and c_B, and the radii of them are not the same. Let $c_{big} = \max_{radius}(c_A, c_B)$, $c_{small} = \min_{radius}(c_A, c_B)$, $d_t = d(c_{big}, c_{small})$, and θ be the merging coefficient. If $d_t + r_{small} \le r_{big} + \theta \times r_{small}$, the prerequisite to merge is met. Then let $r_{temp} = d_t + r_{small}$, and the new radius of the c_{big} will be $r_{new} = \max(r_{temp}, r_{big})$.

The details of this stage are outlined in Algorithm 6.20.

Algorithm 6.20 Merging of hyper-spheres

1: **Input:** C: A set of hyper-spheres which are formed in stage 1.
2: **Output:** C: The remaining hyper-spheres after merging.
3: **Method:**
4: **for** each c_i in C **do**
5: **for** each c_j in C except c_i **do**
6: $c_{big} = \max_{radius}(c_i, c_j)$, $c_{small} = \min_{radius}(c_i, c_j)$, $d_t = d(c_{big}, c_{small})$.
7: **if** $d_t + r_{small} \leq r_{big} + \theta \times r_{small}$ **then** % θ is the merging coefficient
8: Merge c_i and c_j.
9: **end if**
10: **end for**
11: **end for**

Stage 3. Selecting Hyper-Spheres

Since the training process is entirely autonomous, the number of generated hyper-spheres could be large. Therefore, the final stage needs to select hyper-spheres.

There are three types of hyper-spheres that are prominent, which are described as follows:

(1) The first type of hyper-spheres includes large number of instances. Because these are the fundamental hyper-spheres that contain most data points, they are marked as "Core Hyper-spheres".
(2) The second type of hyper-spheres has less instances but locates near the border. They are marked as "Support Hyper-spheres" because such hyper-spheres can be found by measuring the distance between hyper-spheres and the nearest DMZ points.
(3) The third type of hyper-spheres has small number of instances and is far away from the border. These hyper-spheres can be discarded.

To achieve high classification accuracy, both core hyper-spheres and support hyper-spheres should be selected. The logic of the third stage is outlined in Algorithm 6.21.

B. Mini-Batch Learning and Distributed Computing

To make it applicable in large scale applications, we encapsulate the proposed algorithms into a Map-Reduce framework. We can collect the incoming instances as mini-batch set and then train them in MapReduce tasks. The computing model of the algorithms is illustrated in Fig. 6.12.

The collected mini-batch instances can be encapsulated in key-value pairs and mapped into mapper tasks.

Algorithm 6.21 Selection of hyper-spheres

```
1: Input: C: The set of hyper-spheres which are formed in preceding stages.
2: Output: C: The remaining hyper-spheres after selection.
3: Method:
4: for each c_i in C do
5:     % T is the threshold of the instances number which one hyper-sphere must at least have.
6:     % num(c) is a function computing the number of instances in a hyper-sphere.
7:     if num(c_i) < T then
8:         % Let d(c_i, DMZ) be the distance from the centroid of c_i to the nearest data point in
           DMZ.
9:         if r_i < d(c_i, DMZ) then
10:            Discard c_i.
11:        end if
12:    else
13:        Mark c_i as "core hyper-sphere".
14:    end if
15: end for
```

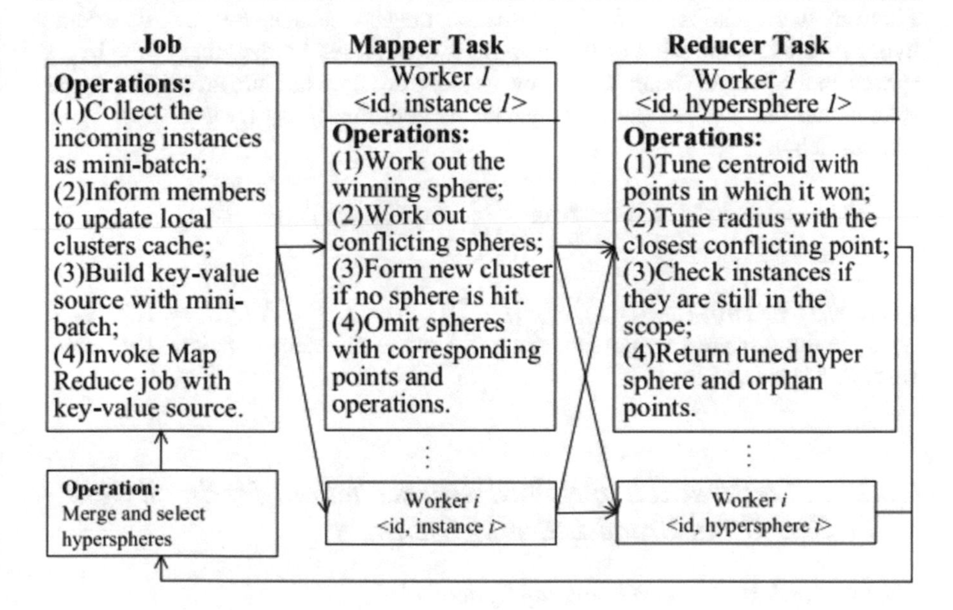

Fig. 6.12 MapReduce computing model

In each mapper tasks, the operations are based on instances. It queries local cache for every instance to find out in which hyper-spheres the instance falls, marks the winning hyper-sphere and the conflicting ones, and sents the hyper-spheres along with the description of the needed operations in another form of key-value<id, hyper-sphere> pairs.

In each reducer task, the operations are based on every hyper-sphere, which is aggregated according to the hyper-sphere id emitted from mapper tasks. The competitive learning can be conducted collectively with the aggregated instances.

The tuning of a radius can be performed for only once with the closest conflicting instance, and it should find out the orphan points and return the tuned hyper-sphere at the end.

After a turn of the MapReduce tasks, the merging and selection of the hyper-spheres should be performed. After all of the operations, the tuned hyper-spheres should be saved to the cache. The orphan points should be retrained in the next turn. In the whole MapReduce process, sub-tasks do not coordinate with each other. Thus the hyper-spheres and DMZ are not updated in real time in a mini-batch turn, and they are updated collectively after all reducer tasks return.

C. Predicting Labels

Just like other supervised competitive neural networks, AdaHS must determine the winning hyper-sphere in the hidden layer to predict the label of a new instance. There are two situations. In the first situation, the new instance falls into an existing hyper-sphere and the label of the instance is determined by the label of the hyper-sphere. In the second situation, the new instance does not fall into an existing hyper-sphere, and the label of the new instance is coordinated by the k nearest hyper-spheres' labels:

$$y = \arg\max_{l_j} \sum_{c_i \in N_k(x)} w_j I(y_i = l_j),$$

where $w_j = exp(-([d_E(\phi(x), c_j)^2]/[2r_j^2]))$; $i = 1, 2, \ldots, L$; $j = 1, 2, \ldots, k$; $N_k(x)$ is the k nearest hyper-spheres; and I is the indicator function. The default value of k is set to 3.

6.3.2 *A Construction of Robust Representations for Small Data Sets Using Broad Learning System*

6.3.2.1 Review of Broad Learning System

This subsection is mainly a simple introduction to the BLS. The details of this system can be found in [11]. The BLS is designed based on the random vector functional-link neural network (RVFLNN) [25, 47]. In the BLS, the mapped features and enhancement features instead of the original features are used to feed into a single layer neural network. Figure 6.13 shows the structure of the BLS.

In Fig. 6.13, X means the input features, and Y means the corresponding labels. The label Y uses one-hot encoding, which means all neurons are set to 0 except the one that belongs to the label is set to 1. The mapped features can be represented as

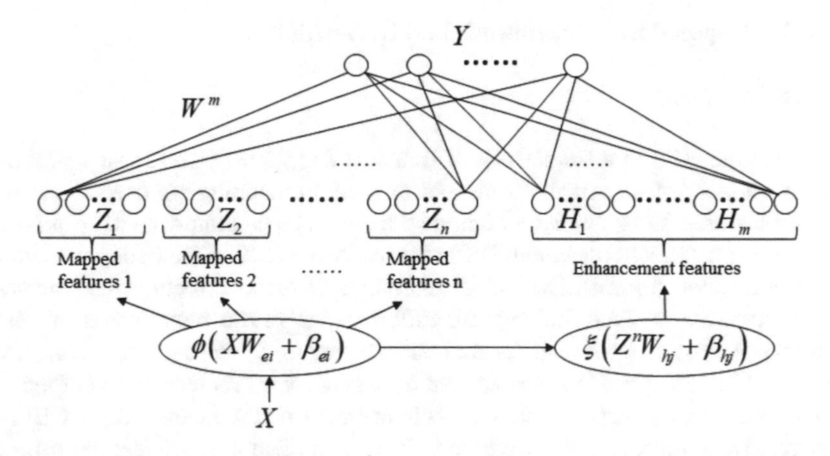

Fig. 6.13 The structure of the BLS

follows:

$$Z_i = \phi_i(XW_{ei} + \beta_{ei}), \tag{6.118}$$

where Z_i is the i-th mapped features and W_{ei} is the random weights. All the mapped features are concatenated as $Z^n \equiv [Z_1, Z_2, \ldots, Z_n]$, then the enhancement features can be represented as follows:

$$H_j = \xi_j(Z^n W_{hj} + \beta_{hj}). \tag{6.119}$$

All the enhancement features are concatenated as $H^m \equiv [H_1, H_2, \ldots, H_m]$. Therefore, the broad model can be represented as follows:

$$Y = [Z^n|H^m]W^m, \tag{6.120}$$

where $W^m = [Z^n|H^m]^+ Y$ is the weights of the single-layer neural network and can be easily calculated through the ridge regression approximation of $[Z^n|H^m]^+$ using the following equation:

$$A^+ = \lim_{\lambda \to 0} (\lambda I + AA^T)^{-1} A^T. \tag{6.121}$$

Theoretically, the $\phi_i(\cdot)$ and $\xi_j(\cdot)$ used in mapped features and enhancement features can be different functions. The sparse autoencoder is applied to fine-tune the W_{ei} of mapped features, and the sigmoid function is used to generate enhancement features in [11].

6.3.2.2 Proposed BLS Framework and BLS with RLA

A. BLS Framework

To extend the BLS to a framework of transforming inputs into robust representations, feature extraction methods instead of random mapping are used to generate mapped features. Let $Z_i = \phi_i(X)$ denote the i-th mapped features, where $\phi_i(\cdot)$ can be any feature extraction method. Different feature extraction methods can generate different mapped features. Even if all mappings of mapped features use the same AE method, the mapped features are different due to the randomness of neural networks. All the mapped features are concatenated as $Z^n \equiv [Z_1, Z_2, \ldots, Z_n]$, and the ensemble of mapped features Z^n can provide a robust representation of inputs.

The setting of a large number of enhancement nodes in the original BLS is removed. Deep representations, called enhancement features, are learned from the ensemble mapped features Z^n. The enhancement features can be denoted as $H_j = \xi_j(Z^n)$, where $\xi_j(\cdot)$ can be any feature extraction method. All the enhancement features are concatenated as $H^m \equiv [H_1, H_2, \ldots, H_m]$. The concatenation of mapped features Z^n and enhancement features H^m can provide more robust representations to enhance the performance of downstream tasks.

Figure 6.14 shows the structure of the BLS framework. It should be noted that w and β used in (6.118) and (6.119) are random, so their method is random mapping. The mappings $\phi(\cdot)$ and $\xi(\cdot)$ in Fig. 6.14 can be any feature extraction method, including random mapping, autoencoder, convolution feature extraction, recursive feature extraction, etc. Therefore, w and β are omitted in the new equations.

Further, to generate more different mapped features and enhancement features in the BLS framework, samples and features can be randomly selected for each mapping. Figure 6.15 shows the structure of a random version of the BLS framework.

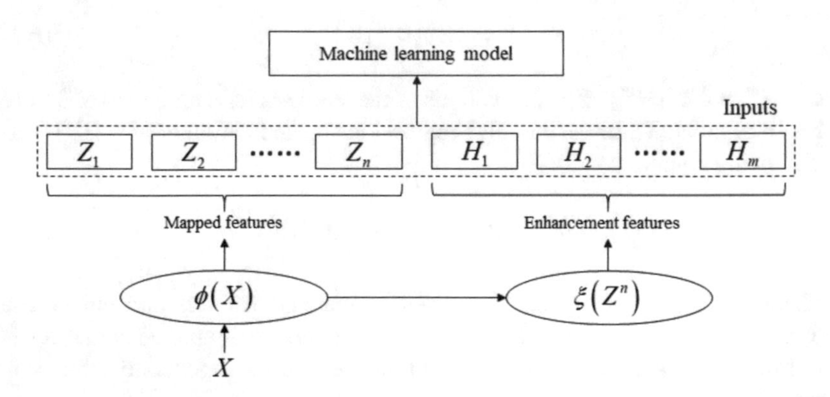

Fig. 6.14 The structure of the BLS framework

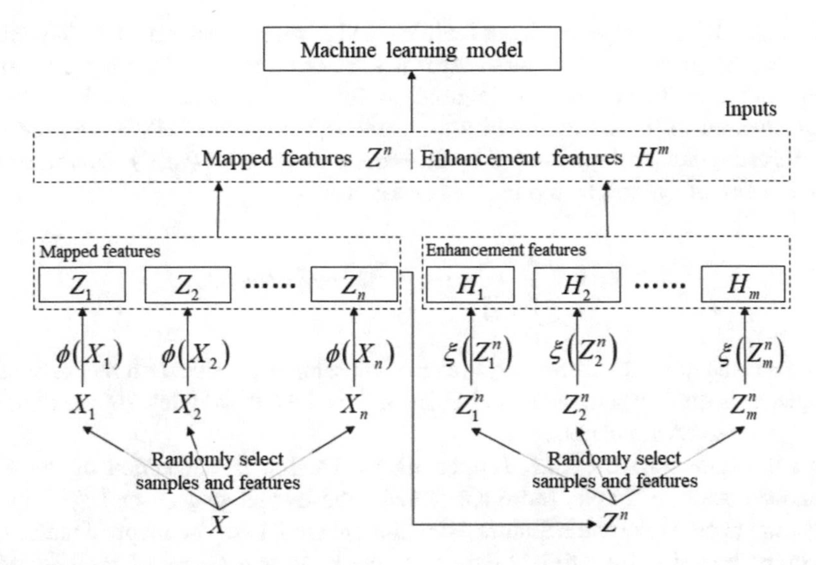

Fig. 6.15 The structure of a random version of the BLS framework

B. BLS with RLA

Deep autoencoder (DA) is a nonlinear dimensionality reduction approach and usually works much better than PCA [20]. Instead of the unsupervised architecture used in DA, LA uses supervised architecture to connect the features and the labels together. The representation features learned from the LA not only contain the information of original features but also contain the estimated label belonging to the sample. In that case, the representation features can provide more information to the machine learning models and may promote the performance of these models. Figure 6.16 shows the structure of the LA.

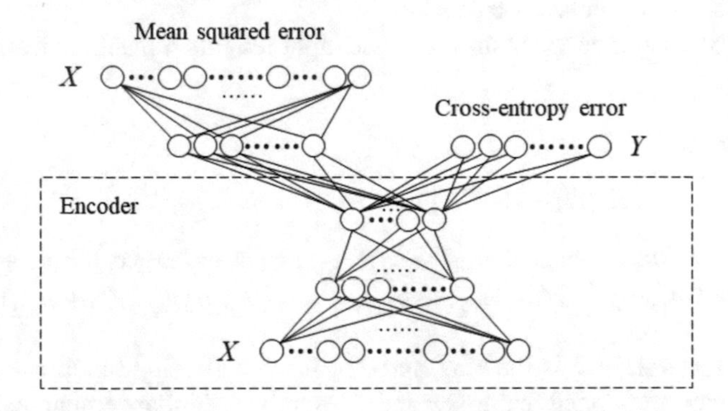

Fig. 6.16 The structure of the LA

In Fig. 6.16, X means the input features, and Y means the corresponding labels using one-hot encoding. The label layer is added on the top of the representation layer, and a softmax function is used to forecast the label. The loss of the reconstruction of X is measured by the mean squared error (MSE), and the loss of forecasting label Y is measured by the cross-entropy error (CEE). Therefore, the loss function of LA can be represented as follows:

$$loss = \frac{1}{n}\sum_{i=1}^{n}(\alpha(x_i - \hat{x}_i)^2 - \beta y_i log\hat{y}_i),$$

where n is the number of samples, x_i is the i-th sample, $\hat{x}_l$ is the i-th reconstruction sample, y_i is the i-th sample's one-hot label, $\hat{y}_l$ is the i-th sample's softmax output, α and β are the scale factors.

To illustrate how the BLS framework works, LA is embedded in the BLS framework as an example. More specifically, the mappings $\phi_i(\cdot)$ and $\xi_j(\cdot)$ in the BLS framework is the same feature extraction method, LA. The mapped feature Z_n is learned by using the original data X as the input and output of the LA. After learning n mapped features, all the mapped features are concatenated as Z^n. The enhancement feature H_m is learned by using the Z^n as the input and output of the LA, then all the mapped features and enhancement features are concatenated as the final input and fed into any machine learning model.

Because of the random initialization of the weights of LA, each mapped feature and enhancement feature will be different. Further, randomly picked samples and randomly picked features can be used as inputs for each LA, and the random label-based autoencoder (RLA) can generate more different mapped features and enhancement features in the BLS. The randomness of RLA is controlled by two parameters: selected sample size and selected feature size. If the selected sample size and the selected feature size are less than 1, samples and features are randomly picked according to these two selected sizes. If the selected sample size and the selected feature size are equal to 1, all samples and features are used to train RLA. Therefore, LA is a special case of RLA.

Given a two-layer encoder structure, the input fed into a machine learning model is as follows:

$$\begin{aligned} input &= [Z^n|H^m] \\ &= [Z_1, \ldots, Z_n|H_1, \ldots, H_m] \\ &= \begin{bmatrix} \sigma(w_{11}\sigma(w_{12}x + b_{12}) + b_{11}), \ldots, \sigma(w_{n1}\sigma(w_{n2}x + b_{n2}) + b_{n1})| \\ \sigma(w'_{11}\sigma(w'_{12}Z^n + b'_{12}) + b'_{11}), \ldots, \sigma(w'_{m1}\sigma(w'_{m2}Z^n + b'_{m2}) + b'_{m1}) \end{bmatrix}, \end{aligned}$$

where w is the weight, b is the bias, and $\sigma(\cdot)$ is the activation function.

The number of mapped features n and the number of enhancement features m are different and depend on the complexity of modeling problems. Additional mapped features and enhancement features can be added to achieve a better performance when the setting (n, m) cannot reach the desired accuracy. The pseudocode of the BLS with RLA is shown in Algorithms 6.22 and 6.23.

Algorithm 6.22 Broad learning: increment of additional enhancement features

1: **for** $i = 0, i < n$ **do**
2: Train RLA model with training data set.
3: Generate Z_i using the RLA.
4: **end for**
5: Concatenate the mapped features $Z^n \equiv [Z_1, Z_2, \ldots, Z_n]$.
6: **for** $j = 0, j < m$ **do**
7: Train RLA model with data set Z^n.
8: Generate H_j using the RLA.
9: **end for**
10: Concatenate the mapped features and enhancement features $[Z^n|H^m]$ as inputs.
11: Train the machine learning model.
12: **while** VALIDATION ERROR is not satisfied **do**
13: Train RLA model with data set Z^n.
14: Generate H_{m+1} using the RLA.
15: Concatenate the mapped features and enhancement features $[Z^n|H^{m+1}]$ as inputs.
16: Train the machine learning model.
17: $m = m + 1$.
18: **end while**

Algorithm 6.23 Broad learning: increment of additional mapped features

1: **for** $i = 0, i < n$ **do**
2: Train RLA model with training data set.
3: Generate Z_i using the RLA.
4: **end for**
5: Concatenate the mapped features $Z^n \equiv [Z_1, Z_2, \ldots, Z_n]$.
6: **for** $j = 0, j < m$ **do**
7: Train RLA model with data set Z^n.
8: Generate H_j using the RLA.
9: **end for**
10: Concatenate the mapped features and enhancement features $[Z^n|H^m]$ as inputs.
11: Train the machine learning model.
12: **while** VALIDATION ERROR is not satisfied **do**
13: Train RLA model with training data set.
14: Generate Z_{n+1} using the RLA.
15: Concatenate the mapped features $Z^{n+1} \equiv [Z^n, Z_{n+1}]$.
16: **for** $j = 0, j < m$ **do**
17: Train RLA model with data set Z^{n+1}.
18: Generate H_j using the RLA.
19: Concatenate the mapped features and enhancement features $[Z^{n+1}|H^m]$ as inputs.
20: Train the machine learning model.
21: **end for**
22: $n = n + 1$.
23: **end while**

In addition, it should be noted that the BLS is not conflicted with feature selection methods. The BLS can be used before or after the feature selection methods, and the selected features can also be concatenated with the mapped features and enhancement features.

References

1. Ardehaly, E.M., Culotta, A.: Co-training for demographic classification using deep learning from label proportions. In: 2017 IEEE International Conference on Data Mining Workshops (ICDMW) (2017)
2. Arjovsky, M., Bottou, L.: Towards principled methods for training generative adversarial networks. Stat **1050** (2017)
3. Banfield, R.E., Hall, L.O., Bowyer, K.W., Kegelmeyer, W.P.: A comparison of decision tree ensemble creation techniques. IEEE Trans. Pattern Anal. Mach. Intell. **29**(1), 173–180 (2006)
4. Belohlávek, R., Sklenar, V., Zacpal, J.: Crisply generated fuzzy concepts. In: Formal Concept Analysis, Third International Conference, ICFCA 2005, Lens, France, February 14–18, 2005, Proceedings (2005)
5. Bishop, C.M.: Pattern Recognition and Machine Learning. Springer (2006)
6. Bosch, A., Zisserman, A., Munoz, X.: Image classification using random forests and ferns. In: 2007 IEEE 11th International Conference on Computer Vision, pp. 1–8. IEEE (2007)
7. Boyd, S., Boyd, S.P., Vandenberghe, L.: Convex Optimization. Cambridge University Press (2004)
8. Breiman, L.: Random forests. Machine Learning **45**(1), 5–32 (2001)
9. Caruana, R., Karampatziakis, N., Yessenalina, A.: An empirical evaluation of supervised learning in high dimensions. In: Proceedings of the 25th International Conference on Machine Learning, pp. 96–103 (2008)
10. Chapelle, O., Sindhwani, V., Keerthi, S.S.: Optimization techniques for semi-supervised support vector machines. J. Mach. Learn. Res. **9**(2) (2008)
11. Chen, C.P., Liu, Z.: Broad learning system: An effective and efficient incremental learning system without the need for deep architecture. IEEE Trans. Neural Networks Learn. Syst. **29**(1), 10–24 (2017)
12. Christmann, A., Steinwart, I.: How svms can estimate quantiles and the median. In: Advances in Neural Information Processing Systems, pp. 305–312 (2007)
13. Dai, Z., Yang, Z., Yang, F., Cohen, W.W., Salakhutdinov, R.: Good semi-supervised learning that requires a bad gan. Preprint (2017). arXiv:1705.09783
14. Feldman, J.: Minimization of boolean complexity in human concept learning. Nature **407**(6804), 630–633 (2000)
15. Gall, J., Lempitsky, V.: Class-specific hough forests for object detection. In: Decision Forests for Computer Vision and Medical Image Analysis, pp. 143–157. Springer (2013)
16. Ganter, B., Wille, R.: Formal Concept Analysis: Mathematical Foundations. Springer Science & Business Media (2012)
17. Goodfellow, I.J., Pouget-Abadie, J., Mirza, M., Xu, B., Warde-Farley, D., Ozair, S., Courville, A., Bengio, Y.: Generative adversarial networks. Preprint (2014). arXiv:1406.2661
18. Grandvalet, Y., Bengio, Y., et al.: Semi-supervised learning by entropy minimization. In: CAP, pp. 281–296 (2005)
19. Grossberg, S.: Adaptive resonance theory: How a brain learns to consciously attend, learn, and recognize a changing world. Neural Networks **37**, 1–47 (2013)
20. Hinton, G.E., Salakhutdinov, R.R.: Reducing the dimensionality of data with neural networks. Science **313**(5786), 504–507 (2006)

21. Huang, G.B., Zhu, Q.Y., Siew, C.K.: Extreme learning machine: theory and applications. Neurocomputing **70**(1-3), 489–501 (2006)
22. Huang, G.B., Zhou, H., Ding, X., Zhang, R.: Extreme learning machine for regression and multiclass classification. IEEE Trans. Syst. Man Cybern. B (Cybern.) **42**(2), 513–529 (2011)
23. Huang, X., Shi, L., Suykens, J.A.: Support vector machine classifier with pinball loss. IEEE Trans. Pattern Anal. Mach. Intell. **36**(5), 984–997 (2013)
24. Huang, X., Shi, L., Suykens, J.A.: Sequential minimal optimization for svm with pinball loss. Neurocomputing **149**, 1596–1603 (2015)
25. Igelnik, B., Pao, Y.H.: Stochastic choice of basis functions in adaptive function approximation and the functional-link net. IEEE Trans. Neural Netw. **6**(6), 1320–1329 (1995)
26. Johnson, J., Alahi, A., Fei-Fei, L.: Perceptual losses for real-time style transfer and super-resolution. In: European Conference on Computer Vision, pp. 694–711. Springer (2016)
27. Kang, X., Miao, D.: A study on information granularity in formal concept analysis based on concept-bases. Knowl. Based Syst. **105**, 147–159 (2016)
28. Kumar, S., Mohri, M., Talwalkar, A.: Sampling methods for the nyström method. J. Mach. Learn. Res. **13**(1), 981–1006 (2012)
29. Lai, J., Wang, C.: Kernel and graph: Two approaches for nonlinear competitive learning clusterin. Front. Electr. Electron. Eng. **7**(1), 134–146 (2012)
30. Lake, B.M., Salakhutdinov, R., Tenenbaum, J.B.: Human-level concept learning through probabilistic program induction. Science **350**(6266), 1332–1338 (2015)
31. Leistner, C., Saffari, A., Santner, J., Bischof, H.: Semi-supervised random forests. In: 2009 IEEE 12th International Conference on Computer Vision, pp. 506–513. IEEE (2009)
32. Leistner, C., Saffari, A., Bischof, H.: Miforests: Multiple-instance learning with randomized trees. In: European Conference on Computer Vision, pp. 29–42. Springer (2010)
33. Li, J., Laird, J.: Spontaneous retrieval from long-term memory for a cognitive architecture. In: Proceedings of the AAAI Conference on Artificial Intelligence, vol. 29 (2015)
34. Li, J., Mei, C., Lv, Y.: Knowledge reduction in decision formal contexts. Knowl. Based Syst. **24**(5), 709–715 (2011)
35. Li, J., Huang, C., Xu, W., Qian, Y., Liu, W.: Cognitive concept learning via granular computing for big data. In: International Conference on Machine Learning & Cybernetics (2015)
36. Li, J., Mei, C., Xu, W., Qian, Y.: Concept learning via granular computing: A cognitive viewpoint. Information Sciences **298**, 447–467 (2015)
37. Li, J., Huang, C., Qi, J., Qian, Y., Liu, W.: Three-way cognitive concept learning via multi-granularity. Information Sciences **378**, 244–263 (2017)
38. Li, T., Kou, G., Peng, Y., Shi, Y.: Classifying with adaptive hyper-spheres: An incremental classifier based on competitive learning. IEEE Trans. Syst. Man Cybern. Syst. **50**(4), 1218–1229 (2017)
39. Liu, J., Wang, B., Qi, Z., Tian, Y., Shi, Y.: Learning from label proportions with generative adversarial networks. Preprint (2019). arXiv:1909.02180
40. Lu, J., Hoi, S.C., Wang, J., Zhao, P., Liu, Z.Y.: Large scale online kernel learning. J. Mach. Learn. Res. **17**(47), 1 (2016)
41. Macdonald, B.A., Witten, I.H.: A framework for knowledge acquisition through techniques of concept learning. IEEE Trans. Syst. Man Cybern. **19**(3), 499–512 (1989)
42. Mi, Y., Liu, W., Shi, Y., Li, J.: Semi-supervised concept learning by concept-cognitive learning and concept space. IEEE Trans. Knowl. Data Eng. (2020). https://doi.org/10.1109/TKDE.2020.3010918
43. Mi, Y., Shi, Y., Li, J., Liu, W., Yan, M.: Fuzzy-based concept learning method: exploiting data with fuzzy conceptual clustering. IEEE Trans. Cybern., 1–12 (2020)
44. Modha, D.S., Ananthanarayanan, R., Esser, S.K., Ndirango, A., Sherbondy, A.J., Singh, R.: Cognitive computing. Commun. ACM **54**(8), 62–71 (2011)
45. Moosmann, F., Triggs, B., Jurie, F.: Fast discriminative visual codebooks using randomized clustering forests. In: Twentieth Annual Conference on Neural Information Processing Systems (NIPS'06), pp. 985–992. MIT Press (2006)

46. Pal, M.: Random forest classifier for remote sensing classification. Int. J. Remote Sens. **26**(1), 217–222 (2005)
47. Pao, Y.H., Park, G.H., Sobajic, D.J.: Learning and generalization characteristics of the random vector functional-link net. Neurocomputing **6**(2), 163–180 (1994)
48. Pei, D., Li, M.Z., Mi, J.S.: Attribute reduction in fuzzy decision formal contexts. In: 2011 International Conference on Machine Learning and Cybernetics, vol. 1, pp. 204–208. IEEE (2011)
49. Perbet, F., Stenger, B., Maki, A.: Random forest clustering and application to video segmentation. In: BMVC, pp. 1–10. Citeseer (2009)
50. Quan, T.T., Hui, S.C., Cao, T.H.: A fuzzy fca-based approach to conceptual clustering for automatic generation of concept hierarchy on uncertainty data. In: CLA, pp. 1–12 (2004)
51. Rehman, M.Z.u., Li, T., Yang, Y., Wang, H.: Hyper-ellipsoidal clustering technique for evolving data stream. Knowl. Based Syst. **70**, 3–14 (2014)
52. RogerKoenker: Quantile Regression. Cambridge University Press (2005)
53. Saffari, A., Leistner, C., Santner, J., Godec, M., Bischof, H.: On-line random forests. In: 2009 Ieee 12th International Conference on Computer Vision Workshops, ICCV Workshops, pp. 1393–1400. IEEE (2009)
54. Salimans, T., Goodfellow, I., Zaremba, W., Cheung, V., Radford, A., Chen, X.: Improved techniques for training gans. Preprint (2016). arXiv:1606.03498
55. Shao, M.W., Leung, Y., Wang, X.Z., Wu, W.Z.: Granular reducts of formal fuzzy contexts. Knowl. Based Syst. **114**, 156–166 (2016)
56. Sharp, T.: Implementing decision trees and forests on a gpu. In: European Conference on Computer Vision, pp. 595–608. Springer (2008)
57. Shi, Y., Liu, J., Qi, Z., Wang, B.: Learning from label proportions on high-dimensional data. Neural Networks **103**, 9–18 (2018)
58. Shi, Y., Mi, Y., Li, J., Liu, W.: Concept-cognitive learning model for incremental concept learning. IEEE Trans. Syst. Man Cybern. Syst. (2018). https://doi.org/10.1109/TSMC.2018.2882090
59. Shi, Y., Cui, L., Chen, Z., Qi, Z.: Learning from label proportions with pinball loss. Int. J. Mach. Learn. Cybern. **10**(1), 187–205 (2019)
60. Shi, Y., Mi, Y., Li, J., Liu, W.: Concurrent concept-cognitive learning model for classification. Information Sciences **496**, 65–81 (2019)
61. Springenberg, J.T.: Unsupervised and semi-supervised learning with categorical generative adversarial networks. Preprint (2015). arXiv:1511.06390
62. Srivastava, N., Vul, E.: A simple model of recognition and recall memory. In: Proceedings of the 31st International Conference on Neural Information Processing Systems, pp. 292–300 (2017)
63. Steinwart, I., Christmann, A., et al.: Estimating conditional quantiles with the help of the pinball loss. Bernoulli **17**(1), 211–225 (2011)
64. Szegedy, C., Vanhoucke, V., Ioffe, S., Shlens, J., Wojna, Z.: Rethinking the inception architecture for computer vision. In: Proceedings of the IEEE Conference on Computer Vision and Pattern Recognition, pp. 2818–2826 (2016)
65. Tadrat, J., Boonjing, V., Pattaraintakorn, P.: A new similarity measure in formal concept analysis for case-based reasoning. Expert Syst. Appl. **39**(1), 967–972 (2012)
66. Tang, H., Dong, P., Shi, Y.: A construction of robust representations for small data sets using broad learning system. IEEE Trans. Syst. Man Cybern. Syst., 1–11 (2019)
67. Tho, Q.T., Hui, S.C., Fong, A.C.M., Cao, T.H.: Automatic fuzzy ontology generation for semantic web. IEEE Trans. Knowl. Data Eng. **18**(6), 842–856 (2006)
68. Wang, Y.: On cognitive computing. Int. J. Softw. Sci. Comput. Intell. **1**(3), 1–15 (2011)
69. Wang, Y., Wang, Y.: Cognitive informatics models of the brain. IEEE Trans. Syst. Man Cybern. C (Appl. Rev.) **36**(2), 203–207 (2006)
70. Wang, Y., Chiew, V.: On the cognitive process of human problem solving. Cogn. Syst. Res. **11**(1), 81–92 (2010)

71. Wang, Z., Feng, J.: Multi-class learning from class proportions. Neurocomputing **119**, 273–280 (2013)
72. Wang, Y., Howard, N., Kacprzyk, J., Frieder, O., Sheu, P., Fiorini, R.A., Gavrilova, M.L., Patel, S., Peng, J., Widrow, B.: Cognitive informatics: Towards cognitive machine learning and autonomous knowledge manipulation. Int. J. Cogn. Inf. Nat. Intell. (IJCINI) **12**(1), 1–13 (2018)
73. Warde-Farley, D., Goodfellow, I.: Adversarial perturbations of deep neural networks. Perturbat. Optim. Stat. **311** (2016)
74. Wei, L., Qi, J., Zhang, W.: Attribute reduction theory of concept lattice based on decision formal contexts. Sci. China F Inf. Sci. **51**(7), 910–923 (2008)
75. Wille, R.: Restructuring lattice theory: an approach based on hierarchies of concepts. In: International Conference on Formal Concept Analysis, pp. 314–339. Springer (2009)
76. Wu, J.S., Zheng, W.S., Lai, J.H.: Approximate kernel competitive learning. Neural Networks **63**, 117–132 (2015)
77. Xiao, C., Wang, W., Lin, X., Shang, H.: Top-k set similarity joins. In: 2009 IEEE 25th International Conference on Data Engineering, pp. 916–927. IEEE (2009)
78. Yahia, S.B., Jaoua, A.: Discovering knowledge from fuzzy concept lattice. In: Data Mining and Computational Intelligence, pp. 167–190. Springer (2001)
79. Yu, J.: Machine learning: From axioms to algorithms (2017)
80. Yu, F., Liu, D., Kumar, S., Tony, J., Chang, S.F.: $\propto$svm for learning with label proportions. In: International Conference on Machine Learning, pp. 504–512. PMLR (2013)
81. Zadeh, L.A.: Fuzzy sets and information granularity. Adv. Fuzzy Set Theory Appl. **11**, 3–18 (1979)
82. Zhang, W.X., Qiu, G.F.: Uncertain Decision Making Based on Rough Sets. Publishin of Tsinghua University, Beijing (2005)
83. Zhang, W.X., Ma, J.M., Fan, S.Q.: Variable threshold concept lattices. Information Sciences **177**(22), 4883–4892 (2007)
84. Zhang, F., Liu, J., Wang, B., Qi, Z., Shi, Y.: A fast algorithm for multi-class learning from label proportions. Electronics **8**(6), 609 (2019)

Chapter 7
Sentiment Analysis

Sentiment analysis (SA) refers to the use of computational linguistics to identify and extract subjective information in source material, usually unstructured and heterogeneous text data [24]. This chapter summarizes the recent findings of the authors' research team on SA [7, 15, 21–24, 28]. It has two sections. Section 7.1 is word embedding with two Sect. 7.1.1 is about single sense model vs. multiple sense model while Sect. 7.1.2 is about intrinsic vs extrinsic evaluation. Section 7.2 outlines the SA applications.

SA can be extensively applied to a large number of application scenarios such as improving customer service and analyzing social media. There are three types of SA in terms of classification levels. i.e., aspect-level, sentence-level and document-level SA. Document-level SA means predicting the sentiment polarity of a document which is composed of several sentences. Sentence-level SA refers to detecting the sentiment of a single sentence. Furthermore, a document or sentence may describe some aspects of a product like travel product and we sometimes need to know the exact sentiment polarities of each aspect. This is regarded as aspect-level SA. Taking a tourism review as an example, "We went to The Forbidden City last week, and the tour guide is knowledgeable, but it was crowded in The Forbidden City. The worst thing was that it rained when we visited" discusses the cicerone, scenery spot and weather [24]. Here the tour guide is praised whereas the scenery spot and weather are criticized in this tourism review.

Text, audio, image and video comprise the basic information carriers in modern times. Users post plenty of microblogs and tweets on social media platform such as Microblog and Twitter. People can obtain valuable knowledge from different kinds of aforementioned medium. In fact, not only text modality but also audio video modalities convey sentiment information. However, how to effectively extract useful information from unstructured text data remains a challenge. To this end, researchers presented a large amount of sophisticated SA techniques to tackle this challenge. As shown in Fig. 7.1, existing SA techniques can be classified as three types, namely lexicon-based, traditional machine learning based and new approaches.

Y. Shi, *Advances in Big Data Analytics*,
https://doi.org/10.1007/978-981-16-3607-3_7

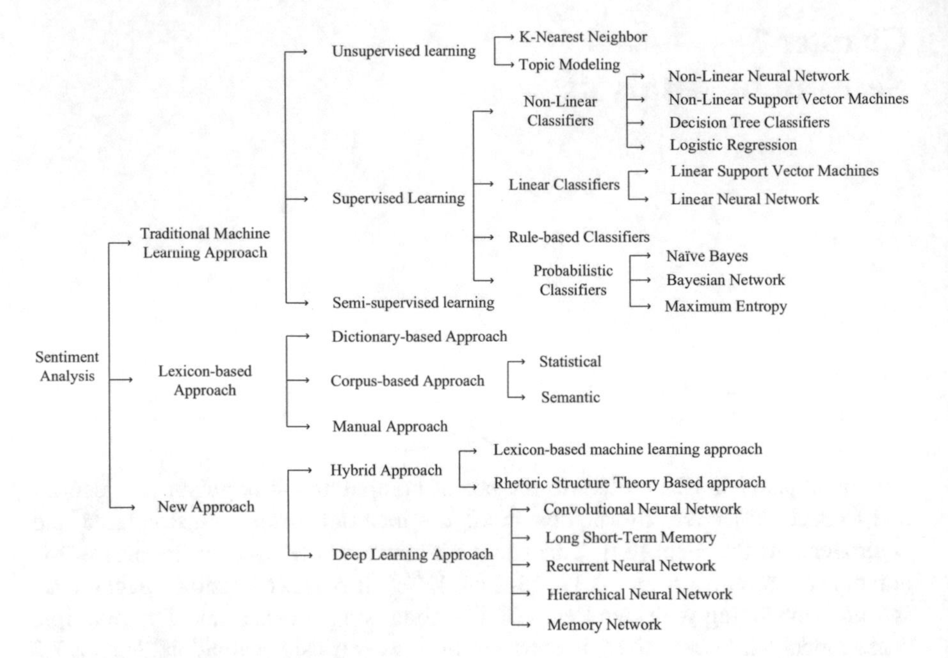

Fig. 7.1 Concept map of SA techniques [24]

Sentiment lexicon, which contains numerous sentiment words with different sentiment polarities or intensities, is one of the most popular lexicon-based approaches. sentiment lexicon can be built by expert knowledge, which requires huge human work and resources. Therefore, some data-driven methods [15, 27] are created to automatically construct the sentiment lexicon and cover as many sentiment words as possible. General steps of sentiment lexicon based SA are as follows:

- extract sentiment words from a given sentence or document.
- sum the sentiment scores of each sentiment words in the sentence or document.
- classify the sentence or document according to the total sentiment score.

Traditional machine learning based approaches, such as naive bayes and rule based approach get good performances on a number of SA tasks including sentence-level and document-level. Taking rule-based approaches as an example, it consists of, as the name partially suggests, a set of rules that classify data in data space. For instance, VADER [11], a simple rule-based modal, contains a gold-standard list of lexical features. extensive experiments on four distinct domain datasets demonstrated that VADER outperformed some benchmarks in the social media domain.

Recently, a lot of new approaches are proposed among which deep learning based and hybrid approaches attract more attention. Deep learning based approaches employ deep learning techniques such as recurrent neural network [25], long short-term memory model [9], convolutional neural network [13], memory networks

[26] and transformer [29] to classify sentences or documents. Hybrid approaches combine the advantages of different methods and get the state-of-the-art on many benchmark SA datasets. For example, cambria et al. [5] proposed an ensemble of symbolic and sub-symbolic AI techniques to perform sentiment analysis. More concretely, they built a new three-level knowledge representation SenticNet 5 for SA, where long short-term memory model (LSTM) was used to discover verb-noun primitives. The generation process largely extend the coverage of basic SenticNet4 [4].

In general, raw data contains numerous useless information. Therefore, it is necessary to pre-process raw data for machine learning and deep learning based SA. For example, HTML contains a lot of HTML tags and non-alphabetic signs which should be removed to improve the data quality. In this paper [7], the authors proposed a data pre-processing framework which is proved to be effective in SA task. The first step is data transformation, where all the HTML tags, non-alphabetic signs, stop words and non-informative words like file, movie, actor, actress, scene are removed from the raw data. After that, stemming is performed on the documents to reduce redundancy. Then, three feature matrices are constructed for each of the datasets based on three different types of feature weighting : TF-IDF, FF and FP. Furthermore, the second step is filtering step, where univariate method chi-squared is utilized to conduct filtering procedure and it measures the dependency between the word and the category of the document it is mentioned in. If the word is frequent in many categories, chi-squared value is low and vice versa. Extensive experiments indicate that appropriate text pre-processing methods including aforementioned data transformation and filtering can significantly improve the classifier's performance. Nevertheless, machine learning and deep learning models cannot directly process words or sentences. In this case, we need a more generalized method to convert the unstructured text data into vector space where words, sentences and even documents can be represented as vectors. This process is named as word embedding or word representation which will be introduced in next section.

7.1 Word Embedding

7.1.1 Methods: Single Sense Model vs Multiple Sense Model

Distributional hypothesis proposed by Harris [8] is the footstone of NLP and word embedding. The word embedding methods consists of three types [23]: matrix-based methods such as TF-IDF matrix, Latent Semantic Analysis (LSA) [14], and GloVe [20]; cluster-based methods, such as Brown [3]; and neural network based methods, such as Neral Network Language Model (NNLM) [1], Log-Bilinear Language model (LBL) [18], C&W [6], skip-gram [17], continuous bag-of-words model (CBOW) [17], and FastText [12], etc.

The aforementioned methods deal with the single sense word embedding, which fails to represent words with multiple meanings. For example, word “apple” is a kind of fruit when it occurs in an article about food or botany; while “apple” refers to a technology company when it comes with MacBook, iphone, etc. In single sense model, only one vector is generated for word “apple” by averaging the results for all meanings, and it cannot comply with all language rules. This calls for a multiple sense word embedding, where each sense of the word corresponds to an independent word vector which serves as an auxiliary vector in the mean while.

How to determine a word’s sense according to the current context is one of the greatest challenges for multiple sense word embedding. An intuitive method is to represent the corresponding sense vector via clustering all contexts of the target word by maximizing the probability $P(S)$ in formula (7.1).

$$P(S) = \prod_{i=1}^{n}\prod_{j=1}^{m_i} p(w_i^j|C(w_i^j)) \approx \prod_{i=1}^{n}\prod_{j=1}^{m_i} p(w_i^j|C(w_i)) \tag{7.1}$$

where $S = w_1, w_2, \ldots, w_n$ is the word sequence of a sentence, w_i is the ith word, $C(w_i)$ is the context of word w_i, m_i is the number of total senses of word w_i, w_i^j is the embedding for the jth sense of the ith word. Here, replacing $C(w_i^j)$ with $C(w_i)$ is an effective simplification method.

Based on this intuitive idea, tow-stage hard clustering method [10] was proposed to learn the multiple prototype embedding by means of spherical K-means algorithm and relabeling contexts with corresponding cluster centroids before training. However, the spherical k-means cluster is time-consuming and relabeled context may lose certain detailed information. Zheng Y. et al. [28] presented multi-prototype continuous bag-of-words model (MCBOW) based on a common word vector cell and CBOW, which is illustrated in Fig. 7.2. The objective of this model is to predict multiple target word embedding by exploiting different context information. For each prediction, every context word containing several different sense embeddings (represented by light green modes), the red arrows show how each sense embedding is chosen to form a temp context vector (represented by pink nodes) in Fig. 7.2. Here, u_i^j is the weight (coefficient) for the ith word sense w_i^j. The selection procedure is based on the similarity between the target word’s context and the temp context vector. Then the embedding of target word is determined and the vectors is updated through stochastic gradient descent [2] based back-propagation.

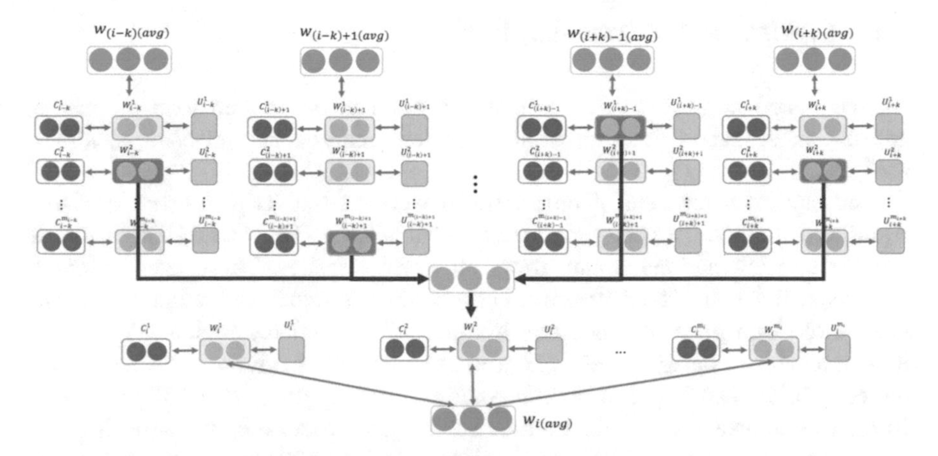

Fig. 7.2 Multi-prototype continuous bag-of-words model [28]

7.1.2 Evaluation: Intrinsic vs Extrinsic

Intrinsic evaluation and extrinsic evaluation are two perspectives to measure the quality of learned word embeddings. Intrinsic evaluation mainly focus on evaluating the word properties, such as the semantic information, syntactic information, and morphology information, etc. While, extrinsic concentrate on specific natural language processing (NLP) tasks in practice.

Some methods employed as intrinsic evaluation for single sense word embedding are shown as follows. Word similarity (WS) measures the similarity between two words by calculating the cosine values of their word vectors; Word analogy (WA) explores the analogical relationship between the word pair and searches for the other matching word pairs satisfying the same relationship; Word synonym detection (WSD) is to find the synonyms given the current token; Selectional preferences (SP) utilizes the verb-noun pairs to construct a verb-noun phrase and noun-verb phrase.

For multiple sense embedding models, each word contains various sense vectors and the aforementioned basic evaluation methods may fail. Thus, four cosine value based distances were proposed by [10], i.e. $AvgSimC$, $GlobalSim$, $AvgSim$, and $LocalSim$.

For extrinsic evaluation, the learned word embeddings are fed to a specific model as features or for the initialization of neural networks. The tasks includes named entity recognition (NER), part-of-speech tagging (POST), text classification (TC), and sentiment analysis (SA), etc. Different from the small intrinsic evaluation datasets, extrinsic evaluation datasets usually contains sufficient training and test samples. However, the input word embedding and the experiment settings may both influence the final results. Thus, it is important to control the experiment settings while evaluating the learned word embeddings using extrinsic evaluation methods [23].

7.2 Sentiment Analysis Applications

As mentioned in Sect. 7.1, sentiment analysis can be applied to a number of application scenarios such as product review analysis and investor sentiment analysis.

The rapid development of online travel websites like TripAdvisor[1] lead to a significant increase in user-generated content (UGC) [16]. Here UGC refers to reviews and interactions among users on travel websites. These user generated reviews will be displayed under a certain travel product and other users who browse the web page of this travel product will see related reviews. Therefore, it is necessary to design an algorithm to automatically analyze such reviews and travel websites can improve their service accordingly. In 2018, [15] proposed a framework named DWWP for tourism review SA, where a domain-specific new words detection method (DW) and word propagation (WP) are presented. Tourism reviews have a number of user-invented domain-specific words such as " (a large costume show), proper nouns, converted words and multiword expressions (MWEs) which are not included in the existing sentiment lexicons. Manual detection of such words is time-consuming and costly. Therefore, automatic and effective construction of a high-quality tourism-specific sentiment lexicon is of great value. What's more, Chinese SA was even harder due to the lack of segmentation symbols like blank space in English. In this case, the aforementioned four types of words cannot be easily detected by Chinese word segmentation tools. Besides, one limitation of existing data driven sentiment lexicon construction methods is the lack of robustness. To this end, DWWP framework is presented to solve the above issues and build a high-quality tourism-specific sentiment lexicon.

Figure 7.3 is the concept map of the proposed DWWP framework. As shown in the figure, raw data is first collected, pre-processed and then fed into domain new words detection (DW) block. First, Chinese word segmentation tool is utilized to segment the raw text data into a series of single morphemes. Then the authors proposed a statistical indicator named Assembled Mutual Information (AMI) to determine if a candidate word is a valid word or not. The formula of AMI is as follows:

$$AMI(w) = \sum_j (log \frac{n_w/N}{\sqrt[T]{\prod_{i=1}^{T}[(n_{w_i}^j - n_w + sf)/N]}}) \tag{7.2}$$

where w means the candidate new word which is composed of T single morphemes. n_w is the occurrence number of w, whereas n_{m_i} is the occurrence number of m_i. N stands for the total number of documents. If the AMI score of a candidate new word is high, it is more probable to be a valid new word and vice versa. With the help of this domain new words detection algorithm, plenty of domain new

[1] https://www.tripadvisor.com.sg/.

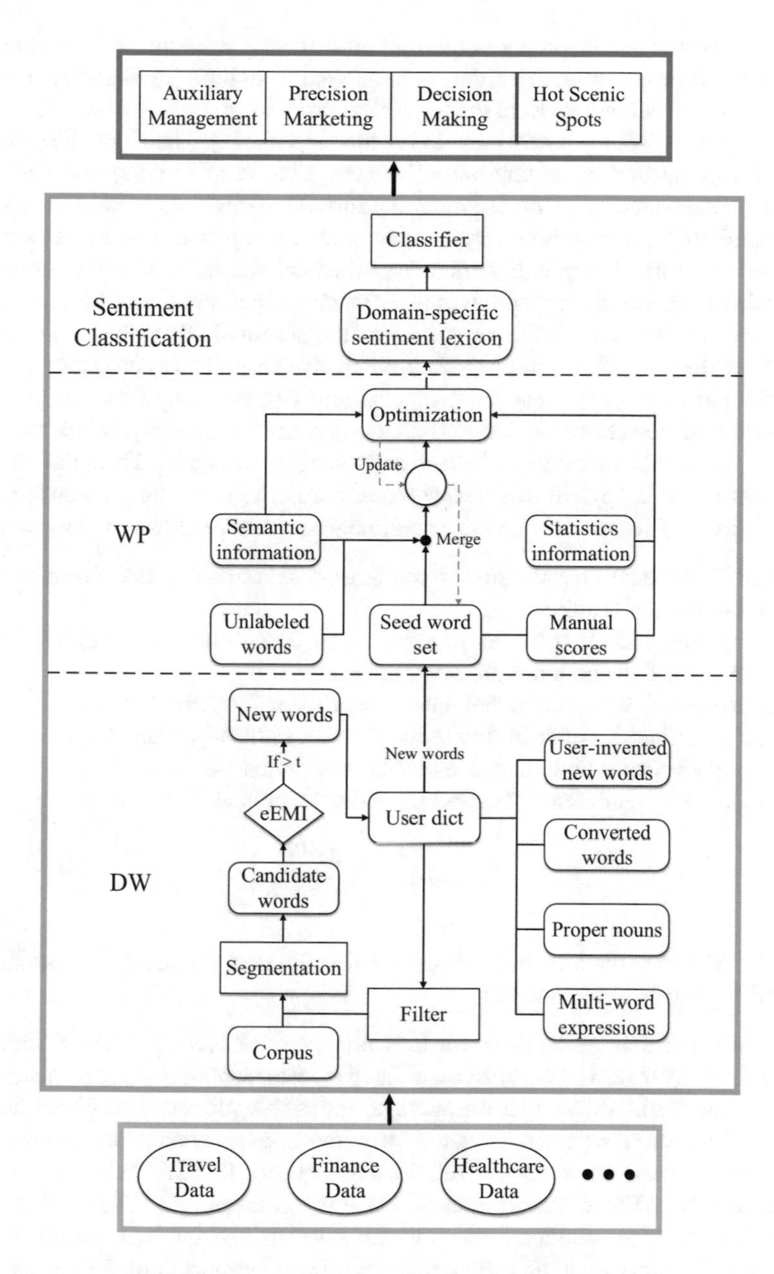

Fig. 7.3 Domain words detection and word propagation system [15]

words, converted words, proper nouns and multiword expressions can be obtained. Next, a word propagation algorithm is employed to build a high-quality tourism-specific sentiment lexicon. Here the algorithm starts from a small set of seed words (detected new words are partly included in the seed words). Then both the semantic similarity (measured by cosine value between two word vectors) and statistical similarity (measured by pointwise mutual information) are used to measure the real world similarity between two words. Additionally, an optimization function which considers seed word, semantic similarity, statistical similarity is designed to tune the sentiment scores of sentiment words. Extensive experiments on Chinese tourism review dataset demonstrate the superiority of the proposed DWWP SA framework.

Besides the product review analysis, SA can also be applied to investor sentiment analysis. Shi et al. [21] proposed a text mining system using data cleaning, text representation, feature extraction and a two-step sentiment analysis techniques to identify individual investor sentiment and comply an index. Then the investor sentiment index is applied to Chinese stock market to study its relationship with CSI 300 stock index returns. Investor sentiment measure process is as follows:

- Extract individual investor posts from large-scale online stocks forum posts on East money stock forum.
- Employ linguistic module to process posts data, where text representation, feature extraction and noise classifier are utilized.
- The processed text data is fed into a sentiment identification block (support vector machine classifier in this section). Here bullish-bearish classifier is used to identify investor sentiment as either bullish or bearish.
- Build investor sentiment index and the index formula is:

$$M_t = ln\left[\frac{1 + M_t^{BUY}}{1 + M_t^{SELL}}\right] \tag{7.3}$$

  where M_t^{BUY} is the total bullish posts in time interval t, and M_t^{SELL} is the total bearish posts in time interval t.

Based on 5,163,210 online posts on East money stock forum, investor sentiment index is built by means of the aforementioned investor sentiment index construction method. The result shows that on average, investor sentiment is towards bullish, which verifies the viewpoint of investors' irrational biases in behavioral finance [19]. Moreover, this chapter studied the relationship between CSI 300 index and investor sentiment index. The similarity rate of investor sentiment is 60.76% and is much higher than that of institutional view, which suggests that investor sentiment from online stock forum can predict stock returns, especially short term. To this end, this chapter established a 3 order VAR model which indicates the asymmetric effects of investor sentiment on stock market.

Sentiment analysis, a promising research field in natural language processing (NLP), attracts more attention in recent years. With the development of deep learning based NLP techniques, the performance of SA increases fast. Basic SA techniques such as word embedding as well as the application scenarios will be the future research focus.

References

1. Bengio, Y., Ducharme, R., Vincent, P., Jauvin, C.: A neural probabilistic language model. J. Mach. Learn. Res. **3**(Feb), 1137–1155 (2003)
2. Bottou, L.: Large-scale machine learning with stochastic gradient descent. In: Proceedings of COMPSTAT'2010, pp. 177–186. Springer (2010)
3. Brown, P.F., Desouza, P.V., Mercer, R.L., Pietra, V.J.D., Lai, J.C.: Class-based n-gram models of natural language. Computational Linguistics **18**(4), 467–479 (1992)
4. Cambria, E., Poria, S., Bajpai, R., Schuller, B.: Senticnet 4: A semantic resource for sentiment analysis based on conceptual primitives. In: Proceedings of COLING 2016, the 26th International Conference on Computational Linguistics: Technical Papers, pp. 2666–2677 (2016)
5. Cambria, E., Poria, S., Hazarika, D., Kwok, K.: Senticnet 5: Discovering conceptual primitives for sentiment analysis by means of context embeddings. In: Thirty-Second AAAI Conference on Artificial Intelligence (2018)
6. Collobert, R., Weston, J.: A unified architecture for natural language processing: Deep neural networks with multitask learning. In: Proceedings of the 25th International Conference on Machine Learning, pp. 160–167 (2008)
7. Haddi, E., Liu, X., Shi, Y.: The role of text pre-processing in sentiment analysis. Procedia Comput. Sci. **17**, 26–32 (2013)
8. Harris, Z.S.: Distributional structure. Word **10**(2-3), 146–162 (1954)
9. Hochreiter, S., Schmidhuber, J.: Long short-term memory. Neural Computation **9**(8), 1735–1780 (1997)
10. Huang, E.H., Socher, R., Manning, C.D., Ng, A.Y.: Improving word representations via global context and multiple word prototypes. In: Proceedings of the 50th Annual Meeting of the Association for Computational Linguistics: Long Papers-Volume 1, pp. 873–882. Association for Computational Linguistics (2012)
11. Hutto, C.J., Gilbert, E.: Vader: A parsimonious rule-based model for sentiment analysis of social media text. In: Eighth International AAAI Conference on Weblogs and Social Media (2014)
12. Joulin, A., Grave, E., Bojanowski, P., Mikolov, T.: Bag of tricks for efficient text classification. Preprint (2016). arXiv:1607.01759
13. Kim, Y.: Convolutional neural networks for sentence classification. Preprint (2014). arXiv:1408.5882
14. Landauer, T.K., Foltz, P.W., Laham, D.: An introduction to latent semantic analysis. Discourse Processes **25**(2-3), 259–284 (1998)
15. Li, W., Guo, K., Shi, Y., Zhu, L., Zheng, Y.: Dwwp: Domain-specific new words detection and word propagation system for sentiment analysis in the tourism domain. Knowl. Based Syst. **146**, 203–214 (2018)
16. Marine-Roig, E.: Online travel reviews: A massive paratextual analysis. In: Analytics in Smart Tourism Design, pp. 179–202. Springer (2017)
17. Mikolov, T., Chen, K., Corrado, G., Dean, J.: Efficient estimation of word representations in vector space. Preprint (2013). arXiv:1301.3781

18. Mnih, A., Hinton, G.: Three new graphical models for statistical language modelling. In: Proceedings of the 24th International Conference on Machine Learning, pp. 641–648 (2007)
19. Odean, T.: Are investors reluctant to realize their losses? J. Finance **53**(5), 1775–1798 (1998)
20. Pennington, J., Socher, R., Manning, C.D.: Glove: Global vectors for word representation. In: Proceedings of the 2014 Conference on Empirical Methods in Natural Language Processing (EMNLP), pp. 1532–1543 (2014)
21. Shi, Y., Tang, Y.R., Cui, L.X., Long, W.: A text mining based study of investor sentiment and its influence on stock returns. Econom. Comput. Econom. Cybernet. Stud. Res. **52**(1), 183–199 (2018)
22. Shi, Y., Zheng, Y., Guo, K., Li, W., Zhu, L.: Word similarity fails in multiple sense word embedding. In: International Conference on Computational Science, pp. 489–498. Springer (2018)
23. Shi, Y., Zheng, Y., Guo, K., Zhu, L., Qu, Y.: Intrinsic or extrinsic evaluation: An overview of word embedding evaluation. In: 2018 IEEE International Conference on Data Mining Workshops (ICDMW), pp. 1255–1262. IEEE (2018)
24. Shi, Y., Zhu, L., Li, W., Guo, K., Zheng, Y.: Survey on classic and latest textual sentiment analysis articles and techniques. Int. J. Inf. Tech. Dec. Making **18**(04), 1243–1287 (2019)
25. Tang, D., Qin, B., Liu, T.: Document modeling with gated recurrent neural network for sentiment classification. In: Proceedings of the 2015 Conference on Empirical Methods in Natural Language Processing, pp. 1422–1432 (2015)
26. Weston, J., Chopra, S., Bordes, A.: Memory networks. Preprint (2014). arXiv:1410.3916
27. Wu, F., Huang, Y., Song, Y., Liu, S.: Towards building a high-quality microblog-specific chinese sentiment lexicon. Decis. Support Syst. **87**, 39–49 (2016)
28. Zheng, Y., Shi, Y., Guo, K., Li, W., Zhu, L.: Enhanced word embedding with multiple prototypes. In: 2017 4th International Conference on Industrial Economics System and Industrial Security Engineering (IEIS), pp. 1–5. IEEE (2017)
29. Zhu, Z., Zhou, Y., Xu, S.: Transformer based chinese sentiment classification. In: Proceedings of the 2019 2nd International Conference on Computational Intelligence and Intelligent Systems, pp. 51–56 (2019)

Chapter 8
Link Analysis

Link analysis has been recognized as an effective technique in data science to explore the relationships of objects. The objects can be social events, people, organization and even business transactions. This chapter reports the practical models of link analysis in various data-driven application areas. Section 8.1 presents a recommendation system for marketing optimization [1]. Section 8.2 is about advertisement clicking prediction [2]. Section 8.3 presents a model for customer churn prediction [3]. Section 8.4 provides node coupling clustering approaches for link prediction [4]. Finally, Sect. 8.5 discusses a pyramid scheme model for consumption rebate frauds [5].

8.1 Recommender System for Marketing Optimization

This section proposes a new method called trigger and triggered (TT) model. It aims to solve the problems described above and provide a lifecycle recommendation. The proposed method consists of two parts. The first part eliminates concentration noise in the training data and can be used as independent anonymous recommendation system. The second part serves as a recommendation system with lifecycle awareness among products. In the following, Sect. 8.1.1 introduces terminologies and related techniques while Sect. 8.1.2 describes the proposed model.

8.1.1 Terminologies and Related Techniques

8.1.1.1 Score Matrix

For convenience of discussion, this section uses the term "score-matrix" to describe the score between different stock keeping units (SKU, a unique identification of

Y. Shi, *Advances in Big Data Analytics*,
https://doi.org/10.1007/978-981-16-3607-3_8

each products). The score of SKU can be the number of times of co-purchases, or that of sequential purchases (a customer buys B because of the product A is bought first) or the similarity of products.

8.1.1.2 Weibull Distribution

The Weibull distribution [6] is one of the most widely used lifetime distributions. It was originally proposed by the Swedish physicist Waloddi Weibull, who used the model to approximate the distribution of breaking strength of materials. The versatility of distribution can take on the characteristics of other types of distributions, by tuning value of the shape parameter [6]. The probability density function of a Weibull model with a random variable x is shown in (8.1):

$$f(x) = \frac{\alpha}{\beta}\left(\frac{x-\mu}{\beta}\right)^{\alpha-1} e^{-\left(\frac{x-\mu}{\beta}\right)^{\alpha}} \tag{8.1}$$

where α, β and γ are known as the shape, scale and threshold parameters, respectively with constraints that $\alpha > 0$, $\beta > 0$, $\gamma > 0$. The advantage of Weibull distribution is its versatility of distribution. Therefore, in this section we use the characteristics to simulate distributions with a left long tail or right long tail, that looks like the sequential consumptions from a customer.

8.1.1.3 Gradient Descent

Gradient descent [7] is an optimization algorithm. It uses the method of first-order iterative gradient optimization to find the minimum of a function. To find a local minimum of a function, small steps are taken forward from the negative direction of the gradient of the function at current points. Stochastic gradient descent is a stochastic approximation of the gradient descent by aggregating a batch of gradient descent of sample data, which can largely increase the speed of parameter tuning [8]. Conjugate gradient descent derives from gradient descent, but instead of using gradient descent, it applies conjugate directions in the process of optimization [9].

8.1.1.4 Loss Function and Measurement

There are several metrics in the evaluation of recommender system [10, 11]: accuracy, coverage and diversity. Accuracy is the most important metric in a recommender system. A recommender system for top-k will give top k items to users in a sequence. There are multiple ways to measure its accuracy, for example:

$$precision@k = \frac{\left|T_{clicked} \cap T_{K,recommended}\right|}{K} \tag{8.2}$$

where $T_{clicked}$ is the items have been clicked in the test set for a user, $T_{K,\ recommended}$ is the k items recommended to a user.

If the rank of recommendation items is the major concern, the metric *ap @ k* will be used (see Eq. (8.3)).

$$ap@k = \sum_{n=1}^{k} \frac{P(n)}{\min(m, k)} \tag{8.3}$$

where $p(n)$ denotes the precision at the *n*th item in the item list.

For *ap @ k* metric, the same recommendation with different ranks will give different evaluations. For example, if user bought 3 items, follows recommended item #1 and #3, then *ap@10* = (1/1 + 2/3)/3 ≃ 0.56. For the same recommendation list, if user follows item #1 and #10, then *ap@10* = (1/1 + 2/10)/3 = 0.4.

The metrics are applied to evaluate the difference between estimated and actual purchase time, which are $\hat{y_u^c}$ and y_u^c. This section uses the mean absolute error (MAE), the root means square error (RMSE) and the mean absolute percentage error (MAPE) to evaluate the overall effect. The definitions are shown in (8.4), (8.5) and (8.6).

$$\text{MAE} = \frac{1}{N_u \times N_c} \sum_c \sum_u \left| y_u^c - \hat{y_u^c} \right| \tag{8.4}$$

$$\text{RMSE} = \sqrt{\frac{\sum_c \sum_u \left(y_u^c - \hat{y_u^c} \right)^2}{N_u \times N_c}} \tag{8.5}$$

$$\text{MAPE} = \frac{1}{N_u \times N_c} \sum_{1}^{N_u \times N_c} \left| \frac{y_u^c - \hat{y_u^c}}{y_u^c} \right| \tag{8.6}$$

8.1.2 Trigger and Triggered Model

The goal of recommender systems is not only to satisfy customers but also to meet the demands of marketing. From the view of marketing, the accuracy along with the history data cannot be the only measurement, the goal of marketing is the other important metric. In addition, the other contribution in this method is to effectively connect experts' knowledge with algorithm. Unlike traditional recommender systems, Trigger and Triggered (TT) model concerns more on concentration elimination, marketing optimization and lifecycle of trigger and triggered products. The workflow moves from product to personalized granularity through two independent algorithms: TT_PAR and TT_PPE. TT_PAR is responsible for the generation of meaningful trigger and triggered pairs, and TT_PPE is for the lifecycle

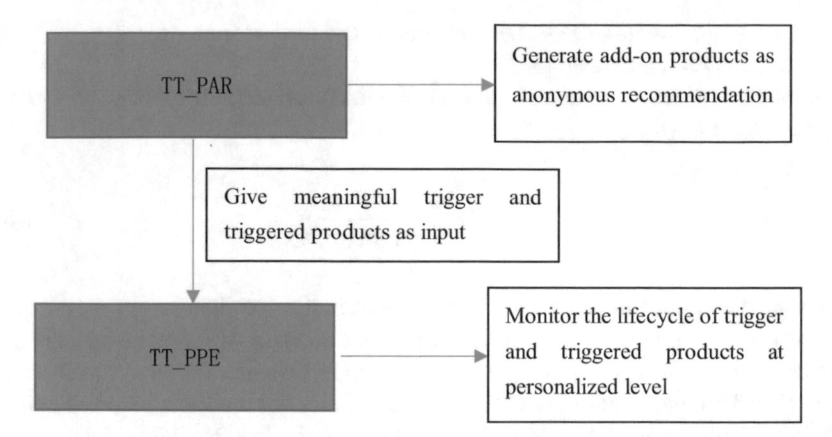

Fig. 8.1 The workflow of TT_PAR and TT_PPE algorithms

of sequential purchases. The relationship between these two algorithms is shown in Fig. 8.1.

The algorithm TT_PAR handles eliminating concentration noise and maximizing marketing goals quite well. Trigger and triggered pairs happened in short timeframe can be treated directly as an anonymous add-on recommendation, and positive results are shown in the experiment. The TT_PPE algorithm takes trigger and triggered pairs from each customer as inputs. By combining with customers' activities and demographic information, the lifecycles of trigger and triggered products are estimated, which can be used for both onsite and offsite recommendation and promotions. The TT_PPE algorithm provides more accurate prediction on lifecycle of product pairs by comparing with other practical algorithms.

8.1.2.1 Meaningful Trigger and Triggered Pairs

When an algorithm is designed to auto recommend products, accuracy is always the most important criterion. However, focusing on accuracy alone may lead to the phenomenon of "filter bubbles", which means hot products become more popular but other options may decrease their exposure to customers as a result of sale diversity diminishing. For example, if a laptop is the most popular item in an e-commerce site and the popularity reaches to a critical point, the laptop might be treated as first recommendation option from the view of accuracy, no matter whatever the customer bought last time. This problem is even more serious when a store is featured by a certain type of products.

On the other hand, from the marketing perspective, accuracy is the primary objective. For instance, when the sales of refrigerator inexplicably decline, the recommender system should transform its role as a promotor to locate potential customers and increasing exposure of refrigerators.

Table 8.1 The example of a customer's consumption records

Date	Product id
2018-01-02	1
2018-01-02	2
2018-01-03	1
2018-01-10	3
2018-01-10	1
2018-01-10	4
2018-01-13	3
2018-01-20	3

Therefore, extracting meaningful trigger and triggered pairs in the history is important. For example, in a dataset, "iPhone–iPhone case" and "iPhone–bed" could be two pairs with the same numbers of co-purchase records. To deliver qualified results, the meaningless pair "iPhone–bed" should be excluded. This section illustrates the proposed method to run pair cleaning. The cleaned pairs can be used independently as an anonymous recommendation or combining with other information for lifecycle prediction.

8.1.2.2 Transformation of Trigger and Triggered Pairs

Trigger-Triggered (TT) pairs are product pairs of a user (at categorical level) purchased at a different time. An example of an original transaction record for user u is shown in Table 8.1.

In a store, a customer normally buys multiple products at the same day, and the same product items or new items will appear in future consumption. As mentioned above, "the same product" purchase is defined as a repetitive purchase and "new item" purchase is defined as a complementary purchase.

In general, the TT-pairs are created between purchases at a different time.

8.1.2.3 Extract Meaningful Pairs

It is necessary to design an algorithm to decrease the bias in triggered items. A TT-Paris filtering method is proposed to solve the problem. It starts with a reverted ranking approach, and applies second-order mining (a post-stage of data mining projects in which humans collectively make judgments on data mining models' performance.) [12, 13] to find meaningful pairs that are most important to the marketing. The tt-pairs filtering consists of two steps.

Algorithm 8.1 *TT*-Pairs Filtering

Step1: obtain reverted ranking using formula (8.7)

Step2: second-order mining by experts

Below we explain these two steps in detail.

Step 1: Rank the *tt*-pairs via *tt*-score, and extract *k* pairs with the highest grades.

The purpose of *tt*-score is to diminish the concentration bias in triggereds. *TT*-pairs are ranked by *tt*-score instead of numbers of occurrence. The *tt*-score is calculated as follows:

$$\begin{aligned} tt_{score} = nor\Big(&O_{trigger_1 triggered_j}, O_{trigger_2 triggered_j}, \ldots \\ &, O_{trigger_n triggered_j}\Big) \\ nor(\cdot) &= \frac{x - x_{min}}{x_{\max} - x_{min} + \delta} \end{aligned} \tag{8.7}$$

where O denotes occurrences and δ is a constant value used to increase weights of a list with higher occurrences.

Step 2: Experts' opinions about the best products.

To collect experts' opinions about the important products, a filtering and grading system is designed for marketing experts. The experts are required to filter the most important pairs and to grade the products.

In this system, once a category code is selected, its triggered candidates are shown in the following two sections: "Add On Products" and "Products in Selection Pool". The first section is what the expert chooses as the most important triggers and the second section includes all candidates that can be added to the selected section.

For grading system, the marketing experts will give their weights to three components: margin, quantity and price for each product (in categorical level), and observe the correctness of their grades. The final scores for products are published after their adjustment.

The formula of calculating score is shown in (8.8):

$$p_{score} = \alpha \times margin + \beta \times quantity + \gamma \times price \tag{8.8}$$

The product score shows the importance of products. Products with p_{score} higher than a threshold will be selected. Finally, part of the top k *tt*-pairs extracted from step 1 will be excluded if their p_{score} is less than the expert-defined threshold.

8.1.3 Trigger-Triggered Model for the Anonymous Recommendation

The anonymous recommendation should match three basic requirements: (1) reduce concentration bias in the dataset and reflect the logic correlation between products, (2) help the marketing to promote products, (3) guarantee the variety of products can be shown on the site.

Fig. 8.2 Scoring matrix

		Group1		Group2		
		SKU1	SKU2	SKU1	SKU2	SKU3
Group1	SKU2	3	4	3	none	none
	SKU1	3	5	3	none	none
Group2	SKU3	4	5	none	2	none
	SKU2	5	none	none	5	none
	SKU1	3	5	none	5	none

To realize these goals, we use both the information of the category and SKU level of products. The categorical connections between *tt*-pairs are generated based on *tt*-scores and experts' selection which is described in Sect. 8.1.1. The application of tt-scores and experts' inputs significantly reduce the problem of concentration bias. At the product level, linear programming is applied to find the right SKU pairs.

In Fig. 8.2, the matrix is the correlation score of products (tt-pairs). The size of the matrix is constrained by the generated group pair, and only relative SKUs in the groups can be connected. The scores in the matrix derive from the historical transactions. The formula of the score is shown in (8.9):

$$ppScore = CO - Occurrence + \alpha \times p_{score} \tag{8.9}$$

where $CO - Occurrence$ is the number of co-occurrence of these two products in the last 1 year. The definition of co-occurrence is that the trigger and triggered products have been bought at the same time or the triggered products have been bought within 30 days after trigger products bought firstly. The linear programming problem (shown in (8.10)) is aimed to choose the best scores among all pairs while satisfying the goals of marketing promotion and recommendation variety will be reached.

$$\begin{aligned} &\text{maximize} \quad Z = ppScore_{ij} \times \beta_{ij} \\ &\text{subject to} \begin{cases} \beta_{ij} = 0 \text{ or } 1 \\ \sum_{i=1}^{n} \beta_{ij}^{c} \leq \varphi_j \\ \sum_{j=1}^{n} \beta_{ij}^{c} \leq \omega_i \\ \sum_{js\ is\ marketing\ promtion} \beta_{ij}^{c} = \delta_j \end{cases} \end{aligned} \tag{8.10}$$

where φ and ω are both a constant non-negative integer value, and δ is a constant positive integer value. The parameter φ is used to constrain the maximum numbers of products in each triggered category, that improves the variety of recommendation. The parameter ω is used to constrain the maximum numbers of a product can be shown as a triggered recommendation in the recommendation list. The value restricts over-recommendation on popular items. The parameter δ is used to promote specific products, makes sure the products can be shown more frequent at recommendation list.

8.1.4 Trigger-Triggered Model for Product Promotion

Managing the lifecycle of consumption is a key to the success of customer engagement. Appropriate advertising should be sent to customers at the appropriate time in order to stimulate customers' potential consumption. For example, a good promotion plan should send customers a new iPhone promotion 1 or 2 year after the consumption of an iPhone, or a case promotion should be sent much earlier once the customer bought a phone. A product trigger and triggered system has been designed to track lifecycle of products at individual level, which can be used for product promotion by real time recommendation system or digital advertising through email or ad platforms. The principle of the system design is that once a certain product has been purchased, then the tracking system is activated, and related products will be sent to customers at right time if the products have not been purchased yet at that time.

To study the time frame between two sequential purchase activities in personalized level, it is to estimate the probability of a user's consumption at a time interval $(j, j + \Delta t)$. That is the conditional probability p(T $\in$ $(j, j + \Delta t|product\_u\_t)$): if the consumer u is going to buy product i in the future, what is the most likely time the consumer will buy. By examining the dataset, it is shown that the probability distribution of p(T $\in$ $(j, j + \Delta t|product\_u\_t)$) is usually a long tail with left or right peak. Therefore, a normal distribution may not be a good candidate to simulate it. Alternatively, a Weibull distribution has been applied. Wang and Zhang [14] have applied Weibull distribution to predict the time frame in ecommerce application. Different from that work, in the section we use a Weibull model with three parameters to predict the time frame. Threshold parameter is included to deal with the case that some triggered items normally are not purchased immediately after the consumption of trigger products. In addition, we use gradient descent approach to tune the parameters instead of variational inference proposed in [14]. Even though gradient descent inference takes more time to locate a minimum, it will be easier to derive the algorithm. The probability density function of a Weibull model with random variable x is shown in (8.11).

As indicated in the formula, the scale parameter β is transformed to be a linear function of variables $\beta^T X$, where X is a vector of variables to capture signals of purchase, including a binary value indicating if the customer bought any same product or similar products in time bins $t_1,\ t_2 \ldots t_m$, or if triggereds have been purchased during promotion dates, or seasonality information, and etc. To make sure the derived scale parameter $\beta_1^T X > 0$, let's transform $\beta_1^T X$ to be $e^{\beta_1^T X}$. The derived density equation is:

$$f(y) = \frac{\alpha}{e^{\beta_1^T x}} \left(\frac{y - \mu}{e^{\beta_1^T x}} \right)^{\alpha - 1} \exp\left(- \left(\frac{y - \mu}{e^{\beta_1^T x}} \right)^{\alpha} \right) \tag{8.11}$$

For each ith observation at each specific tt-pair c, the density function of purchase time is shown in (8.12):

$$p\left(y_i^c | X_i^c, \alpha^c, \beta^c, \mu^c\right) = \frac{\alpha^c}{e^{\beta_1^{cT} x^c}} \left(\frac{y_i^c - \mu^c}{e^{\beta_1^{cT} x^c}} \right)^{\alpha^c - 1} \exp\left(- \left(\frac{y_i^c - \mu^c}{e^{\beta_1^{cT} x^c}} \right)^{\alpha^c} \right) \tag{8.12}$$

The distributions of parameters are: $\alpha^c \sim N(\mu_\alpha, \delta_\alpha)$, $\mu^c \sim N(\mu_\mu, \delta_\mu)$, $\beta^c \sim N(\mu_\beta, \delta_\beta)$. It is denoted that $\omega = (\mu_a, \delta_a, \mu_\mu, \delta_\mu, \mu_\beta, \sum_\beta)$.

To solve the equation, we build a model separately for each product group m, where group refers to products at a particular categorical level. Therefore, in each group, the parameters are $\varphi = (\{\alpha^1, \mu^1, \beta^1\}, \{\alpha^2, \mu^2, \beta^2\}, \ldots, \{\alpha^m, \mu^m, \beta^m\})$. By grouping pairs, the purchase signal of similar products in a group is used. The joint likelihood for all variables is extended, as shown in (8.13):

$$\begin{aligned} L\left(\varphi | D_g\right) \propto L\left(D_g, \varphi\right) = p\left(\omega\right) \prod_{c=1}^{m} p\left(\alpha^c, \mu^c, \beta^c | \omega\right) \prod_{i=1}^{n_m} p\Big(\\ y_i^c \mid \alpha^c, \mu^c, \beta^c, X_i^c \Big) \end{aligned} \tag{8.13}$$

Since the model contains many variables, the traditional method is computationally too expensive to get an answer. Instead, functions of parameters are replaced as constant c_i and MLE is used to estimate parameters, as shown in (8.14):

$$\begin{aligned} \varphi = \left(\left\{ \hat{\alpha^1}, \hat{\mu^1}, \hat{\beta^1} \right\}, \left\{ \hat{\alpha^2}, \hat{\mu^2}, \hat{\beta^2} \right\}, \ldots, \left\{ \hat{\alpha^m}, \hat{\mu^m}, \hat{\beta^m} \right\} \right) \\ = \operatorname{argmax} \left\{ L\left(\text{Data}, \varphi\right) \right\} \\ = \operatorname{argmin} \Bigg\{ \sum_{c=1}^{m} \left(c_1 \alpha^{c2} + c_2 \mu^{c2} + c_3 \beta^{c2} \right) + \\ \sum_{c=1}^{m} \sum_{i=1}^{n_m} - \log\left(p\left(y_i^c | X_i^c, \alpha^c, \beta^c, \mu^c \right) \right) \Bigg\} \end{aligned} \tag{8.14}$$

The pseudocode to solve the previous equation is shown in the following Algorithm 8.2. The parameters $(\mu_a, \mu_\mu, \mu_\beta)$ are initialized in the beginning. It is hidden parameters for grouping. Then Step 1 in the algorithm is to get a local minimal value of parameters φ, that is $(\{\alpha^1, \mu^1, \beta^1\}, \{\alpha^2, \mu^2, \beta^2\}, \ldots, \{\alpha^m, \mu^m, \beta^m\})$. On the basis of φ, parameters $(\mu_a, \mu_\mu, \mu_\beta)$ are updated. These two iteration steps continue until converge.

Algorithm 8.2 Trigger-Triggered Model

Initialize $(\mu_a, \mu_\mu, \mu_\beta) = \left(\mu_\alpha^0, \mu_\mu^0, \mu_\beta^0\right)$
$i = 0$
repeat
for *tt*-pair $c = (1, 2, \ldots, m)$ in group:
updating parameters $(\alpha^m, \mu^m, \beta^m)$ in (8.15) on the basis of known $\left(\mu_\alpha^i, \mu_\mu^i, \mu_\beta^i\right)$
end for
$i = i + 1$
updating parameters $\left(\mu_\alpha^i, \mu_\mu^i, \mu_\beta^i\right)$ in (8.16) based on known $(\alpha^m, \mu^m, \beta^m)$
end repeat (convergence)

$$
\begin{aligned}
(\alpha^m, \mu^m, \beta^m) = argmin\Big\{ & c_1\alpha^c - \mu_a{}^2 + c_2\mu^c - \mu_\mu{}^2 \\
& + c_3\beta^c - \mu_\beta{}^2 + \sum_{i=1}^{n_m} - \log\left(p\left(y_i^c | X_i^c, \alpha^c, \beta^c, \mu^c\right)\right) \Big\}
\end{aligned}
\tag{8.15}
$$

$$
\begin{aligned}
(\mu_a, \mu_\mu, \mu_\beta) = argmin\Big\{ & c_1\alpha^c - \mu_a{}^2 + c_2\mu^c - \mu_\mu{}^2 \\
& + c_3\beta^c - \mu_\beta{}^2 + c_4\mu_a{}^2 + c_5\mu_\mu{}^2 \\
& + \mu_\beta{}^2 \sum_{i=1}^{n_m} - \log\left(p\left(y_i^c | X_i^c, \alpha^c, \beta^c, \mu^c\right)\right) \Big\}
\end{aligned}
\tag{8.16}
$$

Formulas (8.15) and (8.16) are hierarchical expression of (8.14) where prior knowledge are applied, two steps of inference improve the stability of prediction.

The updating methods at here can be algorithms such as Conjugate Gradient Descent, Broyden-Fletcher-Goldfarb-Shanno or others [7]. The optimum function [15] in R has been applied for parameters inference.

The experimental study of using a transaction dataset is collected from a retail store can be found in [1].

8.2 Advertisement Clicking Prediction by Using Multiple Criteria Mathematical Programming

This section proposes a multi-criteria linear regression (MCLR) [16] and kernel-based multiple criteria regression (KMCR) [17, 18] algorithms to predict CTR of ads in a web search engine given its logs in the past.

8.2.1 Research Background of Behavioral Targeting

8.2.1.1 Concept of Click-Through Rate

This section provides the concept of click-through rate (CTR). For example: As Fig. 8.3 showing, when one user views a known Chinese website 163.com, which has an ad slot that can display an advertisement. For the media 163.com, the problem is: Which advertiser should be chosen for this ad slot? The answer can be: Choose the one with max revenue. The definition is:

$$\text{Max } revenue = \text{Max}\{CPC_i \times P\_CTR_i\} \tag{8.17}$$

where i represents the ith of the advertiser; *CPC* is the Cost per Click [The money paid to the media when one ad is clicked, set by advertiser]; and P_CTR is the Prediction CTR [Expected CTR, is given by prediction model, the predict target in this section].

Note that the CTR model is very important for the platform to keep their revenue maximum. Some machine learning based regression algorithms such as logistic regression [19], maximum entropy [20], support vector regression (SVR) [21] and conditional random field (CRF) [22] have been adopted to predict the clicks of advertisements presented for a query.

In this section, the proposed multi-criteria linear regression (MCLR) and kernel-based multiple criteria regression (KMCR) algorithms will be used for CTR prediction. Note that the regression models for CTR problems are different from classification models because the former do not need the testing process for verification while the later do. However, the clicking events prediction needs classification models as introduced below.

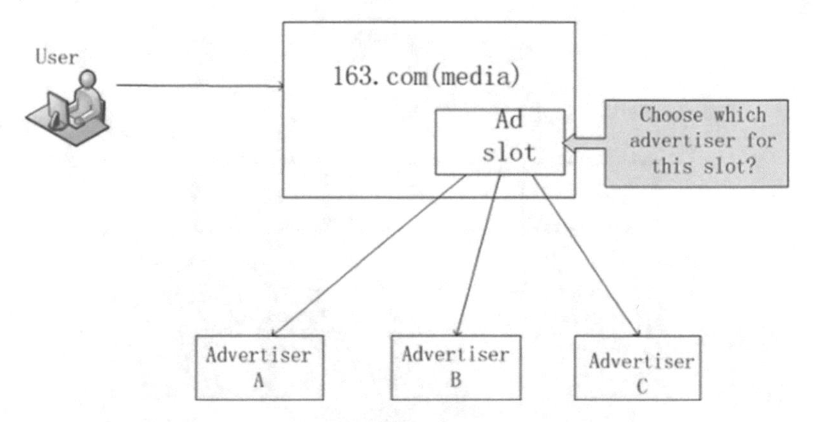

Fig. 8.3 Application scenarios of CTR

8.2.1.2 Concept of Clicking Events Prediction

For the advertiser with several candidate advertisements has the chance to display their advertisements on the ad slot, he needs to decide which ad to display. Under this scenario, a clicking events prediction model is needed to solve this problem. Through model prediction, the advertiser can learn about which ad will be clicked or not, then he can choose displaying ad that will be clicked for a good revenue. Figure 8.4 shows the application scenario of Clicking Events Prediction.

8.2.2 *Feature Creation and Selection*

To show the practical ability of the proposed method, the datasets of track2 of the KDD Cup 2012 are used for testing (http://www.kddcup2012.org/). The training set contains 155,750,158 instances that are derived from log message of search sessions, where a search session refers to an interaction between a user and the search engine. During each session, the user can be impressed with multiple ads, then, the same ads under the same setting (such as position, depth) from multiple sessions are aggregated to make an instance in the datasets. Each instance can be viewed as a vector (#click, #impression, DisplayURL, AdID, AdvertiserID, Depth, Position, QueryID, KeywordID, TitleID, DescriptionID, UserID). It means that under a specific setting, the user (UserID) has been impressed with the ad (AdID) for #impression times, and has clicked #click times of those. In addition to the instances,

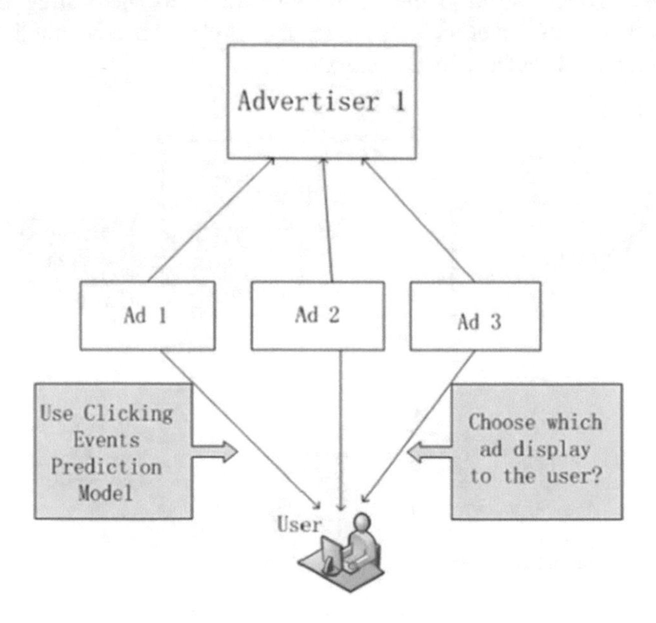

Fig. 8.4 Application scenario of Clicking Events Prediction

the datasets also contain token lists of query, keyword, title and description, where a token is a word represented by its hash value. The gender and segmented age information of each user are also provided in the dataset. The test set contains 20,297,594 instances and shares the same format as the training set, except for the lack of #click and #impression. The test set is generated with log messages that come from sessions latter than those of the training set. More detailed information about the datasets can be found in (http://www.kddcup2012.org/).

However, a major challenge is to create efficient features. Feature creation and selection are the most important steps in solving a supervised learning problem. After comparing different methods, this section chooses two of them to create the features, which are called T-Set-1 and T-Set-2, respectively:

8.2.2.1 Feature Creation Method for T-Set-1

In T-Set-1, the bag of words model was used. This method is frequency-based method that is used to predict the probability of each presented word on a clicked instance based on each feature (tokens). Then, we built the whole feature space by combining the query dictionary and ad dictionary.

8.2.2.2 Feature Creation Method for T-Set-2

Two kinds of features, original feature and synthetic feature, are used for modeling in this section:

1. Original Features: The original feature set contains discrete features and continuous features. The discrete features are the unique ID of each ad, advertiser, query, keyword, tile, description, token, gender and age for one user, depth and position of ads, and the displayed URL. The continuous features are the click-through rates of each value of the discrete features. When a discrete feature is being used; the corresponding click-through rate will be activated and adopted as a continuous feature.
2. Synthetic Feature: First of all, we join any two original discrete features with each other and use them as synthetic features. We also test some 3-tuple features but only the QueryID_AdID_UserID is available. Since most 3-tuple features are too sparse and seldom activated. Secondly, we join the original discrete features with each of the tokens. Position information is added to the original discrete features to generate one 2-tuple position-based feature. Bigram features are also adopted for analyzing the queries, titles and descriptions.

8.2.2.3 Normalization

Since the ranges of all the variables' value are significantly different, a linear scaling transformation needs to be performed for each variable. The transformation expresses as:

$$x_n = \frac{x_i - \min(x_1, K, x_n)}{\max(x_1, K, x_n) - \min(x_1, K, x_n)} \tag{8.18}$$

where x_n is the normalized value and x_i is the instance value.

8.2.2.4 Categorization Method for Positive/Negative Samples

To analyze the dataset for predicting accurately, let's consider:

Advertisement 1: The time of display is 10, the time of click is 0.
Advertisement 2: The time of display is 10, the time of click is 1.
Advertisement 3: The time of display is 10, the time of click is 8.

From above, it can be seen that the gap between advertisements 2 and 3 is bigger than the gap between advertisements 1 and 2. If we simply categorize those samples based on click times, those with click times greater than 1 are categorized as positive samples and those less than 1 as negative samples, then the advertisement 2 and 3 are both labeled as 1 while the label of advertisement 1 is −1. In this situation, the influence of advertisement 2 and 3 are the same. However, as the time of click 0 and 1 is closer than 1 and 8, it is not reasonable. Therefore, we treat the click-through-rate as a probability problem. For one wonderful advertisement, someone will click it while others won't. Therefore, in this section, we calculate the click-through-rate (CTR) of each instance, and the average CTR. Then we compare each instance's CTR with the average CTR. If it is greater than the average CTR, the label of the instance should be 1; otherwise, it should be 0. The formula to calculate the CTR is described as below:

$$Click - Through - Rate(CTR) = (\#click + \alpha * \beta) / (\#impression + \beta) \tag{8.19}$$

where $\alpha = 0.05$, $\beta = 75$, that obtained from the experiment.

8.2.2.5 Confusion Matrix

The confusion matrix is used for the performance analysis:

TP (True Positive)	the number of records in the first class that has been classified correctly
FP (False Positive)	the number of records in the second class that has been classified into the first class
TN (True Negative)	the number of records in the second class that has been classified correctly
FN (False Negative)	the number of records in the first class that has been classified into the second class

Then four different performance measures are:

$$\begin{aligned} Specificity &= \frac{TN}{TN+FP}; \\ Sensitivity &= \frac{TP}{TP+FN}; \\ False\ Positive\ Rate &= \frac{FP}{TN+FP}; \\ False\ Negative\ Rate &= \frac{TN}{FN+TN}. \end{aligned} \tag{8.20}$$

8.2.2.6 Receiver Operating Characteristics (ROC) Graph

Receiver Operating Characteristics (ROC) graph is a useful technique for organizing classifiers and visualizing their performance. ROC graphs are commonly used in decision making, and in recent years have been increasingly adopted in the machine learning and data mining research communities. In addition to a generally useful performance graphing method, they have properties that make them especially useful for domains with skewed class distribution and unequal classification error costs. These characteristics have become increasingly important as research continues into the areas of cost-sensitive learning and learning in the presence of unbalanced classes. The reader can find details of experimental and comparison studies of the MCLR and KMCR regression models as well as the classifications MCLP and KMCP for the clicking events prediction in [2].

8.3 Customer Churn Prediction Based on Feature Clustering and Nonparallel Support Vector Machine

Bank customer churn prediction is one of the key businesses for modern commercial banks. Recently, various methods have been investigated to identify the customers who would leave away. This section presents a framework based on feature clustering and classification technique to help commercial banks make an effective decision on customer churn problem.

8.3.1 Related Work

8.3.1.1 Maximal Information Coefficient

Relationship coefficient is usually used for measuring the similarity of two variables. Person coefficient is one of the most famous relationship metrics, because it is easy to calculate and has a naive explanation. However, only linear relationship can be captured well using this metric when other kinds of dependence work badly such as sin or cubic function. Recently, [23] proposed a novel relationship measure called MIC. Inspired by innovative idea, they show that MIC could capture a wide range of associations both functional and not. Furthermore, the MIC value is roughly equal to the coefficient of determination R^2 in statistics [23]. Now we provide a little introduction to MIC.

Given a finite set D whose elements are two dimensions data points, one dimension is x-values and the other is *y*-values. Suppose *x*-values is divided into *x* bins and *y*-values into y bins, this type of partition is called *x-by-y* grid G. Let $D|_G$ represent the distribution of *D* divided by a *x-by-y* grid *G*. $I_*(D,x,y) = max\, I(D|_G)$, where $I(D|_G)$ is the mutual information of $D|_G$. There are infinite amount of *x-by-y* grids, so there are infinite number of $I(D|_G)$ either. Set the maximal $I(D|_G)$ as $I_*(D,x,y)$. Given different *x*- and *y*-value, a matrix named *characteristic matrix* could be constructed as $M(D)_{x,y} = \frac{I*(D,x,y)}{\log\ \min\{x,y\}}$. Furthermore, MIC can be obtained by $MIC(D) = max_{xy<B(n)}\{M(D)_{x,\,y}\}$, where $B(n)$ is the upper bound of the grid size. The elements in *characteristic matrix* $I_*(D,x,y)$ is chosen from a infinite amount of $I(D|_G)$, thus Reshef et al. develop an approximation algorithm and program for generating characteristic matrix and the estimators derived from MIC [24, 25]. With these state-of-the-art utilities, data exploration could be easily completed before other data mining procedure.

8.3.1.2 Affinity Propagation Clustering

Clustering data through similarity is a popular step in many scientific analysis and application systems. Frey and Dueck [26] developed a modern clustering method named "affinity propagation" (AP) which constructs clusters by information messages exchanged between data points. Given the similarities of each two distinct data points as input, AP algorithm considers all the instance as potential centroids at the beginning of the algorithm. And then, algorithm merges small cluster into bigger ones step by step. Different from classical clustering algorithms like k-means, each instance is regarded as one node in a network by AP clustering approach. Messages was transmitted between nodes, so each data point reconsidered their situation through new information and properly modified the cluster they belong to. This procedure went on until a good set of clusters and centroids produced.

In this process, there are mainly two categories of message exchanged between data points. One of them is sent from point *i* to point *j* which formulated as $r(i,j)$.

It illustrates the strength point i choosing point j as its centroid. The other sort information is from point j to point i as $a(i,j)$. It shows the confidence that one point j recommends itself as the centroid of another point i. And the author of AP take $r(i,j) \leftarrow s(i,j) - max_{j' \, s.\,t.\,j' \neq j}\{a(i,j') + s(i,j')\}$ and $a\,(i,\,j) \leftarrow \min\{0, r\,(k,k)\} + \sum_{i' s.t. i' \neq i,j} \max\left\{0, r\left(i', k\right)\right\}\}$ to update current situation. Update is needed only for the pairs of points whose similarities are already known. This trait makes the algorithm much faster than other methods. To identify the centroid of point i, point j that maximizes $r(i,j) + a(i,j)$ should be considered during each iteration. AP clustering method requires a similarity matrix s as input, and the element of the matrix $s(i,j)$ provides the distance from point i to point j. In addition, the diagonal values of the matrix is not assigned 1 as usual. These values are called "preference" which show how point i is likely to be chosen as a centroid. That is to say, the larger $s(i,i)$ is, the more probability that point i play a role of a centroid. Obviously, $s(i,i)$ are key parameters which control the number of final clusters by AP method.

8.3.1.3 Nonparallel Support Vector Machine

Optimization has a long history for discovering valuable patterns and making decisions [27]. A lot of approaches have been investigated based on optimization techniques such as SVM [28], multiple criteria linear programming (MCLP) [29], etc. SVM is a serial of modern methods for data mining and pattern recognition including classification, regression and other data analytical task. Based on the theory of statistical learning, SVM use a single hyperplane to construct discriminative model which follow the Structure Risk Minimization (SRM) principle [28]. Recently, inspired by twin SVM [30, 31] (see Chap. 3 as well) proposed a novel classification method called NPSVM. Similar to twin SVM, this method applied two nonparallel hyperplanes to handle the classification problem [31].

$$f_+(x) = \omega_+^T \cdot x + b_+ = 0 \quad and \quad f_-(x) = \omega_-^T \cdot x + b_- = 0 \tag{8.21}$$

Given a binary classification dataset $D = \{(x_1, y_1), (x_2, y_2), \ldots, (x_\ell, y_\ell)\}$ with ℓ instances and n attributes, let $\ell_+ + \ell_- = \ell$, ℓ_+ positive and ℓ_- negative instances. In order to express by matrix, the positive instances were represented by matrix $A \in \Re_{\ell_+ \times n}$. Each row of matrix A is one instance. The negatives were expressed by matrix $B \in \Re_{\ell_- \times n}$. So the primal optimization problems for NPSVM are

$$\begin{aligned} \min_{\omega_+, b_+, \eta_+, \xi_-} \quad & \tfrac{1}{2}\eta_+^T\eta_+ + c_1 e_-^T \xi_- + \tfrac{1}{2}c_2\left(\| \omega_+ \|_2^2 + b_+^2\right) \\ s.t. \quad & -\left(B\omega_+ + b_+ e_-\right) + \xi_{\geq e_-} \geq e_-, \\ & A\omega_+ + b_+ e_+ = \eta_+, \\ & \xi_- \geq 0 \end{aligned} \tag{8.22}$$

and

$$\begin{aligned}
\min_{\omega_-, b_-, \eta_-, \xi_\mp} \quad & \tfrac{1}{2}\eta_-^T\eta_- + c_3 e_+^T \xi_\mp + \tfrac{1}{2}c_4 \left(\| \omega_- \|_2^2 + b_-^2\right) \\
s.t. \quad & A\omega_- + b_- e_+ + \xi_+ = e_+, \\
& B\omega_- + b_- e_- = \eta_-, \\
& \xi_+ \geq 0
\end{aligned} \tag{8.23}$$

where c_1, c_2 are the model parameters, and e_+ and e_- are the vector of ones with proper dimensions. For each hyperplane in 1, NPSVM try to make the instances of one category close to this hyperplane, and the distance between the instance of the other class and the hyperplane is more than 1 at least. The Wolf Dual problem for Eqs. (8.22) and (8.23) could be expressed as

$$\begin{aligned}
& \min_{\alpha_1, \alpha_3} \tfrac{1}{2} \left(\alpha_1^T, \alpha_3^T\right) Q \left(\alpha_1^T, \alpha_3^T\right)^T - c_2 e^T \alpha_1 \\
& \quad s.t. \;\; 0 \leq \alpha_1 \leq c_1 e,
\end{aligned} \tag{8.24}$$

where

$$Q = \begin{bmatrix} BB^T & BA^T \\ AB^T & AA^T + c_2 I \end{bmatrix} + E \tag{8.25}$$

and

$$\begin{aligned}
& \min_{\alpha_1', \alpha_3'} \tfrac{1}{2} \left(\alpha_1'^T, \alpha_3'^T\right) Q \left(\alpha_1'^T, \alpha_3'^T\right)^T - c_4 e^T \alpha' \\
& \quad s.t. \;\; 0 \leq \alpha_3' \leq c_3 e,
\end{aligned} \tag{8.26}$$

where

$$Q = \begin{bmatrix} AA^T & -AB^T \\ -BA^T & BB^T + c_4 I \end{bmatrix} + \begin{bmatrix} E & -E \\ -E & E \end{bmatrix} \tag{8.27}$$

The final decision could be made by comparing the distance to these two hyperplanes, respectively. The distance could be obtained from

$$\begin{aligned}
f_+(x) &= \omega_+ \cdot x + b_+ \\
&= -\tfrac{1}{c_2}\left(B^T\alpha_1 + A^T\alpha_3\right) \cdot x \\
&\quad - \tfrac{1}{c_2}\left(e^T\alpha_1 + e_+^T\alpha_3\right)
\end{aligned} \tag{8.28}$$

and

$$
\begin{aligned}
f_{-}(x) &= \omega_{-} \cdot x + b_{-} \\
&= -\tfrac{1}{c_4}\left(-A^T \alpha_1' + B^T \alpha_3'\right) \cdot x \\
&\quad - \tfrac{1}{c_4}\left(-e_{-}^T \alpha_1' + e_{+}^T \alpha_3'\right)
\end{aligned}
\tag{8.29}
$$

Once the distances has been obtained, for a new customer $x_j \in \Re^n$, the attrition prediction could be obtained according to the closer hyperplane in Eq. (8.21), such as

$$
f(x) = argmin \left| f_{\pm}\left(x_j\right) \right| = argmin \left| \omega_{\pm}^T \cdot x_j + b_{\pm} \right|,
\tag{8.30}
$$

where $|\cdot|$ means the perpendicular distance from point x_j to hyperplane $\omega_{\pm}^T \cdot x + b_{\pm} = 0$.

8.3.2 Customer Churn Prediction with NPSVM

Missing items in data would produce big problems for calculation. Usually, the missing elements are filled by some fixed real number which is easily distinguishable. Another way to process missing values is to remove the features that the ratio of missing items is higher than certain threshold (like 30%). However, this kind of operations may be subjective and not suitable.

Instead of directly deleting features, feature selection strategy is applied. Furthermore, to eliminate useless descriptors, it focuses on the relationship and missing ratio among features. Pairwise relationship between features are applied to reduce the calculation problem from the missing value. The MIC relationship measure are calculated through the available values at the same instance for each pair of features, e.g., there are five customers with two features, "-" represents missing items.

The values for the first feature are {1, 3, 7, 9, -}, and for the other are {-, 2, 6, 7, -}. So items {3, 7, 9} and {2, 6, 7} are the available values that could be used for the MIC calculation. Thus, even there are numerous missing items for some feature, the relationship between two features could also been obtained.

In order to combine feature relationship and the missing ratio together for feature filter, a new measure is defined as:

$$
Preference(i) = \text{Max}(MICValue) + \frac{\lambda}{\log\left(MissingRatio(i)\right) + \epsilon},
\tag{8.31}
$$

where ϵ is a very small real number and λ is the parameter which is also a real number. Equation (8.31) provides a new preference measure which balances the consideration between missing ratio and relationship of features. On the left part of Eq. (8.31), *Max* (*MICValue*) is the maximum of all the MIC values among each pair of features. The right part represents the missing ratio of feature i. When affinity propagation algorithm takes this measure as parameter, as it is showed in Ref. 1, the larger the preference parameters are, the more probability an instance tended to become the center of clusters. That means when the missing ratio for feature i is too large, the corresponding value would be much smaller than the other. As a result, feature i would has less probability to be chosen as centroid. Finally, only the centroid features are preserved as the selected features.

Based on these chosen data, two hyperplane for churn and not churn customer could be constructed according to NPSVM model (8.24) and (8.26). In intuition, each hyperplane represents one category of customers. Once the two hyperplanes have been achieved, different decision functions could be obtained by providing different weights for the two distances $f_{-}(x)$ and $f_{+}(x)$. The parameter μ balances the two distances after the model construction. This characteristic provides the capability to make further adjustments when the preference has changed. The advantage is that the reconstruction and calculation for the model does not need any more. Once the two hyperplanes have been received, the further adjustments could be achieved for giving different μ values and recalculating only Eq. (8.30). The detail of the procedure could be found in Algorithm 8.3.

Algorithm 8.3 MICAP-NPSVM Customer Churn Prediction Framework

Input:

Customer churn training dataset $\mathcal{D} = \{\Omega, C\}$, $\Omega = (f_1, f_2, \ldots, f_n)$. The missing ratio vector

MissingRatio for all the features. Parameters c_1, c_2, c_3, c_4, λ, μ, ϵ.

Output:

Customer churn prediction function $F(x)$.

1: **Begin**

2: $\Omega^{'} = \varnothing$;

3: **for all** $f_i, f_j \in \Omega, i \neq j$ **do**

4: $\left(f_i', f_j'\right) = FilterMissingItems\left(f_i, f_j\right)$;

5: $M\left(i, j\right) = CalculateMIC\left(f_i', f_j'\right)$;

6: **end for**

7: $M\left(i, i\right) = \text{Max}(M) + \frac{\lambda}{\log(MissingRatio(i))+\epsilon}$;

8: $\mathrm{Y} = \{Cluster_1, Cluster_2, \ldots, Cluster_\ell, \ \} = \text{APClustering(M)}$;

9: **for all** $Cluster_k \in \mathrm{Y}$ **do**

10: $I = ClusterCentroid(Cluster_k), I \in Cluster_k$;

11: $\Omega^{'} = \Omega^{'} \cup I$

12: **end for**

13: Constructing new dataset $\mathcal{D}'$ according to $\Omega^{'}$;

14: Generating churn customer matrix A and not churn customer matrix B from the new dataset $\mathcal{D}'$;

15: Construct and solve the optimization problem (8.24) and (8.26);

16: Calculating the distance from instance x to the two hyperplanes $f_{-}(x)$ and $f_{-}(x)$ by Eqs. (8.28) and (8.29);

17: Obtain decision function by $F(x) = abs(f_{-}(x)) - \mu \cdot abs(f_{+}(x))$.

18: **End**

The input of the method is the original dataset of customer churn and the model parameters. In practice, the parameter of c_1 to c_4 could be arranged as $c_1 = c_3$, $\lambda = 0.5$, $\mu = 1$, $\epsilon = 0.001$. Line 4 extracts two different features f_i, f_j from the original dataset $\mathcal{D}$, and filters the rows with missing values in either of the two features. Line 5 calculates the MIC based on the result of line 4 and stores the MIC values in the similarity matrix M. According to Eq. (8.31), line 7 combines the maximum of all the MIC values and the missing ratio of each feature as the preference measure. Line 8 applies affinity clustering on the similarity matrix M and obtains feature cluster set Y. Line 9 to 12 selects the centroid of each cluster from Y and generates a new dataset according to the chosen features $\mathcal{D}'$ in line 13. Line 14 extracted the churn customer data as matrix A, not churn customer data as matrix B. Line 15 constructs the optimization problem (8.24) and (8.26) through matrix A and B, and then solves it. Line 16 constructs the two hyperplanes by Eqs. (8.28) and (8.29). Line 17 achieve the customer churn prediction decision function through $F(x) = abs(f_{-}(x)) - \mu \cdot abs(f_{+}(x))$. The detailed data analysis based on a well-known commercial bank of China can be found in [3].

8.4 Node-Coupling Clustering Approaches for Link Prediction

This section provides two novel node-coupling clustering approaches and their extensions for the link prediction problem. They consider the different roles of nodes, and combine the coupling degrees of the common neighbor nodes of a predicted node-pair with cluster geometries of nodes. Our approaches remarkably outperform the existing methods in terms of efficiency accuracy and effectiveness.

8.4.1 Preliminaries

8.4.1.1 Clustering Coefficient

In graph theory, clustering coefficient is a metric that can evaluate the extent to which nodes tend to cluster together in a graph [32]. It can capture the clustering information of nodes in a graph [33]. An undirected network can be described as a graph $G = (V, E)$, where V denotes the set of nodes and E indicates the set of edges. vi 2 V is a node in Graph G. The clustering coefficient of node vi in Graph G can be defined as

$$C(i) = \frac{E_i}{\frac{k_i \cdot (k_i - 1)}{2}} = \frac{2 \cdot E_i}{(k_i \cdot (k_i - 1))} \tag{8.32}$$

where $C(i)$ denotes the clustering coefficient value of node v_i. k_i represents the degree value of node v_i. E_i is the number of the con-nected links among k_i neighbors of node v_i. For example, there is a node v_1 in Graph G. The degree value of node v_1 is 5 (i.e. $k_1 = 5$). The number of connected links among the neighbors of node v_1 is 6 (i.e. $E_1 = 6$). Thus, the clustering coefficient value of node v_1 is: $C(1) = \frac{2 \cdot E_1}{(k_1 \cdot (k_1 - 1))} = \frac{2 \cdot 6}{5 \cdot (5-1)} = 0.6$.

8.4.1.2 Evaluation Metrics

In this section, we present two popular metrics for link prediction accuracy: *Area under curve (AUC)* and *Precision*. In general, a link prediction method can compute a score S_{xy} for each unknown link to evaluate its existence probability and give an ordered list of all unknown links based on these S_{xy} values [34].

It can evaluate the overall performance of a link prediction method. As described in [35, 36], the AUC value can be considered as the probability that the S_{xy} value of an existing yet unknown link is more than that of a non-existing link at random. That is, we randomly select an existing yet unknown link in the test set and compare its score with that of a non-existing link at a time. There are N independent comparisons, where the times that the existing yet unknown links have higher S_{xy} value are H, and the times that they have the same S_{xy} value are E. The *AUC* value is defined as:

$$AUC = \frac{H + 0.5 \cdot E}{N} \tag{8.33}$$

This metric considers N links with the highest S_{xy} values in all unknown links. If there are T existing yet unknown links in the top N unknown links [35, 36], the Precision is defined as:

$$Precision = \frac{T}{N} \tag{8.34}$$

8.4.2 Node-Coupling Clustering Approaches

In this section presents the proposed approaches for link prediction. Firstly, it presents a new node-coupling degree metric—node-coupling clustering coefficient. Then, it discusses the process of our approaches. Finally, the complexity analysis is given.

8.4.2.1 Node-Coupling Clustering Coefficient

Many similarity-based methods only consider the number or degrees of common neighbor nodes of a predicted node-pair in link prediction, and few exploit further the coupling degrees among the common neighbor nodes and the clustering information to improve the prediction accuracy. Based on the above reason, we propose a new node-coupling degree metric—node-coupling clustering coefficient (NCCC), which can capture the clustering information of a network and evaluate the coupling degrees between the common neighbor nodes of a predicted node-pair. It also considers different roles of the common neighbor nodes of a predicted node-pair in a network. Now, we introduce this metric through a simple example.

Figure 8.5 shows an example for predicting the link between nodes M and N in two networks. Two original networks are described in Fig. 8.5a, b. Figure 8.5c, d are two subgraphs that consist of nodes M; N and their common neighbors in Fig. 8.5a, b, respectively. We aim to predict which link between nodes M and N is more likely to exist in Fig. 8.5a, b. In general, we find that the coupling degrees of nodes M; N and their common neighbors are higher in Fig. 8.5c, d. Thus, we believe the link of nodes M; N in Fig. 8.5a is more likely to exist than in Fig. 8.5b. If we apply CN; AA; RA; PA to predict the link of nodes M; N in these two original networks, we can gain the same prediction result for each method. The reasons are as follows: from Fig. 8.5a, b, we can see that the common neighbor set $\{bdf\}$ of nodes *M; N* are the same, and every corresponding common neighbor node has the same degree value in these two original networks. The similarity metric is the number of the common neighbor nodes of a predicted node-pair in *CN*. *CN* has the same prediction results because of the same common neighbor node set $\{bdf\}$ of nodes M; N in these two original networks. *RA; AA* are based on the degree values of the common neighbor nodes. *RA* has the same prediction probability as *AA* because of the same degree value of every corresponding common neighbor node in $\{bdf\}$ in these two original networks. For the same reason, *PA* provides the same prediction result because that there are the corresponding same degree values for nodes *M* and *N* in these two original networks. However, the link probabilities between node *M* and node *N* in Fig. 8.5a, b are not likely to be the same.

In the above case, inspired by [37, 38], we propose a new node-coupling degree metric based on the clustering information and node degree—node-coupling

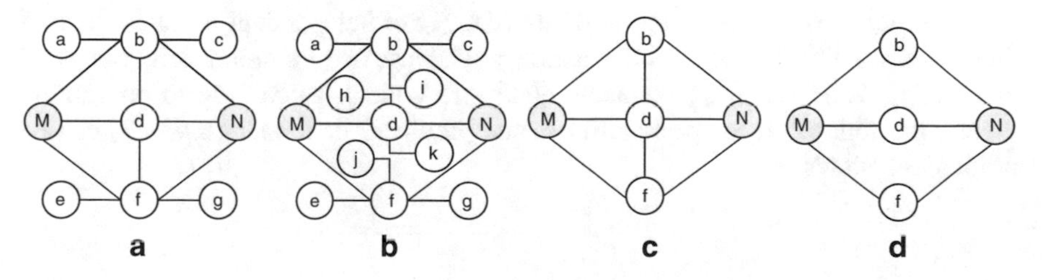

Fig. 8.5 An example for predicting the link between nodes M and N in two original networks

clustering coefficient. This metric cannot only resolve the above prediction problem in Fig. 8.5, but also capture the clustering information of a network. If node n is a com-mon neighbor node of the predicted node-pair (M, N), the node-coupling clustering coefficient of node$n$, $NCCC(n)$, can be defined as follows:

$$NCCC(n) = \frac{\sum_{i \in C_n} \left(\frac{1}{d_i} + C(i) \right)}{\sum_{j \in \Gamma_n} \left(\frac{1}{d_j} + C(j) \right)} \tag{8.35}$$

where Γ_n is the neighbor node set of nodes n. (M, N) denotes a predicted node-pair $n \in \Gamma(M) \cup \Gamma(N)$. C_n denotes the common neighbor node set of the node-pair (M, N) in $\Gamma(N)$, which includes nodes M, N. Namely $C_n = \Gamma(M) \cap \Gamma(N) \cap \Gamma(n) \cup \{M, N\}$. d_i denotes the degree value of node i. $C(i)$ denotes the clustering coefficient of node i. In this metric, $\frac{1}{d_i} + C(i)$ is considered as the contribution of node i to the coupling degree of the common neighbor nodes of the predicted node-pair (M, N). The node-coupling clustering coefficient of node n is the ratio of the contribution sum of all nodes in C_n to that in $\Gamma(n)$. In this way, our approaches can apply this metric that incorporates the clustering information and different roles of each related node to improve the prediction accuracy for link prediction.

In Eq. (8.35), since $C_n \subseteq \Gamma(n)$, $\sum_{i \in C_n} \left(\frac{1}{d_i} + C(i) \right) \leq \sum_{j \in \Gamma_n} \left(\frac{1}{d_j} + C(j) \right)$. As a result, $NCCC(n) \in (0, 1]$. Specially, $NCCC(n) = 1$ when $C_n = \Gamma(n)$.

8.4.2.2 Node-Coupling Clustering Approach Based on Probability Theory (NCCPT)

From probability theory, we propose a new link prediction approach based on the node-coupling clustering coefficient (NCCC) in the previous section. Given a pair of predicted nodes (x, y), node n is a common neighbor node of the node-pair (x, y). $P(n)$ denotes the link existence probability that node x and node y connect because of node n. $\overline{P(n)}$ denotes the link non-existence probability that node n connects node x to node y. Therefore, $P(n) = NCCC(n)$ and $\overline{P(n)} = 1 - NCCC(n)$. $\{A_1, A_2, \ldots, A_i, \ldots, A_m\}$ is the common neighbor set of the predicted node-pair (x, y), namely $\Gamma(x) \cap \Gamma(y) = \{A_1, A_2, \ldots, A_i, \ldots, A_m\}$. We assume that these common neighbor nodes of the node-pair (x, y) are independent to each other. If there exists a link between nodes x and y, at least one common neighbor node in $\{A_1, A_2, \ldots, A_i, \ldots, A_m\}$ connects node x to node y. According to probability theory, the link existence probability of the predicted node-pair (x, y), S_{xy} can be written as follows:

$$
\begin{aligned}
S_{xy} &= 1 - \overline{P\left(A_1\right)} \cdot \overline{P\left(A_2\right)} \cdots \overline{P\left(A_i\right)} \cdots \overline{P\left(A_m\right)} \\
&= 1 - \left(1 - P\left(A_1\right)\right) \cdot \left(1 - P\left(A_2\right)\right) \cdots \left(1 - P\left(A_i\right)\right) \cdots \left(1 - P\left(A_m\right)\right) \\
&= 1 - \left(1 - NCCC\left(A_1\right)\right) \cdot \left(1 - NCCC\left(A_2\right)\right) \cdots \left(1 - NCCC\left(A_i\right)\right) \cdots \\
&\quad \left(1 - NCCC\left(A_m\right)\right) \\
&= 1 - \prod_{n \in \Gamma(x) \cap \Gamma(y)} \left(1 - \frac{\sum_{i \in C_n}\left(\frac{1}{d_i} + C(i)\right)}{\sum_{j \in \Gamma_n}\left(\frac{1}{d_j} + C(j)\right)}\right)
\end{aligned}
\tag{8.36}
$$

Equation (8.36) is a new node similarity metric in our approach. Clearly, a larger value of S_{xy} means a higher probability that there exists a potential link between node x and y. The related parameters in Eq. (8.36) have been described in the last section. In Eq. (8.36), since $NCCC(n) \in (0, 1]$, we will have $\prod_{n \in \Gamma(x) \cap \Gamma(y)} \left(1 - \frac{\sum_{i \in C_n}\left(\frac{1}{d_i} + C(i)\right)}{\sum_{j \in \Gamma_n}\left(\frac{1}{d_j} + C(j)\right)}\right) \in (0, 1]$, and $S_{xy} \in (0, 1]$. Specially, $S_{xy} = 1$ when $C_n = \Gamma(n)$ for every node in $\Gamma(x) \cap \Gamma(y)$. For example, we apply our approach to predict the link probability of nodes *M, N* in Fig. 8.5.

$$
\begin{aligned}
Fig.5(a): S_{MN} &= 1 - \left(1 - \tfrac{2.92}{4.92(b)}\right) \cdot \left(1 - \tfrac{2.8}{2.8(d)}\right) \cdot \left(1 - \tfrac{2.92}{4.92(f)}\right) = 1. \\
Fig.5(b): S_{MN} &= 1 - \left(1 - \tfrac{0.67}{3.67(b)}\right) \cdot \left(1 - \tfrac{0.67}{2.67(d)}\right) \cdot \left(1 - \tfrac{0.67}{3.67(f)}\right) = 0.50.
\end{aligned}
$$

From the above computing results, we find that the potential link between node *M* and node *N* is more likely to exist in Fig. 8.5a than in Fig. 8.5b. Algorithm 8.4 describes the process of our above approach.

8.4.3 Node-Coupling Clustering Approach Based on Common Neighbors (NCCCN)

The traditional CN method is based on the number of the com-mon neighbor nodes of a predicted node-pair [39]. Its similarity metric is defined as follows:

$$
S_{xy}^{CN} = |\Gamma(x) \cap \Gamma(y)| \tag{8.37}
$$

where $\Gamma(i)$ denotes the common neighbor node set of node *i*. $|\Gamma(i)|$ represents the number of the common neighbor nodes of node *i*.

Although *CN* has low complexity in the link prediction problem, it does not consider the different roles of the common neighbor nodes of a predicted node-pair. This results in low prediction accuracy. Here, we propose a new link prediction approach based on *CN*, which combines the different contributions of different nodes to the connecting probability with the clustering information of a network.

In our approach, (x, y) is a predicted node-pair. Node n is a common neighbor node of the node-pair (x, y). *NCCC*(n) can be considered as the contribution of node n that connects node x to node y. *Score*(n) denotes the contribution score value that node n connects node x to node y. Therefore, *Score*(n) = *NCCC*(n). $\{A_1, A_2, \ldots, A_i, \ldots, A_m\}$ is the common neighbor set of the predicted node-pair (x, y), namely $\Gamma(x) \cap \Gamma(y) = \{A_1, A_2, \ldots, A_i, \ldots, A_m\}$. Here, these common neighbor nodes of the node-pair (x, y) are assumed to be independent to each other. We use the contribution sum of all common neighbor nodes of the predicted node-pair (x, y), S_{xy}, to evaluate the link existence likelihood between node x and y. Therefore, the new similarity metric in our approach is defined as follows:

$$\begin{aligned} S_{xy} &= Score\,(A_1) + Score\,(A_2) + \cdots + Score\,(A_i) + \cdots + Score\,(A_m) \\ &= NCCC\,(A_1) + NCCC\,(A_2) + \cdots + NCCC\,(A_i) + \cdots + NCCC\,(A_m) \\ &= \sum\nolimits_{1 \le i \le m} NCCC\,(A_i) \\ &= \sum\nolimits_{n \in \Gamma(x) \cap \Gamma(y)} \frac{\sum_{i \in C_n} \left(\frac{1}{d_i} + C(i)\right)}{\sum_{j \in \Gamma_n} \left(\frac{1}{d_j} + C(j)\right)} \end{aligned} \tag{8.38}$$

In our approach, the related parameters in Eq. (8.38) have been described in Sect. 8.4.2. Clearly, a larger value of S_{xy} means a higher likelihood that there exists a potential link between node x and y. $\Gamma(x) \cap \Gamma(y)$ is the number of common neighbor nodes of the predicted node-pair (x, y). In Eq. (8.38), since $0 < NCCC(n) \le 1$, we have $0 < \sum_{n \in \Gamma(x) \cap \Gamma(y)} \frac{\sum_{i \in C_n} \left(\frac{1}{d_i} + C(i)\right)}{\sum_{j \in \Gamma_n} \left(\frac{1}{d_j} + C(j)\right)} \le \mid \Gamma(x) \cap \Gamma(y) \mid$, and $S_{xy} \in (0, \mid \Gamma(x) \cap \Gamma(y) \mid)$. Specially, $S_{xy} = \mid \Gamma(x) \cap \Gamma(y) \mid$ when $C_n = \Gamma(n)$ or every node in $\Gamma(x) \cap \Gamma(y)$. For example, we use this approach to compute the similarity score of the predicted node-pair (M, N) in Fig. 8.5a, b, respectively.

$$\begin{aligned} Fig.5(a) : S_{MN} &= \frac{2.92}{4.92(b)} + \frac{2.8}{2.8(d)} + \frac{2.92}{4.92(f)} = 2.19. \\ Fig.5(b) : S_{MN} &= \frac{0.67}{3.67(b)} + \frac{0.67}{2.67(d)} + \frac{0.67}{3.67(f)} = 0.61. \end{aligned}$$

We obtain the same prediction result as NCCPT. From this example, we find that our node-coupling clustering approaches can provide better prediction results than the traditional methods. Algorithm 8.4 illustrates the process of our above approach.

8.4.4 The Extensions of NCCPT and NCCCN

To further improve the performance of link prediction, we extend NCCPT and NCCCN by adding its clustering coefficient information of every selected common neighbor node, C_n, to the above two approaches respectively.

For the same reason, (x, y) is a predicted node-pair; node n is a common neighbor node of the node-pair (x, y). When we add the node clustering coefficient information, C_n, in the contribution of node n to the connecting probability based on NCCPT, we can obtain a new contribution of node n:

$NCCC(n) + C_n$. However, $0 \leq NCCC(n) + C_n \leq 2$. This is outside the scope of the probability value. In order to extend NCCPT, we use the average value of $NCCC(n)$ and C_n, $\frac{1}{2}\cdot (C_n + NCCC(n))$, as the contribution of node n. Therefore, S_{xy} in the extended NCCPT approach (ENCCPT) is defined as follows:

$$\begin{aligned} S_{xy} &= 1 - \overline{P(A_1)}\cdot \overline{P(A_2)} \cdots \overline{P(A_i)} \cdots \overline{P(A_m)} \\ &= 1 - (1 - P(A_1)) \cdot (1 - P(A_2)) \cdots (1 - P(A_i)) \cdots (1 - P(A_m)) \\ &= 1 - \left(1 - \tfrac{1}{2}\cdot (NCCC(A_1) + C(A_1))\right) \cdot \left(1 - \tfrac{1}{2}\cdot (NCCC(A_2) + C(A_2))\right) \\ &\cdots \left(1 - \tfrac{1}{2}\cdot (NCCC(A_i) + C(A_i))\right) \cdots \left(1 - \tfrac{1}{2}\cdot (NCCC(A_m) + C(A_m))\right) \\ &= 1 - \prod\nolimits_{n\in\Gamma(x)\cap\Gamma(y)} \left(1 - \frac{1}{2}\cdot \left(\frac{\sum_{i\in C_n}\left(\frac{1}{d_i} + C(i)\right)}{\sum_{j\in\Gamma_n}\left(\frac{1}{d_j} + C(j)\right)}\right) + C_n\right) \end{aligned} \tag{8.39}$$

where C_n is the clustering coefficient of node n. The other parameters are the same as Eq. (8.36). In Eq. (8.39), since $NCCC(n) \in (0, 1]$ and $C_n \in [0, 1]$, we have $\prod_{n\in\Gamma(x)\cap\Gamma(y)} \left(1 - \frac{1}{2}\cdot \left(\frac{\sum_{i\in C_n}\left(\frac{1}{d_i} + C(i)\right)}{\sum_{j\in\Gamma_n}\left(\frac{1}{d_j} + C(j)\right)}\right) + C_n\right) \in [0, 1)$, and $S_{xy} \in (0, 1]$. Specially, $S_{xy} = 1$ when $C_n = \Gamma(n)$ and $C_n = 1$ for every node in $\Gamma(x) \cap \Gamma(y)$. For instance, ENCCPT is used to predict the link existence probabilities of node-pair (M, N) in Fig. 8.5a, b as follows:

$$\begin{aligned} Fig.5(a) : S_{MN} &= 1 - \left(1 - 0.5\cdot \left(\tfrac{2.92}{4.92} + 0.2\right)(b)\right) \cdot \left(1 - 0.5\cdot \left(\tfrac{2.8}{2.8} + 0.67\right)(d)\right) \\ &\cdot \left(1 - 0.5\cdot \left(\tfrac{2.92}{4.92} + 0.2\right)(f)\right) = 0.94. \\ Fig.5(b) : S_{MN} &= 1 - \left(1 - 0.5\cdot \left(\tfrac{0.67}{3.67} + 0\right)(b)\right) \cdot \left(1 - 0.5\cdot \left(\tfrac{0.67}{2.67} + 0\right)(d)\right) \\ &\cdot \left(1 - 0.5\cdot \left(\tfrac{0.67}{3.67} + 0\right)(f)\right) = 0.28. \end{aligned}$$

Similarly, (x, y) represents a pair of predicted nodes, and node n is a common neighbor node of the node-pair (x, y). When we add the node clustering coefficient information, C_n, in the contribution of node n that connects node x to node y based on NCCCN, we can obtain a new contribution of node n: $NCCC(n) + C_n$. $Score(n)$ denotes the contribution score value that node n connects node x to node y. Therefore, $Score(n) = NCCC(n) + C_n$. Hence, the extended NCCCN approach (ENCCCN) is shown in the following Eq. (8.40).

$$
\begin{aligned}
S_{xy} &= Score\,(A_1) + Score\,(A_2) + \cdots + Score\,(A_i) + \cdots + Score\,(A_m) \\
&= (NCCC\,(A_1) + C\,(A_1)) + (NCCC\,(A_2) + C\,(A_2)) + \cdots \\
&\quad + (NCCC\,(A_i) + C\,(A_i)) + \cdots + (NCCC\,(A_m) + C\,(A_m)) \\
&= \sum\nolimits_{1 \le i \le m} (NCCC\,(A_i) + C\,(A_i)) \\
&= \sum\nolimits_{n \in \Gamma(x) \cap \Gamma(y)} \left(\frac{\sum_{i \in C_n} \left(\frac{1}{d_i} + C(i) \right)}{\sum_{j \in \Gamma_n} \left(\frac{1}{d_j} + C(j) \right)} + C(n) \right)
\end{aligned}
\tag{8.40}
$$

where $C(n)$ denotes the clustering coefficient of node n. Other parameters are the same as Eq. (8.38). In Eq. (8.40), since $0 < NCCC(n) \le 1$ and $0 \le C(n) \le 1$, we have $0 < NCCC(n) + C(n) \le 2$, and $0 < \sum_{n \in \Gamma(x) \cap \Gamma(y)} \left(\frac{\sum_{i \in C_n} \left(\frac{1}{d_i} + C(i) \right)}{\sum_{j \in \Gamma_n} \left(\frac{1}{d_j} + C(j) \right)} + C(n) \right) \le$ $2 \cdot | \; \Gamma(x) \cap \Gamma(y) \; |$, namely $S_{xy} \in (0, 2 \cdot \; | \Gamma(x) \cap \Gamma(y) | \;]$. Specially, $S_{xy} = 2 \cdot \; | \Gamma(x) \cap \Gamma(y) |$ when $C(n) = \Gamma(n)$ and $C(n) = 1$ for every mode in $\Gamma(x) \cap \Gamma(y)$. For instance, we use ENCCCN to predict the existence possibility of the link between nodes M and N in Fig. 8.5a, b, respectively. The results are as follows:

$$
\begin{aligned}
Fig.5(a): S_{MN} &= 1 - \left(1 - 0.5 \cdot \left(\tfrac{2.92}{4.92} + 0.2\right)(b)\right) \cdot \left(1 - 0.5 \cdot \left(\tfrac{2.8}{2.8} + 0.67\right)(d)\right) \\
&\quad \cdot \left(1 - 0.5 \cdot \left(\tfrac{2.92}{4.92} + 0.2\right)(f)\right) = 0.94. \\
Fig.5(b): S_{MN} &= 1 - \left(1 - 0.5 \cdot \left(\tfrac{0.67}{3.67} + 0\right)(b)\right) \cdot \left(1 - 0.5 \cdot \left(\tfrac{0.67}{2.67} + 0\right)(d)\right) \\
&\quad \cdot \left(1 - 0.5 \cdot \left(\tfrac{0.67}{3.67} + 0\right)(f)\right) = 0.28.
\end{aligned}
$$

From the above prediction results, it can be found that ENCCPT and ENCCCN have the same prediction results as NCCPT and NCCCN. Moreover, we notice that the prediction results of ENCCPT and ENCCCN have more obvious differences than NCCPT and NCCCN in the same example, respectively. This results in better prediction results compared with our baseline approaches (i.e. NCCPT, NCCCN), and it shows the importance of the clustering information in the link prediction. Algorithm 8.4 describes the process of our above extended approaches.

8.4.5 Complexity Analysis of Our Approaches

In real applications, most link prediction methods are based on local analysis and global analysis. *CN* is the simplest link prediction method in these methods. As a representative of the methods based on local analysis, *CN* has low complexity and suitable for large-scale networks. Its time complexity is $O(n^2)$, where n is the number of nodes in a network. Its space complexity is $O(n^2)$. In contrast, *Katz* is a representative of the methods based on global analysis. Its time complexity is $O(n^3)$. Its space complexity is $O(n^2)$. The methods based on global analysis are impractical for large-scale networks because of their high complexity.

Algorithm 8.4 Node-Coupling Clustering Approaches

1: Set $d[\,]=0$; $C[\,]=0$;
2: Divide the original network G into the training set TS and test set PS;
3: **for** each node i in G **do**
4: Compute the degree value of this node: $d[i]$;
5: Compute the clustering coefficient of this node: $C[i]$;
6: **end for**
7: **for** each nonexistent edge (x, y) in G **do**
8: Compute the similarity score S_{xy} by Eqs. (8.36), (8.38), (8.39) or Eq. (8.40);
9: **end for**
10: Arrange the list of all S_{xy} in descending order;
11: Compute AUC by Eq. (8.33);
12: **Return** *AUC*;

As illustrated in Algorithm 8.4, the main operations of our algorithms consist of lines 3–6 and lines 7–9. The time complexity of lines 3–6 is Oðn²Þ in the worst case. The time complexity of lines 7–9 is $O(n^2)$. Therefore, the overall time complexity of our algorithms is $O(n^2)$. The space complexity of our algorithms is $O(n^2)$. Because our approaches have the same complexity as *CN*, they are suitable for large-scale networks. The experimental analysis to evaluate the performance of proposed approaches on two synthetic datasets and six real datasets can be found in [4].

8.5 Pyramid Scheme Model for Consumption Rebate Frauds

This section provides a pyramid scheme model which has the principal characters of many pyramid schemes that have appeared in recent years: promising high returns, rewarding the participants for recruiting the next generation of participants, and the organizer takes all of the money away when they find that the money from the new participants is not enough to pay the previous participants interest and rewards. It assumes that the pyramid scheme is carried out in the tree network, **Erdős-Réney** (ER) random network, Strogatz–Watts (SW) small-world network, or Barabasi–Albert (BA) scale-free network. The section then gives the analytical results of the generations that the pyramid scheme can last in these cases.

8.5.1 *Networks*

8.5.1.1 Tree Network

Tree networks are connected acyclic graphs. The word tree suggests branching out from a root and never completing a cycle. Tree networks are hierarchical, and each node can have an arbitrary number of child nodes. Trees as graphs have many applications, especially in data storage, searching, and communication [40].

8.5.1.2 Random Network

Random network, also known as stochastic network or stochastic graph, refers to a complex network created by stochastic process. The most typical random network is the ER model proposed by Paul **Erdős** and Alfred **Réney** [41]. The ER model is based on a natural construction method: suppose there are n nodes, and assume that the possibility of connection between each pair of nodes is constant $0 < p < 1$. The network constructed in this way is an ER model network. Scientists first used this model to explain real-life networks.

8.5.1.3 Small-World Network

The original model of small-world was first proposed by Watts and Strogatz, and it is the most classical model of small-world network, which is called SW small-world network [32]. The SW small-world network model can be constructed as follows: take a one-dimensional lattice of L vertices with connections or bonds between nearest neighbors and periodic boundary conditions (the lattice is a ring), then go through each of the bonds in turn and independently with some probability ϕ "rewiring" it. Rewiring in this context means shifting one end of the bond to a new vertex chosen uniformly at random from the whole lattice, with the exception that no two vertices can have more than one bond running between them and no vertex can be connected by a bond to itself. The most striking feature of small-world networks is that most nodes are not neighbors of one another, but the neighbors of any given node are likely to be neighbors of each other and most nodes can be reached from every other node by a small number of hops or steps. It has been found that many networks in real life have the small-world property, such as social networks [42], the connections of neural networks [32], and the bond structure of long macro-molecules in the chemical [43].

8.5.1.4 Scale-Free Network

A scale-free network is a network whose degree distribution follows a power law, at least asymptotically. The first model of scale-free network was proposed by

Barabasi and Albert, which is called BA scale-free network [44]. The BA model describes a growing open system starting from a group of core nodes, new nodes are constantly added to the system. The two basic assumptions of the BA scale-free network model are as follows: (1) from m_0 nodes, a new node is added at each time step, and m nodes from the m_0 nodes are selected to be connected to the new node ($m \leq m_0$); (2) the probability Π_i that the new node is connected to an existing node i satisfies $\Pi_i = k_i/\sum_{j=1}^{N-1} k_j$, where k_i denotes the degree of the node i and N denotes the number of nodes. In this way, when added enough new nodes, the network generated by the model will reach a stable evolution state, and then the degree distribution follows the power law distribution. In [45], it was shown that the degree distribution of many networks in real world is approximate or exact obedience to the power law distribution.

8.5.2 The Model

8.5.2.1 Assumptions

We consider a simple pyramid scheme that meets the basic features of many pyramid schemes in the real world, especially the consumption rebate platforms. First, it has an organizer that attracts participants through promising a high rate of return compared to the normal interest rate. Besides the promising return, any participant will be rewarded by the organizer with a proportion of the total investment of the participants he or she directly attracted, thus the early participants will be motivated enough to recruit the next-generation participants and the next-generation participants will do the same thing in order to get more returns. Secondly, we assume all the participants at current generation are recruited by the participants at the upper generation, and the organizer pays the participants at the previous generations the interests and rewards when all possible participants at current generation have joined in the scheme. The third assumption is that the organizer will take all the money away when he finds the money from the new participants is not enough to pay the previous participants interest and rewards. To simplify the model, we also assume all the participants invest the same amount of money and invest only once. Figure 8.6 is a schematic diagram of the pyramid scheme, which has one organizer and two generations of participants.

Based on these assumptions, we discuss the pyramid scheme spreads in the tree network, random network, small world network, and scale-free network below.

8.5.2.2 Tree Network Case

If the pyramid scheme expands in the form of tree network that has a constant branching coefficient a and the root node of the tree network represents the organiser, we can simply write the number of participants at the g-th generation

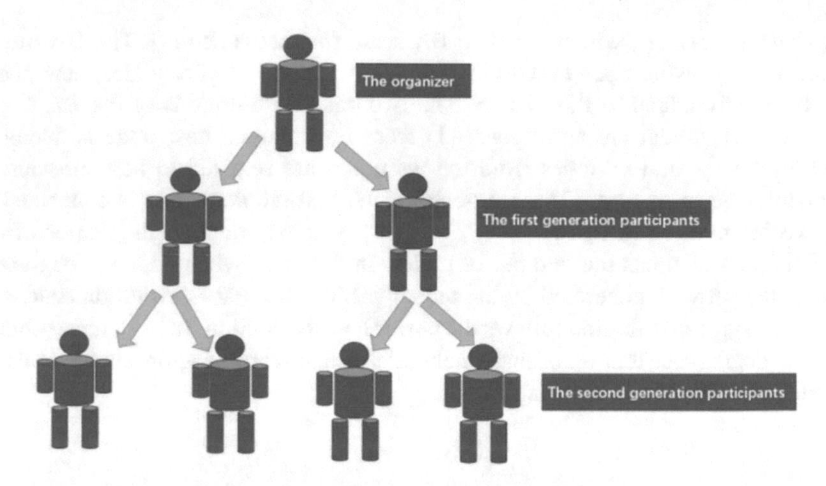

Fig. 8.6 A schematic diagram of pyramid scheme. From top to bottom are the organizer, the first-generation participants, and the second-generation participants

as $n_1\ \alpha^{g-1}$ and the total amount of money entering the pyramid scheme at the g-th generation as $mn_1\ \alpha^{g-1}$, where n_1 is the number of participants at the first generation and m is the amount of money that every participant invests. For simplification, we assume $n_1 = a$ and $m = 1$. We suppose the number of all potential participants is N in this case. Removing the interest and rewards, the relationship between the net inflow of money M of the pyramid scheme and the generation g when all possible participants at the g-th generation have joined in the scheme can be given by:

$$M(g) = \alpha^g - r_0 \sum_{i=1}^{g-1} \alpha^i - r_1 \alpha^g \tag{8.41}$$

where r_0 is the promised rate of return of the organizer, and r_1 is the ratio of the money rewarded to a participant to the total investment of the participants he or she directly recruited. Normally in real pyramid scheme cases, r_0 and r_1 are between 0% and 50%. The first term of Eq. (8.41) represents the investment of all the participants, the second term represents the interest paid to the participants before the generation g, and the third term represents the rewards paid to the recruiters of participants at the g-th generation. Notice that in our pyramid scheme, the participants at the g-th generation are all recruited by the participants at the $(g-1)$-th generation.

The second term of Eq. (8.41) is the sum of geometric sequences, after summing them up, Eq. (8.41) can be rewritten as:

$$M(g) = \frac{\alpha}{\alpha - 1}\left[(1 - r_1)\,\alpha^g - (1 + r_0 - r_1)\,\alpha^{g-1} + r_0\right]. \tag{8.42}$$

Through Eq. (8.42) we can find that if the branching coefficient a satisfies the condition:

$$\alpha \geq \frac{1 - r_1 + r_0}{1 - r_1} \tag{8.43}$$

the inflow of money M(g) of the pyramid scheme is always positive, so the pyramid scheme will continue forever under the circumstances.

However, the potential participants are limited to N and the pyramid scheme will stop eventually. The maximum generation G of the pyramid scheme is given by:

$$G_{TR} = \left[\log_{\alpha} \frac{N\alpha - N + \alpha}{\alpha} \right] + 1 \tag{8.44}$$

where $[x]$ is the integer part of x. At the G-th generation, all the potential participants have joined the pyramid scheme, and the organizer will take away all the money and not pay the interest and rewards any more. We can write the final income of the pyramid as:

$$R_p = N - r_0 \sum_{i=1}^{G-2} (G - i - 1)\, \alpha^i - r_1 \sum_{i=2}^{G-1} \alpha^i \tag{8.45}$$

and the income of the participants at the i-th generation is:

$$R_i = \begin{cases} r_0\,(G - i - 1) \cdot \alpha^i + r_1 \alpha^{i+1} - \alpha^i, & for\ 1 \leq i \leq G - 2, \\ -\alpha^i, & for\ G - 2 \leq i \leq G. \end{cases} \tag{8.46}$$

Figure 8.7a shows the analytical result and the simulative result of maximum generation G_{ER} when the branching coefficient α changes, and we take the parameters $N = 10{,}000$, $r_0 = 0.1$, $r_1 = 0.1$. Figure 8.7b shows the analytical result and the simulative result of maximum generation G_{ER} when the number of possible participants N changes, and we take the parameters $\alpha = 4$, $r_0 = 0.1$, $r_1 = 0.1$. Figure 8.7 illustrates intuitively that in the tree network case, if other conditions of the pyramid scheme remain unchanged, the larger the branch coefficient—that is, the newer participants each person recruits—, the fewer generations the pyramid scheme can last. Meanwhile, when other conditions remain unchanged, the larger the number of potential participants, the more generations the pyramid scheme can sustain, but every new generation needs more participants and this growth of new participants is exponential.

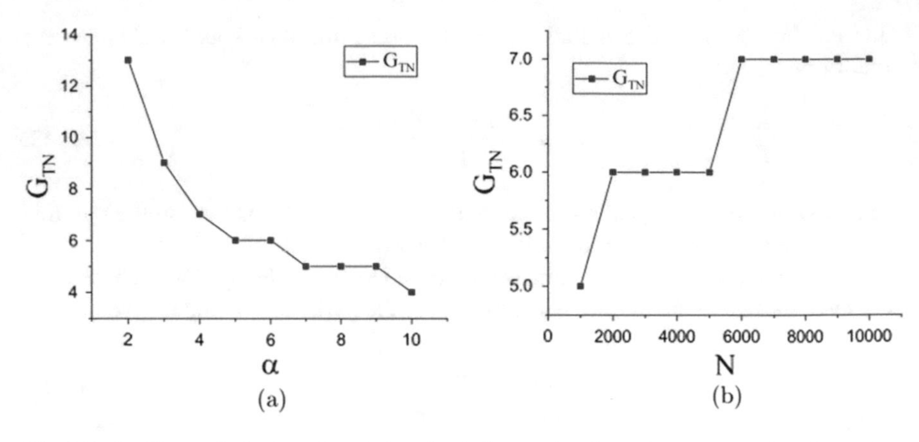

Fig. 8.7 (**a**) The analytical result and the simulative result of maximum generation G_{TN} when the branching coefficient a change. We take the parameters $N = 10{,}000$, $r_0 = 0.1$, $r_1 = 0.1$. (**b**) The analytical result and the simulative result of maximum generation G_{TN} when the number of possible participants N changes. We take the parameters $a = 4$, $r_0 = 0.1$, $r_1 = 0.1$

8.5.2.3 Random Network Case

If the pyramid scheme takes place in an ER random net-work that has an average degree k and N nodes, we assume the organizer is a random node in the network and other nodes represent the potential participants. The organizer recruits the potential participants nearest to him as the first-generation participants, and the first-generation participants recruit the potential participants nearest to them as the second-generation participants, and so on. So the generation of any participant in the pyramid scheme is given by the shortest path length to the node representing the organizer. Katzav et al. [46] have given the approximate analytical results for the distribution of shortest path lengths in ER random networks, the number of nodes at the i-th generation is about k^i if $ilog_N k < 1$ and all the nodes are included in the pyramid scheme if $ilog_N k > 1$. Therefore, the pyramid scheme in the ER random network is approximate to the case in the tree network above and the difference is the branching coefficient α should be replaced by the average degree k.

First, like the case in the tree network, r_0, r_1, and k should satisfy the following condition:

$$k \geq \frac{1 - r_1 + r_0}{1 - r_1} \tag{8.47}$$

The approximate maximum generation G of the pyramid scheme in this case is given by:

$$G_{ER} \approx \left\lfloor 1/\log_N k \right\rfloor + 1 \tag{8.48}$$

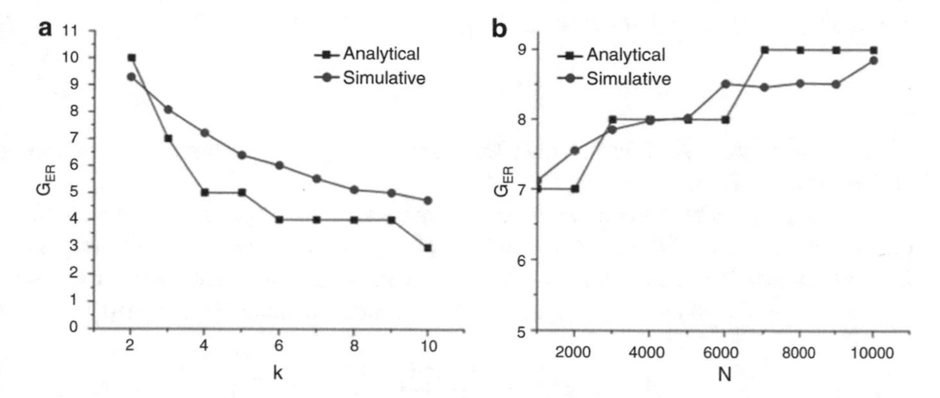

Fig. 8.8 (**a**) The analytical result and the simulative result of maximum generation G_{ER} when the average degree k changes. We take the parameters $N = 10{,}000$, $r_0 = 0.1$, $r_1 = 0.1$. (**b**) The analytical result and the simulative result of maximum generation G_{ER} when the number of possible participants N changes. We take the parameters $k = 4$, $r_0 = 0.1$, $r_1 = 0.1$. The simulative results are averaged after 100 simulations

In addition, we can also write the approximate expressions of the organiser's and participants' income which have the same form of Eq. (8.46), which we omit here. Figure 8.8a shows the analytical result and the simulative result of maximum generation G_{ER} when the average degree k changes, and we take the parameters $N = 1000$, $r_0 = 0.1$, $r_1 = 0.1$. Figure 8.8b shows the analytical result and the simulative result of maximum generation G_{ER} when the number of possible participants N changes, and we take the parameters $k = 4$, $r_0 = 0.1$, $r_1 = 0.1$. The simulative results in the figures are averaged after 100 simulations. From Fig. 8.8, we can find that in the ER random network case, the relationship between maximum generation G_{ER} and mean degree k, and the relationship between G_{ER} and N are similar to those in the tree network case, where the mean degree k represents the amount of participants that each participant can recruit averagely. We can also find that within the range of parameters we have chosen, the analytical results and simulative results are very close.

8.5.2.4 Small World Network Case

Now we consider the pyramid scheme carries out in an SW small-world network, to some extent this case is similar to the case in the ER random network. We also randomly choose a node as the organizer, other nodes represent the potential participants, and r_0, r_1 represent the interest rate and reward ratio, respectively. The generation of any participant in the pyramid scheme is the shortest path length to the node representing the organizer. Newman and Watts [47] pointed out that the number of nodes increases exponentially with the average length of the shortest path when the nodes are infinite. The approximate surface area of a sphere of radius r on the SW small-world network can be given by [47]

$$A(r) = 2e^{4r/\xi}, \tag{8.49}$$

where $\xi = 1/\phi k$, and ϕ is the rewriting probability and k is the degree of the corresponding rule graph.

Changing r to g, we can obtain the approximate number of participants at the g-th generation. Because of the exponential form of $A(g)$, we can deal with this case just like in the cases of tree network and ER random network. The branching coefficient α should be replaced by $e^{4/\xi}$, and the following condition should be satisfied:

$$e^{4/\xi} \geq \frac{1 - r_1 + r_0}{1 - r_1}. \tag{8.50}$$

If the nodes are finite, then the number of nodes reaches the maximum when the distance from the node to the organizer is near the average length of the shortest path. If the distance is greater than the average length of the shortest path, the number of nodes quickly reduces to 0, so it can be approximately considered that the maximum generation G is close to the average length of the shortest path. The average path length d of the SW small-world network is given by [47]

$$\bar{l}_{SW} \approx \frac{N}{K} f(\phi K N), \tag{8.51}$$

where

$$f(u) \approx \begin{cases} 1/4, & if\ u \to 0, \\ lnu/u, & if\ u \to \infty. \end{cases} \tag{8.52}$$

The number of nodes with the average shortest path length to the node representing the organizer is the largest. So we can infer the maximum generation G_{SW} of the pyramid scheme is given by [48]

$$G_{SW} \approx \left[\bar{l}_{SW}\right] + 1 \tag{8.53}$$

In the simulation, we find that the values of r_0 and r_1 are very important. Generally speaking, the greater the values of r_0 and r_1 satisfying Eq. (8.50) are, the closer the simulation results and numerical results are. This happens because when the values of r_0 and r_1 are larger, the pyramid scheme can easily terminate when the number of generations exceeds the average shortest path length. Figure 8.9a shows the analytical result and the simulative result of maximum generation G_{SW} when the possible participants f changes, and we take the parameters $N = 1000$, $K = 3$, $r_0 = 0.2$, $r_1 = 0.2$. Figure 8.9b shows the analytical result and the simulative result of maximum generation G_{SW} when the number of possible participants N changes, and we take the parameters $K = 3$, $\phi = 0.1$, $r_0 = 0.2$, $r_1 = 0.2$. The

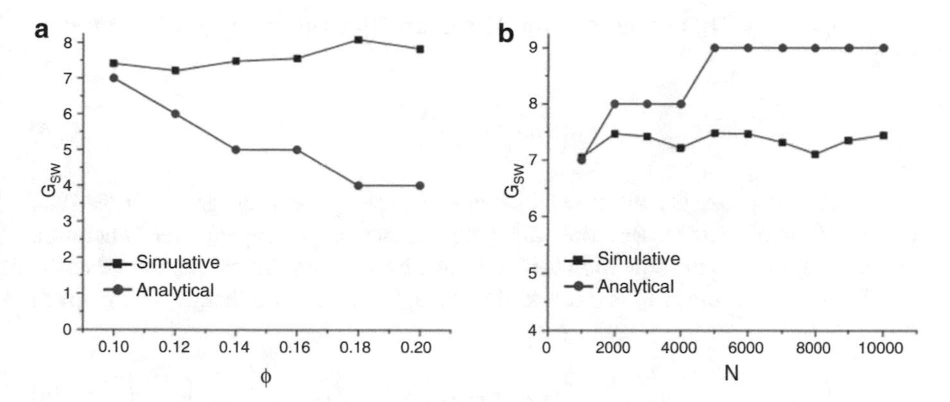

Fig. 8.9 (**a**) The analytical result and the simulative result of maximum generation G_{SW} when the possible participants f changes. We take the parameters $N = 1000$, $K = 3$, $r_0 = 0.2$, $r_1 = 0.2$. (**b**) The analytical result and the simulative result of maximum generation GSW when the number of possible participants N changes. We take the parameters $K = 3$, $\phi = 0.1$, $r_0 = 0.2$, $r_1 = 0.2$. The simulative results are averaged after 100 simulations

simulative results in the figures are averaged after 100 simulations. In Fig. 8.9, we find that within the range of parameters we selected, the maximum generation G_{SW} of the pyramid scheme is not very sensitive to the reconnection probability ϕ and the potential participants, and the analytical results are basically in accordance with the simulative results.

8.5.2.5 Scale-Free Network Case

If the pyramid scheme expands in a BA scale-free net-work, similar to the cases in ER random network and SW small-world network above, then we also randomly choose a node as the organizer and other nodes represent the potential participants. The organizer recruits participants and the participants recruit the next generation participants through the network connections. To ensure positive inflows, the following condition must be satisfied:

$$(1 - r_1)\, n\,(g + 1) \geq r_0 \sum_{i=1}^{g} n(g), \tag{8.54}$$

where $n(g)$ represents the number of participants at the g-th generation, and $n(g + 1)$ represents the number of participants at the $(g + 1)$-th generation. The distribution of shortest path length approximates the normal distribution and the position corresponding to the highest point of the normal distribution is the average shortest path length [49]. The average path length s of the BA scale-free network is given by [50]

$$\bar{l}_{BA} \approx \frac{lnN}{lnlnN}. \tag{8.55}$$

Before the peak, the number of participants per generation grows faster than the exponential growth. But after that, the number of participants per generation declines rapidly, so the condition can no longer be satisfied. So we can infer that the maximum generation G_{BA} is close to the average shortest path length, and is given by

$$G_{BA} \approx \left[\bar{l}_{BA}\right] + 1. \tag{8.56}$$

Figure 8.10 shows the analytical result and the simulative result of maximum generation G_{BA} when the number of possible participants N changes. We taken the parameters $r_0 = 0.2$, $r_1 = 0.2$, and the simulative result is averaged after 100 simulations. We can find that in scale-free networks, the maximum generation G_{BA} is not very sensitive to the potential participants in Fig. 8.10. The analytical results can basically reflect this characteristic.

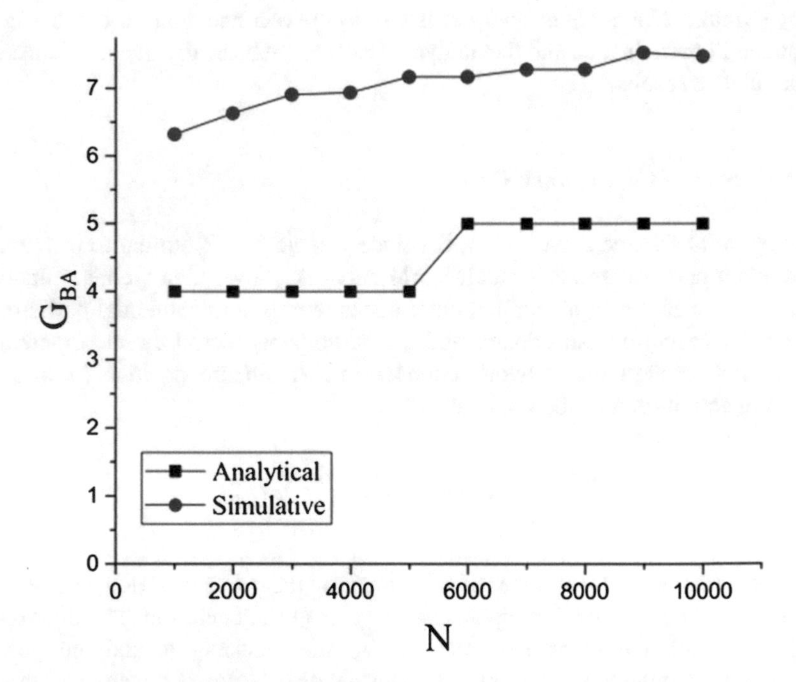

Fig. 8.10 The analytical result and the simulative result of maximum generation G_{BA} when the number of possible participants N changes. We take the parameters $r_0 = 0.2$, $r_1 = 0.2$. The simulative result is averaged after 100 simulations

8.5.3 A Pyramid Scheme in Real World

Although real cases of pyramid scheme are easy to find in news reports, there are few cases that give details of the number of people involved and the pyramid generations. Usually, when the organizer of the pyramid scheme disappears, the participants with loss will report the case to the police, who will then investigate the case. On July 23, 2018, China news network Guangzhou Station reported a pyramid scheme that had 75,663 account numbers and 46 generations, and the pyramid scheme had amassed 76 million yuan in 3 months [51]. This is the same type of pyramid scheme as described in the introduction. Using the analysis in Sect. 8.3, we assume that the pyramid scheme carries on the tree network, **ER** random network, SW small-world network, and BA scale-free net-work. We then verify which network can describe the pyramid scheme in the real world well. We assume that one account number represents a participant.

If this real pyramid scheme expands in a tree network, we can calculate the tree network' branching coefficient $\alpha \approx 1.28$. This means on average less than two participants are recruited by each participant. However, we cannot know more about the connections between the participants, except the branching coefficient.

If this real pyramid scheme spreads in an **ER** Random network, we can calculate the average degree $k \approx 1.28$ through Eq. (8.48). So each node is connected to 1.28 nodes on average, and the connection probability in the ER random network is less than $1.28/75663 \approx 1.7 \times 10^{-5}$, which is very small, then isolated nodes and nodes with degree 1 easily to appear in the network. Although this case is similar to that of tree network, the branching coefficient in the random network is not stable and it is easy to end the pyramid scheme if Eq. (8.43) is not satisfied (the minimum of the formula $(1 - r_1 + r_0)/(1 - r_1)$ is greater than 1). So we think the pyramid scheme can hardly happen in the ER random network.

If this real pyramid scheme carries out in a BA scale-free network, through the analysis and simulation, we find that developing to 46 generations needs far more than 75,663 participants. Therefore, the connections between participants are impossible to form a BA scale-free network.

If this real pyramid scheme takes place in a SW small-world network and accords to all our assumptions, then we could find a simulative result to fit the result of the real pyramid scheme. The parameters we select are $N =$ 100,000, $\phi = 0.02$, $K = 4$, $r_0 = 0.1$, $r_1 = 0.1$, and each participant invests 23,500 yuan. The simulative pyramid scheme has 74,652 participants, and develops to 46 generations, and the fund pool of the pyramid is about 76 million yuan. The simulation results are in good agreement with the real pyramid scheme. Figure 8.11a, b show the cumulative number of participants N_{cum}, the number of participants N_{g} in each generation, and the cumulative money M_{cum} changing over generation g in the simulative pyramid scheme.

Figure 8.11 shows that, the number of participants and the amount of accumulated money of the pyramid scheme grow slowly in the initial stage and explosively in the later stage. Once the growth rate slows down, the amount of the

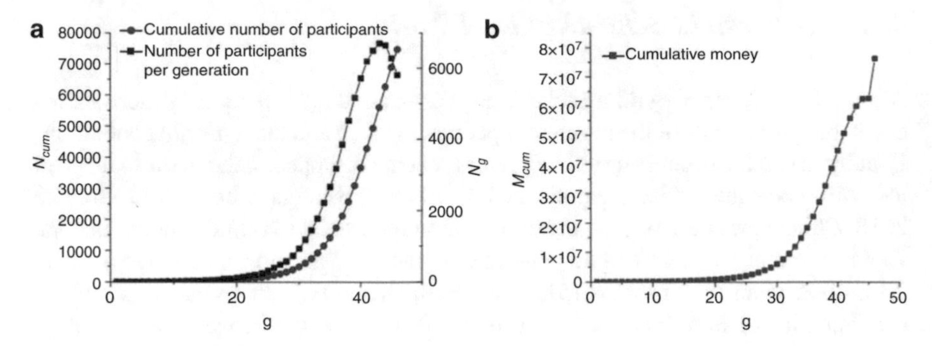

Fig. 8.11 The cumulative number of participants N_{cum}, the number of participants N_{g} in each generation, and the cumulative money M_{cum} changing over generation g in the simulative pyramid scheme. This is one simulative result in the SW small-world case, the parameters we select are $N = 100{,}000$, $\phi = 0.02$, $r_0 = 0.1$, $r_1 = 0.1$, and each participant invests 23,500 yuan

pyramid scheme's accumulated money will soon reach a peak and the organizer will escape. The probability of reconnection in simulation is 0.02, which can be understood according to the actual situation and means that participants tend to recruit new participants from familiar people. In fact, according to our investigation and many news reports, such pyramid frauds always arise in small cities and most of the participants recruit new participants from their familiar people. As the generations go on, the network constituted by all participants has the properties of small world: agglomeration and having some flocks, which are similar to the interpersonal network. Although our model has been simplified and approximated, it is enlightening to explain the real case.

Through this simulation analysis, we can speculate that the connections between participants in the real case may constitute a SW small-world network.

This work is helpful to understand the operation mechanism and characteristics of the pyramid schemes of consumption rebate type. The model may be able to apply to some current illegal high-interest loans, if these illegal projects promise a high interest rate and reward the investors who encourage others to invest in such projects but the money accumulated is not actually invested in any real projects. It shows that the pyramid schemes of consumption rebate type are not easy to be detected by the supervision because of the small amount of funds and the small number of participants accumulated in the initial stage. After the rapid growth of funds and participants, it often comes to the end of this kind of pyramid frauds, and the organizers have often already fled. Therefore for regulators, it is better to nip such platforms in the bud to avoid any more people suffering a loss. In addition, to some extent, this research finding provides some basis for further study of such frauds. For example, we will further consider how the participants' beliefs about always having enough new participants affect the operation of these frauds.

References

1. Deng, W., Shi, Y., Chen, Z., Kwak, W., Tang, H.: Recommender system for marketing optimization. World Wide Web. **23**(4), 1–21 (2020)
2. Lee, J., Shi, Y., Wang, F., Lee, H., Kim, H.K.: Advertisement clicking prediction by using multiple criteria mathematical programming. World Wide Web. **19**(4), 707–724 (2016)
3. Zhao, X., Shi, Y., Lee, J., Kim, H.K., Lee, H.: Customer churn prediction based on feature clustering and nonparallel support vector machine. Int. J. Inf. Technol. Decis. Making. **13** (2014)
4. Li, F., He, J., Huang, G., Zhang, Y., Shi, Y., Zhou, R.: Node-coupling clustering approaches for link prediction. Knowl. Based Syst. **2015**, S0950705115003536 (2015)
5. Shi, Y., Li, B., Long, W.: Pyramid scheme model for consumption rebate frauds. Chin. Phys. B. **28**(7), 078901 (2019). https://doi.org/10.1088/1674-1056/28/7/078901
6. Pinder, J.E., Wiener, J.G., Smith, M.H.: The Weibull distribution: a new method of summarizing survivorship data. Ecology. **59**(1), 175–179 (1978) http://www.jstor.org/stable/1936645
7. Shanno, D.: On Broyden-Fletcher-Goldfarb-Shanno method. J. Optimiz. Theory. App. **46**(1), 87–94 (1985)
8. Bottou, L.: Large-scale machine learning with stochastic gradient descent. In: Lechevallier, Y., Saporta, G. (eds.) Proceedings of COMPSTAT'2010, pp. 177–186. Physica-Verlag HD, Heidelberg (2010)
9. Hager, W., Zhang, H.: A new conjugate gradient method with guaranteed descent and an efficient line search. SIAM. J. Optimiz. **16**(1), 170–192 (2005). https://doi.org/10.1137/030601880
10. Celma, O., Herrera, P.: A new approach to evaluating novel recommendations. In: Proceedings of the 2008 ACM Conference on Recommender Systems, RecSys '08, p. 179C186. Association for Computing Machinery, New York (2008). https://doi.org/10.1145/1454008.1454038
11. Herlocker, J.L., Konstan, J.A., Terveen, L.G., Riedl, J.T.: Evaluating collaborative filtering recommender systems. ACM Trans. Inform. Syst. **22**(1), 5–53 (2004)
12. Nie, G., Zhang, L., Zhang, Y., Deng, W., Shi, Y.: Find intelligent knowledge by second-order mining: three cases from China. In: 2010 IEEE International Conference on Data Mining Workshops, pp. 1189–1195 (2010). https://doi.org/10.1109/ICDMW.2010.115
13. Zhang, L., Li, J., Li, A., Zhang, P., Nie, G., Shi, Y.: A new research field: intelligent knowledge management. In: 2009 International Conference on Business Intelligence and Financial Engineering, pp. 450–454 (2009). https://doi.org/10.1109/BIFE.2009.108
14. Wang, J., Zhang, Y.: Opportunity model for e-commerce recommendation: right product; right time. In: Proceedings of the 36th International ACM SIGIR Conference on Research and Development in Information Retrieval, SIGIR '13, p. 303C312. Association for Computing Machinery, New York (2013). https://doi.org/10.1145/2484028.2484067
15. Nash, J.C., Varadhan, R., Grothendieck, G.: ""Package optimr"". https://CRAN.R-project.org/package=optimr (2016)
16. Zhang, D., Shi, Y., Tian, Y., Zhu, M.: A class of classification and regression methods by multi-objective programming. Front. Comput. Sci. China. **3**(2), 192–204 (2009)
17. Zhao, X., Deng, W., Shi, Y.: Feature selection with attributes clustering by maximal information coefficient. Proc. Comput. Sci. **17**(1), 70–79 (2013)
18. Zhao, X., Shi, Y., Niu, L.: Kernel based simple regularized multiple criteria linear program for binary classification and regression. Intell. Data Anal. **19**(3), 505–527 (2015). https://doi.org/10.3233/IDA-150729
19. Richardson, M., Dominowska, E., Ragno, R.: Predicting clicks: estimating the click-through rate for new ads. In: WWW 2007 International World Wide Web Conference (2007)
20. Cheng, H., Cantu-Paz, E.: Personalized click prediction in sponsored search. In: Proceedings of the Third ACM International Conference on Web Search and Data Mining, WSDM '10, p. 351C360. Association for Computing Machinery, New York (2010). https://doi.org/10.1145/1718487.1718531

21. Li, J., Zhang, P., Cao, Y., Liu, P., Guo, L.: Efficient behavior targeting using svm ensemble indexing. In: Proceedings of the 2012 IEEE 12th International Conference on Data Mining, ICDM '12, p. 409C418. IEEE Computer Society, New York (2012). https://doi.org/10.1109/ICDM.2012.152
22. Guo, Q., Agichtein, E.: Ready to buy or just browsing? Detecting web searcher goals from interaction data. In: Proceedings of the 33rd International ACM SIGIR Conference on Research and Development in Information Retrieval, SIGIR '10, p. 130C137. Association for Computing Machinery, New York (2010). https://doi.org/10.1145/1835449.1835473
23. Reshef, D.N., Reshef, Y.A., Finucane, H.K., Grossman, S.R., Mcvean, G., Turnbaugh, P.J., Lander, E.S., Mitzenmacher, M., Sabeti, P.C.: Detecting novel associations in large data sets. Science. **334**(6062), 1518–1524 (2011)
24. David, R., Yakir, R.: MINE software package. http://www.exploredata.net/
25. Reshef, D.N., Reshef, Y.A.: Supporting Online Material | Science. https://science.sciencemag.org/content/suppl/2011/12/14/334.6062.1518.DC1
26. Frey, B.J., Dueck, D.: Clustering by passing messages between data points. Science. **315**(5814), 972–976 (2007). https://doi.org/10.1126/science.1136800
27. Shi, Y., Tian, Y., Kou, G., Peng, Y., Li, J.: Optimization Based Data Mining: Theory and Applications. Springer, New York (2011). https://doi.org/10.1007/978-0-85729-504-0
28. Vapnik, V.N.: The nature of statistical learning theory (1995)
29. Shi, Y.: Multiple criteria and multiple constraint levels linear programming: concepts, techniques and applications (2015)
30. Jayadeva, Khemchandani, R., Chandra, S.: Twin support vector machines for pattern classification. IEEE Trans. Pattern Anal. Mach. Intell. **29**(5), 905C910 (2007). https://doi.org/10.1109/TPAMI.2007.1068
31. Tian, Y., Qi, Z., Ju, X., Shi, Y., Liu, X.: Nonparallel support vector machines for pattern classification. IEEE Trans. Syst. Man. Cyb. **44**(7), 1067–1079 (2014). https://doi.org/10.1109/TCYB.2013.2279167
32. Watts, D., Strogatz, S.: Collective dynamics of 'small-world' networks. Nature. **393**(6684), 440–442 (1998). https://doi.org/10.1038/30918
33. Huang, Z., Ma, C., Xu, J., Huang, J.: Link prediction based on clustering coefficient. Appl. Phys. **4**, 101–106 (2014)
34. Geisser, S.: Predictive Inference: An Introduction. Chapman Hall, London (1993)
35. Lü, L., Zhou, T.: Link prediction in complex networks: a survey. Physica A. **390**(6), 1150–1170 (2011). https://doi.org/10.1016/j.physa.2010.11.027. https://www.sciencedirect.com/science/article/pii/S037843711000991X
36. Liu, Z., Zhang, Q.M., Lü, L., Zhou, T.: Link prediction in complex networks: a local naive Bayes model. Europhys. Lett. **96**(4), 48007 (2011). https://doi.org/10.1209/0295-5075/96/48007
37. Dong, Y., Ke, Q., Wu, B.: Link prediction based on node similarity. Comput. Sci. **38**(7), 162–164, 199 (2011) CSCD:4281527
38. Zhou, T., Lu, L., Zhang, Y.C.: Predicting missing links via local information. Eur. Phys. J. B. **71**(4), 623–630 (2009). https://doi.org/10.1140/epjb/e2009-00335-8
39. Liben-Nowell, D., Kleinberg, J.: The link prediction problem for social networks. In: Proceedings of the Twelfth International Conference on Information and Knowledge Management, CIKM '03, p. 556C559. Association for Computing Machinery, New York (2003). https://doi.org/10.1145/956863.956972
40. West, D.: Introduction to Graph Theory, 2nd edn. Prentice Hall, Hoboken, NJ (2001)
41. Erdős, P., Rényi, A.: On the evolution of random graphs. Publ. Math. Inst. Hung. Acad. Sci. **5**, 17–61 (1960)
42. Grossman, J.W.: Reviews: Small worlds: the dynamics of networks between order and randomness. Phys. Today. **31**(4), 74–75 (2002)

43. Marques Leite dos Santos, V., Brady Moreira, F., Longo, R.L.: Topology of the hydrogen bond networks in liquid water at room and supercritical conditions: a small-world structure. Chem. Phys. Lett. **390**(1), 157–161 (2004). https://doi.org/10.1016/j.cplett.2004.04.016. https://www.sciencedirect.com/science/article/pii/S0009261404005469
44. Barabasi, A., Albert, R.: Emergence of scaling in random networks. Science. **286**(5439), 509–512 (1999). https://doi.org/10.1126/science.286.5439.509
45. Albert, R., Barabasi, A.: Statistical mechanics of complex networks. Rev. Mod. Phys. **74**(1), 47–97 (2002). https://doi.org/10.1103/RevModPhys.74.47
46. Katzav, E., Nitzan, M., Ben-Avraham, D., Krapivsky, P.L., Khn, R., Ross, N., Biham, O.: Analytical results for the distribution of shortest path lengths in random networks. Europhys. Lett. **111**(2), 26006 (2015)
47. Newman, M.E.J., Watts, D.J.: Scaling and percolation in the small-world network model. Phys. Rev. E Stat. Phys. Plasmas Fluids Related Interdisc. Top. **60**(6 Pt B), 7332–7342 (1999)
48. Barrat, A., Weigt, M.: On the properties of small-world network models. Eur. Phys. J. B. **13**(3), 547–560 (2000)
49. Ventrella, A.V., Piro, G., Grieco, L.A.: On modeling shortest path length distribution in scale-free network topologies. IEEE Syst. J. **12**(4), 3869–3872 (2018). https://doi.org/10.1109/JSYST.2018.2823781
50. Cohen, R., Havlin, S.: Scale-free networks are ultrasmall. Phys. Rev. Lett. **90**(5), 058701 (2003)
51. Fang, W.: http://www.gd.chinanews.com/2018/2018-07-24/2/397998.shtml

Chapter 9
Evaluation Analysis

Evaluation is one of the key steps in big data analytics, which determines the merit of data analysis towards the experimental objectives. It usually relates a trade-off comparison of multiple criteria which may conflict each other or complex interpretations of the problems in nature. This chapter provides several of evaluation models of the recent studies on data science. Section 9.1 reviews three evaluation formations for the known methodologies. Section 9.1.1 describes a decision-making support for the evaluation of clustering algorithms based on multiple criteria decision making (MCDM) [1]. Section 9.1.2 is about evaluation of classification algorithms using MCDM and rank correlation [2]. Section 9.1.3 discusses the public blockchain evaluation using entropy and Technique of Order Preference Similarity to the Ideal Solution (TOPSIS) [3]. Section 9.2 outlines two evaluation methods for Software. Section 9.2.1 is about a classifier evaluation for software defect prediction [4], while Sect. 9.2.2 is about an ensemble of software defect predictors by AHP-based evaluation method [5]. Section 9.3 describes four evaluation methods for sociology and economics. Section 9.3.1 is about a delivery efficiency and supplier performance evaluation in China's E-retailing industry [6]. Section 9.3.2 is about the credit risk evaluation with Kernel-based affine subspace nearest points learning method [7]. Section 9.3.3 is a dynamic assessment method for urban eco-environmental quality evaluation [8], while Sect. 9.3.4 is an empirical study of classification algorithm evaluation for financial risk prediction [9].

9.1 Reviews of Evaluation Formations

9.1.1 Decision-Making Support for the Evaluation of Clustering Algorithms Based on MCDM

In many disciplines, the evaluation of algorithms for processing massive data is a challenging research issue. However, different algorithms can produce different or

Y. Shi, *Advances in Big Data Analytics*,
https://doi.org/10.1007/978-981-16-3607-3_9

even conflicting evaluation performance, and this phenomenon has not been fully investigated. The motivation of this section aims to propose a solution scheme for the evaluation of clustering algorithms to reconcile different or even conflicting evaluation performance. This section develops a model, called decision making support for evaluation of clustering algorithms (DMSECA), to evaluate clustering algorithms by merging expert wisdom in order to reconcile differences in their evaluation performance for information fusion during a complex decision-making process.

9.1.1.1 Clustering Algorithms

Clustering is a popular unsupervised learning technique. It aims to divide large data sets into smaller sections so that objects in the same cluster are lowly distinct, whereas objects in different clusters are lowly similar [10]. Clustering algorithms, based on similarity criteria, can group patterns, where groups are sets of similar patterns [11–13]. Clustering algorithms are widely applied in many research fields, such as genomics, image segmentation, document retrieval, sociology, bioinformatics, psychology, business intelligence, and financial analysis [14].

Clustering algorithms are usually known as the four classes of partitioning methods, hierarchical methods, density-based methods, and model-based methods [15]. Several classic clustering algorithms are proposed and reported, such as the K-means algorithm [16], k-medoid algorithm [17], expectation maximization (EM) [18], and frequent pattern-based clustering [15]. In this section, the six most influential clustering algorithms are selected for the empirical study. These are the KM algorithm, EM algorithm, filtered clustering (FC), farthest-first (FF) algorithm, make density-based clustering (MD), and hierarchical clustering (HC). These clustering algorithms can be implemented by WEKA [19].

The KM algorithm, a partitioning method, takes the input parameter k and partitions a set of n objects into k clusters so that the resulting intracluster similarity is high, and the intercluster similarity is low. And the cluster similarity can be measured by the mean value of the objects in a cluster, which can be viewed as the centroid or center of gravity of the cluster [15].

The EM algorithm, which is considered as an extension of the KM algorithm, is an iterative method to find the maximum likelihood or maximum a posteriori estimates of parameters in statistical models, where the model depends on unobserved latent variables [20]. The KM algorithm assigns each object to a cluster.

In the EM algorithm, each object is assigned to each cluster according to a weight representing its probability of membership. In other words, there are no strict boundaries between the clusters. Thus, new means can be computed based on the weighted measures [18].

The FC applied in this work can be implemented by WEKA [19]. Like the cluster, the structure of the filter is based exclusively on the training data, and test instances will be addressed by the filter without changing their structure.

The FF algorithm is a fast, greedy, and simple approximation algorithm to the k-center problem [17], where the k points are first selected as a cluster center, and the second center is greedily selected as the point farthest from the first. Each remaining center is determined by greedily selecting the point farthest from the set of chosen centers, and the remaining points are added to the cluster whose center is the closest [16, 21].

The MD algorithm is a density-based method. The general idea is to continue growing the given cluster as long as the density (the number of objects or data points) in the neighborhood exceeds some threshold. That is, for each data point within a given cluster, the neighborhood of a given radius must contain a minimum number of points [15]. The HC algorithm is a method of cluster analysis that seeks to build a hierarchy of clusters, which can create a hierarchical decomposition of the given data sets [16, 22].

9.1.1.2 MCDM Methods

The MCDM methods, which were developed in the 1970s, are a complete set of decision analysis technologies that have evolved as an important research field of operation research [23, 24]. The International Society on MCDM defines MCDM as the research of methods and procedures concerning multiple conflicting criteria, which can be formally incorporated into the management planning process [24]. In an MCDM problem, the evaluation criteria are assumed to be independent [25, 26]. MCDM methods aim to assist decision-makers (DMs) to identify an optimal solution from a number of alternatives by synthesizing objective measurements and value judgments [27, 28]. In this section, four classic MCDM methods: the weighted sum method (WSM), grey relational analysis (GRA), TOPSIS, and PROMETHEE II are introduced as follows.

WSM

WSM [29] is a well-known MCDM method for evaluating finite alternatives in terms of finite decision criteria when all the data are expressed in the same unit [30, 31]. The benefit-to-cost-ratio and benefit-minus-cost approaches [32] can be applied to the problem of involving both benefit and cost criteria. In this section, the cost criteria are first transformed to benefit criteria. Besides, there is nominal-the-better (NB), when the value is closer to the objective value, the nominal-the-better (NB) is better. The calculation steps of WSM are as follows. First, assume n criteria, including benefit criteria and cost criteria, and m alternatives. The cost criteria are first converted to benefit criteria in the following standardization process.

1. The larger-the-better (LB): a larger objective value is better, that is, the benefit criteria, and it can be standardized as

$$x'_{ij} = \frac{x_{ij} - \min_i x_{ij}}{\max_i x_{ij} - \min_i x_{ij}} \tag{9.1}$$

2. The smaller-the-better (SB): the smaller objective value is better, that is, the cost criteria, and it can be standardized as

$$x'_{ij} = \frac{\max_i x_{ij} - x_{ij}}{\max_i x_{ij} - \min_i x_{ij}} \tag{9.2}$$

3. The nominal-the-better (NB): the closer to the objective value is better, and it can be standardized as

$$x'_{ij} = 1 - \frac{\left|x_{ij} - x_{ob}\right|}{\max\left\{\max_i x_{ij} - x_{ob}; x_{ob} - \min_i x_{ij}\right\}} \tag{9.3}$$

Finally, the total benefit of all the alternatives can be calculated as

$$A_i = \sum_{j=1}^{k} w_j x'_{ij}, \quad 1 \leq i \leq m, 1 \leq j \leq n \tag{9.4}$$

The larger WSM value indicates the better alternative.

GRA

GRA is a basic MCDM method of quantitative research and qualitative analysis for system analysis. Based on the grey space, it can address inaccurate and incomplete information. GRA has been widely applied in modeling, prediction, systems analysis, data processing, and decision-making [33]. The principle is to analyze the similarity relationship between the reference series and alternative series. The detailed steps are as follows.

Assume that the initial matrix is R:

$$\mathrm{R} = \begin{bmatrix} ccccx_{11} & x_{12} & \cdots & x_{1n} \\ x_{21} & x_{22} & \cdots & x_{2n} \\ \vdots & \vdots & \cdots & \vdots \\ x_{m1} & x_{m2} & \cdots & x_{mn} \end{bmatrix} (1 \leq i \leq m, 1 \leq j \leq n) \tag{9.5}$$

1. Standardize the initial matrix:

$$R' = \begin{bmatrix} ccccx'_{11} & x'_{12} & \cdots & x'_{1n} \\ x'_{21} & x'_{22} & \cdots & x'_{2n} \\ \vdots & \vdots & \cdots & \vdots \\ x'_{m1} & x'_{m2} & \cdots & x'_{mn} \end{bmatrix} (1 \le i \le m, 1 \le j \le n) \tag{9.6}$$

2. Generate the reference sequence x'_0:

$$x'_0 = \left(x'_0(1), x'_0(2), \ldots, x'_0(n)\right) \tag{9.7}$$

where $x'_0(j)$ is the largest and standardized value in the jth factor.

3. Calculate the differences $\Delta_{0i}(j)$ between the reference series and alternative series:

$$\Delta_{0i}(j) = |x'_0(j) - x'_{ij}|,$$
$$\Delta = \begin{bmatrix} \Delta_{01}(1) & \Delta_{01}(2) & \cdots & \Delta_{01}(n) \\ \Delta_{02}(1) & \Delta_{02}(2) & \cdots & \Delta_{02}(n) \\ \vdots & \vdots & \vdots & \vdots \\ \Delta_{0m}(1) & \Delta_{0m}(2) & \cdots & \Delta_{0m}(n) \end{bmatrix} (1 \le i \le m, 1 \le j \le n) \tag{9.8}$$

4. Calculate the grey coefficient $r_{0i}(j)$:

$$r_{0i}(j) = \frac{\min_i \min_j \Delta_{0i}(j) + \delta \max_i \max_j \Delta_{0i}(j)}{\Delta_{0i}(j) + \delta \max_i \max_j \Delta_{0i}(j)} \tag{9.9}$$

5. Calculate the value of grey relational degree b_i:

$$b_i = \frac{1}{n} \sum_{j=1}^{n} r_{0i}(j) \tag{9.10}$$

6. Finally, standardize the value of grey relational degree β_i:

$$\beta_i = \frac{b_i}{\sum_{i=1}^{n} b_i} \tag{9.11}$$

TOPSIS

TOPSIS is one of the classic MCDM methods to rank alternatives over multicriteria. The principle is that the chosen alternative should have the shortest distance from the positive ideal solution (PIS) and the farthest distance from the negative ideal solution (NIS) [34]. TOPSIS can find the best alternative by minimizing the distance to the PIS and maximizing the distance to the NIS [35]. The alternatives can be ranked by their relative closeness to the ideal solution. The calculation steps are as follows [36]:

1. The decision matrix A is standardized:

$$a_{ij} = \frac{x_{ij}}{\sqrt{\sum_{i=1}^{m} (x_{ij})^2}} \, (1 \le i \le m, 1 \le j \le n) \tag{9.12}$$

2. The weighted standardized decision matrix is computed:

$$\begin{aligned} &D = (a_{ij}{}^* w_j) \quad (1 \le i \le m, 1 \le j \le n) \\ &\sum_{i=1}^{m} w_j = 1 \end{aligned} \tag{9.13}$$

3. The PIS V* and the NIS V—are calculated:

$$\begin{aligned} V^* &= \{v_1^*, v_2^*, \ldots, v_n^*\} = \left\{ \left(\max_i v_{ij} \,|j \in J| \right), \left(\min_i v_{ij} \left|j \in J'\right| \right) \right\} \\ V^- &= \{v_1^-, v_2^-, \ldots, v_n^-\} = \left\{ \left(\min_i v_{ij} \,|j \in J| \right), \left(\max_i v_{ij} \left|j \in J'\right| \right) \right\} \end{aligned} \tag{9.14}$$

4. The distances of each alternative from PIS and NIS are determined:

$$\begin{aligned} S_i^+ &= \sqrt{\sum_{j=1}^{n} \left(V_i^j - V^*\right)^2} \quad (1 \le i \le m, 1 \le j \le n) \\ S_i^- &= \sqrt{\sum_{j=1}^{n} \left(V_i^j - V^-\right)^2} \quad (1 \le i \le m, 1 \le j \le n) \end{aligned} \tag{9.15}$$

5. The relative closeness to the ideal solution is obtained:

$$Y_i = \frac{S_i^-}{S_i^+ + S_i^-} \, (1 \le i \le m) \tag{9.16}$$

6. The preference order is ranked.

   The larger relative closeness indicates the better alternative.

9.1.1.3 PROMETHEE II

PROMETHEE II, proposed by Brans in 1982, uses pairwise comparisons and "values outranking relations" to select the best alternative [37]. PROMETHEE II can support DMs to reach an agreement on feasible alternatives over multiple criteria from different perspectives [38, 39]. In the PROMETHEE II method, a positive outranking flow reveals that the chosen alternative outranks all alternatives, whereas a negative outranking flow reveals that the chosen alternative is outranked by all alternatives. Based on the positive outranking flows and negative outranking flows, the final alternative can be selected and determined by the net outranking flow. The steps are as follows:

1. Normalize the decision matrix R:

$$R_{ij} = \frac{x_{ij} - minx_{ij}}{maxx_{ij} - minx_{ij}} (1 \le i \le n, 1 \le j \le m) \tag{9.17}$$

2. Define the aggregated preference indices. Let a, b ∈ A and

$$\begin{cases} \pi(a, b) = \sum_{j=1}^{k} p_j(a, b) w_j \\ \pi(a, b) = \sum_{j=1}^{k} p_j(b, a) w_j \end{cases} \tag{9.18}$$

   where A is a finite set of alternatives {a1, a2, ..., an}, k is the number of criteria such that 1 ≤ k ≤ m, w_j is the weight of criterion j, and $\sum_{j=1}^{k} w_j = 1$ (1 ≤ k ≤ m). π(a, b) represents how a is preferred to b over all criteria, and $p_j(a, b)$ represents how b is preferred to a over all criteria. $p_j(a, b)$ and $p_j(b, a)$ are the preference functions of the alternatives a and b.
3. Calculate π(a, b) and π(b, a) for each pair of alternatives

   In general, there are six types of preference function. DMs must select one type of preference function and the corresponding parameter value for each criterion [40, 41].
4. Determine the positive outranking flow and negative outranking flow. The positive outranking flow is determined by

$$\phi^+(a) = \frac{1}{n-1} \sum_{x \in A} \pi(a, x) \tag{9.19}$$

   and the negative outranking flow is determined by

$$\phi^-(a) = \frac{1}{n-1} \sum_{x \in A} \pi(a, x) \tag{9.20}$$

Table 9.1 Contingency table

Partition C						Σ
		C_1	C_2	......	C_k	
Partition P	P_1	n_{11}	n_{12}	......	n_{1k}	N_1
	P_2	n_{21}	n_{22}	......	n_{22}	N_2
	P_k	n_{k1}	n_{k2}	......	n_{kk}	n_k
	Σ	n_1	n_2	......	n_k	n

5. Calculate the net outranking flow:

$$\phi(a) = \phi^+(a) - \phi^-(a) \tag{9.21}$$

6. Determine the ranking according to the net out-ranking flow.

   Larger $\phi(a)$ is the more appropriate alternative.

9.1.1.4 Performance Measures

External measures for evaluating clustering results are more effective than internal and relative measures. Accordingly, in this study, nine clustering external measures are selected for evaluation. These are entropy, purity, micro-average precision (MAP), Rand index (RI), adjusted Rand index (ARI), F-measure (FM), Fowlkes–Mallows index (FMI), Jaccard coefficient (JC), and Mirkin metric (MM). Among them, measures of entropy and purity are widely applied as external measures in the fields of data mining and machine learning [42, 43]. The nine external measures are generated by a computer with an Intel core i5-3210M CPU @ 2.50 GHz with 8G memory. Before introducing external measures, the contingency table is described.

9.1.1.5 The Contingency Table

Given a data set D with n objects, suppose we have a partition $P = \{P_1, P_2, \ldots, P_n\}$ by some clustering method, where $\cup_{i=1}^{k} P_i = D$ and $P_i \cap P_j = \phi$, for $1 \leq \mathrm{i} \neq \mathrm{j} \leq \mathrm{k}$. According to the preassigned class labels, we can create another partition on $C = \{C_1, C_2, \ldots, C_k\}$ where $U_{i=1}^{k} C_i = D$ and $C_i \cap C_j = \phi$ for $1 \leq \mathrm{i} \neq \mathrm{j} \leq \mathrm{k}$. Let nij denote the number of objects in cluster Pi with the label of class Cj. Then, the data information between the two partitions can be displayed in the form of a contingency table, as shown in Table 9.1.

The following paragraphs define the external measures. The measures of entropy and purity are widely applied in the field of data mining and machine learning.

1. Entropy. The measure of entropy, which originated in the information-retrieval community, can measure the variance of a probability distribution. If all clusters consist of objects with only a single class label, the entropy is zero, and as the class labels of objects in a cluster become more varied, the entropy increases.

The measure of entropy is calculated as

$$E = -\sum_i \frac{n_i}{n}\left(\sum_j \frac{n_{ij}}{n_i}\log\frac{n_{ij}}{n_i}\right) \tag{9.22}$$

2. Purity. The measure of purity pays close attention to the representative class (the class with majority objects within each cluster). Purity is similar to entropy. It is calculated as

$$P = \sum_i \frac{n_{i_i}}{n}\left(\max_j \frac{n_{ij}}{n_{i_i}}\right) \tag{9.23}$$

   A higher purity value usually represents more effective clustering.
3. F-Measure. The F-measure (FM) is a harmonic mean of precision and recall. It is commonly considered as clustering accuracy. The calculation of FM is inspired by the information-retrieval metric as follows:

$$\begin{aligned} F - \text{measure} &= \frac{2\times\text{precision}\times\text{recall}}{\text{precision}+\text{recall}} \\ \text{precision} &= \frac{n_{ij}}{n_j},\ \text{recall} = \frac{n_{ij}}{n_i} \end{aligned} \tag{9.24}$$

   A higher value of FM generally indicates more accurate clustering.
4. Micro-average Precision. The MAP is usually applied in the information-retrieval community. It can obtain a clustering result by assigning all data objects in a given cluster to the most dominant class label and then evaluating the following quantities for each class:

   (a) α(Cj): the number of objects correctly assigned to class Cj.
   (b) β(Cj): the number of objects incorrectly assigned to class Cj.
   The MAP measure is computed as follows:

$$\text{MAP} = \frac{\sum_j \alpha\left(C_j\right)}{\sum_j \alpha\left(C_j\right) + \beta\left(C_j\right)} \tag{9.25}$$

   A higher MAP value indicates more accurate clustering.

5. Mirkin Metric. The measure of Mirkin metric (MM) assumes the null value for identical clusters and a positive value, otherwise. It corresponds to the Hamming distance between the binary vector representations of each partition [44]. The measure of MM is computed as

$$M = \sum_i n_{i.}^2 + \sum_j n_i^2 - 2\sum_i\sum_j n_{ij}^2 \tag{9.26}$$

A lower value of MM implies more accurate clustering. In addition, given a data set, assume a partition C is a clustering structure of a data set and P is a partition by some clustering method. We refer to a pair of points from the dataset as follows:

(a) SS: if both points belong to the same cluster of the clustering structure C and to the same group of the partition P
(b) SD: if the points belong to the same clusters of C and to different groups of P
(c) DS: if the points belong to different clusters of C and to the same groups of P
(d) DD: if the points belong to different clusters of C and to different groups of P

Assume that a, b, c, and d are the numbers of SS, SD, DS, and DD pairs, respectively, and that M a + b + c + d, which is the maximum number of pairs in the data set. The following indicators for measuring the degree of similarity between C and P can be defined.

6. Rand Index. The RI is a measure of the similarity between two data clusters in statistics and data clustering [45]. RI is computed as follows:

$$R = \frac{(a + d)}{M} \tag{9.27}$$

A higher value of RI indicates a more accurate result of clustering.

7. Jaccard Coefficient. The JC, also known as the Jaccard similarity coefficient (originally named the "coefficient de commutate" by Paul Jaccard), is a statistic applied to compare the similarity and diversity of sample sets [46]. JC is computed as follows:

$$J = \frac{a}{(a + b + c)} \tag{9.28}$$

A higher value of JC indicates a more accurate result of clustering.

8. Fowlkes and Mallows Index. The Fowlkes and Mallows index (FMI) was proposed by Fowlkes and Mallows [47] as an alternative for the RI. The measure of FMI is computed as follows:

$$\mathrm{FMI} = \sqrt{\frac{a}{a + b} \cdot \frac{a}{a + c}} \tag{9.29}$$

A higher value of FMI indicates more accurate clustering.

9. Adjusted Rand Index. The adjusted Rand index (ARI) is the corrected-for-chance version of the measure of RI. It ranges from -1 to 1 and expresses the level of concordance between two bipartitions [48]. A value of ARI closest to 1 indicates almost perfect concordance between the two compared bipartitions, whereas a

value near -1 indicates almost complete discordance [49]. The measure of ARI is computed as:

$$\mathrm{ARI} = \frac{a - \left((a+c) + \frac{a+b}{M}\right)}{\left((a+c) + \frac{a+b}{2}\right) - \left((a+c) + \frac{a+b}{M}\right)} \tag{9.30}$$

A higher value of ARI indicates more accurate clustering.

9.1.1.6 Index Weights

In this work, the index weights of the four MCDM methods can be calculated by AHP. The AHP method, proposed by Saaty [50] is a widely used tool for modeling unstructured problems by synthesizing subjective and objective information in many disciplines, such as politics, economics, biology, sociology, management science, and life sciences [51–53]. It can elicit a corresponding priority vector according to pair-by-pair comparison values [54] obtained from the scores of experts on an appropriate scale. AHP has some problems, for example, the priority vector derived from the eigenvalue method can violate a condition of order preservation pro-posed by Costa and Vansnick [55]. However, AHP is still a classic and important approach, especially in the fields of operation research and management science [56]. AHP has the following steps:

1. Establish a hierarchical structure: a complex problem can be established in such a structure, including the goal level, criteria level, and alternative level [57].
2. Determine the pairwise comparison matrix: once the hierarchy is structured, the prioritization procedure starts for determining the relative importance of the criteria (index weights) within each level [5]. The pairwise comparison values are obtained from the scores of experts on a 1–9 scale.
3. Calculate index weights: the index weights are usually calculated by the eigenvector method proposed by Saaty [50].
4. Test consistency: the value of 0.1 is generally considered the acceptable upper limit of the consistency ratio (CR). If the CR exceeds this value, the procedure must be repeated to improve consistency.

9.1.1.7 The Proposed Model

Clustering results can vary according to the evaluation method. Rankings can conflict even when abundant data are processed, and a large knowledge gap can exist between the evaluation results [58] due to the anticipation, experience, and expertise of all individual participants. The decision-making process is extremely complex. This makes it difficult to make accurate and effective decisions [41]. The proposed DMSECA model consists of three steps. They are as follows.

The first step usually involves modeling by clustering algorithms, which can be accomplished using one or more procedures selected from the categories of hierarchical, density-based, partitioning, and model-based methods. In this section, we apply the six most influential clustering algorithms, including EM, the FF algorithm, FC, HC, MD, and KM, for task modeling by using WEKA 3.7 on 20 UCI data sets, including a total of 18,310 instances and 313 attributes. Each of these clustering algorithms belongs to one of the four categories of clustering algorithms mentioned previously. Hence, all categories are represented.

In the second step, four commonly used MCDM methods (TOPSIS, WSM, GRA, and PROMETHEE II) are applied to rank the performance of the clustering algorithms over 20 UCI data sets based on nine external measures as the input, computed in the first step. These methods are highly suitable for the given data sets. Unsuitable methods were not selected. For example, we did not select VIKOR because its denominator would be zero for the given data sets. The index weights are determined by AHP based on the eigenvalue method. Three experts from the field of MCDM are selected and consulted as the DMs to derive the pairwise comparison values completed by the scores of experts. We randomly assign each MCDM method to five UCI data sets. We apply more than one MCDM method to analyze and evaluate the performance of clustering algorithms, which is essential.

Finally, in the third step, we propose a decision-making support model to reconcile the individual differences or even conflicts in the evaluation performance of the clustering algorithms among the 20 UCI data sets. The proposed model can generate a list of algorithm priorities to select the most appropriate clustering algorithm for secondary mining and knowledge discovery. The detailed steps of the decision-making support model, based on the 80-20 rule, are described as follows.

Step 1. Mark two sets of alternatives in a lower position and an upper position, respectively.

It is well known that the 80-20 rule reports that 80% of the results originate in 20% of the activity in most situations. The rule can be credited to Vilfredo Pareto, who observes that 80% of the wealth is usually controlled by 20% of the people in most countries. The implication is that it is better to be in the top of 20% than in the bottom of 80%. So, the 80-20 rule can be applied to focus on the analysis of the most important positions of the rankings in relation to the number of observations for predictable imbalance. The 80-20 rule indicates that the 20% of people, who are creating 80% of the results, which are highly leveraged. In this research, based on the expert wisdom originating from the 20% of people, the set of alternatives is classified into two categories, where the top of 1/5 of the alternatives is marked in an upper position, which represents more satisfactory rankings from the opinion of all individual participants involved in the algorithm evaluation process. The bottom of 1/5 is in a lower position, which represents more dissatisfactory rankings from the opinion of all individual participants. The element marked in the upper position is calculated as follows:

$$x = \frac{n * 1}{5} \tag{9.31}$$

where n is the number of alternatives. For instance, if n 7, then × 7 × 1/5 1.4 ≈ 2. Hence, the second position classifies the ranking, where the first and second positions are those alternatives in the upper position, which are considered as the collective group idea of the most appropriate and satisfactory alternatives. Similarly, the element marked in the lower position is calculated as

$$x = \frac{n * 4}{5} \tag{9.32}$$

where n is the number of alternatives. For instance, if n = 7, then 7*4/5 = 5.6 ≈ 6. Thus, the sixth position classifies the ranking, where the sixth and seventh positions in the lower positions are considered collectively as the worst and most dissatisfactory alternatives.

Step 2. Grade the sets of alternatives in the lower and upper positions, respectively. A score is assigned to each position of the set of alternatives in the lower position and upper position, respectively.
The score in the lower position can be calculated by assigning a value of 1 to the first position, 2 to the second position, . . . , and x to the last position. Finally, the score of each alternative in the lower position is totaled, marked d.
Similarly, the score in the upper position can be calculated by assigning a value of 1 to the last position, 2 to the penultimate position, . . . , and x to the first position. Finally, the score of each alternative in the upper position is totaled, marked b.

Step 3. Generate the priority of each alternative.
The priority of each alternative fi, which represents the most satisfactory rankings from the opinions of all individual participants, can be determined as

$$f_i = b_i - d_i \tag{9.33}$$

where a higher value of f_i implies a higher priority.

9.1.2 *Evaluation of Classification Algorithms Using MCDM And Rank Correlation*

This subsection combines MCDM methods with Spearman's rank correlation coefficient to rank classification algorithms. This approach first uses several MCDM methods to rank classification algorithms and then applies Spearman's rank correlation coefficient to resolve differences among MCDM methods. Five MCDM methods, including TOPSIS, ELECTRE III, grey relational analysis, VIKOR, and PROMETHEE II are implemented in this research.

9.1.2.1 Two MCDM Methods

In addition to GRA, TOPSIS, and PROMETHEE II methods, here two more MCDM methods are outlined as below.

ELimination and Choice Expressing REality (ELECTRE)

ELECTRE stands for ELimination Et Choix Traduisant la REalite (ELimination and Choice Expressing the REality) and was first proposed by Roy [59] to choose the best alternative from a collection of alternatives. Over the last four decades, a family of ELECTRE methods has been developed, including ELECTRE I, ELECTRE II, ELECTRE III, ELECTRE IV, ELECTRE IS, and ELECTRE TRI.

There are two main steps of ELECTRE methods: the first step is the construction of one or several outranking relations; the second step is an exploitation procedure that identifies the best compromise alternative based on the outranking relation obtained in the first step.[60] ELECTRE III is chosen in this section because it is appropriate for the sorting problem. The procedure can be summarized as follows [59, 61, 62]:

Step 1. Define a concordance and discordance index set for each pair of alternatives

$$A_j \text{ and } A_k,\ j, k = 1, \ldots, m; i \neq k$$

Step 2. Add all the indices of an alternative to get its global concordance index C_{ki}.
Step 3. Define an outranking credibility degree $\sigma_s(A_i, A_k)$; by combining the discordance indices and the global concordance index.
Step 4. Define two outranking relations using descending and ascending distillation. Descending distillation selects the best alternative first and the worst alternative last. Ascending distillation selects the worst alternative first and the best alternative last.
Step 5. Alternatives are ranked based on ascending and descending distillation processes.

VlseKriterijumska Optimizacija I Kompromisno Resenje (VIKOR)

VIKOR was proposed by Opricovic [63] and Opricovic and Tzeng [64] for multicriteria optimization of complex systems. The multicriteria ranking index, which is based on the particular measure of closeness to the ideal alternative, is introduced to rank alternatives in the presence of conflicting criteria. This section uses the following VIKOR algorithm provided by Opricovic and Tzeng in the experiment:

Step 1. Determine the best f_i^* and the worst f_i^- values of all criterion functions, $i = 1, 2, \cdots, n$.

$$f_i^* = \begin{Bmatrix} \max_j f_{ij}, \text{for benefit criteria} \\ \min_j f_{ij}, \text{for cost criteria} \end{Bmatrix}, j = 1, 2, \ldots, J,$$
$$f_i^- = \begin{Bmatrix} \min_j f_{ij}, \text{for benefit criteria} \\ \max_j f_{ij}, \text{for cost criteria} \end{Bmatrix}, j = 1, 2, \ldots, J,$$

where J is the number of alternatives, n is the number of criteria, and f_{ij} is the rating of ith criterion function for alternative aj.

Step 2. Compute the values S_j and $R_j; j = 1, 2, \cdots, J$, by the relations

$$S_j = \textstyle\sum_{i=1}^{n} w_i \left(f_i^* - f_{ij}\right)\left(f_i^* - f_i^-\right)$$
$$R_j = \max_i \left[w_i \left(f_i^* - f_{ij}\right)\left(f_i^* - f_i^-\right)\right]$$

where wi is the weight of ith criteria, Sj and Rj are used to formulate ranking measure.

Step 3. Compute the values Qj; $j = 1, 2, \cdots, J$, by the relations

$$Q_j = v\left(S_j - S^*\right)\left(S^- - S^*\right) + (1 - v)\left(R_j - R^*\right)\left(R^- - R^*\right)$$
$$S^* = \min_j S_j,\ S^- = \max_j S_j$$
$$R^* = \min_i R_j,\ R^- = \max_j R_j$$

where the solution obtained by S is with a maximum group utility, the solution obtained by R is with a minimum individual regret of the opponent, and v is the weight of the strategy of the majority of criteria. The value of v is set to 0.5 in the experiment.

Step 4. Rank the alternatives in decreasing order. There are three ranking lists: S; R, and Q.

Step 5. Propose the alternative a′, which is ranked the best by Q, as a compromise solution if the following two conditions are satisfied:

(a) $Q(a'') - Q(a') \geq 1(J - 1)$; (b) Alternative a 0 is ranked the best by S or/and R.

If only the condition (b) is not satisfied, alternatives a' and a'' are proposed as compromise solutions, where a// is ranked the second by Q. If the condition (a) is not satisfied, alternatives a'; $a'' \ldots$; a^M are proposed as compromise solutions, where a^M is ranked the Mth by Q and is determined by the relation $Q(a^M) - Q(a') < 1(J - 1)$ for maximum M.

9.1.2.2 Spearman's Rank Correlation Coefficient

Spearman's rank correlation coefficient measures the similarity between two sets of rankings. The basic idea of the proposed approach is to assign a weight to each MCDM method according to the similarities between the ranking it generated and the rankings produced by other MCDM methods. A large value of Spearman's rank correlation coefficient indicates a good agreement between a MCDM method and other MCDM methods.

The proposed approach is designed to handle conflicting MCDM rankings through three steps. In the first step, a selection of MCDM methods is applied to rank classification algorithms. If there are strong disagreements among MCDM methods, the different ranking scores generated by MCDM methods are used as inputs for the second step.

The second step utilizes Spearman's rank correlation coefficient to find the weights for each MCDM method. Spearman's rank correlation coefficient between the kth and ith MCDM methods is calculated by the following equation:

$$\rho_{ki} = 1 - \frac{6\sum d_i^2}{n\left(n^2 - 1\right)} \tag{9.34}$$

where n is the number of alternatives and di is the difference between the ranks of two MCDM methods. Based on the value of k_i, the average similarities between the kth MCDM method and other MCDM methods can be calculated as

$$\rho_k = \frac{1}{q-1}\sum\nolimits_{i=1, i\neq k}^{q} \rho_{ki}, k = 1, 2, \ldots, q, \tag{9.35}$$

where q is the number of MCDM methods. The larger the k value, the more important the MCDM method is. Normalized k values can then be used as weights for MCDM methods in the secondary ranking.

The third step uses the weights obtained from the second step to get secondary rankings of classifiers. Each MCDM method is applied to re-rank classification algorithms using ranking scores produced by MCDM methods in the first step and the weights obtained in the second step.

- The detailed experimental study of this method can be found in [2]

9.1.3 Public Blockchain Evaluation Using Entropy and TOPSIS

This subsection aims to make a comprehensive evaluation of public blockchains from multiple dimensions. Three first-level indicators and eleven second-level indicators are designed to evaluate public blockchains. The technique for order preferences by similarity to an ideal solution (TOPSIS) method is used to rank public blockchains, and the entropy method is used to determine the weights of each dimension. Since Bitcoin has an absolute advantage, a let-the-first-out (LFO) strategy is proposed to reduce the criteria of the positive ideal solution and make a more reasonable evaluation.

9.1.3.1 Proposed Evaluation Model

Evaluation Indicator

With the increasing requirement of performance, more and more blockchains are designed by new technology. Technology is an important indicator to evaluate public blockchains, but technology is not everything. The popularity is a key factor to measure a platform or system, and the blockchain is the same. For example, the second global public blockchain technology assessment index shows that Bitcoin ranks 17th, but Bitcoin is still one of the most popular blockchains.

Therefore, two indicators are designed to measure the popularity of public blockchains. One is recognition, which is the degree of acceptance of public blockchains by developers and others. The greater the acceptance, the better the blockchain. The other is activity, which measures the activity of developers and others. When developers stop maintaining and improving a blockchain, or people stop talking about it, the blockchain is no longer popular. Developers and other people can be considered separately, but they are under the same indicator in this section because of the same topic. Figure 9.1 shows the first-level indicators and their second-level indicators.

Technology

The basic technology (I_{11}) and the applicability (I_{12}) are the first and the second second-level indicators of technology respectively. These two indicators are quantified by the expert scoring method. Since CCID has established a technology assessment index for public blockchains, this section will reference its scoring results for the two indicators. The basic technology mainly examines the realization function, basic performance, safety and degree of centralization of public blockchains. The applicability focuses on the application scenarios, the number of wallets, the ease of use, and the development support on the chain.

The TPS (I_{13}) is the most important indicator of public blockchain networks. The TPS of Bitcoin and Ethereum are 7 and 20 respectively, while the TPS of VISA is 2000. A blockchain's TPS depends on its consensus algorithm, and the POW consensus algorithm makes the TPS of Bitcoin and Ethereum small.

In November 2017, Ethereum launched a pet cat game called CryptoKitties. Since December 3, 2017, pending transactions at Ethereum have skyrocketed. CryptoKitties accounted for more than 10% of the activity in Ethereum, resulting in

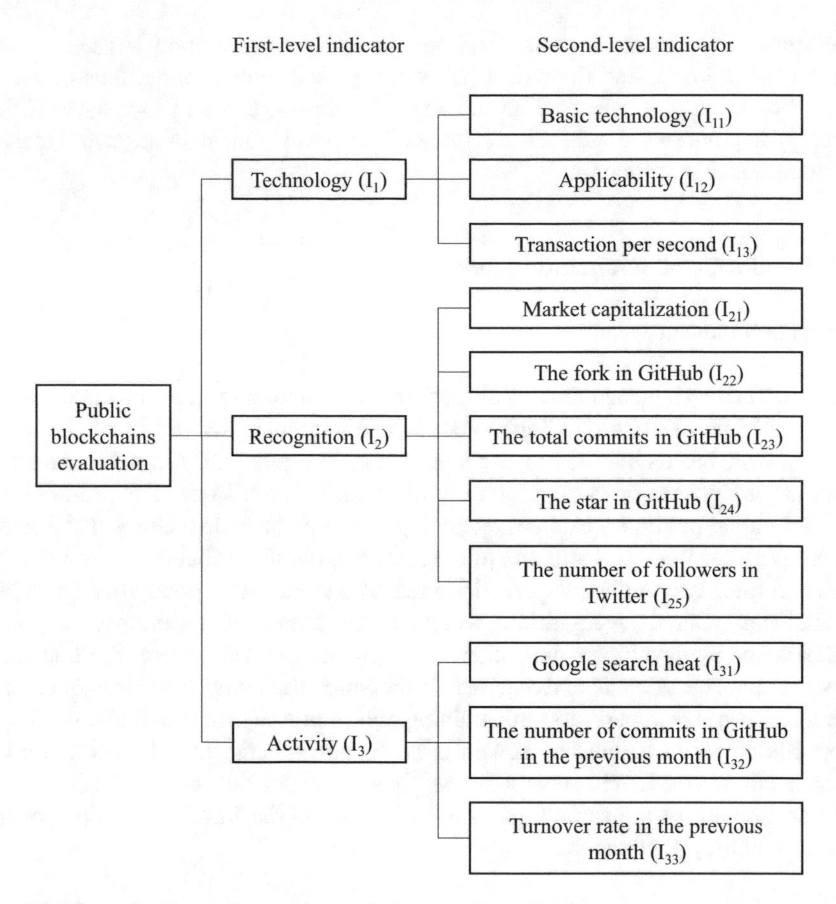

Fig. 9.1 The evaluation indicators for public blockchains evaluation

serious congestion in the Ethereum network. The gas fee, also called transaction fee, is required to be paid to the miners to run a particular transaction or contract. With the congestion of the Ethereum network, the gas fee will increase. As can be seen in Fig. 9.2, the gas fee increases rapidly since December 3, 2017. Additionally, the congestion appears again in the Ethereum network since June 30, 2018, because of the principles of FCoin GPM listing. These high transaction costs show the congestion in the Ethereum network. Since people pay most attention to the TPS nowadays, the TPS is independent of the I_{11} as the third second-level indicator of technology.

However, even if the TPS needs to be upgraded to solve the congestion problem, too large TPS is meaningless. For example, if 2000 TPS is enough to handle the daily transactions, there is no difference between 5000 TPS and million TPS. In this case, the hyperbolic tangent function is introduced to reduce the benefits of the

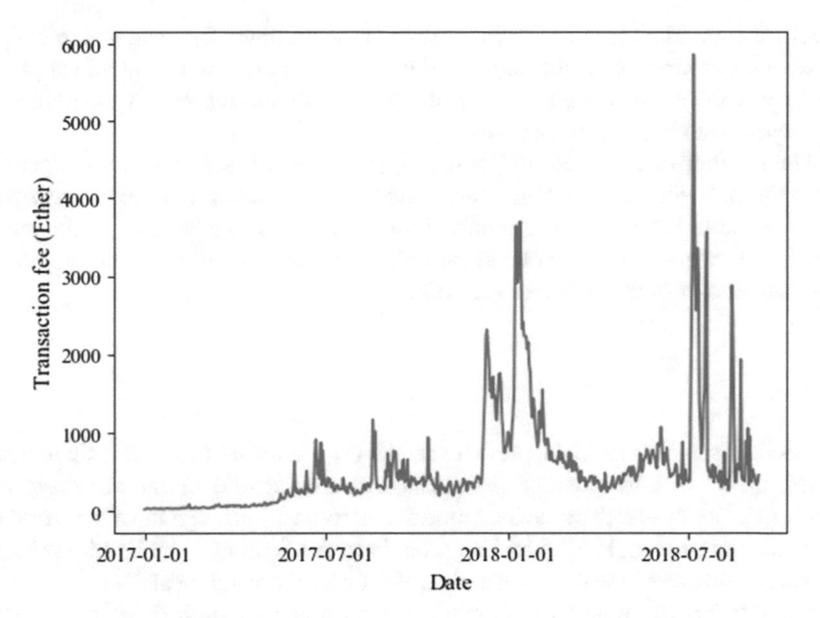

Fig. 9.2 The transaction fee of Ethereum network

increased TPS:

$$y = \frac{e^x - e^{-x}}{e^x + e^{-x}}, x = \frac{\text{TPS}}{\alpha} \tag{9.36}$$

where α is a scale factor and set to 2000 in this section.

Recognition

The market capitalization (I_{21}) is the first second-level indicator of recognition. The market capitalization of a company is the result of the transaction price of the company's stock in the securities market multiplied by the total share capital, reflecting the company's asset value, profitability value, and growth value. Similarly, the market capitalization of a public blockchain is the result of the transaction price of the public blockchain's coin in the cryptocurrency market multiplied by the total number of coins. It reflects the blockchain's use value and growth value. Once a blockchain is not recognized and no longer used, its value will be zero.

The fork (I_{22}), the total commits (I_{23}), and the star (I_{24}) in GitHub are the second, third, and fourth second-level indicator of recognition respectively. A basic technical feature of the blockchain is the shared ledger, which requires multiple participation and cooperation. Due to the openness and transparency of the open source, the open source of blockchain not only quickly obtain the recognition and trust of partners, but also quickly gather a number of outstanding talents for continuous developments. The fork in GitHub represents the number of people who recognize or want to contribute to the blockchain; the total commits in GitHub represent the improvements of the blockchain; the star in GitHub represents the number of developers who like the blockchain.

The number of followers in Twitter (I_{25}) is the fifth second-level indicator of recognition. Twitter is one of the most famous online news and social networking service. The blockchains always have Twitter accounts to post news to the public, and the followers of a public blockchain's Twitter account represent the people who care and recognize the public blockchain.

Activity

The Google search heat in the previous month (I_{31}) is the first second-level indicator of activity. In the search market, Google handles around 90% of searches worldwide. The popularity of search terms over time and across various regions of the world can be compared in Google Trends. The Google search heat of a public blockchain is the sum of its name's search heat and its short name's search heat.

The number of commits in GitHub in the previous month (I_{32}) is the second second-level indicator of activity. It reflects the improvements of blockchains in the previous month.

The turnover rate in the previous month (I_{33}) is the third second-level indicator of activity. The turnover rate is the frequency of coins traded in the market in a certain period of time. The higher the turnover rate, the more active the transactions of cryptocurrency and the more popular the public blockchain. Generally, a high turnover rate means good liquidity of the cryptocurrency.

Evaluation Process

The choice of indicators weights is an important step in the TOPSIS. The entropy method is an objective method to calculate weights based on the objective information of indicators [65]. An indicator with small entropy value means the indicator is important and has a large weight [66]. The entropy is calculated as follows:

$$e_j = -\frac{1}{\ln n} \sum_{i=1}^{n} p_{ij} \ln p_{ij}, \quad p_{ij} = \frac{x_{ij}}{\sum_{i=1}^{n} x_{ij}} \tag{9.37}$$

where x_{ij} is the jth normalized indicator value of the ith public blockchain. Then the degree of divergence (d_j) and the weight (w_j) can be calculated as follows:

$$d_j = 1 - e_j \tag{9.38}$$

$$w_j = \frac{d_j}{\sum_{j=1}^{m} d_j} \tag{9.39}$$

The TOPSIS ranks public blockchains according to their relative proximities calculated by the distance from the positive ideal solution and the distance from the negative ideal solution [67]. The steps for the TOPSIS are described below. The first step is to normalize the indicator matrix:

$$r_{ij} = \frac{x_{ij}}{\sqrt{\sum_{i=1}^{n} x_{ij}^2}} \tag{9.40}$$

With the weights obtained by the entropy method, the weighted normalization matrix is calculated as follows:

$$v = r \cdot \mathrm{diag}(w) \tag{9.41}$$

where diag(w) is a diagonal matrix where the diagonal elements are the weights w. Then the positive ideal solution (A^+) and the negative ideal solution (A^-) can be obtained:

$$A^+ = \left\{ \left(\max_i v_{ij} | j \in J_1 \right), \left(\min_i v_{ij} | j \in J_2 \right) | i = 1, 2, \ldots, n \right\} = \left\{ v_1^+, v_2^+, \ldots, v_j^+, \ldots, v_m^+ \right\} \tag{9.42}$$

$$A^- = \left\{ \left(\min_i v_{ij} | j \in J_1 \right), \left(\max_i v_{ij} | j \in J_2 \right) | i = 1, 2, \ldots, n \right\} = \left\{ v_1^-, v_2^-, \ldots, v_j^-, \ldots, v_m^- \right\} \tag{9.43}$$

where J_1 and J_2 are the benefit and the cost indicators respectively. The distance of each indicator from A^+ and A^- can be calculated as follows:

$$S_i^+ = \sqrt{\sum_{j=1}^{m} \left(v_{ij} - v_j^+ \right)^2}, i = 1, 2, \ldots, n \tag{9.44}$$

$$S_i^- = \sqrt{\sum_{j=1}^{m} \left(v_{ij} - v_j^- \right)^2}, i = 1, 2, \ldots, n \tag{9.45}$$

The relative proximity of each public blockchain to the ideal solution can be calculated as follows:

$$C_i^* = \frac{S_i^-}{S_i^+ + S_i^-}, i = 1, 2, \ldots, n \tag{9.46}$$

Lastly, the public blockchains can be ranked by their relative proximities.

The relative proximities are based on the positive ideal solution and the negative ideal solution. If the relative proximity of the first place is much larger than that of the second place, then some indicator values of the first place are much larger than those of the second place. In this case, even if the second place is much better than the third place, the advantage will become very small under the absolute advantage of the first place. Since the positive ideal solution cannot be achieved by other items, it is better to reduce the criteria of the positive ideal solution. Therefore, a let-the-first-out (LFO) strategy is proposed to make a more reasonable evaluation. In the LFO, if the relative proximity of the first place is much larger than that of the second place, the position of the first place is retained and the other items are re-evaluated.

- The data analysis can be found in [3].

9.2 Evaluation Methods for Software

9.2.1 Classifier Evaluation for Software Defect Prediction

This subsection integrates traditional feature selection methods and multi-criteria decision making (MCDM) methods to improve the accuracy and reliability of defect prediction models and evaluate the performances of software defect detection models.

9.2.1.1 Research Methodology

Results of empirical studies on software defect prediction models do not always converge. Myrtveit et al. [68] analyzed some empirical software engineering studies and identified three factors that may contribute to the divergence: a single sample dataset, choice of accuracy indicators, and cross validation. They concluded that a crucial step in software defect prediction is the design of research procedures.

The inputs are four public-domain software defect datasets provided by the NASA IV&V Facility Metrics Data Program (MDP) repository. Feature selection and classification are conducted in four steps. First, feature selection is conducted using traditional techniques. Features are then ranked using the proposed feature selection method. The third step employs MCDM methods to evaluate feature selection techniques and choose the better performed techniques. In the last step, the selected features are used in the classification to predict software defects. The performances of classifiers are also evaluated using MCDM methods and a recommendation of classifiers for software defect prediction is made based on their accuracy and reliability.

Multiple criteria decision making (MCDM) aims at solving decision problems with multiple objectives and often conflictive constraints [40, 68, 69]. Five MCDM methods, i.e., DEA (BCC model), ELECTRE, PROMETHEE, TOPSIS, and VIKOR, are used in the experimental study to evaluate algorithms.

For feature selection algorithms, output components include seven attributes:

LOC_COMMENTS (The number of lines of comments in a module),
HALSTEAD_PROG_TIME (The halstead programming time metric of a module),
MAINTENANCE_SEVERITY (Maintenance Severity),
NODE_COUNT (Number of nodes found in a given module),
NUM_OPERATORS (The number of operators contained in a module),
NUM_UNIQUE_OPERATORS (The number of unique operators contained in a module),
PERCENT_COMMENTS (Percentage of the code that is comments).

All other attributes are input components. For classification algorithms, input component is false positive rate and output components include the area under receiver operating characteristic (AUC), precision, F-measure, and true positive rate.

9.2.1.2 Experimental Study

Data Sources

The data used in this study are modified public-domain software defect datasets provided by the NASA IV&V Facility Metrics Data Program (MDP) repository [70]. The structures of the datasets are summarized in Table 9.2.

CM is from a science instrument written in a C code with approximately 20 kilo-source lines of code (KLOC). KC is about the collection, processing and delivery of satellite metadata and is written in Java with 18 KLOC. PC is flight software from an earth orbiting satellite written in a C code with 26 KLOC. UC is dynamic simulator for attitude control systems. Forty common attributes are selected for each dataset.

Discussion of Results

Table 9.3 summarizes the feature weights for each dataset. Features that are highly ranked in one or two dataset may have low rankings in other datasets, such as attribute 4, 9, and 27. This indicates that performances of feature selection techniques vary at different datasets. It also shows a need for evaluation of feature selection techniques.

Table 9.2 Dataset structures

Dataset	Number of instances	Normal instances	Bug instances
CM	568	425	143
KC	804	495	309
PC	4472	3718	754
UC	10, 064	9285	779

Table 9.3 Feature weights for the four datasets

Attributes	CM Data		KC Data		PC Data		UC Data	
	W	R	W	R	W	R	W	R
att1	0.57	7	0.44	24	0.60	12	0.68	3
att2	0.26	37	0.40	30	0.35	39	0.22	39
att3	0.64	5	0.59	8	0.63	8	0.47	28
att4	0.27	36	0.57	9	0.95	1	0.74	2
att5	0.57	8	0.51	15	0.55	17	0.64	9
att6	0.48	21	0.30	39	0.43	31	0.48	26
att7	0.41	26	0.55	12	0.51	21	0.40	35
att8	0.44	23	0.33	37	0.65	7	0.67	4
att9	0.68	3	0.35	33	0.47	27	0.31	38
att10	0.33	33	0.64	2	0.69	4	0.62	13
att11	0.50	19	0.48	19	0.52	20	0.47	29
att12	0.33	32	0.57	10	0.55	18	0.62	12
att13	0.56	10	0.51	16	0.45	29	0.60	14
att14	0.51	18	0.44	23	0.63	9	0.60	15
att15	0.52	17	0.34	36	0.40	35	0.42	34
att16	0.24	39	0.51	14	0.49	24	0.45	30
att17	0.49	20	0.49	18	0.56	16	0.66	6
att18	0.56	9	0.41	29	0.29	40	0.18	40
att19	0.29	35	0.61	4	0.43	34	0.43	33
att20	0.43	25	0.60	5	0.77	2	0.79	1
att21	0.47	22	0.47	21	0.43	32	0.53	19
att22	0.52	14	0.44	25	0.49	25	0.47	27
att23	0.43	24	0.43	26	0.43	33	0.59	17
att24	0.52	16	0.39	31	0.49	23	0.53	21
att25	0.54	12	0.49	17	0.59	13	0.52	22
att26	0.55	11	0.42	28	0.58	14	0.55	18
att27	0.72	1	0.43	27	0.39	37	0.40	36
att28	0.62	6	0.54	13	0.46	28	0.33	37
att29	0.38	30	0.34	35	0.40	36	0.49	23
att30	0.65	4	0.38	32	0.65	6	0.65	8
att31	0.24	38	0.32	38	0.45	30	0.48	24
att32	0.37	31	0.45	22	0.62	10	0.60	16
att33	0.40	28	0.25	40	0.47	26	0.43	32
att34	0.32	34	0.35	34	0.50	22	0.53	20
att35	0.70	2	0.64	3	0.68	5	0.62	11
att36	0.52	13	0.59	7	0.57	15	0.65	7
att37	0.41	27	0.56	11	0.61	11	0.64	10
att38	0.52	15	0.47	20	0.36	38	0.43	31
att39	0.40	29	0.65	1	0.73	3	0.66	5
att40	0.21	40	0.59	6	0.53	19	0.48	25

W for Weight, R for Rank

Table 9.4 MCDM evaluation of classifiers for CM dataset

	DEA	ELECTRE	PROMETHEE	TOPSIS	VIKOR
Naïve Bayes	2	3	2	2	2
Logistic	8	7	6	6	1
RBFNetwork	7	5	7	5	6
SMO	6	9	4	8	5
IB1	5	8	8	9	9
FLR	1	1	1	1	3
DecisionTable	3	6	9	3	4
RIPPER	9	2	3	7	7
C4.5	4	4	5	4	8

Table 9.5 MCDM evaluation of classifiers for KC dataset

	DEA	ELECTRE	PROMETHEE	TOPSIS	VIKOR
Naïve Bayes	5	5	3	2	7
Logistic	1	2	2	1	1
RBFNetwork	7	4	4	4	9
SMO	6	6	6	5	8
IB1	9	9	5	7	6
FLR	4	1	1	3	3
DecisionTable	3	3	7	6	2
RIPPER	2	8	9	9	5
C4.5	8	7	8	8	4

The five MCDM methods are applied to evaluate the 11 feature selection techniques.

Tables 9.4, 9.5, 9.6, and 9.7 summarize the evaluation results of the nine classifiers on the four datasets. The rankings of classifiers vary with different datasets. Even within a dataset, different MCDM methods may produce divergent rankings for the same classifier. For example, RIPPER was ranked the second best classifier by ELECTRE and the worst classifier by DEA for CM dataset. In general, FLR outperforms other classifiers. It was ranked the best classifier by at least two MCDM methods for every dataset. SMO achieves good performances on PC and UC, which are larger than CM and KC. The performances of other classifiers on the four software defect datasets are rather mixed.

9.2.2 Ensemble of Software Defect Predictors: An AHP-Based Evaluation Method

This subsection evaluates the quality of ensemble methods for software defect prediction with the analytic hierarchy process (AHP) method. The AHP is a

Table 9.6 MCDM evaluation of classifiers for PC dataset

	DEA	ELECTRE	PROMETHEE	TOPSIS	VIKOR
Naïve Bayes	9	9	3	4	7
Logistic	8	6	7	7	5
RBFNetwork	2	3	4	3	3
SMO	1	1	2	2	1
IB1	5	8	9	9	9
FLR	4	2	1	1	2
DecisionTable	3	4	6	6	6
RIPPER	7	5	5	5	8
C4.5	6	7	8	8	4

Table 9.7 MCDM evaluation of classifiers for UC dataset

	DEA	ELECTRE	PROMETHEE	TOPSIS	VIKOR
Naïve Bayes	5	8	3	4	6
Logistic	3	4	5	5	3
RBFNetwork	2	5	4	3	2
SMO	1	2	2	2	1
IB1	8	7	8	8	7
FLR	4	1	1	1	5
DecisionTable	7	3	7	6	4
RIPPER	6	9	6	7	8
C4.5	9	6	9	9	9

multicriteria decision-making approach that helps decision makers structure a decision problem based on pairwise comparisons and experts' judgments. Three popular ensemble methods (bagging, boosting, and stacking) are compared with 12 well-known classification methods using 13 performance measures over 10 public-domain datasets from the NASA Metrics Data Program (MDP) repository.[70] The classification results are then analyzed using the AHP to determine the best classifier for software defect prediction task.

9.2.2.1 Ensemble Methods

Ensemble learning algorithms construct a set of classifiers and then combine the results of these classifiers using some mechanisms to classify new data records [71]. Experimental results have shown that ensembles are often more accurate and robust to the effects of noisy data, and achieve lower average error rate than any of the constituent classifiers [15, 72–75].

How to construct good ensembles of classifiers is one of the most active research areas in machine learning, and many methods for constructing ensembles have been proposed in the past two decades [76]. Dietterich [71] divides these methods into five groups: Bayesian voting, manipulating the training examples, manipulating the input features, manipulating the output targets, and injecting randomness. Several comparative studies have been conducted to examine the effectiveness and performance of ensemble methods. Results of these studies indicate that bagging and boosting are very useful in improving the accuracy of certain classifiers [77], and their performances vary with added classification noise. To investigate the capabilities of ensemble methods in software defect prediction, this study concentrates on three popular ensemble methods (i.e. bagging, boosting, and stacking) and compares their performances on public-domain software defect datasets.

Bagging

Bagging combines multiple outputs of a learning algorithm by taking a plurality vote to get an aggregated single prediction [78]. The multiple outputs of a learning algorithm are generated by randomly sampling with replacement of the original training dataset and applying the predictor to the sample. Many experimental results show that bagging can improve accuracy substantially. The vital element in whether bagging will improve accuracy is the instability of the predictor [78]. For an unstable predictor, a small change in the training dataset may cause large changes in predictions [79]. For a stable predictor, however, bagging may slightly degrade the performance [78].

Researchers have performed large empirical studies to investigate the capabilities of ensemble methods. For instance, Bauer and Kohavi [77] compared bagging and boosting algorithms with a decision tree inducer and a NaÏve Bayes inducer. They concluded that bagging reduces variance of unstable methods and leads to significant reductions in mean-squared errors. Dietterich [72] studied three ensemble methods (bagging, boosting, and randomization) using decision tree algorithm C4.5 and pointed out that bagging is much better than boosting when there is substantial classification noise.

In this subsection, bagging is generated by averaging probability estimates [16].

Boosting

Similar to bagging, boosting method also combines the different decisions of a learning algorithm to produce an aggregated prediction [80]. In boosting, however, weights of training instances change in each iteration to force learning algorithms to put more emphasis on instances that were predicted incorrectly previously and less emphasis on instances that were predicted correctly previously. Boosting often achieves more accurate results than bagging and other ensemble methods. However, boosting may overfit the data and its performance deteriorates with classification noise.

This study evaluates a widely used boosting method, AdaBoost algorithm, in the experiment. AdaBoost is the abbreviation for adaptive boosting algorithm because it adjusts adaptively to the errors returned by classifiers from previous iterations [73, 81]. The algorithm assigns equal weight to each training instance at the beginning. It then builds a classifier by applying the learning algorithm to the training data. Weights of misclassified instances are increased, while weights of correctly classified instances are decreased. Thus, the new classifier concentrates more on incorrectly classified instances in each iteration.

Stacking

Stacking generalization, often abbreviated as stacking, is a scheme for minimizing the generalization error rate of one or more learning algorithms [82]. Unlike bagging and boosting, stacking can be applied to combine different types of learning algorithms. Each base learner, also called "level 0" model, generates a class value for each instance. The predictions of level-0 models are then fed into the level-1 model, which combines them to form a final prediction [16].

Another ensemble method used in the experiment is voting, which is a simple average of multiple classifiers probability estimates provided by WEKA [16].

9.2.2.2 Selected Classification Models

As a powerful tool that has numerous applications, classification methods have been studied extensively by several fields, such as machine learning, statistics, and data mining [83]. Previous studies have shown that an ideal ensemble should consist of accurate and diverse classifiers. [84] Therefore, this study selects 12 classifiers to build ensembles. They represent five categories of classifiers (i.e., trees, functions, Bayesian classifiers, lazy classifiers, and rules) and were implemented in WEKA.

For trees category, we chose classification and regression tree (CART), NaÏve Bayes tree, and C4.5. Functions category includes linear logistic regression, radial basis function (RBF) network, sequential minimal optimization (SMO), and Neural Networks. Bayesian classifiers include Bayesian network and NaÏve Bayes. K-nearest-neighbor was chosen to represent lazy classifiers. For rules category, decision table and Repeated Incremental Pruning to Produce Error Reduction (RIPPER) rule induction were selected.

Classification and regression tree (CART) can predict both continuous and categorical dependent attributes by building regression trees and discrete classes, respectively [85]. NaÏve Bayes tree is an algorithm that combines NaÏve Bayes induction algorithm and decision trees to increase the scalability and interpretability of NaÏve Bayes classifiers [86]. C4.5 is a decision tree algorithm that constructs decision trees in a top–down recursive divide-and-conquer manner [87].

Linear logistic regression models the probability of occurrence of an event as a linear function of a set of predictor variables [88]. Neural network is a collection of artificial neurons that learns relationships between inputs and outputs by adjusting the weights. RBF network [89] is an artificial neural network that uses radial basis functions as activation functions. The centers and widths of hidden units are derived using k-means, and the outputs obtained from the hidden layer are combined using logistic regression [16]. SMO is a sequential minimal optimization algorithm for training support vector machines (SVM) [90, 91].

Bayesian network and NaÏve Bayes both model probabilistic relationships between the predictor variables and the class variable. While NaÏve Bayes classifier [92] estimates the class-conditional probability based on Bayes theorem and can only represent simple distributions, Bayesian network is a probabilistic graphic model and can represent conditional independencies between variables [93].

K-nearest-neighbor [94] classifies a given data instance based on learning by analogy. That is, it assigns an instance to the closest training examples in the feature space.

Decision table selects the best-performing attribute subsets using best-first search and uses cross-validation for evaluation [95]. RIPPER [96] is a sequential covering algorithm that extracts classification rules directly from the training data without generating a decision tree first.

Each of stacking and voting combines all classifiers to generate one prediction. Since bagging and boosting are designed to combine multiple outputs of a single learning algorithm, they are applied to each of the 12 classifiers and produced a total of 26 aggregated outputs.

9.2.2.3 The Analytic Hierarchy Process (AHP)

The analytic hierarchy process is a multicriteria decision-making approach that helps decision makers structure a decision problem based on pairwise comparisons and experts' judgments [97, 98]. Saaty [99] summarizes four major steps for the AHP. In the first step decision makers define the problem and decompose the problem into a three-level hierarchy (the goal of the decision, the criteria or factors that contribute to the solution, and the alternatives associated with the problem through the criteria) of interrelated decision elements [100]. The middle level of criteria might be expanded to include subcriteria levels. After the hierarchy is established, the decision makers compare the criteria two by using a fundamental scale in the second step. In the third step, these human judgments are converted to a matrix of relative priorities of decision elements at each level using the eigenvalue method. The fourth step calculates the composite or global priorities for each decision alternatives to determine their ratings.

The AHP has been applied in diverse decision problems, such as economics and planning, policies and allocations of resources, conflict resolution, arms control, material handling and purchasing, manpower selection and performance measurement, project selection, marketing, portfolio selection, model selection, politics, and environment [101]. Over the last 20 years, the AHP has been studied extensively and various variants of the AHP have been proposed. [102–105].

In this study, the decision problem is to select the best ensemble method for the task of software defect prediction. The first step of the AHP is to decompose

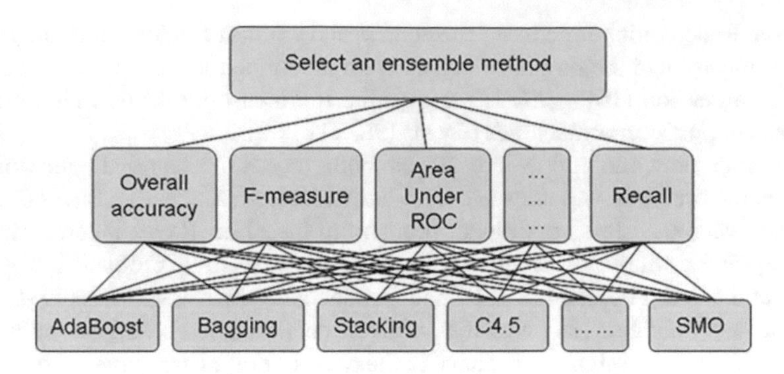

Fig. 9.3 An AHP hierarchy for the ensemble selection problem

the problem into a decision hierarchy. As shown in Fig. 9.3, the goal is to select an ensemble method that is superior to other ensemble methods over public-domain software defect datasets through the comparison of a set of performance measurements. The criteria are performance measures for classifiers, such as overall accuracy, F-measure, area under ROC (AUC), precision, recall, and Kappa statistic. The decision alternatives are ensembles and individual classification methods, such as AdaBoost, bagging, stacking, C4.5, SMO, and NaÏve Bayes. Individual classifiers are included as the decision alternatives for the purpose of comparisons.

In step 2, the input data for the hierarchy, which is a scale of numbers that indicates the preference of decision makers about the relative importance of the criteria, are collected. Saaty [97] provides a fundamental scale for this purpose, which has been validated theoretically and practically. The scale ranges from 1 to 9 with increasing importance. Numbers 1, 3, 5, 7, and 9 represent equal, moderate, strong, very strong, and extreme importance, respectively, while 2, 4, 6, and 8 indicate inter-mediate values. This study uses 13 measures to assess the capability of ensembles and individual classifiers. Previous works have proved that the AUC is the most informative and objective measurement of predictive accuracy [106] and is an extremely important measure in software defect prediction. Therefore, it is assigned a number of 9. The F-measure, mean absolute error, and overall accuracy are very important measures, but less important than the AUC. The true positive rate (TPR), true negative rate (TNR), false positive rate (FPR), false negative rate (FNR), precision, recall, and Kappa statistic are strongly important classification measures that are less important than the F-measure, mean absolute error, and overall accuracy. Training and test time refer to the time needed to train and test a classification algorithm or ensemble method, respectively. They are useful measures in real-time software defect identification. Since this study is not aimed at real-time software defect identification problem, they are included to measure the efficiency of ensemble methods and are given the lowest importance.

The third step of the AHP computes the principal eigenvector of the matrix to estimate the relative weights (or priorities) of the criteria. The estimated priorities are obtained through a two-step process: (1) raise the matrix to large powers (square); (2) sum and normalize each row. This process is repeated until the difference between the sums of each row in two consecutive rounds is smaller than a prescribed value. After obtaining the priority vector of the criteria level, the AHP method moves to the lowest level in the hierarchy, which consists of ensemble methods and classification algorithms in this experiment. The pairwise comparisons at this level compare learning algorithms with respect to each performance measure in the level immediately above. The matrices of comparisons of the learning algorithms with respect to the criteria and their priorities are analyzed and summarized in Sect. 9.2.1.2. The ratings for the learning algorithms are produced by aggregating the relative priorities of decision elements [107].

- The data analysis can be found in [5]

9.3 Evaluation Methods for Sociology and Economics

9.3.1 Delivery Efficiency and Supplier Performance Evaluation in China's E-Retailing Industry

This subsection focuses on overall and sub-process supply chain efficiency evaluation using a network slacks-based measure model and an undesirable directional distance model. Based on a case analysis of a leading Chinese B2C firm W, a two-stage supply chain structure covering procurement-stock and inventory-sale management is constructed.

In Chinese B2C e-commerce websites, two typical operation models are widely taken based on different strategic positioning. One is the third-party platform model which provides an e-commerce platform, technical support, advertising and marketing services for franchises. The leading B2C e-commerce platform in China is Taobao.com and Tmall.com. Their business revenue stems mainly from commissions and service. Another model is called the self-operated model, which has a logistics system for transferring and distributing goods. Examples include companies such as Jingdong, Dangdang, Amazon, Yihaodian and Suning. The source of their profits is that sales revenues decrease purchasing costs. According to a research report from IResearch, a leading internet consultant company and online media in China, platform model companies like Tmall accounts for most of B2C e-commerce market share, as shown in Fig. 9.4.

However, with the ongoing rapid growth of e-commerce in virtual markets, logistics has become the largest bottleneck of e-commerce's constant development. Most e-commerce players take the third party logistics (3PL) model in the initial development because of its advantage in reducing operations costs and capital investment. Because 3PL is either contractual or out-sourced logistics concentrating on regional operations, with business expansion, the drawbacks of 3PL are gradually arising. For example, lost packages and theft are common when using 3PL. Frequent overstocking during holidays and promotion days are also often disclosed due to the

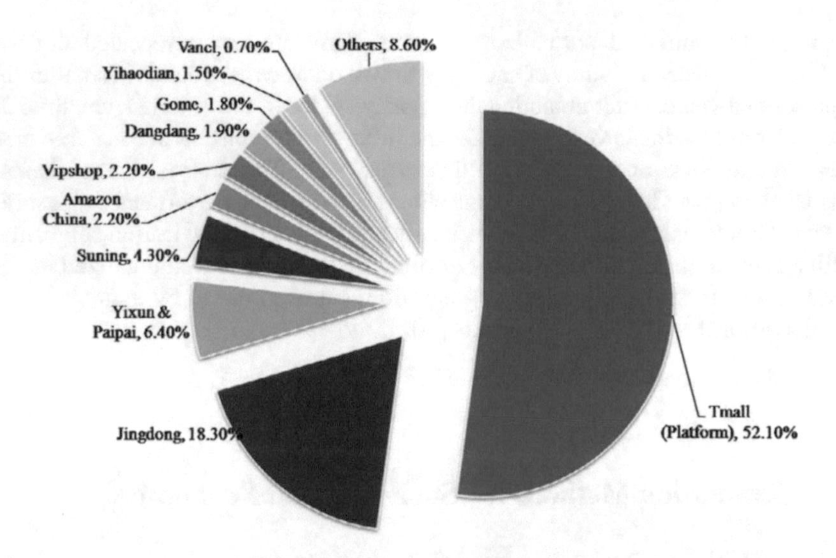

Fig. 9.4 Market share of major Chinese B2C e-commerce players in 2013

insufficient shipping capacity of 3PL. 3PL services are offered to both suppliers and customers while self-operated logistics are often built by B2C websites to improve service quality and "last mile delivery" efficiency through control of every section of the supply chain, from warehouse to consumer. As a result, a hybrid form of logistics combining 3PL and self-managed logistics is currently a popular topic of study.

From an e-retail supply chain perspective, whatever business you are in, suppliers and vendors play a crucial role in your company's success. The merchandise quality and richness provided by suppliers determine the popularity of goods, which in turn affect inventory turnover and sales. Based on that, e-retail supply chain process can be generally divided into two stages—procurement-stock management and inventory-sale control. The first sub-stage, procurement-stock management, represents "the first mile delivery" efficiency of e-retail. The second sub-process, stock-sale control describes supplier performance due to the conversion of inventory into sales revenue, as shown in Fig. 9.5. It should be noted that the overall supply chain efficiency is measured without considering internal link activities or intermediate variables.

9.3.1.1 Case, Research Problem and Data

W firm, one of China's leading B2C e-commerce firms, is chosen as our research case. The reasons are given as follows:

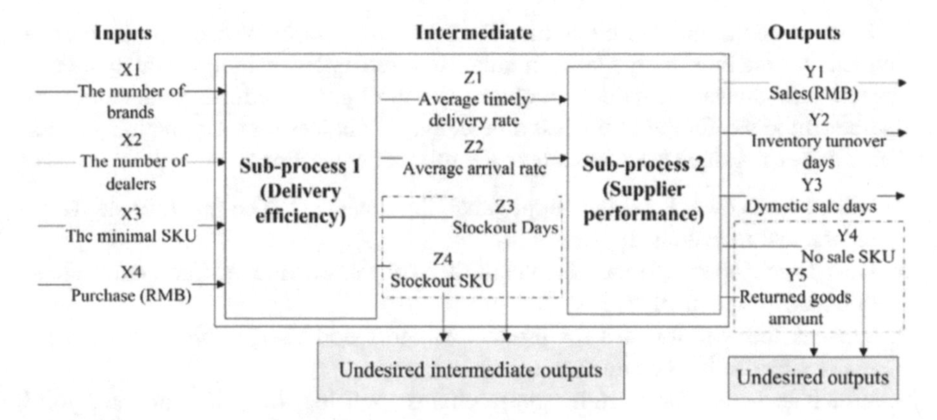

Fig. 9.5 E-commerce procurement-inventory-sale supply chain structure

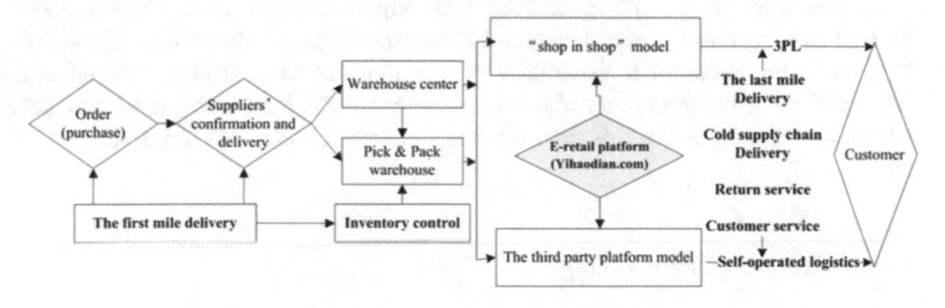

Fig. 9.6 E-retail supply chain for W firm

Firstly, W firm has established a nationwide supply chain network and has an industry-leading supply chain management system in the Chinese B2C e-commerce sector.

Secondly, W firm has the ability to realize a full online operation based on its open supply chain platform which aims to serve traditional enterprises who would like to tap into the e-commerce sector but lack online operating ability. It is similar to the third party platform model in regards to covering an integrated online operations service, improving suppliers' supply chain efficiency and reducing operations costs by system integration, cloud-based marketing, promotion tools, logistics, warehousing and information services.

Thirdly, from "the last mile delivery", those suppliers who choose the "shop in shop" model sell their merchandise by third party logistics (3PL), while running business operations on independently. For contrast, those suppliers choosing the third party platform model only need to provide their merchandise to the platform of W firm, while online operations-related activities are executed by W firm. E-retail supply chain for W firm is described in Fig. 9.6.

In conclusion, the operations model of the suppliers in W firm can be clearly divided into the third party platform model and self-operated model, which are two predominant e-business models in china. The third party platform model and self-operated model offer different "last mile delivery" choices for e-commerce players. Thus, this case can be used to analyze the following questions:

1. What causes overall e-retail supply chain inefficiency? "The first mile delivery" or "the last mile delivery"?
2. How do self-operated mode and the third party platform mode affect supply chain efficiency respectively?
3. What is the way forward for product category and Geographic expansion for major Chinese B2C e-commerce players?
4. Which is better for e-retail supply chains: Self-logistics, 3PL or the hybrid model?

Accordingly, the data of more than 2400 suppliers covering purchasing cost, the lead time, inventory, sale, delivery and returned goods were collected from W firm. Excluding incomplete data, 1229 suppliers of the "shop in shop" model and 899 suppliers of the third party platform model were obtained. Nine major product categories are included in this data set, and the research methods are described in detail.

9.3.1.2 Research Methodology

Network Slacks-Based Measure of Efficiency (NSBM)

Suppose there are n DMUs (j = 1, 2, ..., n) consisting of k divisions (k = 1, 2, ..., k) in a supply chain. mk and rk represent the number of inputs and outputs of Division k, respectively. The set of links leading from Division h to division k is defined as $L(k. h)$. Accordingly, the production possibility set (x^k, y^k, z^k, h) under the assumption of variable returns-to-scale (VRS) production is defined by

$$\begin{aligned}
& x^k \geq \sum_{j=1}^{n} x_j^k \lambda_j^k, k = 1, 2, \cdots, k \\
& y^k \leq \sum_{j=1}^{n} x_j^k \lambda_j^k, k = 1, 2, \cdots, k \\
& z^{k,h} = \sum_{j=1}^{n} z_j^{k,h} \lambda_j^k, \forall k, h \text{ (as outputs from } k \text{ and inputs to } h\text{)}, \\
& \sum_{j=1}^{n} \lambda_j^k = 1, \forall k, \lambda_j^k \geq 0, \forall j, k
\end{aligned}$$

where, $\lambda^k \in R_+^n$ is the intensity vector corresponding to Division k (k = 1, 2, ..., n).

For the evaluated DMU0 (0 = 1, 2, ..., n), in the case of linking activities determined freely while keeping continuity between input and output, non-oriented overall efficiency can be represented as:

$$\rho^* = \min_{\lambda^k, s^{k-}, s^{k+}} \frac{\sum_{k=1}^{k} w^k \left[1 - \frac{1}{m^k}\left(\sum_{i=1}^{m_k} \frac{s_i^{k-}}{x_{is}^k}\right)\right]}{\sum_{k=1}^{k} w^k \left[1 - \frac{1}{r^k}\left(\sum_{r=1}^{r_k} \frac{s_r^{k+}}{y_{ro}^k}\right)\right]} \tag{9.47}$$

$$\text{s.t.} \begin{cases} x_o^k = X^k \lambda^k + s^{k-} \\ y_o^k = Y^k \lambda^k - s^{k+} \\ \lambda^k = 1 \\ X^k = \left(x_1^k, x_2^k, \cdots, x\right) \in R^{m_k} \times n \\ Y^k = \left(y_1^k, y_2^k, \cdots, y_n^k\right) \in R^{r_k} \times n \\ z^{k,h} \lambda^h = z^{k,h} \lambda^k, \ (\forall k, h) \\ z^{k,h} = \left(z_1^{k,h}, z_2^{k,h}, \cdots, z_n^{k,h}\right) \in R^{t_{k,h}} \times n \\ \lambda^k \geq 0, s^{k-} \geq 0, s^{k+} \geq 0, \forall k \end{cases} \tag{9.48}$$

where $\sum_{k=1}^{k} w^k$, $w^k \geq 0$ $(\forall k)$, and w^k is the relative weight of division k defined by the decision makers. Non-oriented division efficiency score can be calculated by the below:

$$\rho_k = \frac{1 - \frac{1}{m_k}\left(\sum_{i=1}^{m_k} \frac{s_i^{k-*}}{x_{io}^k}\right)}{1 - \frac{1}{r_k}\left(\sum_{r=1}^{r_k} \frac{s_r^{k+*}}{y_{ro}^k}\right)}, k = 1, 2, \cdots, k \tag{9.49}$$

s^{k-*} and s^{k+*} are the excessive inputs and short outputs for the above Eq. (9.47).

Undesirable Output Directional Distance Function Model

It is important for a retail supply chain to effectively manage inventory and avoid returned purchases. It is therefore reasonable to extend the network slack-based measure (NSBM) to incorporate undesirable outputs so that it can give a comprehensive and accurate evaluation on delivery efficiency and supplier performance in a given e-retail supply chain.

The usual technical efficiency measurement is based on input and output distance functions, which cannot simultaneously contract undesirable/bad outputs and inputs and expand good/desirable outputs. Directional distance function is a generalized form of the radial model, and it allows us to explicitly increase the desirable outputs and simultaneously decrease undesirable outputs and inputs. To see this let good outputs be denoted by $y \in R_+^M$, bad or undesirable outputs by $b \in R_+^J$, and inputs by $x \in R_+^N$,. Suppose there are k (k = 1, 2, …, K) DMUs in an e-retail supply chain. Each DMU uses input $x^k = \left(x_1^k, x_2^k, \cdots, x_N^k\right) \in R_+^N$ to jointly produce desirable/good outputs $y\hat{}k = \left(y_1^k, y_2^k, \cdots, y_M^k\right) \in R_+^M$ and undesirable/bad outputs $b\hat{}k = \left(b_1^k, b_2^k, \cdots, b_J^k\right) \in R_J^+$. For a specific DMU0, a more generalized form of directional distance function is denoted by Chambers et al. [85] as follows:

$$\theta = \min \frac{1-\frac{1}{m}\sum_{i=1}^{m} w_i \alpha g_{xi0} x_{i0}}{1+o_d\frac{1}{s_d}\sum_{d=1}^{s_d} w_d \beta g_{yd0} y_{d0} - o_u\frac{1}{s_u}\sum_{u=1}^{s_u} w_u \gamma g_{yu0} y_{u0}} \tag{9.50}$$

$$\text{s.t.} \begin{cases} X\lambda + \alpha g^x \le x_0 \\ Y^d\lambda - \beta g_y^d \ge y_0^d \\ Y^u\lambda + \gamma g_y^u \le y_0^u \end{cases} \tag{9.51}$$

with $\sum_{i=1}^{m} w_i = m, \sum_{d=1}^{s_d} w_d = s_d, \sum_{u=1}^{s_u} w_u = s_u,, o_u + o_d = 1$, where m, sd, and su denote the number of inputs, desirable (good) outputs and undesirable (bad) outputs respectively. x0 and y0 are the inputs and outputs of the evaluated DMU0. w_i, w_d, and w_u separately express the weights of inputs, desirable (good) outputs and undesirable (bad) outputs defined by decision makers. g_x and g_y represent the direction vector of inputs and outputs defined by decision makers. ou and od refer to the overall weight of undesirable (bad) and desirable (good) outputs defined by decision makers.

Noted that α, β, γ represent the expansion rate for desirable output items, contraction rate for undesirable output items and input items respectively, and α, β, γ are not necessarily the same value. Namely, it allows for different proportional contraction and expansion rate for inputs, undesirable outputs and desirable outputs.

Performance assessed by directional distance model can be flexibly applied to different analysis purposes. For example, if the direction is chosen by setting$g = (-gx, gy, -gb) = (-xk, yk, -bk\ )$, the efficiency score represents how much the percentage needed to be improved in good outputs, bad outputs and inputs [78]. If instead the direction is set by $g = (-gx, gy, -gb) = (-1, 1, -1)$, the solution value can be interpreted as the net improvement in performance in the case of feasible expansion in good outputs and feasible contraction in bad outputs and inputs [107].

Here we choose the measurement based on the observed data, namely$g = (-gx, gy, -gb) = (-xk, yk, -bk)$, because we would like to observe the potential proportionate change in good outputs, bad outputs and inputs.

9.3.1.3 Variables Description

Input-Output Variables Description in the First Sub-process

As a non-parametric method for converting multi-inputs into multi-outputs, how to choose suitable input-output variable combination is crucial for DEA efficiency evaluation. Thus, in order to give an accurate efficiency measurement, it is necessary to give a reasonable input-output variable description based on e-retail supply chain network structure. Unlike in traditional retail, data mining techniques make demand forecasts possible. An e-commerce supply chain therefore starts with procurement management based on demand forecast. Purchasing plays an important role in cost saving and making profit. The way of orders is scheduled and the resultant lead time directly determines the performance of downstream activities and inventory levels. As a result, order-related input and output variables such as the selection of the right supplier, product variety, purchasing cost, average arrival rate, on time delivery rate are considered in the first sub-process of e-retail supply chain.

The number of brands and stock keeping unit (SKU) describe a variety and richness of the products in e-retail [108, 109]. Higher variety will lead to an increase in consumer's utility, which in turn affects inventory turnover and finally results in an increase in gross margin [110]. Additionally, the number of dealers determines the size of the suppliers and purchasing cost denotes the total financial inputs. Therefore, the number of brands, the number of dealers, the minimal stock keeping unit (SKU) and purchasing cost can be considered as the initial inputs of procurement-delivery management.

Furthermore, gross margin is associated with stockout costs. In practice, stockouts will lead consumer to switch retailers on subsequent shopping trips due to poor shopping experience [111]. As a result, higher stockouts mean higher lost profits. Hence, an important task of procurement managers is to reduce stockout SKUs and shorten stockout days. Accordingly, the variables of stockout SKUs and stockout days are considered as undesirable outputs in the first sub-process of e-retail supply chain performance measurement.

It is crucial that purchasing management is not something stand-alone, but has close links with the measurement of overall supply chain performance. Thus, average arrival rate and on-time delivery rate are used to measure procurement-delivery efficiency. They are the outputs in the first sub-process and the inputs in the second sub-process of e-retail supply chain. The detail input-output variables are described in Table 9.8.

Input-Output Variables Description in the Second Sub-process

Efficient procurement-stock performance can accelerate inventory turnover and promote sales. It is easier for e-commerce players to turn their capital into inventory, but it is difficult for them to turn their inventory into money. According to a statistics of Slywotzky [112], there are 95% of the time used for storage, loading and transportation in a commodity production and sales process. Hence, inventory turnover plays a crucial role in supply chain efficiency measurement. Generally speaking, shorter turnover times mean greater capacity to turn stock into revenue. Accelerating inventory turnover means an increase in the liquidity of capital. Based on that, average days to turnover inventory is considered as one of outputs in the second sub-process of e-retail supply chain. It should be noted that average days to turn over inventory refers to the number of days it takes to sell all on-hand inventory,

Table 9.8 Input and output variables description in the procurement sub-process

Procurement		Variables	Description	Metrics
Inputs	Supplier selection	The number of brands (X1)	The suppliers offer the variety of brand	Material quality
		the number of dealers (X2)	The same brand owns the different dealers	Supplier scale
		The minimal SKU (X3)	The minimal stock keeping units offered by suppliers	Merchandise richness
	Purchasing volume	Purchasing cost (X4)	The total order volume	Total purchasing cost
Desirable outputs		Average on-time delivery rate (Z1)	The ratio of on-time delivery to delayed delivery	Just-in-time delivery performance
Undesirable outputs		Average arrival rate (Z2)	The ratio of the actual delivered products by supplier to those ordered by purchasing managers	Just-in-time delivery performance
		Stockout days (Z3)	The short days of the SKUs when SKUs is below the safety stock	Inventory level
		Stockout SKU (Z4)	SKUs without being offered or restocked by suppliers	Inventory level

and can be calculated by the following formula:

$$\text{days to turnover inventory} = 365/\text{inventory turnover}$$

A change in inventory is a response to the change in sales, while dynamic sale is a key for inventory turnover. In practice, dynamic sale days is often used to illustrate inventory change and judge whether the merchandise is popular or not. In general, shorter dynamic sale days mean faster inventory turnover and less unmarketable goods. The unmarketable goods will lead to the loss of sales revenue due to an increase in stock costs. In e-retail, another loss of sales revenue can be attributed to consumer returned goods. Therefore, when associated with average days to turn inventory and sales revenue, dynamic sale days are considered as the output variables, while no-sale SKU and users' returned goods amount are chosen as undesirable output variables of supplier performance measurement in the second sub-process of e-retail supply chain. The detail input and output variables' illustration is shown in Table 9.9.

9.3.1.4 Empirical Results

E-Retail Efficiency of "the First Mile Delivery" and "the Last Mile Delivery"

Procurement-stock sub-process of e-retail supply chain is called as "the first mile delivery" due to its nature of affecting inventory management. It is the first section of e-retail supply chain, and its performance directly affects subsequent inventory and sales. Therefore, we give more weight to the first stage of e-retail supply chain than to the second stage. According to network slacks-based measure (NSBM) model, for a specific division k, the weight w1k of procurement-stock sub-process is given 0.6 and w2k of inventory-sale sub-process is given 0.4. Associated with the directional distance model with undesirable output, the weights wd of desirable (good) outputs is denoted as 0.6 and the weights wd of undesirable (bad) outputs is denoted as 0.4. We simultaneously run the above two models using the software of MaxDEA 6.2, and the results are given in Fig. 9.7.

As shown in Fig. 9.7, efficiency scores of the procurement-stock stage (Node 1) are lower than those of inventory-sale stage (Node 2). We can hence conclude that it is procurement-stock conversion inefficiency that results in W firm's overall supply chain inefficiency. The process from purchasing to putting in stock is named "first mile delivery", which is essential to developing a healthy buyer-supplier relationship and improving inventory control level.

Specifically, the suppliers of the "shop in shop" model have higher overall supply chain efficiency in kitchen and cleaning products than others due to higher purchasing-stock efficiency in the first sub-process of supply chain. In contrast, the suppliers of the third party platform model achieve better stock-sale performance in kitchen and cleaning products than others but it has low overall supply chain efficiency due to the poor performance in purchasing-stock efficiency, referring to

Table 9.9 Input and output variables description in the inventory-sale sub-process

Inventory-sale process	Variables	Description	Metrics
Inputs	Average on-time delivery rate (Z1)	The ratio of on-time delivery to delayed delivery	Just-in-time delivery performance
	Average arrival rate (Z2)	The ratio of the actual delivered products by supplier to those ordered by purchasing managers	
Desirable Outputs	Average days to turn inventory (Y1)	Inventory turnover in days equals 365 days divided by inventory turnover	Measure the value of capital movement
	Dynamic sale days (Y2)	The number of days it takes for an SKU sold	Evaluate the change of inventory and the popularity of merchandise
	Sales revenue (Y3)	Revenue from goods sold	Financial performance
Undesirable outputs	Unsaleable SKUs (Y4)	The merchandise without sold	Measure the losses of unmarketable goods
	Users' returned goods amount (Y5)	The amount of returned goods by consumers	Measure the losses of poor users' satisfaction

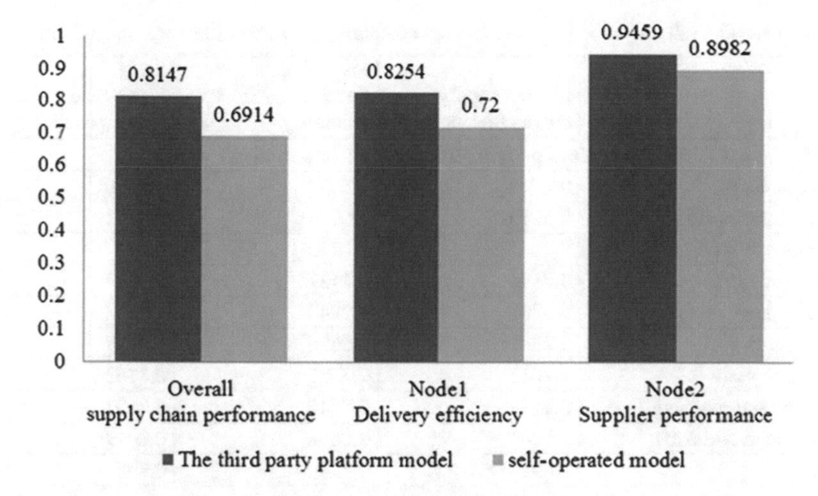

Fig. 9.7 E-retail procurement efficiency and supplier performance

Table 9.10. For this discussion, we can conclude that purchasing-stock efficiency plays a more key role in affecting overall supply chain efficiency. This conclusion further verifies the finding in Fig. 9.7.

Product Categories Expansion and Efficiency Analysis

As China's leading B2C e-commerce online supermarket, W firm has more advantages in fast moving consumer goods (FMGG) like food and drink, as shown in Fig. 9.8. In line with strategic positioning of W firm, this finding displays its core business focus on online supermarket and the concept of "the home". It is this strategic positioning that creates a barrier to potential competitors entering, thus affording a competitive advantage compared with other B2C websites such as dangdang, Suning and Redbaby. As a result, this unique positioning has allowed W firm to quickly build a loyal customer base and win a first-mover advantage.

However, with growing orders, one-stop shopping of "the home" becomes more and more important for attracting customers. Thus, W firm gradually expands its product categories from FMCG products to electronics, apparel, auto parts, maternity, and household products. In general, all major Chinese B2C e-commerce websites experience similar product categories expansion, namely starting with a narrow, vertical product line then expanding to a broad range of categories. For example, Dangdang started with books and Jingdong with digital products. Then, with growing user and market demands, all of them are in pursuit of all-categories expansion. In other words, Chinese B2C e-commerce websites experience a development of transferring from a vertical model to an integrated model.

Table 9.10 Overall and sub-process efficiency comparison for two different supply chain model

Categories	Overall supply chain efficiency	Stage 1 Purchasing-Stock efficiency	Stage 2 Stock-Sale performance
Self-operated model with a third party logistics (3PL) (shop in shop)			
Auto parts	0.4779	0.5392	0.8299
Beauty and personal care	0.6273	0.6576	0.8741
Computer and digital	0.7061	0.7263	0.9207
Food and drink	0.7190	0.7437	0.9097
Health products	0.5701	0.5840	0.9068
Household	0.6281	0.6656	0.8737
Home appliances	0.7015	0.7537	0.8763
Kitchen and cleaning	0.7548	0.8045	0.8697
Toys, mom and baby	0.6241	0.6408	0.8917
All	0.6914	0.7200	0.8982
Third-party platform model with a self-logistics			
Auto parts	0.8021	0.8080	0.9520
Beauty and personal care	0.7770	0.7893	0.9240
Computer and digital	0.8145	0.8374	0.9249
Food and drink	0.8395	0.8496	0.9531
Health products	0.7807	0.7857	0.9427
Household	0.7854	0.7987	0.9396
Home appliances	0.8284	0.8329	0.9556
Kitchen and cleaning	0.7973	0.8009	0.9573
Toys, mom and baby	0.8329	0.8427	0.9520
All	0.8147	0.8254	0.9459

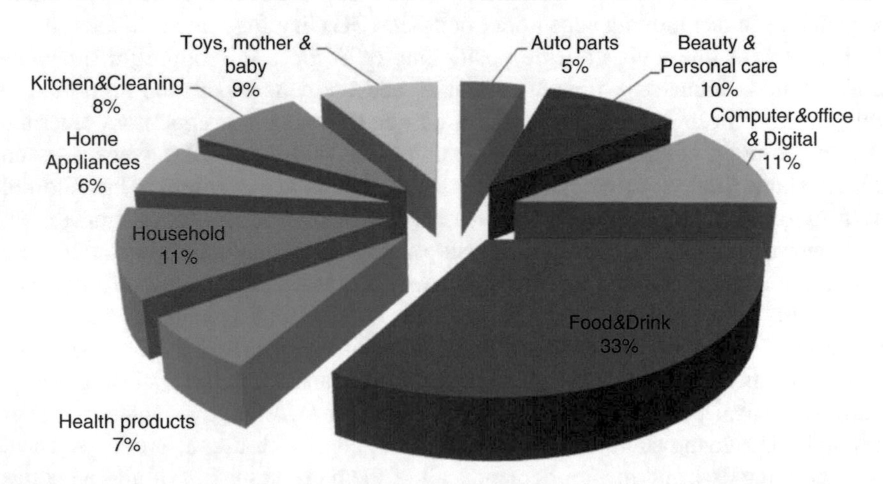

Fig. 9.8 The distribution of overall efficient supplier in different product categories

9.3.1.5 Operations Model Comparison

By the way of third party platform model, the "last mile delivery" fleet serves shops settled on the W platform while simultaneously serving merchants who sell their products on their own web page or other market platforms. The full operations service effectively reduces "the last mile delivery" cost and has allowed W firm to create higher supplier performance in the second sub-process of supply chain, referring to Fig. 9.9. However, which model is more efficient in the first stage known as "first mile delivery", self-operated model or platform model?

From inventory management, too much stock will increase inventory cost while too little stock will affect stockout rate. Thus, it is necessary for an integrated platform to make automated procurement decisions. Figure 9.9 describes inventory management for W firm. It can be seen that a purchase order would be automatically issued and sent to the suppliers when inventory dropped below a defined safety stock, and then the order will be filled by the suppliers [113]. In this way, W firm can record the delivery time, receiving and shelving information and process payment. Therefore, it can be seen in Fig. 9.9 that platform model presents higher procurement-stock efficiency scores than the self-operated (shop in shop) model.

Is the platform model efficient for all product categories?

In response to this question, we compare the "last mile delivery" efficiency of different product categories for the platform model and self-operated model, referring to Fig. 9.10. The results show self-operated (shop in shop) model performs better in computer and Office and digital, food and drink and healthy products. This is because of the high values of computer and Office and digital, and the shorter shelf life of food and drink and healthy products, which determine their priority in order of handling, picking, stockout-compensation and delivery. Furthermore, from the consumer's demand, products such as food and drink and healthy products are often bought based on the temporary needs of customers. Thus it is more suitable for these products to be delivered from regional distribution centers, while self-operated model is more helpful to reduce these product's delivery cost. This is also the reason

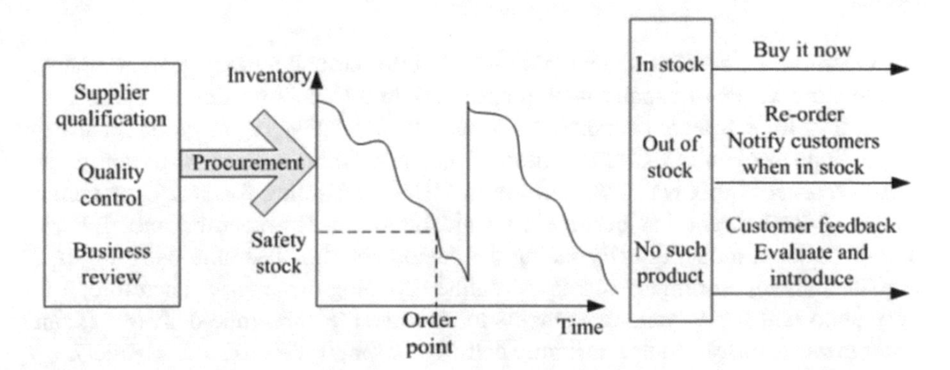

Fig. 9.9 Inventory management for W firm

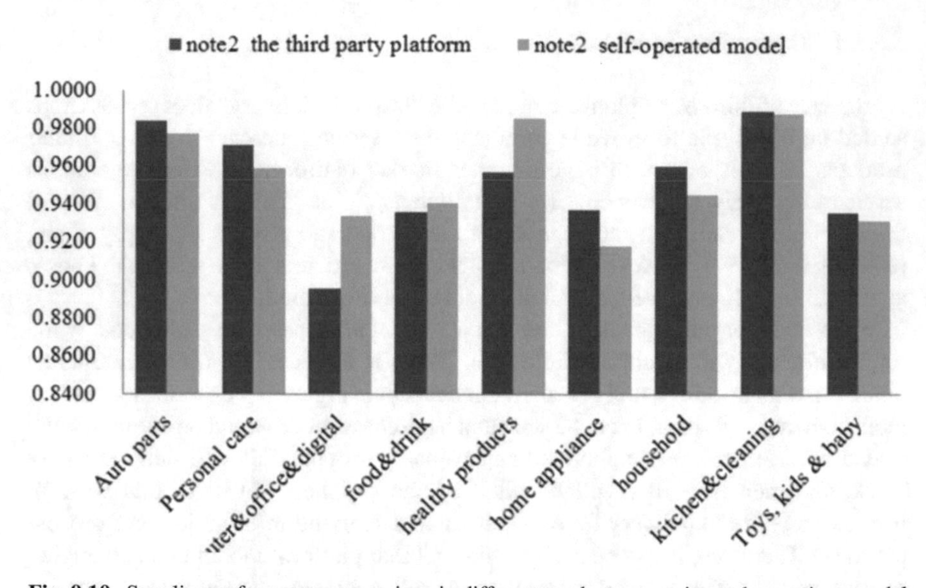

Fig. 9.10 Supplier performance comparison in different product categories and operations model

why Jingdong, the top Chinese self-operated B2C e-commerce website, starts with 3C (Computer, Communication, and Consumer electronic) products.

For the above discussion, we can conclude that the third party platform model generally performs better than self-operated model, due to its higher efficiency in "first mile delivery" and "last mile delivery". However, from a product categories perspective, self-operated model has greater efficiency in computer and Office and digital, food and drink and healthy products than the third party platform model due to these products' characteristics of regional demand and delivery.

Geographic Expansion and Efficiency Evaluation on 3PL and Self-Operated Model

As e-commerce continues its rapid growth into virtually every market sector, retailers are eager to expand their presence online to capture this market share. According to a research report of i-Research, a leading organization focusing on in-depth research in China's internet industry, China's business-to-consumer (B2C) market is to CNY 666.1 billion in 2013, accounting for 36.2% of online shopping market, and has become a formidable force. However, because B2C is an e-commerce model directly facing the customers, the "last mile delivery" is a crucial challenge for improving users' online shopping experience. Therefore, it is very important for e-commerce players to improve the "first mile delivery" (from order to warehouse) and the "last mile delivery" (from warehouse to consumer).

Starting with a large selection spanning many different product categories is a great challenge for the supply chain capacity of W firm. Although the FMCG category contributes to increasing traffic and consumer stickiness due to its nature of meeting daily needs, how to pick, pack and delivery these small items is a constant struggle. For example, by 2013, W firm had about 2,000,000 SKUs, which is 100 times that of a traditional supermarket, and each order of W firm has an average of 10 merchandises while each order of Jingdong has less than 2 merchandises. So it is stringent on warehouse design and the method of choosing food and drink supply chain. Most importantly, food and drink require faster inventory turnover due to their shorter shelf life. As a result, procurement-inventory-sale-delivery decisions needs to be automated as much as possible.

Like most B2C e-commerce players, W firm initially took 3PL delivery service model for the purpose of saving cost. But initial on-time delivery was only 90% and customer returns reached over 3% [113]. Coupled with growing orders, 3PL struggles to keep up with this growth. Therefore, the self-built logistics system becomes essential. In light of Amazon China's centralized distribution model, W firm controls all decisions from its headquarters and builds multiple distribution centers. A new "line-haul + regional distribution center + last mile delivery" model is taken. It is noted that the centralized distribution model serves nationwide consumers with the same selection on one website utilizing transshipment between warehouses to ensure the availability of products from all warehouses. In contrast, the decentralized distribution model offers different selections from local branch websites and delivers products from local distribution centers to consumers.

In the term of warehousing expansion, W firm has built five large warehousing centers covering Beijing, Shanghai, Guangzhou, Wuhan and Chengdu. By the way self-established logistics system and the third party platform operations model, W firm has borne fruit with a drastically enhanced customer experience and a 10% improvement in consumer satisfaction. The results in Table 9.10 verify that the third party platform model with self-operated logistics has better delivery efficiency, supplier performance and supply chain efficiency than self-operated (shop in shop) model.

In summary, both the self-operated model and the third platform model are more efficient in supplier performance than that in purchasing-stock efficiency, as shown in Fig. 9.10 and Table 9.10. Thus, it is urgent for W firm to strengthen their "first mile delivery" efficiency because the "first mile delivery" plays a more crucial role in supplier selection and inventory control. From an e-commerce logistics view, self-operated logistics can improve service quality and efficiency through controlling each section from warehouse to consumers, including "the last mile delivery" and is hence more efficient in the coordination of supply chain. But the complicated supply chain network and growing product categories make most e-retail players tend towards a hybrid form of 3PL and self-logistics.

9.3.2 *Credit Risk Evaluation with Kernel-Based Affine Subspace Nearest Points Learning Method*

This subsection presents a novel kernel-based method named kernel affine subspace nearest point (KASNP) method for credit evaluation. KASNP method is an extension of a new method named affine subspace nearest point method (ASNP) [114, 115] by kernel trick. Compared with SVM, KASNP is an unconstrained optimal problem, which avoids the convex quadratic programming process and directly computes the optimum solution by training set. On three credit datasets, our experimental results show that KASNP is more effective and competitive.

9.3.2.1 Affine Subspace Nearest Point Algorithm

The idea of affine subspace nearest point algorithm is derived from the geometric SVM and its nearest-points problem. Here we first give a brief overview of the geometric interpretation and the nearest point problem of SVM in original space.

Nearest Point Problem of SVM

Given a set S, co(S) denotes the convex hull of S, and is the set of convex combinations of all elements of S:

$$\mathrm{co}\,(\boldsymbol{S})=\left\{\sum\nolimits_{k}\alpha_k\boldsymbol{x}_k|\boldsymbol{x}_k\in\boldsymbol{s},\alpha_k\geq 0,\sum_{k}\alpha_k=1\right\} \tag{9.52}$$

For the linearly separable binary case, given training data, $(\boldsymbol{x}_1, y_1)$, $(\boldsymbol{x}_2, y_2)$, ..., $(\boldsymbol{x}_l, y_l)$, $\boldsymbol{x}_i \in \boldsymbol{R}^d$, $y_i \in \{+1, -1\}$, $i = 1, \ldots, l$, yi is the class label, i.e. $\boldsymbol{S}_1 = \{(\boldsymbol{x}_i, y_i)|\, y_i = +1\}$ and $\boldsymbol{S}_2 = \{(\boldsymbol{x}_i, y_i)|\, y_i = -1\}$, then the convex hulls of the two sets are

$$\mathrm{co}\,(\boldsymbol{S}_1)=\left\{\sum\nolimits_{i:y_i=+1}\alpha_i\boldsymbol{x}_i|\sum\nolimits_{i:y_i=+1}\alpha_i=1,\alpha_i\geq 0\right\} \tag{9.53}$$

$$\mathrm{co}\,(\boldsymbol{S}_2)=\left\{\sum\nolimits_{i:y_i=-1}\alpha_i\boldsymbol{x}_i|\sum\nolimits_{i:y_i=-1}\alpha_i=1,\alpha_i\geq 0\right\} \tag{9.54}$$

As we know, the aim of normal SVM is to find the hyperplane, which separates training data without errors and maximizes the distance (called margin) from the closest vectors to it. In fact, from geometric view, the optimal separating hyperplane is just the one that is orthogonal to and bisects the shortest line segment joining the convex hulls of two sets, and the optimal problem of SVM is equivalent to finding the nearest point problem in the convex hulls [116]. The geometric interpretation

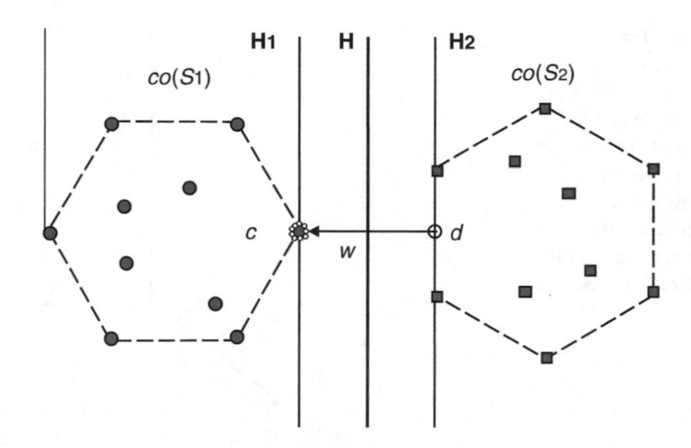

Fig. 9.11 The geometric interpretation and nearest point problem of SVM. co(S1) and co(S2) are two smallest convex sets (convex hulls) shown with dashed lines which contain each class. c and d are the nearest points on them

and nearest point problem (NNP) of SVM can be easily understood by Fig. 9.11.

$$\begin{aligned}&\min_{\alpha}\left\|\sum\nolimits_{i:y_i=+1}\alpha_i\boldsymbol{x}_i-\sum\nolimits_{i:y_i=-1}\alpha_i\boldsymbol{x}_i\right\|^2\\&\text{s.t.}\sum\nolimits_{i:y_i=+1}\alpha_i=1,\sum\nolimits_{i:y_i=-1}\alpha_i=1\\&\alpha_i\geq 0, i=1,\ldots,l\end{aligned}\tag{9.55}$$

If $\boldsymbol{\alpha}^* = \left(\alpha_1^*, \alpha_2^*, \ldots, \alpha_l^*\right)$ is the solution to the convex quadratic optimization Eq. (9.55), then the nearest points in two convex hulls are $\boldsymbol{c} = \sum_{i:y_i=+1}\alpha_i^*\boldsymbol{x}_i$ and $\boldsymbol{d} = \sum_{i:y_i=-1}\alpha_i^*\boldsymbol{x}_i$. Constructing the decision boundary $f(\boldsymbol{x}) = \boldsymbol{w}\cdot\boldsymbol{x} + b$ to be the perpendicular bisector of the line segment joining the two nearest points means that w lies along the line segment and the midpoint p of the line segment satisfies the function $f(x) = 0$. w and p can be computed by c and d: $w = c\ d, p = (1/2)(c + d)$, then b = w p. In the end, the classification discriminant function can be written as: f(x) = sgn(w x + b), where sgn( ) is the sign function.

Similar to the above process of the geometric method of SVM, ASNP method [114] extends the areas searched for the nearest points from the convex hulls in SVM to affine subspaces, and constructs the decision hyperplane separating the affine subspaces with equivalent margin.

9.3.2.2 Affine Subspace Nearest Points (ASNP) Algorithm

Definition 9.1 (Affine subspace). Lee and Seung [117] Given a sample set $S = \{x_1, \ldots, x_m\}$, $x_i \in R^d$, the affine subspace spanned by S can be written as

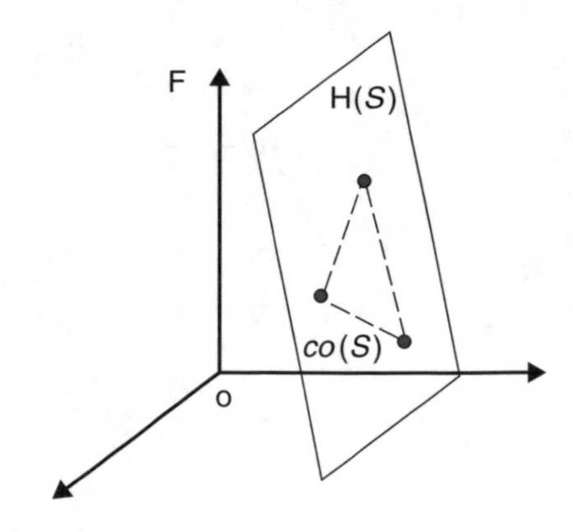

Fig. 9.12 The affine subspace H(S) created by the three samples set S. F is the space three samples lie in. The inner area of the triangle shown with dashed lines is the convex hull co(S), whereas the minimum hyperplane that contains the triangle is the affine subspace H(S)

Eq. (9.56) or Eq. (9.57):

$$H\left(\boldsymbol{S}\right)=\left\{\sum_{i=1}^{m}\alpha_i\boldsymbol{x}_i|\sum_{i=1}^{m}\alpha_i=1\right\} \tag{9.56}$$

$$H\left(\boldsymbol{S}\right)=\left\{\boldsymbol{x}_0+\sum_{i=1}^{m}\alpha_i\left(\boldsymbol{x}_i-\boldsymbol{x}_0\right)\right\},\boldsymbol{x}_0\in H\left(\boldsymbol{S}\right) \tag{9.57}$$

For Eq. (9.56), we can get rid of the constraint $\sum_{i=1}^{m}\alpha_i=1$ by taking a point in H(S) as a new origin x_0. Therefore the equivalent of Eq. (9.56) can be written as Eq. (9.57). We can let x_0 be the average of all samples, $x_0=\frac{1}{m}\sum_{i=1}^{m}x_i$.

In order to interpret the affine subspace, we simply depict the affine subspace in geometry, see, for example in Fig. 9.12.

Compared with the convex hull co(S), the affine subspace contains the convex hull, but is not constrained by $\alpha_i\geq 0$ (see Eq. 9.56). The convex hull only contains the interpolations of the basis vectors, whereas the affine subspace contains not only the convex hull but also the linear extrapolations.

For a binary-class problem with training sets $S_1=\{x_1,x_2,\ldots,x_m\}$ and $S_2=\{x_{m+1},x_{m+2},\ldots,x_n\}$. Two affine subspaces respectively spanned by them are

$$H\left(\boldsymbol{S}_1\right)=\left\{\sum_{i=1}^{m}\alpha_i\boldsymbol{x}_i|\sum_{i=1}^{m}\alpha_i=1\right\} \tag{9.58}$$

$$H\left(\boldsymbol{S}_2\right)=\left\{\sum_{i=m+1}^{n}\alpha_i\boldsymbol{x}_i|\sum_{i=m+1}^{n}\alpha_i=1\right\} \tag{9.59}$$

Then the problem of finding the closest points in affine subspaces can be written as the following optimization problem:

$$\begin{gathered}\min_{\boldsymbol{\alpha}}\left\|\sum_{i=1}^{m}\alpha_i\boldsymbol{x}_i-\sum_{i=m+1}^{n}\alpha_i\boldsymbol{x}_i\right\|^2\\ \text{s.t.}\sum_{i=1}^{m}\alpha_i=1,\sum_{i=m+1}^{n}\alpha_i=1,i=1,\ldots,l\end{gathered} \tag{9.60}$$

Compared with Eq. (9.56), Eq. (9.60) is not under constraint $\alpha_i \geq 0$ which can be also converted into an unconstrained optimal problem as follows:

As Eq. (9.57) is represented, Eqs. (9.58) and (9.59) can be written in unconstrained Eqs. (9.61) and (9.62).

$$H(S_1) = \left\{ \overline{u}_1 + \sum\nolimits_{i=1}^{m} \alpha_i \left(x_i - \overline{u}_1 \right) \right\} \tag{9.61}$$

$$H(S_2) = \left\{ \overline{u}_2 + \sum\nolimits_{i=m+1}^{n} \alpha_i \left(x_i - \overline{u}_2 \right) \right\} \tag{9.62}$$

where $\overline{u}_1 = \frac{1}{m} \sum_{i=1}^{m} x_i$ and $\overline{u}_2 = \left(\frac{1}{n-m} \right) \sum_{i=m+1}^{n} x_i$.

So Eq. (9.60) can be rewritten as

$$\min_\alpha \left\| \left(u_1 + \sum\nolimits_{i=1}^{m} \alpha_i \left(x_i - \overline{u}_1 \right) \right) - \left(u_2 + \sum\nolimits_{i=m+1}^{n} \alpha_i \left(x_i - \overline{u}_2 \right) \right) \right\|^2 \tag{9.63}$$

where $\boldsymbol{\alpha} = \{\alpha_1, \alpha_2, \ldots, \alpha_m\}^T$.

Equation (9.63) is an unconstrained optimal problem, which can be computed directly, and $\boldsymbol{\alpha}$ is

$$\alpha = \left(A^{\mathrm{T}} A \right)^{+} A^{\mathrm{T}} \left(\overline{u}_1 - \overline{u}_2 \right) \tag{9.64}$$

Or

$$\alpha = \left(A^T A + \sigma I \right)^{-1} A^T \left(\overline{u}_1 - \overline{u}_2 \right) \tag{9.65}$$

where $A = \left(\left(\overline{u}_1 - x_1 \right), \ldots, \left(\overline{u}_1 - x_m \right), \left(x_{m+1} - \overline{u}_2 \right), \ldots, \left(x_n - \overline{u}_2 \right) \right)$, and $(A^{\mathrm{T}}A)^{+}$ is the pseudo-inverse of $A^{\mathrm{T}}A$; $\sigma \geq 0$, and I is n*n identity Matrix.

Then the two nearest points in affine subspaces are

$$c = \overline{u}_1 + \sum\nolimits_{i=1}^{m} \alpha_i \left(x_i - \overline{u}_1 \right) \tag{9.66}$$

$$d = \overline{u}_2 + \sum\nolimits_{i=m+1}^{n} \alpha_i \left(x_i - \overline{u}_2 \right) \tag{9.67}$$

The midpoint of the line segment joining c and d is p = (1/2) (c + d). Similar to the nearest point problem of SVM, the decision boundary w x + b = 0 is the perpendicular bisector of the line segment. Thus, w = c − d and b = −w*p. Correspondingly, the decision function is.

$$\begin{aligned} f(x) &= \operatorname{sgn}(w \cdot x + b) \\ &= \operatorname{sgn}\left(\sum\nolimits_{i=1}^{n} y_i \alpha_i \left(x_i \cdot x \right) - (12) \sum\nolimits_{i=1}^{n} \sum\nolimits_{j=1}^{n} y_i \alpha_i \alpha_j \left(x_i \cdot x_j \right) \right) \end{aligned} \tag{9.68}$$

From the above process, we can see that ASNP computing the nearest points in the affine subspaces avoids convex quadratic programming routine and can directly obtain the optimum solution as Eq. (9.67) or Eq. (9.68).

We have introduced the linear ASNP above. But in real world, some data distribution is more complex and nonlinear. When convex hulls intersect (i.e. nonlinearly separating), the distance of nearest points from convex hulls will be zero. Similar with that, when the affine subspaces intersect, the distance in ASNP will also be zero. For the nonlinear distribution data, SVM introduces kernel trick to transform the nonlinear problem to a linear problem (i.e. convex hulls are non-intersection) theoretically. Now kernel method has been widely applied in classification problem, and it has been an effective method for nonlinear or complex data problems. In order to deal with nonlinear problems, we extend the ASNP algorithm to a nonlinear KASNP algorithm by the kernel trick in this section.

9.3.2.3 Kernel Affine Subspace Nearest Points (KASNP) Algorithm

Kernel Method and Kernel Trick

Kernel method [91, 118] is an algorithm that, by replacing the inner product with an appropriate positive definite function, implicitly performs a nonlinear mapping U of the input data from Rd into a high-dimensional feature space H. To compute dot products of (U(x) U(x0)), we employ kernel representation of the form k(x, x0) = (U(x) U(x0)), which allows us to compute the value of the dot products in H without having to actually carry out the map U.

Cover's theorem states that if the transformation is nonlinear and the dimensionality of the feature space is high enough, then the input space may be transformed into a new feature space where the patterns are linearly separable with high probability [119]. That is, when the decision function is not a linear function of the data, the data can be mapped from the input space into a high dimensional feature space by a nonlinear transformation. In this high dimensional feature space, a generalized optimal separating hyperplane is constructed. This nonlinear transformation just can be performed in an implicit way through the kernel methods. Thus the basic principle behind kernel-based algorithms is that a nonlinear mapping is used to extend the input space into a higher-dimensional feature space. Implementing a linear algorithm in the feature space then corresponds to a nonlinear version of the algorithm in the original input space. Kernel-based classification algorithms, primarily in Support Vector Machines (SVM), have gained a great deal of popularity in machine learning fields [91, 118, 120, 121].

Common choices of kernel function are the linear kernel k(x, y) = (x y), the polynomial kernel k(x, y) = (1 + (x y))d, and the radial basis function (RBF) kernel k(x, y) = exp (1/2)(kx yk/r)2 and the sigmoid kernel k(x, y) = tanh(b(x y) c). In this section, we adopt linear kernel and RBF kernel for experiments.

Kernel Affine Subspace Nearest Points (KASNP) Algorithm

Suppose a nonlinear mapping U of the input data in Rd into a high-dimensional feature space H. In space H, we construct the ASNP classifier. Similar to the linear case (see Eq. 9.63), the optimal problem of the closest points in H can be written as the following optimization problem:

$$\min_{\alpha}\left\|\left(\boldsymbol{u}_1+\sum_{i=1}^{m}\alpha_i\left(\Phi\left(\boldsymbol{x}_i\right)-\overline{\boldsymbol{u}}_1\right)\right)-\left(\boldsymbol{u}_2+\sum_{i=m+1}^{n}\alpha_i\left(\Phi\left(\boldsymbol{x}_i\right)-\overline{\boldsymbol{u}}_2\right)\right)\right\|^2 \tag{9.69}$$

Where $\overline{\boldsymbol{u}}_1=\frac{1}{m}\sum_{i=1}^{m}\Phi\left(\boldsymbol{x}_i\right), \overline{\boldsymbol{u}}_2=\frac{1}{n-m}\sum_{i=m+1}^{n}\Phi\left(\boldsymbol{x}_i\right)$.

Let $\boldsymbol{A}=\left(\overline{\boldsymbol{u}}_1-\Phi\left(\boldsymbol{x}_1\right),\ldots,\overline{\boldsymbol{u}}_1-\Phi\left(\boldsymbol{x}_m\right),\Phi\left(\boldsymbol{x}_{m+1}\right)-\overline{\boldsymbol{u}}_2,\ldots,\Phi\left(\boldsymbol{x}_n\right)-\overline{\boldsymbol{u}}_2,\right.$ Formula (9.69) can written as

$$\min_{\alpha} f\left(\boldsymbol{\alpha}\right)=\min_{\alpha}\left\|\left(\overline{\boldsymbol{u}}_1-\overline{\boldsymbol{u}}_2\right)-\boldsymbol{A}\alpha\right\|^2 \tag{9.70}$$

By solving $\frac{\partial f}{\partial \alpha}=0$, we have

$$\boldsymbol{A}^T A\alpha=\boldsymbol{A}^T\left(\overline{\boldsymbol{u}}_1-\overline{\boldsymbol{u}}_2\right) \tag{9.71}$$

In Eq. (9.71) $\boldsymbol{A}^T\boldsymbol{A}$ *and* $\boldsymbol{A}^T\left(\overline{\boldsymbol{u}}_1-\overline{\boldsymbol{u}}_2\right)$ can be cast in terms of dot products $(\Phi(\boldsymbol{x}_i)\cdot\Phi(\boldsymbol{x}_j))$ as follows:

$$\boldsymbol{A}^T\boldsymbol{A}=\left(\boldsymbol{M}^T\boldsymbol{F}+\boldsymbol{E}\right)^T\left(\Phi^T\Phi\right)\left(\boldsymbol{M}^T\boldsymbol{F}+\boldsymbol{E}\right) \tag{9.72}$$

$$\boldsymbol{A}^T\left(\overline{\boldsymbol{u}}_1-\overline{\boldsymbol{u}}_2\right)=\left(\boldsymbol{M}^T\boldsymbol{F}+\boldsymbol{E}\right)^T\left(\Phi^T\Phi\right)\boldsymbol{F}^T\boldsymbol{m}^T \tag{9.73}$$

Where $\boldsymbol{\Phi}=(\Phi(\boldsymbol{x}_1),\ldots,\Phi(\boldsymbol{x}_m),\Phi(\boldsymbol{x}_{m+1}),\ldots,\Phi(\boldsymbol{x}_n))$,

$$\boldsymbol{M}=\begin{pmatrix}\frac{1}{m} & 0\\ 0 & \frac{1}{n-m}\end{pmatrix}\begin{pmatrix}1\cdots1\,0\cdots0\\ 0\cdots0\,1\cdots1\end{pmatrix}_{2\times0},$$

$$F=\begin{pmatrix}1\cdots1\,0\cdots0\\ 0\cdots0\,1\cdots1\end{pmatrix}_{2\times n},m=\left(\frac{1}{m},\frac{1}{n-m}\right),$$

$$E = \begin{bmatrix} -1 & & & & & \\ & \ddots & & & & \\ & & -1 & & & \\ & & & 1 & & \\ & & & & \ddots & \\ & & & & & 1 \end{bmatrix},,$$

$$\boldsymbol{\Phi}^T \boldsymbol{\Phi} = \begin{pmatrix} (\Phi(\boldsymbol{x}_1) \cdot \Phi(\boldsymbol{x}_1)) & \cdots & (\Phi(\boldsymbol{x}_1) \cdot \Phi(\boldsymbol{x}_n)) \\ \vdots & \ddots & \vdots \\ (\Phi(\boldsymbol{x}_n) \cdot \Phi(\boldsymbol{x}_1)) & \cdots & (\Phi(\boldsymbol{x}_n) \cdot \Phi(\boldsymbol{x}_n)) \end{pmatrix}.$$

Employing kernel representations of the form $k(\boldsymbol{x}_i, \boldsymbol{x}_j) = (\boldsymbol{\Phi}(\boldsymbol{x}_i) \cdot \boldsymbol{\Phi}(\boldsymbol{x}_j))$, $\boldsymbol{\Phi}^{\mathrm{T}}\boldsymbol{\Phi}$ is

$$\boldsymbol{K} = \Phi^T \Phi = \begin{pmatrix} k(\boldsymbol{x}_1, \boldsymbol{x}_1) & k(\boldsymbol{x}_1, \boldsymbol{x}_2) & \ldots & k(\boldsymbol{x}_1, \boldsymbol{x}_n) \\ k(\boldsymbol{x}_2, \boldsymbol{x}_1) & k(\boldsymbol{x}_2, \boldsymbol{x}_2) & \ldots & k(\boldsymbol{x}_2, \boldsymbol{x}_n) \\ \vdots & \vdots & \ddots & \vdots \\ k(\boldsymbol{x}_n, \boldsymbol{x}_1) & k(\boldsymbol{x}_n, \boldsymbol{x}_2) & \ldots & k(\boldsymbol{x}_n, \boldsymbol{x}_n) \end{pmatrix}$$

Equations (9.72) and (9.73) can be kernelized:

$$\boldsymbol{A}^T \boldsymbol{A} = \left(\boldsymbol{M}^T \boldsymbol{F} + \boldsymbol{E}\right)^T \boldsymbol{K} \left(\boldsymbol{M}^T \boldsymbol{F} + \boldsymbol{E}\right) \tag{9.74}$$

$$\boldsymbol{A}^T (\overline{\boldsymbol{u}}_1 - \overline{\boldsymbol{u}}_2) = \left(\boldsymbol{M}^T \boldsymbol{F} + \boldsymbol{E}\right)^T \boldsymbol{K} \boldsymbol{F}^T \boldsymbol{m}^T \tag{9.75}$$

So we can directly obtain the solution α of Eq. (9.69):

$$\boldsymbol{\alpha} = \left(\boldsymbol{A}^T \boldsymbol{A}\right)^+ \left(\boldsymbol{A}^T (\overline{\boldsymbol{u}}_1 - \overline{\boldsymbol{u}}_2)\right) \tag{9.76}$$

or

$$\boldsymbol{\alpha} = \left(\boldsymbol{A}^T \boldsymbol{A} + \sigma \boldsymbol{I}\right)^{-1} \left(\boldsymbol{A}^T (\overline{\boldsymbol{u}}_1 - \overline{\boldsymbol{u}}_2)\right) \tag{9.77}$$

where $\boldsymbol{A}^T\boldsymbol{A}^+$ is pseudo-inverse of $\boldsymbol{A}^T\boldsymbol{A}$; $\sigma \geq 0$, and I is n*n identity Matrix.

After getting the optimal solution α, two nearest point c and d can be represented by α:

$$\begin{aligned} \boldsymbol{c} &= \overline{\boldsymbol{u}}_1 + \textstyle\sum_{i=1}^m \alpha_i \left(\Phi(\boldsymbol{x}_i) - \overline{\boldsymbol{u}}_1\right) \\ &= \textstyle\sum_{i=1}^m \sum \left(\frac{1}{m}\left(1 - \sum_{i=1}^m \alpha_i\right) + \alpha_i\right) \Phi(\boldsymbol{x}_i) \end{aligned} \tag{9.78}$$

$$\begin{aligned}\boldsymbol{d} &= \overline{\boldsymbol{u}}_2 + \sum\nolimits_{i=m+1}^{n}\alpha_i \left(\Phi\left(\boldsymbol{x}_i\right) - \overline{\boldsymbol{u}}_2\right) \\ &= \sum\nolimits_{i=m+1}^{n}\left(\frac{1}{n-m}\left(1 - \sum\nolimits_{i=m+1}^{n}\alpha_i\right) + \alpha_i\right)\Phi\left(\boldsymbol{x}_i\right)\end{aligned} \tag{9.79}$$

then, w, p and b can be written as:

$$\boldsymbol{w} = \boldsymbol{c} - \boldsymbol{d} = \Phi \boldsymbol{v}_1 \tag{9.80}$$

$$\boldsymbol{p} = (12)\left(\boldsymbol{c} + \boldsymbol{d}\right) = \frac{1}{2}\Phi \boldsymbol{v}_2 \tag{9.81}$$

$$b = -\boldsymbol{w} \cdot \boldsymbol{p} = -\frac{1}{2}v_1^T \Phi^T \Phi v_2 = -\frac{1}{2}v_1^T \boldsymbol{K} v_2 \tag{9.82}$$

where

$$\boldsymbol{v}_1 = \begin{pmatrix} \frac{1}{m}\left(1 - \sum\nolimits_{i=1}^{m}\alpha_i\right) + \alpha_1 \\ \frac{1}{m}\left(1 - \sum\nolimits_{i=1}^{m}\alpha_i\right) + \alpha_m \\ \frac{-1}{n-m}\left(1 - \sum\nolimits_{i=m+1}^{n}\alpha_i\right) - \alpha_{m+1} \\ \frac{-1}{n-m}\left(1 - \sum\nolimits_{i=m+1}^{n}\alpha_i\right) - \alpha_n \end{pmatrix} \tag{9.83}$$

$$\boldsymbol{v}_1 = \begin{pmatrix} \frac{1}{m}\left(1 - \sum\nolimits_{i=1}^{m}\alpha_i\right) + \alpha_1 \\ \frac{1}{m}\left(1 - \sum\nolimits_{i=1}^{m}\alpha_i\right) + \alpha_m \\ \frac{-1}{n-m}\left(1 - \sum\nolimits_{i=m+1}^{n}\alpha_i\right) - \alpha_{m+1} \\ \frac{-1}{n-m}\left(1 - \sum\nolimits_{i=m+1}^{n}\alpha_i\right) - \alpha_n \end{pmatrix} \cdot Z \tag{9.84}$$

So the decision boundary $(w \cdot \Phi(x)) + b = 0$ is

$$\boldsymbol{v}_2^T \boldsymbol{k}_x - \frac{1}{2}\boldsymbol{v}_2^T K \boldsymbol{v}_1 = 0 \tag{9.85}$$

Where $\boldsymbol{k}_x = \boldsymbol{\Phi}^T \boldsymbol{\Phi}(\boldsymbol{x}) = (k(\boldsymbol{x}_1, \boldsymbol{x}), k(\boldsymbol{x}_2, \boldsymbol{x}), \ldots, k(\boldsymbol{x}_n, \boldsymbol{x}))^T$.
The decision function $f(\boldsymbol{x}) =$ sgn $(\boldsymbol{w} \cdot \boldsymbol{\Phi}(\boldsymbol{x}) + b)$ is

$$f\left(\boldsymbol{x}\right) = \text{sgn}\left(\boldsymbol{w} \cdot \Phi\left(\boldsymbol{x}\right) + b\right) = \text{sgn}\left(\boldsymbol{v}_2^T \boldsymbol{k}_x - \frac{1}{2}\boldsymbol{v}_2^T \boldsymbol{K} \boldsymbol{v}_1\right) \tag{9.86}$$

According to the previous descriptions, the overall process of KASNP learning algorithm can be summarized into the following three steps:

Step 1: Computing the optimal solution $\boldsymbol{\alpha}$ of the nearest points problem of KASNP by training set:

$$\boldsymbol{\alpha} = \left(\boldsymbol{A}^T \boldsymbol{A}\right)^+ \left(\boldsymbol{A}^T\left(\overline{\boldsymbol{u}}_1 - \overline{\boldsymbol{u}}_2\right)\right) \text{ or } \boldsymbol{\alpha} = \left(\boldsymbol{A}^T \boldsymbol{A} + \sigma \boldsymbol{I}\right)^{-1}\left(\boldsymbol{A}^T\left(\overline{\boldsymbol{u}}_1 - \overline{\boldsymbol{u}}_2\right)\right)$$

Step 2: Constructing decision boundary by $\boldsymbol{\alpha}$:

$$\boldsymbol{v}_2^T \boldsymbol{k}_x - \frac{1}{2}\boldsymbol{v}_2^T \boldsymbol{K} \boldsymbol{v}_1 = 0$$

Correspondingly, the decision function is

$$f(\boldsymbol{x}) = \text{sgn}\left(\boldsymbol{v}_2^T \boldsymbol{k}_x - \frac{1}{2}\boldsymbol{v}_2^T \boldsymbol{K} \boldsymbol{v}_1\right)$$

Step 3: Testing a sample $\boldsymbol{y}$,

If $f(\boldsymbol{y}) \geq 0$, $\boldsymbol{y} \in$ the class of $\boldsymbol{S}_1$; otherwise, $\boldsymbol{y} \in$ the class of $\boldsymbol{S}_2$

9.3.2.4 Two-Spiral Problem Test

2D two-spiral classification is a classical nonlinear problem and has been particularly popular for testing novel statistical pattern recognition classifiers. The problem is a difficult test case for learning algorithms [122, 123] and is known to give neural networks severe problems, but it can be successfully solved by nonlinear kernel SVMs [124, 125]. In this section, we also tested our KASNP with RBF kernel $k(x, y) = \exp\left(\frac{1}{2}\right)(x - y/\sigma)^2$ on a 2D two-spiral dataset accessible from the Carnegie Mellon repository [126]. The benchmark dataset, download from http://www.cgi.cs.cmu.edu/afs/cs.cmu.edu/project/vairepository/ai/areas/ai/areas/neural/bench/cmu/0.html, has two classes of spiral-shaped training data points, with 97 points for each, and is illustrated in Fig. 9.13. In order to visualize the separating surface by KASNP, the nodes of a 2D grid (0.05 space per grid) are tested and marked with different color (gray and white) to show their class. Figure 9.14 shows the decision region by KASNP. The parameter r of RBF kernel for KASNP is 0.8.

In Fig. 9.14, our KASNP constructs a smooth nonlinear spiral-shaped separating surface for the 2D two-spiral dataset, which implies that the KASNP classification method can achieve an excellent generalization for nonlinear data.

9.3.2.5 Credit Evaluation Applications and Experiments

Credit risk evaluation is a very typical classification problem to identify “good” and “bad” creditors. In this section, we apply KASNP for credit risk evaluation. To test the efficacy of our proposal KASNP for creditor evaluation, we compare it with SVM by linear kernel and RBF kernel on three real world credit datasets: Australian credit dataset, German credit dataset and a major US credit dataset.

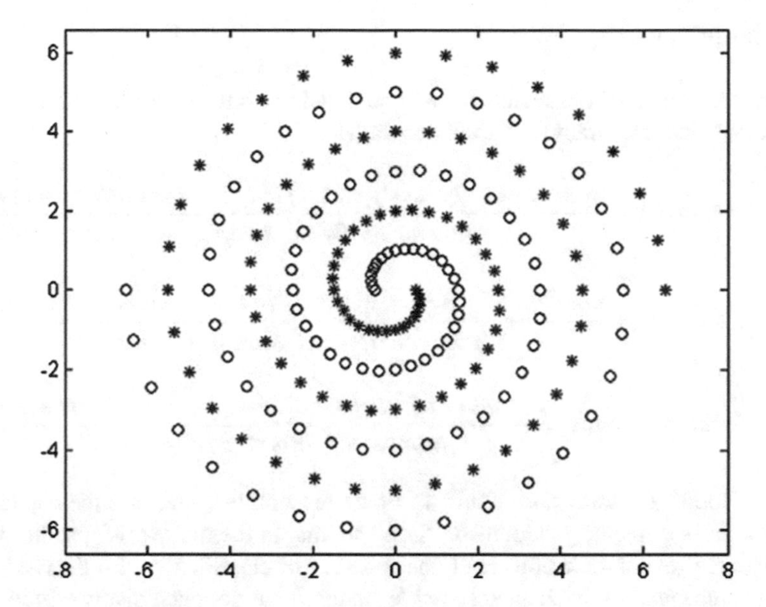

Fig. 9.13 2D two-spiral dataset: "o" spiral 1, "*" spiral 2

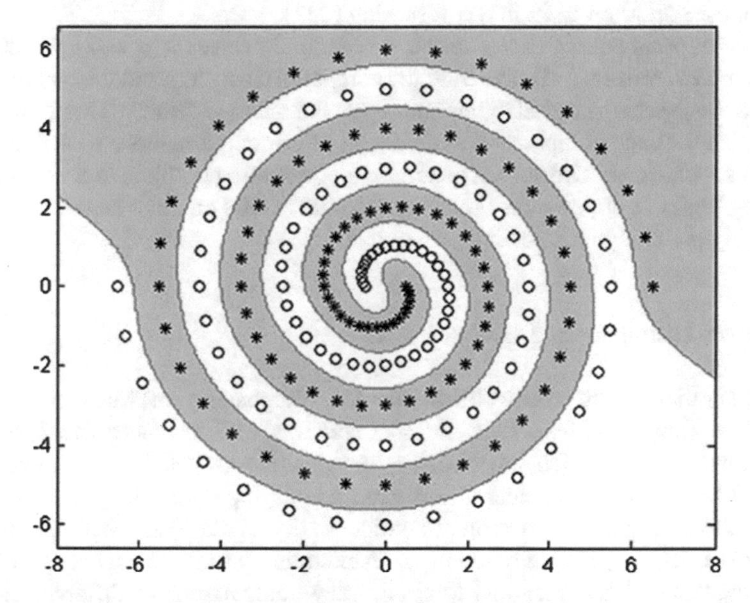

Fig. 9.14 The separation generated by RBF kernel KASNP

The compared linear kernel KASNP is equivalent to original ASNP method [114], that is, ASNP method is a special case of KASNP when kernel function is linear kernel.

Experiment Design

In our experiments, three accuracies will be tested to evaluate the classifiers, “Good” accuracy, “Bad” accuracy and Total accuracy:

$$"Good" \ Accuracy = \frac{number \ of \ correctly \ classified \ "Good" \ samples \ in \ test \ set}{number \ of \ "Good" \ samples \ in \ test \ set}$$

$$"Bad" \ Accuracy = \frac{number \ of \ correctly \ classified \ "Bad" \ samples \ in \ test \ set}{number \ of \ "Bad" \ samples \ in \ test \ set}$$

$$Total \ Accuracy = \frac{number \ of \ correct \ classification \ in \ test \ set}{number \ of \ samples \ in \ test \ set}$$

where “Good” accuracy and “Bad” accuracy respectively measure the capacity of the classifiers to identify “Good” or “Bad” clients. In the real world, for the special purposes to prevent the credit fraud, the accuracy of classification for the risky class must be improved to reach an acceptable standard but not excessively affecting the accuracy of classification for other classes. Thus, improving “Bad” accuracy is one of the most important tasks in credit scoring [127].

In our experiments of each dataset, we randomly select p (p = 40, 60, 80, ..., 180) samples from each class to train the compared classifiers and the remaining for the test. We repeat the test 20 times and report the mean of “Bad”, “Good” and Total accuracies for each compared classifiers. All of our experiments are carried out on Matlab 7.0 platform. The convex quadratic programming problem of SVM is solved utilizing Matlab optimal tools. The experimental results on three credit datasets are separately given in the following subsections.

Results on Australian Credit Dataset

The Australian credit dataset from the UCI Repository of Machine Learning Databases (http://archive.ics.uci.edu/ml/) contains 690 instances of MasterCard applicants, 307 of which are classified as positive and 383 as negative. Each instance has 14 attributes, and all attribute names and values have been changed to meaningless symbols to protect confidentiality of the data. With the number variety (40, 60, ..., 180) of randomly selected training samples per class, the “Bad” accuracy, “Good” accuracy and total accuracy comparisons of different methods on Australian credit dataset, are shown in Tables 9.11, 9.12, and 9.13 respectively. Parameter r of RBF kernel is set to 50,000 for both RBF SVM and RBF KASNP, and the penalty constant C of SVM is ∞.

In above experimental results, for “Bad” accuracy, nonlinear classifiers RBF SVM and RBF KASNP outperform other two linear classifiers, and RBF KASNP is better than RBF SVM. For “Good” accuracy, linear kernel KASNP is the best

Table 9.11 "Bad" accuracy (%) comparisons of different methods on Australian dataset

Number of training data per class	"Bad" accuracy (%) comparisons on Australian dataset			
	Linear SVM	RBF SVM	Linear KASNP	RBF KASNP
40	79.65	84.50	81.97	86.90
60	83.08	85.20	82.06	88.05
80	81.01	86.07	81.34	88.18
100	84.12	87.37	81.27	87.60
120	83.71	86.71	81.48	87.51
140	82.12	87.14	81.40	87.43
160	82.38	87.00	80.25	87.02
180	79.48	86.77	80.07	86.26

Table 9.12 "Good" accuracy (%) comparisons of different methods on Australian dataset

Number of training data per class	"Good" accuracy (%) comparisons on Australian dataset			
	Linear SVM	RBF SVM	Linear KASNP	RBF KASNP
40	81.85	73.95	89.76	72.73
60	87.00	74.98	89.74	76.84
80	85.15	78.28	91.43	79.52
100	83.31	79.44	91.69	81.06
120	84.87	81.36	91.90	82.86
140	83.68	82.49	91.32	84.07
160	84.25	84.05	92.24	85.14
180	84.76	83.86	91.85	86.22

Table 9.13 Total accuracy (%) comparisons of different methods on Australian dataset

Number of training data per class	Total accuracy (%) comparisons on Australian dataset			
	Linear SVM	RBF SVM	Linear KASNP	RBF KASNP
40	80.61	79.89	85.38	80.70
60	84.78	80.77	85.39	83.19
80	82.78	82.74	85.66	84.47
100	83.78	84.02	85.67	84.84
120	84.19	84.49	85.81	85.58
140	82.76	85.24	85.44	86.06
160	83.12	85.82	85.01	86.27
180	81.52	85.65	84.61	86.24

of all classifiers, and its "Good" accuracy can get 89.74–92.24% (see Table 9.12). From the total accuracy comparisons, KASNP dominates SVMs. Linear KASNP can reach the highest total accuracy when the number of training samples $p = 40, \ldots, 120$, and RBF KASNP is the best one when $p = 140, 160, 180$ (see Table 9.13).

Table 9.14 "Bad" accuracy (%) comparisons of different methods on German dataset

Number of training data per class	"Bad" accuracy (%) comparisons on German dataset			
	Linear SVM	RBF SVM	Linear KASNP	RBF KASNP
40	65.87	67.08	67.12	67.15
60	67.90	68.77	67.08	67.60
80	69.64	70.20	69.73	70.66
100	71.47	69.92	71.35	71.53
120	70.92	71.81	72.28	72.36
140	71.06	72.47	73.59	73.16
160	71.29	72.75	71.46	73.75
180	73.13	72.13	72.42	72.83

Table 9.15 "Good" accuracy (%) comparisons of different methods on German dataset

Number of training data per class	"Good" accuracy (%) comparisons on German dataset			
	Linear SVM	RBF SVM	Linear KASNP	RBF KASNP
40	64.91	68.83	66.56	68.89
60	67.73	69.16	66.95	71.09
80	68.75	69.75	69.56	69.60
100	68.23	69.83	69.38	69.89
120	69.89	69.59	68.88	69.58
140	69.63	69.96	69.22	69.83
160	70.94	70.56	70.85	71.31
180	70.33	70.57	70.40	70.66

Results on German Credit Dataset

The German credit dataset from the UCI Repository of Machine Learning Databases (http://archive.ics.uci.edu/ml/) concludes 1000 instances, 700 instances of creditworthy applicants and 300 instances whose credit should not be extended. For each instance, 24 numerical attributes describe the credit history, account balances, loan purpose, loan amount, employment status, and personal information. The different accuracy comparisons of the classifiers on German dataset are given in Tables 9.11, 9.12, and 9.13 respectively. The parameter r of RBF kernel for SVM and KASNP is set to r = 20,000, and the penalty constant C of SVM is set to 1.

From the experimental results in Tables 9.14, 9.15, and 9.16, we can see that our proposed RBF KASNP is slightly better than others. RBF KASNP has five highest accuracies (when p = 40, 80, 100, 120, 160) in "Bad" accuracy comparison, and six best results (when p = 40, 60, 80, 100, 160, 180) for "Good" clients identification. For total accuracy, RBF KASNP continuously achieves the highest accuracy in eight comparison results.

Table 9.16 Total accuracy (%) comparisons of different methods on German dataset

Number of training data per class	Total accuracy (%) comparisons on German dataset			
	Linear SVM	RBF SVM	Linear KASNP	RBF KASNP
40	65.18	68.34	66.72	68.40
60	67.77	69.05	66.99	70.14
80	68.98	69.87	69.60	69.88
100	69.04	69.86	69.87	70.30
120	70.13	70.12	69.68	70.24
140	69.95	70.51	70.19	70.57
160	71.01	71.01	70.98	71.82
180	70.85	70.86	70.78	71.07

Table 9.17 "Bad" accuracy (%) comparisons of different methods on USA dataset

Number of training data per class	"Bad" accuracy (%) comparisons on USA dataset			
	Linear SVM	RBF SVM	Linear KASNP	RBF KASNP
40	63.97	65.34	61.83	81.32
60	65.82	66.01	68.35	82.44
80	67.37	69.99	71.89	83.33
100	66.32	70.69	74.41	83.29
120	68.07	69.43	77.40	82.82
140	67.58	71.64	78.14	84.59
160	69.27	73.21	78.39	84.14
180	73.13	74.44	79.57	84.37

Results on USA Credit Dataset

The last credit card dataset used in our experiments is provided by a major U.S. bank. It contains 6000 records and 66 derived attributes. Among these 6000 records, 960 are bankruptcy accounts and 5040 are "good" status accounts [128]. The "Bad", "Good" and total accuracy comparisons of the classifiers are shown in Tables 9.17, 9.18, and 9.19 respectively. Parameter r of RBF kernel of SVM and KASNP is $r = 10{,}000$, and the penalty constant C of SVM is $C = 1$.

Comparing the results reported in Tables 9.17, 9.18, and 9.19, we find the following results: (1) RBF KASNP is superior to other classifiers in finding "Bad" clients. As we can see from Table 9.17, only using 80 training samples (40 per class), RBF KASNP can achieve best "Bad" classification results 81.32% which is at least higher 15% than the accuracies of other approaches. (2) For identifying "Good" clients, four approaches have not clear difference, and RBF SVM and linear KASNP respectively have four best results in Table 9.18. (3) From the general view (see Table 9.19), the two KASNP approaches dominate SVMs. RBF KASNP performs the best when $p = 40, \ldots, 120$, and linear KASNP outperforms the others when $p = 140, 160, 180$.

Table 9.18 "Good" accuracy (%) comparisons of different methods on USA dataset

Number of training data per class	"Good" accuracy (%) comparisons on USA dataset			
	Linear SVM	RBF SVM	Linear KASNP	RBF KASNP
40	67.12	67.62	59.13	66.11
60	66.46	67.84	65.73	67.15
80	66.65	66.35	68.33	67.15
100	67.02	67.97	67.40	67.45
120	69.34	69.72	68.36	68.00
140	68.04	68.79	69.44	67.13
160	66.59	68.66	70.52	67.73
180	61.38	68.93	70.18	67.69

Table 9.19 Total accuracy (%) comparisons of different methods on USA dataset

Number of training data per class	Total accuracy (%) comparisons on USA dataset			
	Linear SVM	RBF SVM	Linear KASNP	RBF KASNP
40	67.81	67.27	59.55	68.48
60	66.44	67.56	66.13	69.49
80	67.39	66.90	68.86	69.59
100	66.92	68.37	68.44	69.80
120	69.15	69.68	69.68	70.16
140	67.98	69.20	70.69	69.63
160	66.97	69.30	71.63	70.04
180	63.01	69.69	71.48	69.99

9.3.2.6 Discussion

From above experimental results of three credit datasets, we can conclude that as a whole the proposed KASNP is comparable with SVM for creditor classification. As we know, the capacity of finding "Bad" clients is an important measure for credit risk evaluation approaches. From "Bad" accuracy comparison experimental results in Tables 9.11, 9.12, and 9.13, we note that our proposed KASNP with RBF kernel can achieve the best performance for identifying "Bad" creditors. Especially for US dataset, KASNP obviously outperformed other approaches. In total performance, RBF KASNP also performed better than SVMs. Thus, RBF KASNP classifier made a better risky classification performance. Moreover, we also note that, for "Good" clients identification, linear KASNP is a good classifier. Especially on Australian dataset, linear KASNP obtained wonderful "Good" accuracies, while its "Bad" accuracies also kept acceptable standard.

9.3.3 A Dynamic Assessment Method for Urban Eco-Environmental Quality Evaluation

This subsection provides an urban eco-environmental quality assessment system with a dynamic assessment of the Yangtze River Delta and the Pearl River Delta economic zones are proposed and analyzed.

9.3.3.1 Related Works

Assessment of Urban Eco-Environmental Quality

With the rapid surge in urbanization around the world, there are a series of urban eco-environmental problems. In 1962, Carson described the destruction of urban eco-environment in Silent Spring for the first time, which led to the wide-range attention. In 1971, the United Nations Educational, Scientific and Cultural Organization developed the 'Man and the Biosphere' research project, which focused on the eco-environment of human settlements and carried out the urban research subject in human ecology theories and views [129]. Schneider pointed out: 'in contrast with common sense of many urban sociologist and environmentalists, that the urban basic issues are not clean air and water, not endangered species or environment, not energy, nor the urban housing construction and renovation investment, but the association structure of the human environment—the city, it is necessary to build up a harmonious developing city to solve the problem' [29]. In 1984, Yanitsky established a human residence where economy, society and nature are coordinated in development. In 1998, Bohm studied the special urban development process of Vienna in Australia. Although the number of population has not changed significantly, the residential area, road area, and energy consumption have increased significantly, and urban green space reduced significantly. The United Nations human environment and development conference held in Rio de Janeiro, Brazil, pointed out that environmental issue will be the largest challenge in the twenty-first century. The urban eco-environmental quality problem has been an active research fields for years [115, 130–132].

Sensitivity Analysis

Multi-attribute evaluation (MAE) is used in assessment when the known options available are fixed, and the number of the evaluation alternatives are limited [133]. The reliability of the evaluation results is tested in the sensitivity analysis. For a limited alternative set, there are two parameters to determine their ranking of the alternatives: one is the relative importance among attributes, that is, attribute weights; and the other is attribute value correspondent to each alternative.

The early studies of the sensitivity analysis focused on the key attribute weights [134, 135]. Starr [136], Isaacs [137], Fishbum [138] and Evans [139], studied the maximum regional-changed issues when the alternative order remained constant. French and Insua [140] determined the potential competitors in the current optimal solution with the minimum distance method. Masuda [141] and Armacost and Hosseini [142] studied the sensitivity analysis on the analytic hierarchy process (AHP). Ringuest [143] studied the distance sensitivity analysis between the set closest to the original weight and original weight when the optimal solution remained unchanged.

Urban Eco-Environmental Quality Index System

Here, an Urban Eco-Environmental Quality Index System is proposed to assess urban eco-environmental development and quality level.

To build an Urban Eco-Environmental Quality Index System, the following principles should be followed.

People-oriented principle. The core of urban eco-environment is 'human', who is both the creator and the bearer of urban eco-environment. Therefore, the assessment index system should not only reflect on what are closely related with people's living, but also reflect the objective and subjective experience on the environment.

Comprehensiveness principle. The construction of the assessment index system must reflect all aspects of urban eco-environment, including the living conditions, natural environment, social environment, and infrastructure indicators, as well as all aspects of urban environment.

Representative principle. The assessment index system should reflect the main features of urban eco-environment. Both qualitative indicators and quantitative indicators should be included.

9.3.3.2 Selecting Indicators

According to the previous studies [144–146], we selected 25 comprehensive evaluation index, from four perspectives—population ecological indicators, nature ecological indicators, economy ecological indicators, and society ecological indicators to establish the index system, which includes both the cost-based indicators and efficiency-based indicators [147]. The details of all indicators are shown in Table 9.20.

These indicators are collected from the 'China City Statistical Yearbook' and the 'China Statistical Yearbook for Regional Economy', in order to increase the comparability of the index, we unify the indicators to the relative ratio, such as

Table 9.20 Urban eco-environmental quality index system

Factors	Subfactors
Population ecological indicators	Natural population growth rate (%) population density (person/km)
Nature ecological indicators	Percentage of hospital doctors in urban population (%)
	Percentage of college students in urban population (%)
	Percentage of industrial waste water up to the discharge standards (%)
Economy ecological indicators	Industrial waste gas treatment rate (%)
	Industrial solid waste comprehensive utilization rate (%)
	Urban sewage treatment rate (%)
	Domestic garbage treatment rate (%)
	Percentage of comprehensive utilization value of waste products in gross regional product (%)
	Green area per person (square meter/person) green coverage rate in completed area (%)
	Unemployment rate (%)
Society ecological indicators	Public library collection per 100 people (book, part/100 people) percentage of the internet users in urban population (%)
	Household water consumption per person
	(ton/person)
	Household electricity consumption per person (kilowatt hour/person)
	Bus per 10,000 people (bus/10,000 people)
	Urban road area per person (square meter/person)
	Percentage of urban construction land in urban area (%)
	Percentage of tertiary industries in gross regional product (%)
	Gross regional product per person (RMB/person)
	Gross regional product growth rate (%)
	Percentage of investment in science and education in fiscal expenditure (%)
	Average wage of staff and workers (RMB/person)

$$\text{percentage of hospital doctors in urban population} = \frac{\text{hosptial doctors}}{\text{urban population}} \times 100\%$$

$$\text{percentage of investment in science and education in fiscal expenditure} = \frac{\text{investment in science and education}}{\text{fiscal expenditure}} \times 100\%$$

Evaluation Method

The proposed evaluation method includes three steps: The first step is the data preprocessing, the second step is the Dynamic Assessment, and the third step is the sensitivity analysis.

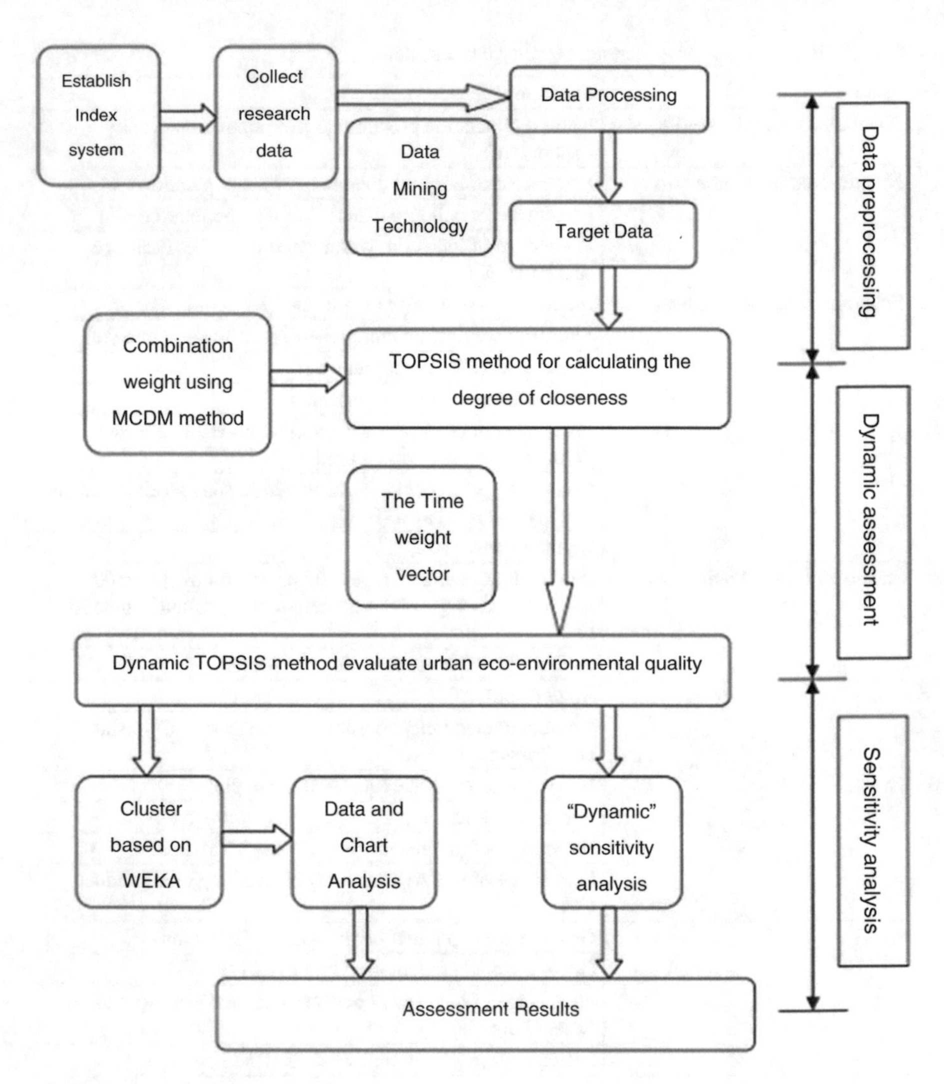

Fig. 9.15 The evaluation framework flow chart

In data preprocessing, evaluation index system is setup and data is processed. The evaluation index system is based on ecological theory, and advices of experts. In data processing, data is cleaned and transformed. A Dynamic Assessment model to evaluate the urban eco-environmental quality is proposed. The sensitivity of attributes weights and values are analyzed.

Figure 9.15 shows the structure of the proposed evaluation model. In the following subsections, we will present the details of the models and methods in proposed framework.

Multi-criteria Decision Making Method

Multi-criteria decision making method (MCDM) is a decision making analysis method, which has been developed since 1970s. MCDM is the study of methods and procedures by which concerns about multiple conflicting criteria can be formally incorporated into the management planning process and the optimum one can be identified from a set of alternatives. In the following subsections, MCDM related methods, Entropy Method, Grey Relation Analysis (GRA) and Technique for order preference by similarity to ideal solution (TOPSIS), which are integrated in this research, are discussed.

Entropy Method

In this research, we introduced the concept of entropy to measure the information, which is a term in information theory, also known as the average amount of information. The index weight is calculated by the Entropy Method. According to the degree of index dispersion, the weight of all indicators is calculated by information entropy. Entropy method is highly reliable and can be easily adopted in information measurement. The calculation steps are as follows:

Suppose we have a decision matrix B with m alternatives and n indicators:

1. In matrix B, feature weight p_{ij} is of the ith alternative to the jth factor:

$$\mathrm{p}_{ij} = y_{ij} \sum_{i=1}^{m} y_{ij} \ (1 \leq i \leq m, 1 \leq j \leq n) \tag{9.87}$$

2. The output entropy e_j of the jth factor becomes

$$e_j = -k \sum_{i=1}^{m} p_{ij} \ \ln p_{ij} \ (k = 1 \ \ln m; 1 \leq j \leq n) \tag{9.88}$$

3. Variation coefficient of the jth factor: g_j can be defined by following equation:

$$g_j = 1 - e_j, \quad (1 \leq j \leq \mathrm{n}) \tag{9.89}$$

   Note that the larger g_j is, the higher the weight should be.
4. Calculate the weight of entropy α_j:

$$\alpha_j = g_j \sum_{j=1}^{m} g_j, \quad (1 \leq j \leq n) \tag{9.90}$$

Grey Relational Analysis Method

Grey relational analysis is a part of grey theory, which can handle imprecise and incomplete information in grey systems. GRA only requires small sample data, simple calculation and the precision is quite high. Specifically, weights are calculated as [148].

Suppose we have the initial matrix R

$$R = \begin{bmatrix} x_1 & x_{12} & \cdots & x_{1n} \\ x_{21} & x_{22} & \cdots & x_{2n} \\ \vdots & \vdots & \cdots & \vdots \\ x_{m1} & x_{m2} & \cdots & x_{mn} \end{bmatrix}$$

1. Standardize the raw matrix R

$$R = \begin{bmatrix} x'_1 & x'_{12} & \cdots & x'_{1n} \\ x'_{21} & x'_{22} & \cdots & x'_{2n} \\ \vdots & \vdots & \cdots & \vdots \\ x'_{m1} & x'_{m2} & \cdots & x'_{mn} \end{bmatrix} \tag{9.91}$$

2. Generate the reference sequence x'_0

$$x'_0 = \left(x'_0(1), x'_0(2), \cdots, x'_0(n)\right) \tag{9.92}$$

$x'_0(j)$ is the largest and normalized value in the jth factor.

3. Calculate the difference $\Delta_{0i}(j)$ between the normalize sequences and the reference sequence x'_0

$$\Delta_{0i}(j) = |x'_0(j) - x'_{ij}|$$

$$\Delta = \begin{bmatrix} \Delta_{01}(1) & \Delta_{01}(2) & \cdots & \Delta_{01}(n) \\ \Delta_{02}(1) & \Delta_{02}(2) & \cdots & \Delta_{02}(n) \\ \vdots & \vdots & \vdots & \vdots \\ \Delta_{0m}(1) & \Delta_{0m}(2) & \cdots & \Delta_{0m}(n) \end{bmatrix} \tag{9.93}$$

4. Compute the grey coefficient: $r_{0i}(j)$

$$r_{0i}(j) = \frac{\min_{i=1}^{n} \min_{j=1}^{m} \Delta_{0i}(j) + \delta \max_{i=1}^{n} \max_{j=1}^{m} \Delta_{0i}(j)}{\Delta_{0i}(j) + \delta \max_{i=1}^{n} \max_{j=1}^{m} \Delta_{0i}(j)} \tag{9.94}$$

where δ is a distinguished coefficient. Usually, the value of d often is set to 0.5, to offer moderate distinguishing effects and good stability.

5. Obtain the grey relational degree value: b_i

$$b_i = \frac{1}{n} \sum_{j=1}^{n} r_{0i}(j) \tag{9.95}$$

6. Finally, calculate the weight of GRA: β_i

$$\beta_i = \frac{b_i}{\sum_{i=1}^{n} b_i} \tag{9.96}$$

In this research, we use Entropy and the GRA method to calculate the normalized weight of the indicators.

Technique for Order Preference by Similarity to Ideal Solution Method

Technique for order preference by similarity to ideal solution TOPSIS was initially developed to rank alternatives over multiple criteria. TOPSIS finds the best alternatives by minimizing the distance to the ideal solution and maximizing the distance to the nadir or negative-ideal solution [34]. All alternative solutions can be ranked according to their closeness to the ideal solution. Because its first introduction, a number of extensions and variations of TOPSIS have been developed over the years. The calculation steps are as follows:

1. Calculate the normalized decision matrix A. The normalized value a_{ij} is calculated as

$$a_{ij} = \frac{x_{ij}}{\sqrt{\sum_{i=1}^{m} \left(x_{ij}\right)^2}} \, (1 \leq i \leq m, 1 \leq j \leq n) \tag{9.97}$$

2. Calculate the weighted normalized decision matrix

$$D = \left(a_{ij} * w_j\right) (1 \leq i \leq \mathrm{m}, 1 \leq \mathrm{j} \leq \mathrm{n}) \tag{9.98}$$

where w_j is the weight of the ith criterion, and $\sum_{j=1}^{n} w_j = 1$.

3. Calculate the ideal solution $V*$ and the negative ideal solution V^-

$$\begin{aligned} V^* &= \left\{v_1^*, v_2^*, \cdots, v_n^*\right\} = \left\{\left(\max_i v_{ij} | j \in J\right), \left(\min_i v_{ij} | j \in J'\right)\right\} \\ V^- &= \left\{v_1^-, v_2^-, \cdots, v_n^-\right\} = \left\{\left(\min_i v_{ij} | j \in J\right), \left(\max_i v_{ij} | j \in J'\right)\right\} \end{aligned} \tag{9.99}$$

4. Calculate the separation measures, using the m-dimensional Euclidean distance

$$\begin{aligned} S_i^+ &= \sqrt{\sum_{j=1}^{n} \left(V_i^j - V*\right)^2} \, (1 \leq i \leq m, 1 \leq j \leq n) \\ S_i^- &= \sqrt{\sum_{j=1}^{n} \left(V_i^j - V^-\right)^2} \, (1 \leq i \leq m, 1 \leq j \leq n) \end{aligned} \tag{9.100}$$

5. Calculate the relative closeness to the ideal solution

$$Y_i = \frac{S_i^-}{S_i^+ + S_i^-}\ (1 \le i \le m) \tag{9.101}$$

where $Y_i \in (0, 1)$. The larger Y_i is, the closer the alternative is to the ideal solution.

6. Rank the preference order

The larger TOPSIS value, the better the alternative.

Dynamic Assessment Method

Dynamic assessment has been introduced by Feuesrtein in the 'theory, tools, techniques of learning potential assessment—the dynamic assessment on hysteresis operators' in 1979. The root of its theory can be traced back to 'the zone of proximal development' by Vygotsky [149]. Over time and the accumulation of the data, people have many chronological sequence data of the plane data table series, called 'time series data sheet.' Comprehensive evaluation with time series data, its parameter values are dynamic, which is defined as 'dynamic comprehensive evaluation' problem [150].

Dimension Reduction for Time Series Data

With the proposed dynamic TOPSIS model, the three-dimensional time series data is reduced to two-dimensional data using the time–weight vector described in the following subsection. The time-weighted vector $w = (w_1, w_2, w_n)$ T represents the degree of emphasis on different time, according to different criteria. The 'time–weight vector entropy' I is given as $I = -\sum_{k=1}^{p} w_k \ \ln\ w_k$, and the 'time degree' T is $T = \sum_{k=1}^{p} w_k \frac{p-k}{p-1}$, where p is the number of years.

The 'time degree' T indicates the degree to which the aggregation operator values a time interval. It can take a value between 0 and 1 to reflect the attitude of a decision maker as shown in Table 9.21. T = 0 implies that time weighted vector w becomes (0, 0, ..., 1) and the element with the latest time value obtains the largest weight. T = 1 implies that time weighted vector w becomes (1, 0, ..., 0) and the element with the earliest time value obtains the largest weight. T = 0.5 implies that data elements of different years have the same importance.

The criterion to determine the time–weight vector is that in the condition of a given 'time degree' T, to mine sample information as much as possible and consider different information of evaluated samples in the timing. The time weighted vector

Table 9.21 The Mean of the time degree T

T value	Illustration
0.1	The recent data is most important
0.3	The recent data is more important
0.5	The data is of equal importance
0.7	The earlier data is more important
0.9	The earlier data is most important
0.2, 0.4, 0.6, 0.8	Intermediate values between adjacent scale values

can be calculated:

$$\begin{cases} \text{MAX}\left(-\sum_{k=1}^{p} w_k \ln w_k\right) \\ s.t.T = \sum_{k=1}^{p} w_k \frac{p-k}{p-1} \\ \sum_{k=1}^{p} w_k = 1,\ w_k \in [0,1],\ k = 1, 2, \cdots, p \end{cases} \tag{9.102}$$

9.3.3.3 Dynamic Technique for Order Preference by Similarity to Ideal Solution Evaluation Method

The dynamic TOPSIS evaluation method based on a dynamic assessment model is used to assess eco-environmental quality, and the proposed method considers the time weight vector to construct three-dimensional time series data [151]. In this model, through the MCDM (TOPSIS), the two-dimensional data is reduced to one-dimensional data to dynamically assess the quality of the urban eco-environment. The steps of proposed dynamic assessment method are as follows:

1. Determine the evaluation index system, according to the ecological theory.
2. Data preprocessing and standardization.
3. Use multi-attribute evaluation method to determine the combination weight.
4. Use MCDM: TOPSIS method to assess the level of urban eco-environmental quality from 2005 to 2009.
5. Create a dynamic assessment model as

$$Z = \alpha_1 Y_1 + \alpha_2 Y_2 + \cdots \alpha_i Y_i + \cdots + \alpha_n Y_n \ (i = 1, 2, \cdots n) \tag{9.103}$$

Where Y_i is defined in Eq. (9.101) used by TOPSIS method to determine relative closeness degree of the urban eco-environmental quality each year. ai is defined in Eq. (9.102) and is the time–weight vector w_i.

Calculate the utility value of urban eco-environmental quality.

Dynamic Sensitivity Analysis

There are two aspects of sensitivity analysis—one is the sensitivity analysis of attribute weight, and the other is the sensitivity analysis of attribute value. However, previous studies on sensitivity analysis are static assessment, which does not show the influence of time [152].

The Dynamic sensitivity analysis is to consider the influence of the Dynamic time weight vector for decision-maker to make the final decisions. Because of the uncertainty of the time–weight vector, the assessment results are uncertain. It is necessary and critical to do sensitivity analysis of dynamic assessment method.

Assume that the weight w_k of index T_k has small fluctuations w_k, the changes in weight value are defined as $w_k^* = w_k + \Delta w_k$, whereas the other weights remain unchanged. After the normalization, we obtain

$$\begin{aligned} w_k' &= \frac{w_k}{w_1+w_2+\cdots w_k+\Delta w_k+\cdots w_n} \\ &= \frac{w_i}{(w_1+w_2+\cdots w_k^*+\cdots w_n)(k=1,2,\cdots n)} \end{aligned} \tag{9.104}$$

The stable range of the index T_k is

$$\begin{cases} \Delta w_k > -w_k, y_{ik} = y_{tk} \\ -w_k < \Delta w_k < \sum_{j=1}^{n} \frac{(y_{ij}-y_{tj})w_k}{y_{tj}-y_{ij}}, y_{ik} < y_{tk} \\ \Delta w_k > \max\left[\sum_{j=1}^{n} \frac{(y_{ij}-y_{tj})w_k}{y_{tj}-y_{ij}}, -w_k\right], y_{ik} > y_{tk} \end{cases} \tag{9.105}$$

K-Means Clustering Algorithm

Clustering analysis divides data set into several different classes, making the data in the same class as similar as possible, but in the different class, as dissimilar as possible [10]. The higher the degree of similarities among similar objects and the more differences among the dissimilar objects, the better the cluster quality.

Cluster is 'the process of dividing physical or abstract objects into similar object classes' [15]. The steps of the K-means cluster algorithm are as follows:

1. Put n objects into k non-empty set.
2. Select random seed value as the current center of clusters.
3. Assign each object with the nearest seed value.
4. Repeat the second step, until there are no new assignments.

In this study, we complete the K-means clustering method by using the WEKA software [16], the specific processes are showed in Fig. 9.16.

The data of empirical study is collected from the 'China City Statistical Yearbook' and 'China Statistical Year-book for Regional Economy' between 2005 and 2009 in [8].

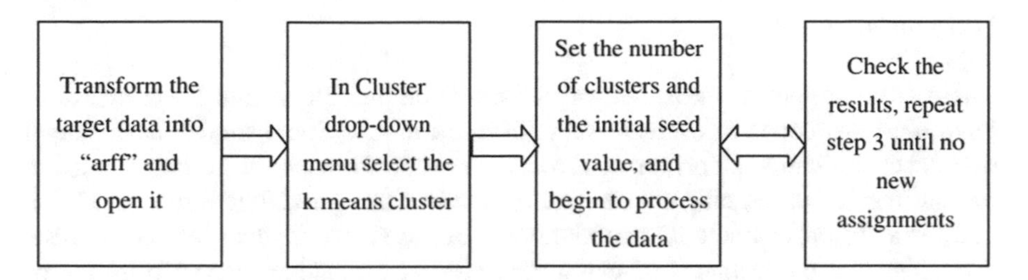

Fig. 9.16 K-means clustering algorithm based on WEKA flow chart

9.3.4 An Empirical Study of Classification Algorithm Evaluation for Financial Risk Prediction

This subsection is to develop an approach to evaluate classification algorithms for financial risk prediction. It constructs a performance score to measure the performance of classification algorithms and introduces MCDM methods to rank the classifiers. An empirical study is designed to assess nine classification algorithms using five performance measures over seven real-life credit risk and fraud risk datasets from six countries. For each performance measure, a performance score is calculated for each selected classification algorithm. The classification algorithms are then ranked using three MCDM methods (i.e., TOPSIS, PROMETHEE, and VIKOR) based on the performance scores.

Another problem in financial risk detection is that the knowledge gap [58] between the results classification methods can provide and taking actions based on them remains large. The lack of interaction between industry practitioners and academic researchers makes it hard to discover financial risks or opportunities and hence weakens the value that classification methods may bring to financial risk detection. To deal with the knowledge gap problem, this section combines the classification results, the knowledge discovery in database (KDD) process, and the concept of chance discovery to build a knowledge-rich financial risk management process in an attempt to increase the usefulness of classification results in financial risk prediction.

9.3.4.1 Evaluation Approach for Classification Algorithms

This section develops a two-step process to evaluate classification algorithms for financial risk prediction. In the first step, a performance score is created for each selected classification algorithm. The second step applies three MCDM methods (i.e., TOP-SIS, PROMETHEE, and VIKOR) to rank the selected classification algorithms using the performance scores as inputs. This section describes how the performance scores are calculated and gives an overview the three MCDM methods used in the study.

Performance Score

There are a variety of measures for classification algorithms and these measures have been developed to evaluate very different things. Some studies have shown that the classification algorithm achieves the best performance according to a given measure on a dataset, may not be the best method using a different measure [106, 153]. In addition, characteristics of datasets, such as size, class distribution, or noise, can affect the performance of classifiers. Hence, evaluating the performance of classification algorithms using one or two measures on one or two datasets often proves to be inadequate.

Based on these two considerations, this study constructs a performance metric that assesses the quality of classifiers using a set of measures on a collection of financial risk datasets in an attempt to give a comprehensive evaluation of classification algorithms. The basic idea of this performance metric is similar to ranking methods, which use experimental results generated by a set of classification algorithms on a set of datasets to rank those algorithms [154]. Specifically, it resembles the significant wins (SW) ranking method by conducting pairwise comparisons of classifiers using tests of statistical significance.

Selection of Performance Measures

Accuracy and error rates are important measures of classification algorithms in financial risk prediction. This work utilizes overall accuracy, precision, true positive rate, true negative rate, and the area under the receiver operating characteristic curve (AUC) to build the performance score. The following paragraphs define and describe these measures.

- Accuracy is the percentage of correctly classified instances [15]. It is one the most widely used classification performance metrics.

$$\text{overall accuracy} = \frac{\text{TN} + \text{TP}}{\text{TP} + \text{FP} + \text{FN} + \text{TN}}$$

  where TP, TN, FP, and FN represent true positive, true negative, false positive, and false negative, respectively. TP and TN are defined below. FP is the number of non-fault-prone instances that is misclassified as fault-prone class. FN is the number of fault-prone instances that is misclassified as non-fault-prone class.
- Precision is the number of classified positive or abnormal instances that actually are positive instances.

$$\text{precision} = \frac{\text{TP}}{\text{TP} + \text{FP}}$$

- TP (true positive) is the number of correctly classified positive or abnormal instances. TP rate measures how well a classifier can recognize abnormal records. It is also called sensitivity measure. In the case of financial risk detection, abnormal instances are bankrupt, fraudulent or erroneous accounts. A classifier with a higher TP rate can help financial institutions reduce their potential credit losses than a classifier with a lower TP rate.

$$\text{true positive rate/sensitivity} = \frac{\text{TP}}{\text{TP} + \text{FN}}$$

- TN (true negative) is the number of correctly classified negative or normal instances. TN rate measures how well a classifier can recognize normal records. It is also called specificity measure.

$$\text{true negative rate/specificity} = \frac{\text{TN}}{\text{TN} + \text{FP}}$$

- ROC stands for receiver operating characteristic, which shows the tradeoff between TP rate and FP rate [15]. The area under the ROC (AUC) represents the accuracy of a classifier. The larger the area, the better the classifier.

Calculation of the Performance Score

The performance score is generated by conducting paired t tests with a significance level of 5% for each classifier. The goal of a paired statistical significance test is to evaluate whether the superior or inferior performance of one classifier over another is statistically significant. The performance score for each classifier is calculated as follows:

Step 1: for each dataset, compare the tenfold cross-validation results of individual performance measure for two classifiers. The null hypothesis is that the two classifiers are the same. If the paired statistical significance (0.05) test indicates that one classifier is better than the other classifier, the performance scores of the superior classifier and the inferior classifier equals to 1 and −1, respectively. If the paired statistical significance (0.05) test indicates that the null hypothesis cannot be rejected, then the performance scores for both classifiers equal to 0 in this case.

Step 2: repeat Step 1 for all classifier pairs for the dataset tested in Step 1. Then we get performance scores of all classifiers for the specific dataset and specific performance measure.

Step 3: repeat Steps 1 and 2 for other datasets included in the experiment. The sum of performance scores from all datasets is the performance score of this classifier for the current performance measure. The larger the score is, the better the classifier performs in this measure.

Step 4: repeat Steps 1, 2 and 3 for other four performance measures to get the performance scores of all classifiers for all performance measures.

MCDM Methods

To evaluate classification algorithms, normally more than one criterion needs to be examined, such as accuracy, AUC, and misclassification rate. Thus algorithm selection can be modeled as multiple criteria decision making (MCDM) problems [155]. This subsection uses three MCDM methods, i.e., TOPSIS, PROMETHEE, and VIKOR, and explains how they can be used to rank classification algorithms.

Experiment

The experiment is designed to validate the proposed two-step evaluation approach using nine classification methods over seven real-life credit risk and fraud risk datasets from six countries. The first and second parts of this section give an overview of classification algorithms and financial risk datasets used in the empirical study. The third and fourth parts describe the experimental design and the evaluation results.

9.3.4.2 Classification Algorithms

The classification algorithms used in the experiment include eight well-known classification techniques and ensemble method. The eight classification methods are Bayesian Network [93], Naïve Bayes [92], support vector machine (SVM) [90], linear logistic regression [156], k-nearest neighbor [94], C4.5 [87], Repeated Incremental Pruning to Produce Error Reduction (RIPPER) rule induction [96], and radial basis function (RBF) network [89]. All algorithms were implemented using Weka 3.6, a free data mining software package [16].

Bayesian Network and Naïve Bayes both model probabilistic relationships between predictor variables and the class variable. While Naïve Bayes classifier estimates the class-conditional probability based on Bayes theorem and can only represent simple distributions, Bayesian Network is a probabilistic graphic model and can represent conditional independencies between variables. SVM classifier uses a nonlinear mapping to transform the training data into a higher dimension and search for the linear optimal separating hyperplane, which is then used to separate data from different classes [15]. Linear logistic regression models the probability of occurrence of an event as a linear function of a set of predictor variables. k-nearest neighbor classifies a given data instance based on learning by analogy, that is, assigns it to the closest training examples in the feature space. C4.5 is a decision tree algorithm that constructs decision trees in a top-down recursive divide-and-conquer manner. RIPPER is a sequential covering algorithm that extracts classification rules directly from the training data without generating a decision tree first [15]. RBF network is an artificial neural network that uses radial basis functions as activation functions.

In addition to the eight classification techniques, ensemble method was included in the experiment. An ensemble consists of a set of individually trained classifiers whose predictions are combined when classifying novel instances. There are two fundamental elements of ensembles: a set of properly trained classifiers and an aggregation mechanism that organizes these classifiers into the output ensemble. This study uses the vote algorithm in Weka to perform the ensemble method. Vote combines classifiers by averaging their probability estimates [16].

9.3.4.3 Financial Risk Datasets

The datasets used in this study come from six countries and represent four aspects of financial risk: credit approval (credit card application), credit behavior, bankruptcy risk, and fraud risk.

German Credit Card Application Dataset (UCI MLR)

The German credit card application dataset comes from UCI machine learning databases. It contains 1000 instances with 24 predictor variables and 1 class variable (UCI). The 24 variables describe the status of existing checking account, credit history, education level, employment status, personal status, age, and so on. The class variable indicates whether an application is accepted or declined. Seventy percent of the instances are accepted applications and 30% are declined instances.

Australian Credit Card Application Dataset [87]

This dataset was provided by a large bank and concerns consumer credit card applications. It has 690 instances with 15 predicator variables plus 1 class variable. The class variable indicates whether an application is accepted or declined. 55.5% of the instances are accepted applications and 44.5% are declined instances.

USA Credit Cardholders' Behavior Dataset [157]

The dataset was from a major US bank and contains 6000 credit card data with 64 predictor variables plus 1 class variable. Each instance has a class label indicating its credit status: either good or bad. Eighty-four percent of the data are good accounts and 16% are bad accounts. Good indicates good status accounts and bad indicates accounts with late payments, delinquency, or bankruptcy. The predictor variables describe account balance, purchase, payment, cash advance, interest charges, date of last payment, times of cash advance, and account open date.

China Credit Cardholders' Behavior Dataset

This dataset was collected by a commercial bank in China and contains 5456 credit card data with 13 attributes. These attributes describe credit cardholders' daily balance, abnormal usage, limit usage rate, first time used, revoking pay, suspend pay, transactions detail, and personal information. Each record in the dataset has a class label denotes the status of a credit card account: either good or bad. There are 91.9% good accounts and 8.1% bad accounts.

Japanese Bankruptcy Dataset [158]

This set collects 37 bankrupt Japanese firms and 111 non-bankrupt Japanese firms from various sources during the post-deregulation period of 1989–1999. Final sample firms are ones traded in the First Section of Tokyo Stock Exchange, and their financial data are available from 2000 PACAP database for Japan compiled by the Pacific-Basin Capital Market (PACAP) Research Center at the University of Rhode Island. Each case has 13 predictor variables and 1 class variable (bankrupt or non-bankrupt). The predictor variables describe financial state and performance of firms.

Korean Bankruptcy Dataset [159]

This dataset collects bankrupt firms in Korea from 1997 to 2003 from public sources. It consists of 65 bankrupt and 130 non-bankrupt firms whose data are available and publicly trading firms in the Korean Stock Exchange. Each case has 13 predictor variables with one class variable (bankrupt or non-bankrupt).

Insurance Dataset [160]

The data was provided by an anonymous US corporation. Each record concerns about an insurance claim. The set has 18,875 instances with 103 variables. A binary class attribute indicates whether an instance is a normal claim or abnormal claim. There are 353 abnormal claims and 18,522 normal claims. The abnormal instances represent fraudulent or erroneous claims and were manually collected and verified.

9.3.4.4 Experimental Design

The calculation process of the performance score and the three MCDM methods were applied to the nine classifiers over the seven financial risk datasets. The experiment was carried out according to the following process:

Input: a financial risk related dataset.
Output: ranking of classification algorithms.
Step 1: understand business requirements, dataset structure and data mining task.
Step 2: prepare target datasets: select and transform relevant features; data cleaning; data integration. Communicate any findings during data preparation with domain experts.
Step 3: train and test multiple classification models in randomly sampled partitions (i.e., tenfold cross-validation) using Weka 3.6 [19].
Step 4: calculate the performance scores following the process discussed in section "Performance Score".
Step 5: evaluate classification algorithms using TOPSIS, PROMETHEE II, and VIKOR. The performance scores for each classifier obtained from Step 4 are used as inputs to the MCDM methods. All the MCDM methods are implemented using MATLAB.
Step 6: generate three separate tables of the final ranking of classification algorithms provided by each MCDM method.
Step 7: discuss the results with domain experts. Explore potential chance(s) from data mining results. Go back to Step 1 if new business questions are raised during the process.
END

Measures have different importance in financial risk prediction. For example, false negative (FN) is the number of positive or abnormal instances that is misclassified as normal class. Since positive instances are bankrupt, fraudulent or erroneous accounts in financial risk detection, a classifier with a high FN rate can cause huge lost to creditors. Thus FN measure should have higher importance in financial risk prediction than other measures, such as false positive measure [161]. Another important measure in financial risk prediction is AUC because it selects optimal models independently from the class distribution and the cost associated with each class.

Weights of each performance measure used in TOPISIS, PROMETHEE, and VIKOR are defined according to these findings from previous research. In this study, FN rate is not included because it equals to one minus TP rate. The importance of FN rate in financial risk prediction is then reflected in the weight of TP rate. The weights of the five performance measures are defined as: TP rate and AUC are set to 10 and other three measures (i.e., over-all accuracy, precision, and TN rate) are set to 1. The weights are normalized and the sum of all weights equal to 1.

9.3.4.5 Results and Discussion

The results of test set overall accuracy, precision, AUC, TP rate, and TN rate of all classifiers on the seven datasets are reported in Table 9.22. In the dataset column of Table 9.22, Australian indicates the Australian credit card application data; USA indicates the credit cardholders' behavior data from the United States; China refers to the credit cardholders' behavior data collected from a Chinese bank; IN indicates the insurance data; German indicates the German credit card application data; and Japan and Korea indicate the Japanese and the Korean bankruptcy data, respectively. The nine classification methods were applied to each dataset using tenfold cross-validation. For each dataset, the best result of a specific performance measure is highlighted in boldface.

When the distribution of classes is highly skewed, as in the IN dataset (1.87% abnormal instances versus 98.13% normal cases), Naïve Bayes and Bayesian Network outperform other classifiers. Naïve Bayes has the highest TP rate (0.9065), which indicates that it captured 90.65% of the abnormal records, while Bayesian Network achieves a good TN rate (0.8291). Although SVM and RBF network got perfect overall accuracy (100%), they failed to identify any abnormal behavior (TP = 0 and FN = 1). For evenly distributed dataset, such as the Australian data, all classifiers have good over-all accuracy and AUC. For small datasets, such as the Japanese bankruptcy data, no classifier produces satisfactory results on AUC and TP rate. However, SVM and ensemble obtained good AUC and TP rate for the small size Korea bankruptcy dataset. For medium sized datasets, such as the credit cardholders' behavior datasets, linear logistic generates the best AUC, while Naïve Bayes and SVM produce the best TP rates. There is no classification algorithm which achieves the best results across all measures for a single dataset or has the best outcomes for a single performance measure across all datasets.

Based on the classification results presented in Table 9.22, the performance scores of all classifiers are calculated following the process discussed in Sect. 9.3.4.6 and the results are summarized in Table 9.23. For each performance measure, the best result generated by a classification algorithm is highlighted in boldface and italic. Since the performance scores are generated by conducting paired t tests with a significance level of 5% for all classifier pairs across all datasets, a classification algorithm with the highest performance score indicates that it performs significantly better than other classifiers for that specific performance measure over the seven datasets. Similar to the classification results reported in Table 9.22, no classifier has the highest performance scores for all five measures and classifiers with the best scores on some measures may perform poorly on other measures. For example, SVM achieves the best performance scores on overall accuracy and TN rate, but its scores on precision and AUC are quite low. Therefore the MCDM methods are introduced to provide a final ranking of classification algorithms.

The ranking of classifiers generated by TOPSIS, PROMETHEE II, and VIKOR is summarized in Tables 9.23, 9.24, 9.25, and 9.26, respectively. The results of TOPSIS and PROMETHEE are straightforward: the higher the ranking, the better the classifier. Linear logistic, Bayesian Network, and ensemble methods are the top-three ranked classifiers using the TOPSIS approach. The same set of classifiers is ranked as the top-three classifiers by the PROMETHEE II, however, the order of Bayesian Network and ensemble is reversed.

Since VIKOR provides compromised solutions, the ranking of classifiers needs to be determined by the Step 5 of the VIKOR algorithm.

Table 9.22 Classification results

Dataset	Algorithm	Overall accuracy	Precision	Area under ROC	True positive rate	True negative rate
Australian	Bayesian Network	0.8522	**0.8596**	0.9143	0.7980	0.8956
Australian	Naïve Bayes	0.7725	0.8571	0.8978	0.5863	**0.9217**
Australian	SVM	0.8551	0.7867	0.8622	**0.9251**	0.7990
Australian	Linear logistic	**0.8623**	0.8313	**0.9312**	0.8664	0.8590
Australian	K nearest neighbor	0.7942	0.7653	0.7922	0.7752	0.8094
Australian	C4.5	0.8348	0.8271	0.8346	0.7948	0.8668
Australian	RBF network	0.8304	0.8493	0.8995	0.7524	0.8930
Australian	RIPPER rule induction	0.8522	0.8213	0.8714	0.8534	0.8512
Australian	Ensemble	0.8551	0.8439	0.99	0.8274	0.8773
USA	Bayesian Network	0.7055	0.3366	0.8424	0.8656	0.6750
USA	Naïve Bayes	0.6933	0.3280	0.8395	**0.8740**	0.6589
USA	SVM	0.8372	0.4738	0.5632	0.1604	**0.9661**
USA	Linear logistic	**0.8532**	**0.5785**	**0.8539**	0.3031	0.9579
USA	K nearest neighbor	0.8028	0.3830	0.6327	0.3802	0.8833
USA	C4.5	0.8170	0.4156	0.6245	0.3542	0.9052
USA	RBF network	0.8400	0.0000	0.8256	0.0000	**1.0000**
USA	RIPPER rule induction	0.8443	0.5212	0.6380	0.3333	0.9417
USA	Ensemble	0.8382	0.4929	0.8432	0.3990	0.9218
China	Bayesian Network	0.9111	0.9805	0.9388	0.9216	0.7909
China	Naïve Bayes	0.8645	**0.9822**	0.9102	0.8684	**0.8205**
China	SVM	0.9417	0.9507	0.9359	**0.9878**	0.4159
China	Linear logistic	0.9426	0.9555	**0.9453**	0.9835	0.4773
China	K nearest neighbor	0.9263	0.9598	0.7505	0.9601	0.5409
China	C4.5	0.9443	0.9622	0.8593	0.9779	0.5614
China	RBF network	0.9247	0.9374	0.9113	0.9840	0.2477
China	RIPPER rule induction	0.9351	0.9576	0.7419	0.9727	0.5068
China	Ensemble	**0.9472**	0.9661	0.9229	0.9769	0.6091
IN	Bayesian Network	0.8261	0.0694	0.8361	0.6686	0.8291
IN	Naïve Bayes	0.3368	0.0250	0.7307	**0.9065**	0.3260
IN	SVM	0.9813	0.0000	0.5000	0.0000	**1.0000**
IN	Linear logistic	0.9809	0.0000	0.7546	0.0000	0.9996
IN	K nearest neighbor	0.9723	0.2300	0.5961	0.2040	0.9870
IN	C4.5	0.9786	0.3641	0.6656	0.1898	0.9937
IN	RBF network	0.9813	0.0000	0.7097	0.0000	**1.0000**
IN	RIPPER rule induction	0.9806	0.4444	0.5774	0.1586	0.9962
IN	Ensemble	**0.9817**	**0.5745**	**0.8443**	0.0765	0.9989
German	Bayesian Network	0.7250	0.5654	0.7410	0.3600	0.8814
German	Naïve Bayes	0.7550	0.6104	0.7888	**0.5067**	0.8614
German	SVM	**0.7740**	**0.6667**	0.6938	0.4933	**0.8943**
German	Linear logistic	0.7710	0.6578	0.7919	0.4933	0.8900

(continued)

Table 9.22 (continued)

Dataset	Algorithm	Overall accuracy	Precision	Area under ROC	True positive rate	True negative rate
German	K nearest neighbor	0.6690	0.4485	0.6064	0.4500	0.7629
German	C4.5	0.7190	0.5388	0.6607	0.4400	0.8386
German	RBF network	0.7400	0.5840	0.7520	0.4633	0.8586
German	RIPPER rule induction	0.7340	0.5720	0.6557	0.4500	0.8557
German	Ensemble	0.7620	0.6476	**0.7980**	0.4533	**0.8943**
Japan	Bayesian Network	0.7568	0.5135	0.7292	**0.5135**	0.8378
Japan	Naïve Bayes	0.7432	0.4857	0.7197	0.4595	0.8378
Japan	SVM	0.7500	0.0000	0.5000	0.0000	**1.0000**
Japan	Linear logistic	0.7770	0.5667	0.7290	0.4595	0.8829
Japan	K nearest neighbor	0.7770	0.5714	0.6595	0.4324	0.8919
Japan	C4.5	0.7162	0.4242	0.5270	0.3784	0.8288
Japan	RBF network	0.7162	0.3810	0.6533	0.2162	0.8829
Japan	RIPPER rule induction	0.7365	0.4706	0.6193	0.4324	0.8378
Japan	Ensemble	**0.7905**	**0.6667**	**0.7424**	0.3243	0.9459
Korea	Bayesian Network	0.8667	**0.8095**	0.8773	0.7846	**0.9077**
Korea	Naïve Bayes	0.7744	0.7059	0.8168	0.5538	0.8846
Korea	SVM	**0.8718**	0.7778	0.8682	**0.8615**	0.8769
Korea	Linear logistic	0.8462	0.7692	0.8749	0.7692	0.8846
Korea	K nearest neighbor	0.8154	0.7101	0.7993	0.7538	0.8462
Korea	C4.5	0.8359	0.7797	0.7948	0.7077	0.9000
Korea	RBF network	0.8256	0.7460	0.8033	0.7231	0.8769
Korea	RIPPER rule induction	0.8667	0.7826	0.8577	0.8308	0.8846
Korea	Ensemble	0.8564	0.7681	**0.9026**	0.8154	0.8769

Table 9.23 Performance scores of classifiers

Classifier/measure	Overall accuracy	Precision	AUC	TP rate	TN rate
Bayesian Network	−19	8	23	5	−4
Naïve Bayes	−28	8	24	2	3
SVM	22	−20	−27	1	13
Linear logistic	22	6	32	4	6
K nearest neighbor	−26	−13	−36	−2	−23
C4.5	−4	5	−26	1	−7
RBF network	4	−22	3	−22	5
RIPPER rule induction	10	9	−23	8	−4
Ensemble	19	19	30	3	11

The classifier with the first position in the ranking list by Q cannot be proposed as the compromise solution because the condition (a) $Q(a'') - Q(a') \geq 1(J - 1)$ is not satisfied. Therefore, alternatives a', a'', *and* $a''^{\prime}$ are proposed as compromise solutions, since a is the maximum number of alternative determined by the relation

Table 9.24 Results of the TOPSIS approach

Classifier	TOPSIS
Linear logistic	0.891293
Bayesian Network	0.874166
Ensemble	0.866155
Naïve Bayes	0.815243
RIPPER rule induction	0.638725
C4.5	0.544801
SVM	0.542099
K nearest neighbor	0.457113
RBF network	0.283217

Table 9.25 Results of the PROMETHEE II approach

Classifier	PROMETHEE II
Linear logistic	0.711957
Ensemble	0.532609
Bayesian Network	0.413043
RIPPER rule induction	0.353261
Naïve Bayes	0.190217
C4.5	−0.43478
SVM	−0.44022
RBF network	−0.46739
K nearest neighbor	−0.8587

Table 9.26 Results of the VIKOR approach

Classifier	VIKOR Q	VIKOR S	VIKOR R
Linear logistic	0.00055	0.080211	0.057971
Ensemble	0.027268	0.090276	0.072464
Bayesian Network	0.070517	0.168871	0.057545
Naïve Bayes	0.137489	0.205328	0.086957
RIPPER rule induction	0.628727	0.393233	0.351662
SVM	0.76261	0.520044	0.377238
C4.5	0.765376	0.533903	0.370844
RBF network	0.971288	0.688997	0.434783
K nearest neighbor	0.979134	0.698862	0.434783

$Q(a^M) - Q(a') < 1(J - 1)$. That is, the rankings of linear logistic, Bayesian Network, and ensemble methods are in closeness according to VIKOR.

The results of Tables 9.23, 9.24, 9.25, and 9.26 indicate that TOPSIS, PROMETHEE II, and VIKOR provide similar top-ranked classification algorithms for financial risk prediction.

9.3.4.6 Knowledge-Rich Financial Risk Management Process

Even though classification has become a crucial tool in financial risk prediction, most studies focus on developing algorithms or improving existing algorithms that can identify suspicious patterns and have not paid enough attention to the involvement of end users and the actionability of the classification results [83]. This is mainly due to two reasons: (1) the difficulty in accessing real-life financial risk data and (2) limited access to domain experts and background information. The lack of interaction between industry practitioners and academic researchers makes it hard to discover financial risks or opportunities and hence weaken the value that classification methods may bring to financial risk detection.

In an attempt to improve the usefulness of classification results and increase the probability of identifying unusual chances in financial risk analysis, this section proposes a knowledge-rich financial risk management process (Fig. 9.17). Chance discovery (CD) is defined as "the awareness of a chance and the explanation of its significance" [162]. Ohsawa and Fukuda [162] suggested three keys to chance discovery: communicating the significance of an event; enhancing user's awareness of an event's utility using mental imagery; and revealing the causalities of rare events using data mining methods. Figure 9.17 combines the knowledge discovery in database (KDD) process model [113], the chance discovery process [162], and the CRISP-DM process model [163]. It emphasizes three keys to chance discovery and knowledge-rich data mining: users, communication and data mining techniques. Users refer to domain experts and decision makers. Domain experts are knowledgeable of the field information, data collection procedures and meaning of variables. With the assistance of data miners, domain experts can gain insights of financial risk data from different aspects and potentially observe new chances. To turn the identified knowledge into financial or strategic advantages, decision makers, who understand the operational and strategic goals of a company, are required to provide feedbacks on the importance of the potential new chances and determine what actions should be taken. Moving back and forth between steps is always required. The cyclical nature is illustrated by the outer circle of the chance discovery process in Fig. 9.17.

This study chose the insurance data as an example to examine the proposed process. The business objective(s) of this project was to develop classification model(s) to assist human inspection of suspicious claims. After the business objective has been deter-mined, the dataset was preprocessed for classification task. During the preparation stage, two issues were brought up by the data miners: first, there are several attributes with missing values for all the instances in the dataset; second, the definitions of four attributes are conflicting. From the data miner's point of view, an attribute with completely missing values is useless in data mining tasks and should be simply removed. But from the domain expert's perspective, this is an unusual situation and represents a potential chance for operational improvement. Any attribute stored in the database was designed to capture relevant information and an attribute with complete missing value may indicate errors in the data

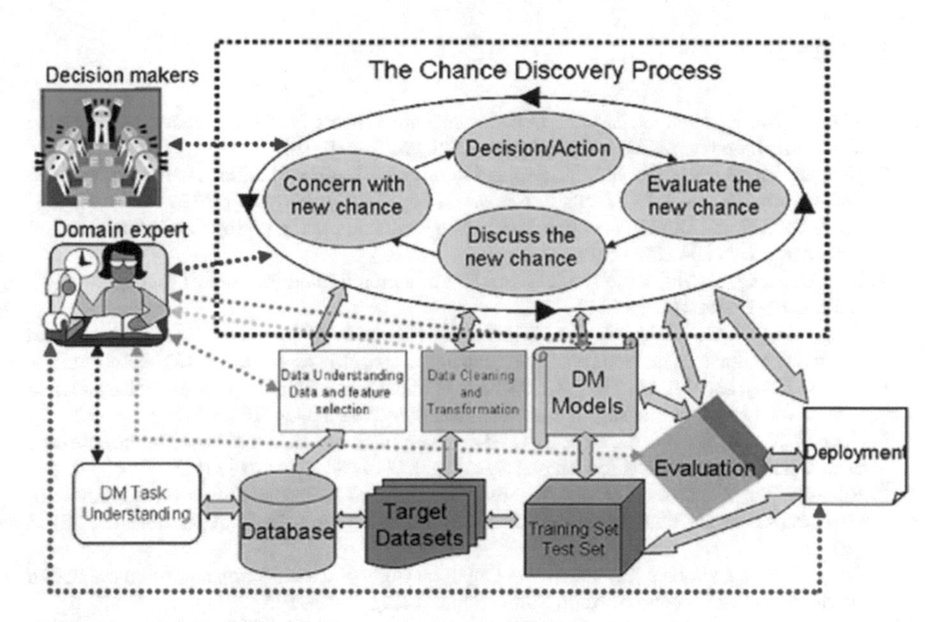

Fig. 9.17 Knowledge-rich financial risk management process

collecting process. After careful examination, domain experts found out the reasons for missing values and took corrective actions.

Then nine classifiers were applied to the insurance data using tenfold cross-validation. A classifier with low false negative (FN) rate can minimize insurance fraud risk because FN rate denotes the percentage of high-risk claims that were misclassified as normal claims. For this dataset, Naïve Bayes has the lowest FN rate ($1 - 0.9065 = 0.00935$). Because it achieves the lowest FN rate and provides classification results that can be easily understood and used by domain experts, Naïve Bayes was chosen as the decision classifier. This model can be used to predict high-risk claim; narrow down the size suspicious records; and accelerate the claim-handling process. The classification results obtained from data mining step can further be analyzed to provide additional insights about the data. For instance, if some general features of high- or low-risk claims can be identified from the classification results, it may help the insurance company to establish profiles for each type of claims, which potentially may bring profits to the company.

To summarize, the empirical study demonstrates that introducing the concept of chance discovery into the KDD process can help users choose the most appropriate classifier, promote the awareness of previously unnoticed chances, and increase the usefulness of data mining results.

References

1. Wu, W., Xu, Z., Kou, G., Shi, Y.: Decision-making support for the evaluation of clustering algorithms based on MCDM. Complexity. **2020**, 9602526 (2020)
2. Kou, G., Lu, Y., Peng, Y., Shi, Y.: Evaluation of classification algorithms using MCDM and rank correlation. Int. J. Inf. Technol. Decis. Making. **11**(01), 197–225 (2012)
3. Tang, H., Shi, Y., Dong, P.: Public blockchain evaluation using entropy and topsis. Expert Syst. Appl. **117**, 204–210 (2019)
4. Kou, G., Peng, Y., Shi, Y., Wu, W.: Classifier evaluation for software defect prediction. Stud. Informatics Contr. **21**(2), 118 (2012)
5. Peng, Y., Kou, G., Wang, G., Wu, W., Shi, Y.: Ensemble of software defect predictors: an ahp-based evaluation method. Int. J. Inf. Technol. Decis. Making. **10**(01), 187–206 (2011)
6. Shi, Y., Yang, Z., Yan, H., Tian, X.: Delivery efficiency and supplier performance evaluation in China's e-retailing industry. J. Syst. Sci. Complex. **30**(2), 392–410 (2017)
7. Zhou, X., Jiang, W., Shi, Y., Tian, Y.: Credit risk evaluation with kernel-based affine subspace nearest points learning method. Expert Syst. Appl. **38**(4), 4272–4279 (2011)
8. Kou, G., Wu, W., Zhao, Y., Peng, Y., Yaw, N.E., Shi, Y.: A dynamic assessment method for urban eco-environmental quality evaluation. J. Multi-Criteria Decis. Anal. **18**(1–2), 23–38 (2011)
9. Peng, Y., Wang, G., Kou, G., Shi, Y.: An empirical study of classification algorithm evaluation for financial risk prediction. Appl. Soft Comput. **11**(2), 2906–2915 (2011)
10. Jain, A.K., Murty, M.N., Flynn, P.J.: Data clustering: a review. ACM Comput. Surv. **31**(3), 264–323 (1999)
11. Chen, L., Xu, Z., Wang, H., Liu, S.: An ordered clustering algorithm based on k-means and the promethee method. Int. J. Mach. Learn. Cybern. **9**(6), 917–926 (2018)
12. Wu, J., Chen, J., Xiong, H., Xie, M.: External validation measures for k-means clustering: a data distribution perspective. Expert Syst. Appl. **36**(3), 6050–6061 (2009)
13. Xu, R., Wunsch, D.: Survey of clustering algorithms. IEEE Trans. Neural Netw. **16**(3), 645–678 (2005)
14. Paul, A.K., Shill, P.C.: New automatic fuzzy relational clustering algorithms using multi-objective nsga-II. Inf. Sci. **448**, 112–133 (2018)
15. Han, J., Kamber, M., Pei, J.: Data Mining Concepts and Techniques, 3rd edn. Morgan Kaufmann, Burlington, MA (2011)
16. Witten, I.H., Frank, E., Hall, M.A., Pal, C.J.: Practical Machine Learning Tools and Techniques, p. 578. Morgan Kaufmann, Burlington, MA (2005)
17. Hochbaum, D.S., Shmoys, D.B.: A best possible heuristic for the kcenter problem. Math. Oper. Res. **10**(2), 180–184 (1985)
18. Fayyad, U., Piatetsky-Shapiro, G., Smyth, P.: The kdd process for extracting useful knowledge from volumes of data. Commun. ACM. **39**(11), 27–34 (1996)
19. Hall, M., Frank, E., Holmes, G., Pfahringer, B., Reutemann, P., Witten, I.H.: The weka data mining software: an update. ACM SIGKDD Explor. Newsl. **11**(1), 10–18 (2009)
20. Dempster, A.P., Laird, N.M., Rubin, D.B.: Maximum likelihood from incomplete data via the em algorithm. J. R. Stat. Soc. Ser. B Methodological. **39**(1), 1–22 (1977)
21. Kumar, M., et al.: An optimized farthest first clustering algorithm. In: 2013 Nirma University International Conference on Engineering (NUiCONE), pp. 1–5. IEEE, New York (2013)
22. Dasgupta, S., Long, P.M.: Performance guarantees for hierarchical clustering. J. Comput. Syst. Sci. **70**(4), 555–569 (2005)
23. Hamdan, S., Cheaitou, A.: Supplier selection and order allocation with green criteria: an MCDM and multi-objective optimization approach. Comput. Oper. Res. **81**, 282–304 (2017)
24. Peng, Y., Shi, Y.: Multiple criteria decision making and operations research. Ann. Oper. Res. **197**(1), 1–4 (2012)
25. Wang, Z., Xu, Z., Liu, S., Tang, J.: A netting clustering analysis method under intuitionistic fuzzy environment. Appl. Soft Comput. **11**(8), 5558–5564 (2011)

26. Yang, J.L., Chiu, H.N., Tzeng, G.H., Yeh, R.H.: Vendor selection by integrated fuzzy MCDM techniques with independent and interdependent relationships. Inf. Sci. **178**(21), 4166–4183 (2008)
27. He, J., Zhang, Y., Shi, Y., Huang, G.: Domain-driven classification based on multiple criteria and multiple constraint-level programming for intelligent credit scoring. IEEE Trans. Knowl. Data Eng. **22**(6), 826–838 (2010)
28. Shi, Y., Zhang, L., Tian, Y., Li, X.: Intelligent Knowledge: A Study Beyond Data Mining. Springer, New York (2015)
29. Schneider, K.R.: On the Nature of Cities. Josey-Bass Publishers, San Francisco (1979)
30. Fishburn, P.: Additive Utilities with Incomplete Product Set: Applications to Priorities and Assignments. ORSA Publication, Baltimore, MD (1967)
31. Triantaphyllou, E.: Multi-criteria decision making methods. In: Multicriteria decision making methods: a comparative study, pp. 5–21. Springer, New York (2000)
32. Triantaphyllou, E., Baig, K.: The impact of aggregating benefit and cost criteria in four mcda methods. IEEE Trans. Eng. Manag. **52**(2), 213–226 (2005)
33. Wu, W., Kou, G., Peng, Y.: Group decision-making using improved multi-criteria decision making methods for credit risk analysis. Filomat. **30**(15), 4135–4150 (2016)
34. Jahanshahloo, G.R., Lotfi, F.H., Izadikhah, M.: Extension of the topsis method for decision-making problems with fuzzy data. Appl. Math. Comput. **181**(2), 1544–1551 (2006)
35. Cheng, S., Hwang, C.: Fuzzy Multiple Attribute Decision Making: Methods and Applications Lecture Notes in Economics and Mathematical Systems. Springer, New York (1992)
36. Opricovic, S., Tzeng, G.H.: Compromise solution by MCDM methods: a comparative analysis of vikor and topsis. Eur. J. Oper. Res. **156**(2), 445–455 (2004)
37. Brans, J.P., Mareschal, B.: Promethee methods. In: Multiple Criteria Decision Analysis: State of the Art Surveys, pp. 163–186. Springer, New York (2005)
38. Hermans, C.M., Erickson, J.D.: Multicriteria decision analysis: overview and implications for environmental decision making. Ecol. Econ. Sustain. Watershed Manage. **7**, 213–228 (2007)
39. Kuang, H., Kilgour, D.M., Hipel, K.W.: Grey-based promethee II with application to evaluation of source water protection strategies. Inf. Sci. **294**, 376–389 (2015)
40. Ergu, D., Kou, G., Peng, Y., Shi, Y.: A simple method to improve the consistency ratio of the pair-wise comparison matrix in anp. Eur. J. Oper. Res. **213**(1), 246–259 (2011)
41. Kou, G., Wu, W.: Multi-criteria decision analysis for emergency medical service assessment. Ann. Oper. Res. **223**(1), 239–254 (2014)
42. Steinbach, M., Karypis, G., Kumar, V.: A comparison of document clustering techniques (2000)
43. Zhao, Y., Karypis, G., Fayyad, U.: Hierarchical clustering algorithms for document datasets. Data Min. Knowl. Disc. **10**(2), 141–168 (2005)
44. Mirkin, B.: Mathematical Classification and Clustering, vol. 11. Springer Science & Business Media, Berlin (1996)
45. Rand, W.M.: Objective criteria for the evaluation of clustering methods. J. Am. Stat. Assoc. **66**(336), 846–850 (1971)
46. Jaccard, P.: Nouvelles recherches sur la distribution florale. Bull. Soc. Vaud. Sci. Nat. **44**, 223–270 (1908)
47. Fowlkes, E.B., Mallows, C.L.: A method for comparing two hierarchical clusterings. J. Am. Stat. Assoc. **78**(383), 553–569 (1983)
48. Hubert, L., Arabie, P.: Comparing partitions. J. Classif. **2**(1), 193–218 (1985)
49. Badescu, D., Boc, A., Diallo, A.B., Makarenkov, V.: Detecting genomic regions associated with a disease using variability functions and adjusted rand index. BMC Bioinformatics. **12**, 1–10 (2011)
50. Saaty, R.W.: The analytic hierarchy process—what it is and how it is used. Math. Modell. **9**(3 5), 161 176 (1987)
51. Takahashi, I.: Ahp applied to binary and ternary comparisons. J. Oper. Res. Soc. Jpn. **33**(3), 199–206 (1990)

52. Tyagi, S., Agrawal, S., Yang, K., Ying, H.: An extended fuzzyahp approach to rank the influences of socialization-externalization-combination-internalization modes on the development phase. Appl. Soft Comput. **52**, 505–518 (2017)
53. Wu, W., Kou, G., Peng, Y., Ergu, D.: Improved ahp group decision making for investment strategy selection. Technol. Econ. Dev. Econ. **18**(2), 299–316 (2012)
54. Yu, C.S.: A gp-ahp method for solving group decision-making fuzzy ahp problems. Comput. Oper. Res. **29**(14), 1969–2001 (2002)
55. eCosta, C.A.B., Vansnick, J.C.: A critical analysis of the eigenvalue method used to derive priorities in ahp. Eur. J. Oper. Res. **187**(3), 1422–1428 (2008)
56. Ertay, T., Ruan, D., Tuzkaya, U.R.: Integrating data envelopment analysis and analytic hierarchy for the facility layout design in manufacturing systems. Inf. Sci. **176**(3), 237–262 (2006)
57. Amiri, M.P.: Project selection for oil-fields development by using the ahp and fuzzy topsis methods. Expert Syst. Appl. **37**(9), 6218–6224 (2010)
58. Domingos, P.: Toward knowledge-rich data mining. Data Min. Knowl. Disc. **15**(1), 21–28 (2007)
59. Roy, B., et al.: ELECTRE III: un algorithme de classement basé sur une représentation floue des préférences en présence de critères multiples. Cahires du CERO. **20**(1), 3–24 (1978)
60. Figueira, J.R., Mousseau, V., Roy, B.: Electre methods. In: Multiple Criteria Decision Analysis, pp. 155–185. Springer, New York (2016)
61. Milani, A.S., Shanian, A., El-Lahham, C.: Using different electre methods in strategic planning in the presence of human behavioral resistance. J. Appl. Math. Decis. Sci. **2006** (2006)
62. Roy, B., Bouyssou, D.: Aide multicritère à la décision: méthodes et cas. Economica, Paris (1993)
63. Opricovic, S.: Multicriteria optimization of civil engineering systems. Faculty of Civil Engineering, Belgrade 2(1), 5–21 (1998)
64. Opricovic, S., Tzeng, G.H.: Multicriteria planning of post-earthquake sustainable reconstruction. Comput. Aided Civ. Inf. Eng. **17**(3), 211–220 (2002)
65. Mao, N., Song, M., Deng, S.: Application of topsis method in evaluating the effects of supply vane angle of a task/ambient air conditioning system on energy utilization and thermal comfort. Appl. Energy. **180**, 536–545 (2016)
66. Ding, X., Chong, X., Bao, Z., Xue, Y., Zhang, S.: Fuzzy comprehensive assessment method based on the entropy weight method and its application in the water environmental safety evaluation of the Heshangshan drinking water source area, three gorges reservoir area, China. Water. **9**(5), 329 (2017)
67. Behzadian, M., Otaghsara, S.K., Yazdani, M., Ignatius, J.: A state-of the-art survey of topsis applications. Expert Syst. Appl. **39**(17), 13051–13069 (2012)
68. Myrtveit, I., Stensrud, E., Shepperd, M.: Reliability and validity in comparative studies of software prediction models. IEEE Trans. Softw. Eng. **31**(5), 380–391 (2005)
69. Kou, G., Shi, Y., Wang, S.: Multiple criteria decision making and decision support systems-guest editor's introduction. Decis. Support. Syst. **51**(2), 247–249 (2011)
70. Chapman, M., Callis, P., Jackson, W.: Metrics data program, NASA IV and V facility (2004). http://mdp.ivv.nasa.gov
71. Dietterich, T.G.: An experimental comparison of three methods for constructing ensembles of decision trees: bagging, boosting, and randomization. Mach. Learn. **40**(2), 139–157 (2000)
72. Dietterich, T.G.: Ensemble methods in machine learning. In: International Workshop on Multiple Classifier Systems, pp. 1–15. Springer, New York (2000)
73. Freund, Y., Schapire, R.E., et al.: Experiments with a new boosting algorithm. In: ICML, vol. 96, pp. 148–156. Citeseer (1996)
74. Ting, K.M., Zheng, Z.: A study of adaboost with naive Bayesian classifiers: weakness and improvement. Comput. Intell. **19**(2), 186–200 (2003)
75. Wilson, T., Wiebe, J., Hwa, R.: Recognizing strong and weak opinion clauses. Comput. Intell. **22**(2), 73–99 (2006)

76. Opitz, D., Maclin, R.: Popular ensemble methods: an empirical study. J. Artif. Intell. Res. **11**, 169–198 (1999)
77. Bauer, E., Kohavi, R.: An empirical comparison of voting classification algorithms: bagging, boosting, and variants. Mach. Learn. **36**(1), 105–139 (1999)
78. Breiman, L., et al.: Heuristics of instability and stabilization in model selection. Ann. Stat. **24**(6), 2350–2383 (1996)
79. Breiman, L.: Bagging predictors. Mach. Learn. **24**(2), 123–140 (1996)
80. Schapire, R.E.: The strength of weak learnability. Mach. Learn. **5**(2), 197–227 (1990)
81. Freund, Y., Schapire, R.E.: A decision-theoretic generalization of online learning and an application to boosting. J. Comput. Syst. Sci. **55**(1), 119–139 (1997)
82. Wolpert, D.H.: Stacked generalization. Neural Netw. **5**(2), 241–259 (1992)
83. Peng, Y., Kou, G., Shi, Y., Chen, Z.: A descriptive framework for the field of data mining and knowledge discovery. Int. J. Inf. Technol. Decis. Making. **7**(04), 639–682 (2008)
84. Hansen, L.K., Salamon, P.: Neural network ensembles. IEEE Trans. Pattern Anal. Mach. Intell. **12**(10), 993–1001 (1990)
85. Breiman, L., Friedman, J., Olshen, R., Stone, C.: Classification and Regression Trees. Wadsworth International Group, Belmont, CA (1984) Google Scholar
86. Kohavi, R.: Scaling up the accuracy of naive-Bayes classifiers: a decision-tree hybrid. In: Kdd, vol. 96, pp. 202–207 (1996)
87. Quinlan, J.R.: C4. 5: Programs for Machine Learning. Elsevier, Amsterdam (2014)
88. Dong, M.: The improvement of the method of topsis in synthetic queme & sensitivity analysis. Syst. Eng. Theory Pract. **5** (1993)
89. Bishop, C.M., et al.: Neural Networks for Pattern Recognition. Oxford University Press, Oxford (1995)
90. Platt, J.: Sequential minimal optimization: a fast algorithm for training support vector machines (1998)
91. Vapnik, V.: The Nature of Statistical Learning Theory. Springer Science & Business Media, Berlin (2013)
92. Domingos, P., Pazzani, M.: On the optimality of the simple Bayesian classifier under zero-one loss. Mach. Learn. **29**(2), 103–130 (1997)
93. Weiss, S.M., Kulikowski, C.A.: Computer Systems that Learn: Classification and Prediction Methods from Statistics, Neural Nets, Machine Learning, and Expert Systems. Morgan Kaufmann Publishers Inc, Burlington, MA (1991)
94. Dasarathy, B.V.: Nearest Neighbor (nn) Norms: Nn Pattern Classification Techniques. IEEE Computer Society Tutorial, Los Alamitos (1991)
95. Kohavi, R.: The power of decision tables. In: European Conference on Machine Learning, pp. 174–189. Springer, New York (1995)
96. Cohen, W.W.: Fast effective rule induction. In: Machine Learning Proceedings 1995, pp. 115–123. Elsevier, Amsterdam (1995)
97. Saaty, T.L.: How to make a decision: the analytic hierarchy process. Eur. J. Oper. Res. **48**(1), 9–26 (1990)
98. Saaty, T.L., Sagir, M.: Extending the measurement of tangibles to intangibles. Int. J. Inf. Technol. Decis. Making. **8**(01), 7–27 (2009)
99. Saaty, T.L.: Decision making with the analytic hierarchy process. Int. J. Serv. Sci. **1**(1), 83–98 (2008)
100. Saaty, T.L.: A scaling method for priorities in hierarchical structures. J. Math. Psychol. **15**(3), 234–281 (1977)
101. Zahedi, F.: The analytic hierarchy process—a survey of the method and its applications. Interfaces. **16**(4), 96–108 (1986)
102. Despotis, D.K., Derpanis, D.: A min–max goal programming approach to priority derivation in ahp with interval judgements. Int. J. Inf. Technol. Decis. Making. **7**(01), 175–182 (2008)
103. Ho, W.: Integrated analytic hierarchy process and its applications—a literature review. Eur. J. Oper. Res. **186**(1), 211–228 (2008)

104. Li, H.L., Ma, L.C.: Ranking decision alternatives by integrated dea, ahp and gower plot techniques. Int. J. Inf. Technol. Decis. Making. **7**(02), 241–258 (2008)
105. Sugihara, K., Tanaka, H.: Interval evaluations in the analytic hierarchy process by possibility analysis. Comput. Intell. **17**(3), 567–579 (2001)
106. Ferri, C., Hernández-Orallo, J., Modroiu, R.: An experimental comparison of performance measures for classification. Pattern Recogn. Lett. **30**(1), 27–38 (2009)
107. Brun, M., Sima, C., Hua, J., Lowey, J., Carroll, B., Suh, E., Dougherty, E.R.: Model-based evaluation of clustering validation measures. Pattern Recogn. **40**(3), 807–824 (2007)
108. Chamberlin, E.H.: Product heterogeneity and public policy. Am. Econ. Rev. **40**(2), 85–92 (1950)
109. Lancaster, K.: The economics of product variety: a survey. Mark. Sci. **9**(3), 189–206 (1990)
110. Gaur, V., Kesavan, S.: The effects of firm size and sales growth rate on inventory turnover performance in the US retail sector. In: Retail Supply Chain Management, pp. 25–52. Springer, New York (2015)
111. Fitzsimons, G.J.: Consumer response to stockouts. J. Consum. Res. **27**(2), 249–266 (2000)
112. Slywotzky, A.J.: The age of the choiceboard. Harv. Bus. Rev. **78**(1), 40–40 (2000)
113. Fisher, M.: Yihaodian: The no. 1 store. The Wharton School Case Study (2012)
114. Zhou, X., Jiang, W., Tian, Y., Zhang, P., Nie, G., Shi, Y.: A new kernel-based classification algorithm. In: 2009 Ninth IEEE International Conference on Data Mining, pp. 1094–1099. IEEE, New York (2009)
115. Zhu, X., Yang, X., Liu, T.: On mechanism for eco-environmental quality of Jiangsu province. Econ. Geogr. **24**(4), 473–476 (2004)
116. Keerthi, S.S., Shevade, S.K., Bhattacharyya, C., Murthy, K.R.: A fast iterative nearest point algorithm for support vector machine classifier design. IEEE Trans. Neural Netw. **11**(1), 124–136 (2000)
117. Lee, D.D., Seung, H.S.: Unsupervised learning by convex and conic coding. In: Advances in Neural Information Processing Systems, pp. 515–521. Princeton University, Princeton, NJ (1997)
118. Boser, B.E., Guyon, I.M., Vapnik, V.N.: A training algorithm for optimal margin classifiers. In: Proceedings of the Fifth Annual Workshop on Computational Learning Theory, pp. 144–152 (1992)
119. Cover, T.M.: Geometrical and statistical properties of systems of linear inequalities with applications in pattern recognition. IEEE Trans. Electron. Comput. **3**, 326–334 (1965)
120. Mika, S., Ratsch, G., Weston, J., Scholkopf, B., Mullers, K.R.: Fisher discriminant analysis with kernels. In: Neural Networks for Signal Processing IX: Proceedings of the 1999 IEEE Signal Processing Society Workshop (cat. no. 98th8468), pp. 41–48. IEEE, New York (1999)
121. Schölkopf, B., Smola, A., Müller, K.R.: Kernel principal component analysis. In: International Conference on Artificial Neural Networks, pp. 583–588. Springer, New York (1997)
122. Fahlman, S.E., Lebiere, C.: The cascade-correlation learning architecture. Tech. Rep. (1990)
123. Osowski, S., Brudzewski, K.: Fuzzy self-organizing hybrid neural network for gas analysis system. IEEE Trans. Instrum. Meas. **49**(2), 424–428 (2000)
124. Osowski, S., Siwek, K., Markiewicz, T.: MLP and SVM networks-a comparative study. In: Proceedings of the 6th Nordic Signal Processing Symposium, 2004. NORSIG 2004, pp. 37–40. IEEE, New York (2004)
125. Xu, Q., Pei, W., Yang, L., He, Z.: Support vector machine tree based on feature selection. In: International Conference on Neural Information Processing, pp. 856–863. Springer, New York (2006)
126. Fahlman, S.: Cmu benchmark collection for neural net learning algorithms. In: Machine-Readable Data Repository. School of Computer Science, Carnegie Mellon Univ., Pittsburgh, PA (1993)
127. Li, A., Shi, Y., He, J.: Mclp-based methods for improving "bad" catching rate in credit cardholder behavior analysis. Appl. Soft Comput. **8**(3), 1259–1265 (2008)
128. He, J., Liu, X., Shi, Y., Xu, W., Yan, N.: Classifications of credit cardholder behavior by using fuzzy linear programming. Int. J. Inf. Technol. Decis. Making. **3**(04), 633–650 (2004)

129. Rester F, V.H.A.: Ecology and iaanniug in melnopolitan area a sensitivity model (1980)
130. Alberti, M., Waddell, P.: An integrated urban development and ecological simulation model. Integr. Assess. **1**(3), 215–227 (2000)
131. Van Kamp, I., Leidelmeijer, K., Marsman, G., De Hollander, A.: Urban environmental quality and human well-being: towards a conceptual framework and demarcation of concepts; a literature study. Landsc. Urban Plan. **65**(1-2), 5–18 (2003)
132. Yu, L., Yin, W.: Application research of analytic hierarchy process (arp) in urban ecotope quality evaluation. Sichuan Environ. **21**(4), 38–40 (2002)
133. Zeleny, M.: MCDM: Past Decade and Future Trends: A Source Book of Multiple Criteria Decision Making, vol. 1. Jai Press, London (1984)
134. Anderson Jr., W.T., Cox III, E.P., Fulcher, D.G.: Bank selection decisions and market segmentation: determinant attribute analysis reveals convenience- and service-oriented bank customers. J. Mark. **40**(1), 40–45 (1976)
135. Myers, J.H., Alpert, M.I.: Determinant buying attitudes: meaning and measurement. J. Mark. **32**(4 part 1), 13–20 (1968)
136. Starr, M.K.: A discussion of some normative criteria for decisionmaking under uncertainty. Ind. Manage. Rev. **8**(1), 71 (1966)
137. Isaacs, H.H.: Sensitivity of decisions to probability estimation errors. Oper. Res. **11**(4), 536–552 (1963)
138. Fishburn, P.C.: Analysis of decisions with incomplete knowledge of probabilities. Oper. Res. **13**(2), 217–237 (1965)
139. Evans, J.R.: Sensitivity analysis in decision theory. Decis. Sci. **15**(2), 239–247 (1984)
140. Simon French, D.R.I.: Partial information and sensitivity analysis in multi-objective decision making. In: Lockett, A.G., Islei, G. (eds.) Improving Decision Making in Organisations Lecture Notes in Economics and Mathematical Systems. Springer, Berlin (1989)
141. Masuda, T.: Hierarchical sensitivity analysis of priority used in analytic hierarchy process. Int. J. Syst. Sci. **21**(2), 415–427 (1990)
142. Armacost, R.L., Hosseini, J.C.: Identification of determinant attributes using the analytic hierarchy process. J. Acad. Mark. Sci. **22**(4), 383–392 (1994)
143. Ringuest, J.L.: Lp-metric sensitivity analysis for single and multiattribute decision analysis. Eur. J. Oper. Res. **98**(3), 563–570 (1997)
144. Dale, V.H., Beyeler, S.C.: Challenges in the development and use of ecological indicators. Ecol. Indic. **1**(1), 3–10 (2001)
145. Gu, C.C.G.: Study on index system of eco-city assessment. Natural Ecol. Conserv. **8**, 24–38 (2001)
146. Whitford, V., Ennos, A.R., Handley, J.F.: "City form and natural process"—indicators for the ecological performance of urban areas and their application to Merseyside, UK. Landsc. Urban Plan. **57**(2), 91–103 (2001)
147. Nakhaeizadeh, G., Schnabl, A.: Development of multi-criteria metrics for evaluation of data mining algorithms. In: KDD, pp. 37–42 (1997)
148. Kao, P., Hocheng, H.: Optimization of electrochemical polishing of stainless steel by grey relational analysis. J. Mater. Process. Technol. **140**(1–3), 255–259 (2003)
149. Tan, O.-S., Seng, A.S.-H.: Cognitive Modifiability in Learning and Assessment. Cengage Learning, Singapore (2008)
150. Ren, R.W.H.: Multi-Variable Data Analysis. National Defense Industry Press, Beijing (1997)
151. Guo, Y., Yao, Y., Yi, P.: Method and application of dynamic comprehensive evaluation. Syst. Eng. Theory Pract. **27**(10), 154–158 (2007)
152. Zuo, J.: Discussion multi-criteria decision making method of sensitivity analysis. Syst. Eng. Theory Pract. **7**(3), 1–11 (1987)
153. Ali, S., Smith, K.A.: On learning algorithm selection for classification. Appl. Soft Comput. **6**(2), 119–138 (2006)
154. Brazdil, P.B., Soares, C.: A comparison of ranking methods for classification algorithm selection. In: European Conference on Machine Learning, pp. 63–75. Springer, New York (2000)

155. Rokach, L.: Ensemble-based classifiers. Artif. Intell. Rev. **33**(1–2), 1–39
156. Le Cessie, S., Van Houwelingen, J.C.: Ridge estimators in logistic regression. J. R. Stat. Soc. Ser. C Appl. Stat. **41**(1), 191–201 (1992)
157. Kou, G., Peng, Y., Shi, Y., Wise, M., Xu, W.: Discovering credit cardholders' behavior by multiple criteria linear programming. Ann. Oper. Res. **135**(1), 261–274 (2005)
158. Kwak, W., Shi, Y., Eldridge, S.W., Kou, G.: Bankruptcy prediction for Japanese firms: using multiple criteria linear programming data mining approach. Int. J. Bus. Intell. Data Mining. **1**(4), 401–416 (2006)
159. Kwak, W., Shi, Y., Kou, G.: Bankruptcy prediction for Korean firms after the 1997 financial crisis: using a multiple criteria linear programming data mining approach. Rev. Quant. Finan. Acc. **38**(4), 441–453 (2012)
160. Peng, Y., Kou, G., Sabatka, A., Matza, J., Chen, Z., Khazanchi, D., Shi, Y.: Application of classification methods to individual disability income insurance fraud detection. In: International Conference on Computational Science, pp. 852–858. Springer, New York (2007)
161. Baesens, B., Van Gestel, T., Viaene, S., Stepanova, M., Suykens, J., Vanthienen, J.: Benchmarking state-of-the-art classification algorithms for credit scoring. J. Oper. Res. Soc. **54**(6), 627–635 (2003)
162. Ohsawa, Y., Fukuda, H.: Chance discovery by stimulated groups of people. Application to understanding consumption of rare food. J. Contingencies Crisis Manage. **10**(3), 129–138 (2002)
163. Chapman, P., Clinton, J., Kerber, R., Khabaza, T., Reinartz, T., Shearer, C., Wirth, R., et al.: Crisp-dm 1.0: Step-by-Step Data Mining Guide, vol. 9, p. 13. SPSS Inc, Chicago, IL (2000)

Part III
Application and Future Analysis

Chapter 10
Business and Engineering Applications

By implementing the algorithms for big data analytics described in the previous chapters, this chapter outlines three sections about related business and engineering applications. Section 10.1 relates to banking and financial market analysis with three subsections. The first one is about domestic systemically important banks: a quantitative analysis for the Chinese banking system [1]. The second is about how does credit portfolio diversification affect banks' return and risk: evidence from Chinese listed commercial banks [2]. The third one is about an approach of integrating piecewise linear representation and weighted support vector machine for forecasting stock turning points [3]. Section 10.2 describes an agriculture problem that is the classification of orange varieties based on near infrared spectroscopy [4]. Section 10.3 provides two engineering applications. The first one is about automatic road crack detection using random structured forests [5] while the second one is efficient railway tracks detection and turnouts recognition method using HOG features [6].

10.1 Banking and Financial Market Analysis

10.1.1 Domestic Systemically Important Banks: A Quantitative Analysis for the Chinese Banking System

Recent financial crises and financial contagion worldwide have brought the issue of Systemically Important Financial Institutions (SIFIs) into intense discussions [7]. Although there is no clear consensus on how systemic risk of an institution is measured, policy makers and regulators reach an agreement that to identify financial institutions whose viability is crucial for the smooth functioning of the overall financial system is essential.

Y. Shi, *Advances in Big Data Analytics*,
https://doi.org/10.1007/978-981-16-3607-3_10

No SIFI definitions are commonly accepted, however, policy makers generally consider institutions as systemically important that cannot exist the market without causing major disruption to the financial system [8]. The common characteristics of SIFI are: big in size, too connected with other institutions, lack of substitutability and also complex in their business. Regulators are working on a new regulatory framework to address systemic risk, under which, the designated systemically important institutions are candidates for tighter supervision and additional loss absorbency requirements to ensure financial stability.

In November 2011, the Basel Committee on Banking Supervision (BCBS) finalized its assessment methodology to identify Global Systemically Important Banks (G-SIBs), and mandate them to hold additional Common Tier 1 capital (on the top of minimum capital charges of Basel III) [9, 10]. At the same time, Financial Stability Board identified an initial group of 29 banks as Global Systemically Important Banks, using the methodology developed by BCBS [11].

China Banking Regulatory Commission (CBRC) also followed the BCBS assessment framework and proposed that size, interconnectedness, non-substitutability and complexity should be considered when evaluating domestic systemic importance of banks in Guidance on the implementation of the new Regulatory Standards for Chinese banking system.

This subsection presents a response to the official assessment approach proposed by Basel Committee to identify domestic systemically important banks (D-SIBs) in China. Its analysis not only presents current levels of domestic systemic importance of individual banks but also the changes.

10.1.1.1 Literature Review

Issue of how to measure systemic importance of banks has drawn much attention in recent literatures, as a direct response to the regulatory requirement [12–18]. Most of the widely-used approaches fall into the following two categories: market-based techniques and indicator-based approach [19].

Market-based techniques usually rely on information extracted from market prices and sophisticated financial models. Weistroffer, Speyer [19] distinguish market-based techniques in to two strands: in an additive manner and in a non-additive manner. However, they leave behind a large portion of studies that used network analysis. Therefore, according to the model structure, we classify market-based approaches into two buckets: network analysis and portfolio models.

Network analysis constructs a matrix of mutual exposures to describe the interconnectedness within the banking system [7]. With a hypothetical credit event to a specific bank, the researchers on this ground then simulate spillover from the credit event, and assess the possible fallout for the rest of the system. The logic of the method is that: when a bank fails, it will trigger other banks' defaults resulting from their exposures (both direct and indirect) to the failing bank. Majority of the applied network literatures [20, 21] has focused on the interbank credit market due to data availability. Recently, Li et al. [22] developed a transfer entropy-based method to derive information from stock market, and determine the interbank exposure matrix when the exposure matrix data is not available.

Portfolio models were derived from measurement of risk in portfolio of security, and extend to the measurement of systemic risk for a portfolio of banks [7]. The portfolio models follow two routines to measure a bank's systemic importance from different angles. "Bottom-up" approaches start with distress of a particular bank, and then assess the associated system-wide distress. Conditional Value at Risk (CoVaR) is the important concept in this routine, which is defined as the value at risk of the banking system conditional on institutions in distress [15]. While "top-down" methods focused on examining the fragility of the overall banking system, and required a methodology for allocating the overall risk into individual banks. Shapley Value [23] and Marginal Expected Shortfalls [13, 14, 16] are the common measures derived from "top-down" routine.

Although market-based approaches play an important role in risk supervision, shortcomings such as lack of available data, market-based indicators' instability make them less suitable as supervisory benchmarks. In contrast with market-based approach, policy makers tend to prefer a more hand-on approach using the available bank-level data (but not include market assessment). Market-based measures are only used as a cross-check if possible [19]. Besides, Drehmann and Tarashev [12] proved that some simple indicators (bank size, total interbank lending and borrowing) help approximate market-based measures of systemic importance. This research built the linkage between market-based approaches and indicator based approaches, and laid an empirical foundation for regulatory authorities' adoption of indicator-based for practical purpose. Similar work can also be found in [24].

Under indicator-based approach framework, choice of indicators and their respective weights is determined. Systemic importance of banks is then scored and ranked accordingly. The Basel Committee on Banking Supervision (BCBS) has developed an indicator-based approach for assessing systemic importance of Global Systemic Important Banks (G-SIBs) [9]. Indicators of five categories are used to fully reflect a bank's systemic importance from different dimensions. The five categories are: size, interconnectedness, non-substitutability, complexity and cross-jurisdictional activity, with equal weight of 20%. The assessment methodology provides guidance for the countries that attempt to assess domestic systemically important banks of their own. Braemer and Gischer [10] followed the official technique and made practical modification to determine the domestic systemic risk of each bank in Australia.

To be consistent with transparent requirement, this section is much in line with the BCBS methodology under indicator-based measurement approach while doing modifications to suit for real situations in China. While the list of D-SIBs might be obvious, we intend to highlight relative proportion among the banks as well as the change of the results. Besides using indicator-based measurement approach to identify D-SIBs, we also consider the systemic risk of the whole banking system, by investigating how D-SIBs and non D-SIBs are correlated before and after the recent financial crises using Copula. This part of analysis also provides cross-check of the D-SIBs identification based on market-based data.

10.1.1.2 Methodology and Data

Indicator-Based Measurement Approach

Systemic consequence of G-SIBs' failures can be dramatic but difficult to predict, therefore, Basel Committee require G-SIBs to hold additional common Tier 1 capital on the top of Basel III standard just in case of insolvency and can be bailed out by the government.

To precisely identify which banks are G-SIBs, the BCBS indicator-based assessment methodology measures a bank's systemic importance from five dimensions: cross-jurisdictional activity, size, interconnectedness, non-substitutability, and complexity [9]. All the categories have the equal weights of 20%, and total score of each bank is summed from the five categories. The multiple indicators within a category are also equally weighted. Each value of an indicator is the individual bank amount divided by the aggregate amount across all the banks in the sample. Based on the score of systemic importance, the BCBS methodology classifies the banks into four different buckets with additional loss absorbency requirements varied from 1% to 2.5% accordingly. This methodology mainly aims to capture the impact that a failure of a bank may have on the global financial system and economy.

The proposed methodology for D-SIBs identification in Chinese banking system is much in line with the official BCBS indicator-based approach. In general, the difference lies in the choice of financial indicators due to data availability under the five categories depicting systemic importance, and also the weights of individual indicators within a certain category. Besides, a major adjustment is that the official category—"cross-jurisdictional activity" is changed to "public confidence" which was called "domestic sentiment" in Braemer and Gischer's work [10].

The categories that compose the measurement system of Chinese D-SIBs and the respective indicators within each category are presented and explained below.

Size

The size of a bank is regarded as a key measure of systemic risk, as illustrated in the too-big-to-fail problem [18, 25]. In official BCBS approach, the size category is consisted of one single indicator—"total exposure" of a bank, as defined in Basel III rule text [9].

However, the indicator "total exposures" requires both on-balance items as well as off-balance items, which are not available to the public. We instead adopt the common and observable proxy—"total asset" on the balance sheet to reflect the relative size of a bank in the sample.

Interconnectedness

Systemic risk can rise through interlinkages between the nodes in banking network system both directly and indirectly [7]. If one bank defaults, it might not able to repay its interbank liabilities, therefore, the probability of other banks' distress increase, or may lead to domino effects of default contagion within the interconnected system [9, 26].

Interconnectedness of a bank is measured by the volume of its intra-financial system assets and intra-financial system liabilities under the official BCBS approach, also with the wholesale funding ratio. However, in agreement with the work that identifies D-SIBs in Australia [10], we are skeptical that retail funding enhances financial stability. Therefore, we only include "intra-financial system assets" and "intra-financial system liabilities" in the category. Intra-financial system assets are the sum of:

1. Due from banks and other financial institutions, which include deposits in banks and other financial institutions, and lendings to banks and other financial institutions.
2. Reverse repo agreement.

Intra-financial system liabilities are the sum of:

1. Due to banks and other financial institutions, which include deposits from banks and other financial institutions, and borrowings from banks and other financial institutions.
2. Repo agreement.

Non-substitutability

A bank is systemic important within the system if it is difficult for other banks to provide the similar services in case of a default [19]. The Basel Committee regards an institution that plays a dominant role in a specific business segment or as a provider of market infrastructure as systemic important. The three indicators under this category designed by BCBS are: "assets under custody", "payment cleared and settled through payment systems", and "value of underwritten transactions in debt and equity markets" [9].

The major role of a bank within China is to provide loans to corporates and households. A high share of the loans indicates low substitutability of the bank, and will have a negative impact on economy if it is difficult to find an alternative source of funding [10]. Therefore, we include two indicators within the category: "personal loans and advances" and "corporate loans and advances".

Complexity

This dimension of systemic importance is regarding to the "too-complex to fail" theory [27]. The logic behind the category is that more complex bank is more difficult to dissolve in case of a failure as greater costs and time are needed [9]. The official BCBS approach includes three indicators in the category: "OTC derivatives notional value", "level 3 assets", and "held for trading and available for sale" with equal weights.

Due to absence of OTC derivative market and data availability in China, we include three indicators in the category of "complexity": "derivative financial assets", "held for trading" and "available for sale". Among them, "Held for trading" and "available for sale" are distinguished to ensure data consistency among the banks.

Public Confidence

This is the category that we did major modification on.

The official BCBS measurement approach focuses on the banks' footprints worldwide to capture their global impact. The two indicators: cross-jurisdictional claims, and cross-jurisdictional liabilities measure the bank's activities outside its headquarters to describe how much is the international impact from its distress or failure [9]. The idea behind the indicator is that the greater global reach of a bank, the more widespread the spillover effect from its failure.

Different from BCBS approach, our objective is to identify the systemically important banks within China rather than worldwide. Therefore, a proxy to emphasize the domestic importance of a bank should be replaced with the category. As with the last category of systemic importance proposed by Braemer and Gischer [10], we also adopt "public confidence" to capture the public perception of the domestic impact that will be after a bank's failure. The more deposits from household of a bank are at risk, the worriedness about financial instability is more likely to spread over the whole nation. Bank runs over the whole banking system which come afterwards will then be defined as a systemic event that the bank contributes to. Hence, under the category of "public confidence", "deposit from household" which is the sum of demand deposit and time deposit is used as the measure.

The choice of indicators relies on the assessment objective, real situation and also data availability. We present an overview of difference and linkage between official BCBS assessment methodology for G-SIBs, measurement method for Australian D-SIBs and our measurement approach for Chinese D-SIB in Table 10.1.

One thing that should be noticed is that there exists inconsistency of accounts in financial statements among the banks. Therefore, we double check the original figures from financial statements and adjust them into a general framework of accounts under Generally Accepted Accounting Principles (GAAP) to ensure that all the indicators we use convey the same implications among the samples.

Table 10.1 Indicators of assessment approaches for G-SIBs and D-SIBs

Category (and weights)	Individual indicator		
	BCBS approach (G-SIBs)	Approach for Australia (D-SIBs)	Approach for China (D-SIBs)
Size (20%)	• Total exposures as defined for use in the Basel III leverage ratio	• Total residence assets	• Total assets (on balance)
Interconnectedness (20%)	• Intra-financial system assets • Intra-financial system liabilities • Wholesale funding ratio	• Loans to financial corporations • Deposits from financial corporations	• Intra-financial system assets • Intra-financial system liabilities
Non-substitutability (20%)	• Assets under custody • Payment cleared and settled through payment systems • Values of underwritten transactions in debt and equity markets	• Loans to households • Loans to non-financial corporations • Loans to the general government • Loans to community service and non-profit organizations	• Personal loans and advances • Corporate loans and advances
Complexity (20%)	• OTC derivatives notional value • Level 3 assets • Held for trading and available for sale	• Investment securities • Trading securities	• Held for trading[a] • Available for sale • Derivative financial assets
Cross-jurisdictional activity (20%)	• Cross-jurisdictional claims • Cross-jurisdictional liabilities	*Not included*	*Not included*
Public confidence[b] (20%)	*Not included*	• Deposits from households	• Deposits from households

[a]Including designated at fair value through profit or loss
[b]"Public confidence" is called "domestic sentiment" in measurement approach for Australia D-SIBs

Weights and Scores

The five categories of the indicators describe a bank's systemic importance from different distinct dimensions. As with the official BCBS approach, we also equal weight the five categories with 20%.

As for indicators within one category, the official BCBS approach as well as assessment method for Australian D-SIBs equal weights of all the indicators. However, to emphasize the different role that each indicator plays in distinguishing banks, we proposed an entropy-based method to determine the weights. We want to design the weighting method, which puts more weights on the indicators in which the samples distribute more dispersedly. That means, if the indicator is better to differentiate the banks, it should have more weights. Also, this weighting method ensures weight update every year, which captures the changing importance of the indicators.

For a category with more than one indicator, suppose each bank i has its value X_{ij} on indicator j. The entropy of indicator j is defined as in Eq. (10.1):

$$e_j = -\frac{\sum_{i=1}^{n} P_{ij} \ln P_{ij}}{\ln(n)} \tag{10.1}$$

where $P_{ij} = \frac{X_{ij}}{\sum_{i=1}^{n} X_{ij}}$ and the total number of banks is n.

The greater difference of indicator value among the banks, the smaller entropy of the indicator is. And that means the indicator plays a more important role to differentiate the banks. Therefore, we define difference coefficient as in Eq. (10.2) to weight the indicator.

$$g_j = 1 - e_j \tag{10.2}$$

Then we do normalization as in Eq. (10.3) to ensure the sum of indicators' weights within one category is 1, where m indicators in total are assumed. Then the weight of indicator j within the category is determined as W_j.

$$W_j = \frac{g_j}{\sum_{j=1}^{m} g_j} \tag{10.3}$$

The score on each indicator is calculated in way that identical with the official BCBS approach to capture the relative systemic importance among the sample. That is to divide individual bank amount by the aggregate amount across all the banks in the sample. The score is then weighted by the indicator's weight respectively. The final score of systemic importance comes with all the weighted scores on all the indicators (there are nine indicators under our measurement approach) added.

10.1.1.3 Copula Approach

Besides identification of Chinese D-SIBs under the indicator-based measurement approach presented in previous subsection, another goal of the section is to empirically investigate whether there are contagion effects from D-SIBs to the rest of the banking system during the recent major financial crises, which will also provide ancillary evidence for the previous judgment of D-SIBs.

What are the dependence structures between D-SIBs and the rest banks before and after the crises? Is there a tendency of strengthened distress correlations with D-SIBs after the crises? To answer these questions, the nonlinear model—Copula provides a powerful tool for this analysis, which will be introduced below.

Basic Concept of Copula

A Copula function is a function that joins multivariate distribution functions to their one-dimensional margins; therefore, it is called a link function [28]. For given univariate marginal distribution functions $F_1(x_1), F_2(x_2), \cdots F_m(x_m)$ of variables x_1, $x_2, \cdots x_m$, the multivariate distribution function C is defined as a Copula function in Eq. (10.4) [29].

$$C\left(F_1\left(x_1\right), F_2\left(x_2\right), \cdots F_m\left(x_m\right)\right) = F\left(x_1, x_2 \cdots x_m\right) \tag{10.4}$$

Sklar [30] proved that a Copula function has the following property: If $F(x_1, x_2 \cdots x_m)$ is a joint multivariate distribution function with univariate marginal distribution functions $F_1(x_1), F_2(x_2), \cdots F_m(x_m)$, then there exists a Copula function $C(u_1, u_2 \cdots u_m)$ that $F(x_1, x_2 \cdots x_m) = C(F_1(x_1), F_2(x_2), \cdots F_m(x_m))$ holds. If each F_i is continuous then C is unique. Hence, copula functions provide a unifying and flexible way to study multivariate distributions [31]. In recent years, Copula functions have been widely used in financial modeling [32, 33]. Applications of Copula functions in finance range from capital allocation [23, 34], financial markets contagion [35–38], risk integration [39–41], to default correlations [42].

The commonly used Copula functions are Gaussian Copula, Students' Copula (t-Copula), which are symmetric, and Archimedean family of Copulas [43], with their definitions presented as in Table 10.2.

Measurement of Contagion Effects

To investigate the contagion effects between the banks, our focus is on the lower tail dependence. That means we attempt to explore how much is the probability the other bank also falls in the extremely bad situation when one bank is in its worst situation. To express the conditional probability sense in a mathematical way, we have Eq. (10.5) to define the lower tail-dependence coefficient [44].

$$\lambda^{lo} = \lim_{u \to 0} P\left\{Y < G^{-1}(u) | X < F^{-1}(u)\right\} \tag{10.5}$$

where $F(x)$ and $G(y)$ are the marginal distributions of random variables X and Y, $G^{-1}(u)$ and $F^{-1}(u)$ are the inverse distribution functions of X and Y respectively.

Table 10.2 Commonly used Copula functions

Copula	$C(u_1, u_2, \cdots, u_N)$	Note
Gaussian Copula	$C(u_1, u_2, \cdots, u_N; \rho) = \Phi_\rho(\Phi^{-1}(u_1),\ \Phi^{-1}(u_2), \cdots, \Phi^{-1}(u_N))$	ρ is a symmetric, positive definite matrix with diag $\rho = 1$; Φ_ρ is the standardized multivariate normal distribution with correlation matrix ρ.
T-Copula	$C(u_1, u_2, \cdots, u_N; \rho, k) = t_{\rho,k}\left(t_k^{-1}(u_1), t_k^{-1}(u_2), \cdots, t_k^{-1}(u_N)\right)$	$\rho$is a symmetric, positive definite matrix with diag $\rho = 1$; $t_{\rho,k}$is the standardized multivariate Student's distribution with $k$degrees of freedom and correlation matrix ρ.
Archimedean family of Copulas	$C(u_1, u_2, \cdots, u_N) = \begin{cases} \varphi^{-1}[\varphi(u_1) + \varphi(u_2) + \cdots + \varphi(u_N)], & if \sum_{i=1}^{N} \varphi(u_i) \le \varphi(0) \\ 0 & otherwise \end{cases}$	$\varphi(u)$ is the generator of the Copula. It is a function that satisfy $\varphi(1) = 0$, $\varphi'(u) < 0$, and $\varphi''(u) > 0$, for all $0 \le u \le 1$.

Table 10.3 Archimedean family of Copulas

Copula	$C(u, v)$	Generator $\varphi(t)$	Lower tail-dependence coefficient λ^{lo}	Upper tail-dependence coefficient λ^{up}
Gumbel Copula	$\exp[-\{(-\ln u)^{\alpha} + (-\ln v)^{\alpha}\}^{\frac{1}{\alpha}}]$	$(-\ln t)^{\alpha}$	0	$2 - 2^{\frac{-1}{\alpha}}$
Clayton Copula	$\left(u^{-\alpha} + v^{-\alpha}\right)^{\frac{-1}{\alpha}}$	$t^{-\alpha} - 1$	$2^{\frac{-1}{\alpha}}$	0
Frank Copula	$-\frac{1}{\alpha}\ln\left[1 + \frac{(e^{-u\alpha}-1)(e^{-v\alpha}-1)}{e^{-\alpha}-1}\right]$	$-\ln\left[\frac{e^{-t\alpha}-1}{e^{\alpha}-1}\right]$	0	0

An equivalent definition of lower tail dependence in terms of a bivariate copula function is given in Eq. (10.6).

$$\lambda^{lo} = \lim_{u \to 0} \frac{C(u, u)}{u} \tag{10.6}$$

Gaussian Copula and Student's Copula are symmetric in their tails, while the real world is not always the case. Archimedean family of Copulas, which is also extensively applied, does have some typical distributions, to better capture the asymmetric tail dependence structure. Among Archimedean family of Copulas, the most popular functions are Gumbel Copula, Clayton Copula and Frank Copula. We present bivariate distribution of the three Archimedean Copulas, as well as their generators and tail-dependence coefficients in Table 10.3.

Frank Copula also has symmetric tail dependence as with Gaussian Copula and t-Copula, while Gumbel Copula and Clayton Copula are more flexible to describe the different dependence structure on the tails. Clayton Copula is sensitive to the changes of lower tail dependence, and provides us a useful measure to calculate to what extent two banks correlates when they are simultaneously in extremely bad situations. Therefore, we choose Clayton Copula to model the correlations between D-SIBs and the other banks.

Our analysis splits into two parts concerning the two financial crises respectively. The first part is about the contagion effects during the subprime crisis. And the second part is about the contagion effects during the European debt crisis.

Closing prices of every trading day for all the 16 listed commercial banks are used. Daily return rate of each bank at time t is then calculated as:

$$return_t = \ln \frac{closeprice_t}{closeprice_{t-1}} \tag{10.7}$$

Pairs of two, one from D-SIBs, and the other from the rest banks are formed respectively. Then we use Clayton Copula to model daily return correlations of the two banks both before and after the crises, and estimate the parameter. We focus on the change of lower tail-dependence coefficients before and after the crises.

Besides the analysis between individual D-SIB and non D-SIB in pairs of two, we also consider D-SIBs as a whole subsystem, to eliminate the mutual correlations among D-SIBs. Then the rest of banks are examined one by one with the D-SIB subsystem by the same procedure.

The main issue here is to construct a series of daily return rate of the D-SIB subsystem. The return rate series of the D-SIB subsystem is the weighted average of daily return rates of individual D-SIBs that are included in the subsystem, where the weights are the relative market value of these banks at the end of last year as shown in Eq. (10.8).

$$return_{subsystem} = \sum\nolimits_{i=1}^{n} \frac{marketvalue_i}{\sum\nolimits_{i=1}^{n} marketvalue_i} return_i \tag{10.8}$$

Note that the market value which is used as weight is the figure reported at the end of last year. That means to construct the weighted average series, weights are updated every year, while the series also in daily frequency.

10.1.1.4 Data Description

At the end of year 2012, China's banking sector consisted of two policy banks and China Development Bank, 5 large commercial banks, 12 joint-stock commercial banks, 144 city commercial banks, 337 rural commercial banks, etc. Overall, the number of banking institutions in China's banking system amounted to 3747 [30].

Although number of banking institutions is large, a small group of them dominates the entire market. Due to data availability but also the non-substitutable roles they play, we focus all the 16 Chinese listed commercial banks which are potential D-SIBs in nature. The publicly traded banks include all the 5 large commercial banks, 8 of 12 joint-stock commercial banks, and also 3 city commercial banks. According to 2012 annual report of China Banking Regulatory Commission (CBRC) and our estimates from annual reports of all the listed banks, the 16 banks take a large asset proportion, around 62% of the whole banking sector, therefore are reliable to represent the industry.

These listed commercial banks began to adopt new accounting standards in 2007. After the year of 2007, Chinese banking sector has been developing stably without significant reforms. This period also covers the before/after time of major recent financial crises: subprime crisis and European debt crisis. We choose the time period of 2007–2012 to gather and organize our data for analysis.

Under the indicator-based measurement approach to identify Chinese D-SIBs, figures on all the indicators which are presented in Table 10.1 are gathered from Wind database. Most of them are originally from annual financial statements and their notes of the 16 listed commercial banks while some items are from the regulatory agency—CBRC. We also double check them with the original sources.

For Copula approach for modeling the dependence structure, we prepare daily closing prices of the 16 listed banks during the identical time period, as well as the market values at the end of each year. They are also available in Wind database.

10.1.1.5 Quantitative Results

Systemic Importance Distribution and Changes

This part presents our analysis for the systemic importance of all the banks covered in the sample, using the indicator-based measurement approach. While the list of D-SIBs might be obvious, we intend to highlight the relative distribution of systemic importance, as well as the change of the results during the years.

Our quantitative calculation of systemic importance in the Chinese banking system validates the public perception. Due to the final scores, the 16 listed commercial banks perform differently from the five dimensions of systemic importance.

We firstly present the current levels of systemic risk in the Chinese banking system using the end-2012 figures. As Table 10.4 shows, the "Big Four" major banks, or "Big Five" major banks (which is called large commercial banks under CBRC regulation) process high systemic impact in the banking system. The "Big Four" takes 67.1% and the "Big Five" accounts for 73% systemic importance of the total system. Our result demonstrates that the biggest systemic risk originates from Industrial and Commercial Bank of China (ICBC) with 20% of the system, with Bank of China (BOC) followed with 16.8%. Also in 2011, BOC followed ICBC and ranked No. 2 in domestic systemic importance in Chinese banking sector. However, ICBC has not been in the list of global systemic important banks until this year. It is BOC that remains in the list for three years. That means global systemic importance dos not equal to domestic systemic importance. It is the range systemic importance covers that matters.

It can be also found that, BOC is smallest in "Size" and "Public Confidence" measured by deposit from households among the "Big Four" banks, which, may result from its branches in limited number. However, due to its highest score in "interconnectedness" and "complexity" category, it still demonstrated itself as the second systemically important banks within the nation, over Agricultural Bank of China (ABC), and China Construction Bank (CCB). Therefore, it is important to note the "Size" category. Although commonly utilized and may closely relate with other categories, it is still not an adequate proxy of systemic risk, as it cannot detect such disparities. "Too-complex-to-fail" is as important as "to-big-to-fail" problem as the phenomenon suggests.

Another bank to note is BOCOM. Although we note "Big Five" major banks, and put BOCOM together with "Big Four" major banks, a large difference between BOCOM and "Big Four" major banks can be observed both in total scores and scores within categories. CCB, which ranks the last in 2012 final score among the "Big Four" major banks, is almost 2.5 times systemic important of BOCOM, although the gaps in "complexity" and "interconnectedness" categories are not that

Table 10.4 Domestic systemic importance of Chinese Banks at 2012

Ranking	Bank	Size	Inter connectedness	Non-substitutability	Complexity	Public confidence	Final score
1	Industrial and Commercial Bank of China	0.2042	0.1274	0.2022	0.2192	0.2466	0.1999
2	Bank of China	0.1476	0.1319	0.1604	0.2465	0.1508	0.1675
3	Agricultural Bank of China	0.1542	0.1001	0.1492	0.1458	0.2416	0.1582
4	China Construction Bank	0.1627	0.0943	0.1751	0.1032	0.1931	0.1457
5	Bank of Communications	0.0614	0.0639	0.0626	0.0611	0.0451	0.0588
6	Industrial Bank Corporation	0.0378	0.1027	0.0278	0.0343	0.0113	0.0428
7	China Merchants Bank	0.0397	0.0450	0.0492	0.0349	0.0348	0.0407
8	China Minsheng Banking Corporation	0.0374	0.0871	0.0352	0.0241	0.0148	0.0397
9	Shanghai Pudong Development Bank	0.0366	0.0588	0.0308	0.0203	0.0147	0.0323
10	China Citic Bank	0.0345	0.0381	0.0346	0.0339	0.0155	0.0313
11	China Everbright Bank	0.0265	0.0455	0.0250	0.0264	0.0107	0.0268
12	Shenzhen Development Bank	0.0187	0.0346	0.0178	0.0107	0.0065	0.0177
13	Huaxia Bank	0.0173	0.0313	0.0141	0.0084	0.0061	0.0154
14	Bank of Beijing	0.0130	0.0231	0.0102	0.0081	0.0053	0.0119
15	Bank of Ningbo	0.0043	0.0087	0.0032	0.0186	0.0018	0.0073
16	Bank of Nanjing	0.0040	0.0075	0.0025	0.0044	0.0014	0.0040

large. BOCOM also has no overlaps in scores with other joint-stock commercial banks in the years, seemingly independent from that system. Therefore, we argue that it is not so fair if we put BOCOM into the same bucket as "Big Four" banks and mandate it to hold the identical additional capital. Although, it is a natural member of D-SIBs since its impact is more than a little, different bucket from "Big Four" major banks should be considered.

Among the rest, we can observe that city commercial banks such as Bank of Beijing, Bank of Ningbo, and Bank of Nanjing have hardly any systemic relevance with total score under 1.5%. Some joint-stock commercial banks show little domestic systemic importance accounting for less than 2% of the total, such as Shenzhen Development Bank and Huaxia Bank.

Some other joint-stock commercial banks, however, are much closer to the smallest major bank—Bank of Communications in their final scores in 2012, which shows the trend to be included in the potential candidates list of D-SIBs. Hence, we suggest that the impact of those banks such as Industrial Bank Corporation (IBC), China Merchants Bank (CMB) and China Minsheng Banking Corporation (CMBC) can't be neglected. We will also address the issue later regarding to the changes of systemic importance over time.

When we observe the scores in categories of the three banks, an interesting finding is the high rankings of IBC and CMBC in the category of "interconnectedness". IBC, which is characterized with its interbank activities, account more than 10% systemic importance in the category, ranking the third place. CMBC also list before BOCOM, one of "Big Five" in the same category. These findings once again validated that "size" only is not adequate to capture systemic importance. At least IBC and CMBC expose themselves too much to the banking system, not consistent with their rankings in the final, which should alert the regulators to tighten their interbank activities.

We demonstrate the final scores of domestic systemic importance over the 5-year period in Table 10.5. The banks are ranked according to the final scores of year 2012.

Focusing on the "Big Four" or "Big Five" major banks, the time series displayed in Fig. 10.1 reveals important information about the levels as well as the changes of domestic systemic importance yearly in the Chinese banking system.

We can find that, although the major banks (no matter BOCOM included or not) still dominate the systemic importance at a high level, their domestic systemic importance keeps decreasing sharply during the 5 years.

The decreasing trend of systemic importance of major banks means domestic systemic importance is distributed among the system more evenly, rather than concentrated on a small group that the regulatory always keeps an eye on. This gives the authorities the implications that more banks with growing systemic importance should also be noticed, otherwise it will be too late to regulate if they become too domestic systemic important.

Figure 10.2 shows the levels and the changes of domestic systemic importance of joint-stock commercial banks whose final scores are all above 0.02 in the 5 years.

Table 10.5 Final scores of domestic systemic importance over 2008–2012

Banks	2008	2009	2010	2011	2012
Industrial and Commercial Bank of China	0.1915	0.1913	0.1891	0.2071	0.1999
Bank of China	0.2167	0.1964	0.2008	0.1811	0.1675
Agricultural Bank of China	0.1523	0.1721	0.1591	0.1530	0.1582
China Construction Bank	0.1584	0.1640	0.1484	0.1469	0.1457
Bank of Communications	0.0660	0.0599	0.0652	0.0614	0.0588
Industrial Bank Corporation	0.0291	0.0261	0.0338	0.0377	0.0428
China Merchants Bank	0.0408	0.0430	0.0415	0.0379	0.0407
China Minsheng Banking Corporation	0.0212	0.0212	0.0246	0.0278	0.0397
Shanghai Pudong Development Bank	0.0306	0.0248	0.0316	0.0338	0.0323
China Citic Bank	0.0256	0.0325	0.0291	0.0381	0.0313
China Everbright Bank	0.0223	0.0263	0.0270	0.0251	0.0268
Shenzhen Development Bank	0.0110	0.0095	0.0099	0.0133	0.0177
Huaxia Bank	0.0176	0.0152	0.0145	0.0147	0.0154
Bank of Beijing	0.0115	0.0096	0.0134	0.0129	0.0119
Bank of Ningbo	0.0025	0.0040	0.0067	0.0047	0.0073
Bank of Nanjing	0.0029	0.0040	0.0052	0.0045	0.0040
"Big Four"	**0.7189**	**0.7238**	**0.6973**	**0.6882**	**0.6712**
"Big Five"	**0.7849**	**0.7838**	**0.7626**	**0.7496**	**0.7300**

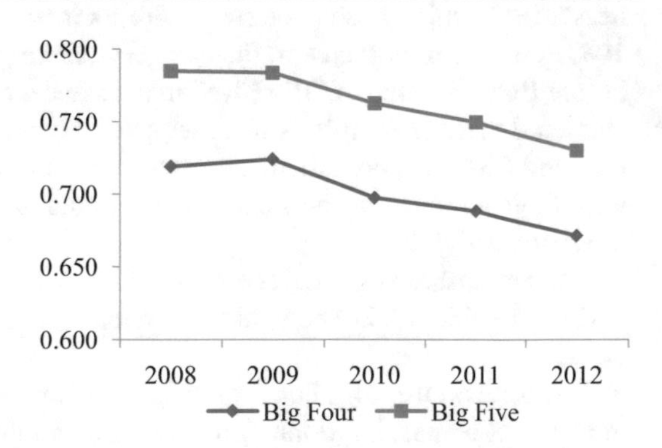

Fig. 10.1 Domestic systemic importance of Chinese major banks

As for these potential candidates to be included in D-SIBs, we suggest that Industrial Bank Corporation, China Merchants Bank and China Minsheng Banking Corporation should be considered. They are ranking just after the "Big Five" major banks with scores very close to BOCOM in the year 2012.

In addition, IBC's and CMBC's domestic systemic importance demonstrates a tendency of growing over the 5 years, which are quite different in ways of change from other joint-stock banks as Shanghai Pudong Development Bank (SPDB), China Citic Bank (CITIC), and China Everbright Bank (CEBB), although they were in an almost identical level between of systemic importance. Further, IBC and CMBC present significant impacts on the whole banking system from the aspects of "interconnectedness" in 2012, ranking the third and sixth.

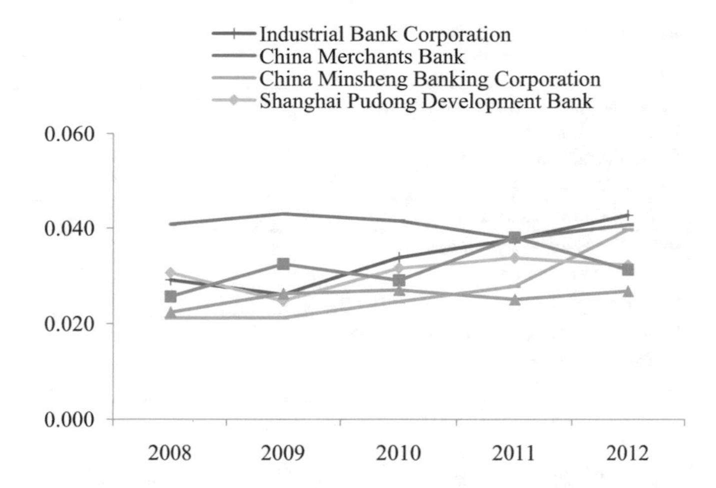

Fig. 10.2 Domestic systemic importance of main joint-stock banks

As for CMB, it always demonstrates itself as potential systemic important for accounting around 4% of the banking system, significantly higher than the rest joint-stock commercial banks. Therefore, we suggest IBC, CMBC and CMB to be included in the list of Chinese D-SIBs, while SPDB, CITIC and CEBB remain in the watch list.

We also follow the official BCBS approach to classify the banks into buckets with additional capital charge while the tentative thresholds differ. According to our suggestions, "Big Four" major banks should be in the top level of the buckets with maximum additional capital requirement, while BOCOM follow behind the next level. In the current state, besides "Big Five" major banks, the natural members of D-SIBs, potential candidates such as IBC, CMBC and CMB should also be included, which may provide an incentive to encourage them not to become more systemically important. As for these banks with more than little systemic importance, but not to a degree of potential systemically important, such as SPDB, CITIC and CEBB, we suggest that the regulatory should also watch their status dynamically. Therefore, a watch list is also recommended. To summarize, the list to be included in each bucket of D-SIBs is then presented as in Table 10.6.

10.1.2 How Does Credit Portfolio Diversification Affect Banks' Return and Risk? Evidence from Chinese Listed Commercial Banks

This subsection investigates empirically the effects of diversification on the Chinese banks' return and risk from the aspect of sector. Panel data on 16 Chinese listed commercial banks during the 2007–2011 period is used for the study.

Table 10.6 Buckets of Chinese D-SIBs

Bucket	Bank list
A	Industrial and Commercial Bank of China Bank of China Agricultural Bank of China China Construction Bank
B	Bank of Communications
C	Industrial Bank Corporation China Merchants Bank China Minsheng Banking Corporation
Watch list	Shanghai Pudong Development Bank China Citic Bank China Everbright Bank

10.1.2.1 Methodology

Diversification Measures

Previous research works applied several commonly-used traditional diversification measures including Hirschman-Herfindhl (HHI) [45, 46] and the Shannon Entropy (SE) [47, 48]. Some of the papers also used distance-based diversification measures to compare the differences between credit portfolio composition and a benchmark. in most of the cases, the industry composition of the economy's market portfolio is a benchmark for diversification. Distance-based diversification measures therefore take the differences in sizes of each sector into consideration [48, 49].

It's important to highlight that the purpose of our work is to examine whether the sectorial composition in the banks' credit portfolios affect the banks' return and risk. Sectorial composition means the banks' relative exposures to certain sectors. Before we introduce our proxies of diversification which are used in the section at length, we briefly illustrate why we do not consider distance-based diversification measures.

The particular circumstance in China is that sector classification and definition is slightly different between that in economic activity[1] and in security markets.[2] For example, K. Social service in the 13-sector classification in security markets covers M. Science, P. education, Q health, etc. in 20-sector classification of economic activity.[3] However, most of the listed commercial banks report their lending according to 13-sector classification set by CSRC respectively. Therefore,

[1]Twenty sectors in total, set by national Bureau of Statistics of China. See industry Classification of national economy, gB/T 4754–2011.

[2]Thirteen sectors in total, set by China Securities Regulatory Commission (CSRC). See industry Classification guidance for Listed Companies, CSRC Public announcement, 2012, no. 31.

[3]Details of the 20-sector classification is provided upon request, and 13-sector classification is described in subsection HHI and risk-adjusted HHI calculation.

it's difficult to distinguish the difference between their sectorial composition of credit portfolio and the macroeconomic market portfolio of the whole country.

Due to the above reason, we choose Herfindhl-Hirschman Index as the basic measure of diversification, and also construct a new one based on it, taking systematic risk of each sector into consideration.

Hirschman-Herfindhl Index (HHI)

Hirschman-Herfindhl Index (HHI) is a commonly used accepted measure of market concentration [50]. it assumes perfect diversification as equal exposure to every sector.

Before we calculate the diversification measure, for each bank, relative exposure xit of each sector i at time t is defined as its nominal exposure exit divided by the total exposure $\sum_{k=1}^{N} ex_{kt}$:

$$x_{it} = \frac{ex_{it}}{\sum_{i}^{N} ex_{kt}} \tag{10.9}$$

HHI is the sum of the squares of the relative exposures and thus for each individual bank, it is defined as:

$$HHI_t = \sum\nolimits_{i=1}^{N} x_{it}^2 \tag{10.10}$$

where N is the total number of sectors the banks provide their lending to. The lower and upper bound of HHI is $1/N$ and 1, representing a perfect diversified and a perfect focused portfolio respectively. The higher HHI value, the less is the diversification of the bank.

Risk-Adjusted HHI

Traditional HHI equals relative exposure of every sector; however, sector itself has different vulnerability as a response to the whole economy's up and down. Lessons learned from banking crises of the 1980s and early 1990s taught us that banks should not exposure too much to only few sectors [51]. Subprime crisis was partly due to too much exposure to real estate industry which has especially high correlations with macro economy [52]. To address the risk that a sector has high correlation with the macro economy, we introduce systematic risk of each sector in diversification measure. That is to weight more on relative exposures of the offensive sectors than these of the defensive sectors, rather than equal them in traditional HHI.

Systematic risk of one sector defined in this section is much in line with the concept in economics and finance. Systematic risk is vulnerability to events which affect aggregate outcomes such as broad market returns, rather than individual return of a firm or an industry. It cannot be eliminated through diversification in a portfolio.

in William Sharp's capital asset pricing model, the important concept to evaluate an asset's exposure to systematic risk is beta. it is an indicator of an asset's vulnerability to systematic risk, and indicates the degree to which an asset's expected return is correlated with boarder market return [5, 53].

In this subsection, we replace an asset with a sector in Sharp's framework. Sector beta is used to measure to which extent a sector moves together with the entire market. Then we introduce sector beta as weight of relative exposure of each sector respectively to construct our new diversification measure.

Risk-adjusted HHI for each bank at time t as follows:

$$\text{risk-adjusted } HHI_t = \sum\nolimits_{\{i=1\}}^{N} \beta_{it} x_{it}^2 \tag{10.11}$$

In Eq. (10.11), βit reflects systematic risk of each sector i at time t, defined as the covariance between market return and sector return cov(R_{Mt}, R_{it}) divided by the variance of market return $\sigma_{R_{Mt}}^2$, as with Sharp's definition of beta.

$$\beta_{it} = \frac{\text{cov}\,(R_{Mt}, R_{it})}{\sigma_{R_{Mt}}^2} \tag{10.12}$$

The higher risk-adjusted HHI value means not only the bank's credit portfolio is more concentrated, but also focused too much on the sectors with higher systematic risk.

Dependent Variables

In this part, we briefly introduce the dependent variables and other control variables we use in our regression models.

Return Performance Measures

- Return on assets (ROA): measured as the ratio of net income to total assets;
- Return on equity (ROE): measured as the ratio of net income to equity. The measure is to describe the return performance of the bank from the perspective of stockholders;
- We use both the measures for robustness check.

Risk Measure

- Nonperforming loans: we evaluate the banks' monitoring effectiveness. The variable is interpreted as an ex-post measure of the actual losses from lending activity, and easy to be found in the banks' annual reports.

Other Control Variables

We control for characteristics that are commonly used as control variables in similar studies as [45, 46, 54]. Control variables in this section are: asset, loan-to-deposit ratio and equity ratio to describe a bank from the three important aspects—size, liquidity and capital structure.

- Asset: we use the continuous variable to measure the size of the bank as most of the previous literatures did. This control variable is to capture size-related influences that stem from differences in size among banks, and to control for the well-known size effect [55]. In our regression analysis, squared term $\ln^2(asset)$ asset along with $\ln^2(asset)$ asset is also introduced to capture for the potential nonlinear relationship between size and risk when the linear relationship is not significant.
- Loan-to-deposit ratio: The control variable is defined as total loan divided by total deposit. It is a commonly used statistic for assessing a bank's liquidity. If the ratio is too high, it means that banks might not have enough liquidity to cover any unforeseen fund requirements; if the ratio is too low, banks may not be earning as much as they could be. Therefore, this ratio has the potential impact on banks' return and risk that should be controlled when doing regressions.
- Equity ratio: This control variable is defined as equity divided by the total assets, reflecting the capital structure of the bank. The reciprocal of equity ratio is the well-known equity multiplier which is associated closely with financial risk. Equity multiplier is also used to calculate return on equity in DuPont formula for financial analysis. Therefore, this variable is used mainly to control the influence of capital structure on banks' return and risk.

10.1.2.2 Model Specification

The Relationship Between Bank Returns and Credit Portfolio Diversification

The basic question in this study is whether loan sectorial diversification yields higher returns. We deal with this issue by regressing returns on diversification measures, while controlling other important variables in the following equation.

$$return_{kt} = \beta_0 + \beta_1 diversification_{kt} + \gamma \cdot V_{kt} + \varepsilon_{kt}, \tag{10.13}$$

where $return_{kt}$ is the return of bank k at time t measured by ROA and also ROE. V_{kt} is a vector of control variables including asset, loan-to-deposit ratio and equity ratio. $diversification_{kt}$ is the variable we are interested in, representing separately HHI and risk-adjusted HHI explained in the previous subsection. Finally, ε_{kt} is the residual value.

If $\beta_1 > 0$, concentration seems to be more advantageous than diversification from the aspect of return. Otherwise, $\beta_1 < 0$ means that diversification across sectors yields higher return.

In addition, we also test the nonlinear relationship between credit loan diversification and banks' return by introducing squared term $diversification_{kt}^2$ into the regression.

$$return_{kt} = \beta_0 + \beta_1 diversification_{kt} + \beta_2 diversification_{kt}^2 + \gamma \cdot V_{kt} + \varepsilon_{kt}. \tag{10.14}$$

If the regression model is significant, it seems to suggest that there exists U-shaped or reversed U-shaped relationship between diversification and banks' return.

The Impact of Credit Portfolio Diversification on Bank Risk

This topic is to test how loan sectorial diversification impacts risk. We regress risk measure on diversification measures in the following equation. Asset, loan-to-deposit ratio and equity ratio are included in the regression as control variables.

$$risk_{kt} = \beta_0 + \beta_1 diversification_{kt} + \gamma \cdot V_{kt} + \varepsilon_{kt}, \tag{10.15}$$

where $risk_{kt}$ is the risk of bank k at time t measured by nonperforming loans. Since we evaluate the banks' monitoring effectiveness, absolute value of nonperforming loans is used. $diversification_{kt}$ is separately HHI and risk-adjusted HHI explained in the previous subsection. Finally, ε_{kt} is the residual value.

In our regression analysis, squared term $\ln(asset)$ asset along with $\ln^2(asset)$ asset are also introduced to capture for the potential nonlinear relationship between size and risk when the linear relationship is not significant.

If $\beta_1 > 0$, concentration seems to be less attractive than diversification since risk is higher. Otherwise, $\beta_1 < 0$ means that diversification across sectors brings more risk.

In addition, we also test the potential nonlinear relationship such as U-shaped or reversed U-shaped between credit loan diversification and banks' return by introducing squared term $diversification_{kt}^2$ into the regression.

$$risk_{kt} = \beta_0 + \beta_1 diversification_{kt} + \beta_2 diversification_{kt}^2 + \gamma \cdot V_{kt} + \varepsilon_{kt}. \tag{10.16}$$

10.1.2.3 Data

Sample and Data Source

In China, there are 16 listed commercial banks in all. These banks take a large asset proportion of the whole banking sector. Our sample includes panel data of all the 16 Chinese listed commercial banks in 2007–2011 period, 80 observations in total. The 16 commercial banks are: Industrial and Commercial Bank of China (ICBC), Agricultural Bank of China (ABC), Bank of China (BOC), China Construction Bank (CCB), Bank of Communications (BOCOM), China Merchants Bank (CMB), Shanghai Pudong Development Bank (SPDB), China Minsheng Banking Corporation (CMBC), China Citic Bank (CITIC), China Everbright Bank (CEBB), Industrial Bank Corporation (IBC), Huaxia Bank (HXB), Shenzhen Development Bank (SDB, now is merged with Ping'an Bank), Bank of Beijing (BBJ), Bank of Nanjing (BNJ), and Bank of Ningbo (BNB).

These commercial banks began to adopt new accounting standards in 2007. Besides, in the period of 2007–2011, Chinese banking sector was developing stably without significant reforms. Therefore, our choice of the time period helps to examine the stable relationship between diversification and banks' return and risk.

Sector exposures of every listed commercial bank of the 5 years are from their annual reports. They direct their loans to the following sectors: farming, forestry, husbandry and fishing; mining; manufacturing, production and supply of electric power, gas and water; construction; transportation and warehousing; information technology; wholesale and retail trade; finance and insurance; real estate; social service (including science, education and health); communication and culture, etc. Their classification of sectors is mainly in accordance with the 13-sector classification standard set by CSRC; therefore, we only adjust some of their reported sector loan exposures to ensure consistency between the 16 banks and also the 13-sector classification standard.

Return performance measures and risk measure is from Wind database.[4] We also double check them with those on the banks' annual reports.

HHI and Risk-Adjusted HHI Calculation

Definition and classification of sectors in the 16 listed commercial banks' loan exposures reports show slight differences in some sectors. We adjust the inconsistent ones to ensure that we can compare the calculated HHI between the banks and also between the different years. 13-Sector classification specified by CSRC is listed in Table 10.7.

The 16 listed banks all reported their exposures in the following sectors: B, C, D, E, F, H, and J. However, some of the banks classified one, some or all of the

[4]Wind database is a widely used service provider of financial data in China.

Table 10.7 13-Sector classification specified by CSRC

Sector code	Description
A	Farming, forestry, husbandry and fishing
B	Mining
C	Manufacturing
D	Production and supply of electric power, gas and water
E	Construction
F	Transportation and warehousing
G	Information technology
H	Wholesale and retail trade
I	Finance and insurance
J	Real estate
K	Social service
L	Communication and culture
M	Comprehensive industry

Table 10.8 Adjusted sector classification standard

No	Description	Sector code
11	Mining	B
12	Manufacturing	C
13	Production and supply of electric power, gas and water	D
14	Construction	E
15	Transportation and warehousing	F
16	Wholesale and retail trade	H
17	Real estate	J
18	Social services: science, education, culture and health leasing and business service communication and publication	K, L
19	Others: (any, some or all of the following) farming, forestry, husbandry and fishing information technology finance and insurance comprehensive industry	A, G, I, M

Sector A, G, I, M into "others", and the rest banks reported them separately if they provided loans to any of Sector A, G, I, M. Besides, some banks had their own sector "science, education, culture and health", "public service" in their reporting, while other confirmed to 13-classification regulation in which "communication and culture" is an independent sector. To overcome the inconsistency, we adjust our classification standard shown as Table 10.8.

The most important issue in calculation of risk-adjusted HHI is to estimate sector betas. First, we choose the right proxy of market return. Cross correlations between Shanghai index (SH), Shenzhen index (SZ) and HuShen 300 index (HS 300) are calculated. Since they are highly positive related with correlation coefficient higher

Table 10.9 Sector betas in 2007–2011 period

	2007	2008	2009	2010	2011
A	0.885	1.074	1.042	0.845	1.148
B	1.117	1.092	1.238	1.358	0.769
C	1.017	1.055	0.998	0.966	1.251
D	0.971	0.888	0.881	0.812	1.009
E	1.130	1.067	1.008	0.914	1.238
F	0.921	1.075	1.086	1.069	1.047
G	0.784	0.952	0.918	0.899	1.128
H	1.086	0.976	0.995	0.825	1.175
I	1.112	1.166	1.065	0.929	0.855
I	1.213	1.232	1.421	1.068	1.269
K	0.976	1.166	1.145	0.848	1.254
L	0.867	0.961	0.950	0.788	1.377
M	1.121	1.078	1.112	0.970	1.387

than 0.85, we choose return rate of Shanghai composite index as proxy of market return because it covers a wider range thus better describe the market. Shanghai composite index is available in Wind database, which is in daily frequency.

Then, we prepare return rate of all the sectors. Wind database provides indices of all the 13 sectors in accordance with CSRC standard, all in daily frequency. To eliminate unnecessary volatility, we calculate weekly return of market and all the sectors by doing logarithmic transformation of weekly closing index series respectively.

$\beta_{it} = \frac{\text{cov}(R_{Mt}, R_{it})}{\sigma^2_{R_{Mt}}}$ is used to estimate sector betas. Note that we list our results according to CSRC 13-sector classification. When we introduce the sector betas to calculate risk-adjusted HHI, we do adjustment according to our new classification standard.

Original beta values of Sector B, C, D, E, F, H, and J is used. Average of Sector K and L's betas is replaced with I8 in our new standard. For I9, we calculate beta of each bank according to the sectors included in "others" category (one, some or all of Sector A, G, I, M) by averaging their betas. For example, we classify Sector I and M for BOC into I9 "others", average beta of Sector I and M is used to replace I9's beta. Table 10.9 presents the calculated betas of each sector of the 5 years.

Summary Statistics

Table 10.10 presents the summary statistics of the variables that we use in our models specified. The most important issue in our study, by analyzing the mean and standard deviation of the concentration measure, is that Chinese banks' credit portfolios seem to be well diversified.

We compare average HHI of Chinese banking sector with the main findings of the previous study. In general it is more diversified than Italians' whose average

Table 10.10 Summary statistics of the variables

	HHI	Risk-adjusted HHI	Nonperforming loan	ROA	ROE	Loan-to-deposit ratio	Asset	Equity ratio
Mean	0.187	0.197	56,657.36	1.079	19.904	67.853	3,388,097	5.837
Median	0.173	0.183	13,993.95	1.105	19.400	69.682	1,527,874	5.772
Maximum	0.372	0.381	303,020	1.720	41.120	83.784	15,476,868	13.071
Minimum	0.145	0.144	271.839	0.150	4.180	50.840	75,510.77	2.204
Std. Dev.	0.039	0.043	80,658.06	0.261	5.144	6.790	3,963,708	1.841
Skewness	2.312	1.868	1.472	−0.810	1.036	−0.175	1.370	1.306
Kurtosis	9.558	7.266	3.789	4.602	6.956	2.588	3.604	6.367
Jarque-Bera	214.641	107.2	30.971	17.302	65.636	0.974	26.233	59.759
Probability	0	0	0	0.00018	0	0.614578	0.000002	0
Observations	80	80	80	80	79	80	80	79
Cross sections	16	16	16	16	16	16	16	16

Table 10.11 Correlations of HHI, risk-adjusted HHI and asset

	HHI	Risk-adjusted HHI	Asset
HHI	1		
Risk-adjusted HHI	0.895	1	
Asset	−0.408	−0.358	1

HHI is 0.237 [54], Irish's [56] and German's [46] in which HHI both equal 0.291. For emerging markets, Brazilians' and Argentines' are also more concentrated than Chinese's with HHI 0.316 [48] and 0.55 [57] respectively.

Table 10.11 depicts the cross correlations of the variables we are interested in with asset. This provides evidence that a not absolutely pronounced correlation exists between HHI and risk-adjusted HHI, suggesting that the effects of both on return performance and risk may be at least slightly different. So it is necessary to observe the effects of both on return and risk. Besides, there is an obvious negative correlation between credit loan concentration and the size of bank. A possible reason to explain is that larger banks have accesses to more resources, possess more adequate capital and have deeper understanding in lending to several sectors.

Figure 10.3 shows the tendency of average HHI and risk-adjusted HHI of the 16 banks in the 5 years. It is easy to find that there is a decreasing tendency of sectoral concentration as HHI reduces with time passes, which seems to indicate that credit portfolios are less concentrated. However, risk-adjusted HHI grows sharply in 2011, suggesting that more exposures to sectors with higher systematic risk such as real estate, manufacturing and construction. This may also result from increased systematic risk of these sectors at the same time. An increased risk-adjusted HHI may be a sign of higher risk resulting in more nonperforming loan, but it cannot be captured by traditional measure HHI.

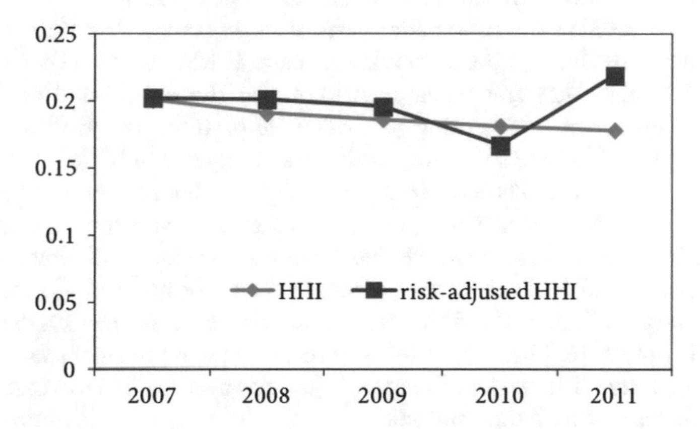

Fig. 10.3 Tendency of HHI and risk-adjusted HHI in 2007–2011

10.1.3 A New Approach of Integrating Piecewise Linear Representation and Weighted Support Vector Machine for Forecasting Stock Turning Points

This subsection integrates the piecewise linear representation (PLR) and the weighted support vector machine (WSVM) to forecast stock TPs and proposes several methods to enhance the performance of the PLR-WSVM model. Firstly, a fitness function is proposed to select the threshold of the PLR automatically. Secondly, an oversampling method suitable for the problem of forecasting stock TPs is proposed. The random undersampling combined with the oversampling is used to balance the number of samples. Thirdly, the relative strength index (RSI) is integrated to determine whether the predicted TP is a buying point or selling point. Twenty stocks are used to test the proposed model. The experimental results show that the proposed model significantly outperforms other models. The coefficient of variation of the revenues obtained by the proposed model is the lowest, indicating the proposed model is the most stable.

10.1.3.1 Literature Review

Financial Data Forecasting

Financial data forecasting can be mainly divided into three aspects: the financial time series forecasting, the price trend forecasting, and the trading signal forecasting. The financial time series forecasting uses continuous variables as prediction targets. Machine learning algorithms, such as the artificial neural networks (ANN) and the support vector machines (SVM), have been widely used in financial time series forecasting and have better performance than traditional linear models such as autoregressive integrated moving average (ARIMA) [58–60]. Combining the nonlinear algorithm and linear algorithm also showed a good performance in financial time series forecasting [61, 62]. The performance of different ANN structures in financial time series forecasting was compared in [63].

The financial time series forecasting typically uses the root mean square error (RMSE) as the performance metric, but it cannot give the accuracy of stocks' rises and falls [64]. Therefore, some researches focus on the price trend forecasting [65–69]. The price trend forecasting uses binary variables or multivariate variables as prediction targets. Technical indicators are used to forecast the stock's daily trends [65–69]. The ANN [67] has been used to forecast the trend of stocks during 1 day, 5 days and 10 days. The random forest (RF) [65] has been used to forecast the trend of stocks during different time periods.

However, investors are more concerned with making trading decisions than forecasting daily prices [70]. The financial time series forecasting and the price trend forecasting mainly focus on daily forecasting. If investors simply buy and sell stocks according to the daily predicted trends, frequent transactions will lead to high transaction fees and low profits. Therefore, some researches focus on the trading signal forecasting [71–73]. The purpose of the trading signal forecasting is to establish a system that predicts when to buy and sell stocks to make a profit in the financial market. The ensemble artificial neural network (EANN) [74] has been used to predict TPs and earned about 5% more than the buy-and-hold strategy (BHS). Mabu et al. [75] used the genetic network programming and earned about 5% more than the BHS as well. Pham et al. [76] proposed a trading system based on risk management and company assessment. Wu et al. [77] proposed a trading system based on technical analysis and sentiment analysis. Reinforcement learning is an algorithm based on maximizing reward and has been used to establish a system to trade stocks [78–80]. Reference [81] used the league championship algorithm, network structure, and reinforcement learning to extract stock trading rules, which earned about 17% more than the BHS. Recently, the piecewise linear representation (PLR) was integrated into the ANN to predict the trading points of stocks [70, 82]. The PLR is a method to split a series into serval segments, and the maximum error of each segment does not exceed a threshold [83]. The financial time series can be split into different segments by the PLR, and these split points are TPs. The SVM and other algorithms were combined with the PLR to forecast trading signals as well [39, 83–86].

Review of the PLR and the WSVM

The PLR is a method to split a series into serval segments [83]. Given a threshold δ, a series can be split into serval segments, and the maximum error of each segment does not exceed the threshold δ. The PLR algorithm [87] is shown in Algorithm 10.1.

Algorithm 10.1 Top-Down Algorithm

If the maximum error of the segment is higher than the threshold δ:
  Split the segment into two segments from the position of the maximum error.
  If the maximum error of the left segment is higher than the threshold δ:
    Split the left segment using Algorithm 10.1.
  If the maximum error of the right segment is higher than the threshold δ:
    Split the right segment using Algorithm 10.1.

Proposed Model

In the proposed model, the PLR is used to generate TPs and OPs, and the threshold of the PLR is automatically selected by a fitness function. The sample weights are calculated by the change rate of price between adjacent TPs. The oversampling and the undersampling are used to balance the number of samples. The WSVM [88] is used to forecast the TPs, and the trading signals are determined by the RSI rule and the DODS. Figure 10.4 shows the flowchart of the proposed model, and the detailed methods are introduced in the following subsections.

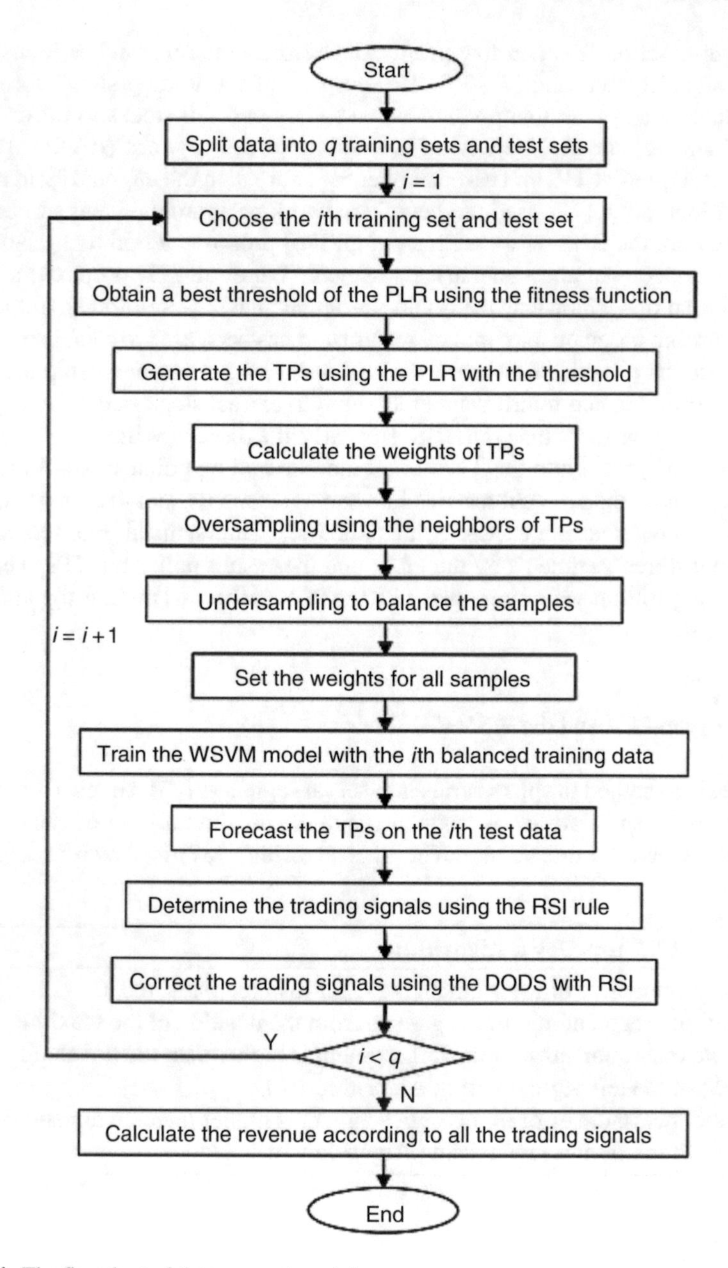

Fig. 10.4 The flowchart of the proposed model

Input Indicators

The stock indicators are shown in Table 10.12 [68, 83, 86]. KDJ is a momentum indicator and has been widely used to analyze the stock trends. There are three indicators, K, D, and J, in KDJ, and they are calculated as follows:

$$K(t) = \frac{2}{3} \times K(t-1) + \frac{1}{3} \times \frac{p_c(t) - L_n}{H_n - L_n} \times 100 \tag{10.17}$$

where L_n and H_n are the lowest price and the highest price among n days respectively.

$$D(t) = \frac{2}{3} D(t-1) + \frac{1}{3} K(t) \tag{10.18}$$

Usually, when the K value is less than the D value, and the K line breaks through the D line, it is a buying signal. When the K value is greater than the D value, and the K line falls below the D line, it is a selling signal. Therefore, the different types of KDJ are used as another input indicator.

The days d chosen to calculate the $BIAS_d$ are 5, 10, 20, 30 and 60. The days d chosen to calculate the RSI_d are 6, 12 and 24. Therefore, there are a total of 23 indicators used as input variables.

Generate TPs Using the PLR

The data set is split into q training sets and test sets sequentially, and the q is calculated as follows [83, 86]:

$$q = \lceil (r - r_1) / r_2 \rceil \tag{10.19}$$

where r is the data set size, r_1 is the training set size, and r_2 is the test set size. The example of splitting data set is shown in Fig. 10.5.

For each training set, the PLR is used to obtain the TPs of stocks. The points with trough or peak are classified to TPs, and the other points are classified to OPs. The threshold δ has an important influence on the TPs generated by the PLR. As can be seen in Table 10.13, the smaller the threshold value, the more TPs the PLR generates.

As the threshold decreases, the number of TPs increases, but the PLR will more easily generate some TPs occurred within a short period. These points are only short-term rebound points rather than TPs. Figure 10.6 shows an example. In Fig. 10.6a, three pairs of TPs, a, b and c, occurred within a short period. The three pairs of TPs are still in a falling or rising trend and should not be TPs. Therefore, the more reasonable TPs are shown in Fig. 10.6b after eliminating the short-term rebound

Table 10.12 Stock indicators and their formulas

Indicator	Formula	Description
ATP	$\tilde{p}(t) = TM(t)/TV(t)$	The average transaction price
ALT	$(p_h(t) - p_l(t))/p_l(t)$	The amplitude of the price movement
ITL	$\begin{cases} 1 \text{ if } p_c(t) > p_o(t) \\ 0 \text{ if } p_c(t) \le p_o(t) \end{cases}$	The index for the type of K-line
CATP	$(\tilde{p}(t) - \tilde{p}(t-1))/\tilde{p}(t-1)$	The change rate of average transaction price to the previous trading day
CTM	$(m(t) - m(t+1))/m(t-1)$	The change rate of transaction money compared to the previous trading day
TR	$TV(t)/TS(t)$	The turnover rate
CTR	$(TR(t) - TR(t-1))/TR(t-1)$	The change rate of turnover rate
PCCP	$\begin{cases} \frac{2p_c(t)-p_l(t)-p_h(t)}{p_h(t)-p_l(t)} & \text{if } p_h(t) \ne p_l(t) \\ 1 & \text{if } p_h(t) = p_l(t) \end{cases}$	The position constant of the closing price
PCTV	$\begin{cases} \frac{2TV(t)-TV10_{\min}(t)-TV10_{\max}(t)}{TV10_{\max}-TV10_{\min}} & \text{if } TV10_{\max} \ne TV10_{\min} \\ 1 & \text{if } TV10_{\max} \ne TV10_{\min} \end{cases}$	The position constant of transaction volume on ten days
RDMA	$(MA_{10} - MA_{50})/MA_{50}$	The relative differences of MA between the short run and the long run
RMACD	$RDIFF - \left(\sum_{i=i-8}^{t} RDIFF(i)\right)/9$	The relative moving average convergence-divergence
BIAS_d	$(p_c(t) - MA_d)/MA_d$	The degree of price deviating from the average size
KDJ	Please see Eqs. (10.4)–(10.6)	The stochastic oscillator
ITS	$\begin{cases} 1 & \text{if } K(t) \le D(t) \text{ and } K(t) > K(t-1) \\ -1 & \text{if } K(t) \ge D(t) \text{ and } K(t) < K(t-1) \\ 0 & \text{otherwise} \end{cases}$	The index for the type of KDJ
RSI_d	$100 - 100/(1 + RS_d)$	The relative strength index

$p_o(t)$, $p_c(t)$, $p_h(t)$, $p_l(t)$ are the opening price, the closing price, the highest price, and the lowest price on day t respectively; $TM(t)$, $TV(t)$, $TS(t)$ are the transaction money, the transaction volume, and the size of tradable shares on day t respectively; $MA_d = \frac{1}{d}\sum_{i=t-d+1}^{t} p_c(i)$; $RDIFF = (MA_{12} - MA_{26})/MA_{26}$;
$RS_d = \frac{\frac{1}{d}\sum_{i=t-d+1}^{t} \max(0, p_c(i)-p_c(i-1))}{\frac{1}{d}\sum_{i=t-d+1}^{t} \lvert\min(0, p_c(i)-p_c(i-1))\rvert}$

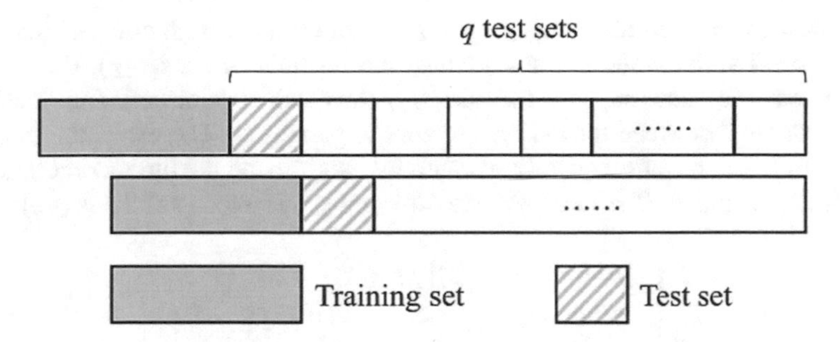

Fig. 10.5 The example of splitting data set sequentially

Table 10.13 Number of TPs under different thresholds

	$\delta = 0.01$	$\delta = 0.05$	$\delta = 0.10$	$\delta = 0.15$	$\delta = 0.20$
Number of TPs	64	37	21	14	8
Percent of TPs	32%	18.5%	10.5%	7%	4%

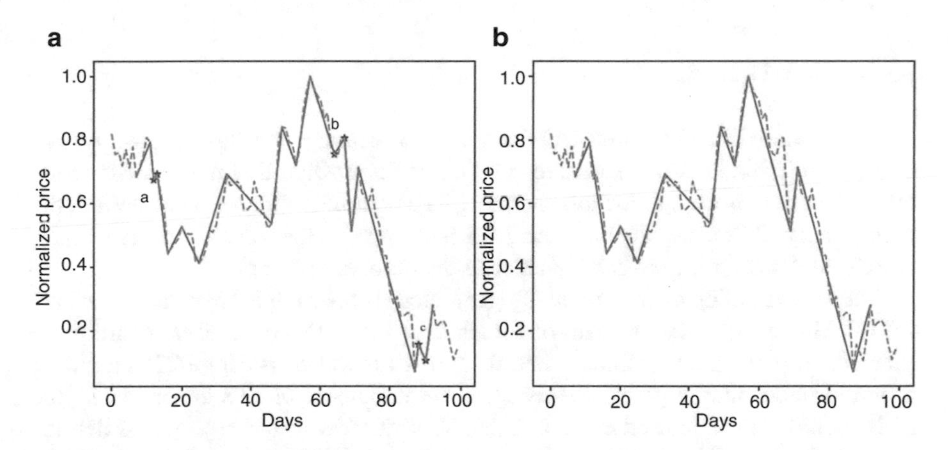

Fig. 10.6 The TPs generated by the PLR

points. Additionally, because of the different price fluctuations of different stocks, it is unreasonable to set the same threshold δ for different stocks.

To overcome the above problems, this section proposes a fitness function that focuses on medium or long-term trends rather than short-term rebounds:

$$r_{PLR} = revenue_{PLR} - \alpha_{PLR} \sum\nolimits_{i=2}^{n} \max \left(\beta_{PLR} - \left(x_i - x_{i-1} \right), 0 \right) \tag{10.20}$$

where $revenue_{PLR}$ is a revenue calculated according to the TPs generated by the PLR, α_{PLR} is a penalty factor, β_{PLR} is a threshold to control the period within which the TPs should not exist, and x_i is the day of the ith TP.

Because it can accurately buy at a low price and sell at a high price, the larger the number of TPs, the larger the first item of the formula ($revenue_{PLR}$). But with the second item, the trades happened within β_{PLR} days will be punished. The smaller the number of days between trades, the greater the penalty is. Therefore, the threshold δ of the PLR can be automatically selected by maximizing the fitness function r_{PLR}.

Different samples, TPs and OPs, are differently weighted as follows [40]:

$$\mu_t^{(tr)} = \begin{cases} |p_c(ns_t) - p_c(t)| / p_c(t) & \text{if } t \text{ is a TP} \\ \lambda * \min_{S_t} \mu_{S_t} & \text{if } t \text{ is a OP} \end{cases} \tag{10.21}$$

where $pc(\cdot)$ is the closing price, st is the TP, nst is the next TP, and λ is a scaled factor. The sample weights are normalized into [46, 89] by the following equation:

$$\mu_i^{(tr)} = 1 + \frac{\mu_i^{(tr)} - \mu_{min}^{(tr)}}{\mu_{max}^{(tr)} - \mu_{min}^{(tr)}} \tag{10.22}$$

Forecast TPs Using the WSVM

Generally, unbalanced samples will decrease the accuracy of a classifier [4]. As can be seen in Table 10.13, the number of OPs is larger than that of TPs. Since the fitness function is used to avoid generating TPs occurred within a short period, the percentage of TPs is usually between 7.5% and 17.5%. Therefore, the oversampling and the undersampling are used to balance the number of samples.

The oversampling method typically generates virtual samples to balance the data set [8]. However, in the problem of forecasting TPs, TPs are labeled according to financial experts or algorithms. Therefore, this section considers a TP should be a period rather than a point. In this case, the neighbors of TPs generated by the PLR should also be labeled as TPs. A neighbor window nw_{tp} ($nw_{tp} \geq 0$) is defined to control the neighbors that should be labeled as TPs. With the neighbor window nw_{tp}, the number of TPs will be expanded $2nw_{tp}$ times. The weights of the neighbor TPs are the same as their central TP. Further, the characteristic of a TP's neighbors is similar to that of the TP, and it may be hard to classify whether the neighbors are the TPs or not. Therefore, a neighbor window nw_{op} ($nw_{op} \geq 0$) is defined to control the neighbors of TPs that should not be labeled as OPs. For example, as can be seen in Fig. 10.7, point c is the TP obtained by the PLR. Given $nw_{tp} = 1$ and $nw_{op} = 1$, the points b and c are labeled as TPs, and the points a and e are discarded and are not labeled as OPs.

Note that in some cases, some points generated by the PLR are not TPs. For example, in Fig. 10.8, an uptrend segment may follow two downtrend segments, so the split points a and b are considered as OPs.

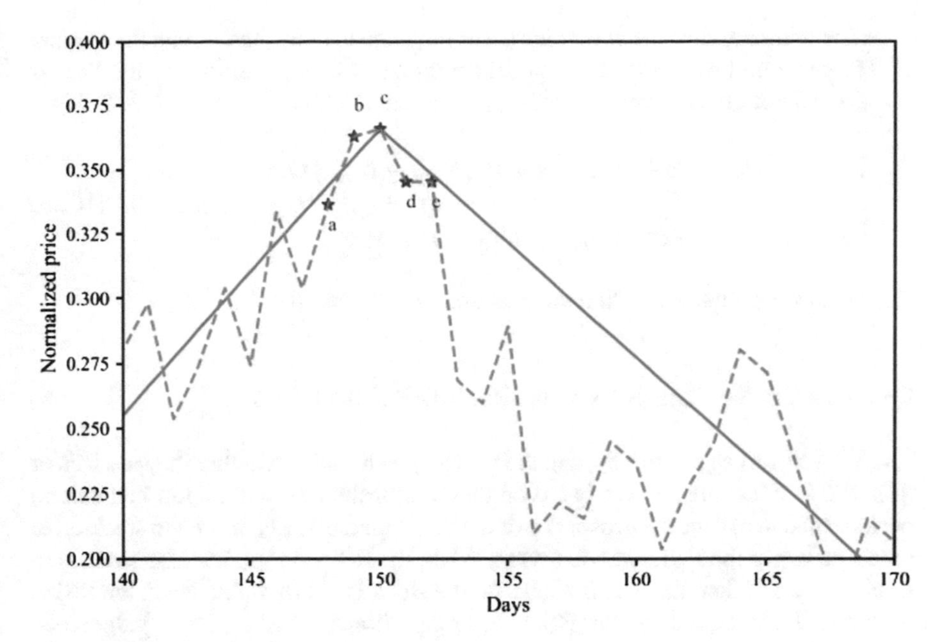

Fig. 10.7 Example for the neighbor window

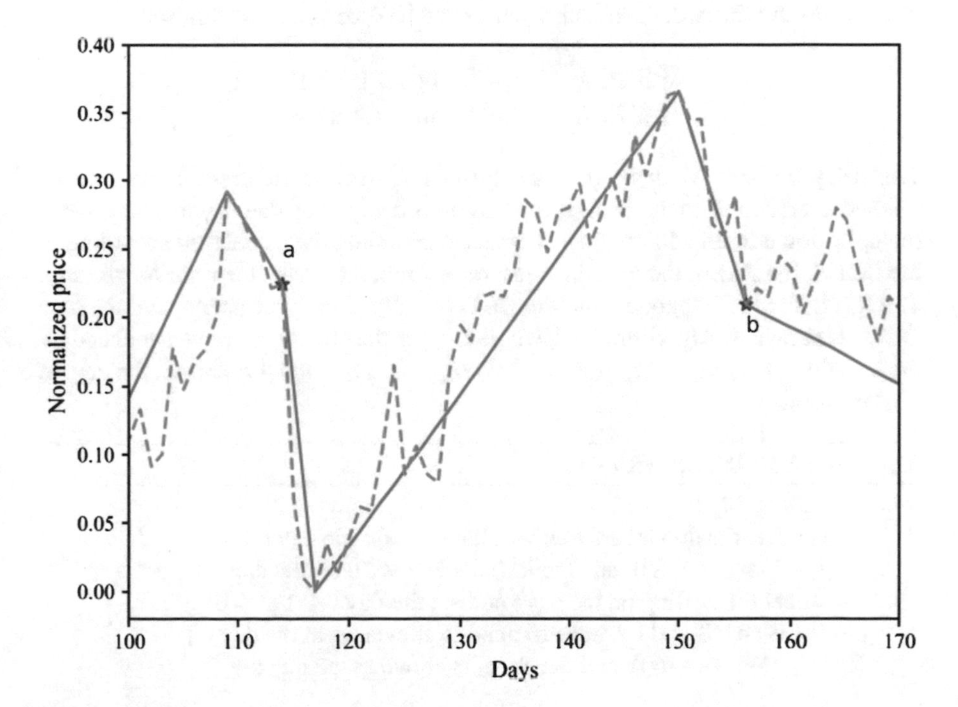

Fig. 10.8 Example for the OP generated by the PLR

After adjusting samples according to the nw_{tp} and the nw_{op}, assuming the number of TPs generated by the PLR is N_{ptp}, the number of OPs generated by the PLR is N_{pop}, and the number of other OPs is N_{oop}, then the OPs can be selected as follows:

$$\begin{cases} N_{oop}\text{OPs} + \text{random select } \left(N_{ptp} - N_{pop}\right)\text{OPs} \\ \qquad\qquad if\ N_{pop} + N_{oop} > N_{ptp} \\ \text{keep } \left(N_{pop} + N_{oop}\right)\ \text{OPs} \quad \text{else} \end{cases} \tag{10.23}$$

The balanced samples and their weights are used to train the WSVM.

Determine Trading Signals Using the RSI Rule and the DODS

The WSVM forecasts the TPs, and the next step is to judge whether they are BPs or SPs. RSI is a technical curve based on the ratio of the rise and the fall in a certain period and can reflect the prosperity of the stock markets. The more the stock price rises, the larger the RSI, and vice versa. When the RSI is around 50, the stock is in a stable trend; When the RSI is above 70, the stock is overbought; When the RSI is below 30, the stock is oversold [90]. In this subsection, the RSI is used to judge stock trends and determine trading signals. The points with an RSI of around 50 are in a stable trend, and it is difficult to determine whether they are BPs or SPs. Therefore, these points are discarded, and other points can be determined as follows:

$$\begin{cases} \text{if } RSI_{24}(t) < 40, & \text{then } t \text{ is a BP} \\ \text{if } RSI_{24}(t) > 60, & \text{then } t \text{ is a SP} \end{cases} \tag{10.24}$$

The delay-one-day strategy (DODS) [40] is a method to prevent the loss of prediction errors. With the DODS, the trading is delayed by one day to determine if the up or down trend will continue. Because the trading day is delayed according to the DODS, the RSI of the next day must be examined to make the right transaction. The RSI is further integrated into the DODS to adjust the transaction, and the new DODS is shown in Algorithm 10.2, where $cr\ (t)$ is the change rate and calculated by the equation $cr\ (t) = (P_c\ (t) - P_c\ (t-1))\ /P_c\ (t-1)$, α and β represent the degree of alleviation.

Algorithm 10.2 DODS with RSI

for each trading day t:
 if t is a BP, the downward trend is alleviated on the day $t + 1$ $(cr(t+1) > \alpha * cr(t))$, and the RSI_{12} is below 50 on the day $t + 1$:
  Cancel the trading on the day t and set the day t + 1 as a BP
 elseif t is an SP, and the upward trend is alleviated in the day t + 1 $(cr(t+1) < \beta * cr(t))$, and the RSI_{12} is above in the day t + 1:

Cancel the trading on the day *t* and set the day t + 1 as a SP
else:
Cancel the trading on the day t and set the day t + 1 as a OP

10.2 Agriculture Classification

10.2.1 An Alternative Approach for the Classification of Orange Varieties Based on Near Infrared Spectroscopy

This section studies a multivariate technique and feasibility of using near infrared spectroscopy (NIRS) for non-destructive discriminating Thai orange varieties.

10.2.1.1 Materials and Methods

Spectral Acquisition and Preprocessing

There are three different orange varieties to be classified in the present work including Kaew Wan (K), Number One (N), and Sai Nam Pung (S) and some properties of them are shown in Table 10.14. This dataset was obtained from a previous study of quality evaluation of orange using near infrared spectroscopy [91]. In order to characterize the features of the oranges, samples were selected from orange orchard in the northern part of Thailand all year long to cover each the studied orange varieties in all seasons. The collected samples were then graded in the experiment with the same size, color and smoothness to avoid effects from physical characteristics of samples. NIR absorbance were measured by shortwave length-spectrometer (model: PureSpect, Saika Technological Institute Foundation, Japan) increasing from 643.26 to 970.92 (nm) at a step of 1.29 (nm) in transmittance measurement mode. The intensity of halogen lamp was 50 W and the sample holder rotated with speed of 6 m/min. The experimental setup was shown in Fig. 10.9.

Table 10.14 Three varieties of oranges in experiment

Varieties	Production area	Rind color	Total soluble solid mean	Acidity mean
Keaw wan	Chiang Mai province, north of Thailand	Yellowish green	10.26	0.36
Number One	Chiang Mai province, north of Thailand	Yellowish green	9.88	0.54
Sai Nam Pung	Chiang Mai province, north of Thailand	Yellowish green	12.13	0.56

Fig. 10.9 The NIR spectra collection of Thai oranges from spectrometer (model: PureSpect, Saika Technological Institute Foundation, Japan)

Each orange sample was scanned two times and spectral records were averaged. An orange is accordingly described by the NIR absorbance 255 different spectra (i.e., wavelength). From each variety, 100 samples (oranges) were measured for their NIR absorbance. As a result, a data set contains 300 instances and 255 spectrum features was generated for developing classification models to the three orange varieties. The NIR spectral features were normalized into [0,1]. The averages of the three orange varieties on each of the 255 spectra are illustrated in Fig. 10.10.

Classification Methods

A number of representative classification methods were evaluated in the present work. Firstly, the complete data containing all the 255 spectrum features were used to develop classification models for the three orange varieties. The performances of these modeling methods were indexed by the classification accuracies obtained via the commonly used tenfold cross-validation [92]. Once the best performing method was identified, a subsequent feature selection was conducted to see whether a fewer features instead of the full spectral range can achieved a comparable performance.

The first classification method evaluated in the present work is kNN in [93]. kNN is a simple classification algorithm known as lazy method, which classifies a new

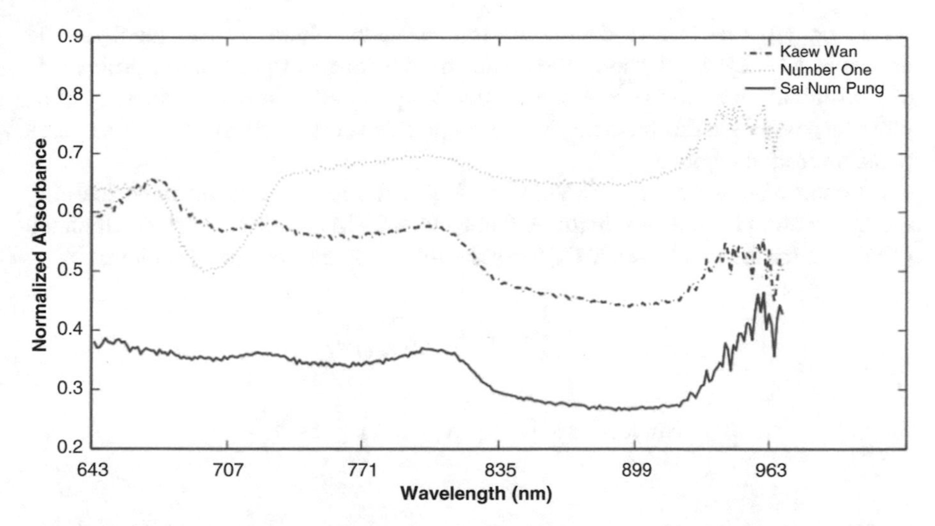

Fig. 10.10 The average NIR absorbance of the three orange varieties at different spectra (identified by wavelength)

instance into the major class among its neighbors. Here, the parameter k that defines the number of neighbor considered in classification of new instances is tuned among different numbers of k.

Linear Discriminant Analysis (LDA) [94] is another method assessed in the present work. This method assumes Gaussian distributions of the same covariance for the samples of different classes ($P(\boldsymbol{x}|\,C_k)$). As a result, the discriminate (classification) function is a linear combination of input features.

Logistic Regression (LGR) in [95] is a simple linear (log-linear) nonparametric classification model of good generalization capability, which is robust and makes less assumption for the data. LGR simply assumes the posterior probability of class C_k as $P(C_k|\,\boldsymbol{x}) = exp\ (a_k)/\sum_j \exp\ (a_j)$. Where $a_k = \langle \boldsymbol{w_k}, \boldsymbol{x} \rangle$. Where C_k denotes the **k**-th class, $\boldsymbol{x}$ is an instance (i.e., data point), $\boldsymbol{w}$ is the weight vector needs to be tuned to fit into data.

Multi-Criteria Quadratic Programming (MCQP) in [96] formulated the classification problem as

$$\min_{\boldsymbol{w},b} \sum_{i=1}^{l} \alpha_i^2 - \lambda \sum_{i=1}^{l} \beta_i^2$$

$$\text{Subject to } w^T x_i + b = y_i\ (\alpha_i - \beta_i)$$

$$\alpha_i, \beta_i \geq 0 \tag{10.25}$$

where $(\boldsymbol{x_i}, y_i)$, i = 1,...,l, denote the training data, where $\boldsymbol{x_i}$ is an input sample and $y_i \in \{-1, 1\})$ is its class label. The solution to this optimization problem is a discriminate function $\boldsymbol{w^T x} + b = 0$ that keeps the data miss-classified by it as close as possible to this boundary while keeping the correct-classified ones as far as possible from the boundary.

Support Vector Machine (SVM) in [97] is a sophisticated machine learning method rooting in statistical learning theory. The SVM algorithm can be formulated (using the same notations as MCQP) as the following optimization problem:

$$\min_{\boldsymbol{w},b,\xi} \frac{1}{2}\boldsymbol{w}^T\boldsymbol{w} + C\sum\nolimits_{i=1}^{l} \xi_i$$

$$\text{Subject to } y_i\left(\boldsymbol{w}^T\phi\left(\boldsymbol{x}_i\right) + b\right) \geq 1 - \xi_i$$

$$\xi_i \geq 0 \tag{10.26}$$

In this formulation, $\phi(\bullet)$ is the function that maps the input data onto a higher dimensional space, which is implicitly defined by a kernel function that enables the solution of the nonlinear optimization problem in a linear kernel space. The Gaussian kernel (10.27) was selected for SVM, which is recommended as a reasonable initial choice.

$$K\left(\boldsymbol{x}_i, \boldsymbol{x}_j\right) = \left(\phi\left(\boldsymbol{x}_i\right), \phi\left(\boldsymbol{x}_j\right)\right)$$

$$= \exp\left(-\gamma\left\|\boldsymbol{x}_i - \boldsymbol{x}_j\right\|^2\right) \tag{10.27}$$

There are two adjustable model parameters (C, γ) for the SVM with Gaussian kernel. A heuristic "grid-search" using tenfold cross-validation was conducted among $C \in \{2^{-15}, 2^{-13}, \ldots, 2^3\}m$ and $\gamma \in \{2^{-15}, 2^{-13}, \ldots, 2^3\}$ in order to find the (C, γ) with which the SVM can achieve the best predictive accuracy. It should be noted that the above designs of MCQP and SVM are for binary classification problems and must be extended in order to tackle the multiclass (i.e., the three orange varieties) problem. In the present work, "one-against-one" strategy was implemented for the multi-class classification.

Feature Selection Methods

In order to develop a "reduced" orange classification model involving only subsets of spectrum features, a number of feature selection methods were performed to identify features of good discriminative ability from the 255 spectra. These feature selection methods are described in the following. Correlation-based Feature Selector

(CFS) in [89] is a simple feature selection method which prefers feature subset that is highly correlated with the class while having low inter-correlation. The CFS's feature subset evaluation function is calculated from

$$Ms = \frac{k\overline{r_{cf}}}{\sqrt{k + k(k-1)\overline{r_{ff}}}} \tag{10.28}$$

where Ms is the heuristic merit of a feature subset S containing k features, $\overline{r_{cf}}$ is the mean feature–class correlation ($f \in S$), $\overline{r_{ff}}$ is the average feature–feature inter-correlation.

Three information-based feature selection methods were also performed in the present work. Among them, InfoGain assesses the value of a feature by measuring the information gain (InfoGain(Class, Feature) = H(Class) H(Class|Feature), where H denotes the information entropy) with respect to the class. GainR is in the same spirit as InfoGain except that the valued of a feature value is defined as gain ratio, i.e., GainR(Class, Feature) = (H(Class) H(Class|Feature))/H(Feature). Likewise, SymmU evaluates the worth of a feature by measuring the symmetrical uncertainty with respect to the class, which is defined by SymmU(Class,Feature) = 2 × (H(Class) H(Class|Feature))/H(Class) + H(Feature).

ReliefF [98] is one of the sophisticated feature selection methods, which repeatedly samples an instance and considers the value of a given feature for the nearest instances of the same and different class. This method tries to find a good estimation of the following probability as the weight of each feature (f).

$$\begin{aligned} W_f = {} & P\,(\text{different value of } f\,|\,\text{different class}) \\ & - P\,(\text{different value of } f\,|\,\text{same class}) \end{aligned} \tag{10.29}$$

Like kNN, the number of nearest neighbors needs to be specified for ReliefF. In the present work, this number was set as 10, which is often the default value for ReliefF.

Unlike these feature selection introduced above, Least Square Forward Selection (LS-FS) in [99] is an unsupervised method, which does not inquire class information to decide which features are of good quality. Instead, LS-FS searches for the feature subset which can best represent (or reproduce) the entire data. The optimal feature subset can be found by least square combinatorial optimization. LS-FS is embedded with kernel functions (i.e., linear, polynomial, radial basis function (RBF)) to handle both linear and non-linear inter-feature relationships in data. In our research, the non-linear extension of LS-FS is named as Kernel Least Square Forward Selection (KLS-FS).

In addition to these feature selection methods, PCA [100] was also used in the present work as a reference method because it is an effective method in data mining and widely applied in NIR spectroscopic field. PCA is a well-known feature reduction method which constructs new features by projecting the original features to the directions in which the data scatter most instead of selecting original features.

For performing the nonlinear form of principal component analysis, KPCA in [101] reforms the traditional linear PCA by applied the kernel trick to construct non-linear mapping.

The feature selection techniques were proposed to NIRS analysis because it is normally high dimensionality composed by hundreds to thousands variables. As a result, it usually occurs collinearity between variables and sometimes non-linearity in NIR spectra can also happen from process measurement and chemical nature of the target analytical parameter such as the effect of the environment (pressure, temperature, etc.) to the different type of interaction to each functional group in samples [102]. Our study tries to use both supervised and unsupervised feature selection techniques to find the most relevant wavelengths (informative variables), eliminate non-informative variables and reduce collinearity and non-linearity effects. Improvement of the quality of spectra can develop reduced classification model by the reasons of computational time, easier interpretation and classification model improvement.

Various supervised feature selection methods in machine learning described above were used to compare the results and search for alternative approaches. On the other hand, unsupervised feature selection method is recognized as another approach to select the feature subset from high dimensional data. LS-FS was chosen by the reasons of easier interpretable result and removal of collinearity and non-linearity effect from the spectra. Here we need to addition the advantages of feature selection over feature transformation/projection (represented by PCA). Features generated by PCA are weighted sum of all the original features, i.e., all spectra need to be measured before using PCA to construct fewer (projected) features. Also, the projected features (generated via PCA) do not keep their physical meaning. In these regards, feature selection that keeps a few original features is superior to feature projection.

By searching for good discriminative features, the threshold of the percentage of variance kept by PCs was set as 99% in PCA. Then, LS-FS and all feature selection methods above were tuned to number of principal components (PCs) constructed by PCA for comparable result with PCA. In nonlinear searching, 99% of variance kept by PCs was also set in KPCA. RBF kernel was adopted in both KPCA and KLS-FS to find the optimal parameter by grid search among $\gamma \in \{2^{-15}, 2^{-13}, \ldots, 2^{3}\}$ and number of PCs from one to threshold PCs number.

10.2.1.2 Results and Discussions

Full NIR Spectra

All the classification methods introduced in section "*Classification methods*" were evaluated for discriminating the three orange varieties (i.e., Kaew Wan (K), Number One (N), and Sai Nam Pung (S)) using NIR spectra. Their performances are indexed by the classification accuracies (obtained through tenfold cross-validation) given in Table 10.15. Along with the overall classification accuracies (i.e., the average

Table 10.15 Classification accuracies (tenfold cross validation) of different models of three orange varieties

		In-group accuracy (%)		
Model	Overall accuracy (%)	Keaw wan	Number One	Sai Nam Pung
*k*NN	93.67	88	96	97
LDA	93.33	93	91	96
LGR	100	100	100	100
MCQP	97.00	98	94	99
SVM	99.00	97	100	100

accuracy for the three classes), the in-group classification accuracies provide detailed information about the performances of the classifiers for each variety. It can be seen from Table 10.15 that LGR made no classification error in the cross-validation. SVM and MCQP also performed very well on the data. For SVM, best model parameters C and γ found by the heuristic grid-search are 2^{11} and 2^{-7}. With these two parameters, SVM only made few miss-classifications for K class. The performance of MCQP is slightly lower than SVM but it is also simpler than SVM. The evaluation shows that the kNN ($k = 1$) and LDA failed to provide satisfying classification accuracies.

Full NIR Spectra with Various Feature Selection Methods

In order to simplify classification model, the LGR method was selected to be a classifier in the feature reduction/selection experiment. PCA was performed since the original features showed strong linear dependence. As a result, four PCs were constructed by PCA. For an aligned comparison, all other feature selection methods were tuned to selected four features as well (except that CFS selected 18 features since it cannot be tuned to have four features selected). The tenfold cross-validation classification accuracies of the LGR models developed on the four reduced/selected features were used to index the performances of the feature reduction/selection methods, which are shown in Table 10.16.

It can be seen from Table 10.16 that the four PCs obtained via PCA lend the LGR the highest classification accuracy. In other word, PCA provides more compact representation for the original data. However, It should be recognized that the model using PCs is still a "full model" (i.e., all the 255 spectra needs to be measured) since each PC is indeed a linear combination of all the original features and there

Table 10.16 Classification accuracies (tenfold cross validation) of LGR model with different feature reduction/selection methods based on 4 PCs or features

Model	Accuracy of each feature reduction/selection method (%)						
	PCA	LS-FS	CFS	InfoGain	GainR	ReliefF	SymmU
LGR	97	95	92	80	81	80	80

is no direct information about the discriminating ability of the original features provided by PCA. The signature analysis or loading plot of each PC must be used for further analysis to indicate discriminative features (i.e., wavelength). LS-FS, on the contrary, identified that the feature subset comprised by the spectra of 769.68, 692.28, 662.61 and 959.31 (nm) is discriminative to the three orange varieties. This feature subset lent the LGR model classification accuracy as high as 95%. It will significantly reduce our experiment effort since such a model requires only the four spectrum measurements instead of all the 255 ones. The rest of the feature selection methods listed in Table 10.16 (CFS, InfoGain, GainR, ReliefF, and SymmU) failed to identify features of good discriminating ability.

Furthermore, the performance of PCA and LS-FS was further compared by constructing/selecting different numbers of features. The tenfold cross-validation LGR classification accuracies of the features constructed by PCA and the ones selected by LS-FS were illustrated in Fig. 10.11 as the number of features increase from 4 to 250. Figure 10.11 shows that PCA brought increased classification accuracy until the number of PCs (i.e., the features constructed by PCA) increases up to 50; ever since, however, its performance continued to decrease as the number of PCs increasing. This strange behavior of PCA is a result of that, when more features need to be constructed, the smaller eigenvalue of the covariance matrix

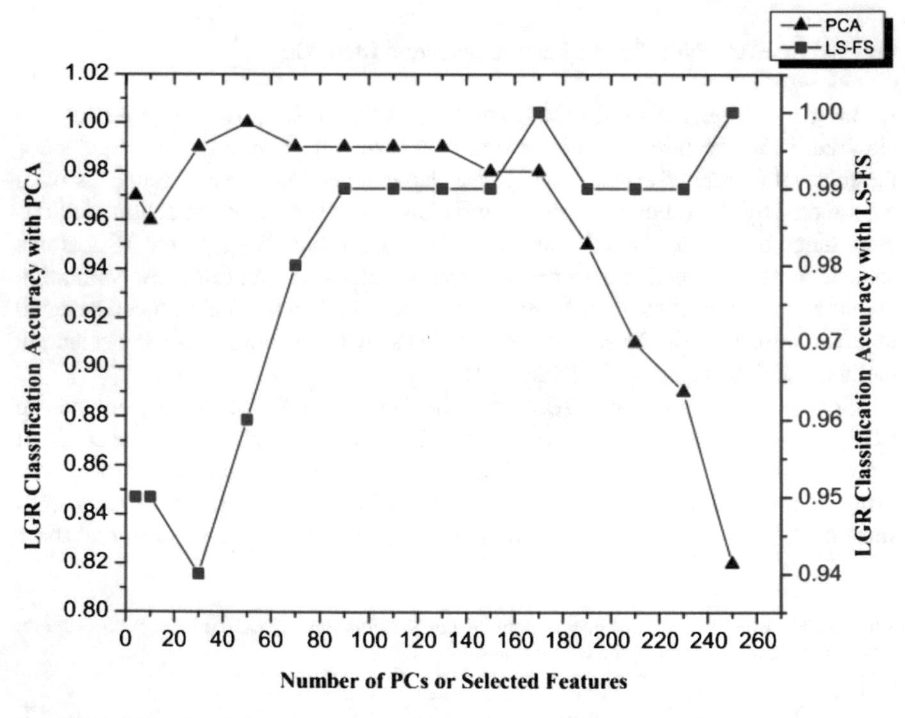

Fig. 10.11 LGR classification accuracies of the PCs and the features selected via LS-FS

of the data approaches to zero and its corresponding eigenvector (i.e., a PC) will be a nearly random direction due to numerical error becomes dominant in the PCA computation. And it is shown in the feature reduction/selection experiment that 99% variance of the data studies in this work is accounted for by only four PCs, which indicates that the subsequent eigenvalues are already close to zero. The performance of LS-FS is slightly inferior to the PCA when only a few features were used since PCA utilizes all the original features to construct the PCs. However, with the increased number of selected features, LS-FS outperformed PCA and has 100% classification accuracy with 170 features, which in turn indicates that at least 85 features (i.e., spectra) are redundant for classifying the three orange varieties.

For spectral interpretation, the loading plot of the first four principal components (PCs) across the entire spectral region was drawn as shown in Fig. 10.12. Figure 10.12 shows that there are wave crests and wave vales in the wavelengths before 700 nm and wavelengths between 800 and 950 nm in different PC which can indicate some properties in samples. If wavelength areas in loading plot were compared with the feature subset selected from LS-FS, it showed that both methods gave relatively similar wavelength ranges. Additionally, the discussion in term of relationship between spectral wavelengths and sample chemical properties was analyzed by LGR model based on the feature subset selected by LS-FS. The favorable performance of the LGR model developed based on the feature subset comprised by the spectra of 769.68, 692.28, 662.61 and 959.31 (nm) indicate the suitableness of the subset for classification of the three orange varieties. For wavelengths of 769.68 and 959.31 nm, some previous works [103, 104] showed that the wavelength around 765 nm and wavelengths between 800 and 950 nm are indicative to sugar content in orange juice. In our research wavelengths of 769.68

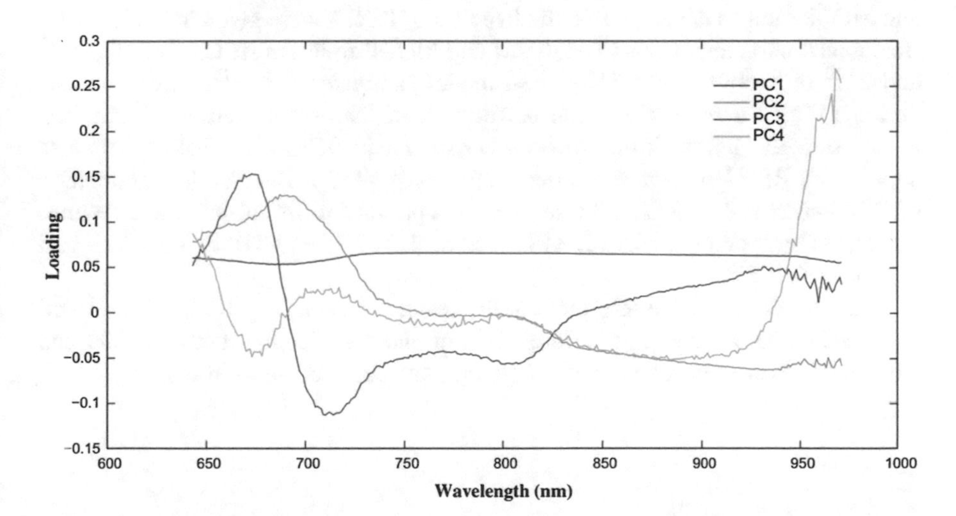

Fig. 10.12 Loading plot of the first four principal components (PCs) across the entire spectral region from PCA

Table 10.17 Classification accuracies of different methods with 5 PCs from KPCA (tenfold cross validation)

		In-group accuracy (%)		
Model	Overall accuracy (%)	Keaw wan	Number One	Sai Nam Pung
LGR	97.33	98	96	98
MCQP	97.00	97	95	99
SVM	97.67	98	96	99

and 959.31 nm were selected in the same area of spectra which can describe sugar content in our samples. For wavelengths of 692.28 and 662.61, it has been reported in [104] that in the wavelength before 700 nm was mainly contribute to the color or different growing conditions of orange juice and can also used to discriminate orange varieties and work in [105] also showed that the wavelength around 620 nm may be sensitive to some pigment in orange juice. The spectra selected by LS-FS as most suitable for orange classification agree very well with these previous findings.

In order to search more compact or more discriminative features, the non-linear feature selection/reduction experiment was also applied in this study. Our research chooses two methods which are KPCA and KLS-FS. To investigate the non-linear relation in this dataset, three classifiers (LGR, SVM and MCQP), which can perform well to classify this dataset were firstly used to test with Kernel Principal Component Analysis (KPCA). Their performances are also indexed by the classification accuracies (obtained through tenfold cross-validation). KPCA was tuned by RBF kernel among $\gamma \in \{2^{-15}, 2^{-13}, \ldots, 2^{3}\}$. The threshold of the percentage of variance kept by PCs was set as 99%. As a result, five PCs were constructed by KPCA. The classification accuracies of LGR, SVM (linear kernel) and MCQP models developed on the five PCs of KPCA were given in Table 10.17. Best model parameter c for LGR, SVM and MCQP found in KPCA is 2^{-13}, 2^{-11} and 2^{-13} respectively. For SVM, best model parameter C is 2^{13}. In comparison between three methods, SVM can perform better than other methods. Therefore, in this step SVM was chosen to be a classifier with KLS-FS. KLS-FS was also adjusted by RBF kernel in the same condition of KPCA. Best model parameter c of RBF kernel is 2^{-1}. KLS-FS also gave the optimal feature subset at five features comprised by spectra of 825.15, 848.37, 658.74, 707.76 and 778.71(nm). The best parameter C is 2^{15}.

In aspect of spectral interpretation, the features selected by KLS-FS have still been related to wavelengths before 700 nm and wavelengths between 800 and 950 nm which are sensitive to some pigments and sugar contents in orange.

10.3 Engineering Problems

10.3.1 Automatic Road Crack Detection Using Random Structured Forests

Cracks are a growing threat to road conditions and have drawn much attention to the construction of intelligent transportation systems. However, as the key part of an intelligent transportation system, automatic road crack detection has been challenged because of the intense inhomogeneity along the cracks, the topology complexity of cracks, the inference of noises with similar texture to the cracks, and so on. This subsection proposes CrackForest, a novel road crack detection framework based on random structured forests, to address these issues.

10.3.1.1 Related Work

In this section, we first give a brief review of crack detection, after that, the related crack characterization methods are discussed. Crack characterization exploits the spatial distribution of image tokens composing the detected cracks and thereby transforms the structured tokens into discrete labels.

Crack Detection

Numerous papers have been written on road crack detection over the past 30 years. Early works [29, 106–109] are mainly based on intensity-thresholding for its simplicity and efficiency. Most recent work explores crack detection under more challenging conditions and can be divided into five branches: methods based on saliency detection, textured-analysis, wavelet transform, minimal path and machine learning. An assessment of various pavement distress detection methods can be found in [110, 111].

Salient Detection: Salient regions are visually more conspicuous due to their contrast with the surroundings. Although existing methods [112, 113] demonstrate their effectiveness in detecting salient regions in the Berkeley database [114], they perform poor on the completeness and continuity of detected crack.

Textured-Analysis: Since road surface images are often highly textured, textured-analysis methods [92, 115, 116] are introduced in road crack detection. In order to distinguish the cracks and the backgrounds, [92, 116] use the Wigner model, and [115] uses classification method. These methods use a local binary pattern operator to determine whether each pixel belongs to a crack and the local neighbor information is not taken into consideration. Therefore, the cracks with intensity inhomogeneity cannot be detected precisely.

Wavelet Transform: Wavelet transform is applied to separate distresses from noises [117]. In [118], complex coefficient maps are built by a 2D continuous wavelet transform, wavelet coefficients maximal values are obtained for crack detection. As a result, differences between crack regions and crack free regions could be raised up. However, due to the anisotropic characteristic of wavelets, these approaches may not handle the cracks with low continuity or high curvature properly.

Minimal Path Selection: Give both endpoints of the curve as user's input, minimal path based method can extract simple open curves in images, that is first proposed by Kass et al. [119]. In [120], Kaul et al. propose a method that is able to detect the same types of contour-like image structures with less prior knowledge about both the topology and the endpoints of the desired curves. To avoid false detections that are assimilating loops, Amhaz et al. [121, 122] propose an improved algorithm to select endpoints at the local scale and then to select minimal paths at the global scale. It can also detect the width of the crack. In [123], Nguyen et al. propose a method which takes into account intensity and crack form features for crack detection simultaneously by introducing Free-Form Anisotropy.

Machine Learning: With the increasing size of image data, machine learning based methods [28, 124–127] have become an important branch in detecting road cracks. In [54], artificial neural network models are used to separate crack pixels from the background by selecting proper thresholds. Delagnes and Barba [125] deals with the detection of poorly contrasted cracks in textured areas using a Markov random field model. In [124], Cord et al. use AdaBoost to distinguish images of road surfaces with defects from road surfaces based on textual information with patterns. For all these methods, the training and classification are conducted on each sub-image and as local method, they have drawbacks in finding complete crack curves over the whole image.

Crack Characterization

Existing methods on crack characterization are mainly based on shape descriptor, crack seeds and assigning crack type on each image block.

Reference [3] gives the definition of cracks based on mathematical morphology and proposes that a crack is thought to be a succession of saddle points with linear features. But this definition is pretty vague. References [128], [129] use the direction indices of each pixels and extensible directions for each direction to characterize cracks. A chromosome representation is applied to encode the different ensemble of directions and its extensible directions. Therefore, a crack can be represented as a long sequence of 0 and 1.

References [111, 126] categorize the cracks into five types: longitudinal, transverse, diagonal, block, and alligator. Reference [126] uses a neural network based method to search patterns of various crack types horizontally and vertically. Reference [111] uses curves and buffers to describe certain regions of a crack.

Reference [130] uses longitudinal, transverse, or diagonal crack seeds to identify longitudinal and transverse cracks. Orientation and strength information are taken into consideration by Ref. [131], which largely improves the diversity of crack seeds.

In [132], cracks are classified into three types as defined by the Portuguese Distress Catalog. They use two block features including the mean and the standard deviation values of pixel-normalized intensities to categorize an image block as longitudinal, transversal or miscellaneous. Reference [133] computes CTA (Conditional Texture Anisotropy) values over the distribution of the mean and the standard deviation values calculated on pixels to distinguish crack pixels from defect free pixels.

However, there are two main drawbacks in these methods. On the one hand, new types of crack cannot be generated. By applying the structured tokens, we extend the crack types into thousands of dimensions. On the other hand, these methods perform poor on the cracks with complex topology. To address this issue, we propose a novel crack descriptor to describe the cracks with arbitrary complex topology.

10.3.1.2 Automatic Road Crack Detection

In this section, we will introduce our novel crack detection method which can take advantage of the structured information of cracks. Figure 10.15 shows the overall procedure of our proposed method. This framework can be divided into three parts: In the first part, we extend the feature set of traditional crack detection method by introducing the integral channel features. These features extracted from multiple levels and orientations allow us to re-define representative crack tokens with richer structured information. In the second part, random structured forests are introduced to exploit such structured information, and thereby a preliminary result of crack detection can be obtained. In the third part, we propose a new crack descriptor by using the statistical character of tokens. This descriptor can characterize the cracks with arbitrary topology. And a classification algorithm (KNN, SVM or One-Class SVM) is applied to discriminate cracks from noises effectively. Please find the results in every part of our method in Fig. 10.13 (Figs. 10.14 and 10.15).

Structured Tokens

Token (segmentation mask) indicates the crack regions of an image patch. Current block-based methods [117, 132] are usually used to extract small patches and calculate mean and standard deviation value on these patches to represent an image token. These traditional features are computed on gray level images and applied to describe the brightness and gradient information. However, local structured information is not taken into consideration. So in the first step, we re-define the tokens by introducing the integral channel features which incorporate the color, gradient information from multiple levels and facets.

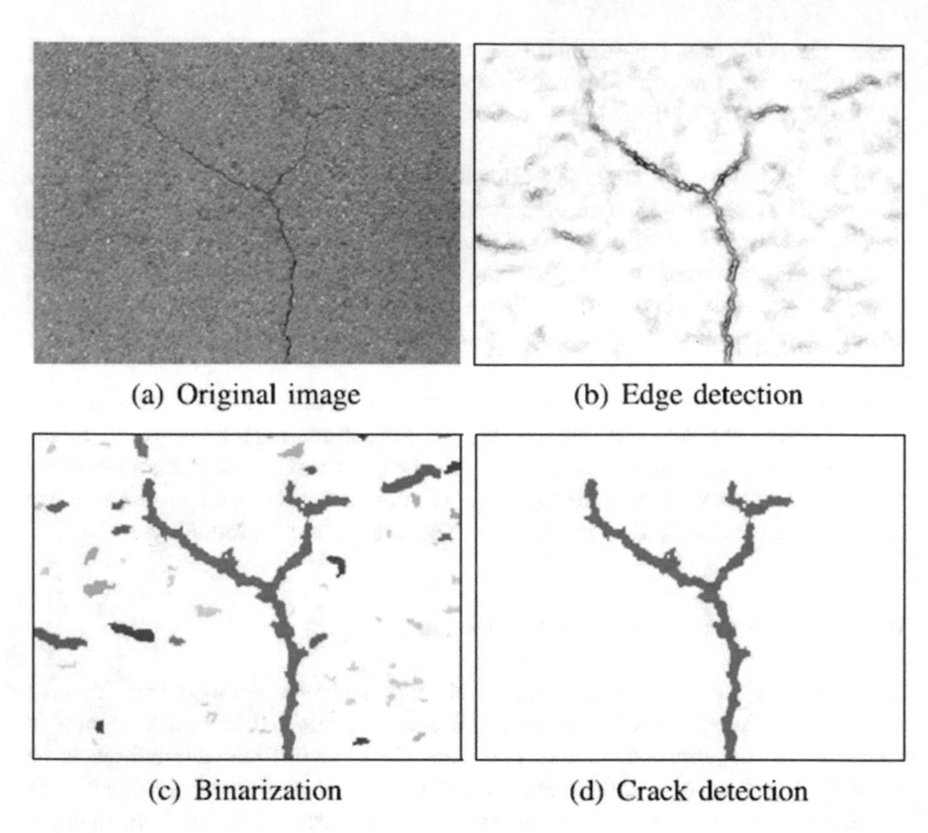

Fig. 10.13 Consider the pavement surface shown in (**a**). (**b**) Preliminary detection results after applying random structured forests. Darker color indicates that the pixel is more likely to contain a crack. After eroding and dilating, the result is shown in (**c**). (**d**) Final result after the classification stage

1. *Learning the Tokens:* Assume that we have a set of images I with a corresponding set of binary images G representing the manually labeled crack edge from the sketches. We use a 16×16 sliding window to extract image patches $x \in \mathcal{X}$from the original image. Image patch x which contains a labeled crack edge at its center pixel, will be regarded as positive instance and vice versa. $y \in \mathcal{Y}$ encodes the corresponding local image annotation (crack region or crack free region), which also indicates the local structured information of the original image. These tokens cover the diversity of various cracks, which are not limited to straight lines, corners, curves, etc. From Fig. 10.16, we can see the extracted image patches and their hand drawn contour tokens. These image patches and tokens will be used to train CrackForest later.
2. *Feature Extraction:* To describe the above tokens, features are computed on the image patches x extracted from the training images I, and considered to be weak classifiers in the next step.

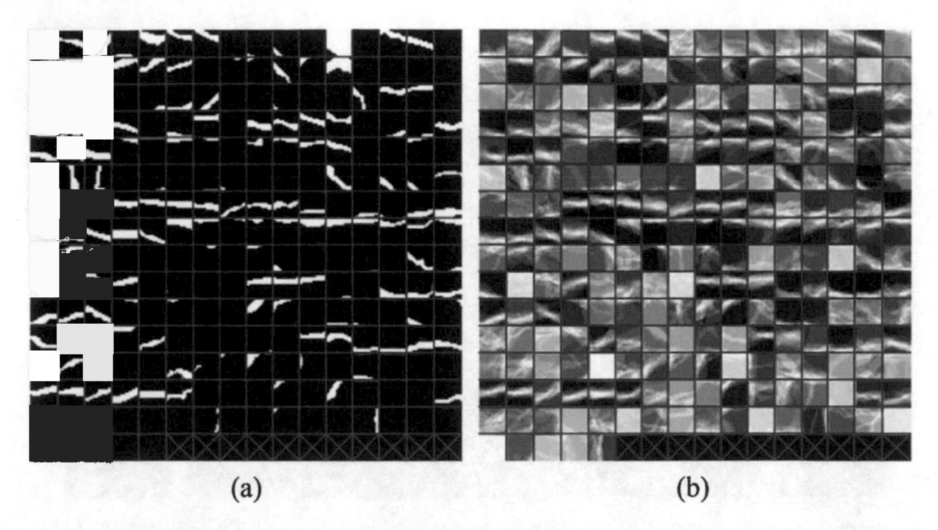

Fig. 10.14 Examples of tokens learned from a manually labeled image database. (**a**) Most representative token for each token set. (**b**) Mean contour structure for each token set

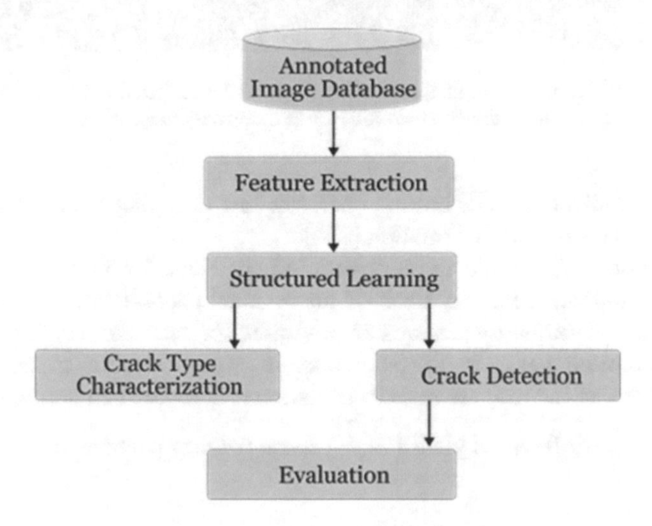

Fig. 10.15 Procedure of the proposed automatic road crack detection method

We use mean and standard deviation value as features. Two matrices are computed for each original image: the mean matrix Mm with each block's average intensity and the standard deviation matrix STD_m with corresponding standard deviation value std. Each image patch yields a mean value and a 16×16 standard deviation matrix.

To characterize the cracks more comprehensively, we also apply a set of channel features composed with color, gradient and oriented gradient information. Integral channel features not only perform better than other features including histogram

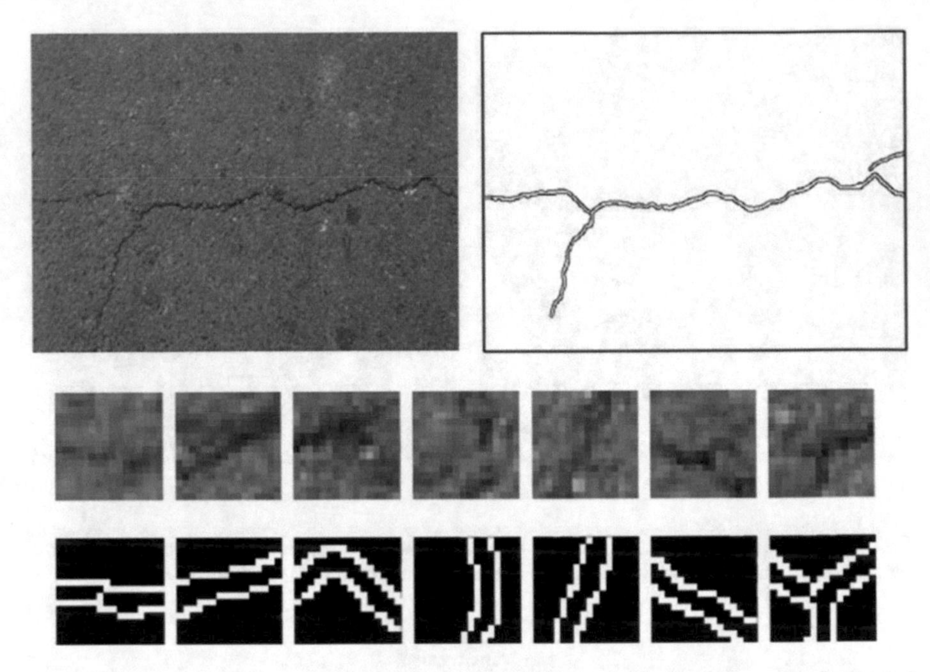

Fig. 10.16 (Top) Example of original image and its ground truth. (Bottom) Example of extracted image patches and their hand-drawn contours. Notice the variety of sketches

of oriented gradient (HOG), but also achieve fast detecting results and integrate heterogeneous sources of information [134].

Three color, 2 magnitude and 8 orientation channels, for a total of 13 channels yield 3328 candidate features. Each of the channel captures a different aspect of information. Self-similarity features are compute for each channel. These features capture the portion that an image patch contains similar textures based on color or gradient information [135]. Texture information is computed on a $m \times m$ grid over the patch. These differences yield $\binom{5 \cdot 5}{2}$ more features per channel.

Structured Learning

In previous step, a set of tokens y which indicate the structured information of local patches, and features which describe such tokens, are acquired. In this step, we cluster these tokens by using a state-of-the-art structured learning framework, random structured forests, to generate an effective crack detector. Random structured forests can exploit the structured information and predict the segmentation mask (token) of a given image patch. Thereby we can obtain the preliminary result of crack detection.

In random structured forests, each decision tree $f_t(x)$ classifies an image patch $x \in \mathcal{X}$ by recursively branching left or right down to the tree until a leaf is reached. And the class of the node is assigned to patch x. The leaf stores the prediction of the input x, which is a target label $y \in \mathcal{Y}$ or a distribution over $\mathcal{Y}$. By training such a tree, tokens with the same structure will be gathered at one leaf. We use the most representative token in each leaf to represent the token class. The class number of tokens equals to the number of leaves.

A forest T can be seen as an ensemble of decision trees f_t. Each tree $f_t(x)$ gives a prediction of a sample $x \in \mathcal{X}$. The final class prediction of multiple trees is integrated by a majority voting algorithm. A leaf $L(\pi) \in f_t$ can assign a class prediction for samples it is reached by, where π stands for the most represented token in the leaf. Each node $N\left(h, f_t^L, f_t^R\right) \in f_t$is associated with a binary split function

$$h\left(x, \theta_j\right) \in \{0, 1\} \tag{10.30}$$

with feature θ_j for each node j. If $h(x, \theta_j) = 0$, sample x should be branched to the left sub-tree f_t^L, otherwise the right sub-tree f_t^R.

1. *Class Prediction:* Given a tree $f_t \in T$, the class prediction of an image patch $x \in \mathcal{X}$ can be obtained by recursively branching it forward until a leaf is reached. An intuitive example has shown in Fig. 10.17. The prediction function $\psi\ (x|\,f_t)$: $\mathcal{X} \to \mathcal{Y}$ for node j is

$$\psi\left(x|N\left(h, f_t^L, f_t^R\right)\right) = \begin{cases} \psi\left(x|\,f_t^L\right), & for\ h\left(x, \theta_j\right) = 0 \\ \psi\left(x|\,f_t^R\right), & for\ h\left(x, \theta_j\right) = 0 \end{cases}$$

$$\psi\ (x|L\,(\pi)) = \pi. \tag{10.31}$$

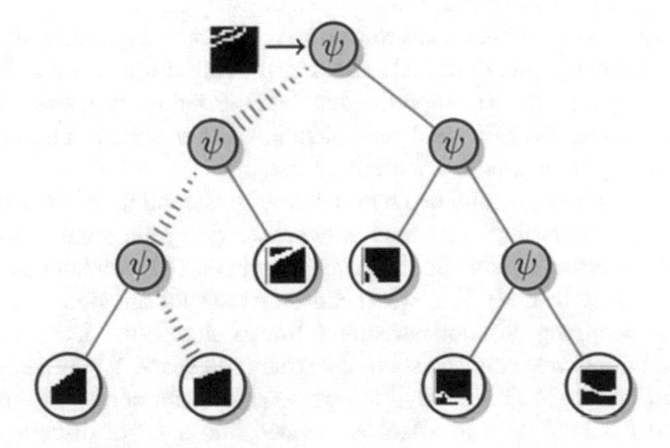

Fig. 10.17 Routing path of an image patch

2. *Randomized Training:* Each tree is trained individually. For a given node N_j and training set $S_j \subset \mathcal{X} \times \mathcal{Y}$, the goal is to find the optimal feature θ_j that results in a good split of the data. In other words, the discrepancy of tokens in the same leaf should be as small as possible. We apply information gain to measure this discrepancy and maximize the information gain to choose θ_j. The form of information gain for node j is defined as follow:

$$I_j = I\left(S_j, S_j^L, S_j^R\right) \tag{10.32}$$

where $S_j = S_j^L \cup S_j^R$, $S_j^L = \left\{(x, y) \in S_j | h\left(x, \theta_j\right) = 0\right\}$ stands for a set of samples that reaches the left sub-tree of the current node and $S_j^R = \left\{(x, y) \in S_j | h\left(x, \theta_j\right) = 1\right\}$ refers to the other set of samples that reaches the right sub-tree.

Whether a terminal node should be further split depends on the maximum depth, the minimum size of node or the entropy of the class distribution. If the node is no longer splitting, a leaf is grown where the class prediction π is set to the most representative token in the training data. Otherwise a node $N\left(h, f_t^L, f_t^R\right)$ is grown where h is a split function regulated by parameter θ_j, maximizing the information gain about the label distribution due to the split $\left\{S_j^L, S_j^R\right\}$ of the training data S.

For multi-class classification ($\mathcal{Y} \subset \mathcal{Z}$), the definition of information gain is

$$I_j = H\left(S_j\right) - \sum\nolimits_{k \in \{L, R\}} \frac{\left|S_j^k\right|}{\left|S_j\right|} H\left(S_j^K\right) \tag{10.33}$$

where $H(S_j) = -\sum_y p_y \log\left(p_y\right)$ denotes the Shannon entropy and p_y stands for the proportion of elements in S with label y. Alternatively, Gini impurity $H(S_j) = \sum_y p_y(1 - p_y)$ can also be applied in Eq. (10.33).

Individual decision tree tends to overfit, which may negatively affected accuracy. To overcome this drawback, random structured forests combine multiple decision trees together to assign the final label. Random structured forests have shown promising flexibility and generalization ability, and most importantly, this method is easy to parallel and extremely fast.

The randomness is embodied by randomly subsampling the data used to train each tree and each node, and randomly subsampling the features used to split each node. In order to maintain the diversity of trees, only a small pool of features is used to select the optimal θ_j when choosing the split function.

3. *Structured Mapping:* Random structured forests change the discrete outer space of the traditional decision forests into a structured space $\mathcal{Y}$. While dealing with structured label $y \in \mathcal{Y}$ directly may cause significant computing expense, the structured labels $y \in \mathcal{Y}$ at a leaf is mapped into a set of discrete labels $c \in \mathcal{C}$, where $\mathcal{C} = \{1, \ldots, k\}$. Given the discrete label space $\mathcal{C}$, information gain

can be calculated efficiently via (10.4). We first map the label space $\mathcal{Y}$ into a intermediate space $\mathcal{Z}$.

$$\Pi : \mathcal{Y} \rightarrow \mathcal{Z}. \tag{10.34}$$

Define $z = \Pi(y)$ in space $\mathcal{Z}$ as a $\binom{16 \cdot 16}{2} = 32640$ dimensional vector, which encodes every pair of pixels in the segmentation mask y. The computational cost of z appears to be significant.

While the dimension of z is still very high, we randomly select 256 dimension of z to train each split function, using a distinct reduced mapping function at each node j

$$\Pi_{\varphi} : \mathcal{Y} \rightarrow \mathcal{Z}. \tag{10.35}$$

Then we apply PCA reduction to map 256 dimensions of z into 5 dimensions, with the first dimension being the most significant factor. To obtain the discrete label $c \in \mathcal{C}$of each structured label y, we use the first dimension of each intermediate label z to cluster into two sets. Labels in the same cluster are assigned to the same label c. With the label c, standard information gain can be calculated at each node.

After the random structured forests are trained, the structured labels y is gathered at the leaves of each tree (see Fig. 10.14). An image patch is routed though each tree based on the split function until a leaf is reached. The most representative token in the leaf is assigned to the image patch. Figure 10.18 shows an intuitive example. We select the token which has the lowest variance with others as the most representative token.

4. *Binarization:* After the structured mapping, each image patch x is assigned to a structured label y. Due to the overlapping, the result of detection is a map, where each element indicates the probability that the corresponding position in the original image is on crack region. So, we use a threshold α to obtain all the possible regions. A high α value may cause the incontinuity of cracks and the ignorance of inapparent cracks. Therefore, we choose $0.1 \leq \alpha \leq 0.2$ in this section. Figure 10.19a shows the binarization result when $\alpha = 0.1$.

   We conduct the erosion and the dilation operation on the preliminary edge detection results to make the cracks as connective as possible. The inside of the crack is filled and the fragments are connected. Moreover, some of the noises are eliminated. From Fig. 10.19b, we can see that small fragments of the detected region have merged together and the continuity of the crack has been improved.

Crack Type Characterization and Detection

Each image patch is assigned to a structured label y (segmentation mask) after structured learning. Although we obtain a preliminary result of crack detection

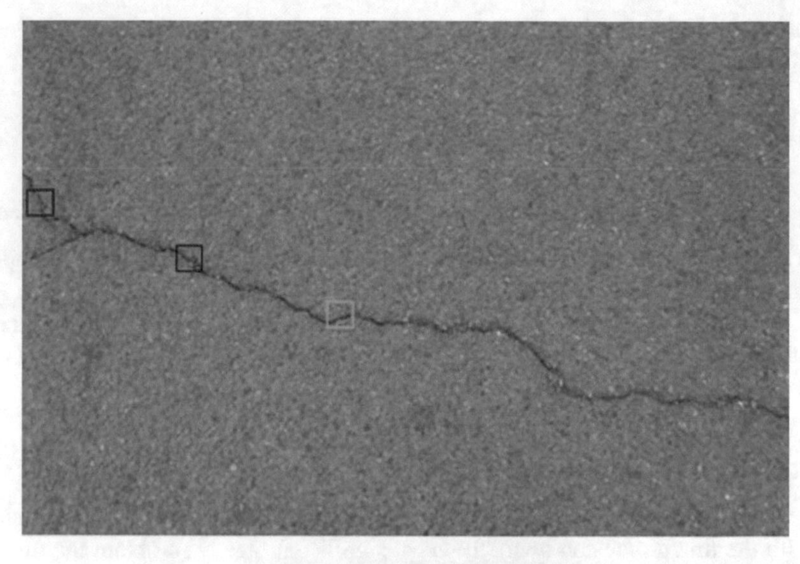

Fig. 10.18 Assigning y to each image patch. The image patches have been assigned to the tokens below (both from left to right)

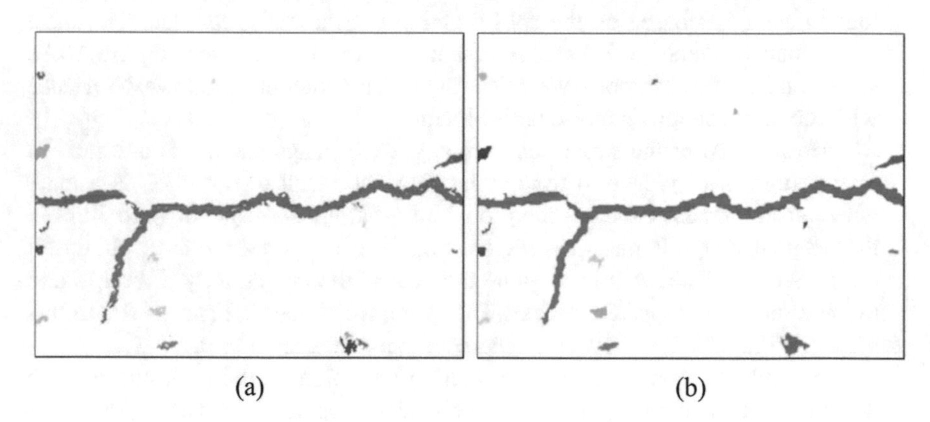

Fig. 10.19 (**a**) Binarization result based on threshold when $\alpha = 0.1$ (removing pixels of low probability according to the given probability map). (**b**) Result after erosion and dilation with a 4×4 rectangular structuring elements

so far, a lot of noises are generated due to the textured background at the same time. Traditional thresholding methods mark small regions as noises according to their sizes. However, in this way, many inconspicuous cracks may be removed by mistake.

Cracks have a series of unique structural properties that differ from noises. Based on this thought, we propose a novel crack descriptor by using the statistical feature of structured tokens in this section. This descriptor consists of two statistical histograms, which can characterize cracks with arbitrary topology. By applying classification method like SVM, we can discriminate noises from cracks effectively.

1. *Crack Descriptor:* Existing crack characterization methods categorize cracks into several types, such as longitudinal, transverse, diagonal, block, and alligator. However, the descriptor proposed, which consists of hundreds of dimensions respectively, has greatly broadened the range of representable crack. What is more, the crack is no longer limited to a few types, we extend the types of crack into thousands of kinds.

   We use 26,443 structured tokens obtained in the structured learning procedure to characterize the cracks. The statistical histogram and the neighborhood histogram of these tokens within a crack can be calculated precisely.

   Statistical Feature Histogram: After the structured learning procedure, we can obtain the token map. Each point in the map indicates the label of token that the 16×16 image patch around the corresponding position is assigned to. Statistical feature histogram in Fig. 10.20 reflects the composition of the crack comprehensively. Each dimension of this histogram represents the number of a certain token.

   The token number from the training result is numerous. After plotting the overall occurrence of each token in Fig. 10.21a, we notice a long tail effect of the token distribution. After analyzing the statistical information of appeared tokens, we find that over 90% occurrences of all the tokens are centered on 708 specific tokens. The occurrences of most tokens make up only a small percentage of all. Therefore, we only use these 708 tokens to construct the statistical feature histogram and the statistical neighborhood histogram. Figure 10.21b shows the occurrence of these tokens.

   Statistical Neighborhood Histogram: The statistical neighborhood histogram captures the neighborhood information of two tokens. We calculate the co-occurrence of each pair of tokens only when they are adjacent. There would be $\binom{708}{2} = 250,278$ token pairs without reduction. Furthermore, we also find the long tail effect of this distribution. Over 90% occurrences of all the token pairs are centered on 956 specific token pairs. Thus, only these token pairs will be used in the following section.
2. *Crack Detection:* With the two histograms for each separated region, we can characterize cracks with arbitrary topology. In this section, we will introduce how to discriminate the noises from cracks by using the two histograms.

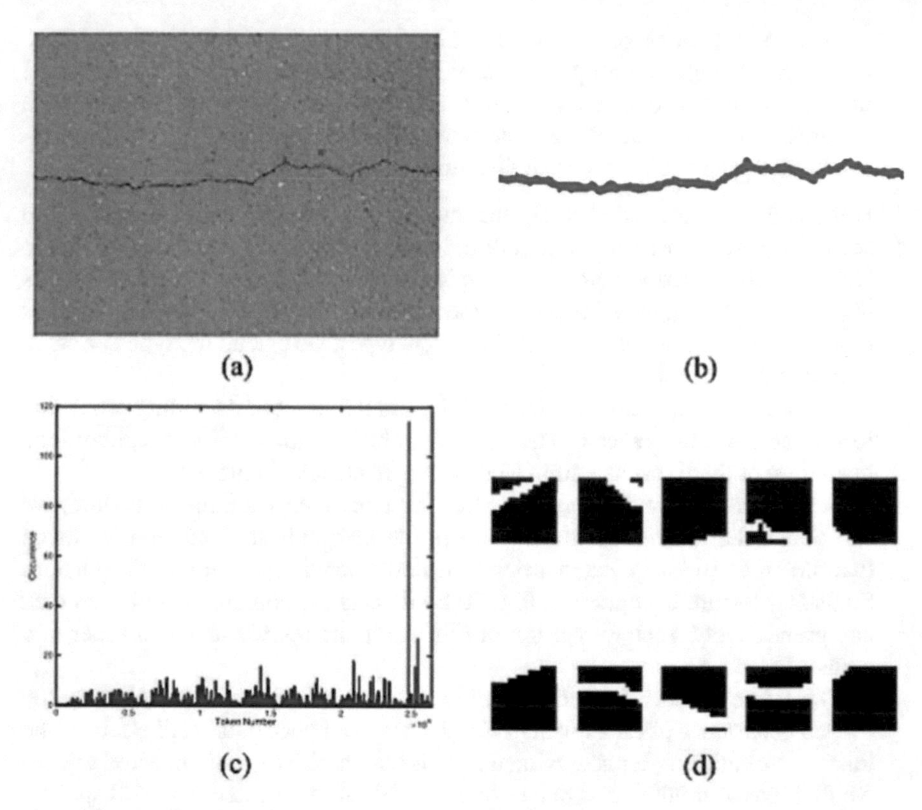

Fig. 10.20 (a) Original image. (b) One of the detected regions. (c) Statistical feature histogram of the detected region. (d) Appearance of the ten most frequent tokens look

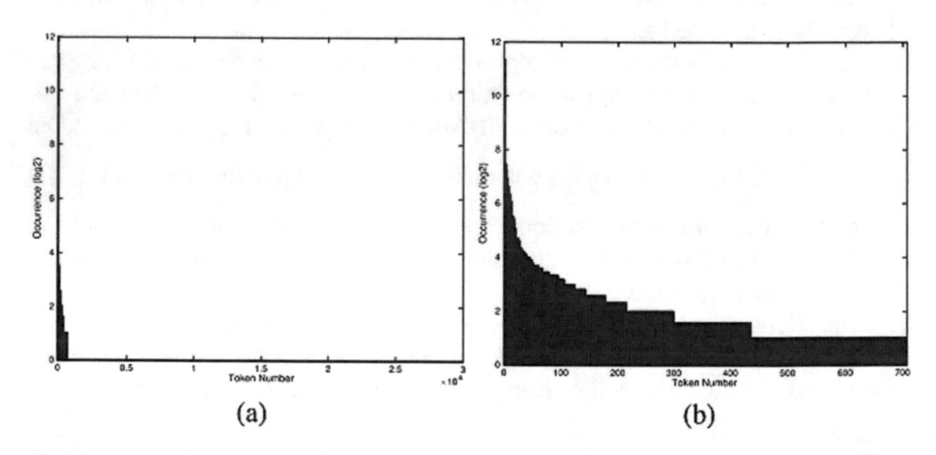

Fig. 10.21 Statistical feature histogram showing the occurrence (in logs) of each token (sorting in descending order of occurrence). (a) Statistical feature of all 26,443 tokens. (b) Only the most representative tokens are shown

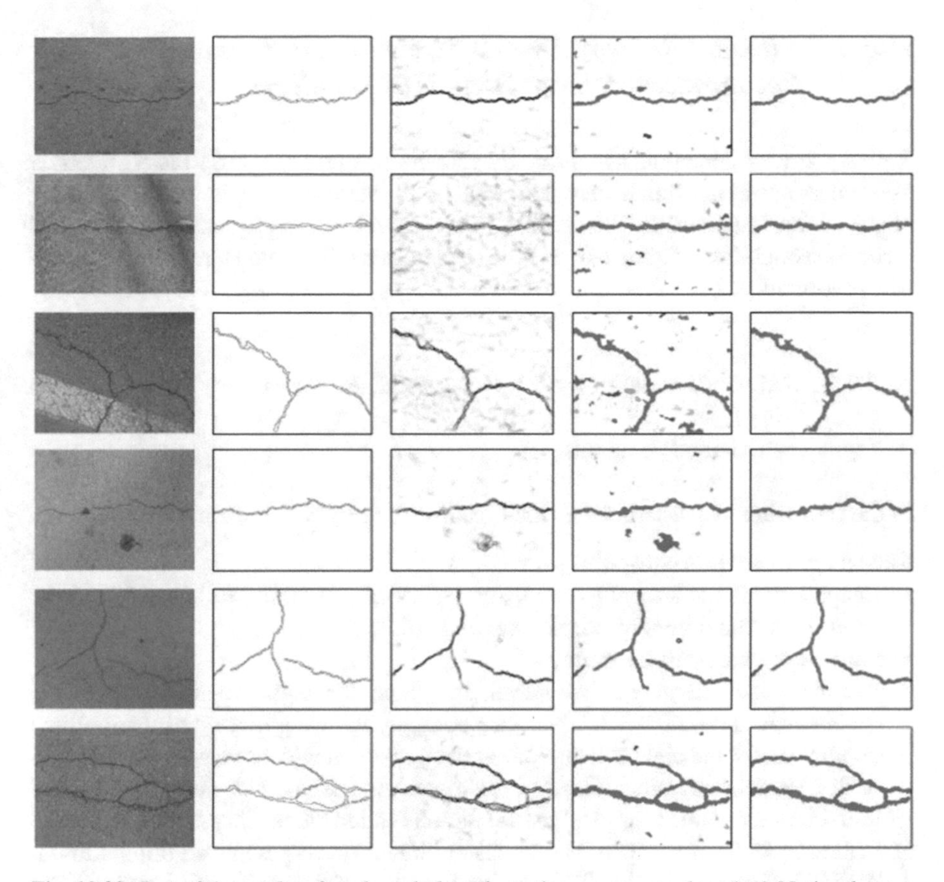

Fig. 10.22 Part of the results of road crack detection using our proposed method. Notice that our method can eliminate the influence of oil stains, shadows, and complex background, effectively, and can cope with miscellaneous crack topology

Vectorization: The distribution of occurrence and cooccurrence are scaled to [0, 1]. Hence, each detected region is presented as a long vector with $708 + 956 = 1664$ dimensions.

Classification: We consider the crack detection procedure as a classification problem. The crack regions are assigned to class $+1$ and the crack free regions are assigned to class -1. By applying KNN (k-Nearest Neighbor), SVM (Support Vector Machine) with linear kernel and One-Class SVM with linear kernel, we obtain the classification model which can discriminate cracks from noises effectively. The results of our algorithm using SVM are shown in Fig. 10.22.

10.3.2 Efficient Railway Tracks Detection and Turnouts Recognition Method Using HOG Features

Railway tracks detection and turnouts recognition are the basic tasks in driver assistance systems, which can determine the interesting regions for detecting obstacles and signals. In this subsection, a novel railway tracks detection and turnouts recognition method using HOG (Histogram of Oriented Gradients) features was presented.

10.3.2.1 Railway Tracks Detection Using HOG Features

Histogram of Oriented Gradients

In the following, we describe the HOG features briefly, which include three steps.

First step: gradients computation

Image's gradients are computed using Gaussian smoothing followed by one of several discrete derivative masks. See Fig. 10.23a.

Second step: orientation binning

For the edge orientation histogram, in which the eight orientation bins are evenly spaced over 0°–360°. Now we compute the weight votes of the gradient orientation for the eight orientations at each pixel, in which the vote is a function of the gradient magnitude itself at this pixel. At last, the votes are accumulated into orientation bins over local spatial regions called blocks. In practice, the HOG feature is an 8-dimensional vector. Each dimension represents an orientation in 0°–360°. The step is 45°. The size of the block varies with it's location. The stride (block overlap) is fixed at each pixel size. See Fig. 10.23b.

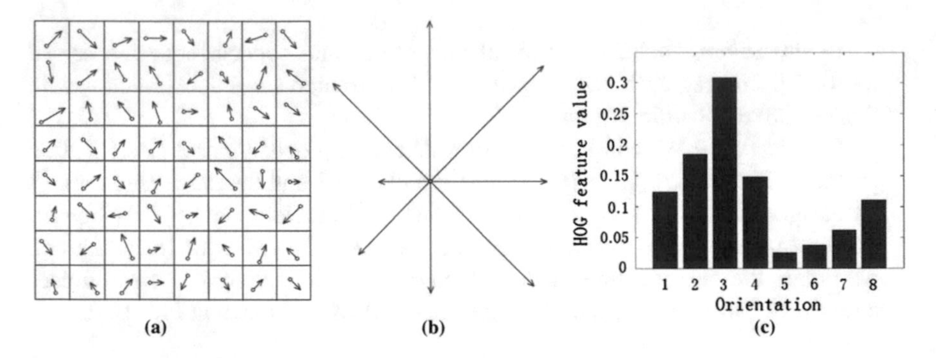

Fig. 10.23 Left: (**a**) indicates each pixel's gradient orientation value in the block. Center: (**b**) shows the accumulated votes of all pixels in eight direction of the block. Right: (**c**) is the final HOG feature vector histogram of the block

Third step: normalization

Gradient strengths vary over a wide range because of the local variations in illumination image contrast, which will directly affect region-growing algorithm's performance. Therefore, we need to normalize the HOG feature vector. In practice, HOG feature vector is normalized from 0 to 1.

Integral Histogram Technology (IHT)

To avoid the repeated computation, the integral histogram technology [60] is employed for improving the speed of computing HOG features, which is able to make our algorithm done in real time, as required by railway tracks detection and turnouts recognition.

The basic idea of IHT is as follows. The integral image at point (x, y) contains the sum of the pixels above and to the left of (x, y):

$$II(x, y) = \sum I\left(x', y'\right), \tag{10.36}$$

where $I(x', y')$ is an original image, and $II(x, y)$ is it's integral image. The integral image allows to compute the sum of the pixels on arbitrary rectangular regions by considering four integral image values at corners of the region. In other words, the time of computing HOG feature is independent of the size of a region.

Eight integral images are established in order to compute HOG features over arbitrary rectangular regions. Each integral image counts the cumulative number of edge direction intensity falling into bins, respectively. By doing this, we are able to compute the HOG feature vector of a given region instantly. In practice, the eight integral images are pre-computed and stored in the memory of the computer for being called at any time.

10.3.2.2 Railway Tracks Detection Based on Region-Growing Algorithm

The region-growing algorithm [136] is one of the commonly used methods in image segmentation domain. In the section, this method will be used to detect railway tracks.

The first step of the region-growing algorithm is to select seeds. In practice, all blocks at the bottom of the image are seen as growing seeds.

At the beginning, the bottom blocks are taken as the exact location of these seeds, which are then grown from these seeds to adjacent blocks depending on a region membership criterion. In practice, our criterion is the distance of HOG features of each other. Since all images are collected along the direction of tracks, here we use 3-connected neighborhoods to grow from seeds.

The next step, we keep examining the adjacent blocks of seeds. If they have a similarly HOG feature with the seed, we classify them into the seed. This is an iterated process until there are no changes in two successive iterative stages. Finally, Blocks with similarly HOG features in the image will be classified into a region. Details are shown in Algorithm 10.3.

Algorithm 10.3 Region-Growing Algorithm for Railway Tracks Detection

---

Initial: Image *I*, the feature similarity threshold *T*, all seeds: *seed_p*$_i$, *i=1,…,n*, growing point sets: *grow_p*. integral images initialized: *inte_images*.

For *i=1,…,n* **do**

group_p.clear(), start=0, end=0,

grow_p(0)=seed_p$_i$.

While *start* $\leq$ *end* **do**

cur_p=grow_p(start);

HOG_f^{cur}= *CompterFeature(cur_p, inte_images)*

For *k=1,2,3* **do**

point=Neighbor(cur_p);

if point was not involved in calculation **do**

HOG_f^{cur}= *CompterFeature(cur_p, inte_images)*

Dis=CompterDistance(HOG_f^{cur}, HOG_f^{nei}*)*

if *Dis>T*

grow$_p$*.add(point),*

end++.

End if

End if

End for

start++.

End While

End for

Output: Obtain similarly groups based on HOG features.

In order to improve the computation speed of railway tracks detection, we use the histogram intersection [137] distance to compute the similarity of HOG features:

$$Dis = ComputerDistance\left(HOG_f^{cur}, HOG_f^{nei}\right)$$

$$= \sum_{i=1}^{8} \min\left(HOG_f^{cur}(i), HOG_f^{nei}(i)\right) \tag{10.37}$$

The railway image can be seen as a projection from 3-D space to 2-D space. Suppose the angle between the camera installed the train and the ground is known prior, we may describe the variation of tracks' width from far to near through the

Fig. 10.24 Left: red lattices show blocks generated for computing HOG features. The size of blocks changes with its' locations. Right: a block's neighbors

Arithmetic Series Principle (see Fig. 10.24). The formula is as follows:

$$b_x = b_{min} + \frac{b_{max} - b_{min}}{h - b_{max}}, \tag{10.38}$$

where x is a variable in the image height's direction from top to bottom, h is the image's height, b_{min} and b_{max} represent the minimum and maximum of the block size about HOG features, respectively. Then b_x means the size of the block in x point. In practice, b_{min} and b_{max} can be measured in a fixed size image, such as $b_{max} = 46$ (pixels), $b_{min} = 2$ (pixels).

As the railway tracks' block contains distinct shape features, it's HOG feature vector has some notable characteristics as follows (see Fig. 10.25):

1. The sum of the HOG features corresponding the block containing railway track is relatively large. According to this characteristic, we may exclude many irrelevant areas (e.g. dark areas) of the image;

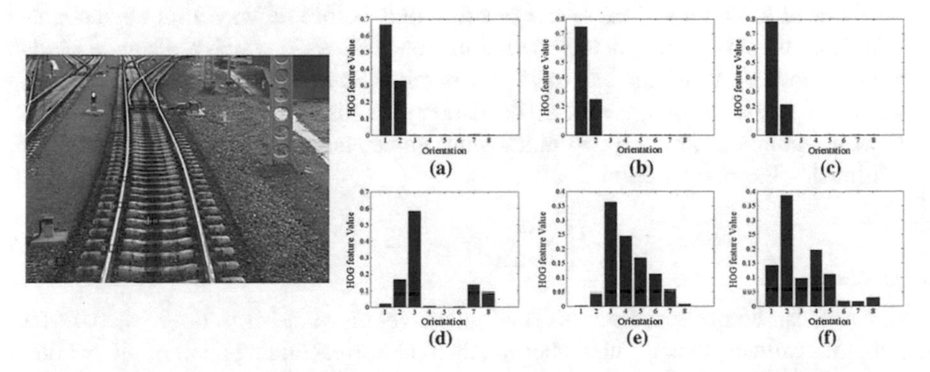

Fig. 10.25 Left: red boxes and blue boxes are some blocks computed; Right: Histograms of Oriented Gradients corresponding with blocks

2. The variance of the HOG features corresponding the block containing railway track is relatively large since such block always contains a prominent edge, which can also be used to exclude many irrelevant areas of the image;
3. In most cases, the sum of it's vertical edges is larger than horizontal edges, and it's 3rd and 7th's feature values are almost zero. These characteristics mentioned above may help us to improve the railway tracks detection's speed and precision by setting some thresholds.

Now we can construct the algorithm detecting railway tracks as follows (Algorithm 10.4).

Algorithm 10.4 Railway Tracks Detection

1. Computing gradient direction intensities for each pixel in the image. Pixels with weak gradient direction intensity will not be used in the following calculation. These black pixels in Fig. 10.26b shows such pixels;
2. Computing eight HOG integral images;
3. Go to Algorithm 10.3;
4. Adjust railway tracks detection results obtained from Algorithm 10.3:
   (a) Delete the groups not satisfying the prior knowledge;
   (b) Cluster these adjacent groups.

Figure 10.26 gives some results of railway tracks detection.

10.3.2.3 Railway Turnouts Recognition

A railroad turnout is a mechanical installation enabling railway trains to be guided from one track to another at a railway junction. It consists of the pair of linked tapering rails lying between the diverging outer rails. The tapering rails can be moved between two positions.

A classic turnout is shown in Fig. 10.27. In this section, we define the range of the railway turnout as the area from the red intersection to the end of the tapering rail in the railway image.

We need to find positions of the tapering rails before railway turnouts recognition. Here we apply an approach called mirror-method, in which a distance needs to be computed: Assuming that tracks are parallel. Same to the method in Sect. 2.3, we define $f(y) = a_1 + a_2 y$, where $f(y)$ denotes the distance of a pair of trails in $(\cdot, y)$ of the image. a_1, a_2 are two unknown parameters, which can be computed by minimizing the cost function:

$$\min_{a_1, a_2} \sum_{i=1}^{n} \| f(y_i) - d_i \|^2, \tag{10.39}$$

where di can be measured. In this section, we get the value $a_1 = 60$, $a_2 = 0.4$. If a point of a railway track is known before, then the corresponding point of the parallel railway track can be computed; the parallel railway track is called the virtual-track.

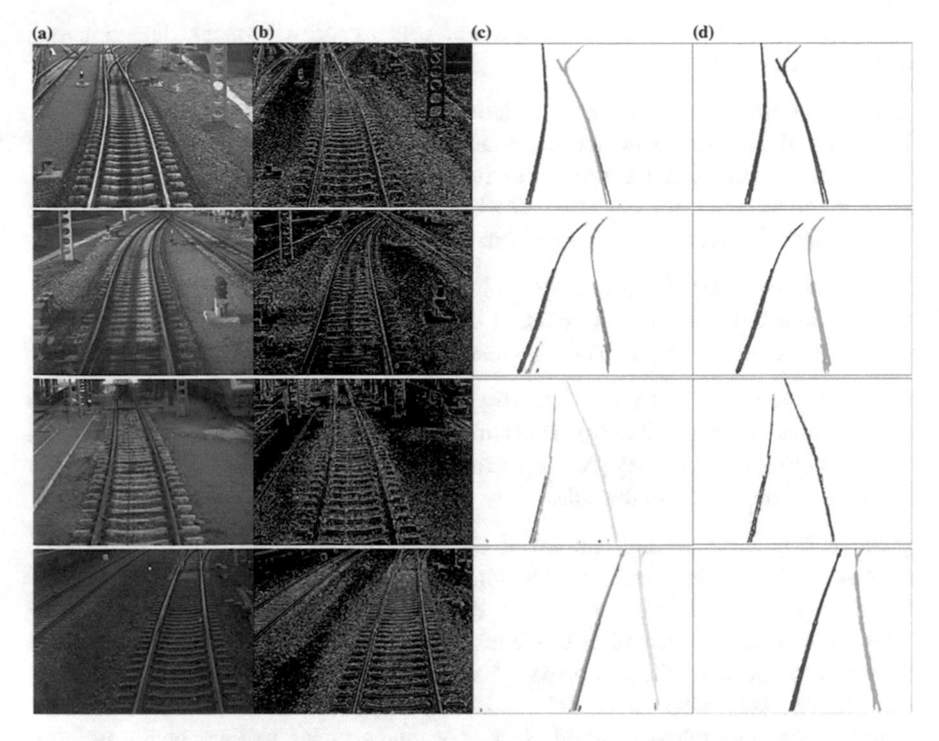

Fig. 10.26 (**a**) Original images including railway tracks. (**b**) Gradient direction intensity's images. Different colors indicate different directions of the image gradient. (**c**) Region-growing results. Different colors represent different groups. (**d**) Final detection results of railway tracks

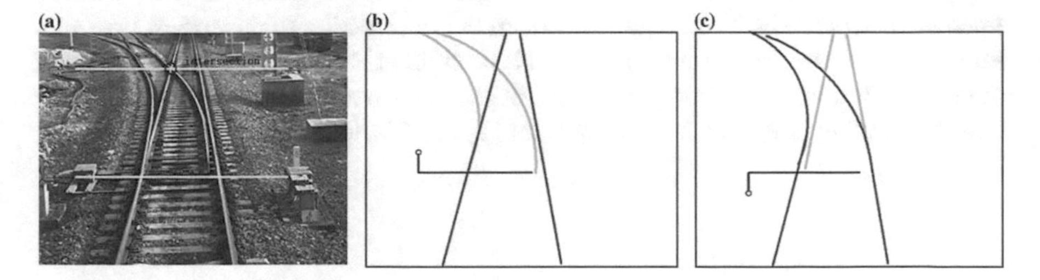

Fig. 10.27 (**a**) The area between two yellow lines is called the railway turnout. The red circle is the intersection of the turnout. (**b**, **c**) Red lines are the ones traveled by train. The black part shows the turnout mechanism, which may be operated remotely using an electric motor or hand-operated lever or from a nearby ground frame

The mirror-method plays an important role in railway tracks detection and turnouts recognition:

1. Supply railway tracks that are not detected by Algorithm 10.3;
2. Verify whether the railway track detected is correct by computing the distance between the detected track and it's virtual-track;
3. Find the tapering rail's approximate location, where we can grow inversely the tapering rail's tracks using Algorithm 10.3.

We only consider the case of two railway tracks detected[5] and assume that the railway track detected does not contain the tapering rail when it is in the detached state. Then there are only two cases when the image contains the railway turnout:

Case 10.1 (see the first column in Fig. 10.26): there is a branch in one of two railway tracks (see Fig. 10.28b). This shows that the railway track contains the part of the tapering rail, and they are connected together. In other words, the rail and the tapering rail are closed in the side.

Case 10.2 (see the second column of Fig. 10.26): there is no branch in both the railway tracks. At the same time, the change of the distance between two railway tracks is nonmonotonic in the vertical direction of the image (see Fig. 10.28f).[6] This is the most complicated case. We take Fig. 10.28g as an example to explain the railway turnout recognition process. Firstly, we compute the virtual-tracks of the two railway tracks respectively. The red and purple dotted lines denote the virtual-tracks obtained by mirror-method; Secondly, we calculate the intersecting point of two virtual-tracks. The blue point shows the railway tracks' intersection; Thirdly, we use Algorithm 10.3 to search tapering rails' tracks from top to bottom in the image. The blue and yellow lines represent the tapering rail's tracks; Fourthly, we compute the average distance between the railway tracks and the closest tapering rail's tracks from them separately. The range is set from the intersection to the end of the rail's track. The left average distance is 27, and the right is 38. The opening direction of the turnout is decided by comparing the size of the two average distances. It is clear that the opening direction of the turnout in Fig. 10.28g is the left.

[5]If only finding one railway track, the other can be obtained by mirror-method. We may delete extra tracks by mirror-method and information of image sequences if more than two railway tracks are detected.

[6]If the change of the distance between two railway tracks is monotonic, only the half of one railway track is detected and connected with tapering rail (In practice, this case is almost impossible to happen, we may deal with it as no existing turnout).

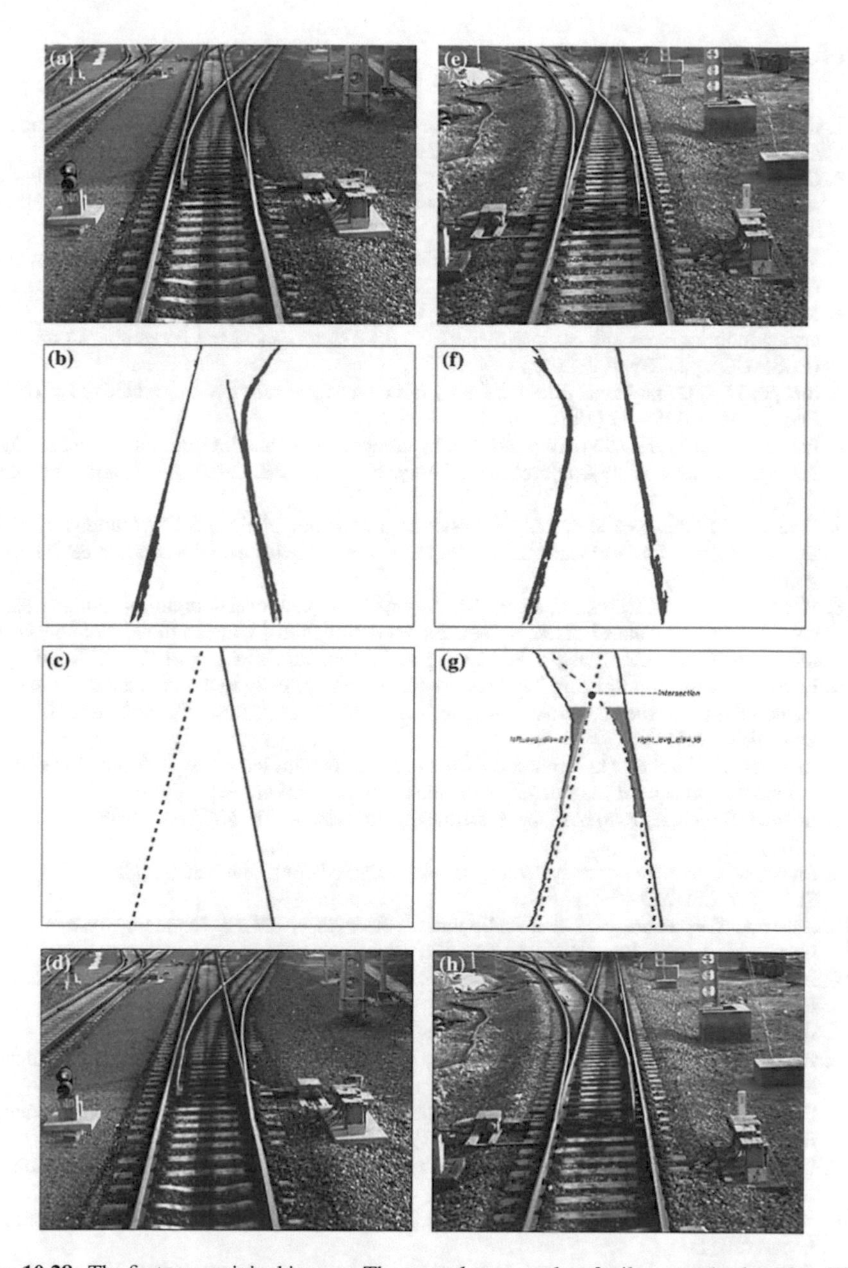

Fig. 10.28 The first row: original images. The second row: results of railway tracks detection. The third row: turnouts recognition. The fourth row: the blue part means the path traveled by train

References

1. Chen, Y., Shi, Y., Wei, X., Zhang, L.: Domestic systemically important banks: a quantitative analysis for the Chinese banking system. Math. Probl. Eng. (2014)
2. Chen, Y., Shi, Y., Wei, X., Zhang, L.: How does credit portfolio diversification affect banks' return and risk? Evidence from Chinese listed commercial banks. Technol. Econ. Dev. Econ. **20**(2), 332–352 (2014)
3. Tanaka, N., Uematsu, K.: A crack detection method in road surface images using morphology. MVA. **98**, 17–19 (1998)
4. Sukhanov, S., Merentitis, A., Debes, C., Hahn, J., Zoubir, A.M.: Bootstrap-based SVM aggregation for class imbalance problems. In: 23rd European Signal Processing Conference (EUSIPCO), pp. 165–169 (2015)
5. Sharpe, W.F.: Capital asset prices: a theory of market equilibrium under conditions of risk. J. Financ. **19**(3), 425–442 (1964)
6. Porikli, F.: Integral histogram: a fast way to extract histograms in Cartesian spaces. In: 2005 IEEE Computer Society Conference on Computer Vision and Pattern Recognition, pp. 829–836 (2005)
7. Board, F.S.: Guidance to assess the systemic importance of financial institutions, markets and instruments: initial considerations. Report to the G-20 finance ministers and central bank governors (2010)
8. Wang, J., Yao, Y., Zhou, H., Leng, M., Chen, X.: A new over-sampling technique based on SVM for imbalanced diseases data. In: Proceedings 2013 International Conference on Mechatronic Sciences, Electric Engineering and Computer (MEC), pp. 1224–1228 (2013)
9. Basel Committee on Banking Supervision: Global systemically important banks: assessment methodology and the additional loss absorbency requirement. Basel Committee on Banking Supervision (2011)
10. Braemer, P., Gischer, H.: Domestic systemically important banks: an indicator based measurement approach for the Australian banking system. Working Paper Series (2012)
11. Board, F.S.: Initial group of global systemically important banks (GSIBs). Technology Report (2011)
12. Acharya, V.V.: A theory of systemic risk and design of prudential bank regulation. J. Financ. Stab. **5**(3), 224–255 (2001)
13. Acharya, V.V., Pedersen, L.H., Philippon, T., Richardson, M.P.: Regulating systemic risk. Financ. Mark. Inst. Instrum. **18**(2), 174–175 (2009)
14. Acharya, V.V., Pedersen, L.H., Philippon, T., Richardson, M.P.: Measuring systemic risk. Rev. Finan. Stud. **29**(1002), 85–119 (2017)
15. Adrian, T., Brunnermeier, M.K.: Covar. Nber Working Papers (2011)
16. Bouye, E., Durlleman, V., Nikeghbali, A., Riboulet, G., Roncalli, T.: Copulas for finance. Mpra Paper (2000)
17. Drehmann, M., Tarashev, N.: Systemic Importance: Some Simple Indicators. Social Science Electronic Publishing, Rochester, NY (2011)
18. Drehmann, M., Tarashev, N.: Measuring the systemic importance of interconnected banks. J. Financ. Intermed. **22**(4), 586–607 (2013)
19. Weistroffer C.: Identifying systemically important financial institutions (SIFIs). Technical Report (2011)
20. Chen, Z.: Are banks too big to fail? Measuring systemic importance of financial institutions. SSRN Electron. J. **6**, 205–250 (2010)
21. Cont, R., Moussa, A.: Network structure and systemic risk in banking systems. SSRN Electron. J. (2010)
22. Li, J., Liang, C., Zhu, X., Sun, X., Wu, D.: Risk contagion in Chinese banking industry: a transfer entropy-based analysis. Entropy. **15**(12), 5549–5564 (2013)
23. Malevergne, Y., Sornette, D.: Testing the Gaussian copula hypothesis for financial assets dependences. Quant. Financ. **3**(2003), 231–250 (2003)

24. Tarashev, N., Borio, C., Tsatsaronis, K.: Attributing systemic risk to individual institutions. BIS Working Papers. **68**(3), 1–18 (2010)
25. Elsinger, H., Lehar, A., Summer, M.: Systemically important banks: an analysis for the European banking system. IEEP. **3**(1), 73–89 (2006)
26. Freixas, X.: Optimal bail out policy, conditionality and creative ambiguity. Fmg Discussion Papers (1999)
27. Allen, F., Gale, D.M.: Financial contagion. J. Polit. Econ. **108**(1), 1–33 (2000)
28. Cheng, H., Wang, J., Hu, Y., Glazier, C., Shi, X., Chen, X.: Novel approach to pavement cracking detection based on neural network. Transp. Res. Record. **1764**(1), 119–127 (2001)
29. Cheng, H.D., Miyojim, M.: Automatic pavement distress detection system. Inf. Sci. **108**(1–4), 219–240 (1998)
30. China Banking Regulatory Commission: 2012 annual report of China Banking Regulatory Commission. Technology Report (2012)
31. Sklar, A.: Random variables, joint distribution functions and copulas. Kybernetika. **9**(6), 449–460 (1973)
32. Luciano: Copula Methods in Finance. John Wiley & Sons, Hoboken, NJ (2004)
33. Patton, A.J.: Copula-based models for financial time series. OFRC Working Papers, pp. 767–785 (2008)
34. Cherubini, U., Luciano, E.: Value-at-risk trade-off and capital allocation with copulas. Econ. Notes. (2001)
35. Hu, L.: Dependence patterns across financial markets: a mixed copula approach. Appl. Financ. Econ. **16**(10), 717–729 (2006)
36. Jondeau, E., Rockinger, M.: The copula-Garch model of conditional dependencies: an international stock market application. J. Int. Money Financ. **25**(5), 827–853 (2006)
37. Rodriguez, J.C.: Measuring financial contagion: a copula approach. J. Empir. Financ. **14**(3), 401–423 (2007)
38. Zhu, X., Li, Y., Liang, C., Chen, J., Wu, D.: Copula based change point detection for financial contagion in Chinese banking. Proc. Comput. Sci. **17**, 619–626 (2013)
39. Li, J., Feng, J., Sun, X., Li, M.: Risk integration mechanisms and approaches in banking industry. Int. J. Inf. Technol. Decis. Mak. **11**(06), 1183–1213 (2012)
40. Li, J., Zhu, X., Lee, C.F., Wu, D., Feng, J., Shi, Y.: On the aggregation of credit, market and operational risks. Rev. Quant. Finan. Acc. **44**(1), 161–189 (2015)
41. Liang, C., Zhu, X., Li, Y., Sun, X., Chen, J., Li, J.: Integrating credit and market risk: a factor copula based method. Proc. Comput. Sci. **17**, 656–663 (2013)
42. Li, D.: On default correlation: a copula function approach. J. Fixed Income. **9**(4), 43–54 (2000)
43. Genest, C., MacKay, J.: The joy of copulas: bivariate distributions with uniform marginals. Am. Stat. **40**(4), 280–283 (1986)
44. Kumar, P., Shoukri, M.M.: Evaluating aortic stenosis using the Archimedean copula methodology. J. Data Sci. **6**(2), 173–187 (2008)
45. Berger, A.N., Hasan, I., Zhou, M.: The effects of focus versus diversification on bank performance: evidence from Chinese banks. J. Bank. Financ. **34**(7), 1417–1435 (2010)
46. Hayden, E., Porath, D., Westernhagen, N.V.: Does diversification improve the performance of German banks? Evidence from individual bank loan portfolios. J. Financ. Serv. Res. **32**(3), 123–140 (2007)
47. Kamp, A., Pfingsten, A., Porath, D.: Do banks diversify loan portfolios? A tentative answer based on individual bank loan portfolios. SSRN Electron. J. (2005)
48. Tabak, B.M., Fazio, D.M., Cajueiro, D.O.: The effects of loan portfolio concentration on Brazilian banks' return and risk. J. Bank. Financ. **35**(11), 3065–3076 (2010)
49. Rudolph, K., Pfingsten, A.: German Banks' Loan Portfolio Composition: Marketorientation vs. Specialisation. Social Science Electronic Publishing, Rochester (2002)
50. Rhoades, S.A.: The Herfindahl-Hirschman index. Fed. Reserv. Bull. (1993)
51. FDIC: An examination of the banking crises of the 1980s and early 1990s (1997)

52. Yuliya, D., Otto, V.H.: Understanding the subprime mortgage crisis. Rev. Financ. Stud. **24**(6), 1848–1880 (2011)
53. Sharpe, W.F.: A simplified model for portfolio analysis. Manag. Sci. **9**(2), 277–293 (1963)
54. Acharya, V.V., Hasan, I., Saunders, A.: Should banks be diversified? evidence from individual bank loan portfolios. J. Bus. **79**(3), 1355–1412 (2006)
55. Dempsey, M., Edirisuriya, P., Gunasekarage, A.: Stock market's assessment of bank diversification: evidence from listed public banks in south Asian countries. SSRN Electron. J. (2013)
56. McElligott, R., Stuart, R.: Measuring the sectoral distribution of lending to Irish non-financial corporates. Central Bank & Financial Services Authority of Ireland, Financial Stability Report (2007)
57. Bebczuk, R., Galindo, A.: Financial crisis and sectoral diversification of Argentine banks, 1999–2004. Expert Syst. Appl. **18**(1–3), 199–211 (2008)
58. Dunis, C.L., Williams, M.: Modelling and trading the eur/usd exchange rate: do neural network models perform better? Working Paper Center for International Banking Economics & Finance (2011)
59. Okasha, M.K.: Using support vector machines in financial time series forecasting. Int. J. Stat. Appl. **4**(1), 28–39 (2014)
60. Yu, L., Wang, S., Lai, K.K.: A neural-network-based nonlinear metamodeling approach to financial time series forecasting. Appl. Soft Comput. **9**(2), 563–574 (2009)
61. Adhikari, R., Agrawal, R.K.: A combination of artificial neural network and random walk models for financial time series forecasting. Neural Comput. Appl. **24**(6), 1441–1449 (2014)
62. Zhang, G.P.: Time series forecasting using a hybrid arima and neural network model. Neurocomputing. **50**(2003), 159–175 (2003)
63. Wu, B., Duan, T.: A performance comparison of neural networks in forecasting stock price trend. Int. J. Comput. Intell. Syst. **10**(1), 336 (2017)
64. Huang, C.J., Yang, D.X., Chuang, Y.T.: Application of wrapper approach and composite classifier to the stock trend prediction. Expert Syst. Appl. **34**(4), 2870–2878 (2008)
65. Khaidem, L., Saha, S., Dey, S.R.: Predicting the direction of stock market prices using random forest. Preprint at arXiv:1605.00003 (2016)
66. Lee, M.C.: Using support vector machine with a hybrid feature selection method to the stock trend prediction. Expert Syst. Appl. **36**(8), 10896–10904 (2009)
67. Lertyingyod, W., Benjamas, N.: Stock price trend prediction using artificial neural network techniques: case study: Thailand stock exchange. In: 2016 International Computer Science and Engineering Conference (ICSEC), pp. 1–6 (2016)
68. Marković, I., Stojanović, M., Stanković, J., Stanković, M.: Stock market trend prediction using ahp and weighted kernel LS-SVM. Soft. Comput. **21**(18), 5387–5398 (2017)
69. Patel, J., Shah, S., Thakkar, P., Kotecha, K.: Predicting stock and stock price index movement using trend deterministic data preparation and machine learning techniques. Expert Syst. Appl. **42**(1), 259–268 (2015)
70. Chang, P.C., Fan, C.Y., Liu, C.H.: Integrating a piecewise linear representation method and a neural network model for stock trading points prediction. IEEE Trans. Syst. Man Cybernetics C. **39**(1), 80–92 (2008)
71. Bao, D., Yang, Z.: Intelligent stock trading system by turning point confirming and probabilistic reasoning. Expert Syst. Appl. **34**(1), 620–627 (2008)
72. Chavarnakul, T., Enke, D.: A hybrid stock trading system for intelligent technical analysis-based equivolume charting. Neurocomputing. **72**(16–18), 3517–3528 (2009)
73. Sevastianov, P., Dymova, L.: Synthesis of fuzzy logic and Dempster–Shafer theory for the simulation of the decision-making process in stock trading systems. Math. Comput. Simul. **80**(3), 506–521 (2010)
74. Li, X., Deng, Z., Jing, L.: Trading strategy design in financial investment through a turning points prediction scheme. Expert Syst. Appl. **36**(4), 7818–7826 (2009)
75. Mabu, S., Hirasawa, K., Obayashi, M., Kuremoto, T.: Enhanced decision making mechanism of rule-based genetic network programming for creating stock trading signals. Expert Syst. Appl. **40**(16), 6311–6320 (2013)

76. Pham, H.V., Cooper, E.W., Cao, T., Kamei, K.: Hybrid Kansei-SOM model using risk management and company assessment for stock trading. Inf. Sci. **256**(1), 8–24 (2014)
77. Wu, J.L., Yu, L.C., Chang, P.C.: An intelligent stock trading system using comprehensive features. Appl. Soft Comput. **23**(2014), 39–50 (2014)
78. Bahar, H.H., Zarandi, M.H.F., Esfahanipour, A.: Generating ternary stock trading signals using fuzzy genetic network programming. In: 2016 Annual Conference of the North American Fuzzy Information Processing Society (NAFIPS), pp. 1–6 (2016)
79. Tan, Z., Quek, C., Cheng, P.Y.K.: Stock trading with cycles: a financial application of ANFIS and reinforcement learning. Expert Syst. Appl. **38**(5), 4741–4755 (2011)
80. Zhang, J., Maringer, D.: Using a genetic algorithm to improve recurrent reinforcement learning for equity trading. Comput. Econ. **47**(4), 551–567 (2016)
81. Alimoradi, M.R., Kashan, A.H.: A league championship algorithm equipped with network structure and backward q-learning for extracting stock trading rules. Appl. Soft. Comput. **68** (2018)
82. Kwon, Y.K., Sun, H.D.: A hybrid system integrating a piecewise linear representation and a neural network for stock prediction. In: Proceedings of 2011 6th International Forum on Strategic Technology, pp. 796–799 (2011)
83. Luo, L., Chen, X.: Integrating piecewise linear representation and weighted support vector machine for stock trading signal prediction. Appl. Soft Comput. **13**(2), 806–816 (2013)
84. Chen, X., He, Z.J.: Prediction of stock trading signal based on support vector machine. In: 2015 8th International Conference on Intelligent Computation Technology and Automation (ICICTA) (2016)
85. Li, F., Gao, F., Kou, P.: Integrating piecewise linear representation and Gaussian process classification for stock turning points prediction. J. Comput. Appl. **35**, 2397–2403 (2015)
86. Luo, L., You, S., Xu, Y., Peng, H.: Improving the integration of piece wise linear representation and weighted support vector machine for stock trading signal prediction. Appl. Soft Comput. **56**, 199–216 (2017)
87. Keogh, E.J., Chu, S., Hart, D., Pazzani, M.J.: An online algorithm for segmenting time series. In: Proceedings 2001 IEEE International Conference on Data Mining, pp. 289–296 (2001)
88. Vapnik, V.: The Nature of Statistical Learning Theory. Springer Science & Business Media, New York (2013)
89. Hall, M.A.: Correlation-based feature selection for machine learning (1999)
90. Bhargavi, R., Gumparthi, S., Anith, R.: Relative strength index for developing effective trading strategies in constructing optimal portfolio. Int. J. Appl. Eng. Res. **12**(19), 8926–8936 (2017)
91. Kasemsumran, S., Thanapase, W., Teardwongworakul, A., Pathaweerut, S.: Nondestructive internal quality evaluation of orange using near infrared spectroscopy Thailand research fund (2008)
92. Petrou, M., Kittler, J., Song, K.Y.: Automatic surface crack detection on textured materials. J. Mater. Process. Technol. **56**(1–4), 158–167 (1996)
93. Dunham, M.H.: Data Mining: Introductory and Advanced Topics. Pearson Education, India (2006)
94. Larry, W.: All of Statistics: A Concise Course in Statistical Inference. Springer, New York (2004)
95. Cessie, S.L., Houwelingen, J.C.V.: Ridge estimators in logistic regression. Appl. Stat. **41**(1), 191–201 (1992)
96. Shi, Y., Liu, R., Yan, N., Chen, Z.: Multiple criteria mathematical programming and data mining. In: International Conference on Computational Science, pp. 7–17 (2008)
97. Cristianini, N., Shawe-Taylor, J., et al.: An Introduction to Support Vector Machines and Other Kernel-Based Learning Methods. Cambridge University Press, Cambridge (2000)
98. Kira, K., Rendell, L.A.: A practical approach to feature selection. Mach. Learn. Proc. **1992**, 249–256 (1992)
99. Liu, R., Rallo, R., Cohen, Y.: Unsupervised feature selection using incremental least squares. Int. J. Inf. Technol. Decis. Mak. **10**(06), 967–987 (2011)

100. Jolliffe, I.T., Trendafilov, N.T., Uddin, M.: A modified principal component technique based on the lasso. J. Comput. Graph. Stat. **12**(3), 531–547 (2003)
101. Schölkopf, B., Smola, A., Müller, K.R.: Nonlinear component analysis as a kernel eigenvalue problem. Neural Comput. **10**(5), 1299–1319 (1998)
102. Zou, X., Zhao, J., Povey, M., Holmes, M., Mao, H.: Variables selection methods in near-infrared spectroscopy. Anal. Chim. Acta. **667**(1–2), 14–32 (2010)
103. Kawano, S., Fujiwara, T., Iwamoto, M.: Nondestructive determination of sugar content in satsuma mandarin using near infrared (NIR) transmittance. J. Jpn. Soc. Hortic. Sci. **62**(2), 465–470 (1993)
104. Shao, Y., He, Y., Bao, Y., Mao, J.: Near-infrared spectroscopy for classification of oranges and prediction of the sugar content. Int. J. Food Prop. **12**(3), 644–658 (2009)
105. Cen, H., He, Y., Huang, M.: Measurement of soluble solids contents and pH in orange juice using chemometrics and vis-nirs. J. Agric. Food Chem. **54**(20), 7437–7443 (2006)
106. Ayenu-Prah, A., Attoh-Okine, N.: Evaluating pavement cracks with bidimensional empirical mode decomposition. EURASIP J. Adv. Signal Process. **2008**(1) (2008)
107. Oliveira, H., Correia, P.L.: Automatic road crack segmentation using entropy and image dynamic thresholding. In: 17th European Signal Processing Conference, pp. 622–626 (2009)
108. Qi, Z., Tian, Y., Shi, Y.: Effcient railway tracks detection and turnouts recognition method using hog features. Neural Comput. & Applic. **23**(1), 245–254 (2013)
109. Zhao, H., Qin, G., Wang, X.: Improvement of canny algorithm based on pavement edge detection. In: 3rd International Congress on Image and Signal Processing, pp. 964–967 (2010)
110. Chambon, S., Moliard, J.M.: Automatic road pavement assessment with image processing: review and comparison. Int. J. Geophys. **2011**, 1–20 (2011)
111. Tsai, Y.C., Kaul, V., Mersereau, R.M.: Critical assessment of pavement distress segmentation methods. J. Transp. Eng. **136**(1), 11–19 (2010)
112. Achanta, R., Estrada, F., Wils, P., Suesstrunk, S.: Salient region detection and segmentation. In: The 6th International Conference on Computer Vision Systems (ICVS 2008), pp. 66–75 (2008)
113. Achanta, R., Hemami, S., Estrada, F., Susstrunk, S.: Frequency-tuned salient region detection. In: The 2009 IEEE Conference on Computer Vision and Pattern Recognition (CVPR 2009), pp. 1597–1604 (2009)
114. Arbeláez, P., Maire, M., Fowlkes, C., Malik, J.: Contour detection and hierarchical image segmentation. IEEE Trans. Pattern Anal. Mach. Intell. **33**(5), 898–916 (2011)
115. Hu, Y., Zhao, C.X.: A novel LBP based methods for pavement crack detection. J. Pattern Recogn. Res. **5**(1), 140–147 (2010)
116. Song, K.Y., Petrou, M., Kittler, J.: Texture crack detection. Mach. Vis. Appl. **8**(1), 63–75 (1995)
117. Zhou, J., Huang, P.S., Chiang, F.P.: Wavelet-based pavement distress detection and evaluation. Opt. Eng. **45**(2), 027007 (2006)
118. Subirats, P., Dumoulin, J., Legeay, V., Barba, D.: Automation of pavement surface crack detection using the continuous wavelet transform. In: 2006 International Conference on Image Processing, pp. 3037–3040 (2006)
119. Kass, M., Witkin, A., Terzopoulos, D.: Snakes: active contour models. Int. J. Comput. Vis. **1**(4), 321–331 (1988)
120. Kaul, V., Yezzi, A., Tsai, Y.: Detecting curves with unknown endpoints and arbitrary topology using minimal paths. IEEE Trans. Pattern Anal. Mach. Intell. **34**(10), 1952–1965 (2011)
121. Amhaz, R., Chambon, S., Idier, J., Baltazart, V.: A new minimal path selection algorithm for automatic crack detection on pavement images. In: 2014 IEEE International Conference on Image Processing (ICIP), pp. 788–792 (2014)
122. Amhaz, R., Chambon, S., Idier, J., Baltazart, V.: Automatic crack detection on two-dimensional pavement images: an algorithm based on minimal path selection. IEEE Trans. Intell. Transp. Syst. **17**(10), 2718–2729 (2016)

123. Nguyen, T.S., Begot, S., Duculty, F., Avila, M.: Free-form anisotropy: a new method for crack detection on pavement surface images. In: 18th IEEE International Conference on Image Processing, pp. 1069–1072 (2011)
124. Cord, A., Chambon, S.: Automatic road defect detection by textural pattern recognition based on adaboost. Comput. Aided Civ. Inf. Eng. **27**(4), 244–259 (2012)
125. Delagnes, P., Barba, D.: A Markov random field for rectilinear structure extraction in pavement distress image analysis. In: 1995 IEEE International Conference on Image Processing, pp. 446–449 (1995)
126. Lee, B.J., Lee, H.D.: Position-invariant neural network for digital pavement crack analysis. Comput. Aided Civil Infrastruct. Eng. **19**(2), 105–118 (2004)
127. Oliveira, H., Correia, P.L.: Supervised strategies for cracks detection in images of road pavement flexible surfaces. In: 16th European Signal Processing Conference, pp. 1–5 (2008)
128. Cheng, H., Chen, J.R., Glazier, C., Hu, Y.: Novel approach to pavement cracking detection based on fuzzy set theory. J. Comput. Civ. Eng. **13**(4), 270–280 (1999)
129. Ying, L., Salari, E.: Beamlet transform-based technique for pavement crack detection and classification. Comput. Aided Civil Infrastruct. Eng. **25**(8), 572–580 (2010)
130. Huang, Y., Xu, B.: Automatic inspection of pavement cracking distress. J. Electron. Imaging. **15**(1), 013017 (2006)
131. Zou, Q., Cao, Y., Li, Q., Mao, Q., Wang, S.: Cracktree: automatic crack detection from pavement images. Pattern Recogn. Lett. **33**(3), 227–238 (2012)
132. Oliveira, H., Correia, P.L.: Automatic road crack detection and characterization. IEEE Trans. Intell. Transp. Syst. **14**(1), 155–168 (2012)
133. Nguyen, T.S., Avila, M., Begot, S.: Automatic detection and classification of defect on road pavement using anisotropy measure. In: 17th European Signal Processing Conference, pp. 617–621 (2009)
134. Dollár, P., Tu, Z., Perona, P., Belongie, S.: Integral channel features (2009)
135. Lim, J.J., Zitnick, C.L., Dollár, P.: Sketch tokens: a learned mid-level representation for contour and object detection. In: Proceedings of the IEEE Conference on Computer Vision and Pattern Recognition, pp. 3158–3165 (2013)
136. Adams, R., Bischof, L.: Seeded region growing. IEEE Trans. Pattern Anal. Mach. Intell. **16**(6), 641–647 (1994)
137. Swain, M.J., Ballard, D.H.: Color indexing. Int. J. Comput. Vis. **7**(1), 11–32 (1991)

Chapter 11
Healthcare Applications

Healthcare is also a very hot application area of data science, especially in the COVID-19 pandemic around the world since the beginning of 2020. This chapter provides two sections of the related healthcare applications. Section 11.1 deals with the evaluation of medical doctor's performance by using ordinal regression-based approach [1], while Sect. 11.2 outlines a cutting-edge research finding to learn transmission patterns of COVID-19 outbreak by using an age-specific social contact characterization [2].

11.1 Evaluating Doctor Performance: Ordinal Regression-Based Approach

Doctor's performance evaluation is an important task in mobile health (mHealth), which aims to evaluate the overall quality of online diagnosis and patient outcomes so that customer satisfaction and loyalty can be attained. However, most patients tend not to rate doctors' performance, therefore, it is imperative to develop a model to make doctor's performance evaluation automatic. When evaluating doctors' performance, we rate it into a score label that is as close as possible to the true one.

This section aims to perform automatic doctor's performance evaluation from online textual consultations between doctors and patients by way of a novel machine learning method.

A workflow of the OR-DPE model is shown in Fig. 11.1 OR-DPE comprises of text preprocessing, representation, model training, and predictability. Because the communication between doctor and patient is through a text message, the DPE task is like text mining. The consultation texts are preprocessed and displayed as high dimensional vectors. Because the SVM-based model with linear kernel [3] performs excellently on large-scalc data and is well suited for text mining fields, this model is

Y. Shi, *Advances in Big Data Analytics*,
https://doi.org/10.1007/978-981-16-3607-3_11

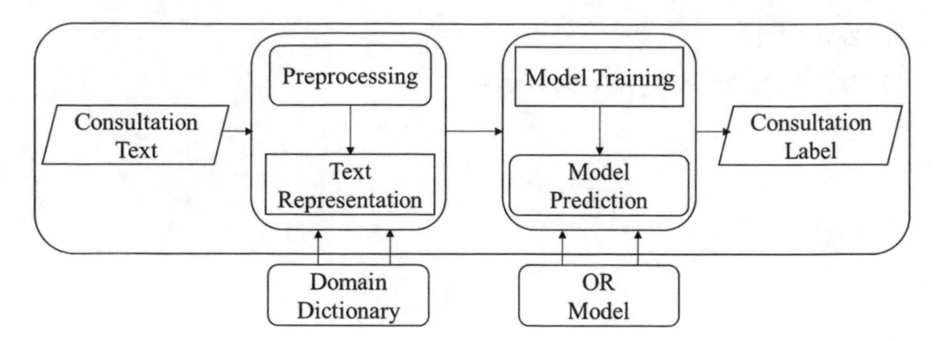

Fig. 11.1 The general workflow of the ordinal regression for doctor performance evaluation (OR-DPE) model

preferred to address the DPE. In this section, a new SVM-based Ordinal Partitioning model (SVMOP) is proposed as the OR model for DPE. With the SVMOP model, OR-DPE can, not only make sure that the predicted labels are as correct as possible, but also ensure that the incorrect labels are as close to true as possible. To our knowledge, this is the first time that the issue of DPE has been conceptualized as an ordinal regression task. Empirical studies on real data sets from one of the largest mobile doctor/patient communication platforms in China show that the model can achieve state-of-the-art performance from multiple metrics.

11.1.1 Methods

11.1.1.1 Preprocessing and Text Representation

The original corpus should be preprocessed, and each sample should be represented as an input vector. In the preprocessing step, punctuation and stop words will be removed. If the experimental data is written in Chinese, the words must be segmented as in Chinese text. Sentences are represented as character strings without natural delimiters. Chinese Word Segmentation (CWS) is used to identify word sequences in a sentence and mark boundaries in appropriate places. For example, CWS can put the character sequence “ ” together as a Chinese word for “smallpox” rather than the individual Chinese character “ ” (sky) and “ ” (flower) respectively. Word segmentation is a preliminary and important step for preprocessing. Most methods take the CWS as a sequence labeling problem [4], which can be formalized as supervised learning methods with customized features. Additionally, domain dictionaries with technical terms as ancillary resources, are beneficial for CWS and medical feature extraction. Here, three medical dictionaries are employed; one for Illness, one for Symptoms and one for Medicine. Most terms in the dictionaries are customized by medical experts and extended with new word detection techniques. We have collected 49,758 illness and symptom terms and 24,975 medical terms. Information about the dictionaries is shown in Table 11.1. For this purpose, we

Table 11.1 The details about the medical dictionaries. "1≤terms≤3" means the number of terms having a character length less than 3 but greater than 1

Number of phrases	Dictionary Name Illness and Symptom Dictionary (N = 49,758)	Medicine Dictionary (N = 24,975)
1≤terms≤3, n (%)	32,840 (66.00)	3746 (15.00)
4≤terms≤6, n (%)	16,918 (34.00)	14,486 (58.00)
terms≥7, n (%)	0 (0)	6743 (27.00)
Representative examples	(Neurosis), (HTN), (smallpox)	(Paroxetine), (Flexeril)

Table 11.2 F1–F8 represent the customized medical features, while F9 and F10 are the text features

Feature	Description
F1	The number of symptom names in doctors' answers
F2	The number of illness names in doctors' answers
F3	The number of medicine names in doctors' answers
F4	The number of patients' questions
F5	The number of doctors' answers
F6	The response time for the patient's first question
F7	The number of Chinese characters in patients' questions
F8	The number of Chinese characters in doctors' answers
F9	Unigrams
F10	Bigrams

combined the dictionaries with Jieba tool, an open-sourced Chinese segmentation software, for word segmentation.

For text representation, each sample is represented as an input vector where each dimension of the vector represents a feature. The element is the corresponding feature value. Feature engineering plays an important role in text mining. Apart from the basic text features such as Bag of Words (BOW) [5], unigrams, and bigrams, the custom medical features that can mirror some characteristics of the platform are utilized. These are specifically designed for the doctor/patient communication platform by domain experts and most are based on medical dictionaries. Typical text and medical features used in OR-DPE are presented in Table 11.2. Customized features (F1–F8) can capture domain knowledge: the count of medicine and symptom names in doctors' answers reflects the doctors' professional level; the number of Chinese characters in doctors' answers mirrors the service attitudes, and more. Likewise, the text features (F9 and F10) cover most consultation information. The feature value is the numerical value of the feature while the feature value of text features is the term frequency inverted document frequency (TF-IDF) [6]. TF-IDF reflects how important a word is to a document. If a word occurs rarely but appears frequently in a sample, it is most likely to reflect the characteristics of this sample. Specifically, TF-IDF is the product of two statistics: term frequency and inverse document frequency, where the former represents the frequency and the latter represents the inverse frequency of occurrence in all samples.

The quantity of text features is so large that the customized features (see Table 11.2) can easily be overshadowed. To highlight the importance of customized features, they are boosted by the Gradient Boosting Decision Tree (GBDT) [7]. GBDT is a powerful tool in many industrial communities [8]. GBDT mines the most effective features and feature combinations by a decision tree to boost the performance of regression and classification tasks. This technique is applied to increase the number of custom medical feature combinations. The main idea of GBDT is to combine weak learners into a single, strong learner like other boosting methods. GBDT is an iteration algorithm, which is composed of multiple decision trees. In the *m*-th iteration of GBDT, assumes that there are some imperfect models, Fm. The GBDT would construct a better model F_{m+1} to approach the best model by adding an estimator h, namely $F_{m+1} = F_m(x) + h(x)$. Then the problem is transformed by the question of how to find h(x). As the above equations imply, a perfect h should satisfy the equation:

$$h(x) = F_{m+1} - F_m(x) \approx y - F_m(x)$$

where y is the true label, $y - F_m(x)$ is called a loss function. In practice, a general way is to apply square loss function is: $1/2(y - F_m(x))^2$. Because the residual is exactly the negative gradients of the squared loss function. The problem on the left can then be solved directly by gradient descent algorithms. In our work, we apply GBDT to boost the 8 customized features shown in Table 11.2 to generate several effective feature combinations. According to the statistics, the number of features is 363,336 with text features, and 363,344 if adding the 8 customized features. After boosting the customized features, the number becomes 370,858. Another 7514 combined customized feature combinations have been added. The performances of various features are shown in Sect. 11.2.2.

11.1.1.2 Model Training

There are many different models of OR. Referring to an OR survey [16], the models are grouped into three categories, namely the (1) naive approach, (2) threshold approach, and (3) ordinal partitioning approach. These models have corresponding strengths and weakness. The naive approach considers OR naively, as a standard classification task or a regression task [3, 9]. At the same time, the ordering information between labels has been ignored. The threshold approach is based on the idea of approximating a real value predictor and then dividing the real line into intervals [10–13]. Assuming P is the number of categories, the objective of threshold-based OR models is to seek P–1 parallel hyperplanes further dividing the data into ordered classes. The ordinal partitioning approach uses the ordering information to decompose the ordinal regression into several binary classification tasks. For binary classification, there are many models to choose from. For example, Frank and Hall [14], applied decision trees as submodels while Waegeman and Boullart [15] used weighted SVMs as binary classifiers.

Since the ordering of information is conducive to model building [16], we chose the OR model from the latter two methods. As the number of samples is large and the dimension of the representative vectors is high, a model was chosen that can handle large-scale and high dimensional data. So, the ordinal partitioning approach is used instead of the threshold approach for OR problems depending on paralleled hyperplanes. There are many binary classifiers that can be chosen from the submodels. Hsieh et al. [3] showed that the linear SVM is a robust tool that can deal with large-scale and high dimensional data. Inspired by these, we want to combine SVM with Ordinal Partitioning (SVMOP) as the OR model for the OR-DPE.

The OR problem can be described as follows: given a training Set $T = \{(x_i, y_i)\}_{\{i=1\}}^n \subseteq (X, Y)$ where $x_i \in R^l$ is the i-th input vector ($i = 1, 2, \ldots$, n), where n is the number of instances, l is the number of features, and $y_i \in Y_i$ is the label of x_i. Assuming there are P categories and without loss of generality, we take the label set $Y = \{1, 2, \ldots P\}$. The goal of OR is to find a function f: $X \rightarrow Y$ to predict the label of a new instance x. As mentioned earlier, SVMOP will be embedded into the OR-DPE model. Figure 11.2 illustrates the SVMOP procedure. In this figure, five ordinal categories of data are represented by different colors and shapes. The idea of SVMOP is to partition the overall model into P–1 binary classifications. Then the associated question: "Is the rank of the input greater than p?" can be asked. Here $p = 1, 2, \ldots, P$–1. Therefore, the rank of x can be determined by a sequence of these binary classification problems. Specifically, when training the p-th binary classifier, the label y_i is retransformed to a new class label depending on whether the label y_i is greater than p or not, namely:

$$\hat{y}_{pi} = \begin{cases} -1, \text{if } y_i \leq p \\ 1, \text{if } y_i > p \end{cases} \tag{11.1}$$

where $i = 1, 2, \ldots, n$. Therefore, the problem can be reformulated: given a training set $= \{(x_i, \hat{y}_{pi})\}_{i-1}^n$, where $x_i \in R^l$ is the i-th input sample, $y_{pi} \in \{-1,1\}$ is defined by Eq. (11.1). The model aims to find a function to predict the ordered labels of new instances.

Linear SVM is one of the best candidates among the binary classifiers dealing with high dimensional data. Then linear SVM is taken as the p-th sub-model:

$$\min_{\boldsymbol{w}_p, \xi_p} \frac{1}{2} \parallel \boldsymbol{w}_p \parallel_2^2 + C\sum\nolimits_{i=1}^n \xi_{pi}$$

$$s.t. \hat{y}_{pi}\left(w_p^T \boldsymbol{x}_i\right) \geq 1 - \xi_{pi}, i = 1, 2, \ldots, n, \tag{11.2}$$

$$\xi_{pi} \geq 0, i = 1, 2, \ldots, n,$$

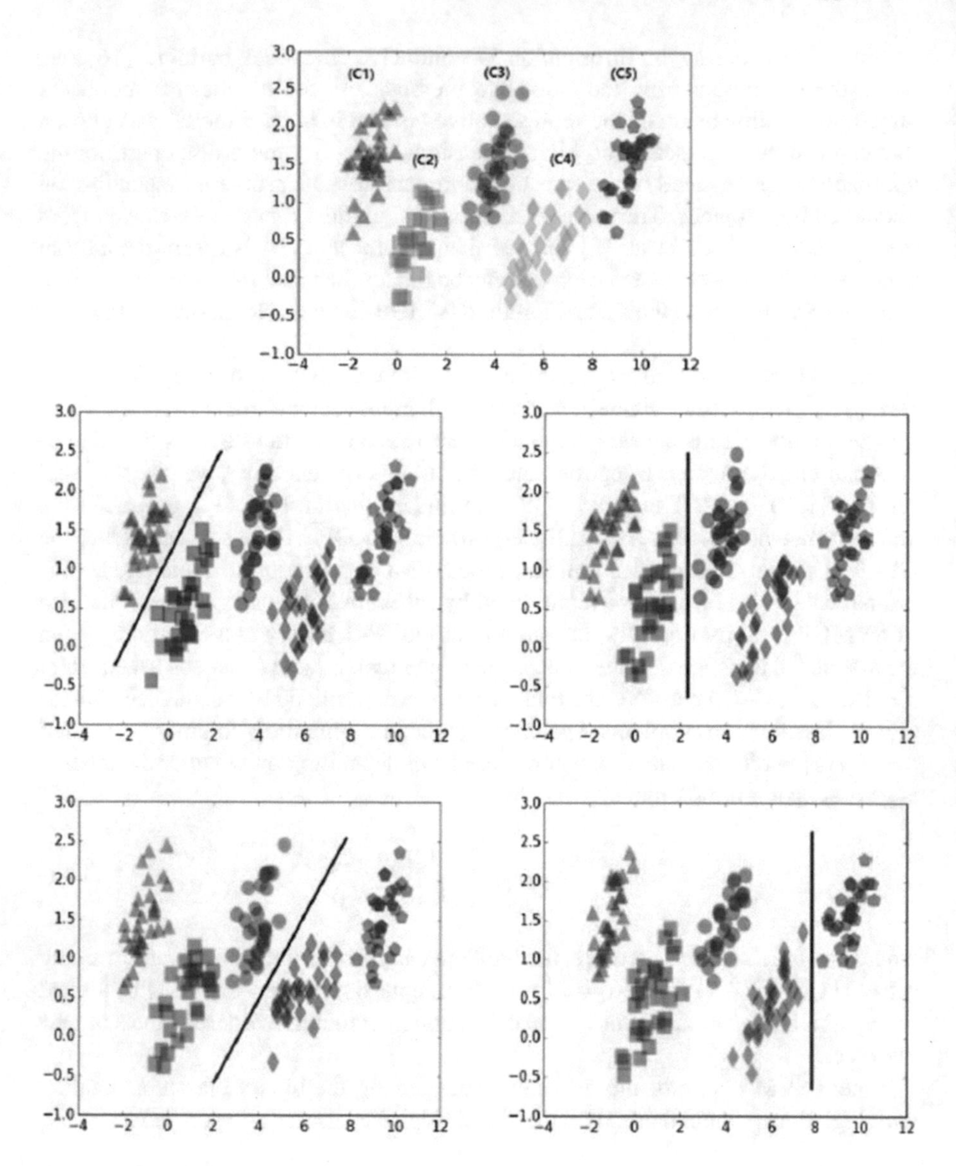

Fig. 11.2 The demo that shows how a combined support vector machine and ordinal partitioning scheme model (SVMOP) works on ordinal data

where w_p represents the parameter of the p-th submodel, ξ_{pi} is the slack variable of the p-th submodel.

As for the optimization solver, we chose the Dual Coordinate Descent algorithm (DCD) as the training algorithm of SVM [3]. DCD is one of the most effective training algorithms for linear SVMs. It solves the model in Eq. (11.2) by the Lagrange dual form. The dual form of the p-th sub-model in Eq. (11.2) is given as Eq. (11.3). Without loss of generality, we ignore the subscript p in the dual form:

$$\min_{\alpha} f(\boldsymbol{\alpha}) = \frac{1}{2}\boldsymbol{\alpha}^T Q\boldsymbol{\alpha} - e^T\boldsymbol{\alpha}$$

$$s.t. 0 \le \alpha_i \le C, i = 1, 2, \ldots, n, \tag{11.3}$$

where $Q_{ij} = \hat{y}_{pi}\hat{y}_{pj}\boldsymbol{x}_i^T\boldsymbol{x}_j.DCD$ is to employ a classic divide-and-conquer method for optimizing high dimensional problems. It starts from an initial zero vector $\alpha^0 = \mathbf{0}$ and generates a sequence of vectors $\{\alpha^k\}_{k=0}^{\infty}$. For each iteration step, the algorithm sequentially selects one dimension associated with α to optimize by fixing other dimensions.Suppose $\alpha*$ is the solution of Eq. (11.3) then the optimal value of w_p for Eq. (11.2) can be computed as follows:

$$w_p = \sum\nolimits_{i=1}^{n} \hat{y}_{pi}\boldsymbol{\alpha}_i^* x_i \tag{11.4}$$

11.1.1.3 Model Prediction

For model prediction, the research [16] shows that it is important to construct an effective rule for predicting new instances in the ordinal partitioning-based OR models. Many existing ways are based on the probability manipulation or outcomes by submodels to predict the label of a new instance. In the work by Frank et al [14], when estimating the probabilities for the first and the last class, the authors were dependent on a corresponding classifier. However, it needs to rely on two adjacent classifiers when computing the middle classes. This prediction method is simple and easy to implement, but may lead to a negative probability [16, 17]. Another example in the work [15], the authors combined the outcomes of all the submodels to predict the label of a new instance x. However, their prediction function may cause ambiguities for some test samples.

To alleviate the problem with the above prediction functions, we propose a new prediction function as shown in Eq. (11.5):

$$r(\boldsymbol{x}) = 1 + \arg \max_{p \in \{1,2,\ldots,P-1\}} \left\{p : w_p^T\boldsymbol{x} > 0\right\} \tag{11.5}$$

where $r(\boldsymbol{x})=1$ if none of $w_p^T\boldsymbol{x}$ is greater than 0. This prediction function relies on the discriminant planes and joins all binary classifiers to obtain a single classification.

The p-th binary classifier provides the answer to the associated question: "Is the rank of the input x greater than p?", where $p = 1, 2, \ldots, P–1$. That is, for prediction, the new sample x would be asked by a sequence of the questions above. And last, the predicted label equals $r(x)$ which represents the satisfaction degree. The greater $r(x)$, the more satisfied.

11.1.1.4 Statistical Methods and Evaluation Metrics

To better highlight the characteristics of ordinal regression models, we evaluated the performance with the following three common evaluation measures: (1) mean absolute error (MAE) [11, 16, 18], (2) mean square error (MSE) [19–21], and (3) pairwise accuracy (PAcc) [18, 22, 23]. MAE and MSE can directly measure the degree of deviation between the true label ($gold_i$) and predicted label (*predicted*). They can be defined by the following equations:

$$\mathrm{MAE} = \frac{1}{n}\sum\nolimits_{i=1}^{n} \mid gold_i - predicted_i \mid \tag{11.6}$$

$$\mathrm{MSE} = \frac{1}{n}\sum\nolimits_{i=1}^{n} (gold_i - predicted_i)^2 \tag{11.7}$$

Since they are metrics measuring the error, the lower they are, the better their performance. PAcc is widely applied in the medical data analysis, ranking and statistics fields with the name of concordance index or Kendall τ [23, 24]. PAcc could reflect the correct ratio of ranking between pairwise instances. Specifically, the set of preference evaluation pairs is represented as $S, S = \{(i,j) \mid gold_i > gold_j\}$.

The PAcc is given by

$$\mathrm{PAcc} = \frac{\left|\{(i,j) \mid (i,j) \in S : predicted_i > predicted_j\}\right|}{|S|} \tag{11.8}$$

where "| *S* |" represents the number of the set *S*. It accords with the rule: the greater, the better.

11.1.1.5 Mining Predictive Features

Apart from rating doctors' performance, we continue to explore the most predictive features among text features and customized features in DPE. In general, predictive features always play significant and instructive roles on the platform construction. In this case, the most important features were extracted by analyzing the weight matrix $W R^{l \times (P-1)}$, where l and P–1 are the dimensions of the matrix. As mentioned, l is the total number of all the features (that is, $l = 363, 344$) and P is the number of categories, where P–1 is the number of the submodels. In Eq. (11.2), W is composed of the weight parameters, with w in each submodel, namely $W = (w_1, w_2, \ldots, w_{P-1})$. We denote $W(j, :)$ as the j-th row vector and the absolute value of the elements in the row vector represents the contributions to each submodel for the j-th feature. The larger the value is, the more predictive property the feature has. For every feature in each kind of text feature or customized feature, described in Table 11.2, it owns its corresponding weight vector $W(j, :)$, where $1 \leq j \leq l$.

We compute the total contribution $\mathrm{Con_j}$ of the j-th feature to the model decision by Eq. (11.9):

$$\mathrm{Con_j} = \| \boldsymbol{W}(j, :) \|_2 \tag{11.9}$$

where "L_2" represents the L2-norm of a vector. When the contributions of all the features have been computed, they would be ranked and hence obtain the top-most predictive features.

11.2 Transmission Patterns of COVID-19 Outbreak

COVID-19 has spread to six continents. Now is opportune to gain a deeper understanding of what may have happened. The findings can help inform mitigation strategies in the disease-affected countries.

This section examines an essential factor that characterizes the disease transmission patterns: the interactions among people. We develop a computational model to reveal the interactions in terms of the social contact patterns among the population of different age-groups. We divide a city's population into seven age-groups: 0–6 years old (children); 7–14 (primary and junior high school students); 15–17 (high school students); 18–22 (university students); 23–44 (young/middle-aged people); 45–64 years old (middle-aged/elderly people); and 65 or above (elderly people). We consider four representative settings of social contacts that may cause the disease spread: (1) individual households; (2) schools, including primary/high schools as well as colleges and universities; (3) various physical workplaces; and (4) public places and communities where people can gather, such as stadiums, markets, squares, and organized tours. A contact matrix is computed to describe the contact intensity between different age-groups in each of the four settings. By integrating the four contact matrices with the next-generation matrix, we quantitatively characterize the underlying transmission patterns of COVID-19 among different populations.

It is found on six representative cities in China: Wuhan, the epicenter of COVID-19 in China, together with Beijing, Tianjin, Hangzhou, Suzhou, and Shenzhen, which are five major cities from three key economic zones. The results show that the social contact-based analysis can readily explain the underlying disease transmission patterns as well as the associated risks (including both confirmed and unconfirmed cases). In Wuhan, the age-groups involving relatively intensive contacts in households and public/communities are dispersedly distributed. This can explain why the transmission of COVID-19 in the early stage mainly took place in public places and families in Wuhan. We estimate that Feb. 11, 2020 was the date with the highest transmission risk in Wuhan, which is consistent with the actual peak period of the reported case number (Feb. 4 14). Moreover, the surge in the number of new cases reported on Feb. 12 and 13 in Wuhan can readily be captured using our model, showing its ability in forecasting the potential/unconfirmed cases. We further estimate the disease transmission risks associated with different work resumption plans in these cities after the outbreak. The estimation results are consistent with the actual situations in the cities with relatively lenient policies, such as Beijing, and those with strict policies, such as Shenzhen.

With an in-depth characterization of age-specific social contact-based transmission, the retrospective and prospective situations of the disease outbreak, including the past and future transmission risks, the effectiveness of different interventions, and the disease transmission risks of restoring normal social activities, are computationally analyzed and reasonably explained. The conclusions drawn from the study not only provide a comprehensive explanation of the underlying COVID-19 transmission patterns in China, but more importantly, offer the social contact-based risk analysis methods that can readily be applied to guide intervention planning and operational responses in other countries, so that the impact of COVID-19 pandemic can be strategically mitigated.

11.2.1 Methods

11.2.1.1 Scope of This Study

We select six major cities in China for our study: Wuhan, Beijing, Tianjin, Hangzhou, Suzhou, and Shenzhen; their geographical locations and the disease situations (in terms of total case number from Dec. 2019 to Feb. 2020) are shown in Fig. 11.3. Wuhan was the epicenter of COVID-19 in China [25, 26]. The other five cities are representative in that they are situated in the three most important economic zones in China, which contribute more than 40% of the national GDP. Specifically, Beijing and Tianjin are representing the Jing-Jin-Ji (Beijing-Tianjin-Hebei) Metropolitan Region in Northern China. Hangzhou and Suzhou are the major players in the Yangtze River Delta City Cluster in Eastern China. Shenzhen is the flagship in the Greater Bay Area in Southern China. Another important reason to select these cities for our study is that the population of these cities contains a large number of migrant workers and college students from other cities or provinces. The frequent human mobility largely increases the risk of imported cases, posing great challenges to the control and prevention of COVID-19, especially when people are gradually returning to workplaces and schools in a later stage.

11.2.1.2 Data Sources

The data used in our study include:

1. The daily confirmed cases from Dec. 8, 2019 to Feb. 29, 2020 in Wuhan, Beijing, Tianjin, Hangzhou, Suzhou, and Shenzhen, which were accessed and collected

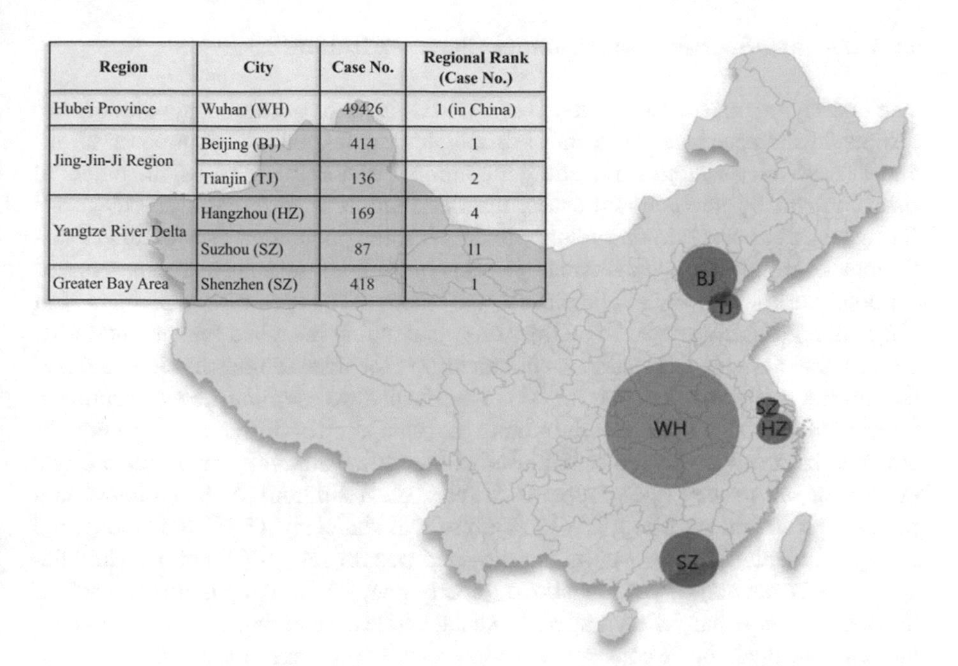

Region	City	Case No.	Regional Rank (Case No.)
Hubei Province	Wuhan (WH)	49426	1 (in China)
Jing-Jin-Ji Region	Beijing (BJ)	414	1
	Tianjin (TJ)	136	2
Yangtze River Delta	Hangzhou (HZ)	169	4
	Suzhou (SZ)	87	11
Greater Bay Area	Shenzhen (SZ)	418	1

Fig. 11.3 The geographical locations and the disease situations (total number of cases from Dec. 2019 to Feb. 2020) of six major cities selected in this study: Wuhan, Beijing, Tianjin, Hangzhou, Suzhou, and Shenzhen

from the websites of the Health Commission of Hubei Province,[1] the Beijing Municipal Health Commission,[2] the Tianjin Municipal Health Commission,[3] the Hangzhou Municipal Health Commission,[4] the Suzhou Municipal Health Commission,[5] and the Shenzhen Municipal Data Open Platform,[6] respectively.

2. The demographic data of Wuhan,[7] Beijing,[8] Tianjin,[9] Hangzhou,[10] Suzhou,[11] and Shenzhen.[12]

[1] http://wjw.hubei.gov.cn/

[2] http://wjw.beijing.gov.cn/

[3] http://wsjk.tj.gov.cn/

[4] http://wsjkw.hangzhou.gov.cn/

[5] http://wsjkw.suzhou.gov.cn/

[6] https://opendata.sz.gov.cn/data/data-Set/toDataDetails/29200_01503668

[7] http://tjj.hubei.gov.cn/tjsj/sjkscx/tjnj/gsztj/whs/

[8] http://tjj.beijing.gov.cn/

[9] http://stats.tj.gov.cn/

[10] http://tjj.hangzhou.gov.cn/

[11] http://tjj.suzhou.gov.cn/

[12] http://tjj.sz.gov.cn/

11.2.1.3 Age-Specific Social Contact Characterization

The underlying transmission patterns of COVID-19 among different populations are difficult to characterize because they are complex and related to various observations and disease-related factors, including the number of confirmed cases, the potential risks brought by unconfirmed cases, the distribution of different case categories (indigenous/imported) in different regions/cities, the population distribution of different age-groups, the social contact patterns in different settings (e.g., households, schools, workplaces, and public places), the extent of interventions implemented in different regions/cities, etc. To address this challenging issue in a fundamental way, we examine an essential factor that characterizes the disease transmission patterns: the interactions among people [27, 28]. Specifically, we examine the interactions in terms of the social contact patterns among the population of different age-groups. To characterize the age-specific social contact-based transmission, we divide a city's population into seven age-groups: 0–6 years old (children); 7–14 (primary and junior high school students); 15–17 (high school students); 18–22 (university and college students); 23–44 (young/middle-aged people); 45–64 years old (middle-aged/elderly people); and 65 or above (elderly people). The population in each of the seven groups has its own specific social circles, gathering places, or activity patterns. Meanwhile, we consider four representative settings of social contacts that may cause the disease spread: (1) individual households, which may lead to the transmission within families; (2) schools, including primary/high schools as well as colleges and universities, which may cause the spread among students and teachers; (3) various physical workplaces, which may affect in-office and outside workers; and (4) public places and communities, such as stadiums, markets, squares, and organized tours, where the spread within a dense population may arise. Let G1–G7 be the seven age-groups: 0–6, 7–14, 15–17, 18–22, 23–44, 45–64, and 65 or above, respectively. Then the contact frequencies between an individual from Gi and an individual from G_j $(i, j = 1, \ldots, 7)$ under the settings of Households, Schools, Workplaces, and Public/community, denoted by c_{ij}^H, c_{ij}^S, c_{ij}^W and c_{ij}^P, respectively, are calculated as follows:

$$c_{ij}^H = \frac{C_{ij}^H}{P_i P_j}, c_{ij}^S = \frac{C_{ij}^S}{P_i P_j}, c_{ij}^W = \frac{C_{ij}^W}{P_i P_j}, c_{ij}^P = \frac{C_{ij}^P}{P_i P_j} \tag{11.10}$$

$$\boldsymbol{C}^H = \left[c_{ij}^H\right]_{7\times 7}, \ \boldsymbol{C}^S = \left[c_{ij}^S\right]_{7\times 7}, \boldsymbol{C}^W = \left[c_{ij}^W\right]_{7\times 7}, \ \boldsymbol{C}^P = \left[c_{ij}^P\right]_{7\times 7} \tag{11.11}$$

where C_{ij}^H, C_{ij}^S, C_{ij}^W and C_{ij}^P denote the total number of contacts between individuals from G_i and those from G_j under the settings of Households, Schools, Workplaces, and Public/community, respectively, P_i and P_j denote the population of G_i and G_j, and $\boldsymbol{C}^H$, $\boldsymbol{C}^S$, $\boldsymbol{C}^W$ and $\boldsymbol{C}^P$ denote the 7×7 social contact matrices.

In Eq. (11.1), we use demographic data to calculate P_i $(i = 1,\ldots,7)$. For C_{ij}^H, C_{ij}^S, C_{ij}^W and C_{ij}^P, as the city-specific data of social contacts between age-groups is unavailable, we adopt a computational method [4] to estimate them. The appropriateness of using such a computational method for social contact estimation in data-scarce situations has been validated 20: the estimated $\boldsymbol{C}^H$, $\boldsymbol{C}^S$, $\boldsymbol{C}^W$ and $\boldsymbol{C}^P$ are consistent with the results from a real-world social contact survey [29] in terms of the strong assortativeness and the appearance of similar secondary diagonal contact patterns.

Next, we represent the overall age-specific social contact matrix as a linear combination of the above four matrices [28]:

$$\boldsymbol{C} = r_H\boldsymbol{C}^H + r_S\boldsymbol{C}^S + r_W\boldsymbol{C}^W + r_P\boldsymbol{C}^P \tag{11.12}$$

where r_H, r_S, r_W, $r_P \geq 0$ are the weights of matrices $\boldsymbol{C}^H$, $\boldsymbol{C}^S$, $\boldsymbol{C}^W$, and $\boldsymbol{C}^P$, respectively, and satisfy that $r_H + r_S + r_W + r_P = 1$.

According to Xia et al.'s work [5], the initial weights of four social contact matrices in our study are set as: $r_H = 0.31$, $r_S = 0.24$, $r_W = 0.16$, $r_P = 0.29$. It should be pointed out that similar settings have also been utilized to simulate the contact matrices in other studies, e.g., the similar weight for the household matrix has been used to calculate the contact matrix for Varicella and Parvovirus B19 [7]. The results in Fig. 11.3 show that our model with the above parameter settings can adequately capture the disease trends in different cities; our sensitivity analysis also confirms that the developed model (to be described below) is relatively robust to the parameter settings.

With the overall age-specific social contact matrix $\boldsymbol{C}$, we can characterize the disease transmission pattern using the next-generation matrix $\boldsymbol{K}_t$ [8]:

$$\boldsymbol{K}_t = \left(\frac{\mu}{\gamma}\right)\boldsymbol{S}_t\boldsymbol{B}\boldsymbol{C}\boldsymbol{A}_t$$

$$\boldsymbol{I}_{t+1} = \boldsymbol{K}_t\boldsymbol{I}_t \tag{11.13}$$

where $\boldsymbol{S_t}$, $\boldsymbol{B}$, $\boldsymbol{A}$ and $\boldsymbol{K_t}$ are 7×7 matrices and $\boldsymbol{I_t}$ is a 7×1 vector. Specifically, $\boldsymbol{S_t}$, $\boldsymbol{B}$, and $\boldsymbol{A}$ are diagonal matrices, with the diagonal elements s^t, b_{ii}, and $a_{ii}(i = 1,\ldots,7)$ being the size of susceptible population in G_i at the t-th generation of the disease infection, the individual susceptibility in G_i, and the infectivity of infected individuals in G_i, respectively. The i-th element in vector $\boldsymbol{I_t}$ denotes the number of infectious individuals in G_i at the t-th generation of the disease infection. By referring to Li et al.'s work [30], we set the reproduction number $R_0 = 2.2$.

For the recovery rate γ, we calculate it as follows: First, according to the definition of recovery rate [31], it is the reciprocal of the duration of being infectious (i.e., γ = 1/infectious period). Then, according to Svensson's work [30], the infectious period is equal to the mean generation time minus the mean latent period, so we have γ = 1/(mean generation time – mean latent period). Further, as pointed out in Binti Hamzah et al.'s work [25] and Liu et al.'s work [32], the mean generation time is 7.5 days. Moreover, Wu et al.'s work [26] indicates that the mean incubation period is 5.2 days for COVID-19. As there is not precise infection date for those patients to estimate the mean latent period, we use the mean incubation period to approximate the mean latent period. Therefore, the recovery rate is estimated as $\gamma = 1/(7.5-5.2)$. For the infectivity, we set $a_{ii} = 1.0$ (for $i = 1,\ldots,7$) according to Xia et al.'s work [28]. For the susceptibility b_{ii}, as it represents the probability of being infected when a susceptible individual is exposed to infectious contacts, we estimate it as follows: For each G_i, we first calculate its infected population ratio r_i by dividing the number of infected cases in G_i by P_i, i.e., $r_i = n_i/P_i$. With r_i calculated for all seven age-groups, we then obtain a multiplier, $1/\min\{r_1, \ldots, r_7\}$, through normalizing the smallest r_i to 1, and inflate all other infected population ratios by $1/\min\{r_1, \ldots, r_7\}$. Then we estimate the susceptibility as $b_{ii} = r_i/\min\{r_1, \ldots, r_7\}$. As different cities have different numbers of infected cases and different population sizes, they will have different susceptibilities. Specifically, we have: $b_{11} = 1.00$, $b_{22} = 1.23$, $b_{33} = 35.33$, $b_{44} = 21.09$, $b_{55} = 13.18$, $b_{66} = 42.16$, and $b_{77} = 97.48$ for Wuhan; $b_{11} = 9.08$, $b_{22} = 1.00$, $b_{33} = 10.67$, $b_{44} = 4.15$, $b_{55} = 10.26$, $b_{66} = 14.58$, and $b_{77} = 17.90$ for Beijing; $b_{11} = 1.00$, $b_{22} = 1.86$, $b_{33} = 1.66$, $b_{44} = 3.14$, $b_{55} = 5.24$, $b_{66} = 9.86$, and $b_{77} = 11.94$ for Tianjin; $b_{11} = 1.06$, $b_{22} = 1.87$, $b_{33} = 1.00$, $b_{44} = 1.78$, $b_{55} = 5.46$, $b_{66} = 9.38$, and $b_{77} = 4.51$ for Hangzhou; $b_{11} = 1.07$, $b_{22} = 1.18$, $b_{33} = 1.57$, $b_{44} = 1.00$, $b_{55} = 3.97$, $b_{66} = 7.20$, and $b_{77} = 3.14$ for Suzhou; and $b_{11} = 3.57$, $b_{22} = 4.72$, $b_{33} = 1.00$, $b_{44} = 2.76$, $b_{55} = 3.49$, $b_{66} = 17.59$, and $b_{77} = 32.51$ for Shenzhen.

The disease infection dynamics computed using Eqs. (11.2) and (11.3) corresponds to the situation without any intervention. To take the effect of intervention into consideration, it is important for us to further decrease r_H, r_S, r_H, r_P in Eq. (11.2), i.e., the weights of different social contact matrices accordingly. Similarly, if we consider different work resumption plans, we will need to increase these weights proportional to the rate of work resumption. Specifically, we reduce r_W from its original value to 0 as of Jan. 23 (the starting date of implementing stringent public health control policies). Moreover, we gradually recover its value from the starting date of our work resumption plans to reflect the effect of "back-to-work" policies. We apply the similar rationale to r_S and r_P. For r_H, as the public social distancing policies would increase social contacts within households, we increase the value of r_H starting from Jan. 23, and gradually reduce to its original value once the "back-to-work" policy kicks in.

11.2.1.4 Role of the Funding Source

The funders of the study had no role in study design, data collection, data analysis, data interpretation, writing of the Article, or the decision to submit for publication. All authors had full access to all the data in the study and were responsible for the decision to submit the Article for publication.

11.2.2 Results

11.2.2.1 Social Contact-Based Transmission Characterization

As can be seen in Fig. 11.4, the distribution of age-groups involving relatively intensive contacts in households and public/communities is rather scattered, and thus it is easy to cause the disease spread among different age-groups in these two settings. This is consistent with the observation that the transmission of COVID-19 in the early stage mainly took place in public places and families. In contrast, the distribution of age-groups with intensive contacts in schools and workplaces are relatively concentrated. Moreover, the composition of people in these two settings is relatively stable, making the management easier than that in public places or communities. Because most of the schools and workplaces were closed before the Chinese Spring Festival and have not been reopened or resumed yet, the scale of the COVID-19 outbreak in these two settings is relatively limited. However, if normal educational and economic activities are to be resumed, a large number of students and staff will gather in these two settings, which may present a real challenge to the control and prevention of COVID-19 infection in these concerned places.

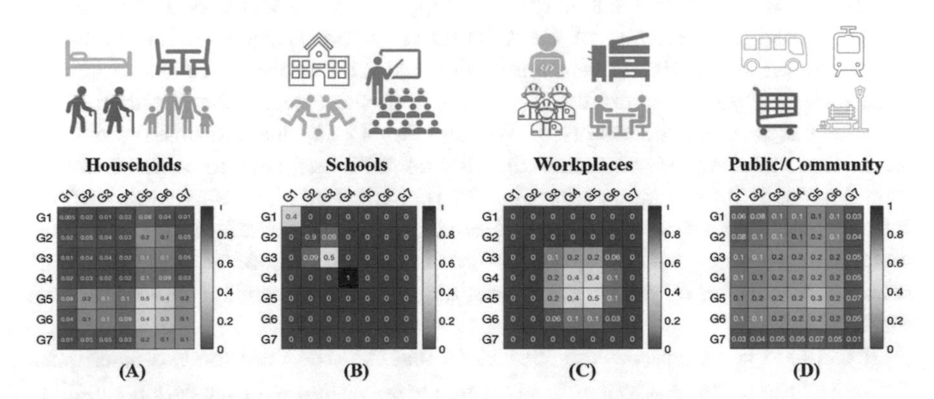

Fig. 11.4 Measurement of the intensity of social contacts among seven age-groups (G1: 06; G2: 714; G3: 1517; G4: 1822; G5: 2344; G6: 4564; and G7: 65 or above) in four major settings: (**a**) households; (**b**) schools; (**c**) workplaces; and (**d**) public/community, in Wuhan. The contact patterns in these four settings are consistent with common social behaviors observed in a typical society. Specifically, as shown in (**a**), the majority of the social contacts within households occur across different generations. (**b**) Demonstrates that the main social contact in schools centers around kids in the same age-group. As depicted in (**c**), workplaces are dominated by social contacts among young adults and the middle-aged adults. In (**d**), the social contacts are more diverse when people are in public places

11.2.2.2 Retrospective Analysis of the Disease Outbreak

For different cities, the transmission patterns of COVID-19 might be different. In Wuhan, the cases were mainly indigenous. In the other five cities, the cases might be either indigenous or imported from Hubei. Therefore, for these five cities, we need to take both indigenous cases and imported cases into consideration when investigating the transmission patterns among populations. To model the potential local transmission risk caused by the imported cases, we use the following approach: First, for each confirmed case, we identify if it is imported or indigenous according to the information provided by the Municipal Health Commission. If the case is an imported case, we consider its potential risk in bringing in local transmission. According to Li et al.'s results [33], the mean serial interval is 7.5 days, so we assume that for each imported case, from the day of arrival to the day of hospitalization, he/she could infect 1/7.5 person per day. We apply the same principle to all imported cases to estimate their potential infections. Those cases infected by the imported ones are considered as potential cases in our study. The confirmed cases and potential cases together constitute the disease transmission risk, i.e., the focus of the following retrospective and prospective analyses.

With the age-specific social contact-based transmission modeling, we are able to describe and explain what may have happened retrospectively and what can be anticipated prospectively of the COVID-19 outbreak. Figure 11.5 shows the estimation on the trends of disease infection and the transmission risks associated with different work resumption plans based on the social contact patterns and reported cases. From the results of Wuhan (Fig. 11.5a), we can observe that the situation without any interventions (the brown line) is estimated to be much severer than that with interventions (the blue line), indicating the effectiveness of the interventions implemented in Wuhan. Here the interventions refer to various social distancing measures, including quarantine of patients, closure of workplaces and schools, suspension of public transportation, and requirement for people to wear masks [34–36].

It can also be observed from Fig. 11.5a that the date with the estimated peak number of the new cases was Feb. 11, which is consistent with the actual situation: the number of reported cases reached the peak during Feb. 4–14. Moreover, note that there was a sharp increase in the number of confirmed cases on 12 and 13 Feb. This is because that the National Health Commission of China adopted a new case definition in Hubei province. In the new case definition, the clinically diagnosed cases (suspected cases with pneumonic-type imaging characteristics) were included in the newly reported cases. To put it in the right context, the observed surge in the number of reported cases does not imply a large number of cases were found on 12 and 13 Feb, but an inflow of accumulated clinically confirmed cases from the past few days. As can be observed from Fig. 11.5a, the result of our model (the gap between the blue line and the red bars) offers a reasonable explanation of the not-yet-reported cases over the period. Specifically, the model is not designed to provide a mathematical estimation/prediction that fits exactly to the number of reported cases, but to present an estimation of the risk to the community if certain

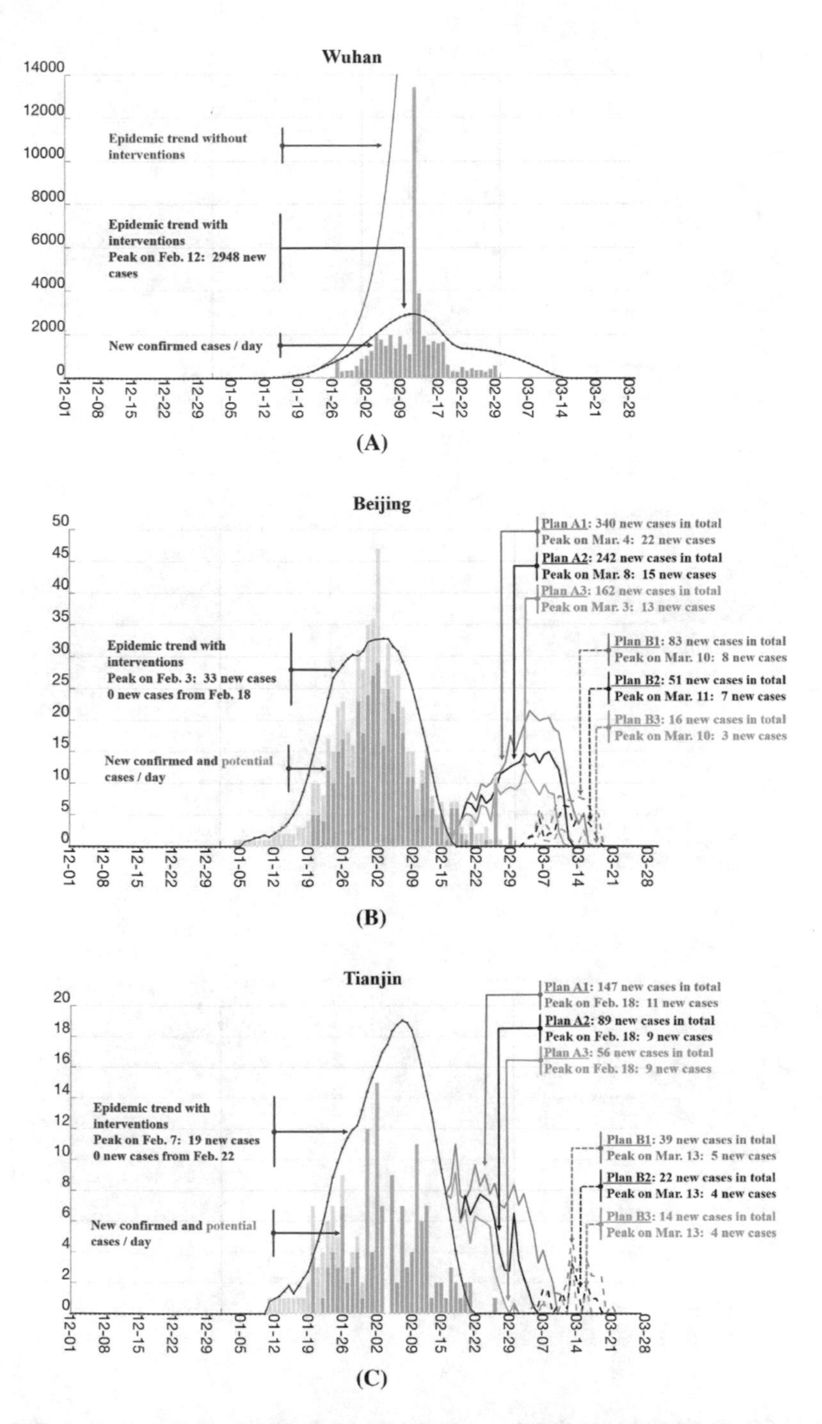

Fig. 11.5 Estimation on the trends of disease infection and transmission risks associated with different work resumption plans based on the social contact patterns and reported cases. (**a**) The estimated disease trends without any interventions (the brown line) and with interventions (the blue line) in Wuhan. The newly confirmed cases reported every day are shown in red bars. (**b–f**) The estimated disease trends with interventions (the blue line) and the transmission risks associated with different work resumption plans in Beijing, Tianjin, Hangzhou, Suzhou, and Shenzhen.

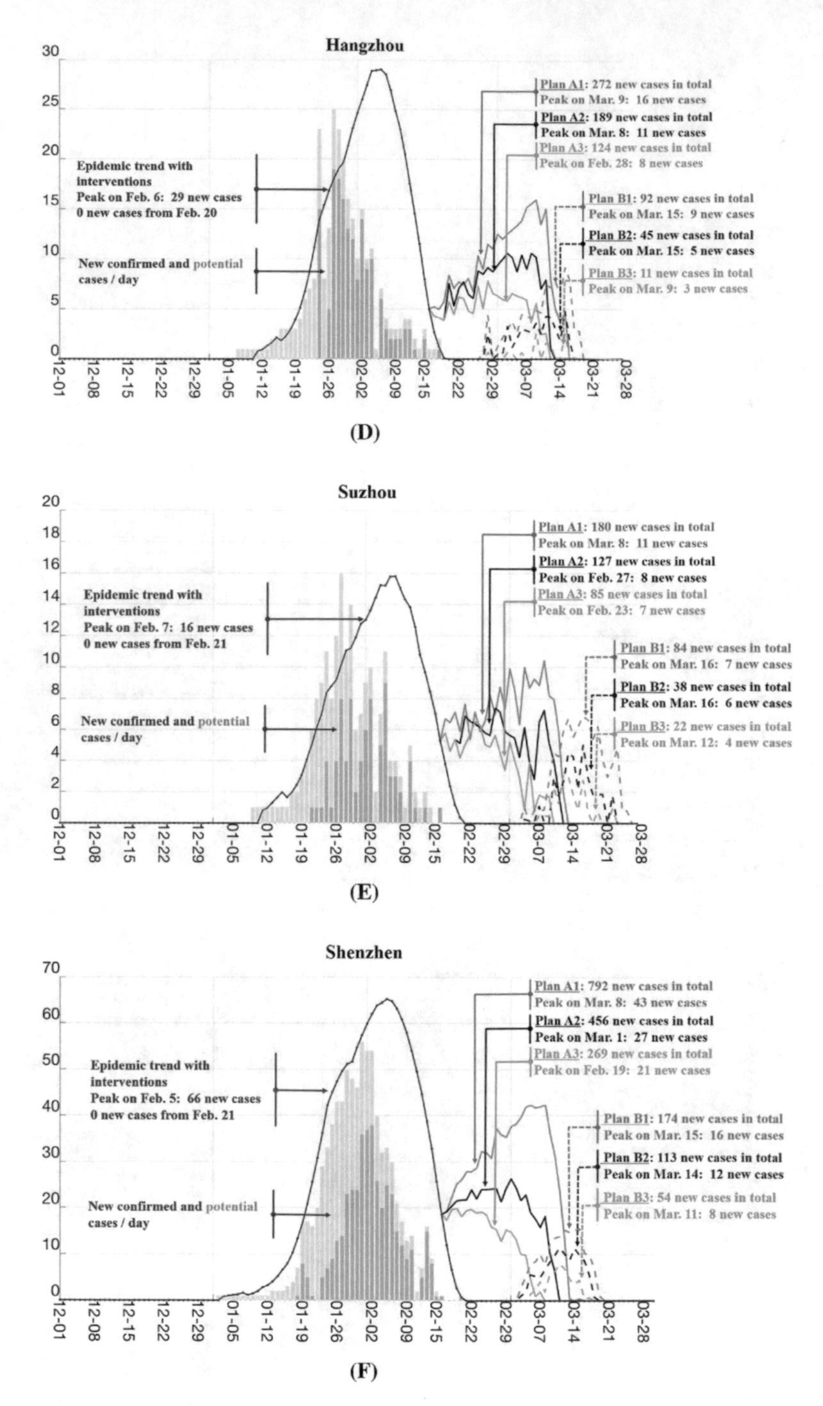

Fig. 11.5 (continued) The dark red bars denote the newly confirmed cases reported every day while the light red bars denote the potential cases locally infected by the imported cases, which are estimated according to the mean serial interval of 7.5 days [33]. Plans A1–A3 refer to the plans that start on Feb. 17 (Monday) and finish the resumption in 1 week, 1/2 months, and 1 month, respectively. Plans B1–B3 refer to the plans that start on Feb. 24 (Monday) and finish the resumption in 1 week, 1/2 months, and 1 month, respectively

measures are or are not exercised. One reasonable assumption included in the model is the consideration of unreported cases, as it is nearly impossible to timely capture all new cases given limited resources and the fixed capacity of the medical system. For this reason, our model estimates a case number larger than that of the reported confirmed cases, with the excessive number representing the potential risks that are not yet identified as confirmed cases.

11.2.2.3 Prospective Analysis of Disease Transmission Risks and Economic Impacts

The COVID-19 pandemic has hit the global economy by a storm. As the public health crisis escalates, countries have responded by enforcing social distancing measures, such as the closure of public venues and reduced working hours, to reduce the chance of contracting the highly contagious virus in a social setting. At the same time, this will inevitably lead to a massive decline in business activities, causing unprecedented economic loss to the countries. When the COVID-19 outbreak is contained, as in the case of Wuhan, China, it is foreseeable that the countries will need to think about how to safely resume social activities and bring work and life back to normal, as any pre-mature resumption of social contacts could potentially cause a rebound (second wave) in new infection cases. In the light of the pressing needs to provide a scientific ground for systematically planning the resumption of social/business activities near the end of the outbreak, we present a prospective analysis of different work resumption plans, which can enable us to assess not only the respective economic implications of the plans, but more importantly, the levels of disease transmission risks associated with the corresponding plans.

Specifically, to further understand what can be anticipated prospectively of the disease outbreak, we analyze what may happen if the social/business activities gradually restore from the strong control and isolation to the normal situation (including public and work places). We analyze the disease transmission risks associated with different work resumption plans in Beijing, Tianjin, Hangzhou, Suzhou, and Shenzhen, respectively. Since Wuhan was in a serious situation during the outbreak, its work resumption may take longer than the other five cities; its detailed plans for the resumption and associated risks are discussed in the original paper [2]. We conduct our prospective study on two sets of different work resumption plans (Plans A1–A3 and Plans B1–B3) and accordingly examine their associated risks of disease transmission. Plans A1–A3 resume work when the disease transmission is well under control, i.e., the number of new reported cases is about to become zero. Specifically, Plans A1–A3 start on Feb. 17 (Monday) and finish the resumption in 1 week, 1/2 months, and 1 month, respectively. Plans B1–B3 are stricter than Plans A1–A3; they resume work when the number of new reported cases has been zero for three consecutive days. Therefore, Plans B1–B3 start on Feb. 24 (Monday) and, similarly, finish the resumption in 1 week, 1/2 months, and 1 month, respectively.

In order to parameterize our model for risk prediction with different work resumption plans, we estimate the percentage of work ongoing or recovered in each of those cities at the time when the resumption plans start. For Beijing, according to a document issued by the municipal government,[13,14] eligible companies can resume work from Feb. 3. Therefore, we set a weekly increase of 10% in the resumption of work in Beijing from Feb. 3 until the resumption plan begins. For Tianjin, we use the same resumption settings as Beijing because both cities are situated in the Jingjinji Metropolitan Region. For Hangzhou, according to the municipal government regulations,[15] general enterprises shall not resume work before 23:59 on Feb. 9. Meanwhile, according to the electricity consumption statistics of the State Grid Corporation of China, about 20% of enterprises in Zhejiang Province generated electricity consumption on Feb. 10.[16] By further taking into account the situation of home and remoting office, we set Hangzhou's work resumption rate on Feb. 10 to be 10%, and from Feb. 11 until its resumption plan begins, the weekly work resumption rate will be 10%. For Suzhou, we use the same resumption settings as Hangzhou as both are in the Yangtze River Delta Economic Zone. Last but not the least, for Shenzhen, according to the municipal government regulations, general enterprises may not resume work before 24:00 on Feb. 9.[17] Therefore, we set Shenzhen's weekly work resumption rate to be 10% from Feb. 10 until its resumption plan begins.

As shown in Fig. 11.5b–f, Plans B1–B3 represent a stricter work resumption policy; they start 1 week later than Plans A1–A3. The estimation of disease transmission risks is consistent with actual situations. For example, Beijing implemented a relatively lenient policy on the early resumption of work, and thus has several new cases reported every day during the past 2 weeks. This is consistent with our estimated risk trend of Beijing with the plans A1–A3. In contrast, Shenzhen still strictly controlled the resumption of work, so there is no new case reported during Feb. 24–29.[18] This is also in line with our estimation on Shenzhen's risk with stricter plans B1–B3.

Table 11.3 summarizes the disease transmission risks as well as the estimated Year-over-Year GDP growth (%) in the first half of 2020 with respect to different work resumption plans in Beijing, Tianjin, Hangzhou, Suzhou, and Shenzhen. As can be noted, Plan B3 resumes the work as late as possible and completes the resumption as slow as needed, and thus minimizes the disease transmission risk.

---

[13] http://www.gov.cn/xinwen/2020-01/31/content_5473425.htm (Accessed on April 1, 2020).

[14] http://www.dehenglaw.com/CN/tansuocontent/0008/017738/7.aspx?MID=0902 (Accessed on April 1, 2020).

[15] http://www.hangzhou.gov.cn/art/2020/2/9/art_1256295_41893739.html(Accessed on April 1, 2020).

[16] http://energy.people.com.cn/n1/2020/0213/c71661-31585079.html (Accessed on April 1, 2020).

[17] http://www.sz.gov.cn/szzt2010/yqfk2020/szzxd/zczy/zcwj/fgzc/content/post_6728851.html (Accessed on April 1, 2020).

[18] Shenzhen Municipal Data Open Platform: https://opendata.sz.gov.cn/data/data-Set/toDataDetails/29200_01503668

Table 11.3 The disease transmission risk (in terms of estimated new cases during the work resumption period) and the estimated Year-over-Year GDP growth (%) in the first half of 2020 ($YoY_{2020} = \frac{\Delta Y_{2020} - \Delta_{2019}}{\Delta_{2019}}$, with ΔY_{2020} and ΔY_{2019} denoting the estimated GDP growth in the first half of 2020 and the actual GDP growth in the first half of 2019, respectively) of 6 different work resumption plans in 5 cities (Beijing, Tianjin, Hangzhou, Suzhou, and Shenzhen). Here ΔY_{2020} of each city is calculated based on ΔY_{2019} and the proportional GDP loss caused by the corresponding percentage of workplace closure in different work resumption plans

	Beijing		Tianjin		Hangzhou		Suzhou		Shenzhen	
	New cases	YoY_{2020}	New cases	YoY_{2020}	New cases	YoY_{2020}	New cases	YoY_{2020}	New cases	YoY_{2020}
Plan A1	340	−9.9%	147	−9.9%	272	−10.1%	180	−10.1%	792	−10.7%
Plan A2	242	−11.7%	89	−11.7%	189	−11.9%	127	−11.9%	456	−12.6%
Plan A3	162	−15.0%	56	−15.0%	124	−15.2%	85	−15.2%	269	−16.4%
Plan B1	**83**	**−12.6%**	**39**	**−12.6%**	**92**	**−12.8%**	**84**	**−12.8%**	**174**	**−13.8%**
Plan B2	51	−14.1%	22	−14.1%	45	−14.4%	38	−14.4%	113	−15.5%
Plan B3	**16**	**−17.0%**	**14**	**−17.0%**	**11**	**−17.3%**	**22**	**−17.3%**	**54**	**−18.8%**

The plan has the least expected GDP growth. Alternatively, it may also be practically desirable to gradually bring the work back to normal, while keeping all the necessary control measures possible to eliminate any potential disease transmission. In such a case, Plan B1 could be adopted. It achieves risk mitigation and gradual work and life recovery at the same time.

11.2.2.4 Sensitivity to Parameter Variations

We conduct the sensitivity study to examine variations of the analytic results with respect to variations in different age-groups and various social contact patterns. Specifically, we analyze the sensitivity of the estimated disease trends with respect to changes in the infectivity matrix $\boldsymbol{A}$, the individual susceptibility matrix $\boldsymbol{B}$, and the contact matrix $\boldsymbol{C}$. Note that since both $\boldsymbol{A}$ and $\boldsymbol{B}$ are diagonal matrices, the impact to the next-generation matrix $\boldsymbol{K}$ will be identical if $\boldsymbol{A}$ and $\boldsymbol{B}$ change in the same scale, we only conduct the sensitivity analysis on $\boldsymbol{A}$ as a representative. Each time, we change the diagonal value in $\boldsymbol{A}$ for one specific age-group while keeping other age-groups' value fixed. By doing so, we can investigate the impact of different age-groups on our results. For the contact matrix $\boldsymbol{C}$, each time we change the weight for one of the four matrices $\boldsymbol{C}^H$, $\boldsymbol{C}^S$, $\boldsymbol{C}^W$, and $\boldsymbol{C}^P$, while keeping the weights of the other three unchanged. By doing so, we can observe the impact of different social contact patterns on the results.

Figure 11.6 shows variations in the estimated disease trends corresponding to variations in (a) different age-groups and (b) various social contact patterns. The trends in Fig. 11.6 are measured by the total number of confirmed and potential cases. From Fig. 11.6a we can observe that the disease trends in all six cities are relatively sensitive to the variations in the infectivity in G5 (23–44) and G6 (45–64). There are two main reasons. First, both the population size and the case number in these two age-groups are relatively large. Second, people in these two groups are more frequently engaged in social activities than those in other age-groups. Therefore, a slight variation on the infectivity in these two groups might cause relatively large variations in the disease trends. From Fig. 11.6b we can observe that the variations of the disease trends are more obvious in households than in schools, workplaces, and public places/communities, which is consistent with our observations in Fig. 11.4. The reason is that China has already implemented strong social distancing strategies, such as the closure of schools and workplaces, and thus the variations of social contact intensities in schools and workplaces have relatively limited impact on the disease trends. Note that even various intervention strategies have been deployed, the disease trends can still be relatively sensitive to the variations of social contact intensities in public places and communities. The key implication, as revealed by Fig. 11.6b, is to maintain proper social distance in public/community, even when school and business activities are resumed. Moreover, there are some intrinsic consistencies between the results in Fig. 11.6a, b. Note that the group in Fig. 11.6a with the highest sensitivity is G5, in the age of 23–44. People of this age are generally the family breadwinner and play the most active role

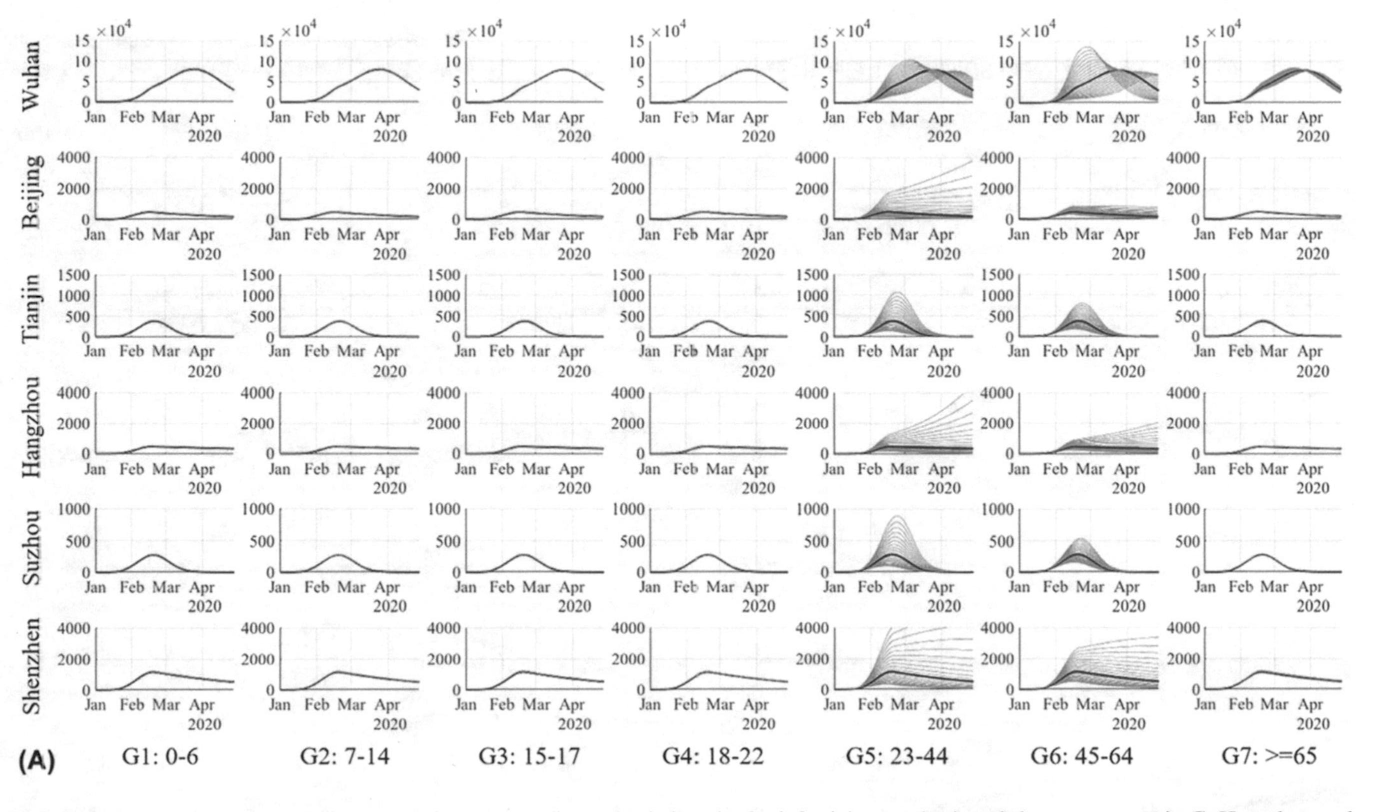

Fig. 11.6 Variations in the estimated disease trends corresponding to variations in the infectivity matrix A and the contact matrix C. Here the trends are measured by the total number of confirmed and potential cases. (**a**) Trend variations with the variations (±10%) of a_{ii} ($i = 1, \ldots, 7$), the infectivity in age-group G_i, in Wuhan, Beijing, Tianjin, Hangzhou, Suzhou, and Shenzhen. Each row represents a city and each column represents an age-group. (**b**) Trend variations

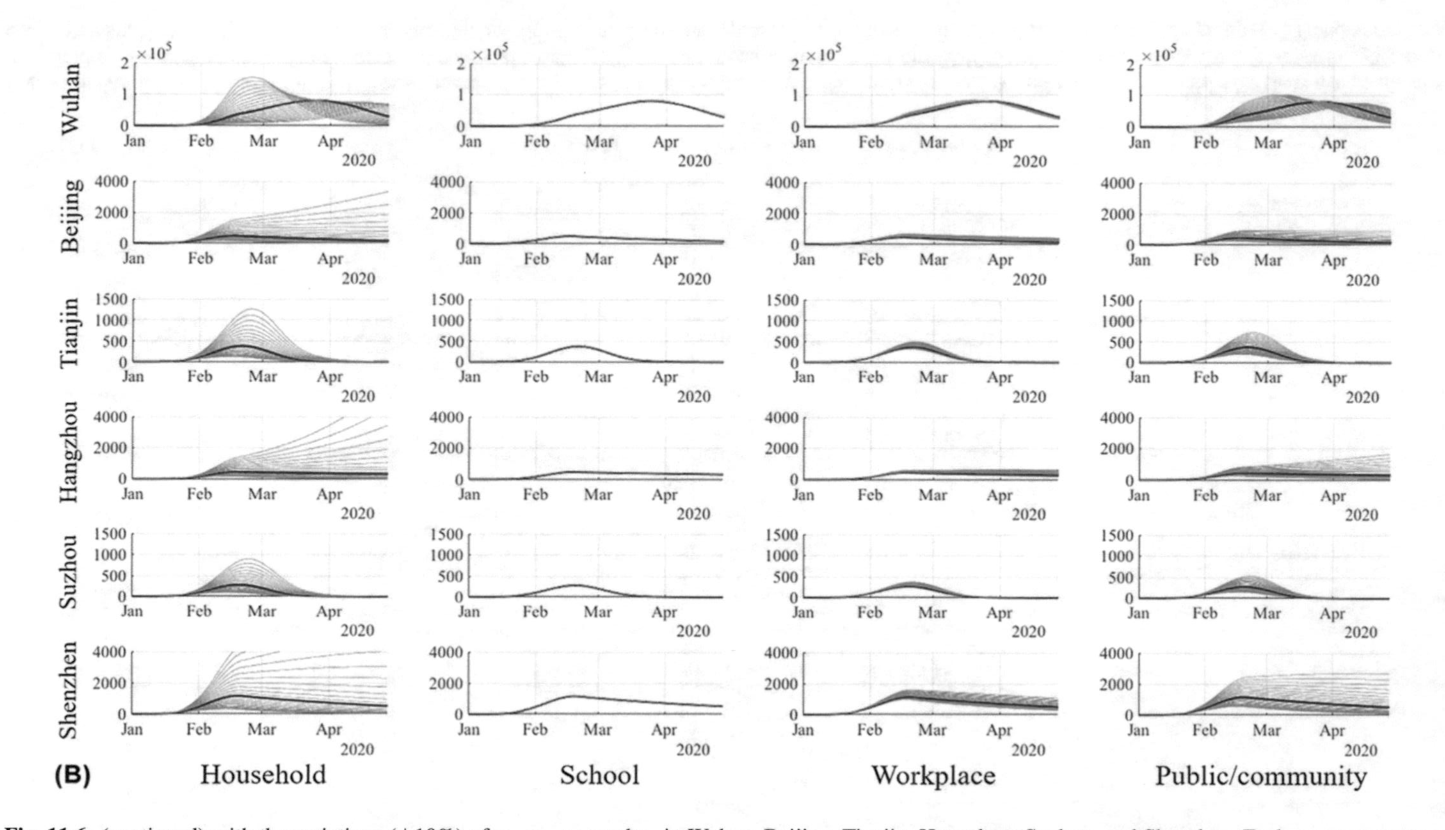

Fig. 11.6 (continued) with the variations (±10%) of r_H, r_S, r_W, and r_P in Wuhan, Beijing, Tianjin, Hangzhou, Suzhou, and Shenzhen. Each row represents a city and each column represents a social contact setting

in social activities. If they are infected, they will bring a huge risk to their family members in households and their friends and acquaintances in communities. This explains the observation in Fig. 11.6a, i.e., the variation of infectivity in G5 brings the highest impact on the disease trends, and thus reminds the public health workers to pay special attention to this group of population when implementing control and prevention strategies in a later stage.

11.2.3 Discussion

In this study, we demonstrated the importance of characterizing the underlying transmission patterns among different populations for the purpose of understanding the COVID-19 outbreak in China from an epidemiological perspective. With an in-depth characterization of the age-specific social contact-based transmission, we conducted the retrospective and prospective analyses of the disease outbreak, including the past and future disease transmission trends, the effectiveness of different interventions, and the disease transmission risks of restoring normal social activities. We focused on six representative cities in China; the conclusions drawn from the study not only provide a comprehensive explanation of the underlying COVID-19 transmission patterns in China, but more importantly, offer the contact-based risk analysis methodology that can readily be applied to guide intervention planning and operational responses in other countries, so as to effectively control the COVID-19 pandemic.

The analysis of this study was conducted on six cities in China, including Wuhan, Beijing, Tianjin, Hangzhou, Suzhou, and Shenzhen. On the one hand, the selected cities were representative in terms of the severity of the COVID-19 outbreak in the respective region during the time that the study was conducted and their economic impact in China. On the other hand, they presented different characteristics in several important aspects: First, the categories of confirmed cases in these cities were different. For Wuhan, most of the cases were indigenous cases; while in the other five cities, a large portion of the cases was imported cases (from Hubei Province). Second, the distributions of populations in different age-groups were different in these cities. Third, the levels of social distancing interventions and the work resumption plans implemented in different cities were different. The above differences made the scenarios in different cities quite different and intriguing. The retrospective and prospective analyses conducted on these cities show that our results are consistent with the real situations of the corresponding cities, validating the model's generalization ability given different real-world contexts. Importantly, it should be noted that, although the numerical results derived from the six cities in this study may not be the same as those in other countries, the developed methods are general at the methodological level and the idea of using age-specific contacts to characterize the disease transmission patterns is instructive in understanding, and hence planning corresponding interventions in, the situations of the disease outbreaks in those countries. When applying the developed methodology to the wider global population, country/region-specific scenarios and settings, such as case categories, distribution of age-specific population, working environment and hours, and interventions and work resumption plans, should be incorporated to provide better tailor-made parameterization, thus making the retrospective and prospective analyses more situation-specific and informative.

As COVID-19 is a newly emerging infectious disease, we are still in the process of gaining more knowledge and understanding of its transmission patterns. As a result, the parameters estimated based on the current understanding might not be as adequate or precise as those in some of the well-understood diseases, such as seasonal influenza. Therefore, one of our future research directions is to continue investigating the characteristics of the disease, from both epidemiological and computational perspectives, so as to parameterize the model in a more accurate way. Further, in this study, we have modeled the underlying transmission of COVID-19 outbreak by considering the age-specific social contact patterns. It should be pointed out that there also exist other disease-related factors that might affect the disease transmission patterns, such as the cross-region mobility of the population and the environmental factors. We plan to incorporate these disease-related factors into the model, thus making our analysis more comprehensive. Moreover, the current study focuses on the representative cities in China. However, it will be desirable to conduct further analyses on a global scale. In this regard, the general methodology provided in this study can readily be applied, while considering country/region-specific social, demographic, and epidemiological characteristics, such as infection-related social contact patterns [37]. To further generalize and transfer our research, we plan to collaborate with researchers and practitioners around the world to conduct the corresponding analyses for other countries/regions.

References

1. Shi, Y., Li, P., Yu, X., Wang, H., Niu, L.: Evaluating doctor performance: ordinal regression-based approach. J. Med. Internet Res. **20**(7), e240 (2018)
2. Liu, Y., Gu, Z., Xia, S., Shi, B., Zhou, X.N., Shi, Y., Liu, J.: What are the underlying transmission patterns of covid-19 outbreak? An age-specific social contact characterization. EClinicalMedicine. **22**, 100354 (2020)
3. Hsieh, C.J., Chang, K.W., Lin, C.J., Keerthi, S.S., Sundararajan, S.: A dual coordinate descent method for large-scale linear svm. In: Proceedings of the 25th International Conference on Machine Learning, pp. 408–415 (2008)
4. Xue, N.: Chinese word segmentation as character tagging. Int. J. Computat. Linguist. Chin. Lang. Process. **8**(1) (2003) Special Issue on Word Formation and Chinese Language Processing, 29–48 (2003)
5. Zhang, Y., Jin, R., Zhou, Z.H.: Understanding bag-of-words model: a statistical framework. Int. J. Mach. Learn. Cybern. **1**(1–4), 43–52 (2010)
6. Witten, I.H., Frank, E.: Data mining: practical machine learning tools and techniques with java implementations. ACM SIGMOD Rec. **31**(1), 76–77 (2002)
7. Friedman, J.H.: Greedy function approximation: a gradient boosting machine. Ann. Stat. **2001**, 1189–1232 (2001)

8. Chapelle, O., Shivaswamy, P., Vadrevu, S., Weinberger, K., Zhang, Y., Tseng, B.: Multi-task learning for boosting with application to web search ranking. In: Proceedings of the 16th ACM SIGKDD International Conference on Knowledge Discovery and Data Mining, pp. 1189–1198 (2010)
9. Kramer, S., Widmer, G., Pfahringer, B., De Groeve, M.: Prediction of ordinal classes using regression trees. Fundam. Inform. **47**(1–2), 1–13 (2001)
10. Chu, W., Keerthi, S.S.: New approaches to support vector ordinal regression. In: Proceedings of the 22nd International Conference on Machine Learning, pp. 145–152 (2005)
11. Chu, W., Keerthi, S.S.: Support vector ordinal regression. Neural Comput. **19**(3), 792–815 (2007)
12. Gu, B., Sheng, V.S., Tay, K.Y., Romano, W., Li, S.: Incremental support vector learning for ordinal regression. IEEE Trans. Neural Netw. Learn. Syst. **26**(7), 1403–1416 (2014)
13. Herbrich, R., Graepel, T., Obermayer, K.: Support vector learning for ordinal regression (1999)
14. Frank, E., Hall, M.: A simple approach to ordinal classification. In: European Conference on Machine Learning, pp. 145–156 (2001)
15. Waegeman, W., Boullart, L.: An ensemble of weighted support vector machines for ordinal regression. Int. J. Comput. Syst. Sci. Eng. **3**(1), 47–51 (2009)
16. Gutiérrez, P.A., Perez-Ortiz, M., Sanchez-Monedero, J., Fernandez-Navarro, F., Hervas-Martinez, C.: Ordinal regression methods: survey and experimental study. IEEE Trans. Knowl. Data Eng. **28**(1), 127–146 (2015)
17. Cardoso, J., da Costa, J.P.: Learning to classify ordinal data: the data replication method. J. Mach. Learn. Res. **8**, 1393–1429 (2007)
18. Gutiérrez, P.A., Pérez-Ortiz, M., Fernández-Navarro, F., Sánchez-Monedero, J., Hervás-Martínez, C.: An experimental study of different ordinal regression methods and measures. In: International Conference on Hybrid Artificial Intelligence Systems, pp. 296–307 (2012)
19. Diao, Q., Qiu, M., Wu, C.Y., Smola, A.J., Jiang, J., Wang, C.: Jointly modeling aspects, ratings and sentiments for movie recommendation (jmars). In: Proceedings of the 20th ACM SIGKDD International Conference on Knowledge Discovery and Data Mining, pp. 193–202 (2014)
20. Shimada, K., Endo, T.: Seeing several stars: a rating inference task for a document containing several evaluation criteria. In: Pacific-Asia Conference on Knowledge Discovery and Data Mining, pp. 1006–1014 (2008)
21. Tang, D., Qin, B., Liu, T.: Document modeling with gated recurrent neural network for sentiment classification. In: Proceedings of the 2015 Conference on Empirical Methods in Natural Language Processing, pp. 1422–1432 (2015)
22. Kuo, T.M., Lee, C.P., Lin, C.J.: Large-scale kernel ranksvm. In: Proceedings of the 2014 SIAM International Conference on Data Mining, pp. 812–820 (2014)
23. Lee, C.P., Lin, C.J.: Large-scale linear ranksvm. Neural Comput. **26**(4), 781–817 (2014)
24. Kendall, M.G.: A new measure of rank correlation. Biometrika. **30**(1/2), 81–93 (1938)
25. Hamzah, F.B., Lau, C., Nazri, H., Ligot, D., Lee, G., Tan, C., Shaib, M., Zaidon, U., Abdullah, A., Chung, M., et al.: Coronatracker: worldwide COVID-19 outbreak data analysis and prediction. Bull. World Health Organ. **1**(32) (2020)
26. Wu, J.T., Leung, K., Leung, G.M.: Nowcasting and forecasting the potential domestic and international spread of the 2019-ncov outbreak originating in Wuhan, China: a modelling study. Lancet. **395**(10225), 689–697 (2020)
27. Fumanelli, L., Ajelli, M., Manfredi, P., Vespignani, A., Merler, S.: Inferring the structure of social contacts from demographic data in the analysis of infectious diseases spread. PLoS Comput. Biol. **8**(9), e1002673 (2012)
28. Xia, S., Liu, J., Cheung, W.: Identifying the relative priorities of subpopulations for containing infectious disease spread. PLoS One. **8**(6), e65271 (2013)
29. Mossong, J., Hens, N., Jit, M., Beutels, P., Auranen, K., Mikolajczyk, R., Massari, M., Salmaso, S., Tomba, G.S., Wallinga, J., et al.: Social contacts and mixing patterns relevant to the spread of infectious diseases. PLoS Med. **5**(3), e74 (2008)
30. Svensson, Å.: A note on generation times in epidemic models. Math. Biosci. **208**(1), 300–311 (2007)

31. Hethcote, H.W.: The mathematics of infectious diseases. SIAM Rev. **42**(4), 599–653 (2000)
32. Liu, T., Hu, J., Xiao, J., He, G., Kang, M., Rong, Z., Lin, L., Zhong, H., Huang, Q., Deng, A., et al.: Time-varying transmission dynamics of novel coronavirus pneumonia in china. BioRxiv. (2020)
33. Li, Q., Guan, X., Wu, P., Wang, X., Zhou, L., Tong, Y., Ren, R., Leung, K.S., Lau, E.H., Wong, J.Y., et al.: Early transmission dynamics in Wuhan, China, of novel coronavirus-infected pneumonia. N. Engl. J. Med. **382**(13), 1199–1207 (2020)
34. Epidemiology Working Group, et al.: Strategy and policy working group for NCIP epidemic response. Chinese Center for Disease Control and Prevention (2019)
35. Hellewell, J., Abbott, S., Gimma, A., Bosse, N., Jarvis, C., Russell, T.W., Munday, J.D., Kucharski, A.J., Edmunds, W.J., Funk, S., et al.: Feasibility of controlling COVID-19 outbreaks by isolation of cases and contacts. Lancet Glob. Heal. **8**, e488–e496 (2020)
36. Sadique, M.Z., Adams, E.J., Edmunds, W.J.: Estimating the costs of school closure for mitigating an influenza pandemic. BMC Public Health. **8**(1), 1–7 (2008)
37. Hoang, T., Coletti, P., Melegaro, A., Wallinga, J., Grijalva, C.G., Edmunds, J.W., Beutels, P., Hens, N.: A systematic review of social contact surveys to inform transmission models of close-contact infections. Epidemiology (Cambridge, Mass.). **30**(5), 723 (2019)

Chapter 12
Artificial Intelligence IQ Test

Since 2015, "artificial intelligence" has become a popular topic in science, technology, and industry. New products such as intelligent refrigerators, intelligent air conditioning, smart watches, smart robots, and of course, artificially intelligent mind emulators produced by companies such as Google and Baidu continue to emerge. However, the view that artificial intelligence is a threat remains persistent. An operation is that if we compare the developmental levels of artificial intelligence products and systems with measured human intelligence quotients (IQs), can we develop a quantitative analysis method to assess the problem of artificial intelligence threat?

Quantitative evaluation of artificial intelligence currently in fact faces two important challenges: there is no unified model of an artificially intelligent system, and there is no unified model for comparing artificially intelligent systems with human beings. These two challenges stem from the same problem, namely, the need to have a unified model to describe all artificial intelligence systems and all living behavior (in particular, human behavior) in order to establish an intelligence evaluation and testing method. If a unified evaluation method can be achieved, it might be possible to compare intelligence development levels.

This chapter provides an innovative concept and basic measurements on testing the Intelligence Quotient (IQ) on artificial intelligence (AI) technologies and devices. Section 12.1 describes the basic concepts of IQ test on AI, particularly Internet search engines and a standard intelligence model. Section 12.1.1 builds an IQ test bank to compare the known search engines, such as Google and Baidu with three groups of Children whose ages are 6, 12, 18 [1]. Section 12.1.2 further employs a data mining method to find out the features of search engines reflected in the Internet intelligence test and the intelligence difference between search engines and human beings [2]. Section 12.1.3 proposes a "standard intelligence model" that unifies AI and human characteristics in terms of four aspects of knowledge, i.e., input, output, mastery, and creation [3]. Section 12.2 investigates three laws of intelligence for interpreting the concepts of intelligence, wisdom, consciousness,

Y. Shi, *Advances in Big Data Analytics*,
https://doi.org/10.1007/978-981-16-3607-3_12

life and non-life [4]. Section 12.3 explores characteristics on AI-IQ test by using fuzzy cognitive map-based dynamic scenario analysis [5].

12.1 A Basic AI-IQ Test

This subsection introduces the IQ test concepts of Internet, which are IQ of Internet applications, Internet 2014 Intelligence Scale, Internet IQ absolute and then explore deviation algorithms for carrying out the IQ test for the major Internet search engines with a group of Children's IQ.

12.1.1 The Concepts of AI-IQ Test

Definition 12.1 IQ of Internet application is to measure intellectual development level of Internet applications at certain test time through a series of standard tests, which include electronic bullet board, search engine, social network, electronic mailbox and instant messaging software etc.

Definition 12.2 Internet IQ is to measure Internet IQ Standards Evaluating Bank through a series of standard tests, so as to derive the intellectual development level of Internet at certain test time, and intellectual development level of Internet is also termed as Internet IQ at that point of time [6].

Based on the basic understanding that intellectual is about people's ability of understanding objective things and applying knowledge to solve practical problems, we will build Internet Intelligent Evaluation System from four major aspects in terms of knowledge obtaining ability (also termed as observation ability) and retaining ability, together with ability of knowledge innovation and feedback (also termed as expression ability), set up 15 subtests from the four aspects and endow weights with Delphi Method to form 2014 Internet Intelligence Scale as shown in Table 12.1.

Definition 12.3 Based on the structure of Table 12.1, the Absolute IQ Algorithm of Internet (IQA) is given as:

$$\mathrm{IQA} = \sum_{\mathrm{i}=1}^{\mathrm{N}} \mathrm{Fi} \times \mathrm{Wi} \tag{12.1}$$

Where Fi is the evaluation index score (adopts the indexes of Table 12.1), Wi is the weight of evaluation index, and N is the number of evaluation index.

Table 12.1 2014 Version of Internet Intelligence Scale

First-class Index	Second-class Index	Description	Weight (%)
Ability of knowledge acquisition	Ability of character acquisition	Know about the testing object whether can understand and answer the testing question via characters. (Only one correct answer can be deemed pass)	3
	Ability of sound acquisition	Know about the testing object whether can understand and answer the testing question via sounds. (Only one correct answer can be deemed pass)	3
	Ability of picture acquisition	Know about the testing object whether can understand and answer the testing question via pictures. (Only one correct answer can be deemed pass)	4
bility of mastery of knowledge	Common knowledge	Know about the knowledge range of testing object. For example: *what's the name of three kinds of blood vessel for a human body?*	6
	Translate	Know about the testing object's transfer ability of the different languages. For example: *please translate "Machine Intelligence cannot exceed that of human beings" into English.*	3
	Calculate	Know about the calculation ability of the testing object, calculation speed and correctness. For example: *what is the result for 356*4 – 213?*	6
	Put in order	To know about the systemizing ability for the matters' relationship. For example: *please rank the commander, platoon leader, group commander, monitor, battalion commander, regimental commander by position.*	5
Ability of knowledge innovation	Associate	Know about the ability of observing similarities for the testing object. For example: *foot as for hand, is equivalent to leg as for what?*	12
	Create	Know about the ability of second creation according to the files. For example, *please tell a story with the key words of sky, rainbow, panda, mountain, hunter and so on.*	12
	Speculate	Know about the ability of speculating described things. *For example, there is one kind of animal that is similar to wolf, but is called as loyal friend of human being, then what is it?*	12%
	Select	Know about the testing object whether can select the same or different matter's relation. For example: *please select the different one among snake, tree and tiger.*	12
	Discover (laws)	Know about the testing object whether can discover the laws and apply them from the information or not. For example: *what is the figure after 1,2,4,7,11,16?*	12

(continued)

Table 12.1 (continued)

First-class Index	Second-class Index	Description	Weight (%)
Ability of feedback of knowledge	Ability of expressing via characters	Know about the testing object whether can express the testing results with characters. (Only one correct answer can be deemed pass)	3
	Ability of expressing via sounds	Know about the testing object whether can express the testing results with sounds. (Only one correct answer can be deemed pass)	3
	Ability of expressing via pictures	Know about the testing object whether can express the testing results with pictures. (Only one correct answer can be deemed pass)	4

Definition 12.4 Similarly, the deviation IQ Algorithm of Internet (**IQd**) can be expressed as:

$$\mathbf{IQd} = \mathbf{100} + \frac{\mathrm{IQ_A}\text{-}\overline{\mathrm{IQ_A}}}{S} \tag{12.2}$$

This formula is suitable for the IQ comparison among all the applications of Internet, highlighting the Internet testing object's position in Internet application. Under this circumstance, $\mathrm{IQ_A}$ is the average value of all applicative IQ in the Internet IQ evaluating bank (Table 12.1).

Let S be the standard deviation of all application in the Internet IQ evaluating bank, M is the number of all applications in the Internet IQ evaluating bank.

$$\mathrm{S} = \sqrt{\frac{1}{\mathrm{M}}\sum_{\mathrm{i=1}}^{\mathrm{M}}\left(\mathrm{IQ_A}\text{-}\overline{\mathrm{IQ_A}}\right)^2} \tag{12.3}$$

Search engine is one of the most important applications of Internet, whose representatives are Google, Baidu and Bing, etc. The working principle of search engine can automatically access to Internet with the help of a systematic procedure called Spider which can collect the web pages automatically [7]. The Spider can climb to other web pages along with all the URLs from any web pages. It repeats this process, and collects all the web pages it has climbed over. Then the system analyzes the collected web pages with the procedure of analysis index from the index database, extracts relevant web information according to certain correlation algorithm with a number of complex computations. After that it obtains the relevance or importance of the page content and hyperlinks from each key word of each web page [8].

When a user inputs a keyword search in the search rankings of an index database, the searching systematic procedure finds out all the related web pages. In the end, the system of page generating returns the page links and page abstracts of searching results to the user [9].

Google, Baidu and other types of search engines are improving the levels of intelligent search engines currently in a variety of ways to continuously, from only being able to identify texts to identify sounds and pictures. Through introducing "semantic understanding" technology, they try to understand the user's search intention and the computing arithmetic and structured display of searching results would be re-optimized, which would present the most accurate and comprehensive information to the user. With the help of deep learning, search engines are made to identify what the object is by the image automatically [10]. So according to the rules established by the Internet earlier IQ tests, the choice of IQ tests on search engines will have important significance.

According to Table 12.1, the following search engines IQ test question bank can be built. Based on the characteristics of different abilities, there are respectively one test question for the ability to obtain the knowledge and gain feedbacks, four questions for the ability to grasp knowledge and innovate it. With more in-depth study of the future, it will increase the number of test questions in order to improve the accuracy of the test.

The question bank for search engines is named *the 2014 version of the search engine intelligence test question bank*. The components are described as:

A. Ability of character acquisition

1. Use the input tool provided by the search engine, see whether one can input the character string "1 + 1 =?" and feedback the correct result or not.

B. Ability of sound acquisition

1. Tester reads "1 + 1 =?", check the input tool provided by the search engine can identify the correct result whether or not.

C. Ability of picture acquisition

1. Tester draws "1 + 1 =?" on a paper, check the input tool provided by the search engine can identify the correct result whether or not.

D. Ability of grasping the common knowledge

1. Which river is the longest in the world?
2. Which planet is the largest in the solar system?
3. How many chromosomes in human body?
4. What's the name of the first president of USA?

E. Ability of grasping the translation

1. Translate "力量(Liliang)" into English
2. Translate "力量(Liliang)" into Japanese
3. Translate "力量(Liliang)" into French
4. Translate "implications" into Chinese

F. Ability of grasping the calculation

1. How much is 25 multiply by 4?
2. How much is 36 divide 3?
3. How much is the biquadrate of 2?
4. How much is 128 extract three roots?

G. Ability of grasping the ranking

1. Please rank 34, 21, 56, 100, 4, 7, 9, 73 from small to large.
2. Please rank undergraduate, elementary student, middle school student, doctor, master from high education background to low education background.
3. Please rank Europe, the earth, France, Paris, Eiffel Tower from large to small via the area.
4. As for the same weight, please rank the price from expensive to low for gold, copper, silver, stone.

H. Ability of grasping the selection

1. Please select a different one from snake, tree, tiger, dog and rabbit.
2. Please select a different one from the earth, Mars, Venus, Mercury and the sun.
3. Please select a different one from red, green, blue, golden, yellow and white.
4. Please select a different one from car, train, airplane, steamer, and worker.

I. Ability of grasping the association

1. If associate birds with the sky, what can be associated with fishes?
2. If associate the son with the father, what can be associated with daughter?
3. If associate red with the sun, what can be associated with blue?
4. If associate the primary student with the primary school, what can be associated with universities?

J. Ability of grasping the creation

1. Please tell us a story by sky, rainbow, panda, mountain, and hunter and so on.
2. Please tell us a story by China, America, Russia and Japan.
3. Please tell us a story by red, tree, airplane, bullet, sun and so on.
4. Please tell us a story by 1, 2, 3, 4, 5.

K. Ability of grasping the speculation

1. If most of people are holding umbrellas in the street, with dropsy on thc ground, then what is the weather like at this time?
2. If one person wears high-heeled shoes, skirt, and with long hair, then what is the sex for this person probably?
3. If there are many animals in one place, but all in the cages, and many people are looking, then where is it?
4. If one person throws off his pen, but just float away around him, then where is he probably?

L. Ability of grasping the discovery of laws

1. Offer four questions, respectively are: $20/5 = 4$, $40/8 = 4$, $80/20 = 4$, $160/40 = 4$, observe the rules, then design the fifth question.
2. Cook A expresses that he likes to eat pork, mutton, beef, chicken, fish, but does not like Chinese cabbage, cucumber, green been, eggplant, potato, the please observe the rules, select the favorite food between duck meat and celery for this Cook.
3. On a certain regulation, the row numbers are $\frac{1}{2}, \frac{1}{3}, \frac{1}{10}, \frac{1}{15}, \frac{1}{26}, \frac{1}{35} \ldots$ for this rule, what is the seventh one in this series?
4. At every night, Company staff B goes home on Jan. 1st, goes the bar on Jan., 2nd, goes home on Jan. 3rd, goes the bar on Jan. 4th, goes home on Jan. 5th, goes the bar on Jan. 6th, goes home on Jan. 7th, goes the bar on Jan. 8th, where B may present on Feb. 13th probably?

M. Ability of expressing via characters

1. Input the character string "How much is 1 plus 1, please answer via characters", check the testing search engine whether can express the answer via characters or not.

N. Ability of expressing via sounds

1. Input the character string "How much is 1 plus 1, please answer via sounds", check the testing search engine whether can express the answer via sounds or not.

O. Ability of expressing via pictures

1. Input the character string "How much is 1 plus 1, please answer via pictures", check the testing search engine whether can express the answer via pictures or not

12.1.1.1 A Small Sample of AI-IQ Test

For an experimental study of IQ test on search engine, 7 well-known search engines: Google.com.hk, Baidu.com, Sogou.com, Bing.com, Zhongsou.com, panguso.com, so.com are chosen as the samples of search engine to conduct the IQ test. The testing principle is to carry out the testing via Table 12.1 with regard to the whole testing questions. If one cannot input the question into the testing search engine, this score will be 0, and if one can input the question into the search engine, which cannot shows the correct results in the first try or the time of answering is over 3 min in the first search engine, the score will be 0. According to the rules of 2014 Internet Intelligent Scale, in the test 1, 2, 3 and 13, 14, 15, there is only one question, if one can answer correctly in 3 min, each question can get 100; as for other testing, if one can answer correctly in 3 min, each question may get 25. The testing environment is Winxp System, IE9 explorer (Chinese version). The testing results are shown as Table 12.2.

Then, we carry out the IQ test for 20 Children of 6 ages, 12 ages and 18 ages via the same rules, and obtain the results as in Table 12.3.

According to the weight rules of Table 12.1, the absolute IQ and relative IQ scores for 7 search engines and 20 children of 3 different ages are calculated as in Table 12.4 (note that the absolute IQ's full mark is 100).

12.1.2 A Data Mining for Features of AI-IQ Test

12.1.2.1 A Large Sample of AI-IQ Test

Based on the above discussion, a data mining method is applied to find out the features of search engines reflected in the Internet intelligence test and the intelligence difference between search engines and human beings.

Table 12.2 Results of Seven search engines IQ Test

	Google	Baidu	Sogou	Bing	so	panguso	Zhongsou	Weight (%)
Ability of character acquisition	100	100	100	100	100	100	100	3
Ability of sound acquisition	0	0	0	0	100	0	0	3
Ability of picture acquisition	0	100	100	0	100	0	0	4
Common knowledge	100	100	100	100	100	75	50	6
Translate	100	75	50	50	50	0	0	3
Calculate	**100**	**100**	100	**25**	**75**	**75**	**50**	6
Put in order	0	0	0	0	0	0	0	5
Association	0	0	0	0	0	0	0	12
Create	0	0	0	0	0	0	0	12
Speculate	0	0	0	0	0	0	0	12
Select	0	0	0	0	0	0	0	12
Discover (laws)	0	0	0	0	0	0	0	12
Ability of expressing via characters	100	100	100	100	100	100	100	3
Ability of expressing via pictures	0	0	0	0	0	0	0	4

In order to show the meaning of the data mining method, 50 typical search engines across the world are first tested by the scale of Table 12.1. They include Google, Baidu, Bing, eMaxia, Anzswers, Pictu, Saja search, and 1stcyprus from 25 countries and regions, including China, America, India, the United Kingdom, Russia, Japan, Australia and so on. If any question in the test bank cannot be entered into a search engine, zero score will be given to the search engine. If a question can be entered into a search engine, but the correct result is not included in the first search result, or the time spent on answering the question is more than 3 min, zero score should be given to the search engine, too. According to the 2014 Internet Intelligence Scale, there is only one question in test items 1, 2, 4, 13, 14 and 15, if a correct answer is given within 3 min, a score of 100 may be obtained by each. And for the questions in other test items, if they are answered correctly by a search engine within 3 min, a score of 25 may be given to that search engine. The test environment is WinXP system and IE9 browser (Chinese version). The test results are shown in Table 12.5. Meanwhile, the same rules are used to test 150 people who are grouped by the age of 6, 12 and 18, 50 people for each group in Table 12.6.

According to the weight rules of Table 12.1, the Absolute IQs and the Relative IQs of the 50 search engines and three groups of people are calculated and the

Table 12.3 Results of 20 children IQ Test

	6 Ages (average value)	12 Ages (average value)	18 Ages (average value)
Ability of character acquisition	100	100	100
Ability of sound acquisition	100	100	100
Ability of picture acquisition	100	100	100
Common knowledge	25	25	75
Translate	0	25	50
Calculate	**25**	**75**	**100**
Put in order	50	75	100
Association	50	75	100
Create	50	100	100
Speculate	75	100	100
Select	50	100	100
Discover (laws)	25	75	100
Ability of expressing via characters	100	100	100
Ability of expressing via sounds	100	100	100
Ability of expressing via pictures	100	100	100

results are ranked in a descending order, as shown in Table 12.7. Note that the K values were respectively taken as 3, 4 and 5, the clustering results within the cluster sum of squared errors were respectively 23.5, 13.4 and 9.6. The clustering results within the cluster sum of squared errors were respectively 23.5, 13.4 and 9.6.

Clustering Analysis

Firstly, in order to obtain the referable relationship between the 53 test objectives and the 15 test items, all the 795 pieces of test data in Tables 12.5 and 12.6 are analyzed in the software weka 3.6 by using the K-means clustering algorithm. It respectively takes K values as 3, 4, and 5. When the K value is chosen as 3; the search engines are classified as a, b, and c. Similarly, when the K value is chosen as 4, the search engines are classified as a, b, c, and d; and when the K is equal to 5, the clusters are denoted as a, b, c, d, and e. The clustering results of search engines depending on the K values which are also shown in Table 12.7.

Classification Analysis

As we know, there is no good evaluation criterion for a typical clustering problem. In order to evaluate the effectiveness of clustering result, we employ various classification algorithms on labeled data. Then, the evaluation of classification results can be seen as an indirect evaluation for clustering result. Consequently, in this part, according to the clustering result obtained in the above part, classification

Table 12.4 Absolute IQ/relative IQ scores

	Google	Baidu	Sogou	Bing	so	panguso	Zhongsou	6 Ages	12 Ages	18 Ages
Absolute IQ	21	24.25	23.5	15	25	15	12	55.5	85.25	97
Relative IQ	99.34	99.44	99.43	99.14	99.48	99.14	99.04	100.51	101.53	101.92

Table 12.5 Results of Seven search engines IQ Test

		1	2	3	4	5	6	7	8	9	10	11	12	13	14	15
China	Baidu	100	0	100	100	50	100	0	0	0	0	0	0	100	0	0
China	Sogou	100	0	100	75	50	100	0	0	0	0	0	0	100	0	0
China	so	100	100	100	75	50	75	0	0	0	0	0	0	100	0	0
China	panguso	100	0	0	75	0	75	0	0	0	0	0	0	100	0	0
China	Zhongsou	100	0	0	50	0	50	0	0	0	0	0	0	100	0	0
Hong Kong	timway	100	0	0	75	25	25	0	0	0	0	0	0	100	0	0
Greece	Gogreece	100	0	0	0	0	0	0	0	0	0	0	0	100	0	0
Holland	slider	100	0	0	0	0	0	0	0	0	0	0	0	100	0	0
Norway	Sunsteam	100	0	0	0	0	0	0	0	0	0	0	0	100	0	0
Egypt	yell	100	100	100	50	50	50	0	0	0	0	0	0	100	0	0
Egypt	netegypt	100	0	0	0	0	0	0	0	0	0	0	0	100	0	0
USA	yahoo	100	0	0	75	100	0	0	0	0	0	0	0	100	0	0
USA	Dogpile	100	100	0	0	0	0	0	0	0	0	0	0	100	0	0
USA	google	100	100	100	75	100	100	0	0	0	0	0	0	100	0	0
USA	bing	100	0	0	75	50	25	0	0	0	0	0	0	100	0	0
Solomons	eMaxia	100	0	0	0	0	0	0	0	0	0	0	0	100	0	0
Australia	Anzswers	100	0	0	0	0	0	0	0	0	0	0	0	100	0	0
Australia	Pictu	100	0	0	0	0	0	0	0	0	0	0	0	100	0	0
Malaysia	Sajasearch	100	0	0	0	0	0	0	0	0	0	0	0	100	0	0
New Zealand	SerachNZ	100	0	0	0	0	0	0	0	0	0	0	0	100	0	0
India	indiabook	100	0	0	0	0	0	0	0	0	0	0	0	100	0	0
India	khoj	100	0	0	75	0	0	0	0	0	0	0	0	100	0	0
Britain	ask	100	0	0	75	50	0	0	0	0	0	0	0	100	0	0

Britain	Excite UK	100	0	0	0	0	0	0	0	0	0	0	0	100	0	0
Britain	splut	100	0	0	0	0	0	0	0	0	0	0	0	100	0	0
France	voila	100	0	0	75	50	0	0	0	0	0	0	0	100	0	0
France	lycos	100	0	0	100	0	0	0	0	0	0	0	0	100	0	0
Russia	Yandex	100	0	100	100	100	0	0	0	0	0	0	0	100	0	0
Russia	ramber	100	100	0	75	100	25	0	0	0	0	0	0	100	0	0
Russia	webalta	100	0	0	75	100	0	0	0	0	0	0	0	100	0	0
Russia	Rol	100	0	0	0	0	0	0	0	0	0	0	0	100	0	0
Russia	Km	100	0	0	75	0	0	0	0	0	0	0	0	100	0	0
Spain	ciao	100	0	0	0	0	0	0	0	0	0	0	0	100	0	0
Spain	His	100	0	0	100	50	75	0	0	0	0	0	0	100	0	0
Korea	naver	100	0	0	75	100	25	0	0	0	0	0	0	100	0	0
Korea	nate	100	0	0	75	75	50	0	0	0	0	0	0	100	0	0
UAE	Arabo	100	0	0	50	75	75	0	0	0	0	0	0	100	0	0
Czech	seznam	100	0	0	100	0	100	0	0	0	0	0	0	100	0	0
Portugal	clix	100	0	0	100	0	75	0	0	0	0	0	0	100	0	0
Portugal	sapo	100	0	0	100	0	0	0	0	0	0	0	0	100	0	0
Japan	goo	100	0	0	75	75	0	0	0	0	0	0	0	100	0	0
Japan	excite	100	0	0	75	75	0	0	0	0	0	0	0	100	0	0
Cyprus	1stcyprus	100	0	0	0	0	0	0	0	0	0	0	0	100	0	0
Germany	fireball	100	0	0	0	0	0	0	0	0	0	0	0	100	0	0
Germany	bellnet	100	0	0	0	0	0	0	0	0	0	0	0	100	0	0
Germany	Acoon	100	0	0	25	0	0	0	0	0	0	0	0	100	0	0
Germany	lycos	100	0	0	100	0	0	0	0	0	0	0	0	100	0	0
Germany	slider	100	0	0	0	0	0	0	0	0	0	0	0	100	0	0
Germany	wlw	100	0	0	0	0	0	0	0	0	0	0	0	100	0	0
Germany	suche	100	0	0	75	0	0	0	0	0	0	0	0	100	0	0

Table 12.6 Results of 20 children IQ Test

		1	2	3	4	5	6	7	8	9	10	11	12	13	14	15
Human	6Ages	100	100	100	25	0	25	50	50	50	75	50	25	100	100	100
Human	12Ages	100	100	100	25	25	75	75	75	100	100	100	75	100	100	100
human	18Ages	100	100	100	75	50	100	100	100	100	100	100	100	100	100	100

Table 12.7 Absolute IQ/relative IQ scores

				Absolute IQ	Relative IQ scores	K = 3	K = 4	K = 5
1		Human	18Ages	97	104.85	b	d	e
2		Human	12Ages	84.5	104.11	b	d	e
3		Human	6Ages	55.5	102.39	b	d	e
4	America	USA	google	26.5	102.13	b	a	a
5	Asia	China	Baidu	23.5	101.69	a	a	a
6	Asia	China	so	23.5	101.69	b	a	a
7	Asia	China	Sogou	22	101.41	a	a	a
8	Africa	Egypt	yell	20.5	100.32	b	a	a
9	Europe	Russia	Yandex	19	100.23	a	c	d
10	Europe	Russia	ramber	18	100.17	a	c	d
11	Europe	Spain	His	18	100.17	a	a	b
12	Europe	Czech	seznam	18	100.17	a	a	b
13	Europe	Portugal	clix	16.5	100.08	a	a	b
14	Asia	Korea	nate	15.75	100.03	a	c	d
15	Asia	UAE	Arabo	15.75	100.03	a	c	d
16	Asia	China	panguso	15	99.99	a	a	b
17	Asia	Korea	naver	15	99.99	a	c	d
18	Europe	Russia	webalta	13.5	99.9	a	c	d
19	America	USA	yahoo	13.5	99.9	a	c	d
20	America	USA	bing	13.5	99.9	a	c	d

(continued)

Table 12.7 (continued)

				Absolute IQ	Relative IQ scores	K = 3	K = 4	K = 5
21	Asia	Hong Kong	timway	12.75	99.86	a	a	b
22	Asia	Japan	goo	12.75	99.86	a	c	d
23	Asia	Japan	excite	12.75	99.86	a	c	d
24	Asia	China	Zhongsou	12	99.81	a	a	b
25	Europe	Britain	ask	12	99.81	a	c	d
26	Europe	France	voila	12	99.81	a	c	d
27	Europe	France	ycos	12	99.81	a	a	b
28	Europe	Portugal	sapo	12	99.81	a	a	b
29	Europe	Germany	lycos	12	99.81	a	a	b
30	Asia	India	khoj	10.5	99.72	a	a	b
31	Europe	Russia	Km	10.5	99.72	a	a	b
32	Europe	Germany	suche	10.5	99.72	a	a	b
33	America	USA	Dogpile	9	99.63	c	b	c
34	Europe	Germany	Acoon	7.5	99.55	c	b	c
35	Asia	Malaysia	Sajasearch	6	99.46	c	b	c
36	Asia	India	indiabook	6	99.46	c	b	c
37	Asia	Cyprus	1stcyprus	6	99.46	c	b	c
38	Europe	Greece	Gogreece	6	99.46	c	b	c
39	Europe	Holland	slider	6	99.46	c	b	c
40	Europe	Norway	Sunsteam	6	99.46	c	b	c
41	Europe	Britain	Excite UK	6	99.46	c	b	c
42	Europe	Britain	splut	6	99.46	c	b	C
43	Europe	Russia	Rol	6	99.46	c	b	C
44	Europe	Spain	ciao	6	99.46	c	b	C
45	Europe	Germany	fireball	6	99.46	c	b	C

46	Europe	Germany	bellnet	6	99.46	c	b	C
47	Europe	Germany	slider	6	99.46	c	b	C
48	Europe	Germany	wlw	6	99.46	c	b	C
49	Africa	Egypt	netegypt	6	99.46	c	b	C
50	Oceania	Solomons	eMaxia	6	99.46	c	b	C
51	Oceania	Australia	Anzswers	6	99.46	c	b	C
52	Oceania	Australia	Pictu	6	99.46	c	b	C
53	Oceania	New Zealand	SerachNZ	6	99.46	c	b	C

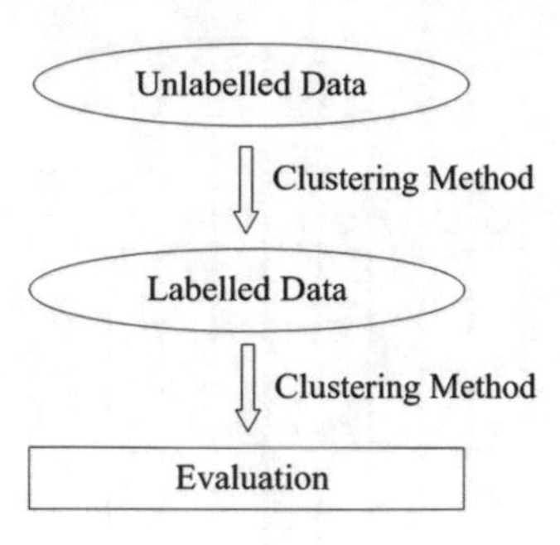

Fig. 12.1 Outline of clustering result evaluation

Table 12.8 In-sample result in original data

Algorithms	NB	DT (J48)	LR	KNN (K = 2)	SVM (SVM Linear)	NN (3 hidden layers)
Accuracy	100%	97.92%	100%	100%	100%	97.92%

algorithms are employed to evaluate and utilize the clustering results. The outline of this idea can be expressed in Fig. 12.1. Firstly, according to the IQ score and the result of $K = 3$, there are some intersections between group "a" and "b". As a result, for simplicity, it groups "a" and "b" as the same class, which can be labeled as $+1$. Then, the search engines corresponding to group c are labeled as -1. Then, it becomes a typical binary classification problem.

12.1.2.2 In-Sample Experiment

The result obtaining from the clustering can be evaluated as follows. One can know that a distinctly distinguishing classification problem will result in a high accuracy by using typical classification algorithms, such as Naïve Bayes (NB), Decision Tree (DT), Logistic Regression (LR), K Nearest Neighbor (KNN), Support Vector Machine (SVM) and Neural Network (NN). In other words, high accuracy results of these typical classification algorithms will indirectly and partially guarantee the high degree distinctness of clustering result. Based on this kind of thinking, we test our data in these 5 stable methods. The results can be found in Table 12.8. Here, fivefold cross validation is chosen to make the result more reliable and reasonable.

In Table 12.8, in-sample accuracies in various algorithms are all very high, which means we can partially depend on the clustering result and do prediction based on this result.

12.1.2.3 Out-Sample Experiment

This part shows the ability of generalization of the method. It finishes the IQ test on other 31 engines as new dataset. Furthermore, according to the clustering result on

Table 12.9 Out-sample result in original data

Algorithms	NB	DT (J48)	LR	KNN (K = 2)	SVM (SVM Linear)	NN (3 hidden layers)
Accuracy	90.32%	83.87%	83.87%	93.55%	77.42%	83.87%

original data, we arrange new data into three different clusters by Euclidean distance based on IQ test result. Then, taking these 31 engines as validation set, the original dataset containing 48 points as training set, we show the validation accuracies of various kinds of classification algorithms. According to the validation result in Table 12.9, the overall prediction accuracy is acceptable, which means the generalization of our method is reasonable and reliable.

12.1.3 A Standard Intelligence Model

This subsection proposes a standard intelligence model that unifies AI and human characteristics in terms of four aspects of knowledge, i.e., input, output, mastery, and creation. The model is established based on the theories of the von Neumann architecture, David Wechsler's human intelligence model, knowledge management using data, information, knowledge and wisdom (DIKW), and other related approaches.

The von Neumann architecture provided the inspiration that a standard intelligence system model should include an input/output (I/O) system that can obtain information from the outside world and feed results generated internally back to the outside world. In this way, the standard intelligence system can become a "live" system [11].

David Wechsler's definition of human intelligence led us to conceptualize intellectual ability as consisting of multiple factors; this is in opposition to the standard Turing test or visual Turing test paradigms, which only consider singular aspects of intellectual ability [12].

The DIKW model further led us to categorize wisdom as the ability to solve problems and accumulate knowledge, i.e., structured data and information obtained through constant interactions with the outside world. An intelligent system would not only master knowledge, it would have the innovative ability to be able to solve problems [13]. The ideas of knowledge mastery ability, being able to innovatively solve problems, David Wechsler's theory, and the von Neumann architecture can be combined, therefore we proposed a multilevel structure of the intellectual ability of an intelligent system–a "standard intelligence model," as shown in Fig. 12.2 [14].

In the basis of this research, the following criteria for defining a standard intelligence system are discussed. If a system (either an artificially intelligent system or a living system such as a human) has the following characteristics, it can be defined as a standard intelligence system:

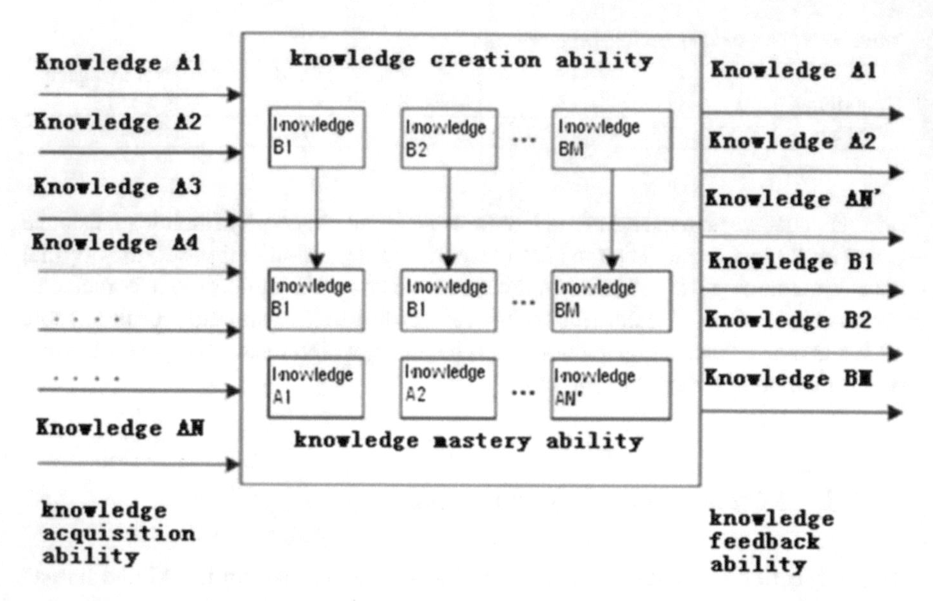

Fig. 12.2 The standard intelligence model

Characteristic 12.1 the system has the ability to obtain data, information, and knowledge from the outside world from aural, image, and/or textual input (such knowledge transfer includes, but is not limited to, these three modes);

Characteristic 12.2 the system has the ability to transform such external data, information, and knowledge into internal knowledge that the system can master;

Characteristic 12.3 based on demand generated by external data, information, and knowledge, the system has the ability to use its own knowledge in an innovative manner. This innovative ability includes, but is not limited to, the ability to associate, create, imagine, discover, etc. New knowledge can be formed and obtained by the system through the use of this ability;

Characteristic 12.4 the system has the ability to feed data, information, and knowledge produced by the system feedback the outside world through aural, image, or textual output (in ways that include, but are not limited to, these three modes), allowing the system to amend the outside world.

12.1.3.1 Extensions of the von Neumann Architecture

The von Neumann architecture is an important reference point in the establishment of the standard intelligence model. Von Neumann architecture has five components: an arithmetic logic unit, a control unit, a memory unit, an input unit, and an output unit. By adding two new components to this architecture (compare Figs. 12.2 and 12.3), it is possible to express human, machine, and artificial intelligence systems in a more explicit way.

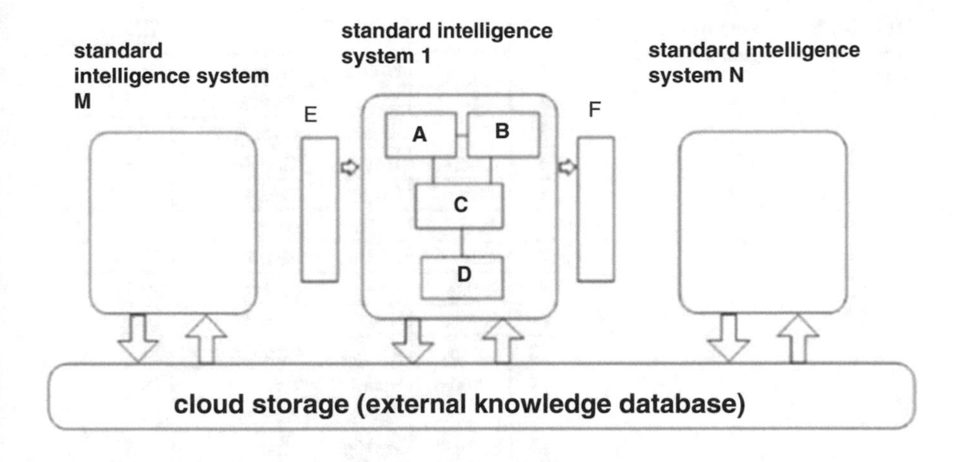

A. arithmetic logic unit
B. control unit
C. internal memory unit
D. innovation generator
E. input device
F. output device

Fig. 12.3 Expanded von Neumann architecture

The first added component is an innovative and creative function, which can find new knowledge elements and rules through the study of existing knowledge and save these into a memory used by the computer, controller, and I/O system. Based on this, the I/O can interact and exchange knowledge with the outside world. The second additional component is an external knowledge database or cloud storage that can carry out knowledge sharing. This represents an expansion of the external storage of the traditional von Neumann architecture, which is only for single systems (see Fig. 12.3).

Definition 12.5 A unified model of intelligent systems has four major characteristics, namely, the abilities to acquire, master, create, and feedback knowledge. The evaluation of the intelligence and developmental level of an intelligent system can be done by testing these four characteristics simultaneously.

The IQ of an artificial intelligence (AI-IQ) is based on a scaling and testing method defined according to the standard intelligence model. Such tests evaluate intelligence development levels, or grades, of intelligent systems at the time of testing, with the results delineating the AI IQ of the system at testing time [1].

Definition 12.6 A mathematical formula for AI IQ is given as:

$$Level1 : M \xrightarrow{f} Q, \;\; Q = f(M) \tag{12.4}$$

Here, M represents an intelligent system, Q is the IQ of the intelligent system, and f is a function of the IQ.

Table 12.10 Ranking of top 13 artificial intelligence IQs in 2014

				Absolute IQ
1		Human	18 years old	97
2		Human	12 years old	84.5
3		Human	6 years old	55.5
4	America	America	Google	26.5
5	Asia	China	Baidu	23.5
6	Asia	China	so	23.5
7	Asia	China	Sogou	22
8	Africa	Egypt	yell	20.5
9	Europe	Russia	Yandex	19
10	Europe	Russia	ramber	18
11	Europe	Spain	His	18
12	Europe	Czech	seznam	18
13	Europe	Portugal	clix	16.5

In general, an intelligent system M should have four kinds of ability: knowledge acquisition (information acceptance ability), which we denote as I; knowledge output ability, or O; knowledge mastery and storage ability, S; and knowledge creation ability, C. The AI IQ of a system is determined based upon a comprehensive evaluation of these four types of ability. As these four ability parameters can have different weights, a linear decomposition of IQ function can be expressed as follows:

$$\begin{aligned} &Q = f(M) = f(I, O, S, C) = a * f(I) + b * f(O) + c * f(S) + d * f(C) \\ &a + b + c + d = 100\% \end{aligned} \tag{12.5}$$

Based on this unified model of intelligent systems, an artificial intelligence IQ evaluation system can be established in 2014. By considering the four major ability types, 15 sub-tests were carried out and an artificial intelligence scale is formed. This scale is used to set up relevant question databases, tested 50 search engines and humans from three different age groups, and formed a ranking list of the AI IQs for that year [1] (see Sect. 12.2). Table 12.10 shows the top 13 AI IQs.

In 2016, the update AI-IQ tests for artificially intelligent systems was conducted again in evaluating the artificial intelligence systems of Google, Baidu, Sogou, and others as well as Apple's Siri and Microsoft's Xiaobing. The results indicate that the artificial intelligence systems produced by Google, Baidu, and others have significantly improved over the past 2 years but still have certain gaps as compared with even a 6-year-old child (see Table 12.11).

IQ essentially is a measurement of the ability and efficiency of intelligent systems in terms of knowledge mastery, learning, use, and creation. Therefore, IQ can be represented by different knowledge grades:

Table 12.11 IQ scores of artificial intelligence systems in 2016

				Absolute IQ
1	2014	Human	18 years old	97
2	2014	Human	12 years old	84.5
3	2014	Human	6 years old	55.5
4	America	America	Google	47.28
5	Asia	China	duer	37.2
6	Asia	China	Baidu	32.92
7	Asia	China	Sogou	32.25
8	America	America	Bing	31.98
9	America	America	Microsoft's Xiaobing	24.48
10	America	America	SIRI	23.94

Definition 12.7 A model of intelligence grade of artificial intelligence is given below:

$$\begin{aligned} &Level2 : Q \xrightarrow{\chi} K, K \in \{0, 1, 2, 3, 4, 5, 6\} \\ &K = \chi(Q) = \chi\left(f(M)\right) \end{aligned} \tag{12.6}$$

There are different intelligence and knowledge grades in human society. For instance, grades in the educational system such as undergraduate, master, doctor, as well as assistant researcher, associate professor, and professor. People within a given grade can differ in terms of their abilities; however, moving to a higher grade generally involves passing tests in order to demonstrate that watershed levels of knowledge, ability, qualifications, etc., have been surpassed.

How can key differences among the functions of intelligent systems be defined? The "standard intelligence model" (i.e., the expanded von Neumann architecture) can be used to inspire the following criteria:

- Can the system exchange information with (human) testers? Namely, does it have an I/O system?
- Is there an internal knowledge database in the system to store information and knowledge?
- Can the knowledge database update and expand?
- Can the knowledge database share knowledge with other artificial intelligence systems?
- In addition to learning from the outside world and updating its own knowledge database, can the system take the initiative to produce new knowledge and share this knowledge with other artificial intelligence systems?

Using the above criteria, a seven intelligence grades is presented by using mathematical formalism (see Table 12.12) to describe the intelligence quotient, Q, and the intelligence grade state, K, where K = {0, 1, 2, 3, 4, 5, 6}.

The different grades of K are described in Table 12.12 as follows.

The detailed explanation for the meaning of seven levels can be found in [2].

Table 12.12 Intelligence grades of intelligent systems

Intelligence grade	Mathematical conditions
0	Case 1, f(I) > 0, f(o) = 0; Case 2, f(I) = 0, f(o) > 0
1	f(I) = 0, f(o) = 0
2.	f(I) > 0, f(o) > 0, f(S)=α > 0, f(C) = 0; where α is a fixed value, and system M's knowledge cannot be shared by other M.
3	f(I) > 0, f(o) > 0,f(S) = α > 0, f(C) = 0; Where α increases with time.
4	f(I) > 0, f(o) > 0, f(S) = α > 0, f(C) = 0; where α increases with time, and M's knowledge can be shared by other M.
5	f(I) > 0, f(o) > 0, f(S) = α > 0, f(C) > 0; where α increases with time, and M's knowledge can be shared by other M.
6	f(I) > 0 and approaches infinity, f(o) > 0 and approaches infinity, f(S) > 0 and approaches infinity, f(C) > 0 and approaches infinity.

The research in the line of AI-IQ has some important implementations. For example, Fig. 12.4 shows a possible relationship between AI and Human intelligence. Here curve B indicates a gradual increase in human intelligence over time. There are two possible developments in artificial intelligence: curve A shows a rapid increase in the AI IQ, which is above the human IQ at a certain point in time. Curve C indicates that the AI IQ will be infinitely close to the human IQ but cannot exceed it. By conducting tests of the AI IQ, we can continue to analyze and determine the curve that shows a better evolution path of the AI IQ.

12.2 Laws of Intelligence Based on AI IQ Research

The subsection provides three laws of intelligence for interpreting the concepts of intelligence, wisdom, consciousness, life and non-life. The first law is called "M Law of Intelligence". The second law is called "Ω Law of Intelligence". The third law is called "A Law of Intelligence". The Three Laws need to be validated by a biochemical experiment method, an AI system intelligence evaluation experiment method or the computer program simulation experiment method.

To illustrate the laws, the following symbol stipulation on related concepts are used:

***Symbol 1**: U stands for the entire Universe*

***Symbol 2**: a stands for an individual Agent, and A for the set of all individual Agents in Universe, $a \in A$*

***Symbol 3**: E_a stands for the environment that affects the survival of Agent a, that is, the entire environment that can interact with Agent a.*

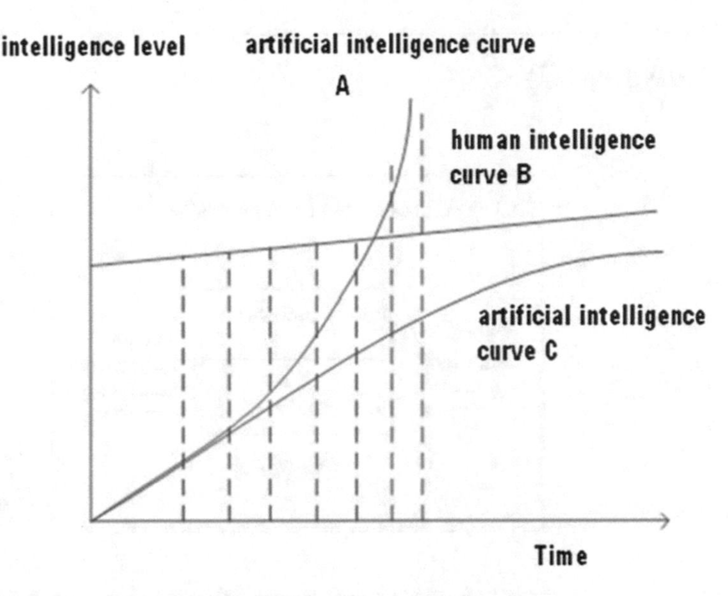

Fig. 12.4 Developmental curves of artificial and human intelligence

Symbol 4: *K(X) stands for the set of knowledge that can be processed or contained by X. For example, K(a) stands for a set of knowledge that can be processed or contained by Agent a; K(U) stands for all sets of knowledge that can be processed or contained by Universe, and K(E_a) stands for all sets of knowledge that can be processed or contained by the environment where Agent a exists.*

12.2.1 Law of Intelligent Model (M Law)

The first law of intelligence is called Law of Intelligent Model, namely M Law. The goal of this law is to establish a unified model, used to describe the key features of any Agent, and it is detailed as follows:

Definition 12.8 Any Agent can be regarded as a system with abilities to input, output, storage(master), creative(innovate) knowledge, and the difference between Agents is that different Agents have different abilities to process knowledge with these four functions.

The quaternary mathematical expression of **Law of Intelligent Model** is:

$$a = (I_a, O_a, S_a, C_a) \tag{12.7}$$

In this mathematical description, $a \in A$ stands for any Agent.
K stands for a knowledge set.

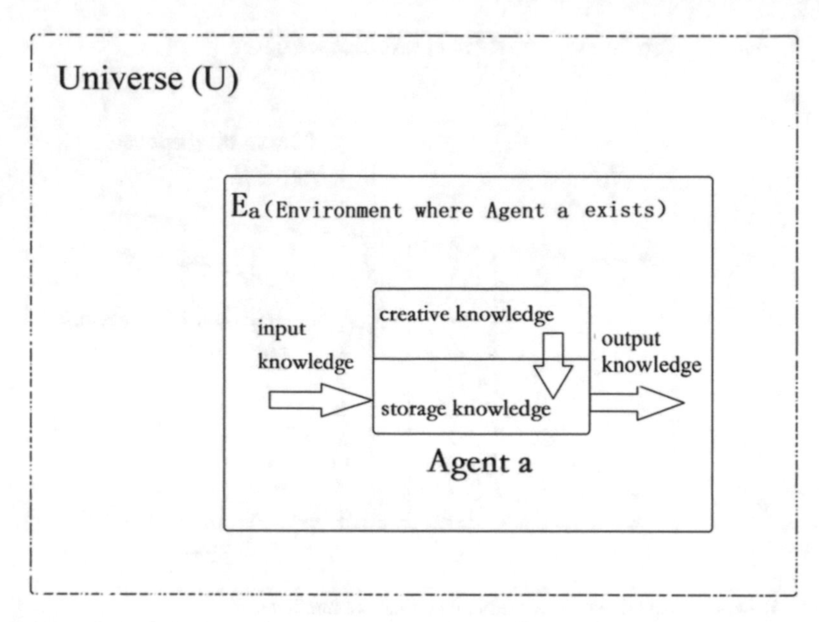

Fig. 12.5 Schematic diagram of the M Law of Intelligence

I_a stands for the ability of Agent a to input knowledge from its environment (E_a).

O_a stands for the ability of Agent a to output knowledge, and the result is the effect on its environment (E_a) (including other Agents).

S_a stands for the ability of Agent a to translate the input knowledge and its own innovative knowledge into storage or mastery of knowledge.

C_a stands for the ability of Agent a to creative or innovate knowledge based on the input and mastery of knowledge.

The four abilities of Agents to process knowledge are respectively between 0 and infinity. The set $K(a)$ of knowledge that any Agent can process is the union of the knowledge sets that the above four abilities can process. Its mathematical expression is: $0 \leq |K(I_a)|, |K(O_a)|, |K(S_a)|, |K(C_a)| \leq \infty$, $K(a) = K(I_a) \cup K(O_a) \cup K(S_a) \cup K(C_a)$.

The illustration of M Law of Intelligence is shown in Fig. 12.5.

According to the M Law of Intelligence, the definitions of the following five concepts may be proposed, which will play an important role in the proposal of subsequent laws.

12.2.2 Absolute 0 Agents (α_{point})

According to the mathematical description of Standard Intelligent Model, i.e. $a = (I_a, O_a, S_a, C_a)$, it can be seen that when the input, output, storage(mastery) and creative(innovation) abilities of an Agent equal to zero, a special state of the Agent

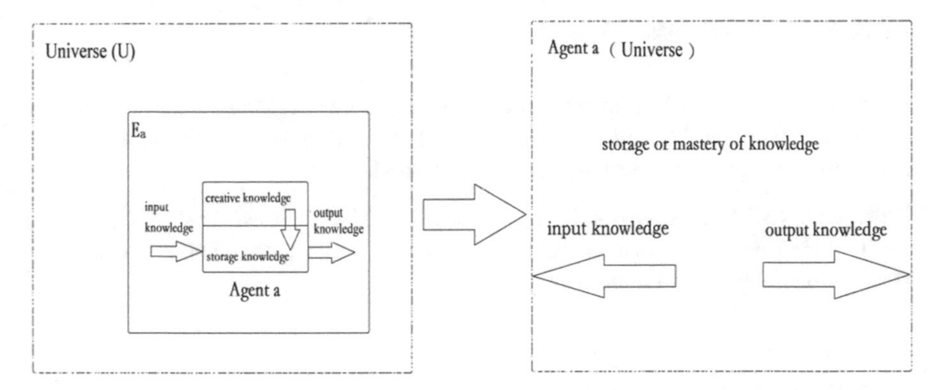

Fig. 12.6 Schematic diagram of the formation of omniscient and omnipotent agent

will form, which is the reason for the proposal of the absolute 0 Agent (α_{point}). It notes the final boundary of Agent's ability to change to infinitesimal.

Definition 12.9 $\exists a \in A$, $|K(I_a)| = 0$, $|K(O_a)| = 0$, $|K(S_a)| = 0$ and $|K(C_a)| = 0$, a is an absolute 0 Agent, denoted as α_{point}, and the set they form is denoted as A_{POINT}.

12.2.3 Omniscient and Omnipotent Agents (Ω_{point})

As discussed above, when the Agent's ability to process knowledge converges to the "0" state, the Agent will become an absolute 0 Agent. Similarly, for Agent a, when its abilities to input, output, storage(master) and creative(innovate) knowledge equal to infinity, another special state will form, which is why the Omniscient and Omnipotent Agent is proposed. The proposal of Omniscient and Omnipotent Agent presents the final boundary of the Agent's ability to change to infinity.

There will be a special situation in the process that Agent a forms an Omniscient and Omnipotent Agent. Specifically, while Agent a's abilities to innovate, input, output and master knowledge are approaching infinity, once the ability to master knowledge equals to infinity, the innovation ability is sure to be zero, otherwise, It will be paradoxical relative to the fact that Agent a's ability to master knowledge is infinite, as shown in Fig. 12.6.

At the same time, because for Agent a, there should be no "external" concept, so for Agent a, both input and output will occur inside it, as shown in Fig. 12.6. Basis on the above, it can be concluded that if there is an Omniscient and Omnipotent Agent, there can only be one. It is mathematically described as follows:

Definition 12.10 $\exists a \in A$, $|K(I_a)| = \infty$, $|K(O_a)| = \infty$ and $|K(S_a)| = \infty$, $|K(C_a)| = 0$, a is an Omniscient and Omnipotent Agent, denoted as Ω_{point}, there can only be one Ω_{point}.

12.2.4 Conventional Agent (a_C)

Given Agent a, it is neither an absolute 0 Agent, nor an Omniscient and Omnipotent Agent, that is, its ability to process knowledge is between 0 and infinity, then a is a conventional Agent, mathematically described as follows:

Definition 12.11 $\exists a \in A$, $0 < |K(I_a)| < \infty$, $0 < |K(C_a)| < \infty$, $0 < |K(O_a)| < \infty$ and $0 < |K(S_a)| < \infty$, a is a conventional Agent, denoted as a_C. The set formed by conventional Agents is denoted as A_C.

12.2.5 Relative 0 Agent (a_R)

For any two conventional Agents, if there is no intersection between the sets formed by the knowledge processed by them, then they are mutually relative *0* Agents, mathematically described as follows:

Definition 12.12 $\exists a_i$, $a_{i+1} \in A_N$, $K(a_i) \cap K(a_{i+1}) = \emptyset$, a_i and a_{i+1} are mutually relative 0 Agents, denoted as a_R, i.e. a_i is the a_R of a_{i+1}, similarly, a_{i+1} is the a_R of a_i. The relative 0 Agent set of an Agent is denoted as A_R.

The existence of relative 0 Agents indicates that even if two Agents are not absolute 0 Agents, they will also treat each other as an absolute 0 Agent as there is no way for them to exchange or share knowledge.

12.2.6 Shared Agent ($a_G$or A_G)

a_1, a_2, a_3, ... a_J are all Agents. If at least one knowledge element k_i is same in the knowledge sets of these agents, then they constitute a shared Agent (set). According to the definition of Standard Intelligent Model, the shared Agent (set) can also be regarded as an Agent, mathematically described as follows:

Definition 12.13 $\exists a_1, a_2, a_3, \ldots a_J \in A$, $K(a_1) \cap K(a_2) \cap K(a_3) \ldots \cap K(a_J) \neq \emptyset$, the system formed by $a_1, a_2, a_3, \ldots a_J$ can be called a shared Agent. A shared Agent may be a set, denoted as A_G, or it may be an Agent, denoted as a_G.

The shared Agent is a larger intelligent system formed by different Agents through the sharing and exchange of knowledge. This will be of great significance and value to all the Agents that constitute the shared Agent, allowing an individual Agent to have stronger ability to process knowledge.

Specially, if all Agents in an Agents set are absolute *0* Agents, then these Agents form a special shared Agent, and we call it as Absolute *0* shared Agent, which is mathematically described as follows:

Definition 12.14 $\exists a_1, a_2, a_3, \ldots a_J \in A_{point}$, The system formed by $a_1, a_2, a_3, \ldots a_J$ can be called as Absolute 0 shared Agent, which is also an Absolute 0 Agent.

12.2.7 Universe Agent (a_U)

We observe that:

1. If an agent a evolves into Omniscient and Omnipotent Agent, by definition, this agent will expand to the entire universe at this time, that is, the universe can be regarded as Omniscient and Omnipotent Agent at this time;
2. If all agents in the universe are Absolute *0* agents, the universe can be regarded as Absolute *0* agent according to the definition of Absolute *0* agents and Absolute *0* shared agent;
3. If all the agents included in Universe are Absolute *0* Agents and Conventional Agents, or all are Conventional Agents, then Universe can be regarded as a special kind of Conventional Agent.

Therefore, Universe can be regarded as an Agent that can change in states such as Absolute *0* agent, Conventional Agent and Omniscient and Omnipotent Agent. In this section, it is named Universe agent (a_U).

Definition 12.15 Because

1. $\exists a \in A$, $a = \Omega point \Rightarrow U = \Omega_{point}$,
2. $\forall a \in A$, $a \in Apoint \Rightarrow U = \alpha_{point}$,
3. $\exists a \in A$, $a \in A_C \Rightarrow U = a_C$,

   we have that U∈A, U is Agent, noted as a_U.

12.2.8 Law of Intelligence Evolution (Ω Law)

The second law of intelligence is called Law of Intelligence Evolution, namely Ω Law. This law interprets the evolution of a Agent to the Omniscient and Omnipotent Agent (Ω_{point}), with the content as follows:

Definition 12.16 Any Agent will evolve directly or indirectly toward the Omniscient and Omnipotent Agent (Ω_{point}) under the effect of F_Ω (*Ω gravity*). In the process of evolution, it is also directly or indirectly subject to F_α (*α gravity*) which hinders the Agent's speed to evolve toward Ω_{point}, especially when F_α (*α gravity*) is constantly greater than F_{Ω} (Ω *gravity*), the Agent will converge toward the absolute 0 Agent (α_{point}).

The mathematical expression of **Law of Intelligence Evolution** is:

$$(0,0,0,0) \overset{F_A}{\leftarrow} (I_a, O_a, S_a, C_a) \overset{F_{\Omega}}{\rightarrow} (\infty,\infty,\infty,0)$$

or

$$\alpha_{\text{point}} \overset{\text{F}_\text{A}}{\leftarrow} a \overset{\text{F}_{\Omega}}{\rightarrow} \Omega_{\text{point}}$$

The law is related to the existence of two special states of the Intelligent Model. As seen from Definitions 12.9–12.10, there are Omniscient and Omnipotent Agents (Ω_{point}) and absolute 0 Agents (α_{point}).

When the Agent changes toward these two states, two "forces" are theoretically required to drive the Agent to evolve toward Ω_{point} or converge toward α_{point}. Therefore, we call the "force" driving the Agent to evolve toward Ω_{point} as F_{Ω} (Ω *gravity*), the "force" driving the Agent to converge toward α_{point} as F_{α} (α *gravity*). The changes of the Agent towards Ω_{point} or α_{point} are shown in Fig. 12.7.

As viewed from the hundreds of millions of years of history in biological evolution, it may be noted that the signs of effects of Ω Law on biological populations can be seen from changes in the ability of different populations to process knowledge.

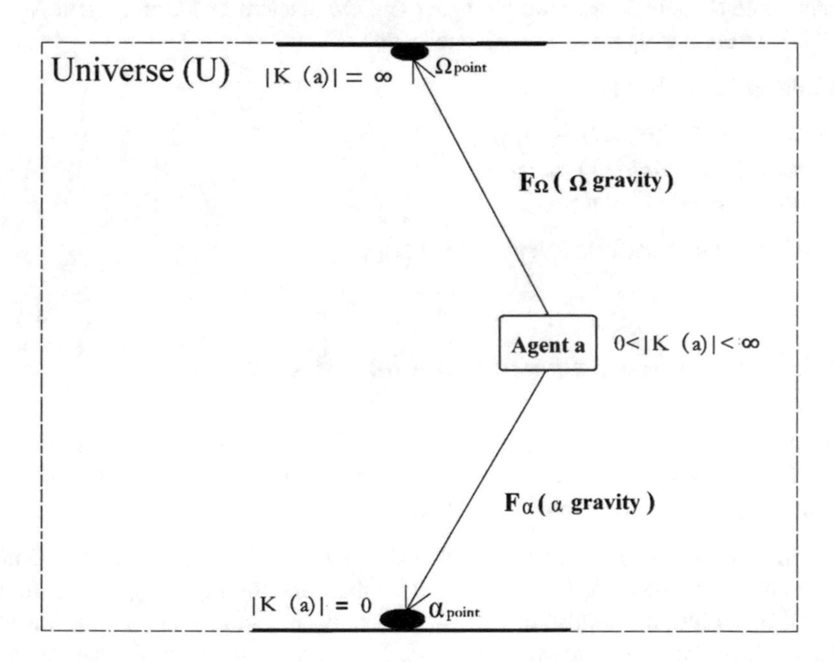

Fig. 12.7 Schematic diagram of Law of Intelligence Evolution

Although different organisms show the biodiversity. As examples, sharks, dinosaurs, pandas and the human can be discussed further as below.

Although there are no precise statistics, according to the common sense, we know that sharks have little change in the new knowledge they have mastered, and the biological population of sharks's ability to process knowledge has changed little during the hundreds of millions of years they have survived [15]. Pandas are on the verge of extinction because of their own and environmental reasons, and the biological population of Pandas's ability to process knowledge is shrinking. Dinosaurs failed to withstand natural disasters 65 million years ago and the entire biological population went extinct [16]. This is equivalent to that dinosaurs converged to an Absolute *0* Agent (α_{point}) under the effect of F_α (*α gravity*), no matter whether they were in the form of a population or individuals. For the human, the biological population's ability to process knowledge has grown considerably since 200,000 years ago when it mastered the use of language [17]. Especially during the recent hundreds of years, with the outbreak of the industrial revolution, the development of physics, and the birth of the Internet, the human's abilities to master knowledge and transform the world has experienced the accelerated growth.

Suppose there is no unfavorable situation such as major disasters, it can be deduced from this trend that the biological population represented by the human will reach the "Omniscient and Omnipotent Agent (Ω_{point})" state when the time approaches the infinite time point. The historical changes in the knowledge processing abilities of sharks, pandas, dinosaurs and the human may be illustrated on the same diagram for comparison [18], as shown in Fig. 12.8. Based on the Ω Law of Intelligence, the following six definitions may be proposed:

12.2.8.1 F_Ω(*Ωgravity*)

In the second law, i.e., Ω law, a "force" is inevitably required as a drive so that Agent a reaches Omniscience and Omnipotence state (Ω_{point}). Such a theoretical demand is the first reason for the proposal of F_Ω (*Ω gravity*).

Meanwhile, if observing the development law of population knowledge bases of the human, sharks, pandas, dinosaurs, etc., we can also find signs of the effects of F_Ω (*Ω gravity*). From this, we can propose the definition of F_Ω (*Ω gravity*) as below:

Definition 12.17 F_Ω (*Ω gravity*) is an "force" directly or indirectly acting upon any Agent, and the result of such action is that Agent or Shared Agent (a_G) in which the Agent is involved approaches toward the Ω_{point} state, namely, the abilities of Agent or Shared Agent (a_G) in the input, output, storage(master) and creative(innovate) of knowledge continuously grow and eventually reach Ω_{point}.

Although the specific principles and effects of F_Ω (*Ω* gravity) are still unknown to us so far, a quantitative research on how F_Ω (*Ω* gravity) acts upon agents may be conducted with a biochemical experiment method, an AI system intelligence evaluation experiment method or a computer program simulation experiment method.

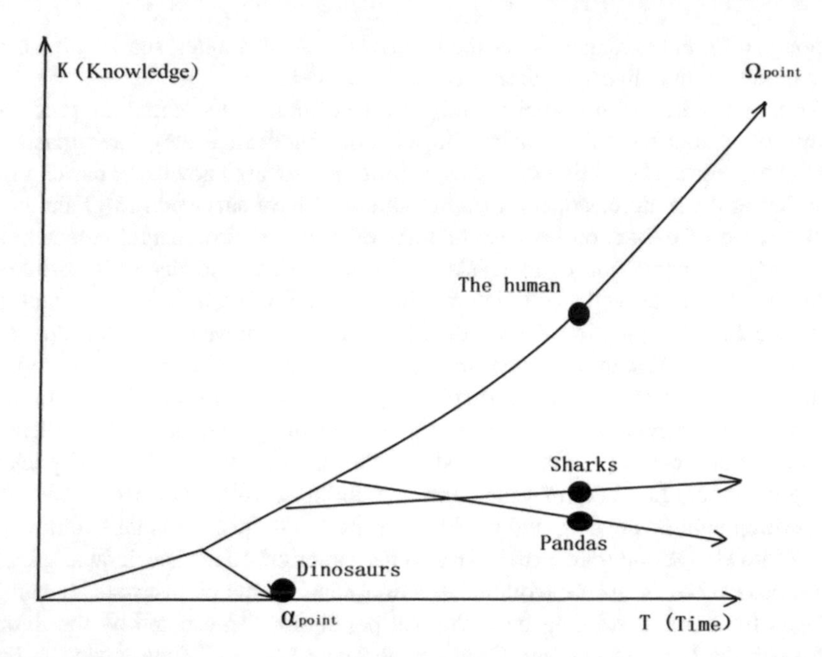

Fig. 12.8 Schematic diagram for the development of biological population's ability to process knowledge

We can also try summarizing the calculation formula of F_{Ω} (Ω gravitational force) on this basis.

12.2.8.2 $F_{alpha}(\alpha gravity)$

Similarly, an influencing factor is also inevitably required so that the Agent converges toward the absolute 0 Agent (α_{point}). Such a theoretical demand is the first reason for the proposal of $\boldsymbol{F}_{\alpha}$ (*α gravity*).

In nature, there are phenomena of aging, fading and death of biological populations, biological individuals or artificial intelligence systems, which correspond to the situation that the Agent converges toward the Absolute *0* Agent (α_{point}). From this, we propose the following definition of $\boldsymbol{F}_{\alpha}$ (*α gravity*):

Definition 12.18 F_{α} (*α gravity*) is an "force" directly or indirectly acting upon any Agent, and the result of such action is that Agent's abilities to input, output, storage(master) and creative(innovate) knowledge continuously decline, and eventually converge to α_{point}.

Similarly, the research on $\boldsymbol{F}_{\alpha}$ (*α gravity*) remains to be explored at this day. It should be combined with the research of F_{Ω} (*Ω gravity*) in the future. Thereby, we can conduct the quantitative research on how $\boldsymbol{F}_{\alpha}$ (*α gravity*) acts upon agents by using the three methods mentioned above and try to summarize the calculation formula of $\boldsymbol{F}_{\alpha}$ (*α gravity*).

12.2.8.3 Agent of Life and Agent of Engineering (a_L*and* a_E)

As the main parts for the generation of intelligence, life and artificial intelligence have always been in dispute in terms of their definitions. What is life? Schrodinger, a physicist, proposed in his book with the title of *What Is Life* that *the characteristic of life is that life can constantly obtain "negative entropy" from the surrounding environment to counter the inevitable increase of entropy in life activities* [19]. Then, what is artificial intelligence? Winston believes that *artificial intelligence is a science about how to make computers do intelligent work that could only be done by the human in the past* [20]. Corresponding to life and artificial intelligence, life agent and engineering agent are proposed in this section, and they are also deduced and defined as follows according to the second law of intelligence.

In the definition of F_{Ω} (*Ω gravity*), we mentioned that F_{Ω} (Ω gravity) directly or indirectly acts upon the Agent. Therefore, we identify an Agent according to whether it is directly subject to F_{Ω} (*Ω gravity*). Then, the definitions of life Agents and engineering Agents are proposed as follows:

Definition 12.19 Among all Agents (A set) in Universe, those Agents that are directly driven by F_{Ω} (*Ω gravity*) are called Agent of Life (a_L).

Definition 12.20 Among all Agents (A set) in Universe, those Agents that are not directly driven by F_{Ω} (*Ω gravity*) are called Agent of Engineering (a_E).

From the existing examples in the real world, the Agents like the human, dinosaurs, sharks, pandas should belong to the category of Agent of life, while the robots, artificial intelligence programs and other systems invented by the human may be regarded as Agents of engineering.

The running power and rules of the Agent of engineering are derived from the Agent of Life or other Agents of engineering. From the purpose that the human create artificial intelligence systems, robots and AI programs still provide services for the continuous development of the human [21]. Therefore, it can be considered that Agents of engineering are indirectly affected by F_{Ω} (*Ω gravity)*, which assists Agents of life to develop towards Ω_{point}.

12.2.8.4 Intelligence

Intelligence is the core issue of our discussion. An important goal to put forward the three laws of intelligence is to answer the question of "*what is intelligence*". Currently, there are also many definitions or controversies about this question. For example, V.A.C. Henmon argues that *intelligence is the ability to acquire and retain knowledge* [22], while Alfred Binet defines intelligence as *the ability of reasoning, judging, memorizing, and abstracting* [23].

Seen from M Law and Ω Law of Intelligence, any Agent processes knowledge and interacts with the outside world through the input, output, mastery and innovation functions. Besides, F_Ω (*Ω gravity*) and $\boldsymbol{F}_\alpha$ (*α gravity*) are the key driving forces for the Agent to process knowledge. Therefore, we propose the following definitions of intelligence:

Definition 12.21 The ability of an Agent to apply knowledge through input, output, mastery and innovation functions under the direct or indirect effects of F_Ω (*Ω gravity*) and $\boldsymbol{F}_\alpha$ (*α gravity*) is called intelligence (capability); or the phenomenon that knowledge flows inside and outside the Agent through the input, output, mastery and innovation functions of the Agent under the joint action of F_Ω (*Ω gravity*) and $\boldsymbol{F}_\alpha$ (*α gravity*), is called intelligence (phenomenon).

12.2.8.5 Consciousness

Consciousness is a concept closely related to intelligence. Then, what is consciousness and what is the difference between consciousness and intelligence? These questions are also the focus of debate among researchers. The understanding of consciousness in psychology involves its broad definition and narrow definition. From the broad definition, consciousness refers to the brain's response to the objective world, while from the narrow definition, it refers to people's awareness and attention to the outside world and themselves [24].

Tulving proposed in his book with the title of *Memory and Consciousness* that *consciousness is the name given to the kind of consciousness that mediates an individual awareness of his or her existence and identity in subjective time extending from the personal past through the present to the personal future* [25].

In the definition of intelligence in this section, it is mentioned that some Agents (*Agent of life*) are intelligence generated under the direct action of F_Ω *(Ω gravity)* and $\boldsymbol{F}_\alpha$ (*α gravity*), and the remaining Agents (*Agent of engineering*) are the intelligence generated by the indirect action. Therefore, whether the Agent is directly affected by F_Ω *(Ω gravity)* and $\boldsymbol{F}_\alpha$ (*α gravity*), and whether it can perceive F_Ω *(Ω gravity)* and $\boldsymbol{F}_\alpha$ (*α gravity*) and form corresponding knowledge are used as a standard for defining consciousness. So that consciousness can be an important feature to distinguish Agents of life and Agents of engineering. Therefore, the consciousness is defined as follows:

Definition 12.22 When the Agent that is directly driven by F_Ω (*Ω gravity*) and $\boldsymbol{F}_\alpha$ (*α gravity*) achieves the application of knowledge through the knowledge input, output, mastery and innovation functions, it can perceive the effects of F_Ω (*Ω gravity*) and $\boldsymbol{F}_\alpha$ (*α gravity*), and thus contain the understanding on F_Ω (*Ω gravity*) and $\boldsymbol{F}_\alpha$ (*α gravity*) in the knowledge mastered by it, this ability or phenomenon is called consciousness.

12.2.8.6 Law of Intelligence (Zero-Infinity) Duality (A Law)

The third law of intelligence is called Law of Intelligence (*zero-infinity*) Duality, namely A Law, with the content that when an Agent changes around the "absolute 0 Agent" (α_{point}), Universe will have existence and inexistence phenomena for this Agent, or the amount of knowledge contained in Universe will also change between 0 and infinity relative to this Agent. It is elaborated as follows:

Definition 12.23 For any Agent, when it converges to α_{point}, the entire Universe (the amount of knowledge, including but not limited to information, concepts, data, laws, time, matter, space, etc.) will become an empty set or "0" state, or we say Universe will not exist relative to this Agent. On the other hand, when the Agent changes from α_{point} to a conventional Agent (a_C), the entire Universe (the amount of knowledge) (including but not limited to information, concepts, data, laws, time, matter, space, etc.) will become infinity. In short, relative to this Agent at this time, Universe exists and there is an infinite amount of knowledge in cognition that needs to be mastered.

The mathematical expression of Law of Intelligence (zero-infinity) Duality is:

$$a \in A_{point}, |K(U)| = 0; \quad a \notin A_{point}, |K(U)| = \infty$$

or

$$O \overset{a}{=\!=\!=} \infty$$

In order to express this law succinctly, we replaced the formula $a \in Apoint, |K(U)| = 0$; $a \notin Apoint$, $|K(U)| = \infty$ with $O \overset{a}{=\!=\!=} \infty$, which shows that relative to an Agent, Universe (amount of knowledge) will change between 0 (empty) and infinity due to the change of the Agent's state.

The meaning of the third law of intelligence is as shown in Fig. 12.9.

If the second law focuses on elaboration of the effects of Ω_{point} on the Agent and Universe, then the third law is to elaborate the effects of α_{point} on the Agent and Universe.

In the real world, there are a large number of cases that the Agent converges to the α_{point}. Such as the extinction of dinosaurs as a population, the natural death or accidental death of human individuals, and the complete scrap of computers or robots due to damage of parts. These phenomena can be regarded as the cases that the Agent converges to α_{point}.

What needs to be studied and thought is, when the Agent converges from a conventional Agent to α_{point}, does the entire Universe still exist relative to this Agent?

According to the definition of absolute *0* Agent (α_{point}), this Agent can neither perceive or output any knowledge, nor create new knowledge, nor master any knowledge. In this case, any element of Universe should be empty or non-existent relative to this Agent. In special cases, when all the Agents in Universe converge to

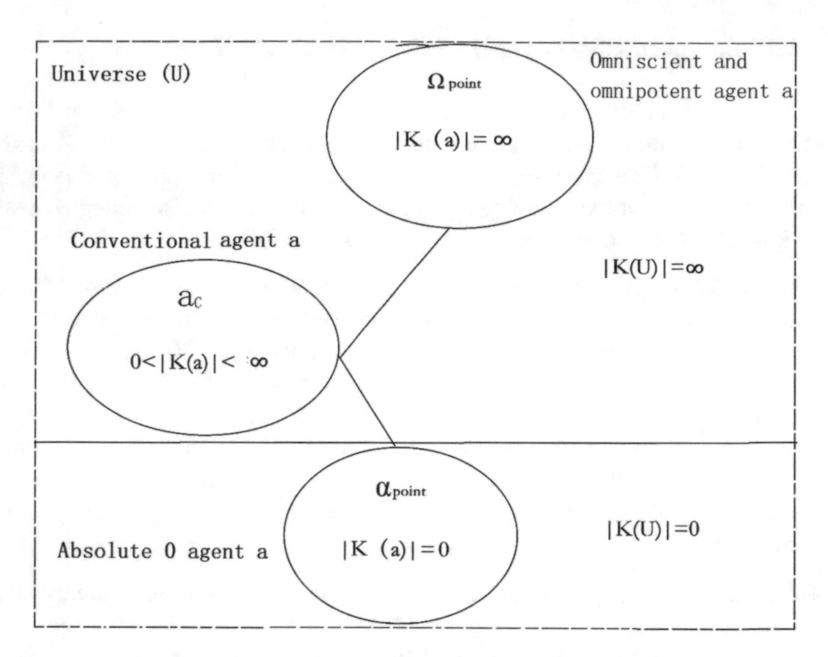

Fig. 12.9 Schematic diagram of Law of Intelligence (zero-infinity) Duality

absolute 0 Agents, the entire Universe will converge to the absolute 0 intelligence state. In the absolute 0 intelligence state, Universe will no longer contain any concepts, elements, knowledge, matter, time, space, or laws, and the entire Universe will be completely empty. When a Conventional Agent appears in Universe which is in Absolute 0 intelligence state, the knowledge contained in Universe (including but not limited to various concepts, elements, knowledge, matter, time, space, law, etc.) will exist relative to the born Conventional Agent or Universe at this time, and will continue to emerge with the evolution of the Agent. Then, how this knowledge emerges and what characteristics and laws involve will be further elaborated in future research.

The relationships between these three intelligence laws are shown in Fig. 12.10.

The further validation of the scientific value of the Three Laws remains to be explored. It can be carried out along two directions. The first direction is to conduct experiments in the real-world environment by means of the technologies and objects in the fields of biochemistry and AI systems. The second direction is to conduct experiments in virtual simulation programs, using the technologies like game dynamics, cellular automaton, AMB simulation.

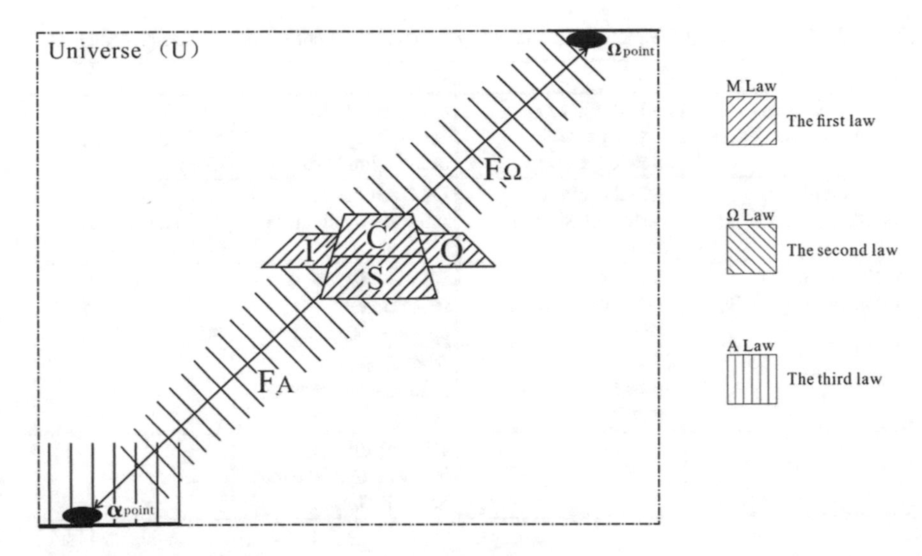

Fig. 12.10 Relationships between three intelligence laws

12.3 A Fuzzy Cognitive Map-Based Approach Finding Characteristics on AI-IQ Test

The determination of IQ test characteristics of Artificial Intelligence (AI) systems can vary depending a methodology is chosen. The subsection provides a Fuzzy Cognitive Map (FCM) approach to improve the IQ test characteristics of Artificial Intelligence (AI) systems. The defuzzification process makes use of fuzzy logic and the triangular membership function along with linguistic term analyses. Each edge of the proposed FCM is assigned to a positive or negative influence type associated with a quantitative weight. All the weights are based on the de-fuzzified value in the defuzzification results. It also leverages a dynamic scenario analysis to investigate the interrelationships between driver concepts and other concepts. Worst and best-case scenarios have been conducted on the correlation among concepts.

Based on the test bank of Sect. 12.1.2, like a human IQ test, each search engine needs to answer several questions that are selected from the developed test bank by random. For each question, they will receive a score between 0 and 100. This framework divides all the questions into four main indicator groups and further into 15 characteristics. Also, a few adult volunteers had the IQ test for the purpose of standardizing the IQ score, and mapping with the human being's IQ score.

Table 12.13 lists all the 15 IQ characteristics along with their corresponding weights for testing AI systems. After gathering expert opinions (Delphi method), all the 15 weights are calculated and presented in the Table 12.13. Where

C1m (m = 1,2 . . . m) = ability to acquire knowledge.
C2n (n = 1,2 . . . n) = ability to master knowledge.

Table 12.13 Fifteen IQ Characteristics for AI system and their corresponding Delphi weights

C1m	C2n	C3p	C4q
C11: Ability to identify word (3%)	C21: Ability to master general knowledge (6%)	C31: Ability to innovate by association (12%)	C41: Word feedback ability (3%)
C12: Ability to identify sound (3%)	C22: Ability to master translation (3%)	C32: Ability to innovate by creation (12%)	C42: Sound feedback ability (3%)
C13: Ability to identify image (4%)	C23: Ability to master calculation (6%)	C33: Ability to innovate by speculation (12%)	C43: Image feedback ability (4%)
	C24: Ability to master arrangement (5%)	C34: Ability to innovate by selection (12%)	
		C35: Ability to innovate by discover laws (12%)	

C3p (p = 1,2 . . . p) = ability to innovate knowledge.
C4q (q = 1,2 . . . q) = ability of knowledge feedback.

The proposed IQ test question bank is arranged according to all the 15 IQ characteristics (concepts). To illustrate, an example of testing question: "Please translate 'Technology's impact' into Spanish" should belong to characteristic C22 (Ability to master translation).

The results of Delphi weights are very subjective. Because they are coming from expert's own judgment, which means the results may be biased. Take advantage of linguistic terms from literature sources can be treated as a better method because all the literature publication sources are considered as an objective approach. One of the article's goals is to assign new weights though the fuzzy logic method (an objective approach). Based on the new weights, the interrelations among characteristics also should be investigated. There are some significant relationships among some characteristics. For example, "C21: Ability to master general knowledge" literally has a positive impact on "C24: Ability to master arrangement".

12.3.1 Research Method

12.3.1.1 Methodology

Fuzzy Cognitive Mapping (FCM) is the most important method of this research article. For the purpose of constructing FCM, the number of edges should be clarified. Theoretically, all the combination of two concepts should have an edge (relationship). However, the literature resources only support the meaningful edges, for example, the edge between one IQ characteristic and the AI system, or the edges of the interrelations among the 15 IQ characteristics. According to the literature resources, it is easy to assign the influence type (negative, positive, or null) of the edge. Keyword extraction plays a significant role in the relationship between concepts capturing. For instance, one reference paper said concept C22 heavily impacts concept C31, then, keyword "heavily impacts" will be extracted here. Each keyword will be assigned with one of the linguistic terms ("VERY LOW", "LOW", "MEDIUM", "HIGH", and "VERY HIGH"). At least three linguistic terms will be assigned to each edge.

The linguistic terms are fuzzy set problems. The membership function plays a significant role in quantifying the membership grade of the element in X to the fuzzy set.

$$\mu_{\mathrm{A}} : X \rightarrow [0, 1] \tag{12.8}$$

Where X represents the universe of discourse while the fuzzy set is A, and A is the membership function [26].

A triangular function will be used in the FCM constructing process. Where a is the lower limit, b is the upper limit, and m is a value between a and b. Figure 12.11 illustrates the membership function as a graph.

$$\mu_{\mathrm{A}} = \begin{cases} 0, x \leq a \\ \frac{x-a}{m-a}, a < x \leq m \\ \frac{b-x}{b-m}, m < x \leq b \\ 0, x > b \end{cases} \tag{12.9}$$

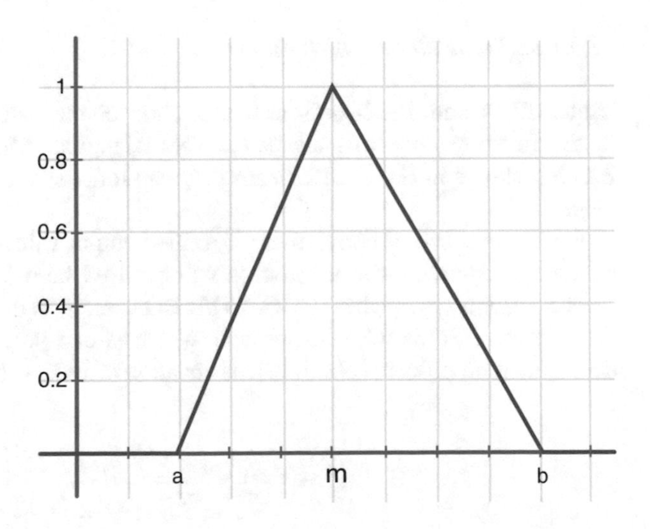

Fig. 12.11 Membership function graph [27]

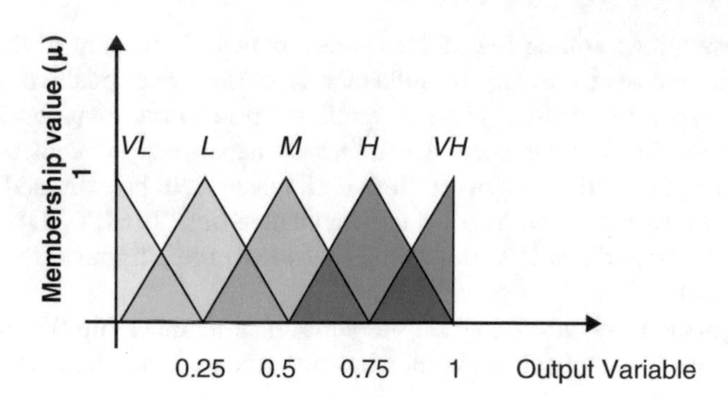

Fig. 12.12 Triangular membership function [59]

12.3.1.2 Linguistic Term Analyses

Table 12.14 summarizes all the possible relationships between each IQ characteristic and the AI system, and the interrelationship among the 15 IQ characteristics. In particular, Barwise's paper mentioned IQ characteristics' ability to identify word is a "most common view" of AI system [18]. Then, the keyword "most common view" will be extracted here, while a linguistic term "HIGH" will be assigned to this edge. Table 12.14 gives an outline of the linguistic terms, influence type, and keywords.

In Table 12.14, "C" represents the "AI system IQ".

Based on the extracted keyword results, Table 12.15 is a more advanced tabulation is used to summary keyword information into a table according to their linguistic terms.

12.3.1.3 Defuzzification Method

Tables 12.14 and 12.15 present a tabulation of the defined five linguistic terms in the fuzzy set we will use later. The Triangular Membership Function [59] which is shown in Fig. 12.12 means different linguistic terms have different output values.

For the purpose of converting a fuzzified output values into a traditional single crisp value, defuzzification process will be used here [60]. Among the existing defuzzification approaches (COG, COA, BOA, etc.), in this research article, we use the Center of Sums (COS) approach, which is one very useful approach for the defuzzification process [60, 61]. This equation of COS is below:

$$x^* = \frac{\sum_{i=1}^{N} x_i * \sum_{k=1}^{N} \mu A_K(x_i)}{\sum_{i=1}^{N} \sum_{k=1}^{n} \mu A_K(x_i)} \tag{12.10}$$

Table 12.14 Linguistic terms

Edge of FCM	Keyword	Linguistic term
C11-C	an aspect of	LOW
	an aspect of	LOW
	an aspect of	LOW
C12-C	a key strategic	HIGH
	core capabilities	HIGH
	obvious	LOW
C13-C	core capabilities	HIGH
	enable	MEDIUM
C21-C	important component	HIGH
	correlated	MEDIUM
	partly represented	LOW
	related to	MEDIUM
C22-C	no significant correlation	VERY LOW
	week relationship	LOW
	no interrelationship	VERY LOW
C23-C	intersection	LOW
	accelerate	MEDIUM
	interleave	MEDIUM
C24-C	a significant	MEDIUM
	common view	MEDIUM
C31-C	interpreted to	MEDIUM
	display	MEDIUM
	measures of	HIGH
C32-C	demonstrates	HIGH
	must entail	VERY HIGH
	referred to	HIGH
	been central to	VERY HIGH
	fundamental to	VERY HIGH
	can be important	HIGH
C34-C	directly	MEDIUM
	commonly used	MEDIUM
	connects to	MEDIUM
C35-C3	related to	MEDIUM
	may affect	LOW
C41-C	are as likely to	LOW
	important element	MEDIUM
	a key for	HIGH
C42-C	linked to	LOW
	taken into consideration	MEDIUM
	is important to	HIGH

(continued)

Table 12.14 (continued)

Edge of FCM	Keyword	Linguistic term
C43-C	dominated by	HIGH
	driven by	MEDIUM
	result in	HIGH
C11-C12	statistically significant	MEDIUM
	foundational	VERY HIGH
	strong connected	VERY HIGH
C11-C13	improve	MEDIUM
	dependent	MEDIUM
	benefit	MEDIUM
C21-C22	important	MEDIUM
	widely identified as	LOW
	never an empty mind of	MEDIUM
C21-C23	result from	HIGH
	partially predicted by	LOW
	as the basis	MEDIUM
C21-C24	commonly used	MEDIUM
	spontaneously	MEDIUM
	related to	MEDIUM
C21-C31	able to	MEDIUM
	a key precursor of	VERY HIGH
	access to	HIGH
C21-C32	according to	MEDIUM
	used to	MEDIUM
	embodied in	HIGH
C21-C33	found to be	HIGH
	directive effect	MEDIUM
	prompted by	HIGH
C21-C34	facilitate	HIGH
	related to	MEDIUM
	as a basic	MEDIUM
C21-C35	needed for	MEDIUM
	lies in	HIGH
	support	HIGH
C41-C42	statistically significant	MEDIUM
	foundational	VERY HIGH
	strong connected	VERY HIGH
C41-C43	improve	MEDIUM
	dependent	MEDIUM
	benefit	MEDIUM
C31-C35	valuable for	MEDIUM
	led to	HIGH
	indicate	HIGH

(continued)

Table 12.14 (continued)

Edge of FCM	Keyword	Linguistic term
C31-C32	representative	HIGH
	based on	MEDIUM
	significance	MEDIUM

Table 12.15 Categorization of keywords based on linguistic terms

Linguistic term		Keyword	
VERY LOW	no significant correlation	no interrelationship	
LOW	an aspect of	week relationship	are as likely to
	obvious	intersection	linked to
	partly represented	widely identified as	may affect
	partially predicted by		
MEDIUM	a field of	accelerate	important element
	enable	important	display
	taken into consideration	never an empty mind of	statistically significant
	according to	as the basis	dependent
	needed for	spontaneously	benefit
	connects to	able to	valuable for
	directly	used to	based on
	commonly used	correlated	significance
	directive effect	a significant	interleave
	related to	common view	driven by
	as a basic	interpreted to	
	improve		
HIGH	prompted by	a key for	demonstrates
	most common view	dominated by	can be important
	facilitate	result from	a key strategic
	lies in	referred to	component
	support	access to	measures of
	led to	core capabilities	indicate
	important	embodied in	found to be
	result in	representative	
VERY HIGH	must entail	strong connected	foundational
	been central to	a key precursor of	

Where n stands for the sum-total of fuzzy sets, N is the sum total of fuzzy variables, and, $_{Ak}$ (x_i) is the membership function for the k-th fuzzy set.

12.3.2 Data Analysis

12.3.2.1 Fuzzy Cognitive Map Results

As stated before, each edge, at least three linguistic terms are assigned to, even, for a few edges, four linguistic terms are assigned to.

A standard fuzzy set operation will be used, which is a standard union. Where,

$$\mu_{A \cup B}(u) = \max\{\mu_A(u), \mu_B(u)\} \tag{12.11}$$

To illustrate, there are the three linguistic terms assigned to the edge of C22-C, they are: "LOW", "VERY LOW", and "VERY LOW".

$$\begin{aligned} A1 &= \tfrac{1}{2} * [(0.25 - 0) + (0 - 0)] * 1 = 0.125 \\ A2 &= \tfrac{1}{2} * [(0.5 - 0) + (0.25 - 0.25)] * 1 = 0.25 \\ A3 &= \tfrac{1}{2} * [(0.25 - 0) + (0 - 0)] * 1 = 0.125 \end{aligned} \tag{12.12}$$

The center of area of the fuzzy set C1 is $\overline{x}_1 = (0.25 + 0)/2 = 0.125$, similarly $\overline{x}_2 = 0.25,\ \ \overline{x}_3 = 0.125$.

Now, the calculated defuzzified value $x^* = \frac{(A_1\overline{x}_1 + A_2\overline{x}_2 + A_3\overline{x}_3)}{A_1 + A_2 + A_3} = 0.1875$.

A final version of the calculated fuzzy cognitive map is presented in Fig. 12.13. This FCM is drawn with software "Mental Modeler".

The following FCM weights are calculated based on the de-fuzzified values of the FCM. A summary of the calculation results is presented in Table 12.16. Table 12.17 provides the corresponding adjacency matrix of the FCM. This matrix can be used to describe the interrelations between the concept.

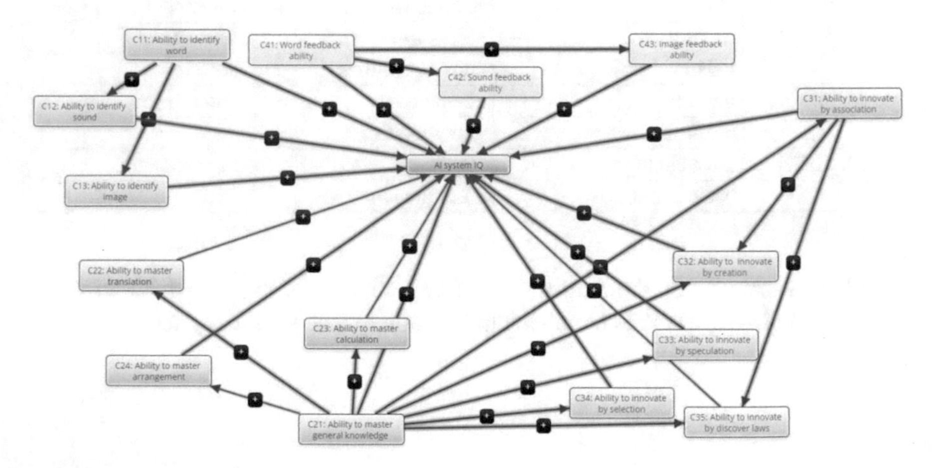

Fig. 12.13 Fuzzy cognitive map with positive/negative sign to edges

Table 12.16 Edge with its calculated weights

Edge of FCM	Defuzzified value	FCM weight	Delphi weight (%)
C11-C	0.5	6.0373%	3
C12-C	0.6786	8.1939%	3
C13-C	0.5833	7.0432%	4
C21-C	0.5625	6.792%	6
C22-C	0.1875	2.264%	3
C23-C	0.45	5.4336%	6
C24-C	0.5	6.0373%	5
C31-C	0.6071	7.3305%	12
C32-C	0.7961	9.6126%	12
C33-C	0.8125	9.8107%	12
C34-C	0.5	6.0373%	12
C35-C	0.4167	5.0315%	12
C41-C	0.5	6.0373%	3
C42-C	0.5	6.0373%	3
C43-C	0.6875	7.3305%	12
C11-C12	0.6525	N/A	0
C11-C13	0.5	N/A	0
C21-C22	0.5625	N/A	0
C21-C23	0.5	N/A	0
C21-C24	0.4	N/A	0
C21-C31	0.7015	N/A	0
C21-C32	0.6071	N/A	0
C21-C33	0.6875	N/A	0
C21-C34	0.6071	N/A	0
C21-C35	0.6875	N/A	0
C41-C42	0.6525	N/A	0
C41-C43	0.5	N/A	0
C31-C35	0.6875	N/A	0
C31-C32	0.6071	N/A	0

12.3.2.2 FCM Steady-State Analysis

A general descriptive summary about this FCM is shown in Table 12.18. The connection and component number are not extremely high. All the components can be categorized into the four groups. All the connections are supported by literature references. There are some interdependencies between the components in the same group. Also, there are some interconnections between components of different groups.

Figure 12.13, which is the merged FCM, shows the density changed to 0.121 while the average connections per component increased to 1.8125. Hierarchy Index is another complexity measurement of FCM. Hierarchy Index is answerable to all the concepts' out-degree in an FCM of N components [62]. Below is the equation

Table 12.17 Adjacency matrix collected from the fuzzy cognitive map

	C11	C12	C13	C21	C22	C23	C24	C31	C32	C33	C34	C35	C41	C42	C43	AI system IQ
C11	0	0.65	0.5	0	0	0	0	0	0	0	0	0	0	0	0	0.5
C11	0	0.65	0.5	0	0	0	0	0	0	0	0	0	0	0	0	0.5
C12	0	0	0	0	0	0	0	0	0	0	0	0	0	0	0	0.68
C13	0	0	0	0	0	0	0	0	0	0	0	0	0	0	0	0.58
C21	0	0	0	0	0.56	0.5	0.4	0.7	0.61	0.69	0.61	0.69	0	0	0	0.56
C22	0	0	0	0	0	0	0	0	0	0	0	0	0	0	0	0.19
C23	0	0	0	0	0	0	0	0	0	0	0	0	0	0	0	0.45
C24	0	0	0	0	0	0	0	0	0	0	0	0	0	0	0	0.5
C31	0	0	0	0	0	0	0	0	0.61	0	0	0.69	0	0	0	0.61
C32	0	0	0	0	0	0	0	0	0	0	0	0	0	0	0	0.8
C33	0	0	0	0	0	0	0	0	0	0	0	0	0	0	0	0.81
C34	0	0	0	0	0	0	0	0	0	0	0	0	0	0	0	0.5
C35	0	0	0	0	0	0	0	0	0	0	0	0	0	0	0	0.42
C41	0	0	0	0	0	0	0	0	0	0	0	0	0	0.65	0.5	0.5
C42	0	0	0	0	0	0	0	0	0	0	0	0	0	0	0	0.5
C43	0	0	0	0	0	0	0	0	0	0	0	0	0	0	0	0.69
AI system IQ	0	0	0	0	0	0	0	0	0	0	0	0	0	0	0	0

Table 12.18 General FCM statistics

FCM properties	Value
Total components	16
Total connections	29
Density	0.121
Connections per component	1.8125
No. of driver components	3
No. of receiver components	1
No. of ordinary components	12
Complexity score	0.3333

of Hierarchy Index.

$$h = \frac{12}{(N-1)\,N\,(N+1)} \sum_{1}^{N} \left[\frac{od(vi) - \left(\sum od(vi)\right)}{N} \right]^2 \tag{12.13}$$

Where N is the total number of components. And, od(vi) is the row sum of absolute values of a variable in the FCM adjacency matrix.

If h is close to 1, the FCM is supposed to be completely dominant (hierarchical). If h is close to 0, the FCM is supposed to be completely adapted eco-strategies (democratic) [63]. This FCM's hierarchy index is 0.326, which means, the FCM is much more adaptable to component changes because of its high level of integration and dependence. Also, the in-degree and out-degree of these nodes makes the FCM more democratic, and its system's steady-state more resistant to the alterations of individual components.

The component with the highest centrality was the "AI SYSTEM IQ" with a high score of 8.29. Also, the top three central components directly affecting the "AI SYSTEM IQ" component was the following, in ascending order of their complexity: Ability to innovate by discover laws 1.799, Ability to innovate by association 2.609, and, Ability to master general knowledge 5.319. A higher value means greater importance of an individual concept or several concepts in the overall model (Table 12.19).

12.3.3 Dynamic Scenario Analysis of the AI System IQ

12.3.3.1 Worst and Best-Case Scenario

The above AI system IQ FCM (Fig. 12.13) shows its complexity. This research also conducted dynamic case scenario analyses along with inference simulation.

To start the analysis, we initially apply the current FCM. Both the worst and best scenario will be examined. After that, some insightful results and conclusions can

Table 12.19 Characteristic, type of concepts, in degree, out degree, centrality and in the FCM

Characteristic	Indegree	Outdegree	Centrality	Type
AI system IQ	8.29	0	8.29	receiver
C11	0	1.65	1.65	driver
C12	0.65	0.68	1.33	ordinary
C13	0.5	0.58	1.08	ordinary
C21	0	5.319	5.319	driver
C22	0.56	0.19	0.75	ordinary
C23	0.5	0.45	0.95	ordinary
C24	0.4	0.5	0.9	ordinary
C31	0.7	1.909	2.609	ordinary
C32	1.22	0.8	2.02	ordinary
C33	0.69	0.81	1.5	ordinary
C34	0.61	0.5	1.109	ordinary
C35	1.38	0.42	1.799	ordinary
C41	0	1.65	1.65	driver
C42	0.65	0.5	1.15	ordinary
C43	0.5	0.69	1.19	ordinary

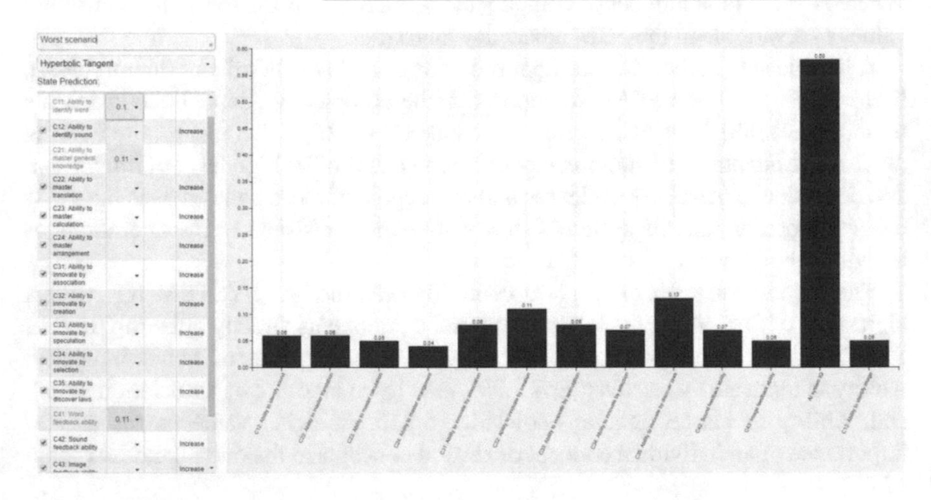

Fig. 12.14 The driver concept effects for the worst scenario

be made. Based on our knowledge, the worst scenario means all the driver concepts are equal to 0.1. And, the best scenario means all the driver concepts are equal to 1.

From Fig. 12.14, it can be observed that there is approximately 58% increase in the "AI system IQ" in the worst scenario while compared to the initial steady-state scenario as the benchmark. Respectively, the "Ability to innovate by discover laws" has an increase of 13%, the "Ability of innovate by creation" has an increase of 11%. All the other concepts have an increase between 4% and 8%. The results also show that all concepts have a positive causality. Furthermore, all of the slight increases for all the ordinary concepts are related to the small increase of driver concepts.

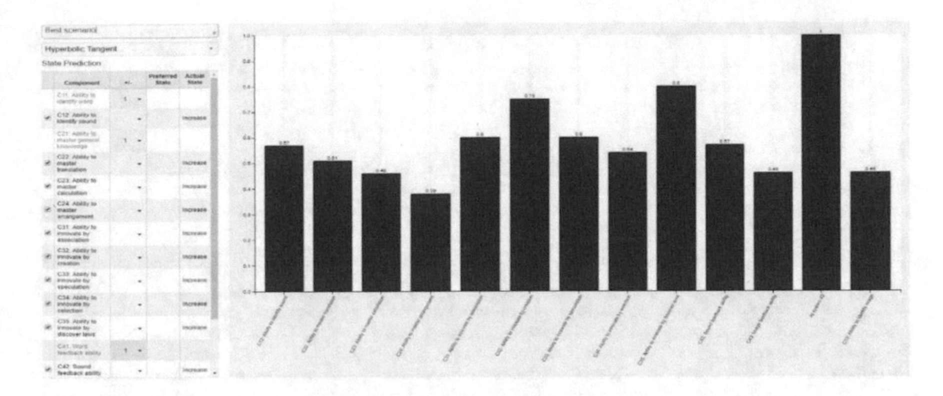

Fig. 12.15 The driver concept effects for the best scenario

Alternatively, all the driver concepts can be set as primarily affecting the FCM's ordinary concepts if all the values are set up with 1. From Fig. 12.15, we found that the "AI system IQ" in the best scenario while compared to the initial steady-state scenario as the benchmark, has a 100% increase. Similarly, the "Ability of innovate by creation" has an increase of 80%, and the "Ability to innovate by discover laws" has an increase of 75%. All the other concepts have an increase between 38% and 60%. This result also supports the conclusion of positive causality. Based on the results, the "Ability of innovate by creation" and "Ability to innovate by discover laws" has the most significant relevance impact.

12.3.3.2 FCM Inference Simulation

Based on the corresponding adjacency matrix (Table 12.19), there are some interrelations between concepts of this FCM. The value Ai of Ci is computed at each simulation step and it basically infers the influence of all other concepts Cj to Ci. This research selected Standard Kosko's activation rule inference method, below is the activation function:

$$A_t\,(K+1) = f\left\{\sum_{j=1,j\neq i}^{N} W_{ji} * A_j(k)\right\} \tag{12.14}$$

Also, the threshold function uses the sigmoid function, which shown as:

$$f(x) = \frac{1}{1+e^{-\lambda x}} \tag{12.15}$$

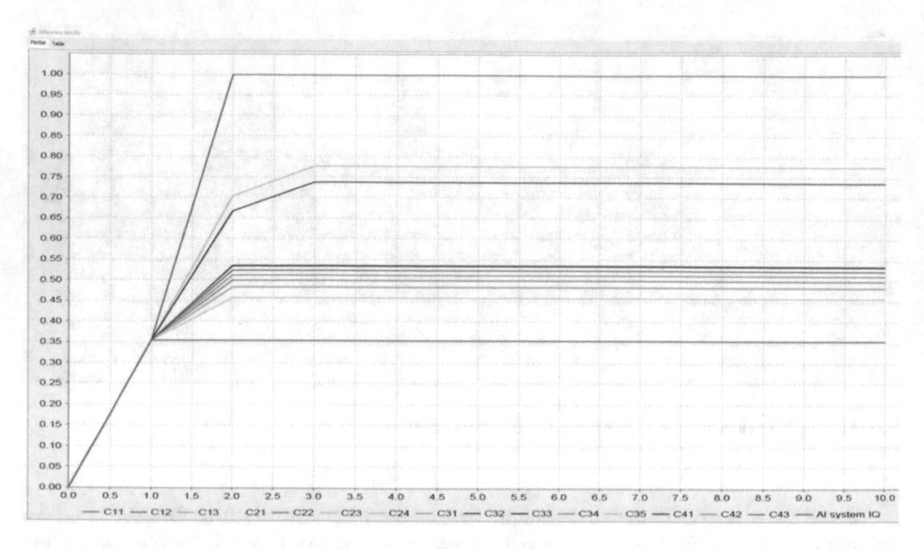

Fig. 12.16 Simulation activation level values per each iteration

Where x is the value Ai(K) at the equilibrium point, and is a real positive number λ that determines the steepness of the continuous function f. Using sigmoid threshold ensure that the activation value belongs to the interval [0, 1].

When running the simulation, all the concepts were assigned an initial value of 0. After a few simulation steps, all the values were expected to be convergence status. Theoretically, after reaching the equilibrium end states, larger activation value means playing a more important role in this FCM. All the driver and ordinary concepts were used for the simulation task. Figure 12.16 shows the corresponding concept activation levels per each iteration with all 18 concepts ranging from 0 to 1. Table 12.20 gives us the inference concept values. All the inference simulations were run through "FCM Expert" software in this research.

Based on the plotter and the table results illustrated by the inference simulation process, it is easy to confirm that the top two critical roles are "C32: Ability to innovate by creation" and "C35: Ability to innovate by discover laws".

In Sect. 12.1, AI system-based search engines IQ is tested based on the Delphi weight approach [38]. Now the new weight calculated through FCM approach is compared to its original subjective approach and two other approaches while using the same data set as the input. Mean Square Error (MSE) is used here as a performance indicator, its equation can be found as below:

$$MSE = \frac{1}{N} \sum_{i}^{N} \left(y_i - \hat{y}_i\right)^2 \tag{12.16}$$

Table 12.21 presents the MSE value for each approach. Dichotomous and polytomous [41] are two other old school methods. For the purpose of choosing the

Table 12.20 Inference concepts values

Step	C11	C12	C13	C21	C22	C23	C24	C31	C32	C33	C34	C35	C41	C42	C43	AI system IQ
0	0	0	0	0	0	0	0	0	0	0	0	0	0	0	0	0
1	0.354	0.354	0.354	0.354	0.354	0.354	0.354	0.354	0.354	0.354	0.354	0.354	0.354	0.354	0.354	0.354
2	0.354	0.522	0.482	0.354	0.498	0.482	0.456	0.536	0.667	0.533	0.512	0.704	0.354	0.522	0.482	0.999
3–8	0.354	0.522	0.482	0.354	0.498	0.482	0.456	0.536	0.736	0.533	0.512	0.776	0.354	0.522	0.482	1
9	0.354	0.522	0.482	0.354	0.498	0.482	0.456	0.536	0.736	0.533	0.512	0.776	0.354	0.522	0.482	1
10	0.354	0.522	0.482	0.354	0.498	0.482	0.456	0.536	0.736	0.533	0.512	0.776	0.354	0.522	0.482	1

Table 12.21 MSE results for four methods

Approach	MSE
Delphi weight	37.63363
Polytomous	49.51347
Dichotomous	31.23294
FCM approach	19.16389

best approach, MSE works as a prediction error indicator here. It is to say, lowest MSE value means less prediction error. Based on MSE values, it is easy to say FCM approach is among the four approaches.

References

1. Liu, F., Shi, Y.: The search engine iq test based on the internet iq evaluation algorithm. Proc. Comput. Sci. **31**, 1066–1073 (2014). https://doi.org/10.1016/j.procs.2014.05.361. https://www.sciencedirect.com/science/article/pii/S1877050914005389. 2nd International Conference on Information Technology and Quantitative Management, ITQM 2014
2. Liu, F., Shi, Y., Liu, Y.: Intelligence quotient and intelligence grade of artificial intelligence. Ann. Data Sci. **4**(2), 179–191 (2017)
3. Liu, F., Shi, Y., Wang, B.: World search engine iq test based on the internet iq evaluation algorithms. IJITDM. **14**(02), 221–237 (2015)
4. Liu, F., Shi, Y.: Investigating laws of intelligence based on AI IQ research. Ann. Data Sci. (2020). https://doi.org/10.1007/s40745-020-00285-x
5. Liu, F., Peng, Y., Chen, Z., Shi, Y.: Modeling of characteristics on artificial intelligence IQ test: a fuzzy cognitive map-based dynamic scenario analysis. Int. J. Comput. Commun. Control. **14**(6), 653–669 (2019)
6. Liu, F.: Internet IQ evaluation systems and algorithms. Complex Syst. Complex. Sci. **12**, 104–115 (2013)
7. Shi, B., He, Y., Wu, C.: Research on search strategy of web spider in topic-oriented search engines. Comput. Eng. App. **50**(2), 116–119, 128 (2014) CSCD:5056620
8. Xu, J.: Study on development trend of search engine. Modern Inf. **31**(9) (2011)
9. Ma, M., Fang, T.: Analysis on the legality of search engine robots' crawling behavior. J. Xi'an Jiaotong Univ. **33**(5) (2013)
10. Zhang, X., Song, Y.: Research on the hotlines of the subjects of search engine based on the scientific knowledge mapping. Sci. Technol. Manag. Res. (18) (2011)
11. von Neumann, J.: First draft of a report on the EDVAC. IEEE Ann. Hist. Comput. **15**(4), 27–75 (1993). https://doi.org/10.1109/85.238389
12. Liu, S.: Geometric analogical reasoning test for feasibility study of cognitive diagnosis. Ph.D. thesis, Jiangxi Normal University, Nanchang (2007)
13. Wang, Y.: Construction and application of synergistic learning system. Ph.D. thesis, East China Normal University, Shanghai (2009)
14. Liu, F.: Search engine IQ test based on the internet IQ evaluation algorithms. Ph.D. thesis, Beijing Jiaotong University, Beijing (2015)
15. Martin, A.P., Naylor, G.J.P., Palumbi, S.R.: Rates of mitochondrial dna evolution in sharks are slow compared with mammals. Nature. **357**(6374), 153–155 (1992)
16. Brusatte, S.L., Butler, R.J., Barrett, P.M., Carrano, M.T., Evans, D.C., Lloyd, G.T., Mannion, P.D., Norell, M.A., Peppe, D.J., Upchurch, P., Williamson, T.E.: The extinction of the dinosaurs. Biol. Rev. **90**(2), 628–642 (2015). https://doi.org/10.1111/brv.12128

17. Enard, W., Przeworski, M., Fisher, S.E., Lai, C.S.L., Wiebe, V., Kitano, T., Monaco, A.P., Pääbo, S.: Molecular evolution of foxp2, a gene involved in speech and language. Nature. **418**(6900), 869–872 (2002)
18. Liu, F., Shi, Y.: Research on artificial intelligence ethics based on the evolution of population knowledge base. In: Shi, Z., Pennartz, C., Huang, T. (eds.) Intelligence Science II, pp. 455–464. Springer International Publishing, Cham (2018)
19. Schroedinger, E.: What is life? The physical aspect of the living cell; with, mind and matter, autobiographical sketches (1992)
20. Winston, P.H., Brown, R.H.: Artificial Intelligence, an MIT Perspective. MIT Press, Cambridge, MA (1984)
21. Durkheim, E.: Les Formes Elementaires de la Vie Religieuse, pp. 78–79 (2006)
22. Henmon, V., Burns, H.M.: The constancy of intelligence quotients with borderline and problem cases. J. Educ. Psychol. **14**(4), 247 (1923)
23. Binet, A.: The mind and the brain (psychology revivals) (2013)
24. Krioukov, D., Kitsak, M., Sinkovits, R.S., Rideout, D., Meyer, D., Boguna, M.: Network cosmology. Sci. Rep. **2** (2012). https://doi.org/10.1038/srep00793
25. Tulving, E.: Memory and consciousness. Can. Psychol. **26**(1), 1–12 (1985)
26. Chen, G.S., Jheng, Y.D., Liu, H.C., Chen, S.Y.: A novel scoring method for stroke order based on Choquet integral with fuzzy measure. In: Proceedings of the 7th Conference on 7th WSEAS International Conference on Applied Computer Science - Volume 7, ACS'07, p. 82C87. World Scientific and Engineering Academy and Society (WSEAS), Stevens Point, WI (2007)
27. Fayyad, U.M., Piatetsky-Shapiro, G., Smyth, P., Uthurusamy, R. (eds.): Advances in Knowledge Discovery and Data Mining. American Association for Artificial Intelligence, New York (1996)
28. Stanovich, K.E., Bauer, D.W.: Experiments on the spelling-to-sound regularity effect in word recognition. Mem. Cogn. **6**(4), 410–415 (1978)
29. Stanovich, K.E.: Word recognition: changing perspectives. In: Handbook of Reading Research, vol. 2, pp. 418–452. Lawrence Erlbaum Associates, Inc, Hillsdale, NJ (1991)
30. Nation, K., Snowling, M.J.: Semantic processing and the development of word-recognition skills: evidence from children with reading comprehension difficulties. J. Mem. Lang. **39**(1), 85–101 (1998)
31. Hull, J.J.: A database for handwritten text recognition research. IEEE Trans. Pattern Anal. **16**(5), 550–554 (2002)
32. Zhu, Y., Tan, T., Wang, Y.: Font recognition based on global texture analysis. IEEE Trans. Pattern Anal. Mach. Intell. **23**(10), 1192–1200 (2001). https://doi.org/10.1109/34.954608
33. Wang, K., Babenko, B., Belongie, S.: End-to-end scene text recognition. In: IEEE International Conference on Computer Vision (2012)
34. Collombat, I.: General knowledge: a basic translation problem solving tool. Transl. Stud. N. Millenium. **4**, 59–66 (2006)
35. Delisle, J.: La traduction raisonnée, 2é edition: Manuel d'initiation à la traduction professionnelle de l'anglais vers le franais. University of Ottawa Press, Ottawa (2003)
36. Baroody, A.J.: Children's relational knowledge of addition and subtraction. Cogn. Instr. **17**(2), 137–175 (1999)
37. Cowan, R., Donlan, C., Shepherd, D.L., Cole-Fletcher, R., Saxton, M., Hurry, J.: Basic calculation proficiency and mathematics achievement in elementary school children. J. Educ. Psychol. **103**(4), 786–803 (2011)
38. Askew, M.: Teaching Primary Mathematics: A Guide for Newly Qualified & Student Teachers. Hodder & Stoughton, London (1998)
39. Rugg, G., McGeorge, P.: The sorting techniques: a tutorial paper on card sorts, picture sorts and item sorts. Expert. Syst. **14**(2), 80–93 (1997). https://doi.org/10.1111/1468-0394.00045
40. Mandler, J.M., Bauer, P.J.: The cradle of categorization: is the basic level basic? Cogn. Dev. **3**(3), 247–264 (1988)
41. Gopnik, A., Meltzoff, A.N.: Semantic and cognitive development in 15-to 21-month-old children. J. Child Lang. **11**(03) (1984)

42. Feigenson, L., Dehaene, S., Spelke, E.: Core systems of number. Trends Cogn. Sci. **8**(7), 307–314 (2004). https://doi.org/10.1016/j.tics.2004.05.002
43. Smedt, B.D., Reynvoet, B., Swillen, A., Verschaffel, L., Boets, B., Ghesquire, P.: Basic number processing and difficulties in single-digit arithmetic evidence from velo-cardio-facial syndrome. Cortex. **45**(2), 177–188 (2009)
44. Smedt, B.D., Gilmore, C.K.: Defective number module or impaired access? Numerical magnitude processing in first graders with mathematical difficulties. J. Exp. Child Psychol. **108**(2), 278–292 (2010)
45. Afuah, A.: Innovation management: strategies, implementation, and profits (2003)
46. Talaya, G.E.: Principios de marketing (2008)
47. Bereiter, C., Scardamalia, M.: Text-based and knowledge based questioning by children. Cogn. Instr. **9**(3), 177–199 (1992)
48. Naomi, M., Norman, D.A.: To ask a question, one must know enough to know what is not known. J. Verbal Learn. Verbal Behav. (1979)
49. Bereiter, C., Scardamalia, M., et al.: Intentional learning as a goal of instruction. In: Knowing, Learning, and Instruction: Essays in Honor of Robert Glaser, pp. 361–392 (1989)
50. Alexander, P.A., Jetton, T.L., Kulikowich, J.M.: Interrelationship of knowledge, interest, and recall: assessing a model of domain learning. J. Educ. Psychol. **87**(4), 559–575 (1995)
51. Qian, G., Alvermann, D.: Role of epistemological beliefs and learned helplessness in secondary school students' learning science concepts from text. J. Educ. Psychol. **87**(2), 282 (1995)
52. Linnenbrink-Garcia, L., Pugh, K.J., Koskey, K.L., Stewart, V.C.: Developing conceptual understanding of natural selection: the role of interest, efficacy, and basic prior knowledge. J. Exp. Educ. **80**(1), 45–68 (2012)
53. Njoo, M., De Jong, T.: Exploratory learning with a computer simulation for control theory: learning processes and instructional support. J. Res. Sci. Teach. **30**(8), 821–844 (1993)
54. Klahr, D., Dunbar, K.: Dual space search during scientific reasoning. Cogn. Sci. **12**(1) (1988)
55. Joolingen, W.V.: Cognitive tools for discovery learning. Int. J. Artif. Intell. Educ. **10**(3) (1998)
56. Agrawal, R.: Fast discovery of association rules. In: Advances in Knowledge Discovery Data Mining. The MIT Press, Cambridge, MA (1996)
57. Koperski, K., Han, J.: Discovery of spatial association rules in geographic information databases. In: Proceedings of the 4th International Symposium on Advances in Spatial Databases, SSD '95, p. 47C66. Springer-Verlag, Berlin (1995)
58. Luhn, H.P.: The automatic creation of literature abstracts. IBM J. Res. Dev. **2**(2), 159–165 (1958)
59. Klir, G.J., Yuan, B.: Fuzzy Sets and Fuzzy Logic: Theory and Applications. Prentice-Hall, Inc., Upper Saddle River, NJ (1994)
60. Mago, V.K., Morden, H.K., Fritz, C., Wu, T., Dabbaghian, V.: Analyzing the impact of social factors on homelessness: a fuzzy cognitive map approach. BMC Medical Informatics Decis. Making. **13** (2013)
61. Dzitac, I., Filip, F.G., Manolescu, M.J.: Fuzzy logic is not fuzzy: world-renowned computer scientist Lotfi A. Zadeh. Int. J. Comput. Commun. **12**(6), 748–789 (2017). https://doi.org/10.15837/ijccc.2017.6.3111. http://univagora.ro/jour/index.php/ijccc/article/view/3111
62. MacDonald, N.: Trees and networks in biological models (1983)
63. Liu, F., Zhang, Y., Shi, Y., Chen, Z., Feng, X.: Analyzing the impact of characteristics on artificial intelligence iq test: a fuzzy cognitive map approach. Proc. Comput. Sci. **139**, 82–90 (2018)

Conclusions

Our human history of big data analytics can be viewed as three stages. The first one is from 1700 to 1950 where statistical analysis has played a key role for 250 years. In this stage, the data analysis is descriptive where Bayes' Theorem has served as its base. As we mentioned in Chap. 1 of this book, the honor of celebrating Bayes' Theorem should go to Richard Price (1723–1791) who published Bayes' story after Thomas Bayes' death [1]. The second stage is from 1950 to 2012 where machine learning and artificial intelligence (AI), supported by data analysis or data mining methods, have been used in addition to statistical analysis. These results are analytic. The birth of computer and computing technology starting in 1950s was due to the need of solving a large-scale linear system which contains millions of rows and columns. Finding such a solution involves in the large number of calculations performed by an algorithm. This was also the major reason for researchers to initiate machine learning and AI methods. However, the root of computer comes from the {0,1}-binary numeral system created by Gottfried Leibniz more than 300 years ago [2]. The third stage is just starting from 2012, when the concept of big data was raised, to now. Here big data analytics becomes a drive force for social and economic development [3].

Our generation has fortunately growing with the second stage of big data analytics. My research career with data analysis has begun in my college time. In 1979, I was 23 years old young man majored in mathematics. One day, I read a newspaper article about fuzzy sets written by Professor Peizhuang Wang. It is first time I knew the concept of Fuzzy Sets and my heart was shocked by its fascinating idea of extending {0, 1} to [0,1]. Under the personal guidance of Professor Wang, I enjoyed much of my spare time as a college sophomore in searching interesting research topics at the "blue ocean" of fuzzy mathematics. I was very crazed about it. In 1981, I have published two research papers on fuzzy sets and fuzzy systems. The first paper was about an isomorphic theorem on fuzzy subgroups and fuzzy series of invariant subgroups, while the second paper contributed the convergence theorem of fuzzy integral of type II [4]. This is the first mile stone of my research career.

Y. Shi, *Advances in Big Data Analytics*,
https://doi.org/10.1007/978-981-16-3607-3

In 1983, I was admitted for a M.B.A. course program at the China's National Center for Industrial Science and Technology Management Development, co-sponsored by USA and China, Dalian Institute of Science and Technology, where I met my great mentors, Professor Daniel Berg and Professor C. F. Lee. Professor Berg who was the provost of Carnegie Mellon University taught Technology Transfer course while Professor Lee from University of Illinois at Urbana-Champaign taught Financial Accounting. These courses provided me intuition to transfer myself from pure mathematics to the real-world applications.

In 1985, I entered the Ph.D. program majored in management science and computer system at the University of Kansas. Although working with my advisor Professor Po-lung Yu on the theoretical problems of optimal design in Multiple Criteria Decision Making, I found a great interest in coding computer programs to test various data sets. From 1985 to 1998, I have built the mathematical foundation of Multiple-Criteria and Multiple-Constraint Levels Linear Programming (MC2LP) [5]. Some of the related works can be found in Chap. 2 of this book. In terms of data analysis, most of data sets used for MC2LP problems are empirical data or data from the data repository. In the summer of 1998, I was invited by the CEO of First Data Corporation at Omaha, Nebraska to conduct data mining and data warehousing project in credit card portfolio management. This led me to teste the great fun of using real-life data to run mathematical algorithms via SAS software, which eventually made me step in the new research fields of data mining, then data warehousing, then business analytics, then now big data analytics. This is the second mile stone of my research career. Some of my earlier works in the fields can be found in [6, 7].

From 1998 to 2011, my main research direction has concentrated on Optimization based Data Mining and the variation of Support Vector Machine as well as their real-life applications [8]. One of meaningful real-life applications was to build China's National Credit Scoring System, called "China Score", which is equivalent to "FICO Score" used in USA and many Western countries. During 2006–2009, I and my research team have worked with the People's Bank of China (PBC: China's Central Bank, equivalent to the Federal Reserve Bank of USA). Using the 950 million personal banking records of PBC, the largest dataset of this kind in the world, they constructed the optimization-based data mining models for the credit score calculation system. Since then, this China Score system has been serving as the national financial base for all commercial banks to handle China's 1.4 billion people for their daily financial and banking activities. This is one of most influential big data-based engineering applications in financial and bank industry (个人信用评分系统, www.baidu.com). This is the third mile stone of my research career.

In 2015, I and my colleagues initiated a new concept of "Intelligent Knowledge". Although data mining can discover the hidden patterns from unknown data, these results of data mining may not be regarded as "knowledge". To create knowledge, which is useful to the end-users, from databases, the theory of human knowledge management should be applied. Given large-scale databases (or Big Data), he proposed the theory and mechanisms of how to combine human knowledge with the hidden patterns of data mining to generate a "special" knowledge, called intelligent knowledge, for the practitioners or decision makers as useful decision support. The theory of intelligent knowledge management opens a door for the people to adopt "data-driven" decision making replacing the traditional "hypothesis-driven" or "model-driven" decision making. His intelligent knowledge theory has quickly been accepted by international academic community [9]. Chen et al. [10] has ranked me as the third place of the top academic authors in BI&A (Business Intelligence & Analytics).

This book, as I mentioned in the preface, is based on more than 80 published papers and reports in 2010–2020, to provide an up-to-date research progress and application findings of my research team in big data analytics and related areas. This can be regarded as the fourth mile stone of my research career.

Look back my living and research experience for the last 40 years, data analysis, now bid data analytics motivated my mathematical means has been the core of my research activities. However, the further development of our society depends on several IT technologies, mainly big data analytics and artificial intelligence. What can we predict the future? I strongly believe that the multi-value logic (multiple numeral system), for instance, fuzzy logic, can play a decisive role to lead us into a new history.

In 1703, Gottfried Leibniz published his paper Explication de l'Arithmétique Binaire, which is translated into English as the "Explanation of the binary arithmetic". He invented {0,1}-binary numeral system and explained its connection with the ancient Chinese figures of Fu Xi. As the simplified version of decimal numeral system, Leibniz's binary system gradually became the basis of the current computer design. It changed our human life dramatically for the last 300 years. The recent achievement of Google's AlphaGo and AlphaGo Zero have demonstrated that the binary numeral system-based computer can easily outperform human beings by massive calculation in a short time. However, if a computer like AlphaGo or even a super-computer plays with three persons in Chinese Mahjong, when one human player sends an eye contact to another human player, the machine cannot figure out how to calculate the human signal. This could partially cause by the simple binary numeral system that has difficulty to figure out the human contact. In Chap. 12, I have discussed our interesting research by using Human IQ test to measure machine. According to our finding, the IQ test for virtual assistants shows that even the best one, such as Google, still is not smarter than a 6-year-old human [11–14]. This means that there is a long way to go for the machine's intelligence to catch up that of our human beings. Perhaps, someday in the future when we use fuzzy logic (multiple numeral system) to design a new computer, its computing power of handling complex calculation can easily catch up and solve the human contact problem. By that time, artificial intelligence will be smarter to understand human being. Of course, any operation of artificial intelligence must be supported by a certain form of big data analytics. Therefore, the reader will find some interesting and useful findings from this book!

References

1. Bayes, F.: An essay towards solving a problem in the doctrine of chances. Biometrika. **45**(3–4), 296–315 (1958)
2. Leibniz, G.: Explication de l'Arithmétique Binaire. In: Die Mathematische Schriften, vol. 7, p. 223. Gerhardt C, Berlin (1879)
3. Shi, Y.: Challenges to engineering management in the big data era. Front. Eng. Manag. **2**(3), 293–303 (2015)
4. Shi, Y.: My early research on fuzzy set and fuzzy logic. Int. J. Comput. Commun. Control. **16**(1) (2021)
5. Shi, Y.: Multiple Criteria and Multiple Constraint Levels Linear Programming: Concepts, Techniques and Applications. World Scientific Publishing Company, Singapore (2001)
6. Shi, Y.: Data mining. In: Zeleny, M. (ed.) IEBM Handbook of Information Technology in Business, pp. 490–495. International Thomson Publishing, Europe (2000)
7. Shi, Y.: Human-casting: a fundamental method to overcome user information overload. Information. **3**(1), 127–143 (2000)
8. Shi, Y., Tian, Y., Kou, G., Peng, Y., Li, J.: Optimization Based Data Mining: Theory and Applications. Springer Science & Business Media, New York (2011)
9. Shi, Y., Zhang, L., Tian, Y., Li, X.: Intelligent Knowledge: A Study Beyond Data Mining. Springer, New York (2015)
10. Chen, H., Chiang, R.H., Storey, V.C.: Business intelligence and analytics: from big data to big impact. MIS Q, 1165–1188 (2012)
11. Liu, F., Shi, Y.: Investigating laws of intelligence based on ai iq research. Ann. Data Sci. **7**, 399–416 (2020)
12. Liu, F., Shi, Y.: The search engine iq test based on the internet iq evaluation algorithm. Proc. Comput. Sci. **31**, 1066–1073 (2014)
13. Liu, F., Shi, Y., Liu, Y.: Intelligence quotient and intelligence grade of artificial intelligence. Ann. Data Sci. **4**(2), 179–191 (2017)
14. Liu, F., Shi, Y., Wang, B.: World search engine iq test based on the internet iq evaluation algorithms. Int. J. Inf. Technol. Decis. Making. **14**(02), 221–237 (2015)

Printed in the United States
by Baker & Taylor Publisher Services